LONDON MATHEMATICAL SOCIETY LECTURE NOT

Managing Editor: Professor N. J. Hitchin, Mathematical Institute,
University of Oxford, 24-29 St Giles, Oxford OX1 3LB, United Kingdom

D1330171

The titles below are available from booksellers, or from Cambridge University P

176   Adams memorial symposium on algebraic topology 2,   N. RAY & G. WALKER (eds)
177   Applications of categories in computer science,   M. FOURMAN, P. JOHNSTONE & A. PITTS (eds)
178   Lower K- and L-theory,   A. RANICKI
179   Complex projective geometry,   G. ELLINGSRUD *et al*
180   Lectures on ergodic theory and Pesin theory on compact manifolds,   M. POLLICOTT
181   Geometric group theory I,   G.A. NIBLO & M.A. ROLLER (eds)
182   Geometric group theory II,   G.A. NIBLO & M.A. ROLLER (eds)
183   Shintani zeta functions,   A. YUKIE
184   Arithmetical functions,   W. SCHWARZ & J. SPILKER
185   Representations of solvable groups,   O. MANZ & T.R. WOLF
186   Complexity: knots, colourings and counting,   D.J.A. WELSH
187   Surveys in combinatorics, 1993,   K. WALKER (ed)
188   Local analysis for the odd order theorem,   H. BENDER & G. GLAUBERMAN
189   Locally presentable and accessible categories,   J. ADAMEK & J. ROSICKY
190   Polynomial invariants of finite groups,   D.J. BENSON
191   Finite geometry and combinatorics,   F. DE CLERCK *et al*
192   Symplectic geometry,   D. SALAMON (ed)
194   Independent random variables and rearrangement invariant spaces,   M. BRAVERMAN
195   Arithmetic of blowup algebras,   W. VASCONCELOS
196   Microlocal analysis for differential operators,   A. GRIGIS & J. SJÖSTRAND
197   Two-dimensional homotopy and combinatorial group theory,   C. HOG-ANGELONI *et al*
198   The algebraic characterization of geometric 4-manifolds,   J.A. HILLMAN
199   Invariant potential theory in the unit ball of C”,   M. STOLL
200   The Grothendieck theory of dessins d'enfant,   L. SCHNEPS (ed)
201   Singularities,   J.-P. BRASSELET (ed)
202   The technique of pseudodifferential operators,   H.O. CORDES
203   Hochschild cohomology of von Neumann algebras,   A. SINCLAIR & R. SMITH
204   Combinatorial and geometric group theory,   A.J. DUNCAN, N.D. GILBERT & J. HOWIE (eds)
205   Ergodic theory and its connections with harmonic analysis,   K. PETERSEN & I. SALAMA (eds)
207   Groups of Lie type and their geometries,   W.M. KANTOR & L. DI MARTINO (eds)
208   Vector bundles in algebraic geometry,   N.J. HITCHIN, P. NEWSTEAD & W.M. OXBURY (eds)
209   Arithmetic of diagonal hypersurfaces over finite fields,   F.Q. GOUVÉA & N. YUI
210   Hilbert C*-modules,   E.C. LANCE
211   Groups 93 Galway / St Andrews I,   C.M. CAMPBELL *et al* (eds)
212   Groups 93 Galway / St Andrews II,   C.M. CAMPBELL *et al* (eds)
214   Generalised Euler-Jacobi inversion formula and asymptotics beyond all orders,   V. KOWALENKO *et al*
215   Number theory 1992–93,   S. DAVID (ed)
216   Stochastic partial differential equations,   A. ETHERIDGE (ed)
217   Quadratic forms with applications to algebraic geometry and topology,   A. PFISTER
218   Surveys in combinatorics, 1995,   P. ROWLINSON (ed)
220   Algebraic set theory,   A. JOYAL & I. MOERDIJK
221   Harmonic approximation.,   S.J. GARDINER
222   Advances in linear logic,   J.-Y. GIRARD, Y. LAFONT & L. REGNIER (eds)
223   Analytic semigroups and semilinear initial boundary value problems,   KAZUAKI TAIRA
224   Computability, enumerability, unsolvability,   S.B. COOPER, T.A. SLAMAN & S.S. WAINER (eds)
225   A mathematical introduction to string theory,   S. ALBEVERIO, *et al*
226   Novikov conjectures, index theorems and rigidity I,   S. FERRY, A. RANICKI & J. ROSENBERG (eds)
227   Novikov conjectures, index theorems and rigidity II,   S. FERRY, A. RANICKI & J. ROSENBERG (eds)
228   Ergodic theory of Z^d actions,   M. POLLICOTT & K. SCHMIDT (eds)
229   Ergodicity for infinite dimensional systems,   G. DA PRATO & J. ZABCZYK
230   Prolegomena to a middlebrow arithmetic of curves of genus 2,   J.W.S. CASSELS & E.V. FLYNN
231   Semigroup theory and its applications,   K.H. HOFMANN & M.W. MISLOVE (eds)
232   The descriptive set theory of Polish group actions,   H. BECKER & A.S. KECHRIS
233   Finite fields and applications,   S. COHEN & H. NIEDERREITER (eds)
234   Introduction to subfactors,   V. JONES & V.S. SUNDER
235   Number theory 1993–94,   S. DAVID (ed)
236   The James forest,   H. FETTER & B. G. DE BUEN
237   Sieve methods, exponential sums, and their applications in number theory,   G.R.H. GREAVES *et al*
238   Representation theory and algebraic geometry,   A. MARTSINKOVSKY & G. TODOROV (eds)
240   Stable groups,   F.O. WAGNER
241   Surveys in combinatorics, 1997,   R.A. BAILEY (ed)
242   Geometric Galois actions I,   L. SCHNEPS & P. LOCHAK (eds)
243   Geometric Galois actions II,   L. SCHNEPS & P. LOCHAK (eds)
244   Model theory of groups and automorphism groups,   D. EVANS (ed)
245   Geometry, combinatorial designs and related structures,   J.W.P. HIRSCHFELD *et al*
246   *p*-Automorphisms of finite *p*-groups,   E.I. KHUKHRO
247   Analytic number theory,   Y. MOTOHASHI (ed)
248   Tame topology and o-minimal structures,   L. VAN DEN DRIES
249   The atlas of finite groups: ten years on,   R. CURTIS & R. WILSON (eds)
250   Characters and blocks of finite groups,   G. NAVARRO
251   Gröbner bases and applications,   B. BUCHBERGER & F. WINKLER (eds)

252	Geometry and cohomology in group theory, P. KROPHOLLER, G. NIBLO & R. STÖHR (eds)
253	The q-Schur algebra, S. DONKIN
254	Galois representations in arithmetic algebraic geometry, A.J. SCHOLL & R.L. TAYLOR (eds)
255	Symmetries and integrability of difference equations, P.A. CLARKSON & F.W. NIJHOFF (eds)
256	Aspects of Galois theory, H. VÖLKLEIN *et al*
257	An introduction to noncommutative differential geometry and its physical applications 2ed, J. MADORE
258	Sets and proofs, S.B. COOPER & J. TRUSS (eds)
259	Models and computability, S.B. COOPER & J. TRUSS (eds)
260	Groups St Andrews 1997 in Bath, I, C.M. CAMPBELL *et al*
261	Groups St Andrews 1997 in Bath, II, C.M. CAMPBELL *et al*
262	Analysis and logic, C.W. HENSON, J. IOVINO, A.S. KECHRIS & E. ODELL
263	Singularity theory, B. BRUCE & D. MOND (eds)
264	New trends in algebraic geometry, K. HULEK, F. CATANESE, C. PETERS & M. REID (eds)
265	Elliptic curves in cryptography, I. BLAKE, G. SEROUSSI & N. SMART
267	Surveys in combinatorics, 1999, J.D. LAMB & D.A. PREECE (eds)
268	Spectral asymptotics in the semi-classical limit, M. DIMASSI & J. SJÖSTRAND
269	Ergodic theory and topological dynamics, M.B. BEKKA & M. MAYER
270	Analysis on Lie groups, N.T. VAROPOULOS & S. MUSTAPHA
271	Singular perturbations of differential operators, S. ALBEVERIO & P. KURASOV
272	Character theory for the odd order theorem, T. PETERFALVI
273	Spectral theory and geometry, E.B. DAVIES & Y. SAFAROV (eds)
274	The Mandlebrot set, theme and variations, TAN LEI (ed)
275	Descriptive set theory and dynamical systems, M. FOREMAN *et al*
276	Singularities of plane curves, E. CASAS-ALVERO
277	Computational and geometric aspects of modern algebra, M.D. ATKINSON *et al*
278	Global attractors in abstract parabolic problems, J.W. CHOLEWA & T. DLOTKO
279	Topics in symbolic dynamics and applications, F. BLANCHARD, A. MAASS & A. NOGUEIRA (eds)
280	Characters and automorphism groups of compact Riemann surfaces, T. BREUER
281	Explicit birational geometry of 3-folds, A. CORTI & M. REID (eds)
282	Auslander-Buchweitz approximations of equivariant modules, M. HASHIMOTO
283	Nonlinear elasticity, Y. FU & R.W. OGDEN (eds)
284	Foundations of computational mathematics, R. DEVORE, A. ISERLES & E. SÜLI (eds)
285	Rational points on curves over finite, fields, H. NIEDERREITER & C. XING
286	Clifford algebras and spinors 2ed, P. LOUNESTO
287	Topics on Riemann surfaces and Fuchsian groups, E. BUJALANCE *et al*
288	Surveys in combinatorics, 2001, J. HIRSCHFELD (ed)
289	Aspects of Sobolev-type inequalities, L. SALOFF-COSTE
290	Quantum groups and Lie theory, A. PRESSLEY (ed)
291	Tits buildings and the model theory of groups, K. TENT (ed)
292	A quantum groups primer, S. MAJID
293	Second order partial differential equations in Hilbert spaces, G. DA PRATO & J. ZABCZYK
294	Introduction to the theory of operator spaces, G. PISIER
295	Geometry and Integrability, L. MASON & YAVUZ NUTKU (eds)
296	Lectures on invariant theory, I. DOLGACHEV
297	The homotopy category of simply connected 4-manifolds, H.-J. BAUES
298	Higher operads, higher categories, T. LEINSTER
299	Kleinian Groups and Hyperbolic 3-Manifolds Y. KOMORI, V. MARKOVIC & C. SERIES (eds)
300	Introduction to Möbius Differential Geometry, U. HERTRICH-JEROMIN
301	Stable Modules and the D(2)-Problem, F.E.A. JOHNSON
302	Discrete and Continuous Nonlinear Schrödinger Systems, M. J. ABLORWITZ, B. PRINARI & A. D. TRUBATCH
303	Number Theory and Algebraic Geometry, M. REID & A. SKOROBOGATOV (eds)
304	Groups St Andrews 2001 in Oxford Vol. 1, C.M. CAMPBELL, E.F. ROBERTSON & G.C. SMITH (eds)
305	Groups St Andrews 2001 in Oxford Vol. 2, C.M. CAMPBELL, E.F. ROBERTSON & G.C. SMITH (eds)
306	Peyresq lectures on geometric mechanics and symmetry, J. MONTALDI & T. RATIU (eds)
307	Surveys in Combinatorics 2003, C. D. WENSLEY (ed.)
308	Topology, geometry and quantum field theory, U. L. TILLMANN (ed)
309	Corings and Comodules, T. BRZEZINSKI & R. WISBAUER
310	Topics in Dynamics and Ergodic Theory, S. BEZUGLYI & S. KOLYADA (eds)
311	Groups: topological, combinatorial and arithmetic aspects, T. W. MÜLLER (ed)
312	Foundations of Computational Mathematics, Minneapolis 2002, FELIPE CUCKER *et al* (eds)
313	Transcendental aspects of algebraic cycles, S. MÜLLER-STACH & C. PETERS (eds)
314	Spectral generalizations of line graphs, D. CVETKOVIC, P. ROWLINSON & S. SIMIC
315	Structured ring spectra, A. BAKER & B. RICHTER (eds)
316	Linear Logic in Computer Science, T. EHRHARD *et al* (eds)
317	Advances in elliptic curve cryptography, I. F. BLAKE, G. SEROUSSI & N. SMART
318	Perturbation of the boundary in boundary-value problems of Partial Differential Equations, D. HENRY
319	Double Affine Hecke Algebras, I. CHEREDNIK
321	Surveys in Modern Mathematics, V. PRASOLOV & Y. ILYASHENKO (eds)
322	Recent perspectives in random matrix theory and number theory, F. MEZZADRI & N. C. SNAITH (eds)
323	Poisson geometry, deformation quantisation and group representations, S. GUTT, J. RAWNSLEY & D. STERNHEIMER (eds)
324	Singularities and Computer Algebra, C. LOSSEN & G. PFISTER (eds)
326	Modular Representations of Finite Groups of Lie Type, J. E. HUMPHREYS
328	Fundamentals of Hyperbolic Manifolds, R. D. CANARY, A. MARDEN & D. B. A. EPSTEIN (eds)
329	Spaces of Kleinian Groups, Y. MINSKY, M. SAKUMA & C. SERIES (eds)
330	Noncommutative Localization in Algebra and Topology, A. RANICKI (ed)
334	The Navier-Stokes Equations, P. G. DRAZIN & N. RILEY
335	Lectures on the Combinatorics of Free Probability, A. NICA & R. SPEICHER
337	Methods in Banach Space Theory, J. M. F. CASTILLO & W. B. JOHNSON (eds)

London Mathematical Society Lecture Note Series. 336

Integral Closure of Ideals, Rings, and Modules

Craig Huneke
University of Kansas

Irena Swanson
Reed College, Portland

CAMBRIDGE
UNIVERSITY PRESS

CAMBRIDGE UNIVERSITY PRESS
Cambridge, New York, Melbourne, Madrid, Cape Town, Singapore, São Paulo

Cambridge University Press
The Edinburgh Building, Cambridge CB2 2RU, UK

Published in the United States of America by Cambridge University Press, New York

www.cambridge.org
Information on this title: www.cambridge.org/9780521688604

First published 2006

Printed in the United Kingdom at the University Press, Cambridge

A catalog record for this publication is available from the British Library

ISBN-13   978-0-521-68860-4 paperback
ISBN-10   0-521-68860-4 paperback

Contents

Contents   v

Table of basic properties   ix

Notation and basic definitions   xi

Preface   xiii

1. What is integral closure of ideals?   1
     1.1. Basic properties   2
     1.2. Integral closure via reductions   5
     1.3. Integral closure of an ideal is an ideal   6
     1.4. Monomial ideals   9
     1.5. Integral closure of rings   13
     1.6. How integral closure arises   14
     1.7. Dedekind–Mertens formula   17
     1.8. Exercises   20

2. Integral closure of rings   23
     2.1. Basic facts   23
     2.2. Lying–Over, Incomparability, Going–Up, Going–Down   30
     2.3. Integral closure and grading   34
     2.4. Rings of homomorphisms of ideals   39
     2.5. Exercises   42

3. Separability   47
     3.1. Algebraic separability   47
     3.2. General separability   48
     3.3. Relative algebraic closure   52
     3.4. Exercises   54

4. Noetherian rings   56
     4.1. Principal ideals   56
     4.2. Normalization theorems   57
     4.3. Complete rings   60
     4.4. Jacobian ideals   63
     4.5. Serre's conditions   70
     4.6. Affine and $\mathbb{Z}$–algebras   73
     4.7. Absolute integral closure   77
     4.8. Finite Lying–Over and height   79
     4.9. Dimension one   83
     4.10. Krull domains   85
     4.11. Exercises   89

5. Rees algebras   93
     5.1. Rees algebra constructions   93
     5.2. Integral closure of Rees algebras   95
     5.3. Integral closure of powers of an ideal   97

5.4. Powers and formal equidimensionality 100
5.5. Defining equations of Rees algebras 104
5.6. Blowing up 108
5.7. Exercises 109

6. Valuations 113
6.1. Valuations 113
6.2. Value groups and valuation rings 115
6.3. Existence of valuation rings 117
6.4. More properties of valuation rings 119
6.5. Valuation rings and completion 121
6.6. Some invariants 124
6.7. Examples of valuations 130
6.8. Valuations and the integral closure of ideals 133
6.9. The asymptotic Samuel function 138
6.10. Exercises 139

7. Derivations 143
7.1. Analytic approach 143
7.2. Derivations and differentials 147
7.3. Exercises 149

8. Reductions 150
8.1. Basic properties and examples 150
8.2. Connections with Rees algebras 154
8.3. Minimal reductions 155
8.4. Reducing to infinite residue fields 159
8.5. Superficial elements 160
8.6. Superficial sequences and reductions 165
8.7. Non–local rings 169
8.8. Sally's theorem on extensions 171
8.9. Exercises 173

9. Analytically unramified rings 177
9.1. Rees's characterization 178
9.2. Module–finite integral closures 180
9.3. Divisorial valuations 182
9.4. Exercises 185

10. Rees valuations 187
10.1. Uniqueness of Rees valuations 187
10.2. A construction of Rees valuations 191
10.3. Examples 196
10.4. Properties of Rees valuations 201
10.5. Rational powers of ideals 205
10.6. Exercises 208

11. Multiplicity and integral closure 212
11.1. Hilbert–Samuel polynomials 212
11.2. Multiplicity 217

11.3. Rees's theorem   222
11.4. Equimultiple families of ideals   225
11.5. Exercises   232

12. The conductor   234
12.1. A classical formula   235
12.2. One–dimensional rings   235
12.3. The Lipman–Sathaye theorem   237
12.4. Exercises   242

13. The Briançon–Skoda Theorem   244
13.1. Tight closure   245
13.2. Briançon–Skoda via tight closure   248
13.3. The Lipman–Sathaye version   250
13.4. General version   253
13.5. Exercises   256

14. Two–dimensional regular local rings   257
14.1. Full ideals   258
14.2. Quadratic transformations   263
14.3. The transform of an ideal   266
14.4. Zariski's theorems   268
14.5. A formula of Hoskin and Deligne   274
14.6. Simple integrally closed ideals   277
14.7. Exercises   279

15. Computing integral closure   281
15.1. Method of Stolzenberg   282
15.2. Some computations   286
15.3. General algorithms   292
15.4. Monomial ideals   295
15.5. Exercises   297

16. Integral dependence of modules   302
16.1. Definitions   302
16.2. Using symmetric algebras   304
16.3. Using exterior algebras   307
16.4. Properties of integral closure of modules   309
16.5. Buchsbaum–Rim multiplicity   313
16.6. Height sensitivity of Koszul complexes   319
16.7. Absolute integral closures   321
16.8. Complexes acyclic up to integral closure   325
16.9. Exercises   327

17. Joint reductions   331
17.1. Definition of joint reductions   331
17.2. Superficial elements   333
17.3. Existence of joint reductions   335
17.4. Mixed multiplicities   338
17.5. More manipulations of mixed multiplicities   344

17.6. Converse of Rees's multiplicity theorem 348
17.7. Minkowski inequality 350
17.8. The Rees–Sally formulation and the core 353
17.9. Exercises 358
18. Adjoints of ideals 360
18.1. Basic facts about adjoints 360
18.2. Adjoints and the Briançon–Skoda Theorem 362
18.3. Background for computation of adjoints 364
18.4. Adjoints of monomial ideals 366
18.5. Adjoints in two–dimensional regular rings 369
18.6. Mapping cones 372
18.7. Analogs of adjoint ideals 375
18.8. Exercises 376
19. Normal homomorphisms 378
19.1. Normal homomorphisms 379
19.2. Locally analytically unramified rings 381
19.3. Inductive limits of normal rings 383
19.4. Base change and normal rings 384
19.5. Integral closure and normal maps 388
19.6. Exercises 390
Appendix A. Some background material 392
A.1. Some forms of Prime Avoidance 392
A.2. Carathéodory's theorem 392
A.3. Grading 393
A.4. Complexes 394
A.5. Macaulay representation of numbers 396
Appendix B. Height and dimension formulas 397
B.1. Going–Down, Lying–Over, flatness 397
B.2. Dimension and height inequalities 398
B.3. Dimension formula 399
B.4. Formal equidimensionality 401
B.5. Dimension Formula 403
References 405
Index 422

Table of basic properties

In this table, R is an arbitrary Noetherian ring, and I, J ideals in R. The overlines $\overline{}$ denote the integral closure in the ambient ring.

(1) $I \subseteq \overline{I} \subseteq \sqrt{I}$ and $\sqrt{0} \subseteq \overline{I}$. (Page 2.)

(2) $\overline{I} = \overline{\overline{I}}$. (Corollary 1.3.1.)

(3) Whenever $I \subseteq J$, then $\overline{I} \subseteq \overline{J}$. (Page 2.)

(4) For any I, J, $\overline{I : J} \subseteq \overline{I} : J$. (Page 7.)

(5) An intersection of integrally closed ideals is an integrally closed ideal. (Corollary 1.3.1.)

(6) **Persistence**: if $R \xrightarrow{\varphi} S$ is a ring homomorphism, then $\varphi(\overline{I}) \subseteq \overline{\varphi(I)S}$. (Page 2.)

(7) If W is a multiplicatively closed set in R, then $\overline{I}W^{-1}R = \overline{IW^{-1}R}$. (Proposition 1.1.4.)

(8) An element $r \in R$ is in the integral closure of I if and only if for every minimal prime ideal P in R, the image of r in R/P is in the integral closure of $(I + P)/P$. (Proposition 1.1.5.)

(9) **Reduction criterion**: $J \subseteq \overline{I}$ if and only if there exists $l \in \mathbb{N}_{>0}$ such that $(I + J)^l I = (I + J)^{l+1}$. (Corollary 1.2.5.)

(10) **Valuative criterion**: $J \subseteq \overline{I}$ if and only if for every (Noetherian) valuation domain V which is an R-algebra, $JV \subseteq IV$. When R is an integral domain, the V need only vary over valuation domains in the field of fractions of R. Furthermore, the V need only vary over valuation domains centered on maximal ideals. (Theorem 6.8.3 and Proposition 6.8.4.)

(11) $\overline{I} \cdot \overline{J} \subseteq \overline{IJ}$. (Corollary 6.8.6.)

(12) Let $R \subseteq S$ be an integral extension of rings. Then $\overline{IS} \cap R = \overline{I}$. (Proposition 1.6.1.)

(13) If S is a faithfully flat R-algebra, then $\overline{IS} \cap R = \overline{I}$. (Proposition 1.6.2.)

(14) The integral closure of a $\mathbb{Z}^n \times \mathbb{N}^m$-graded ideal is $\mathbb{Z}^n \times \mathbb{N}^m$-graded. (Corollary 5.2.3.)

(15) If R is a polynomial ring over a field and I is a monomial ideal, then $\overline{I}$ is also a monomial ideal. The monomials in $\overline{I}$ are exactly those for which the exponent vectors lie in the Newton polyhedron of I. (Proposition 1.4.6, more general version in Theorem 18.4.2.)

(16) Let R be the ring of convergent power series in d variables $X_1, \ldots, X_d$ over $\mathbb{C}$, or a formal power series ring in $X_1, \ldots, X_d$ over field of characteristic zero. If $f \in R$ with $f(\underline{0}) = 0$, then

$$f \in \overline{\left(X_1 \frac{\partial f}{\partial X_1}, \ldots, X_d \frac{\partial f}{\partial X_d} \right)}.$$

(Corollary 7.1.4 and Theorem 7.1.5.)

(17) Let $(R, \mathfrak{m})$ be a Noetherian local ring which is not regular and I an ideal of finite projective dimension. Then $\mathfrak{m}(I : \mathfrak{m}) = \mathfrak{m}I$ and $I : \mathfrak{m}$ is integral over I. (Proposition 1.6.5.)

(18) $x \in \bar{I}$ if and only if there exists $c \in R$ not in any minimal prime ideal such that for all sufficiently large n, $cx^n \in I^n$. (Corollary 6.8.12.)

(19) If R is a Noetherian local ring, then $\operatorname{ht}(I) \le \ell(I) \le \dim R$. (Corollary 8.3.9. The notation $\ell()$ stands for analytic spread.)

(20) If R is local with infinite residue field, then I has a minimal reduction, and every minimal reduction of I is generated by $\ell(I)$ elements. (Proposition 8.3.7.)

(21) If I or J is not nilpotent, then $\ell(IJ) < \ell(I) + \ell(J)$. (Proposition 8.4.4.)

(22) $x \in \overline{I^n}$ if and only if

$$\lim_{i \to \infty} \frac{\operatorname{ord}_I(x^i)}{i} \ge n.$$

(Corollary 6.9.1 and Lemma 6.9.2.)

(23) If $R \to S$ is a normal ring homomorphism of Noetherian rings, then for any ideal I in R, $\bar{I}S = \overline{IS}$. (Corollary 19.5.2.)

Notation and basic definitions

Except where otherwise noted, all rings in this book are commutative with identity and most are Noetherian. An ideal in a ring R generated by $a_1, \ldots, a_n$ is denoted by $(a_1, \ldots, a_n)$ or $(a_1, \ldots, a_n)R$. If $n = 1$, it is also written as $a_1 R$.

A ring is **local** if it has only one maximal ideal. We write $(R, \mathfrak{m})$ to denote a local ring with maximal ideal $\mathfrak{m}$, and $(R, \mathfrak{m}, k)$ to denote $(R, \mathfrak{m})$ with residue field k. An **overring** S of R is a ring that contains R as a subring. An R–algebra S is **essentially of finite type** over R if S is a localization of a finitely generated R–algebra. The **total ring of fractions** is the localization of R inverting all non–zerodivisors. If $f : R \to S$ is a ring homomorphism, I an ideal in R and J an ideal in S, then IS is the ideal in S generated by $f(I)$; and $J \cap R$ is the ideal $f^{-1}(J)$.

An R–module M is **faithful** if $\operatorname{ann}(M) = 0$. M is **torsion–free** if for any non–zerodivisor r of R and any $m \in M$, $rm = 0$ implies that $m = 0$. If I is an ideal in R, M is said to be **separated** in the I–adic topology if $\cap_n I^n M = 0$.

$A \setminus B$	set difference A minus B
$\mathbb{N}$	natural numbers starting from 0
$\mathfrak{m}_R$	the maximal ideal of a local ring R
$\operatorname{Min}(R)$	the set of minimal prime ideals of R
$\operatorname{Max}(R)$	the set of maximal prime ideals of R
$\operatorname{Ass}(M)$	set of associated prime ideals of M
R°	$R \setminus \bigcup_{P \in \operatorname{Min}(R)} P$
R_{red}	ring R modulo its nilradical
$\widehat{R}$	completion of $(R, \mathfrak{m})$ in the $\mathfrak{m}$–adic topology
$R(X)$	faithfully flat extension of R, Section 8.4
$Q(R)$	field of fractions of R
ω_R	canonical module of R
$\dim(\ )$	Krull dimension; vector–space dimension
$\lambda(\ )$	length
$T(R),\ T_k(R)$	Definition 13.4.3
$(R_i),\ (S_i)$	Serre's conditions, Definition 4.5.1
$J_{S/R}$	Jacobian ideal, Definition 4.4.1
K^*	the set of units in K
$\overline{K}$	algebraic closure of field
R^+	absolute integral closure, Section 4.7
$\overline{I}, \overline{R}$	integral closure of ideal, ring
$\operatorname{tr.deg}_F(K)$	transcendence degree of K over F
$[K : F]$	F–vector space dimension of a field K
Tr	trace
$D(R), \widetilde{D}(R)$	set of divisorial valuation(ring)s with respect to R, Definition 9.3.1, Section 18.1 resp.

R_v	the valuation ring of valuation v, Section 6.2
$\mathfrak{m}_v$	the maximal ideal of R_v
κ_v	the residue field of R_v
$\mathrm{rk}(v)$, $\mathrm{rat.rk}(v)$	rank and rational rank of v, Definition 6.6.1
Γ_v	value group of valuation v, Definition 6.2.1,
Γ_V	value group of valuation ring V, page 116
$R[It]$, $R[It, t^{-1}]$	Rees and extended Rees algebras, Chapter 5
$\mathrm{gr}_I(R)$	associated graded ring, Definition 5.1.5
$\mathcal{F}_I(R), \mathcal{F}_I$	fiber cone of I, Definition 5.1.5
$\kappa(P)$	field of fractions of R/P, residue field of P
$\mathrm{adj}(\ )$	adjoint, Chapter 18
$\mathrm{ann}(\ )$	annihilator ideal
$\mathrm{depth}(\ )$	depth of a module or ring, $\mathrm{depth}(R) = \infty$
$\sqrt{I}$	radical of I
$V(I)$	the set of prime ideals in R containing I
$\mathrm{ht}(\ )$	height of an ideal, $\mathrm{ht}(R) = \infty$
$\ell(\ )$	analytic spread, Definition 5.1.5
$\mu(\ )$	minimal number of generators
$\mathrm{ord}_I(\ )$	order function, Definition 6.7.7,
	$\mathrm{ord}_R(\ ) = \mathrm{ord}_{\mathfrak{m}}(\ )$ if $(R, \mathfrak{m})$ is a local ring
$c(I)$	content of an ideal, 14.1.1
$c(f)$	content of a polynomial, 1.7.1
$\mathcal{RV}(\ )$	set of Rees valuations or Rees valuation rings
I^{-1}	$\mathrm{Hom}_R(I, R)$, Definition 2.4.5
I^0, I^{-1}, I^{-2}, etc.	R (powers of an ideal I in a ring R)
$I \mid J$	ideal I divides ideal J, Definition 14.0.1
$I : J^\infty$	the union $\bigcup_n (I : J^n)$, where I and J are ideals
$r_J(I)$	reduction number, Definition 8.2.3
$\mathrm{rk}(M)$	rank of a module
	if R domain: $\mathrm{rk}\,M = \dim_{Q(R)}(M \otimes_R Q(R))$
$e_R(I; M)$, $e(I; M)$, $e(R)$	multiplicity, Chapter 11
$e_R(I_1, \ldots, I_d; M)$,	
$\quad e_R(I_1^{[d_1]}, \ldots, I_k^{[d_k]}; M)$	mixed multiplicity, Chapter 17
$P_{I,M}(n)$	Hilbert–Samuel polynomial
$\det_f(M)$	determinant of a module, Proposition 16.3.2
$I_t(\varphi)$	ideal of $t \times t$ minors of a matrix φ
$\mathcal{E}_R(M)$	page 308
$\mathfrak{R}_F(M)$	a Rees algebra of a module, Definition 16.2.1
$S_F(M)$	Definition 16.2.1
$\mathrm{Sym}_R(M)$	symmetric algebra of M, Chapter 16
$T \succ R$	infinitely near, Definition 14.5.1
$\overline{v}_I$	asymptotic Samuel function, Definition 6.9.3
$\Delta(v)$	Lipman's reciprocity, 14.6.2

Preface

Integral closure has played a role in number theory and algebraic geometry since the nineteenth century, and a modern formulation of the concept for ideals perhaps began with the work of Krull and Zariski in the 1930s. This book is on the integral closure of ideals, rings, and modules over commutative Noetherian rings with identity. Our goal in writing this book was to collect material scattered through many papers, and to present it with a unified treatment. To make the book self-contained, we begin with basic material, and develop most of what is needed to read the book. We hope the presentation makes the book friendly to a beginner. The reader should have basic knowledge of commutative algebra, such as modules, Noetherian chain conditions, prime ideals, polynomial rings, Krull dimension, height, primary decompositions, regular sequences, homomorphisms, regular and Cohen–Macaulay rings, and the process of completion. One exception is the need to have prior knowledge of the integral closure of rings, as this is worked out carefully in Chapter 2.

Integral closures of ideals, rings and modules overlap many important topics, including the core of ideals, Hilbert functions, homological algebra, multiplicities (mixed and otherwise), singularity theory, the theory of Rees algebras, resolution of singularities, tight closure, and valuation theory. While all of these topics are touched on in the book, we ask the reader to forgive us if we didn't fully explore their own favorite from among this list. After over 425 pages, we had to put some limits on what we could cover.

The book is written to be read in linear order. The first eleven chapters are linearly dependent upon each other to some extent, but the last several chapters are largely independent of each other, and the reader or instructor can easily pick and choose which they prefer to cover. In a few places in later chapters more commutative algebra techniques are used without background explanation. Exercises are included at the end of each chapter. Some exercises are easier than others; exercises labelled with a star are hard.

A seminar, beginning in 1984, with William Heinzer, Sam Huckaba, Jee Hub Koh, Bernard Johnston, and Jugal Verma, and later Judith Sally, was instrumental in encouraging the second author's interest in this material. The subject of the seminar, integral closure, became the basis of courses given by the second author at Purdue University. An impetus in writing the book was provided by a grant to the first author from the NSF POWRE program that enabled us to work together during 2000/01 at the University of Kansas. The first author taught there from the very first version of the book. We thank the students in that course for helping us smooth the presentation: Giulio Caviglia, Cătălin Ciupercă, Bahman Engheta, Glenn Rice, Janet Striuli, Emanoil Theodorescu, and Yongwei Yao. Yongwei Yao provided several results and exercises for the book.

Marie Vitulli taught from a second preliminary version at the University of Oregon. We thank Marie Vitulli for her support and feedback, and to Aaron Tresham for working out solutions to the exercises. Several people read parts of earlier versions of the book and provided valuable feedback on the content and on the presentation: Lionel Alberti, Jean Chan, Alberto Corso, Dale Cutkosky, Clare D'Cruz, Trung Dinh, Neil Epstein, Sara Faridi, Terence Gaffney, Daniel Grayson, William Heinzer, Melvin Hochster, Reinhold Hübl, Eero Hyry, Mark Johnson, Olga Kashcheyeva, Daniel Katz, Franz-Viktor Kuhlmann, Monique Lejeune-Jalabert, Joseph Lipman, Thomas Marley, Stephen McAdam, Patrick Morandi, Liam O'Carroll, Bruce Olberding, Greg Piepmeyer, Claudia Polini, Christel Rotthaus, Mark Spivakovsky, Branden Stone, Steven Swanson, Amelia Taylor, Bernd Ulrich, Wolmer Vasconcelos, Janet Vassilev, Jugal Verma, and Cornelia Yuen. We are indebted to all for improving the book and for reducing the number of errors. Of course, the remaining errors are all our own. Terence Gaffney and Joseph Lipman introduced us to material that we had not previously considered, and we thank them for their help.

William Heinzer, Liam O'Carroll, Claudia Polini, Christel Rotthaus, Amelia Taylor, Bernd Ulrich, and Jugal Verma gave us crucial detailed comments that helped us clarify the presentation. Daniel Katz went beyond any call of interest or friendship; he read and corrected the whole book, and often suggested clearer and better proofs. We thank them profusely.

We gratefully acknowledge the partial support of NSF while the book was in progress; the second author was supported in part by NSF grant 024405, while the first author acknowledges support on NSF grants 9970566, 0200420, on the POWRE NSF grant 0073140, and on the ADVANCE Institutional Transformation Program at New Mexico State University, fund NSF0123690, during Spring 2005.

We thank Cambridge University Press, and in particular Roger Astley, for encouragement and help over the years it took us to write this book.

The book was typeset with TeX, and the few pictures were made with Timothy van Zandt's pstricks. Occasionally, to explain some side remark or a well-known point, we use the Lipman-style small-print paragraphs.

Our families put up with us over our years of effort on this book, even when things were not working as we hoped. Nothing works without family support, and our heartfelt gratitude go to Steve, Simon, Edie, Sam, and Ned.

Ultimately, our thanks go to the many researchers who developed integral closure; we ourselves owe a special debt to Joseph Lipman and David Rees, who inspired and taught us, and this book is dedicated to both of them.

1
What is integral closure of ideals?

The main goal of this chapter is to introduce the integral closure of ideals. We give basic definitions, show some elementary manipulations, give a flavor of the theory for monomial ideals, and show how integral closure arises in various contexts.

Why and how did the integral closure of ideals arise in the first place? Much of commutative algebra is dependent, in various guises, upon understanding growth of ideals. A classic example of this phenomenon is the Hilbert–Samuel polynomial for an $\mathfrak{m}$–primary ideal I in a local Noetherian ring $(R, \mathfrak{m})$. For large n, the length of R/I^n, as a function of n, equals an integer–valued polynomial. The coefficients of this polynomial are necessarily rational, and are numerical invariants which are important for study of the pair (R, I). For example, the degree of this polynomial equals $d = \dim R$, and the normalized leading coefficient of the polynomial, namely the leading coefficient times $d!$, is an invariant carrying much information about I and even about R. It is called the multiplicity of I on R (which is the multiplicity of R if $I = \mathfrak{m}$). A key question is what elements r can be added to an $\mathfrak{m}$–primary ideal I so that I and $I + rR$ exhibit the same power–growth, or more specifically, have the same multiplicity?

A first try might be to say that "powers of r grow as powers of I", i.e., if there is an n such that $r^n \in I^n$. This has merit, but allows too few elements r.

A next attempt can be an asymptotic version. Let $v_n(r)$ be the least power of r in I^n. If the limit of $\frac{v_n(r)}{n}$ exists and is at least 1, we can intuitively think that r "grows" at least as fast as I. This approach was taken by David Rees in the 1950s, and successfully picks elements r.

Another approach is to use valuations to measure the relative "sizes" of I and r. One can simply require that $v(r) \geq v(I)$ for every valuation v. This *a priori* leads to a new and different idea of growth, but it turns out to be the same as Rees's asymptotic approach above.

There are other natural approaches. One approach that has become more important through its relationship to tight closure is to ask that there be an element c such that for all large n, $cr^n \in I^n$. Taking nth roots and letting n go to infinity somehow captures the sense that r is almost in I.

All of these different approaches lead to the same concept, which is called the integral closure of the ideal I. Moreover, integral closure occurs naturally in many contexts. Even more remarkably, all of these notions can be subsumed into a single equational definition (see Section 1.1).

The goal of this chapter is to introduce the integral closure of ideals, show some elementary manipulations, and motivate the study of integral closure.

We present many examples of integral closure constructions. In Section 1.4 we illustrate the theory on monomial ideals. In the last two sections we describe how integral closure arises naturally in many contexts. We start with the equational definition, and in Section 1.2 we characterize integral closure with reductions. We expand more on the theory of reductions in Chapter 8. The valuative approach to integral closure is taken up in Chapter 6, Rees's asymptotic approach is in Chapter 10, and the connection with the Hilbert–Samuel polynomial and multiplicity is in Chapter 11.

1.1. Basic properties

Definition 1.1.1 *Let I be an ideal in a ring R. An element $r \in R$ is said to be* **integral over I** *if there exist an integer n and elements $a_i \in I^i$, $i = 1, \ldots, n$, such that*
$$r^n + a_1 r^{n-1} + a_2 r^{n-2} + \cdots + a_{n-1} r + a_n = 0.$$

Such an equation is called **an equation of integral dependence of r over I (of degree n)**.

The set of all elements that are integral over I is called **the integral closure** *of I, and is denoted $\bar{I}$. If $I = \bar{I}$, then I is called* **integrally closed**. *If $I \subseteq J$ are ideals, we say that J is* **integral** *over I if $J \subseteq \bar{I}$.*

If I is an ideal such that for all positive integers n, I^n is integrally closed, then I is called **a normal ideal**.

A basic example is the following:

Example 1.1.2 For arbitrary elements x and $y \in R$, xy is in the integral closure $\overline{(x^2, y^2)}$ of the ideal (x^2, y^2). Namely, with $n = 2$, $a_1 = 0 \in (x^2, y^2)$ and $a_2 = -x^2 y^2 \in (x^2, y^2)^2$, $(xy)^2 + a_1(xy) + a_2 = 0$ is an equation of integral dependence of xy over (x^2, y^2).

Similarly, for any non–negative integer $i \le d$, $x^i y^{d-i}$ is in $\overline{(x^d, y^d)}$.

Remark 1.1.3
(1) $I \subseteq \bar{I}$, as for each $r \in I$, $n = 1$ and $a_1 = -r$ give an equation of integral dependence of r over I.
(2) If $I \subseteq J$ are ideals, then $\bar{I} \subseteq \bar{J}$, as every equation of integral dependence of r over I is also an equation of integral dependence of r over J.
(3) $\bar{I} \subseteq \sqrt{I}$, as from the equation of integral dependence of r over I of degree n as above, $r^n \in (a_1, \ldots, a_n) \subseteq I$.
(4) Radical, hence prime, ideals are integrally closed.
(5) The nilradical $\sqrt{0}$ of the ring is contained in $\bar{I}$ for every ideal I because for each nilpotent element r there exists an integer n such that $r^n = 0$, and this is an equation of integral dependence of r over I.
(6) Intersections of integrally closed ideals are integrally closed.
(7) The following property is called **persistence**: if $R \xrightarrow{\varphi} S$ is a ring homomorphism, then $\varphi(\bar{I}) \subseteq \overline{\varphi(I)S}$. This follows as by applying φ to an

equation of integral dependence of an element r over I to obtain an equation of integral dependence of $\varphi(r)$ over $\varphi(I)S$.

(8) Another important property is **contraction**: if $R \xrightarrow{\varphi} S$ is a ring homomorphism and I an integrally closed ideal of S, then $\varphi^{-1}(I)$ is integrally closed in R. Namely, if r is integral over $\varphi^{-1}(I)$, then applying φ to an equation of integral dependence of r over $\varphi^{-1}(I)$ gives an equation of integral dependence of $\varphi(r)$ over I, so that $\varphi(r) \in I$, whence $r \in \varphi^{-1}(I)$.

(9) In particular, if R is a subring of S, and I an integrally closed ideal of S, then $I \cap R$ is an integrally closed ideal in R.

Integral closure behaves well under localization:

Proposition 1.1.4 *Let R be a ring and I an ideal in R. For any multiplicatively closed subset W of R, $W^{-1}\bar{I} = \overline{W^{-1}I}$.*

Furthermore, the following are equivalent:

(1) $I = \bar{I}$.

(2) For all multiplicatively closed subsets W of R, $W^{-1}I = \overline{W^{-1}I}$.

(3) For all prime ideals P of R, $I_P = \overline{I_P}$.

(4) For all maximal ideals M of R, $I_M = \overline{I_M}$.

Proof: By persistence of integral closure, $W^{-1}\bar{I} \subseteq \overline{W^{-1}I} = \overline{IW^{-1}R}$. Let $r \in \overline{W^{-1}I}$. Write $r^n + a_1 r^{n-1} + \cdots + a_n = 0$ for some positive integer n and some $a_i \in W^{-1}I^i$. There exists $w \in W$ such that $wr \in R$ and for all $i = 1, \ldots, n$, $wa_i \in I^i$. Multiplying the integral equation by w^n yields

$$(wr)^n + a_1 w(wr)^{n-1} + \cdots + a_{n-1} w^{n-1}(wr) + a_n w^n = 0.$$

All the summands are in R, but the equality holds in $W^{-1}R$. Multiplying through by the nth power of some $w' \in W$ gives equality in R:

$$(ww'r)^n + a_1 ww'(ww'r)^{n-1} + \cdots + a_{n-1}(ww')^{n-1}(ww'r) + a_n(ww')^n = 0.$$

This is an integral equation of $ww'r \in R$ over I. Thus $r \in W^{-1}\bar{I}$, which proves the first part.

By the first part, (1) implies (2), and clearly (2) implies (3) and (3) implies (4). Now assume that (4) holds. Let $r \in \bar{I}$. Then for all maximal ideals M in R, $r \in I_M$, hence $r \in I$. This proves (1) and finishes the proof of the proposition. $\square$

The following proposition reduces questions about of integral closure to questions about integral closure in integral domains:

Proposition 1.1.5 *Let R be a ring, not necessarily Noetherian. Let I be an ideal in R.*

(1) The image of the integral closure of I in R_{red} is the integral closure of the image of I in R_{red}: $\bar{I}R_{red} = \overline{IR_{red}}$. Thus $\bar{I}$ equals the natural lift to R of the integral closure of I in the reduced ring R_{red}.

(2) An element $r \in R$ is in the integral closure of I if and only if for every minimal prime ideal P in R, the image of r in R/P is in the integral closure of $(I + P)/P$.

Proof: By persistence of integral closure, $\bar{I}R_{red}$ is contained in $\overline{IR_{red}}$. For the other inclusion, let $r \in R$ such that $r + \sqrt{0} \in \overline{IR_{red}}$. Write $f = r^n + a_1 r^{n-1} + \cdots + a_{n-1}r + a_n \in \sqrt{0}$ for some $a_i \in I^i$. Some power of f is zero. But $f^k = 0$ gives an equation of integral dependence of r over I of degree kn, which finishes the proof of (1).

We prove (2). By persistence of integral closure, the image of $\bar{I}$ in R/P is contained in the integral closure of $(I + P)/P$ for every P. Conversely, let W be the set $\{r^n + a_1 r^{n-1} + \cdots + a_n \mid n \in \mathbb{N}_{>0}, a_i \in I^i\}$. Note that W is a subset of R that is closed under multiplication. If W contains 0, r is integral over I; otherwise there is a prime ideal Q in R disjoint from W. As minimal prime ideals exist in every ring R, there is a minimal prime ideal P contained in Q. By assumption on Q, $W \cap P \subseteq W \cap Q = \emptyset$, but by the assumption on all the minimal prime ideals, $W \cap P$ is not empty, contradicting the assumption that 0 is not in W. $\qquad\square$

When the set of minimal prime ideals of R is finite, an equation of integral dependence of r over I can be constructed from the equations of integral dependence of r over I modulo each of the minimal prime ideals $P_1, \ldots, P_m$ as follows: there exist $a_{ji} \in I^i$, $i = 1, \ldots, n_j$, such that $f_j = r^{n_j} + a_{j1}r^{n_j-1} + \cdots + a_{j,n_j-1}r + a_{j,n_j} \in P_j$. Then $f = f_1 f_2 \cdots f_m \in \sqrt{0}$, so that for some positive integer k, $f^k = 0$, and this gives an equation of integral dependence of r over I.

Remark 1.1.6 Even if there are finitely many minimal prime ideals and the ideal I is contained in none of them, it may happen that the integral closure of I is determined by going modulo only a proper subset of the minimal prime ideals. For example, let $R = k[X, Y, Z]/(XY)$, with X, Y, Z variables over k. For $I = (Y, Z)R$, the integral closure of I modulo the minimal prime ideal (X) lifts to $(X, Y, Z)R$ and the integral closure of I modulo the minimal prime (Y) lifts to $(Y, Z)R$, making the first calculation redundant.

We next rephrase integral closure with ideal equalities:

Proposition 1.1.7 *Let R be a ring, not necessarily Noetherian. For any element $r \in R$ and ideal $I \subseteq R$, $r \in \bar{I}$ if and only if there exists an integer n such that $(I + (r))^n = I(I + (r))^{n-1}$.*

Proof: First suppose that $r \in \bar{I}$. Then an equation of integral dependence of r over I of degree n shows that $r^n \in I(I+(r))^{n-1}$ and hence that $(I+(r))^n = I(I + (r))^{n-1}$. Conversely, if $(I + (r))^n = I(I + (r))^{n-1}$ then $r^n = b_1 r^{n-1} + b_2 r^{n-2} + \cdots + b_{n-1}r + b_n$ for some $b_i \in I^i$, which can be easily rewritten into an equation of integral dependence of r over I. $\qquad\square$

One can also use modules to express integral dependence:

Corollary 1.1.8 (Determinantal trick, cf. Lemma 2.1.8) *Let I be an ideal in R and $r \in R$. Then the following are equivalent:*

(1) r is integral over I.

(2) There exists a finitely generated R–module M such that $rM \subseteq IM$ and such that whenever $aM = 0$ for some $a \in R$, then $ar \in \sqrt{0}$.

Moreover, if I is finitely generated and contains a non–zerodivisor, r is integral over I if and only if there exists a finitely generated faithful R–module M such that $IM = (I + (r))M$.

Proof: Let $r^n + a_1 r^{n-1} + \cdots + a_n = 0$ be an equation of integral dependence of r over I. There exists a finitely generated ideal $J \subseteq I$ such that $a_i \in J^i$ for all $i = 1, \ldots, n$. Thus r is integral over J, and by Proposition 1.1.7, there exists an integer n such that $J(J + (r))^{n-1} = (J + (r))^n$. Then $M = (J + (r))^{n-1}$ is finitely generated, and $rM = r(J + (r))^{n-1} \subseteq (J + (r))^n = J(J + (r))^{n-1} = JM \subseteq IM$. Also, if $aM = 0$, then $(ar)^{n-1} = 0$.

When I is finitely generated, we may take $J = I$, so that $IM = (I + (r))M$. If I contains a non–zerodivisor, M is faithful.

Conversely, assume (2). Let $M = Rb_1 + \cdots + Rb_m$ be an R–module such that $rM \subseteq IM$. For each $i = 1, \ldots, m$, write $rb_i = \sum_{j=1}^{m} a_{ij} b_j$ for some $a_{ij} \in I$. Let A be the matrix $(\delta_{ij} r - a_{ij})$, where δ_{ij} is the Kronecker delta function. Let b be the vector $(b_1, \ldots, b_m)^T$. By construction $Ab = 0$, so that $\det(A)b = \mathrm{adj}(A)Ab = 0$. Hence for all i, $\det(A)b_i = 0$, so that $\det(A)M = 0$. By assumption $(\det(A)r)^k = 0$ for some integer k, and an expansion of $(\det(A)r)^k = 0$ yields an equation of integral dependence of r over I. The last statement now also follows. $\qquad\square$

The first case of Proposition 1.1.7, for special rings, without using the term "integral", is in Prüfer's 1932 paper [224], on pages 14–16. In the same paper, Prüfer used the determinantal trick as in the proof above.

1.2. Integral closure via reductions

We introduce reductions in this section. Reductions are an extremely useful tool for integral closure in general, and we expand on them in Chapter 8.

Definition 1.2.1 *Let $J \subseteq I$ be ideals. J is said to be a **reduction** of I if there exists a non–negative integer n such that $I^{n+1} = JI^n$.*

Proposition 1.1.7 proved the following:

Corollary 1.2.2 *An element $r \in R$ is integral over J if and only if J is a reduction of $J + (r)$.*

Remark 1.2.3 Note that if $JI^n = I^{n+1}$, then for all positive integers m, $I^{m+n} = JI^{m+n-1} = \cdots = J^m I^n$. In particular, if $J \subseteq I$ is a reduction, there exists an integer n such that for all $m \geq 1$, $I^{m+n} \subseteq J^m$.

The reduction property is transitive:

Proposition 1.2.4 *Let $K \subseteq J \subseteq I$ be ideals in R.*

(1) If K is a reduction of J and J is a reduction of I, then K is a reduction of I.

(2) If K is a reduction of I, then J is a reduction of I.

(3) If I is finitely generated, $J = K + (r_1, \ldots, r_k)$, and K is a reduction of I, then K is a reduction of J.

Proof: First we assume that $K \subseteq J$ and $J \subseteq I$ are reductions. Then there exist integers n and m such that $KJ^n = J^{n+1}$ and $JI^m = I^{m+1}$. By Remark 1.2.3 it follows that $I^{m+n+1} = J^{n+1}I^m = KJ^nI^m \subseteq KI^{m+n} \subseteq I^{m+n+1}$, so that equality holds throughout and K is a reduction of I. This proves (1).

Assume that $K \subseteq I$ is a reduction. Then there exists an integer n such that $I^{n+1} = KI^n \subseteq JI^n \subseteq I^{n+1}$, so equality holds throughout and J is a reduction of I. This proves (2).

Assume that I is finitely generated, $J = K + (r_1, \ldots, r_k)$, and K is a reduction of I. Then there exists an integer n such that $KI^n = I^{n+1}$. By (2), for all $i = 0, \ldots, k$, $K + (r_1, \ldots, r_{i-1})$ is a reduction of I. (When $i = 0$, $(r_1, \ldots, r_{i-1})$ is interpreted as the zero ideal.) As $r_i \in I$, by the choice of n it follows that $r_iI^n \subseteq KI^n \subseteq (K + (r_1, \ldots, r_{i-1}))I^n$. If $aI^n = 0$ for some $a \in R$, then as $r_i \in I$, also $ar_i^n = 0$. By assumption I^n is finitely generated. Thus by Corollary 1.1.8, r_i is integral over $K + (r_1, \ldots, r_{i-1})$, so that by Proposition 1.1.7, $K + (r_1, \ldots, r_{i-1})$ is a reduction of $K + (r_1, \ldots, r_i)$. Hence by (1) and induction on k, $K \subseteq K + (r_1, \ldots, r_k) = J$ is a reduction. $\qquad \square$

In case ideals are not finitely generated, the conclusion of Proposition 1.2.4 need not hold (see Exercise 1.8).

Corollary 1.2.5 *Let $K \subseteq I$ be ideals. Assume that I is finitely generated. Then K is a reduction of I if and only if $I \subseteq \overline{K}$.*

Proof: If K is a reduction of I, then by Proposition 1.2.4 (3) for every $r \in I$, K is a reduction of $K + (r)$, so that by Proposition 1.1.7, $r \in \overline{K}$. Thus $I \subseteq \overline{K}$.

To prove the converse, suppose that $I = (r_1, \ldots, r_n) \subseteq \overline{K}$. Then for $j = 1, \ldots, n$, r_j is integral over K and hence over $K + (r_1, \ldots, r_{j-1})$. Then by Proposition 1.1.7, each immediate inclusion in the chain $K \subseteq K + (r_1) \subseteq K + (r_1, r_2) \subseteq \cdots \subseteq K + (r_1, \ldots, r_n) = I$ is a reduction, so that by Proposition 1.2.4 (1), $K \subseteq I$ is a reduction. $\qquad \square$

1.3. Integral closure of an ideal is an ideal

Reductions allow an easy proof of the fact that $\overline{I}$ is an ideal.

Corollary 1.3.1 *The integral closure of an ideal in a ring is an integrally closed ideal (in the same ring).*

Proof: Let K be an ideal in a ring R. Certainly $\overline{K}$ is closed under multiplication by elements of R. It remains to prove that $\overline{K}$ is closed under addition. Let

$r, s \in \overline{K}$. Write an integral equation for r over K: $r^n + k_1 r^{n-1} + \cdots + k_n = 0$ for some $k_i \in K^i$. There exists a finitely generated ideal K' contained in K such that $k_i \in (K')^i$. Thus $r \in \overline{K'}$, and similarly, by possibly enlarging K', $s \in \overline{K'}$. Let $J = K' + (r)$, $I = K' + (r, s) = J + (s)$. By Proposition 1.1.7, K' is a reduction of J, and J is a reduction of I. Thus by Proposition 1.2.4, K' is a reduction of I. As K', J and I are finitely generated, by Proposition 1.2.4 then $K' \subseteq K' + (r + s) \subseteq I$ are reductions. Thus by Proposition 1.1.7, $r + s$ is integral over K' and hence over K, so that $\overline{K}$ is an ideal.

To prove that the integral closure of an ideal is integrally closed, let I be an ideal of R, and $r \in \overline{\overline{I}}$. Then there exists a finitely generated subideal J in $\overline{I}$ such that $r \in \overline{J}$. Write $J = (j_1, \ldots, j_k)$. Similarly there exists a finitely generated ideal $K \subseteq I$ such that each j_i is integral over K. By Proposition 1.2.4, K is a reduction of $K + J$ and $K + J$ is a reduction of $K + J + (r)$, hence K is a reduction of $K + (r)$. Thus r is integral over K and hence over I. This proves the corollary. □

An alternate proof of this corollary is by Proposition 5.2.1 which shows that the integral closure of an ideal is a graded component of the integral closure of a special Rees ring in an overring. Integral closures of rings are discussed in the next chapter and Rees rings in Chapter 5.

The fact that the integral closure is an ideal is very useful in computations and constructions. We use this property in the rest of this section.

Remark 1.3.2
(1) The following are equivalent for an ideal I and an element r in a ring R (cf. Proposition 1.1.4):
   (i) $r \in \overline{I}$.
   (ii) For all multiplicatively closed subsets W of R, $\frac{r}{1} \in \overline{W^{-1}I}$.
   (iii) For all prime ideals P of R, $\frac{r}{1} \in \overline{I_P}$.
   (iv) For all maximal ideals M of R, $\frac{r}{1} \in \overline{I_M}$.
(2) If I and J are ideals in R, then $\overline{I} : J \subseteq \overline{I} : J$. Namely, it is enough to prove that $\overline{I} : J$ is integrally closed. Let r be integral over $\overline{I} : J$. Then r satisfies an equation of integral dependence over $\overline{I} : J$. Say that the degree of this equation is n. For any $a \in J$, multiply the equation by a^n to get an equation of integral dependence of ra over $\overline{I}$. It follows that $ra \in \overline{I}$. Hence $rJ \subseteq \overline{I}$, which means that $\overline{I} : J$ is integrally closed.
(3) If $I \subseteq J$ and $J \subseteq K$ are integral extensions, so is $I \subseteq K$ ($K \subseteq \overline{J} \subseteq \overline{\overline{I}} = \overline{I}$).
(4) If $I \subseteq I'$ and $J \subseteq J'$ are integral extensions of ideals, so are $I + J \subseteq I' + J'$ and $IJ \subseteq I'J'$. Namely, elements of I' and of J' are clearly integral over $I + J$, hence the ideal $I' + J'$ is integral over $I + J$. Also, for any $a \in I$, $b' \in J'$, ab' is integral over IJ (write out the equation of integral dependence for b' over J and multiply by an appropriate power of a), so that the ideal IJ' is integral over IJ. Similarly, $I'J'$ is integral over IJ', so that $I'J' \subseteq \overline{IJ'} \subseteq \overline{IJ}$.

Example 1.3.3   Let $R = k[X, Y]$ be a polynomial ring in X and Y over a field k. The ideal $I = (X^2 + Y^3, XY^3, Y^4)$ is integrally closed.

Proof: Observe that I is homogeneous under the weighted grading $\deg(X) = 3$, $\deg(Y) = 2$. The ideal of R consisting of all elements of degree 8 or higher is generated by X^3, X^2Y, XY^3 and Y^4, and is contained in I. Let $r \in \bar{I} \setminus I$ and having components of degree strictly smaller than 8. Since $\bar{I}$ is an ideal, by possibly subtracting elements of $I \subseteq \bar{I}$ from r, without loss of generality $\deg_X(r) \le 1$ and $\deg_Y(r) \le 3$. An integral equation of an element r over I is of the form $r^n + a_1 r^{n-1} + \cdots + a_n = 0$ with a_i in I^i. The lowest degree component in r^n has to cancel with some components in $a_i r^{n-i}$, or in other words, there exists an integer i such that the lowest degree term r_0 appearing in r has degree equal to the degree of a term appearing in a_i divided by i. But a_i is in I^i, so its degree is at least $6i$, so that $\deg(r_0)$ is either 6 or 7, and so the lowest degree term in a_i has to have degree at most $7i$. Necessarily the lowest degree term in a_i is a multiple of the only generator in I of degree strictly smaller than 8, namely of $X^2 + Y^3$. But then $r_0 \in \sqrt{(X^2 + Y^3)} = (X^2 + Y^3)$, which contradicts the assumption that $\deg_X(r) < 2$. Thus $\bar{I} = I$.

The sum of integrally closed ideals need not be integrally closed:

Example 1.3.4   Let k be a field, X and Y variables over k, and $R = k[X, Y]$. Let $\mathfrak{m} = (X, Y)R$. By Example 1.3.3, $I = (X^2 + Y^3) + \mathfrak{m}^4$ is integrally closed. By similar degree arguments or by Exercise 1.18, $J = (Y^3) + \mathfrak{m}^4$ is also integrally closed. However, the sum of these two integrally closed ideals is not integrally closed. Namely, XY^2 is integral over $I + J = (X^2, Y^3) + \mathfrak{m}^4$ as it satisfies the monic polynomial $f(T) = T^3 - X^3Y^6$ with $X^3Y^6 \in (I + J)^3$, but XY^2 is not in $I + J$.

In rare cases the sum of two integrally closed ideals is still integrally closed:

Proposition 1.3.5   *Let I be an integrally closed ideal in a ring R. Let Z be a variable over R. Let $S = R[Z]$. Then $IS + ZS$ is integrally closed.*

Proof: Let $r \in \overline{IS + ZS}$. We have to prove that $r \in IS + ZS$. Without loss of generality no Z appears in r. Then in an equation of integral dependence of r over $IS + ZS$ we collect all terms of Z–degree zero to obtain an equation of integral dependence of r over I, so that as I is integrally closed, $r \in IS$. $\square$

There is a trick to translate a counterexample for $\overline{I + J} = \bar{I} + \bar{J}$ (as above) to a counterexample for $\overline{IJ} = \bar{I} \cdot \bar{J}$, which goes as follows: given ideals I, J in R and $r \in \overline{I + J} \setminus (\bar{I} + \bar{J})$, let Z be a variable over R. Set $S = R[Z]$, $I' = \bar{I}S + ZS$ and $J' = \bar{J}S + ZS$. Then I' and J' are integrally closed by Proposition 1.3.5. However, $I'J' = \bar{I}\bar{J}S + (\bar{I} + \bar{J})ZS + Z^2S$ is not integrally closed as rZ is integral over $(I + J)ZS$ and hence over $I'J'$, yet if rZ were in $I'J'$, it would have to be in $(\bar{I}\bar{J}S + (\bar{I} + \bar{J})ZS) \cap ZS = (\bar{I} + \bar{J})ZS$, which is a contradiction.

1.4. Monomial ideals

Integral closure of monomial ideals is especially simple and illustrative of the theory in general.

Definition 1.4.1 *Let k be a field, $X_1, \ldots, X_d$ variables over k. A **monomial** in the polynomial ring $k[X_1, \ldots, X_d]$ (or alternatively in the convergent power series ring $\mathbb{C}\{X_1, \ldots, X_d\}$ or in the formal power series ring $k[[X_1, \ldots, X_d]]$) is an element of the form $X_1^{n_1} X_2^{n_2} \cdots X_d^{n_d}$ for some non–negative integers $n_1, \ldots, n_d$. An ideal is said to be **monomial** if it is generated by monomials.*

The polynomial ring $k[X_1, \ldots, X_d]$ has a natural $\mathbb{N}^d$ grading as follows: $\deg(X_i) = (0, \ldots, 0, 1, 0, \ldots, 0) \in \mathbb{N}^d$ with 1 in the ith spot and 0 elsewhere. Under this grading, monomial ideals are homogeneous, and in fact monomial ideals are the only homogeneous ideals.

Let I be a monomial ideal and $r = X_1^{n_1} X_2^{n_2} \cdots X_d^{n_d}$ a monomial in the integral closure of $\overline{I}$. Let $r^n + a_1 r^{n-1} + \cdots + a_{n-1} r + a_n = 0$ be an equation of integral dependence of r over I. As I is a monomial ideal, homogeneous under the natural $\mathbb{N}^d$–grading on $k[X_1, \ldots, X_d]$, each graded piece of each a_i is also an element of I^i. In particular, the graded piece of the equation above of degree $n(n_1, \ldots, n_d)$ is an equation of integral dependence of r over I. So if b_i is the homogeneous component of a_i of degree $i(n_1, \ldots, n_d)$, then $r^n + b_1 r^{n-1} + \cdots + b_{n-1} r + b_n = 0$ is also an equation of integral dependence of r over I. Let i be such that $b_i r^{n-i}$ is non–zero. Note that both r^n and $b_i r^{n-i}$ are elements of degree $n(n_1, \ldots, n_d)$. Since the graded component of $k[X_1, \ldots, X_d]$ of degree $n(n_1, \ldots, n_d)$ is a one–dimensional k–vector space, there exists a unit u in k such that $r^n + u b_i r^{n-i} = 0$. By dividing through by r^{n-i} we get an equation of integral dependence of r over I of the form $r^i - c_i = 0$ for some $c_i \in I^i$ that is a product of i monomials in I.

Thus the problem of finding an equation of integral dependence of a monomial r over a monomial ideal I reduces to finding an integer i and monomials $m_1, \ldots, m_i$ in I such that

$$r^i - m_1 \cdots m_i = 0.$$

With this we can prove that the integral closure of a monomial ideal is a monomial ideal (an alternative proof is in Corollary 5.2.3):

Proposition 1.4.2 *The integral closure of a monomial ideal I in a polynomial ring $k[X_1, \ldots, X_d]$ is a monomial ideal.*

Proof: For contradiction suppose that $f \in \overline{I}$ is not a monomial and that no homogeneous component of f is in $\overline{I}$. Write $f = \sum_{l \in \Lambda} f_l$, where Λ is a finite subset of $\mathbb{N}^d$ and f_l is the component of f of degree l. Let $L \in \Lambda$ with $f_L \neq 0$.

First assume that k is algebraically closed. Any ring automorphism φ of $k[X_1, \ldots, X_d]$ maps an integral equation of f over I into an integral equation of $\varphi(f)$ over $\varphi(I)$. In particular, for any units $u_1, \ldots, u_d$ in k, if φ_u is the ring automorphism taking X_i to $u_i X_i$, then $\varphi_u(I) = I$ and $\varphi_u(f)$ is integral over I. Furthermore, each φ_u has the property that f is non–zero in degree $l \in \mathbb{N}^d$ if and only if $\varphi_u(f)$ is non–zero in degree l. As k is algebraically closed and f is not a scalar multiple of a monomial, there exist $u_1, \ldots, u_d \in k$ such that $\varphi_u(f)$ is not a polynomial multiple of f. As both f and $\varphi_u(f)$ are integral over I, so is $g = u_1^{L_1} \cdots u_d^{L_d} f - \varphi_u(f)$. As $\varphi_u(f)$ is not a multiple of f, then g is not zero. Note that the component of g in degree L is 0 and that whenever g is non–zero in degree l, then f is also non–zero in degree l, so that $l \in \Lambda$. Thus g has strictly fewer non–zero homogeneous components than f, so by induction on the number of components, each component of g lies in the integral closure of I. But each component of g is a scalar multiple of a component of f, which proves that some homogeneous components of f are in the integral closure of I, contradicting the choice of f.

Now let k be an arbitrary field and $\bar{k}$ its algebraic closure. By the previous case we know that each monomial appearing in f with a non–zero coefficient is integral over $I\bar{k}[X_1, \ldots, X_d]$. Thus by the derivation prior to this proposition, each monomial r appearing in f satisfies an integral equation of the form $r^i - a_i = 0$ for some a_i that is a product of i monomials of $I\bar{k}[X_1, \ldots, X_d]$. Hence a_i is also a product of i monomials of I, so that r is integral over I. $\square$

Definition 1.4.3 *Let R be the polynomial ring $k[X_1, \ldots, X_d]$. For any monomial $m = X_1^{n_1} X_2^{n_2} \cdots X_d^{n_d}$, its **exponent vector** is $(n_1, \ldots, n_d) \in \mathbb{N}^d$. For any monomial ideal I, the set of all exponent vectors of all the monomials in I is called **the exponent set** of I.*

Example 1.4.4 Let (X^4, XY^2, Y^3) be a monomial ideal in $\mathbb{C}[X, Y]$. Its exponent set consists of all integer lattice points touching or in the shaded gray area below:

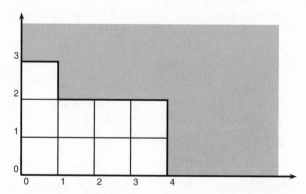

If G is a monomial generating set of I, the exponent set of I consists of all those points of $\mathbb{N}^d$ that are componentwise greater than or equal to the exponent vector of one of the exponent vectors of an element of G. In other words, a monomial m is in a monomial ideal I if and only if m is a multiple of one of the monomial generators of I.

In general, let $k[X_1, \ldots, X_d]$ be a polynomial ring and let I be a monomial ideal. Let its generators be $m_j = X_1^{n_{j1}} X_2^{n_{j2}} \cdots X_d^{n_{jd}}$, $j = 1, \ldots, s$. Let a monomial $r = X_1^{n_1} X_2^{n_2} \cdots X_d^{n_d}$ be integral over I. We have proved that there exist a positive integer i and a product a_i of i monomials of I such that $r^i - a_i = 0$. An arbitrary product a_i of i monomials in I is of the form $bm_1^{k_1} m_2^{k_2} \cdots m_s^{k_s}$, where b is another monomial, and the k_j are non–negative integers which sum up to i. Then $r^i = bm_1^{k_1} m_2^{k_2} \cdots m_s^{k_s}$, and for each coordinate $l = 1, \ldots, d$, $i \cdot n_l \geq \sum_j k_j n_{jl}$, or in other words, $n_l \geq \sum_j \frac{k_j}{i} n_{jl}$.

Thus the problem of finding monomials $r = X_1^{n_1} X_2^{n_2} \cdots X_d^{n_d}$ that are integral over I reduces to finding rational non–negative numbers $c_1, \ldots, c_d$ which add up to 1, such that componentwise
$$(n_1, n_2, \ldots, n_d) \geq \sum_j c_j (n_{j1}, n_{j2}, \ldots, n_{jd}).$$
Conversely, suppose we are given rational non–negative numbers $c_1, \ldots, c_d$ such that $\sum c_j = 1$ and such that the inequality in the last display holds componentwise. Write $c_j = k_j / i$ for some $k_j, i \in \mathbb{N}$, $i \neq 0$. Then $r^i = bm_1^{k_1} \cdots m_d^{k_d}$ for some monomial $b \in k[X_1, \ldots, X_d]$, so that $r \in \bar{I}$. This proves that the problem of finding monomials $r = X_1^{n_1} X_2^{n_2} \cdots X_d^{n_d}$ that are integral over I is equivalent finding rational non–negative rational numbers $c_1, \ldots, c_d$ such that
$$(n_1, n_2, \ldots, n_d) \geq \sum_j c_j (n_{j1}, n_{j2}, \ldots, n_{jd}), \quad \sum_j c_j = 1. \qquad (1.4.5)$$

Geometrically the construction of $(n_1, n_2, \ldots, n_d)$ satisfying the inequality above is the same as finding the integer lattice points in the convex hull of the exponent set of the ideal I. This proves the following:

Proposition 1.4.6 *The exponent set of the integral closure of a monomial ideal I equals all the integer lattice points in the convex hull of the exponent set of I.*

Thus for example the integral closure of (X^4, XY^2, Y^3) can be read off from the convex hull of the exponent set below, proving that $\overline{(X^4, XY^2, Y^3)}$ equals (X^4, X^3Y, XY^2, Y^3):

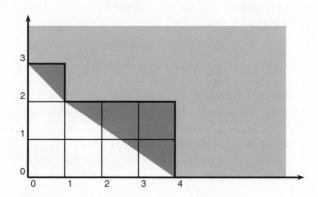

This convex as in the statement of Proposition 1.4.6 has a classical name:

Definition 1.4.7 *For any monomial ideal I in $k[X_1, \ldots, X_d]$, the convex hull in $\mathbb{R}^d$ of the exponent set of I is called the* **Newton polyhedron** *of I.*

Example 1.4.8 Let $I = (X^3, X^2Y, Y^4, Y^2Z, Z^3) \subseteq \mathbb{C}[X, Y, Z]$. The lower boundary of its exponent set is given below. We leave it to the reader to compute the integral closure of I.

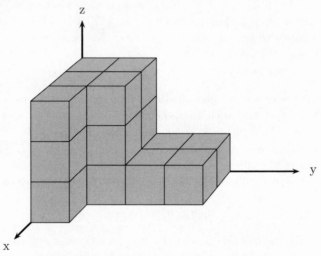

An analysis of the proof above of Proposition 1.4.6 shows even more:

Proposition 1.4.9 *Let I be a monomial ideal in $k[X_1, \ldots, X_d]$. Let N be an upper bound on the degrees of the minimal monomial generators of I. Then the generators of the integral closure of I have degree at most $N + d - 1$.*

Proof: Let $m = X_1^{n_1} \cdots X_d^{n_d}$ be a generator of $\bar{I}$. Then there exist non–negative rational numbers $c_1, \ldots, c_d$ that satisfy equation (1.4.5). Suppose that for some $i \in \{1, \ldots, d\}$, $n_i \geq 1 + \sum_j c_j n_{ji}$. Then the exponent vector of m/X_i also satisfies equation (1.4.5), so that m/X_i is in $\bar{I}$. Hence m is not a generator of $\bar{I}$. This proves that for all i, $n_i < 1 + \sum_j c_j n_{ji}$. Thus the degree of any generator is at most $\sum_{i=1}^d n_i < \sum_{i=1}^d (1 + \sum_j c_j n_{ji}) = d + N$. $\qquad\square$

This makes the computation of the integral closure of monomial ideals feasible; see more in Section 15.4.

We conclude that the integral closure of monomial ideals is special: the equations of integral dependence are simple and the integral dependence relations have a combinatorial/geometric aspect. Nevertheless, the theory of integral closure is sufficiently interesting even for monomial ideals. For example, the power of an integrally closed monomial ideal need not be integrally closed (see Exercises 1.14 and 1.13).

However, if the first few powers of a monomial ideal are integrally closed, then all the powers are integrally closed:

Theorem 1.4.10   (Reid, Roberts and Vitulli [243]) *Let I be a monomial ideal in the polynomial ring $k[X_1, \ldots, X_d]$ such that $I, I^2, \ldots, I^{d-1}$ are integrally closed. Then all the powers of I are integrally closed, i.e., I is normal.*

Proof: Let $n \geq d$. It suffices to prove that I^n is integrally closed under the assumption that $I, I^2, \ldots, I^{n-1}$ are integrally closed. For this it suffices to prove that every monomial integral over I^n lies in I^n. Let $X_1^{a_1} \cdots X_d^{a_d} \in \overline{I^n}$. By the form of the integral equation of a monomial over a monomial ideal there exists an integer r such that $X_1^{a_1 r} \cdots X_d^{a_d r} \in I^{rn}$. By Carathéodory's theorem (Theorem A.2.1), there exist monomial generators $m_1, \ldots, m_d$ of I whose exponent vectors $v_1, \ldots, v_d$, respectively, are linearly independent in $\mathbb{R}$, such that $X_1^{a_1 r} \cdots X_d^{a_d r} \in (m_1, \ldots, m_d)^{rn}$. In other words, there exists non–negative rational numbers $r_1, \ldots, r_d$ such that $\sum_i r_i = n$ and $(a_1, \ldots, a_d) \geq \sum_i r_i v_i$ (componentwise). As $n \geq d$, there exists $j \in \{1, \ldots, d\}$ such that $r_j \geq 1$. Then $(a_1, \ldots, a_d) - v_j \geq \sum_i (r_i - \delta_{ij}) v_i$ says that the monomial corresponding to the exponent vector $(a_1, \ldots, a_d) - v_j$ is integral over I^{n-1}, so that by assumption the monomial corresponding to $(a_1, \ldots, a_d) - v_j$ is in I^{n-1}. Thus $X_1^{a_1} \cdots X_d^{a_d} \in I^{n-1} m_j \subseteq I^n$.   □

In particular, in a polynomial ring in two variables over a field, the power of an integrally closed monomial ideal is integrally closed. (This holds more generally for arbitrary integrally closed ideals in two–dimensional regular rings, by Zariski's theory. See Chapter 14.) A strengthening of the theorem above is in Singla [271] (one needs to test the integral closedness of only the first $l - 1$ powers of I, where l is the analytic spread of I.)

1.5. Integral closure of rings

The integral closure of the ideal R equals R, as by Remark 1.1.3 (1), R is contained in its own integral closure and by the definition, the integral closure of any ideal is contained in R.

Compare this with the definition of the **integral closure of rings**:

Definition 1.5.1 *(A more general definition is in 2.1.1.) If R is a reduced ring with total ring of fractions K, an element $r \in K$ is* **integral** *over the ring R if*

$$r^n + a_1 r^{n-1} + a_2 r^{n-2} + \cdots + a_{n-1} r + a_n = 0$$

for some $a_i \in R$. The set of all elements of K that are integral over R is called the **integral closure** *of R. If R equals its integral closure, then R is called* **integrally closed**.

Beware of the terminology: the integral closure of the ideal R in the ring R is always R, and the integral closure of the ring R may be larger than R.

In Chapters 2 and 4 we will see more on the integral closure of rings and on the connections between the integral closure of ideals and integral closure of rings. Here is a first example, showing that in integrally closed rings every principal ideal is integrally closed (as an ideal).

Proposition 1.5.2 *Let R be a ring, not necessarily Noetherian, and integrally closed in its total ring of fractions. Then for any ideal I and any non–zerodivisor f in R, $\overline{fI} = f \cdot \overline{I}$. In particular, every principal ideal generated by a non–zerodivisor in R is integrally closed.*

As a partial converse, suppose that R is an arbitrary Noetherian ring such that no prime ideal is both minimal and maximal and assume that every principal ideal of height one is integrally closed. Then R is reduced and integrally closed in its total ring of fractions.

Proof: Assume that r is integral over fI. Then each equation of integral dependence of r over fI is of the form $r^n + b_1 f r^{n-1} + \cdots + b_n f^n = 0$ for some $b_i \in I^i$. Dividing through by f^n yields the following equation of integral dependence of the element $\frac{r}{f}$ over the ring R: $(\frac{r}{f})^n + b_1(\frac{r}{f})^{n-1} + \cdots + b_{n-1}(\frac{r}{f}) + b_n = 0$. As $\frac{r}{f}$ lies in the total ring of fractions of R and R is integrally closed, $\frac{r}{f}$ is an element of R, so that $r \in fR$. Furthermore, from the equation, $r/f \in \overline{I}$, so $r \in f \cdot \overline{I}$, proving that $\overline{fI} \subseteq f \cdot \overline{I}$.

Let $r \in f \cdot \overline{I}$ and write $r = sf$ for some $s \in \overline{I}$. Multiplying an equation of integral dependence of s over I of degree n through by f^n yields an equation of integral dependence of $sf = r$ over fI, so that $f \cdot \overline{I} \subseteq \overline{fI}$.

We prove the converse statement. Every integrally closed ideal contains the nilradical N. For every $P \in \operatorname{Min} R$ by assumption there exists a prime ideal Q_P properly containing P. Choose an element $c_P \in Q_P$ that is not any minimal prime ideal. This is possible by Prime Avoidance. The product c of all c_P is not in any minimal prime ideal. Then $N \subseteq \cap_i \overline{(c^i)} = \cap_i (c^i)$. By the Krull Intersection Theorem, $N(1 - rc) = 0$ for some $r \in R$. If $b = 1 - rc$ is contained in a minimal prime ideal P, it is also contained in Q_P, hence as $c \in Q_P$, necessarily $1 \in Q_P$, which gives a contradiction. Thus $b = 1 - rc$ is not contained in any minimal prime ideal P. The same argument as for c shows that $N(1 - sb) = 0$ for some s. Hence $N = N(1 - s(1 - rc)) = N(1 - sb) = 0$. Thus R is reduced. Let $\frac{a}{b}$ be in the total ring of fractions of R ($a, b \in R$, b a non–zerodivisor) and assume that a/b is integral over R. Then a is integral over bR, so by assumption $a \in bR$, and it follows that $\frac{a}{b} \in R$. $\square$

1.6. How integral closure arises

There are many contexts in which integral closure and integral dependence of ideals arise naturally. This section and the following one illustrate this. One of the illustrations is Ratliff's theorem, whose proof we only give in Chapter 5 (Theorem 5.4.1). We prove all the other results in this section.

Proposition 1.6.1 *Let $R \subseteq S$ be an integral extension of rings. Let I be an ideal in R. Then $\overline{IS} \cap R = \overline{I}$.*

Proof: By persistence $\overline{I}$ is contained in $\overline{IS} \cap R$. Let $r \in \overline{IS} \cap R$. This means that r satisfies an equation of integral dependence over IS and thus there exists a finitely generated R–algebra $T \subseteq S$ and a finitely generated ideal $J \subseteq I$ in R such that r satisfies an equation of integral dependence over JT. It suffices to prove that $\overline{JT} \cap R \subseteq \overline{J}$. Thus by replacing S by T and I by J we may assume that I is a finitely generated ideal and S a module–finite extension of R.

Then $r \in \overline{IS} \cap R$ means that there exists an integer n such that $I(I + (r))^n S = (I+(r))^{n+1} S$. Let M be the finitely generated R–module $(I+(r))^n S$. Then $rM \subseteq IM$, and if $aM = 0$ for some $a \in R$, then $(ar)^n = 0$. Thus by Proposition 1.1.8, $r \in \overline{I}$. $\qquad\square$

This proposition says that integral closure extends and contracts from integral extensions. The same holds for faithfully flat extensions:

Proposition 1.6.2 *Let R be a ring and S a faithfully flat R–algebra. For any ideal I of R, $\overline{IS} \cap R = \overline{I}$.*

In particular, if R is Noetherian local with maximal ideal $\mathfrak{m}$ and $\widehat{R}$ is its $\mathfrak{m}$–adic completion, then for any ideal I in R, $\overline{I\widehat{R}} \cap R = \overline{I}$.

Proof: By the persistence property of integral closure, $\overline{I} \subseteq \overline{IS} \cap R$. Now let $r \in \overline{IS} \cap R$. By Proposition 1.1.7 there exists an integer n such that $r^{n+1} \in I(I + (r))^n S$. Thus $r^{n+1} \in I(I + (r))^n S \cap R = I(I + (r))^n$, so that again by Proposition 1.1.7, $r \in \overline{I}$. $\qquad\square$

Discussion 1.6.3 Notice that by combining Propositions 1.6.2, 1.6.1, 1.1.4, and 1.1.5, it follows that the integral closure of an ideal I is completely determined by the integral closure of the images of I in the following rings: in rings which arise by localizing R at prime ideals containing I, those obtained by completing, those obtained by going modulo minimal prime ideals, and those obtained by passing to integral extensions of the rings. In other words, $\overline{I}$ is determined by the integral closure of the image of I in (some) complete local integrally closed domains. This is not only an important theoretical reduction, but it also tells us what difficulties we can expect in dealing with integral closures. The transfer of properties of R to properties of the completion of R, to going modulo minimal prime ideals and passing to integral closures will all be important, and are manifested in various critical definitions and assumptions which appear in this book, including those of analytically unramified and formally equidimensional rings.

Proposition 1.6.4 *Let R be an $\mathbb{N}$–graded ring, generated over R_0 by R_1. Assume that R_0 is reduced. Let $F_1, \ldots, F_m$ be homogeneous elements of degree 1 in R. If $\sqrt{(F_1, \ldots, F_m)} = R_1 R$, then $\overline{(F_1, \ldots, F_m)} = R_1 R$.*

Proof: By assumption there exists n such that $R_1^n \subseteq (F_1, \ldots, F_m)$. By homogeneity, $R_1^n R \subseteq R_1^{n-1}(F_1, \ldots, F_m)$. This proves that $(F_1, \ldots, F_m)$ is a reduction of $R_1 R$, and then by Corollary 1.2.5 that $R_1 R \subseteq \overline{(F_1, \ldots, F_m)}$. But $R_1 R$ is a radical ideal, so $R_1 R \subseteq \overline{(F_1, \ldots, F_m)} \subseteq \sqrt{(F_1, \ldots, F_m)} \subseteq R_1 R$, which finishes the proof. $\qquad\square$

Socle elements modulo an ideal are sometimes integral over the ideal:

Proposition 1.6.5 (Burch [33]) *Let $(R, \mathfrak{m})$ be a Noetherian local ring that is not regular, i.e., $\mu(\mathfrak{m}) > \dim R$, and let I be an ideal of finite projective dimension. Then $\mathfrak{m}(I : \mathfrak{m}) = \mathfrak{m}I$ and $I : \mathfrak{m}$ is integral over I.*

Proof: Without loss of generality $I \neq R$. Let $\mathbb{F}$ be a minimal free resolution of $R/\mathfrak{m}$:

$$\cdots \to F_{i+3} \xrightarrow{\delta_{i+3}} F_{i+2} \xrightarrow{\delta_{i+2}} F_{i+1} \xrightarrow{\delta_{i+1}} F_i \to \cdots \to F_0 \to 0.$$

Let $(\_)'$ denote images after tensoring with R/I over R. Let d be the dimension of R, and consider the following part:

$$F'_{d+3} \xrightarrow{\delta'_{d+3}} F'_{d+2} \xrightarrow{\delta'_{d+2}} F'_{d+1} \xrightarrow{\delta'_{d+1}} F'_d.$$

As R is not regular, these modules are non–zero, but the homologies are $\mathrm{Tor}^R_{d+2}(R/\mathfrak{m}, R/I) = \mathrm{Tor}^R_{d+1}(R/\mathfrak{m}, R/I) = 0$ as R/I has finite projective dimension. Let $r \in I : \mathfrak{m}$. Then as rF'_{d+1} maps to zero under δ'_{d+1}, there exists $v \in F_{d+2}$ such that $\delta'_{d+2}(v') = (r', \ldots, r')$. Then there exists $a \in IF_{d+1}$ such that $\delta_{d+2}(v) = (r, \ldots, r) + a$. Similarly, as $\delta'_{d+2}(\mathfrak{m}v') = 0$, for every $x \in \mathfrak{m}$ there exists $w \in F_{d+3}$ such that $\delta'_{d+3}(w') = xv'$. Then $xv - \delta_{d+3}(w) \in IF_{d+2}$, so that

$$x\left((r, \ldots, r) + a\right) = x\delta_{d+2}(v) = \delta_{d+2}(xv)$$
$$= \delta_{d+2}(xv - \delta_{d+3}(w)) \in \delta_{d+2}(IF_{d+2}) \subseteq IF_{d+1}.$$

Thus $x(r, \ldots, r) \in xaR + I\delta_{d+2}(F_{d+2}) \subseteq xIF_{d+1} + I\delta_{d+2}(F_{d+2})$. From this we read off that that $xr \in \mathfrak{m}I$, whence $\mathfrak{m}(I : \mathfrak{m}) \subseteq \mathfrak{m}I$, as was to be proved.

The last statement follows from Corollary 1.1.8: $M = \mathfrak{m}$ and if $aM = 0$ for some $a \in R$, then $ar = 0$ for all $r \in I : \mathfrak{m} \subseteq \mathfrak{m}$. $\qquad\square$

It is not always true that socle elements are integral over an ideal. For example, let $R = k[[X]]$ be the power series ring in one variable X over a field k. Then with $I = X^2 R$, $I : \mathfrak{m} = (X)$ is not integral over I. Thus the non–regular assumption is necessary in the previous proposition. Also, the finite projective dimension assumption is needed: let R be the non–regular ring $k[[X^2, X^3]]$, $I = (X^3, X^4)$, and $\mathfrak{m} = (X^2, X^3)$. Then $\mathfrak{m} = I : \mathfrak{m}$ but X^2 is not integral over I (by degree count). For a closely related idea, see Goto [94]. See Choi [37] for an interesting application of integral closure to the growth of Betti numbers.

Here is a final example of how integral closure of ideals arises naturally (see Theorem 5.4.1 for a proof).

Theorem 1.6.6 (Ratliff [228]) *Let R be a locally formally equidimensional Noetherian ring and let $(x_1, \ldots, x_n)$ be a parameter ideal, i.e., the height of $(x_1, \ldots, x_n)$ is at least n. For all $m \geq 1$,*

$$(x_1, \ldots, x_{n-1})^m : x_n \subseteq \overline{(x_1, \ldots, x_{n-1})^m} : x_n = \overline{(x_1, \ldots, x_{n-1})^m}.$$

In particular, for all $m \geq 1$, the integral closure of $(x_1, \ldots, x_n)^m$ has no embedded associated prime ideals. This result is proved in Theorems 5.4.1 and 5.4.5 below.

1.7. Dedekind–Mertens formula

Another example of how integral closure arises is taken from a classical formula of Dedekind and Mertens. The treatment here is as in Heinzer and Huneke [111]. One of the corollaries of this formula is that the product of an n–generated ideal with an m generated ideal is always integral over an ideal generated by at most $n + m - 1$ elements. As we will see in Chapter 8, it is desirable and useful to find ideals with the same integral closure and fewer generators.

Definition 1.7.1 *Let R be a commutative ring and let t be a variable over R. The **content** $c(f)$ of a polynomial $f \in R[t]$ is the ideal of R generated by the coefficients of f.*

The classical Dedekind–Mertens Lemma states that if $f = a_0 + a_1 t + \cdots + a_m t^m$ and $g = b_0 + b_1 t + \cdots + b_n t^n$ are polynomials in $R[t]$, then

$$c(f)^n c(f) c(g) = c(f)^n c(fg).$$

See Theorem 1.7.3 below for a stronger statement and a proof.

Since $c(f)c(g) \supseteq c(fg)$, the right–hand side of the above equation is always included in the left–hand side. Provided that $c(f)$ contains a non–zerodivisor, this equation together with Corollary 1.1.8 shows that $c(f)c(g)$ is integral over $c(fg)$. However, since the equation holds in the generic situation in which the coefficients of f and g are allowed to be variables and in this case $c(f)$ obviously contains a non–zerodivisor, it follows that $c(f)c(g)$ is always integral over $c(fg)$, providing a type of "generic" equations for integral closure. For example, in the simple case in which $f = a_0 + a_1 t$ and $g = b_0 + b_1 t$, the product of the contents is generated by the product pairs $a_i b_j$, while the content of the product is generated by $a_0 b_0, a_0 b_1 + a_1 b_0, a_1 b_1$, and an equation of integral dependence of $a_1 b_0$ over the latter ideal is given by $X^2 - (a_0 b_1 + a_1 b_0)X + (a_0 b_0)(a_1 b_1) = 0$.

The fact that $c(f)c(g)$ is always integral over $c(fg)$ can be seen as a generalization of Gauss's Lemma. Gauss's Lemma is stated in many different forms, but one statement of it says that the product of primitive polynomials is primitive, where a polynomial F is **primitive** if $c(F) = 1$. If both f and g

are primitive, then $c(f)c(g)$ is the unit ideal, and the fact that it is integral over $c(fg)$ immediately implies that $c(fg)$ is also the unit ideal, giving Gauss's Lemma in this special case.

Note that in the display above, the exponent n is exactly the degree of g and is independent of f. The exponent can even be smaller:

Definition 1.7.2 *The* **Dedekind–Mertens number** *of a polynomial $g \in R[t]$ is the smallest positive integer k such that*

$$c(f)^{k-1}c(f)c(g) = c(f)^{k-1}c(fg)$$

for every polynomial $f \in R[t]$.

The Dedekind–Mertens number of a polynomial $g(t)$ very much depends upon the coefficient ring R. It is not invariant under base change.

The classical Dedekind–Mertens Lemma says that the Dedekind–Mertens number of g is at most $\deg(g) + 1$, the maximal number of coefficients of g. The next theorem sharpens this inequality:

Theorem 1.7.3 (Heinzer and Huneke [111, Theorem 2.1]) *Let R be a commutative ring, let $g \in R[t]$ be a polynomial, and let $c(g)$ denote the content ideal of g. If for each maximal ideal $\mathfrak{m}$ of R, $c(g)R_\mathfrak{m}$ is generated in $R_\mathfrak{m}$ by k elements, then the Dedekind–Mertens number of g is at most k.*

Proof: Since $c(f)^n c(f)c(g) = c(f)^n c(fg)$ holds in R if and only if for each maximal ideal $\mathfrak{m}$ of R, $c(f)^n c(f)c(g)R_\mathfrak{m} = c(f)^n c(fg)R_\mathfrak{m}$, the Dedekind–Mertens number of g is the maximum of the Dedekind–Mertens numbers of the images of g in $R_\mathfrak{m}[t]$ as $\mathfrak{m}$ varies over the maximal ideals of R. Thus we may assume that the ring R is local (but not necessarily Noetherian), and it suffices to prove the following:

Theorem 1.7.4 *Let $(R, \mathfrak{m})$ be a local ring and let $g \in R[t]$ be a polynomial. If the content ideal $c(g)$ of g is minimally generated by k elements, then the Dedekind–Mertens number of g is at most k, i.e., for every polynomial $f \in R[t]$ we have*

$$c(f)^{k-1}c(f)c(g) \subseteq c(f)^{k-1}c(fg).$$

We prove Theorem 1.7.4 by induction on k. The case $k = 1$ follows by factoring out the principal content of g and using the lemma of Gauss (see Exercise 1.7) that says that a polynomial g with unit content is Gaussian, i.e., $c(fg) = c(f)$ for all f. To continue the induction, we use the following lemma which implies that we may assume that every non–zero coefficient of g is a minimal generator of $c(g)$.

Lemma 1.7.5 *Let $(R, \mathfrak{m})$ be a local ring and let $g \in R[t]$ be a polynomial. Suppose $b \in R$ is such that $b \in \mathfrak{m}c(g)$. Let i be a non–negative integer and set $h = g + bt^i$. Let J be a finitely generated ideal of R and $f \in R[t]$ a polynomial. If $Jc(f)c(h) = Jc(fh)$, then also $Jc(f)c(g) = Jc(fg)$. Therefore g and h have the same Dedekind–Mertens number. More generally, if g^* and h^* are* polynomials in $R[t]$ and if $c(g^* - h^*) \subseteq \mathfrak{m}c(g^*)$, then g^* and h^* have the same Dedekind–Mertens number.

Proof: It is clear that $Jc(f)c(g) \supseteq Jc(fg)$. For the reverse inclusion, since $b \in \mathfrak{m}c(g)$ and $h = g + bt^i$, we have $c(g) = c(h)$. Thus

$$Jc(f)c(g) = Jc(f)c(h) = Jc(fh)$$
$$= Jc(f(g + bt^i)) \subseteq J(c(fg) + bc(f)) \subseteq Jc(fg) + \mathfrak{m}Jc(f)c(g).$$

Nakayama's Lemma implies $Jc(f)c(g) = Jc(fg)$. Letting $J = c(f)^k$, we see that the Dedekind–Mertens number of g is at most the Dedekind–Mertens number of h. Since $g = h + (-b)t^i$, we also have the opposite inequality. If $c(g^* - h^*) \subseteq \mathfrak{m}c(g^*)$, then h^* is obtained from g^* by a finite sequence of operations $h_j = g_j + b_j t^{i_j}$, where $b_j \in \mathfrak{m}c(g_j) = \mathfrak{m}c(g^*)$. Therefore g^* and h^* have the same Dedekind–Mertens number. $\qquad\square$

We now continue our proof of Theorem 1.7.4. Assume that $c(g)$ is minimally generated by $k \geq 2$ elements and that for every polynomial $h \in R[t]$ whose content is minimally generated by fewer than k elements, we have for every polynomial $f \in R[t]$ that

$$c(f)^{k-2}c(f)c(h) \subseteq c(f)^{k-2}c(fh).$$

Let $g = b_m t^m + \cdots + b_1 t + b_0$. By Lemma 1.7.5 we may assume that b_m is a minimal generator of $c(g)$. Write $g = b_m h(t) + g_1(t)$, where $c(h) = R$ and $c(g_1)$ is generated by fewer than k elements. Also write $f(t) = a_n t^n + f_1(t)$, where $\deg(f_1) < \deg(f) = n$. By induction on $\deg(f)$, we may assume $c(f_1)^k c(g) = c(f_1)^{k-1} c(f_1 g)$.

We claim that $c(fg_1) \subseteq c(fg) + b_m c(f_1)$. To prove this claim, note that $c(fg_1) = c(f(g - b_m h)) \subseteq c(fg) + c(b_m fh) = c(fg) + b_m c(fh)$. As $c(h) = R$, the last ideal equals $c(fg) + b_m c(f) = c(fg) + b_m c(a_n t^n + f_1(t)) \subseteq c(fg) + a_n b_m R + b_m c(f_1)$. Since $a_n b_m \in c(fg)$, the last ideal equals $c(fg) + b_m c(f_1)$. In summary, $c(fg_1) \subseteq c(fg) + b_m c(f_1)$.

Claim: $c(f_1 g) \subseteq c(fg) + a_n c(g_1)$. We have $c(f_1 g) = c((f - a_n t^n)g) \subseteq c(fg) + a_n c(t^n g) \subseteq c(fg) + a_n c(g) \subseteq c(fg) + a_n c(b_m h(t) + g_1(t)) \subseteq c(fg) + a_n b_m R + a_n c(g_1) = c(fg) + a_n c(g_1)$, the last equality since $a_n b_m \in c(fg)$.

It suffices to show each term in $c(f)^{k-1}c(f)c(g) = c(f)^k c(g)$ of the form $\theta = a_0^{v_0} \cdots a_n^{v_n} b_j$, where $\sum v_i = k$, is in $c(f)^{k-1}c(fg)$. Since $g = b_m h(t) + g_1(t)$, we can write $b_j = b_m e_j + b_{1j}$, where e_j is the coefficient of t^j in $h(t)$ and b_{1j} is the coefficient of t^j in $g_1(t)$.

Consider the following cases:

Case 1: Suppose that $v_n \neq 0$ and $j = m$. Then $\theta = a_0^{v_0} \cdots a_n^{v_n - 1} a_n b_m \in c(f)^{k-1}c(fg)$.

Case 2: Suppose that $v_n \neq 0$ and $j < m$. Then $\theta = a_0^{v_0} \cdots a_n^{v_n} b_j = a_0^{v_0} \cdots a_n^{v_n}(b_m e_j + b_{1j}) = a_0^{v_0} \cdots a_n^{v_n - 1} a_n b_m e_j + a_0^{v_0} \cdots a_n^{v_n - 1} a_n b_{1j}$, which lies in $c(f)^{k-1}c(fg) + c(f)^{k-1}a_n c(g_1)$.

Case 3: Suppose that $v_n = 0$. Then $\theta \in c(f_1)^k c(g) = c(f_1)^{k-1} c(f_1 g)$ by induction on the degree of f.

Combining these three cases, we have

$$c(f)^k c(g) \subseteq c(f)^{k-1} c(fg) + c(f)^{k-1} a_n c(g_1) + c(f_1)^{k-1} c(f_1 g)$$
$$\subseteq c(f)^{k-1} c(fg) + c(f)^{k-1} a_n c(g_1) + c(f_1)^{k-1} (c(fg) + a_n c(g_1))$$
$$\subseteq c(f)^{k-1} c(fg) + c(f)^{k-1} a_n c(g_1)$$

(since $c(f_1) \subseteq c(f)$). As $c(g_1)$ is generated by fewer than k elements, we have $c(f)^{k-1} c(g_1) = c(f)^{k-2} c(fg_1)$ by induction on k. Therefore

$$c(f)^k c(g) \subseteq c(f)^{k-1} c(fg) + a_n c(f)^{k-2} c(fg_1)$$
$$\subseteq c(f)^{k-1} c(fg) + a_n c(f)^{k-2} (c(fg) + b_m c(f_1))$$
$$\subseteq c(f)^{k-1} c(fg). \qquad \qquad \square$$

Corollary 1.7.6 *Let R be a commutative ring and let I' and J' be ideals of R. Suppose that I' is integral over an ideal $I = (a_0, \ldots, a_{n-1})$ generated by n elements and that J' is integral over an ideal $J = (b_0, \ldots, b_{m-1})$ generated by m elements. Then $I'J'$ is integral over an ideal generated by at most $n + m - 1$ elements.*

Proof: Let t be a variable, and set $f(t) = a_0 + a_1 t + \cdots + a_{n-1} t^{n-1}$, and $g(t) = b_0 + b_1 t + \cdots + b_{m-1} t^{m-1}$. Since $I'J'$ is integral over IJ, it suffices to prove that IJ is integral over an ideal generated by at most $n + m - 1$ elements. By Theorem 1.7.3, $IJ = c(f)c(g)$ is integral over $c(fg)$. Since fg has degree at most $n + m - 2$, the corollary follows. $\qquad \square$

1.8. Exercises

1.1   Let I and J be ideals in a Noetherian ring R. Prove that $\overline{I\overline{J}} = \overline{IJ}$.

1.2   (Cancellation theorem) Let I, J and K be ideals in a Noetherian ring R, I not consisting of zero divisors. Assume that $\overline{IJ} = \overline{IK}$. Prove that $\overline{J} = \overline{K}$. More generally prove that if the height of I is positive and $\overline{IJ} = \overline{IK}$ then $\overline{J} = \overline{K}$.

1.3   Let R be a Noetherian ring, P a prime ideal and q a P–primary ideal. Prove that q is integrally closed if and only if qR_P is.

1.4   Let R be a ring, I an integrally closed ideal and J an arbitrary ideal. Prove that $I : J$ is integrally closed.

1.5   Let R be an integral domain and I an ideal in R. An element $r \in R$ is said to be **almost integral** over I if there exists a non–zero element $c \in R$ such that for all $n \geq 1$, $cr^n \in I^n$. (Cf. Exercise 2.26 in Chapter 2 for the analogous notion for rings in place of ideals.)

(i) Prove that every element of $\overline{I}$ is almost integral over I.

(ii) Prove that the set of all elements in R that are almost integral over I forms an ideal. This ideal is called the **complete integral closure** of I.

1.6   Let R be a polynomial ring in d variables over a field. Let I be a monomial ideal in R. Prove that for all $n \geq d$, $\overline{I^n} = I\overline{I^{n-1}}$.

1.7   (Gauss's Lemma) Let R be a Noetherian ring, X a variable over R, and $f, g \in R[X]$ such that the content $c(g)$ of g is locally generated by one element. Prove that $c(fg) = c(f)c(g)$.

1.8   Let k be a field, X and Y variables over k, and R a direct product or a direct sum of countably infinitely many copies of $k[X, Y]$. Let K be the ideal whose component in the ith piece of R is (X^i, Y^i), J the ideal whose ith component is (X^i, Y^i, XY^{i-1}), and I the ideal whose ith component is $(X, Y)^i$. Certainly $K \subseteq J \subseteq I$. Prove that K is a reduction of I but that K is not a reduction of J.

1.9   (Corso, Huneke, Katz and Vasconcelos [44, Corollary 3.3]) Let $(R, \mathfrak{m})$ be a Noetherian local ring, I an integrally closed $\mathfrak{m}$–primary ideal and M a finitely generated R–module. Prove that if $\mathrm{Tor}_k^R(R/I, M) = 0$, then the projective dimension of M is strictly smaller than k. (Hint: generalize the proof of Proposition 1.6.5.)

1.10   ([44, Corollary 2.5]) Let $(R, \mathfrak{m})$ be a Noetherian local ring and I an $\mathfrak{m}$–primary integrally closed ideal. Let $H_1(I)$ be the first Koszul homology on a system of generators of I. Prove that $\mathrm{ann}\, H_1(I) = I$.

Hint for the exercises below: as in Example 1.3.3, use (various) gradings on the rings to compute the integral closure of ideals.

1.11   Let k be a field, X, Y variables over k, $R = k[X, Y]$, $I = (X^2, Y^2)R$, $J = (X, Y)R$. Prove that $\overline{(I : J)} \neq \overline{I} : J$.

1.12   Let $I = (X^2 + Y^3, XY^3, Y^4) \subseteq k[X, Y]$ from Example 1.3.4. Prove that I is not an intersection of integrally closed irreducible ideals.

1.13   (Faridi) Let R be the polynomial ring $k[X, Y, Z]$, and let X, Y, Z have weights 12, 15, and 20, respectively. Consider the ideal $I = R_{\geq 60}$ (note that 60 here is the least common multiple of the three weights). Prove that I is integrally closed but that I^2 is not.

1.14   (Jockusch and Swanson, unpublished) Let $k[X, Y, Z]$ be the polynomial ring over a field k. Let $I = (X^2, Y^3, Z^7)$.
   (i) Prove that $(\overline{I})^2$ is not integrally closed.
   (ii) Prove that $\overline{I^3} \neq (X^2, Y^3, Z^7)(\overline{I})^2$.

1.15   Show that for $I = (X^{12}, Y^5 Z^7)$ in the polynomial ring $k[X, Y, Z]$, $(X, Y, Z)\overline{I} \neq \overline{(X, Y, Z)I}$.

1.16   Let $k[X_1, \ldots, X_d]$ be the polynomial ring in $X_1, \ldots, X_d$ over a field k, and let $F_1, \ldots, F_m$ be forms of degree n. Assume that $(X_1, \ldots, X_d) = \sqrt{(F_1, \ldots, F_m)}$. Prove that $\overline{(F_1, \ldots, F_m)} = (X_1, \ldots, X_d)^n$.

1.17   (Huckaba and Huneke [128]) Let R be an $\mathbb{N}$–graded Noetherian ring such that R_0 is a field. Let $\mathfrak{m}$ be the homogeneous maximal ideal of R, and I an ideal in R generated by homogeneous elements of the same degree d. Suppose that I is integrally closed (resp. normal). Prove that $I + \mathfrak{m}^{d+1}$ is integrally closed (resp. normal).

1.18 Let $(R, \mathfrak{m})$ be a regular local ring, i.e., $\mu(\mathfrak{m}) = \dim R$. Prove that for any $n \in \mathbb{N}$, the ideal $fR + \mathfrak{m}^n$ is integrally closed if $f \in \mathfrak{m}^{n-1}$ or if $f \in \mathfrak{m} \setminus \mathfrak{m}^2$.

1.19 Let R be a polynomial ring over a regular local ring. Let $\mathfrak{m}$ be the homogeneous maximal ideal of R. Prove that $fR + \mathfrak{m}^n$ is integrally closed for any homogeneous $f \in \mathfrak{m} \setminus \mathfrak{m}^2$ and for any $f \in \mathfrak{m}^{n-1}$.

1.20 Give counterexamples to each of the following:
   (i) $\overline{I + J} = \overline{I} + \overline{J}$.
   (ii) $\overline{IJ} = \overline{I} \cdot \overline{J}$.
   (iii) If $\overline{I} = \overline{J}$, then $I = J$.
   (iv) For any ideal $J \subseteq I$, $\overline{I}/J = \overline{I/J}$ in the ring R/J.

1.21 (Jarrah [150]) Let $I = (X^n Y^n, X^n Z^n, Y^n Z^n) \subseteq k[X, Y, Z]$. (Here k is a field, X, Y, Z variables over k.) Prove that I has no embedded prime ideals but that $\overline{I}$ has embedded prime ideals.

1.22 (Huneke) Let k be a field of characteristic 2, X, Y, Z variables over k, $R = k[X, Y, Z]$, and $I = (X^6 - Y^2 Z^2, Y^4 - X^2 Z^2, Z^4 - X^4 Y^2)$. (One can verify that I is the Frobenius power of the kernel of the natural map $k[X, Y, Z] \to k[t^3, t^4, t^5]$.) Prove that I has no embedded prime ideals but that $\overline{I}$ has embedded prime ideals. (This exercise is computationally challenging.)

2
Integral closure of rings

Integral closure of ideals is intricately connected to, and to a large extent depends on, the notion of the integral closure of rings. For example, the integral closure of ideals can be characterized via integrally closed rings, such as valuation rings. (See Proposition 6.8.2.) In this chapter we present the basic background on the integral closure of rings that is needed in the rest of this book.

The notion of the integral closure of a ring R in an overring S is analogous to the notion of the algebraic closure of a field in an overfield. In fact, the algebraic closure of a field is a special case of the integral closure of rings. Under the operations of integral or algebraic closures, the ring is enlarged to a larger one in which many more equations have solutions. This property is perhaps the most fundamental one of integral closure.

2.1. Basic facts

The basic ingredient in the theory of integral closures, as well as in the theory of algebraic closures, are monic polynomials and their zeros:

Definition 2.1.1 *Let R be a ring and S an R–algebra containing R. An element $x \in S$ is said to be **integral over R** if there exists an integer n and elements $r_1, \ldots, r_n$ in R such that*

$$x^n + r_1 x^{n-1} + \cdots + r_{n-1} x + r_n = 0.$$

This equation is called an **equation of integral dependence of x over R** *(of degree n).*

Equations of integral dependence are not unique, not even if their degrees are minimal possible. For example, let S be the ring $\mathbb{Z}[t]/(t^2 - t^3)$, where t is a variable over $\mathbb{Z}$. Let R be the subring of S generated over $\mathbb{Z}$ by t^2. Then $t \in S$ is integral over R and it satisfies two distinct quadratic equations $x^2 - t^2 = 0 = x^2 - xt^2$ in x. As t is not in R, there can not be equations of integral dependence of degree 1, which shows that equations of integral dependence of minimal degree need not be unique.

However, there are many cases in which equations of minimal degree are unique (see comment after Theorem 2.1.17). Over integral domains, any equation $x^n + r_{n-1} x^{n-1} + \cdots + r_n = 0$ of integral dependence of a non–zero element x, if it is of minimal degree, then r_n is non–zero, for otherwise $x(x^{n-1} + r_1 x^{n-2} + \cdots + r_{n-1}) = 0$, so that necessarily $x^{n-1} + r_1 x^{n-2} + \cdots + r_{n-1} = 0$, which is then an equation of integral dependence of x of strictly smaller degree, contradicting the assumption.

The integral property is preserved under some standard ring operations:

Proposition 2.1.2 *Let $R \subseteq S$ be an extension of rings, and let x be an element of S that is integral over R. Then*

(1) For any R–algebra T, $x \otimes 1 \in S \otimes_R T$ is integral over the image of $T = R \otimes_R T$ in $S \otimes_R T$.

(2) For any ideal I in S, $x + I \in S/I$ is integral over $R/(I \cap R)$.

(3) For any multiplicatively closed subset W of R, $\frac{x}{1} \in W^{-1}S$ is integral over $W^{-1}R$.

Proof: This is straightforward by the observation that an equation of integral dependence of x over R gives an equation of integral dependence under tensor products, quotients, and localization. $\square$

In contrast to the proposition above, the notion of integral non–dependence is not preserved under tensoring, passing to quotients, or localization. For example, $\frac{1}{2}$ is not integral over $\mathbb{Z}$, but is integral over $\mathbb{Z}$ after inverting 2. However, the following is easy to prove:

Proposition 2.1.3 *Let $R \subseteq S$ be an extension of rings, $x \in S$. Then the following are equivalent:*

(1) x is integral over R.

(2) For all multiplicatively closed subsets W of R, x is integral over $W^{-1}R$.

(3) For all prime ideals P of R, x is integral over R_P.

(4) For all maximal ideals M of R, x is integral over R_M.

Definition 2.1.4 *Let $R \subseteq S$ be an inclusion of rings. The set of all elements of S that are integral over R is called **the integral closure of R in S**. If every element of S is integral over R, we say that S is **integral over R**.*

*When S is the total ring of fractions of a reduced ring R, the integral closure of R in S is also called **the integral closure of R**. A reduced ring R is said to be **integrally closed** if the integral closure of R equals R.*

We start with a basic and omnipresent example of integrally closed rings:

Proposition 2.1.5 *A unique factorization domain R is integrally closed.*

Proof: Let $a, b \in R$, with a/b in the field of fractions integral over R. By possibly dividing a and b we may assume that a and b have no non–unit factors in common. Let $x^n + r_1 x^{n-1} + \cdots + r_n = 0$ be an equation of integral dependence of a/b over R. Then $a^n + r_1 b a^{n-1} + \cdots + r_n b^n = 0$, so that $a^n \in (b)$. But by unique factorization, b must be a unit, so that $a/b \in R$. $\square$

In particular, the ring of integers $\mathbb{Z}$, polynomial rings over fields (in arbitrary number of variables), power series rings, regular rings, etc., are all integrally closed.

Here is an example of a ring that is not integrally closed. Let k be a field, t a variable, and set $R = k[t^2, t^3]$. This is the subring of the polynomial ring

$k[t]$ consisting of polynomials without the linear term. As $t = \frac{t^3}{t^2}$ is in the field of fractions but not in R, and as t satisfies the integral equation $x^2 - t^2 = 0$, R is not integrally closed. It is not difficult to see that the integral closure of $k[t^2, t^3]$ is $k[t]$.

The following is proved easily.

Proposition 2.1.6 *Let $R \subseteq S \subseteq T$ be an extension of rings. The following are equivalent:*

(1) S is the integral closure of R in T.

(2) S is the integral closure of R in T after localizing at every multiplicatively closed subset of R.

(3) S is the integral closure of R in T after localizing at the complement of every prime ideal in R.

(4) S is the integral closure of R in T after localizing at the complement of every maximal ideal in R.

Integral dependence of integral domains and fields is a special case:

Lemma 2.1.7 *If $R \subseteq S$ is an integral extension of integral domains, then R is a field if and only if S is a field. Thus $Q \in \operatorname{Spec} S$ is maximal in S if and only if $Q \cap R$ is maximal in R.*

Proof: Assume that R is a field. Let x be a non–zero element of S. Then for some $r_i \in R$, $x^n + r_1 x^{n-1} + \cdots + r_{n-1} x + r_n = 0$. As R is an integral domain, we may assume that r_n is non–zero. By dividing this equation by $r_n x$ one gets $r_n^{-1} x^{n-1} + r_1 r_n^{-1} x^{n-2} + \cdots + r_{n-1} r_n^{-1} + x^{-1} = 0$, so that $x^{-1} \in S$.

Conversely, assume that S is a field. If x is a non–zero element of R, then x^{-1} is an element of S, and is thus integral over R. Hence for some $r_i \in R$, $x^{-n} + r_1 x^{-n+1} + \cdots + r_{n-1} x^{-1} + r_n = 0$. Thus multiplying this equation through by x^{n-1} yields that $x^{-1} \in R$.

The second part follows as $R/(Q \cap R) \subseteq S/Q$ is integral. $\qquad \square$

A weak version of the following lemma was used in Corollary 1.1.8:

Lemma 2.1.8 (Determinantal trick) *Let R be a ring, M an R–module, $\varphi : M \to M$ an R–module homomorphism, and I an ideal of R such that $\varphi(M) \subseteq IM$. Then for some r_i in I^i,*

$$\varphi^n + r_1 \varphi^{n-1} + \cdots + r_n \varphi^0 = 0.$$

In particular, if M is a finitely generated faithful R–module, and if $xM \subseteq M$ for some x in an extension algebra containing R, then x is integral over R.

Proof: The second part follows from the first part: set I to be R, φ to be multiplication by x, and observe that the conclusion that $x^n + r_1 x^{n-1} + \cdots + r_n$ annihilates a faithful module implies that $x^n + r_1 x^{n-1} + \cdots + r_n = 0$, whence x is integral over R. Thus it suffices to prove the first part.

Let $\{m_1, \ldots, m_n\}$ be a generating set of M. Write $\varphi(m_i) = \sum_{j=1}^n a_{ij} m_j$ for some $a_{ij} \in I$. Let A be the matrix whose entry (i, j) equals $\delta_{ij} \varphi -$

$a_{ij}Id$. Then A multiplies the column vector $(m_1, \ldots, m_n) \in M^n$ to zero, i.e., $A[m_1, \ldots, m_n]^T = 0$. Left–multiplication of both sides of this equation by the classical adjoint of A shows that $\det A$ annihilates M. But $\det A$ is a function of the form $\varphi^n + r_1 \varphi^{n-1} + \cdots + r_n \varphi^0$, with each $r_i \in I^i$. $\qquad\square$

The next lemma is a generalization (to rings) of the fact from field theory that a finitely generated algebraic extension of a field is a finite–dimensional vector space over that field:

Lemma 2.1.9   *Let $R \subseteq S$ be an inclusion of rings, and let $x_1, \ldots, x_n \in S$. The following are equivalent:*
(1) For all $1 \le i \le n$, x_i is integral over R.
(2) $R[x_1, \ldots, x_n]$ is a finitely generated R–submodule of S.
(3) There is a non–zero finitely generated faithful R–module M contained in S such that $x_i M \subseteq M$ for each $1 \le i \le n$.

Proof: Assume (1). We use induction on n to prove (2). By induction we may assume that $R[x_1, \ldots, x_{n-1}]$ is a finitely generated R–submodule of S. We can then replace R by $R[x_1, \ldots, x_{n-1}]$; it suffices to prove this implication when $n = 1$. Let $x = x_1$. There exists a positive integer m such that x satisfies an equation of integral dependence of degree m over R. Thus $R[x] = R + Rx + Rx^2 + \cdots + Rx^{m-1}$, which proves (2).

Assume that (2) holds. Assertion (3) follows by setting $M = R[x_1, \ldots, x_n]$ and observing that M is a faithful R-module as it contains the identity of R.

The determinantal trick (Lemma 2.1.8) proves that (3) implies (1). $\qquad\square$

Proposition 2.1.10   *Let $R \subseteq S$ be an inclusion of rings, S generated over R by the elements s_λ, as λ varies over a variable set Λ. Then S is integral over R if and only if each s_λ is integral over R.*

Proof: It suffices to prove that if each s_λ is integral over R, then S is integral over R. Let s be an arbitrary element of S. Then there exists a finite subset Λ_0 of Λ such that $s \in S_0 = R[s_\lambda \,|\, \lambda \in \Lambda_0]$. As all rings contain an identity, S_0 is a faithful R–module. By Lemma 2.1.9, S_0 is a finitely generated R–module. But $sS_0 \subseteq S_0$ implies by Lemma 2.1.9 that s is integral over R. $\qquad\square$

(A version of this for the integral dependence of ideals is in Corollary 5.2.2.)

This further implies that the integral closure of a ring in another is also a ring, so that the integral closure is an operation on the category of rings:

Corollary 2.1.11   *The integral closure of R in an overring S is a ring.*

Proof: Assume that x and y in S are integral over R. By Lemma 2.1.9, $R[x, y]$ is a finitely generated faithful R–module. Note that xy and $x - y$ multiply $R[x, y]$ to $R[x, y]$, so that xy and $x - y$ are both integral over R. $\qquad\square$

Furthermore, the integral closure operation is an operation on the category of fields, i.e., if $R \subseteq S$ are fields, then the integral closure of R in S is also a field. Namely, if x is a non–zero element of the integral closure of R in S, then $R \subseteq R[x]$ is an integral extension of integral domains, so by Lemma 2.1.7, $R[x]$ is a field. Thus $R(\frac{1}{x}) = R(x) = R[x]$. By Lemma 2.1.9, $R[x]$ is module–finite over R. Thus $\frac{1}{x}$ is is contained in a module–finite extension, so by Lemma 2.1.9, $\frac{1}{x}$ is integral (algebraic) over R. This proves that the algebraic closure of a field in an overfield is a field.

Here is another analogy with fields: just as algebraic dependence is a transitive operation, so is integral dependence:

Corollary 2.1.12 *Let* $R \subseteq S \subseteq T$ *be inclusions of rings. Then* S *is integral over* R *and* T *is integral over* S *if and only if* T *is integral over* R.

Proof: If S is integral over R and T is integral over S, then for any element t of T, let $t^n + s_1 t^{n-1} + \cdots + s_n = 0$ be an equation of integral dependence of t over S, with each $s_i \in S$. Then t is integral over $R[s_1, \ldots, s_n]$. By Lemma 2.1.9, $R[s_1, \ldots, s_n, t]$ is module–finite over $R[s_1, \ldots, s_n]$. By assumption, each s_i is integral over R, so that $R[s_1, \ldots, s_n]$ is module–finite over R. Thus $R[s_1, \ldots, s_n, t]$ is module–finite over R, so that by condition (3) of Lemma 2.1.9, t is integral over R. Thus T is integral over R.

The converse is clear. $\square$

Finding the integral closure reduces to integral domains:

Corollary 2.1.13 *Let* R *be a reduced ring. Let* $P_1, \ldots, P_s$ *be all the minimal prime ideals of* R. *The integral closure of* R *in its total ring of fractions is* $\overline{R/P_1} \times \cdots \times \overline{R/P_s}$, *where* $\overline{R/P_i}$ *is the integral closure of* R/P_i *in its field of fractions* $\kappa(P_i)$.

Proof: The total ring of fractions K of R is the zero–dimensional ring obtained from R by inverting all elements of R that are not in any minimal prime ideal. Thus K is the direct product of the $\kappa(P_i)$. Observe that

$$R \subseteq R/P_1 \times \cdots \times R/P_s$$

is a module–finite faithful extension contained in K, which is integral over R by Lemma 2.1.9. As $\overline{R/P_1} \times \cdots \times \overline{R/P_s}$ is integral over $(R/P_1) \times \cdots \times (R/P_s)$ (see Exercise 2.1), it follows that $\overline{R/P_1} \times \cdots \times \overline{R/P_s}$ is integral over R. Thus $\overline{R/P_1} \times \cdots \times \overline{R/P_s}$ is contained in the integral closure of R. But $\overline{R/P_1} \times \cdots \times \overline{R/P_s}$ is integrally closed, for if $(k_1, \ldots, k_s) \in K = \kappa(P_1) \times \cdots \times \kappa(P_s)$, then an equation of integral dependence of $(k_1, \ldots, k_s)$ over $(R/P_1) \times \cdots \times (R/P_s)$ is the product of equations of integral dependence of each k_i over R/P_i. Thus for all $i = 1, \ldots, s$, $k_i \in \overline{R/P_i}$. It follows that $\overline{R/P_1} \times \cdots \times \overline{R/P_s}$ is integrally closed. Thus $\overline{R/P_1} \times \cdots \times \overline{R/P_s}$ is the integral closure of R. $\square$

Observe that a finite direct product of integrally closed domains is locally an integral domain.

Definition 2.1.14 *A ring* R *is said to be* **normal** *if for every prime ideal* P *of* R, R_P *is an integrally closed integral domain.*

Every normal ring is locally an integral domain, thus globally it is reduced. By Corollary 2.1.13, a Noetherian reduced ring is integrally closed if and only if it is normal.

The following lemma helps identify when an intersection of normal rings is normal.

Lemma 2.1.15 *Let R be a reduced ring whose total ring of fractions K is a finite direct product of fields. Then*
(1) R is normal if and only if R is integrally closed in K.
(2) Let L be a ring containing K. For each i in some index set Λ let R_i be an integrally closed ring whose total ring of fractions contains K. Suppose that $R = \bigcap R_i$. Then R is normal.

Proof: Let P be a prime ideal of R. The total ring of fractions of R_P is $K_{R \setminus P}$, which is a direct product of fields. If R is integrally closed in K then R_P is also integrally closed in $K_{R \setminus P}$. By Corollary 2.1.13, by the structure of prime ideals in a direct product of rings, R_P is a domain. The idempotents one obtains by decomposing $K_{R \setminus P}$ as a product of fields are integral over R_P, hence are in R_P, and hence $K_{R \setminus P}$ is a field, which implies that R_P is an integrally closed domain, and so R is normal. Conversely, assume that R is normal. Then R_P is an integrally closed domain. If $t \in K$ is integral over R, then $\frac{t}{1}$ is integral over R_P, and so $\frac{t}{1} \in R_P$ for all prime ideals P, implying that $t \in R$.

To prove the second statement, it is enough to prove that R is integrally closed in K, using the first part of the lemma. Let $x \in K$ be integral over R. For each i, x is in the total ring of fractions of R_i and still integral over R_i as the equation of integral dependence over R is an equation of integral dependence over R_i. By assumption, $x \in R_i$ for each i, hence $x \in R$. This proves that R is integrally closed. $\qquad\square$

Reduction to integral domains is helpful even without the assumption that there be only finitely many minimal prime ideals:

Proposition 2.1.16 *Let R be a ring, not necessarily Noetherian, and S an overring of R. Let $x \in S$. Then x is integral over R if and only if the image of x in S/PS is integral over R/P, as P varies over all the minimal prime ideals of R.*

Proof: Clearly if x is integral over R, then as the image of an integral equation of x over R passes to an integral equation of the same degree in all quotients, it follows that the image of x in S/PS is integral over R/P for every minimal prime ideal P of R.

For the converse, let U be the subset of S consisting of all elements of the form $\{x^n + r_1 x^{n-1} + \cdots + r_n \mid n \in \mathbb{N}_{>0}, r_i \in R\}$. Then U is a subset of S that is closed under multiplication and that by assumption intersects with PS for each $P \in \mathrm{Min}(R)$. If U does not contain 0, then S can be localized at U. If

Q is a prime ideal in $U^{-1}S$, let q denote the contraction of Q in R. Since U intersects qS and qS is contained in Q, it follows that Q intersects U, which is a contradiction. Thus $U^{-1}S$ has no prime ideals, which contradicts the assumption that 0 is not in U. So necessarily $0 \in U$, which gives an equation of integral dependence of x over R. □

It follows that to find the integral closure of R in S it suffices to find the integral closure of R in S modulo the nilradical of R. In other words, the nilpotent elements behave trivially under the integral closure operation.

By Proposition 2.1.16 it is often no loss of generality if in the study of the integral closure and dependence we only consider integral domains. There are a few more tools available for integral closure of integral domains.

None of the theory developed so far gives a good clue towards deciding when an element in an extension is integral over the base ring, or towards finding equations of integral dependence. In Chapter 15 we discuss some of the computational difficulties: while there is a general algorithm for computing the integral closure of an integral domain, in practice it is often unmanageable. Some help in this direction is provided by the following:

Theorem 2.1.17 *Let R be an integral domain, K its field of fractions, and L a field extension of K. Then for every element $s \in L$, s is integral over R if and only if it is algebraic over K and its minimal (monic) polynomial over K has all its coefficients in the integral closure of R.*

Proof: If s is algebraic over K and its minimal monic polynomial over K has all its coefficients in the integral closure of R, then s is integral over the integral closure of R, so s is integral over R by Corollary 2.1.12.

If s is integral over R, it is clearly integral also over K, and satisfies a monic polynomial $f(x)$ all of whose coefficients are in R. This polynomial is a multiple in $K[x]$ of the minimal polynomial of s over K. Let L' be a splitting field of $f(x)$ over K. Let τ be a K–automorphism of L'. Then by applying τ to the given integral equation of s over R, $\tau(s) \in L'$ is also integral over R (and K). From field theory one knows that the minimal polynomial for s over K is $\prod_\tau (x - \tau(s))^n$, where n is either some power of the characteristic of K or is 1 if the characteristic of K is 0. In the expansion of this polynomial, all the coefficients are in K, are symmetric polynomials in the $\tau(s)$, and are thus integral over R. □

Thus in particular, if R is integrally closed in its field of fractions, each element in an algebraic closure of this field that is integral over R satisfies a unique equation of integral dependence of minimal degree over R, namely its minimal polynomial. It is easy to prove the following weaker version:

Proposition 2.1.18 *Let R be an integral domain, K its field of fractions, and L a field extension of K. For any element $s \in L$, s is integral over K if and only if for some non–zero $r \in R$, rs is integral over R.*

The next lemma will be useful later.

Lemma 2.1.19   *Let R be an integral domain, K its field of fractions, and X a variable. Let $f(X)$ be a monic polynomial in $R[X]$, and $g(X), h(X)$ monic polynomials in $K[X]$ such that $f(X) = g(X)h(X)$. Then the coefficients of g and h lie in the integral closure of R.*

Proof: Let L be an algebraic closure of K. Any root of g is also a root of f, so as f is monic with coefficients in R, this root is integral over R. Since the coefficients of $g(X)$ are sums of products of the roots, each coefficient is both integral over R and in K.   □

2.2. Lying–Over, Incomparability, Going–Up, Going–Down

The four most basic theorems concerning the behavior of prime ideals under integral ring extensions have names: Lying–Over, Incomparability, Going–Up, and Going–Down. These were first proved by Krull for integral domains in [175], and in greater generality by Cohen and Seidenberg in [41].

> **Definition 2.2.1** *Let $f : R \to S$ be a ring homomorphism. We say that f satisfies* **Going–Down** *if whenever $P_1 \subseteq P_2$ are prime ideals in R and Q_2 is a prime ideal in S such that $f^{-1}(Q_2) = P_2$, there exists a prime ideal Q_1 in S contained in Q_2 such that $f^{-1}(Q_1) = P_1$.*
>
> *We say that $f : R \to S$ satisfies* **Going–Up** *if whenever $P_1 \subseteq P_2$ are prime ideals in R and Q_1 is a prime ideal in S such that $f^{-1}(Q_1) = P_1$, there exists a prime ideal Q_2 in S containing Q_1 such that $f^{-1}(Q_2) = P_2$.*
>
> *Also, f satisfies* **Lying–Over** *if for any $P \in \operatorname{Spec} R$ there exists $Q \in \operatorname{Spec} S$ such that $f^{-1}(Q) = P$.*

Theorem 2.2.2   (Lying–Over) *Let $R \subseteq S$ be an integral extension of rings. Then for any prime ideal P of R there exists a prime ideal Q of S such that $Q \cap R = P$.*

Proof: We may replace R by R_P and S by $S_{R \setminus P}$: this is still an integral extension of rings, and if the conclusion holds after localization, it holds in the original set–up as well. Thus without loss of generality R is a local ring with maximal ideal P. Let Q be a maximal ideal in S. By Lemma 2.1.7, $Q \cap R$ is a maximal ideal of R, so that $Q \cap R = P$.   □

This implies that $PS \cap R = P$ for every prime ideal P of R. Furthermore, it implies that the natural map $\operatorname{Spec} S \to \operatorname{Spec} R$ is surjective.

Theorem 2.2.3   (Incomparability) *Let $R \subseteq S$ be an integral extension of rings and $P \subseteq Q$ prime ideals of S. If $P \cap R = Q \cap R$, then $P = Q$.*

Proof: Without loss of generality we may localize both R and S at the multiplicatively closed set $R \setminus (Q \cap R)$. By Lemma 2.1.7, P is a maximal ideal of S since $Q \cap R = P \cap R$ is a maximal ideal in R. Thus $P = Q$.   □

A consequence of the Incomparability Theorem is that the Krull dimension satisfies the inequality $\dim S \le \dim R$ whenever $R \subseteq S$ is an integral extension.

Theorem 2.2.4 (Going–Up) *Let $R \subseteq S$ be an integral extension of rings. Then for any chain of prime ideals $P_1 \subseteq P_2 \subseteq \ldots \subseteq P_n$ of R and for any prime Q_1 in S such that $P_1 = Q_1 \cap R$, there exists a chain of prime ideals $Q_1 \subseteq Q_2 \subseteq \ldots \subseteq Q_n$ of S such that $Q_i \cap R = P_i$ for all $1 \le i \le n$.*

Proof: By a straightforward induction, it suffices to prove the case $n = 2$. Without loss of generality we may localize both R and S at the multiplicatively closed set $R \setminus P_2$. Also, we may replace R by R/P_1 and S by S/Q_1: this is still an integral extension of rings, with the images of the P_i and of Q_i prime ideals. The advantage is that now every ideal in S contains $Q_1 = 0$, so it suffices to find a prime ideal in S that contracts to P_2 in R. But this follows by the Lying–Over theorem. $\square$

This implies that whenever $R \subseteq S$ is integral, $\dim S \ge \dim R$. Thus the last two theorems give:

Theorem 2.2.5 *Let $R \subseteq S$ be an integral extension of rings. Then $\dim R = \dim S$.*

Whereas integral extensions preserve dimension, the height of a prime ideal in an integral extension need not be the same as the height of its contraction. For example, if R is a ring and P is a prime ideal in R of positive height, then $R \subseteq R \oplus R/P$ is a module–finite extension of rings, thus an integral extension by Lemma 2.1.9, but the prime ideal P in R of positive height is the contraction of the prime ideal $R \oplus 0 \subseteq R \oplus R/P$ of height 0.

In Corollary 2.2.8 we give some general conditions on R and S so that height is preserved for arbitrary ideals. However, even for a finitely generated integral extension $R \subseteq S$ of Noetherian integral domains it is possible that for some prime ideal Q of S, $\operatorname{ht} Q \ne \operatorname{ht}(Q \cap R)$. Here is an example:

Example 2.2.6 (Nagata [213, E2.1]) Let $(A, \mathfrak{m})$ be a Noetherian local domain and S a Noetherian domain containing A. Assume that S has only finitely many maximal ideals, and that for each maximal ideal Q in S, $Q \cap A = \mathfrak{m}$ and S/Q is a finite algebraic extension of $A/\mathfrak{m}$. Let J be the Jacobson radical of S, i.e., the intersection of all the maximal ideals of S. Then S/J is a module–finite extension of $A/\mathfrak{m}$. Set $R = A + J$. It is easy to verify that R is a commutative domain with identity. Note that J is a maximal ideal of R and $A/\mathfrak{m} = R/J \subseteq S/J$ is a module–finite extension. But as J is an ideal in both R and S and since it is the maximal ideal of R, Nakayama's Lemma gives that $R \subseteq S$ is a module–finite extension. Every element of $R \setminus J$ is a unit in S, thus a unit in R, so that R has a unique maximal ideal. We claim that R is Noetherian. By Cohen's Theorem R is Noetherian provided that every prime ideal P is finitely generated. For any prime ideal P of R, there exists a prime ideal Q in S contracting to P. Then $P = Q \cap R = Q \cap J$. The $A/\mathfrak{m} = R/J$–module P/JP is contained in the $A/\mathfrak{m}$–module PS/JP, which is finitely generated. Thus P/JP is finitely generated as an R–module. Let $p_1, \ldots, p_k$ be elements of P whose images generate the R–module P/JP. By possibly adding elements, we may also assume that $p_1, \ldots, p_k$ generate PS. Now let $p \in P$. Then $p \in (p_1, \ldots, p_k) + JP \subseteq (p_1, \ldots, p_k) + JPS$, so we can write $p = \sum_i r_i p_i + \sum_i j_i p_i$ for some $r_i \in R$ and $j_i \in JS \subseteq R$. This proves that $P = (p_1, \ldots, p_k)$. Thus P is finitely generated as an R–module. Hence R is Noetherian.

Clearly each of the maximal ideals in S contracts to the maximal ideal in R. If we start with S whose maximal ideals have distinct heights, then this yields an example of an integral extension $R \subseteq S$ of Noetherian integral domains such that for some prime ideal Q of S, $\operatorname{ht} Q \neq \operatorname{ht}(Q \cap R)$.

Here is an example of such a ring S, namely, of a ring S that is an overring of a Noetherian local ring $(A, \mathfrak{m})$, such that S has only finitely many maximal ideals, these maximal ideals have different heights, and the residue fields at these maximal ideals are finite algebraic extensions of $A/\mathfrak{m}$. The example below is again due to Nagata. Let $A = k$ be a field, $X, Y_1, \ldots, Y_m$ variables over k, r a positive integer and for $i = 1, \ldots, r$, let $z_i = \sum_{j=0}^{\infty} a_{ij} X^j$ be algebraically independent elements over $k(X)$ for some $a_{ij} \in k$. Set $z_{ik} = X \sum_{j=k}^{\infty} a_{ij} X^{j-k}$. Let $S' = k[X, z_{ik} \,|\, i, k]$. Then S' is not Noetherian. As $X z_{i,k+1} = z_{ik} - a_{ik} X$, it follows that XS' is a prime ideal and that $S'_{XS'}$ is a subring of $k[[X]]$. Any non–zero prime ideal in $S'_{XS'}$ either contains X, in which case it is the maximal ideal, or it contains a power series in X that is not a multiple of X, and then the ideal is the whole ring. Thus $S'_{XS'}$ is a principal ideal domain, so Noetherian. Now let $S'' = k[X, z_{ik}, Y_1, \ldots, Y_m \,|\, i, k]$. Let $\mathfrak{m} = (X, Y_1, \ldots, Y_m)S''$ and $\mathfrak{n} = (X-1, z_i, Y_1, \ldots, Y_m \,|\, i)S''$. By the previous work, $S''_{\mathfrak{m}}$ is a Noetherian local ring of dimension $m + 1$. As $S'[1/X] = k[X, 1/X, z_1, \ldots, z_r]$, it follows that $S''_{\mathfrak{n}}$ is a Noetherian local ring of dimension $m + r + 1$. Let S be the ring obtained from S'' by localizing at the multiplicatively closed set $S'' \setminus \mathfrak{m} \cup \mathfrak{n}$. Then S is Noetherian with exactly two maximal ideals, one of height $m + 1$ and the other of height $m + r + 1$. Also, the two residue fields equal k, so that this S indeed works in the previous example.

Another result regarding expected heights is in Proposition 4.8.6.

Theorem 2.2.7 (Going–Down) *Let $R \subseteq S$ be an integral extension of rings. Assume that R is an integrally closed domain. Further assume that S is torsion–free over R, i.e., every non–zero element of R is regular on S. For any chain of prime ideals $P_1 \subseteq P_2 \subseteq \ldots \subseteq P_n$ of R and Q_n prime in S such that $P_n = Q_n \cap R$, there exists a chain of prime ideals $Q_1 \subseteq \ldots \subseteq Q_n$ of S such that $Q_i \cap R = P_i$ for all $1 \leq i \leq n$.*

Proof: By induction on n it suffices to prove the case $n = 2$. By localizing both R and S at the multiplicatively closed set $R \setminus P_2$, without loss of generality R

is a local integrally closed domain with maximal ideal P_2. It suffices to prove that P_1 contracts from a prime ideal contained in Q_2, or that P_1 contracts from a prime ideal in the ring extension S_{Q_2}. By Exercise 2.4 it suffices to prove that $P_1 S_{Q_2} \cap R = P_1$.

Let $r \in P_1 S_{Q_2} \cap R$. Then $r = \frac{x}{s}$ for some $x \in P_1 S$ and $s \in S \setminus Q_2$. There exists a finitely generated R–subalgebra T of S such that $x \in P_1 T$. Necessarily $R \subseteq T$ is integral, so the finite generation implies that T is module–finite over R. Note that $xT \subseteq P_1 TT = P_1 T$. Thus by Lemma 2.1.8, x satisfies an equation $x^n + a_1 x^{n-1} + \cdots + a_n = 0$ with $a_i \in P_1$ (actually in P_1^i, but we may ignore the powers).

Set $f(X) = X^n + a_1 X^{n-1} + \cdots + a_n$. If $f(X)$ factors into monic polynomials over $K[X]$, then by Lemma 2.1.19 it factors into monic polynomials over $R[X]$. Clearly $R[X]/P_1 R[X]$ is a domain and the image of $f(X)$ in this ring is X^n. This forces each of the factors of $f(X)$ to have all the non–leading coefficients in P_1. This proves that the minimal integral equation for x over K has all the non–leading coefficients in P_1. By changing notation we may assume that this minimal equation is $x^n + a_1 x^{n-1} + \cdots + a_n = 0$, with each $a_i \in P_1$.

It follows that $\left(\frac{x}{r}\right)^n + \frac{a_1}{r}\left(\frac{x}{r}\right)^{n-1} + \cdots + \frac{a_n}{r^n} = 0$. Since $\frac{x}{r} = s$ and $r \in R$, this equation is an integral equation for s over K. By Theorem 2.1.17, the minimality of n implies that this is a minimal equation for $s = \frac{x}{r}$ over K. Thus by Theorem 2.1.17 all the coefficients are in R. Then for $i = 1, \ldots, n$, $\frac{a_i}{r^i}$ is in R, and so $a_i \in r^i R \cap P_1$ for $i = 1, \ldots, n$. If $r \notin P_1$, then $r^i R \cap P_1 = r^i P_1$, so $a_i \in r^i P_1$ and the minimal equation for s over K (and over R) is $s^n + b_1 s^{n-1} + b_2 s^{n-2} + \cdots + b_n = 0$ for some $b_i \in P_1$. In this case $s \in \sqrt{P_1 S} \subseteq Q_2 S$, which is a contradiction. So necessarily $r \in P_1$. $\qquad\square$

Corollary 2.2.8 *Let $R \subseteq S$ be an integral extension of rings. Assume that R is an integrally closed domain. Further assume that S is torsion–free over R, i.e., every non–zero element of R is regular on S. Then for every ideal I of R, $\operatorname{ht} I = \operatorname{ht}(IS)$.*

Proof: By Theorem 2.2.2, $R \subseteq S$ satisfies the Lying–Over condition, and by Theorem 2.2.7, $R \subseteq S$ satisfies the Going–Down condition. Proposition B.2.4 then finishes the proof in the Noetherian case.

The rest covers the general case. By Lemma B.1.3, every prime ideal in R that is minimal over I contracts from a prime ideal in S that is minimal over IS and every prime ideal in S that is minimal over IS contracts to a prime ideal in R that is minimal over I. Hence it suffices to prove the corollary in the case where I is a prime ideal.

Let $Q \in \operatorname{Spec} S$ be minimal over IS and of height equal to $\operatorname{ht}(IS)$. Then $Q \cap R = I$. As $R_I \subseteq S_I$ is integral, by Theorem 2.2.5, $\operatorname{ht} I = \dim(R_I) = \dim(S_I) \geq \dim(S_Q) = \operatorname{ht} Q$, whence $\operatorname{ht} I \geq \operatorname{ht}(IS)$.

Let $P_0 \subsetneq P_1 \subsetneq \cdots \subsetneq P_h = I$ be a chain of prime ideals in R with $h = \operatorname{ht} I$. By Going-Down there exist $Q_0, \ldots, Q_h \in \operatorname{Spec} S$ such that $Q_h = Q$, $Q_0 \subseteq$

$Q_1 \subseteq \cdots \subseteq Q_h$, and $Q_j \cap R = P_j$ for $j = 0, \ldots, n$. By the Incomparability Theorem 2.2.3, $Q_0 \subsetneq Q_1 \subsetneq \cdots \subsetneq Q_h$ is a saturated chain, whence $\operatorname{ht} I = h \leq \operatorname{ht} Q = \operatorname{ht}(IS)$. Thus $\operatorname{ht} I = \operatorname{ht}(IS)$. $\square$

2.3. Integral closure and grading

Recall that a monoid is a non–empty set with an associative binary operation and a unit. A typical example is $\mathbb{N}^d \times \mathbb{Z}^e$ for some noni–negative integers d, e. If a ring is graded by a totally ordered abelian monoid, then so are all of its minimal prime ideals (see for example Section A.3 in the Appendix).

Definition 2.3.1 *Let G be an abelian monoid. A ring R is said to be **G-graded** if the following conditions are satisfied:*
(1) $R = \oplus_{g \in G} R_g$, where R_g is a subset of R,
(2) each R_g is a group under addition,
(3) for each $g, g' \in G$, $R_g R_{g'} \subseteq R_{g+g'}$.
*When R is G–graded, an R–module M is called **G-graded** if*
(1) $M = \oplus_{g \in G} M_g$, where M_g is an R_0–submodule of M,
(2) for each $g, g' \in G$, $R_g M_{g'} \subseteq M_{g+g'}$.
*An element of R (respectively M) is said to be **homogeneous** if it is an element of some R_g (respectively M_g).*

Theorem 2.3.2 *Let $G = \mathbb{N}^d \times \mathbb{Z}^e$, and let $R \subseteq S$ be G–graded and not necessarily Noetherian rings. Then the integral closure of R in S is G–graded.*

Proof: We first prove the case $d + e = 1$. Let $s = \sum_{j=j_0}^{j_1} s_j$, $s_j \in S_j$, be integral over R. We have to show that each s_j is integral over R.

Let r be an arbitrary unit of R. Then the map $\varphi_r : S \to S$ that multiplies elements of S_i by r^i is a graded automorphism of S that restricts to a graded automorphism of R and is identity on S_0. Thus $\varphi_r(s) = \sum_{j=j_0}^{j_1} r^j s_j$ is an element of S that is integral over R.

Assume that R has $n = j_1 - j_0 + 1$ distinct units r_i all of whose differences are also units in R. Define $b_i = \varphi_{r_i}(s)$. Each b_i is integral over R. Let A be the $n \times n$ matrix whose (i, j) entry is $r_i^{j+j_0-1}$. Then

$$A \begin{bmatrix} s_{j_0} \\ s_{j_0+1} \\ \vdots \\ s_{j_1} \end{bmatrix} = \begin{bmatrix} b_{j_0} \\ b_{j_0+1} \\ \vdots \\ b_{j_1} \end{bmatrix}.$$

As A is a Vandermonde matrix, by the choice of the r_i, A is invertible, so that

$$\begin{bmatrix} s_{j_0} \\ s_{j_0+1} \\ \vdots \\ s_{j_1} \end{bmatrix} = A^{-1} \begin{bmatrix} b_{j_0} \\ b_{j_0+1} \\ \vdots \\ b_{j_1} \end{bmatrix}.$$

Thus each s_j is an R–linear combination of the b_i, whence each s_j is integral over R, as was to be proved.

Finally, we reduce to the case when R has $n = j_1 - j_0 + 1$ distinct units r_i all of whose differences are also units in R. Let $t_{j_0}, \ldots, t_{j_1}$ be variables over R. Define $R' = R[t_j, t_j^{-1}, (t_j - t_i)^{-1} \mid i, j = j_0, \ldots, j_1]$ and $S' = S[t_j, t_j^{-1}, (t_j - t_i)^{-1} \mid i, j = j_0, \ldots, j_1]$. We extend the G–grading on R and S to R' and S' by setting the degree of each t_i to be 0. Then $R' \subseteq S'$ are G–graded rings, R' contains at least n distinct units $r_i = t_i$ all of whose differences are also units in R'. By the previous case, each $s_j \in S$ is integral over R'. Consider an equation of integral dependence of s_j over R', say of degree n. Clear the denominators in this equation to get an equation E over $R[t_i \mid i = j_0, \ldots, j_1]$. The coefficient of s_j^n in E is a polynomial in $R[t_i \mid i = j_0, \ldots, j_1]$, with at least one coefficient of this polynomial being a unit of R. Picking out the appropriate multi t_i–degree of E yields an integral equation of s_j over R. Thus s_j is integral over R. This finishes the proof of the case $d + e = 1$.

Now we proceed by induction on $d + e$. Let T be the integral closure of R in S. If $e = 0$ set $G' = \mathbb{N}^{d-1}$ and if $e > 0$ set $G' = \mathbb{N}^d \times \mathbb{Z}^{e-1}$. We impose a G'–grading on $R \subseteq S$ by forgetting about the last component. By induction, $T = \sum_{\nu \in G'} T_\nu$, where T_ν is the homogeneous part of T consisting of elements of degree ν. Now let $s \in T_\nu$. As $s \in S$ and S is G–graded, we may write $s = \sum_{j=j_0}^{j_1} s_j$, where each $s_j \in S_{(\nu, j)}$. Thus by the case $d + e = 1$, each s_j is integral over R. □

It is not true that a result of this type holds for an arbitrary monoid G, as the following example shows:

Example 2.3.3 Let $R = k[X] \subseteq S = k[X, Y, Z]/(X^3 + Y^2 + Z^2)$, where X, Y and Z are variables over a field k of characteristic 2. Then R and S are $(\mathbb{Z}/2\mathbb{Z})$–graded domains as follows: each element in S can be represented uniquely as a polynomial in Y of degree at most 1. With such representation, $S = S_0 \oplus S_1$, $R = R_0 \oplus R_1 = R_0$, each S_i is closed under addition, $S_0 \times S_i \subseteq S_i$, $S_1 \times S_1 \subseteq S_0$. However, the non–homogeneous element $Y + Z$ is integral over R, yet neither of its homogeneous components is integral over R. Thus in this case the integral closure of a G–graded ring in a larger G–graded ring is not G–graded.

However, there is another grading on this particular $R \subseteq S$ under which the integral closure of R in S is graded. (See Exercise 2.21.)

Thus the relative integral closure of graded rings is graded for "good" gradings. We next examine conditions that guarantee that the (non–relative) integral closure be graded as well. The problem with considering the integral closure of a graded ring is that the integral closure is taken in the total ring of fractions, which is a non–graded ring. Consequently, Theorem 2.3.2 cannot be applied directly. In fact, the integral closure of a graded ring need not be graded. To see what can happen, consider the following examples.

Example 2.3.4 It need not be the case that the integral closure of a reduced $(\mathbb{N}^d \times \mathbb{Z}^e)$–graded ring is graded. Let $R = \mathbb{Q}[X, Y]/(XY)$, with X, Y variables over $\mathbb{Q}$. We can impose any of the following gradings on R:
(1) $\mathbb{N}$–grading $\deg X = 0$, $\deg Y = 1$,
(2) $\mathbb{Z}$–grading $\deg X = 1$, $\deg Y = -1$,
(3) $\mathbb{N}^2$–grading $\deg X = (1, 0)$, $\deg Y = (0, 1)$.
Under any of these gradings, $X/(X + Y)$ is an element of the total ring of fractions of R and it satisfies the integral equation $T^2 - T = 0$. Thus $X/(X + Y)$ is integral over R. However, $X/(X + Y)$ cannot be written as a sum of homogeneous components under the given gradings. In fact, there are no homogeneous non–zerodivisors except for units.

For R as in the example above but with the $\mathbb{N}$–grading $\deg X = m$, $\deg Y = n$, with $m, n \in \mathbb{N}_{>0}$, $X/(X + Y)$ can be written as the (clearly) homogeneous element $X^n/(X^n + Y^m)$.

The next proposition gives necessary and sufficient conditions for the integral closure to be graded.

Proposition 2.3.5 *Let $G = \mathbb{N}^d \times \mathbb{Z}^e$, and let R be a G–graded reduced ring with only finitely many minimal prime ideals $P_1, \ldots, P_s$. Let K be the total ring of fractions R, let W be the multiplicatively closed set consisting of homogeneous non–zerodivisors, and let $S = W^{-1}R$. The following are equivalent:*
(1) The ring S is integrally closed.
(2) The integral closure $\overline{R}$ of R is G–graded.
(3) The idempotents of $\overline{R}$ are homogeneous of degree 0.
(4) For $i = 1, \ldots, s$, $P_i + \cap_{j \neq i} P_j$ contains a homogeneous non–zerodivisor. (In case $s = 1$, this condition is vacuously satisfied.)

Proof: Assume condition (1). Since $R \subseteq S \subseteq K$, the integral closure of R is contained in the integral closure of S, which is just S. Hence $\overline{R}$ is the integral closure of R in S. By Theorem 2.3.2 it follows that $\overline{R}$ is $\mathbb{Z}^{d+e}_-$ graded. However, $\overline{R}$ must in fact be G–graded since the equations of integral dependence show that new negative terms cannot appear. This proves (2).

Assume (2). Since the idempotents of $\overline{R}$ are in K and satisfy the monic equation $X^2 - X = 0$, they are integral over R and hence are sums of homogeneous elements in K. Write an idempotent e as $e = \sum_{i=1}^n e_i$, where each e_i is homogeneous. Impose a lexicographic order on $\mathbb{Z}^{d+e}$. Without loss of generality $\deg(e_1) > \cdots > \deg(e_n)$. If $\deg(e_1) > 0$, the homogeneous part of $e^2 - e$ of degree $2\deg(e_1)$ is exactly e_1^2, and since $e^2 - e = 0$ and R is reduced, we get a contradiction. Similarly if $\deg(e_n) < 0$ we get a contradiction. This forces $n = 1$ and e to be homogeneous of degree 0, whence proving (3).

Assume (3). To prove (4), by symmetry it suffices to prove (4) for $i = 1$. Consider the idempotent $e = (1, 0, \ldots, 0) \in \overline{R} = \overline{R/P_1} \times \cdots \overline{R/P_s}$ (using Corollary 2.1.13). By assumption e is homogeneous. There exist homogeneous elements $a, b \in R$, with b a non–zerodivisor in R such that $e = a/b$. This means that the image of a in the direct product is $(b, 0, \ldots, 0)$, which says that $b - a \in P_1$ and that $a \in P_2 \cap \cdots \cap P_s$. Hence $b = (b - a) + a \in P_1 + P_2 \cap \cdots \cap P_s$ is a homogeneous non–zerodivisor. This proves (4).

Finally, assume (4). Let h_i be a homogeneous non–zerodivisor in $P_i + \cap_{j \neq i} P_j$. Set $h = h_1 \cdots h_s$. Then $h \in R$ is a homogeneous non–zerodivisor, and for each i, $h = p_i + r_i$ for some $p_i \in P_i$ and some $r_i \in \cap_{j \neq i} P_j$. Since all the P_i are homogeneous by Corollary A.3.2, we may take r_i and p_i to be homogeneous of the same degree as h. Set $e_i = \frac{p_i}{h}$. As $p_i r_i \in \cap_j P_j = 0$, it follows that $e_i = e_i \frac{(p_i + r_i)}{h} = e_i^2$, so that e_i is a homogeneous idempotent of degree 0. If $i \neq j$, then $p_i p_j$ is in the intersection of all the minimal prime ideals, hence is zero, so that $e_i e_j = 0$. Since $\sum_i p_i$ is not in any minimal prime ideal, it follows that $\sum_i e_i$ is an idempotent that is a unit in K, and thus $\sum_i e_i = 1$. Since S contains the orthogonal idempotents e_i, the isomorphisms $S/P_i \cong Se_i$ and $S \cong (S/P_1) \times \cdots \times (S/P_s)$ preserve the grading, and S has no non–minimal homogeneous prime ideals. We will prove that each S/P_i is integrally closed, whence S is integrally closed by Corollary 2.1.13.

Set $T = S/P_i$. This is a $\mathbb{Z}^{d+e}$–graded domain in which there are no non–zero homogeneous prime ideals. In particular, T_0 is a field, and every non–zero homogeneous element is a unit in T. Let M be the $\mathbb{Z}$–module generated by the degrees of the homogeneous non–zero elements of T. Then $M \subseteq \mathbb{Z}^{d+e}$. Every non–zero element of M is the degree of some non–zero homogeneous element of T (for this we use that T is a domain). In particular, since M is a finitely generated free $\mathbb{Z}$–module, there exist non–zero homogeneous elements $t_1, \ldots, t_c$ in T such that if $\nu_i = \deg(t_i)$, then $\{\nu_1, \ldots, \nu_c\}$ is a $\mathbb{Z}$–basis of M. Since the ν_i are linearly independent over $\mathbb{Q}$, the t_i are algebraically independent over the field T_0.

Clearly $T_0[t_1, \ldots, t_c, t_1^{-1}, \ldots, t_c^{-1}] \subseteq T$. Let $x \in T$ be a homogeneous element. Write $\deg(x) = \sum_i m_i \nu_i$ for some $m_i \in \mathbb{Z}$. Then x and $t_1^{m_1} \cdots t_c^{m_c} \in T$ are homogeneous elements of the same degree, whence their quotient is a non–zero element of T_0. Hence $x \in T_0[t_1, \ldots, t_c, t_1^{-1}, \ldots, t_c^{-1}]$. This proves that $T = T_0[t_1, \ldots, t_c, t_1^{-1}, \ldots, t_c^{-1}]$, with $t_1, \ldots, t_c$ variables over the field T_0. In particular, T is (a localization) of a unique factorization domain, hence T is integrally closed. This proves (1). $\qquad\square$

Corollary 2.3.6 *Let R be a reduced $\mathbb{N}$–graded ring, possibly non–Noetherian, such that the non–zero elements of R_0 are non–zerodivisors in R. Then the integral closure of R in K is $\mathbb{N}$–graded.*

Proof: Let $\alpha \in \overline{R}$. By writing α as a quotient of two elements of R and by collecting all the homogeneous parts of the two elements and of the coefficients of an equation of integral dependence, we see that there exist finitely many homogeneous elements $x_1, \ldots, x_n$ in R such that if R' is the subalgebra of R generated over the primitive subring A of R_0 by the x_i, then α is in the total ring of fractions of R' and is integral over R'. It suffices to prove that the integral closure of R' is G–graded. Observe that the non–zero elements of R'_0 are non–zero divisors in R and hence also in R'. By switching notation, without loss of generality we may assume that $R = A[x_1, \ldots, x_n]$. In particular, we may assume that R is Noetherian.

Localize R at $R_0 \setminus \{0\}$: the localization is still $\mathbb{N}$–graded, and by Theorem 2.3.2 it suffices to prove that the localized R has an $\mathbb{N}$–graded integral closure. Thus without loss of generality we may assume that all non–zero elements of R_0 are units, so that R_0 is a field.

As R is Noetherian, it has only finitely many minimal prime ideals and they are all homogeneous. By assumption on R_0 being a field, these prime ideals are generated in positive degrees. Let $\operatorname{Min} R = \{P_1, \ldots, P_s\}$. For each $i \in \{1, \ldots, s\}$, $P_i + \cap_{j \neq i} P_j$ is a homogeneous ideal generated in positive degrees, and not contained in $P_1 \cup \cdots \cup P_s$. Thus by homogeneous Prime Avoidance, $P_i + \cap_{j \neq i} P_j$ contains a homogeneous non–zerodivisor. Then by Proposition 2.3.5, the integral closure of R is $\mathbb{N}$–graded. $\square$

There are many other examples of graded rings whose integral closures are graded. We present the case of monomial algebras.

Corollary 2.3.7 *Let k be a field, $X_1, \ldots, X_d$ variables over k, and R a subalgebra of $k[X_1, \ldots, X_d]$ generated by monomials in the given variables. Then the integral closure of R is generated by monomials.*

Furthermore, let $E \in \mathbb{N}^d$ be such that a monomial $\underline{X}^{\underline{m}}$ is a monomial generator of R if and only if $\underline{m} \in E$. Then $\underline{m} \in \mathbb{Q}_{\geq 0} E \cap \mathbb{Z} E$ if and only if $\underline{X}^{\underline{m}} \in \overline{R}$.

Proof: Since $\overline{R} \subseteq k[X_1, \ldots, X_d]$, the first part is an immediate corollary of Corollary 2.3.6, applied with $G = \mathbb{N}^d$. If $\underline{m} \in \mathbb{N}^d$, then $\underline{X}^{\underline{m}}$ is in the field of fractions of R if and only if $\underline{m} \in \mathbb{Z} E$, and by the form of the integral equations for monomials, $\underline{X}^{\underline{m}}$ is integral over R if and only if $\underline{m} \in \mathbb{Q}_{\geq 0} E$. $\square$

The main point of the proof above is that the G–graded ring (with $G = \mathbb{N}^d$) is contained in a natural G–graded integrally closed ring.

We finish this section by proving that the integral dependence among $\mathbb{N}$–graded algebras is related to the integral dependence of ideals:

Proposition 2.3.8 *Let $R \subseteq S$ be an inclusion of $\mathbb{N}$–graded rings. Then $R \subseteq S$ is an integral extension if and only if the ring inclusion $R_0 \subseteq S_0$ is an integral extension and for each $n \in \mathbb{N}$, the S–ideal $S_n S$ is integral over the ideal $R_n S$.*

If $R_0 = S_0$, $R = R_0[R_1]$, $S = S_0[S_1]$, then S is integral over R if and only if $S_1 S$ is integral over $R_1 S$.

Proof: If S is integral over R, then any element $s \in S_n$ is integral over R. There exists a homogeneous equation of integral dependence of s over R, which shows that s is integral over R_nS. Conversely, assume that for all $n \in \mathbb{N}$, S_nS is integral over R_nS. Let $s \in S_n$. Then s satisfies an equation of integral dependence over the ideal R_nS, with all the coefficients in $R[S_0]$. Thus the ring S is integral over $R[S_0]$. But S_0 is integral over R_0, so S is integral over R. This proves the first part. The proof of the second part follows easily. $\square$

2.4. Rings of homomorphisms of ideals

We show in this section that whether an integral domain is integrally closed can be detected by computing infinitely many rings of homomorphisms of finitely generated fractional ideals. This is, of course, uncheckable in general, nevertheless, this new characterization does have checkable corollaries, see Section 15.3. This section is only used in later chapters; a reader may want to skip it on first reading.

Rings of homomorphisms of ideals appear in all general algorithms for computing the integral closure, as explained in Chapter 15. They are "computable" over "computable" rings. Other examples of manipulations of rings of homomorphisms of ideals are for example in Katz [163].

Lemma 2.4.1 *Let R be a reduced ring with total ring of fractions K. If K is a direct product of finitely many fields and I and J are R–submodules of K, then every R–homomorphism $I \to J$ is multiplication by an element of K.*

Proof: Write $K = K_1 \times \cdots \times K_r$, where each K_i is a field. Let W be the set of all non–zerodivisors in R. Then W is a multiplicatively closed subset of R, and $W^{-1}I$ is a $W^{-1}R$–module, i.e., a K–module, contained in K. After possibly reindexing, $W^{-1}I = K_1 \times \cdots \times K_s$ for some $s \leq r$. Let e_i be the idempotent of K such that $Ke_i = K_i$. Then $1 = \sum_{i=1}^{r} e_i$. Let $w \in W$ such that for $i = 1, \ldots, r$, $we_i \in R$ and such that for $i = 1, \ldots, s$, $we_i \in I$. Observe that $(e_{s+1}, \ldots, e_r)I = 0$. We claim that φ is multiplication by $\varphi(\sum_{i=1}^{s} we_i)/w$. Let $x \in I$. Then $w\varphi(x) = \varphi(wx) = \varphi(\sum_{i=1}^{r} wxe_i) = \varphi(\sum_{i=1}^{s} wxe_i) = x\varphi(\sum_{i=1}^{s} we_i)$, which proves the claim and the lemma. $\square$

We explicitly write the multiplication element as in the lemma in the case where I is a torsion–free ideal. In this case, with notation as in the proof, $W^{-1}I = K$ and $w = \sum_{i=1}^{r} we_i \in I$. Thus any $\varphi \in \operatorname{Hom}_R(I, J)$ is multiplication by $\varphi(w)/w$. In fact, whenever $y \in I$ is a non–zerodivisor, φ is multiplication by $\varphi(y)/y$ because for any $x \in I$, $y\varphi(x) = \varphi(yx) = x\varphi(y)$.

Lemma 2.4.2 *Let R be a reduced ring with total ring of fractions K. Assume that K is a direct product of finitely many fields, and that I and J are R–submodules of K. Then the natural map*

$$(J :_K I) \longrightarrow \operatorname{Hom}_R(I, J)$$

is a surjective R–module homomorphism with kernel $(0 :_K I)$.

Proof: It suffices to prove that the kernel is as specified. If $k \in K$ and if multiplication by k is the zero function in $\text{Hom}_R(I, J)$, then $kI = 0$, so $k \in (0 :_K I)$. $\square$

If R and K are as above, then by Lemma 2.4.2, for any R–module $I \subseteq K$, $\text{Hom}_R(I, I)$ is naturally identified as an R–submodule of $K/(0 :_K I)$. Furthermore, $\text{Hom}_R(I, I)$ is a commutative subring of $K/(0 :_K I)$ containing the identity. If φ is multiplication by x and ψ is multiplication by y, then $xy = yx$ implies that $\varphi \circ \psi = \psi \circ \varphi$. If in addition I is torsion–free, then $\text{Hom}_R(I, I)$ is a subring of K, and if I is finitely generated, then $\text{Hom}_R(I, I)$ is even contained in the integral closure of R in K: for if $xI \subseteq I$, then by the determinantal trick (Lemma 2.1.8), $x \in \overline{R}$.

The following makes this representation as a submodule of K transparent:

Lemma 2.4.3 *Let R be a reduced ring with only finitely many minimal prime ideals. Let I and J be R–submodules of the total ring of fractions K of R, and suppose that I contains an element x that is a non–zerodivisor on R. Then $\text{Hom}_R(I, J)$ can be identified (as in Lemma 2.4.1) with the R–submodule $\frac{1}{x}(xJ :_R I)$ of K.*

Proof: Certainly multiplication by any element of $\frac{1}{x}(xJ :_R I)$ takes I to J, and conversely if $k \in K$ and $kI \subseteq J$, then in particular kx is in R, so that $kx \in (xJ :_R I)$, whence $k \in \frac{1}{x}(xJ :_R I)$. $\square$

This characterization makes it clearer how $\text{Hom}_R(I, J)$ is a submodule and how $\text{Hom}_R(I, I)$ is a subring of the total ring of fractions of R.

We now restrict out attention to the case where R is a domain.

Definition 2.4.4 *Let R be an integral domain. A **fractional ideal** I of R is a submodule of the field of fractions K of R for which there exists a non–zero element k in K such that $kI \subseteq R$.*

We show that fractional ideals play a big role in the computation of integral closures. In particular, we prove that $\overline{R} = \cup_I \text{Hom}_R(I, I)$, where I varies over fractional ideals of R.

Definition 2.4.5 *If R is a domain with field of fractions K, for any non–zero fractional ideal I, define $I^{-1} = \text{Hom}_R(I, R)$.*

Note that $II^{-1} = I^{-1}I \subseteq R$. Whenever equality holds, both I and I^{-1} are finitely generated: write $1 \in R$ as a finite linear combination $1 = \sum_{i=1}^{n} a_i b_i$ with $a_i \in I$ and $b_i \in I^{-1}$. Then the a_i generate I and the b_i generate I^{-1}: for any $r \in I$, $r = \sum a_i(rb_i) \in \sum_i a_i R$, and for any $s \in I^{-1}$, $s = \sum(sa_i)b_i \in \sum_i b_i R$. Fractional ideals I for which $II^{-1} = R$ are called **invertible fractional ideals**.

It is not true that all finitely generated fractional ideals are invertible. An example is the ideal $I = (X, Y)$ in $k[X, Y]$, where k is a field and X and Y variables over it. Note that here $I^{-1} = k[X, Y]$, for no other element of the field of fraction multiplies both X and Y back into the ring.

By definition $I \subseteq (I^{-1})^{-1}$.

The fractional ideal I^{-1} is in a sense more useful than the original I:

Lemma 2.4.6 *If R is a domain with field of fractions K and I is a non–zero fractional ideal, then as submodules (and even subrings) of K,*

$$(I \cdot I^{-1})^{-1} = \mathrm{Hom}_R(I^{-1}, I^{-1}) = \mathrm{Hom}_R((I^{-1})^{-1}, (I^{-1})^{-1}).$$

Proof: Let $k \in (I \cdot I^{-1})^{-1}$. Thus $k \in K$ and $kI \cdot I^{-1} \subseteq R$, whence $kI^{-1} \subseteq I^{-1}$, so that $k \in \mathrm{Hom}_R(I^{-1}, I^{-1})$.

Next let $k \in \mathrm{Hom}_R(I^{-1}, I^{-1})$. Then $kI^{-1} \subseteq I^{-1}$. If $r \in (I^{-1})^{-1}$, then $rkI^{-1} \subseteq I^{-1}(I^{-1})^{-1} \subseteq R$, so that $rk \in (I^{-1})^{-1}$. This proves that $k \in \mathrm{Hom}_R((I^{-1})^{-1}, (I^{-1})^{-1})$.

Let $k \in \mathrm{Hom}_R((I^{-1})^{-1}, (I^{-1})^{-1})$. Then $k(I^{-1})^{-1} \subseteq (I^{-1})^{-1}$, so that $kI \cdot I^{-1} \subseteq k(I^{-1})^{-1} \cdot I^{-1} \subseteq (I^{-1})^{-1} \cdot I^{-1} \subseteq R$, and $k \in (I \cdot I^{-1})^{-1}$. $\qquad\square$

Discussion 2.4.7 It is worth noting that the lemma above follows from the Hom–tensor adjointness. Namely, the first equality in the display in the lemma follows due to

$$\mathrm{Hom}_R(I^{-1}I, R) \cong \mathrm{Hom}_R(I^{-1} \otimes_R I, R),$$

since the kernel of the natural map from $I^{-1} \otimes_R I$ onto $I^{-1}I$ is a torsion module, and any homomorphism of $I \otimes_R I^{-1}$ to R automatically sends torsion elements to zero. Hence $\mathrm{Hom}_R(I^{-1}I, R) \cong \mathrm{Hom}_R(I^{-1} \otimes_R I, R) \cong \mathrm{Hom}_R(I^{-1}, I^{-1})$, the last isomorphism coming from the Hom–tensor adjointness. Similarly one can prove the other parts of Lemma 2.4.6 as well as other formulas involving I^{-1}.

In general it is not true that $\mathrm{Hom}_R(I, I) = (I \cdot I^{-1})^{-1}$ ("justifying" the claim that I^{-1} is more useful than I). For example, let $R = k[t^3, t^5, t^7]$, where t is a variable over a field k, and let $I = t^3R$. Then $I^{-1} = 1R + t^2R + t^4R$, $I \cdot I^{-1} = (t^3, t^5, t^7)R$, $(I \cdot I^{-1})^{-1} = 1R + t^2R + t^4R$, and $\mathrm{Hom}_R(I, I) = R$. Nevertheless, in a certain sense, I is no less useful than I^{-1}:

Proposition 2.4.8 *Let R be a Noetherian domain. If I varies over non–zero (finitely generated fractional) ideals, then*

$$\overline{R} = \bigcup_I \mathrm{Hom}_R(I, I) = \bigcup_I \mathrm{Hom}_R(I^{-1}, I^{-1}).$$

Proof: By Determinantal trick 2.1.8, $\mathrm{Hom}_R(I^{-1}, I^{-1}), \mathrm{Hom}_R(I, I) \subseteq \overline{R}$, giving one pair of inclusions. Let $s \in \overline{R} \setminus R$. Set $J = R :_R R[s]$. Then J is a non–zero finitely generated ideal in R satisfying the property that $sJ \subseteq J$. This proves that $\overline{R} \subseteq \cup_I \mathrm{Hom}_R(I, I)$. Similarly, $sJ^{-1}J = sJJ^{-1} \subseteq JJ^{-1} \subseteq R$, so that $sJ^{-1} \subseteq J^{-1}$, whence $\overline{R} \subseteq \cup_I \mathrm{Hom}_R(I^{-1}, I^{-1})$. These give the reverse pair of inclusions. $\qquad\square$

This gives a new normality criterion for a domain:

Proposition 2.4.9 *Let R be a Noetherian integral domain with field of fractions K. Then the following are equivalent:*

(1) R is integrally closed.

(2) For all non–zero fractional ideals I, $(I \cdot I^{-1})^{-1} = R$.

(3) For all non–zero ideals I, $(I \cdot I^{-1})^{-1} = R$.

(4) For all non–zero ideals I and all prime ideals P in R of grade one (i.e., the maximal length of a regular sequence contained in P is one), $(I \cdot I^{-1})^{-1} R_P = R_P$.

Proof: For all fractional ideals I, $R \subseteq (I \cdot I^{-1})^{-1}$, so the first three statements are equivalent by Proposition 2.4.8 and Lemma 2.4.6. Clearly (3) implies (4). Now assume (4). Let $x \in (I \cdot I^{-1})^{-1}$. Then for all P of grade 1, $x \in R_P$. But $\cap_P R_P = R$, as P varies over ideals of grade 1. $\qquad\square$

2.5. Exercises

2.1   Let $R_i \subseteq S_i$ be ring inclusions, with i varying in some index set I. Prove that $\oplus_{i \in I} S_i$ is integral over $\oplus_{i \in I} R_i$ if and only if for each i, S_i is integral over R_i. Furthermore, $\oplus_{i \in I} R_i$ is integrally closed in $\oplus_{i \in I} S_i$ if and only if for each i, R_i is integrally closed in S_i.

2.2   Let $A \subseteq B \subseteq C$ be rings such that B is the integral closure of A in C. Let X be a variable over C. Prove that $B[X]$ is the integral closure of $A[X]$ in $C[X]$.

2.3   Let R be an integrally closed domain. Let X be a variable over R.

(i) Prove that $A[X]$ is integrally closed.

(ii) Prove that $R[[X]]$ is integrally closed if R is Noetherian.

2.4   Let $\varphi : R \to S$ be a ring homomorphism and P a prime ideal in R. Prove that $\varphi^{-1}(\varphi(P)S) = P$ if and only if there exists a prime ideal Q in S such that $\varphi^{-1}(Q) = P$.

2.5   Let $(R, \mathfrak{m})$ be an integrally closed local domain and x a non–zero element in $Q(R)$. Prove that $\mathfrak{m}R[x] \cap \mathfrak{m}R[1/x] \subseteq \mathfrak{m}$.

2.6   Let $R \subseteq S$ be an integral extension of rings such that for every $s \in S$, some power of s is in R.

(i) Prove that for every prime ideal P of R there exists a unique prime ideal in S lying over P.

(ii) Prove that there is a one–to–one natural correspondence between prime ideals in R and prime ideals in S. Prove that under this correspondence height is preserved.

2.7   Let R be a ring and G a finite group of automorphisms of R. Let A be the ring of invariants of G, i.e., $A = \{r \in R \mid$ for all $g \in G, g(r) = r\}$.

(i) Prove that $A \subseteq R$ is an integral extension.

(ii) Prove that if Q_1, Q_2 are prime ideals in R such that $Q_1 \cap A = Q_2 \cap A$, then there exists $g \in G$ such that $Q_2 = gQ_1$.

(iii) Prove that $A \subseteq R$ satisfies the Going–Down property.

2.8   Let $R \subseteq S$ be an integral extension of integral domains. Assume that R is integrally closed in its field of fractions K and that the field of fractions L of S is a finite Galois extension of K. Prove that for each prime ideal P of R, the number of prime ideals in S lying over P is at most the degree $[L:K]$. (Hint: see Exercise 2.7.)

2.9   Let k be a field of characteristic different from 2, and X, Y variables over k. Let $A = k[X^2 - 1, X^3 - X, Y] \subseteq R = k[X, Y]$. Prove that $A \subseteq R$ is an integral extension that does not satisfy the Going–Down property.

2.10   Let $R \subseteq S$ be reduced rings satisfying one of the following:
   (i) S is contained in the total ring of fractions of R.
   (ii) S is integral over R and S is an integral domain.
Prove that for every $s \in S$, $(R :_S s)$ is a R–submodule of S not contained in any minimal prime ideal of S.

2.11   Let $R \subseteq S$ be an extension of reduced rings, and let S be a subset of the total ring of fractions T of R. If S is module–finite over R, prove that $R :_T S$ is an ideal in R and in S that contains a non–zerodivisor. This ideal $R :_T S$ is called the **conductor**. Prove that $R :_T S$ equals $R :_R S$ and that $R :_T S$ is the largest ideal that R and S have in common. (More on conductors is in Chapter 12.)

2.12   Let $R \subseteq S$ be a module–finite extension of reduced Noetherian rings. Let P be a prime ideal in R that does not contain the ideal $R :_R S$.
   (i) Prove that $S_{R \setminus P} = R_P$.
   (ii) Prove that $R :_R S$ is not contained in any minimal prime ideal of R if and only if R and S have the same total ring of fractions.

2.13   Let R be an integral domain, let $\overline{R}$ be its integral closure in the field of fractions of R, and let $q \subseteq p$ be prime ideals in R. Suppose that for every prime ideal P in $\overline{R}$ with $p = P \cap R$ there exists $Q \in \operatorname{Spec} \overline{R}$ such that $Q \cap R = q$ and $Q \subseteq P$. Prove that for every torsion–free integral extension S of R and for every $P \in \operatorname{Spec} S$ with $p = P \cap R$ there exists $Q \in \operatorname{Spec} S$ such that $Q \cap R = q$ and $Q \subseteq P$.

2.14   Let d be a square–free integer. Prove that $\mathbb{Z}[\sqrt{d}]$ is not integrally closed if and only if d is congruent to 1 modulo 4.

2.15   Let d be an integer that is not a square and is not square–free. Prove that $\mathbb{Z}[\sqrt{d}]$ is never integrally closed.

2.16   Let D be a non–zero integer. Factor $D = D_0 n^2$, where D_0 is a square–free integer. Let R be the integral closure of $\mathbb{Z}[\sqrt{D}]$. Prove that the elements of R are of the form $a + bw$, with $a, b \in \mathbb{Z}$, where
   (i) $w = \frac{1+\sqrt{D}}{2}$ if $D_0 \equiv D \equiv 1 \bmod 4$,
   (ii) $w = \frac{\sqrt{D}}{2}$ if $D \equiv 0, D_0 \equiv 1 \bmod 4$,
   (iii) $w = \sqrt{D}$ if $D_0 \not\equiv 1 \bmod 4$.

2.17   Let R be a ring that is a direct summand of a ring S as an R–module. Prove that if S is integrally closed, so is R.

2.18   Let k be a field, $X_1, \ldots, X_d$ variables over k, and R a subalgebra of $k[X_1, \ldots, X_d]$ that is generated by finitely many monomials. The goal of this exercise is to prove that $\overline{R}$ is a direct summand (as a module over $\overline{R}$) of a polynomial ring (not necessarily of $k[X_1, \ldots, X_d]$).

   (i) Let E be the set of exponent vectors of monomials in R. Prove that whenever there exists a rational matrix A such that $E = (\ker A) \cap \mathbb{N}^d$, then R is a direct summand of $k[X_1, \ldots, X_d]$ and is integrally closed.

   (ii) Let S be the subalgebra of $k[X_1, \ldots, X_d]$ generated by monomials whose exponents are in $(\mathbb{Q}_{\geq 0}E) \cap \mathbb{N}^d$. Prove that $\overline{R}$ is a direct summand of S.

   (iii) Prove that there exist $b_1, \ldots, b_m \in \mathbb{Z}^d$ such that $e \in \mathbb{Q}_{\geq 0}E$ if and only if for all $i = 1, \ldots, m$, $b_i \cdot e \geq 0$ (dot product).

   (iv) Let B be the $(m+d) \times d$ matrix whose first d rows are identity, and the last m rows are $b_1, \ldots, b_m$. Prove that B defines an injective map $\mathbb{Q}^d \to \mathbb{Q}^{m+d}$ that takes F to $\mathbb{N}^{d+n}$. Let A be the matrix of all relations on the image of F under B. Prove that $(\ker A) \cap \mathbb{N}^{m+d}$ is the image of F under B.

   (v) Prove that $\overline{R}$ is a direct summand of $k[X_1, \ldots, X_{m+d}]$. (More general results of this type are in [121].)

2.19   Let k be a field, and X, Y, t, s variables over k. Prove that the integrally closed ring $k[X, Y, X^2t, XYt, Y^2t]$ is not a direct summand of the polynomial ring $k[X, Y, t]$ but is a direct summand of the polynomial ring $k[X, Y, t, s]$.

2.20   Let k be a field, $X_1, \ldots, X_d$ variables over k, and consider the ring inclusion $k[\underline{X}^{\alpha_1}, \cdots, \underline{X}^{\alpha_n}] \subseteq k[X_1, \cdots, X_n]$, where $\alpha_i = (\alpha_{i1}, \cdots, \alpha_{id}) \in \mathbb{N}^d$. Prove that these rings have the same field of fractions if and only if the determinant of the $d \times d$ matrix $[\alpha_{ij}]$ is ± 1.

2.21   Let $R = k[X] \subseteq S = k[X, Y, Z]/(X^3 + Y^2 + Z^2)$, where X, Y and Z are variables over a field k. Find a monoid G such that $R \subseteq S$ and the integral closure of R in S are all G-graded. (Cf. Example 2.3.3.)

2.22   (Y. Yao) Let $(G, +)$ be a commutative monoid such that

   (i) for all $x, y \in G$ and $n \in \mathbb{N}_{>0}$, $nx = ny$ implies that $x = y$,

   (ii) for all x, y and z in G, $x + z = y + z$ implies that $x = y$.

Prove that whenever $R \subseteq S$ are G-graded rings, the integral closure of R in S is also G-graded.

2.23   (Seidenberg [262]) Let R be a Noetherian integral domain containing $\mathbb{Q}$ and K the field of fractions of R. Let $D : K \to K$ be a derivation such that $D(R) \subseteq R$. Prove that D takes almost integral elements to almost integral elements. (Hint: Let t be a variable. Prove that $1 + tD + (t^2/2!)D^2 + (t^3/3!)D^3 + \cdots$ is a ring endomorphism on $K[[t]]$.)

2.24   ([262]) Let R be an integral domain containing $\mathbb{Q}$ and let t be a variable over R.

   (i) Prove that if R is integrally closed and Noetherian, so is $R[[t]]$.

(ii) If R is Noetherian, prove that if the integral closure $\overline{R}$ of R is Noetherian and module–finite over R, then the integral closure of $R[[t]]$ is $\overline{R}[[t]]$.

(iii) Assume that R is integrally closed (and that R contains a field). Suppose that there exists a non–unit $x \in R$ such that $\cap(x^n) \neq 0$. Prove that $R[[t]]$ is not integrally closed.

2.25   Let k be a field, t an indeterminate over k, and R the ring of **formal Puiseux series** in t over k, i.e., R is the ring consisting of all elements of the form $\sum_{i \geq 0} a_i t^{i/n}$, where $a_i \in k$ and n is a positive integer.

(i) Prove that $R = \cup_{n \geq 1} k[[t^{\frac{1}{n}}]]$ and that the field of fractions of R equals $\cup_{n \geq 1} k((t^{\frac{1}{n}}))$.

(ii) Prove that R has only one maximal ideal and is integrally closed.

(iii) Prove that R is integral over $k[[t]]$.

(iv) If k is $\mathbb{R}$ or $\mathbb{C}$, and A is the ring of all convergent Puiseux series in t over k, prove that A is integrally closed and integral over the ring of all convergent power series $k\{t\}$ in t over k.

2.26   Let $R \subseteq S$ be domains with the same field of fractions. An element $x \in S$ is said to be **almost integral** over R if there exists a non–zero element $c \in R$ such that for all sufficiently large integers n, $cx^n \in R$.

(i) Prove that $x \in S$ is almost integral over R if and only if there exists $c \in R \setminus \{0\}$ such that for all integers $n \geq 1$, $cx^n \in R$.

(ii) Prove that the set of all elements of S almost integral over R forms a ring between R and S. This ring is called the **complete integral closure** of R in S.

(iii) Prove that every element of S that is integral over R is almost integral over R.

(iv) Assume that R is Noetherian and $x \in S$ almost integral over R. Prove that x is integral over R.

(v) Prove that $R + XK[X]$ is integrally closed and that the complete integral closure of $R + XK[X]$ is $K[X]$ if R is integrally closed in its field of fractions K and X is a variable over K.

(vi) Prove that a completely integrally closed domain is integrally closed.

(vii) Prove that a unique factorization domain is completely integrally closed.

(viii) Let X be a variable over R. Prove that R is completely integrally closed if and only if $R[X]$ is completely integrally closed, and that holds if and only if $R[[X]]$ is completely integrally closed.

(ix) Let k be a field, X, Y variables over k, and $R = k[X^n, Y^{n^2} \mid n = 1, 2, \ldots]$. Prove that $S = R[XY^n \mid n = 1, 2, \ldots]$ is integral over R, and that Y is almost integral over S but not over R. Conclude that the complete integral closure of R is not completely integrally closed.

2.27   Let R be a Noetherian domain whose integral closure $\overline{R}$ is a module finite extension strictly containing R. Let $P \in \mathrm{Ass}(\overline{R}/R)$. Prove that depth $R_P = 1$.

2.28   Let k be a field, $X_1, \ldots, X_n$ variables over k and $\mathfrak{m}$ a maximal ideal in $k[X_1, \ldots, X_n]$. Prove that $\mathfrak{m}$ is generated by elements $f_1, \ldots, f_n$, where $f_i \in k[X_1, \ldots, X_i]$ is monic in X_i.

2.29   An integral domain R is said to be **seminormal** if for each x in the field of fractions with $x^2, x^3 \in R$, actually $x \in R$. Prove that an integral domain R is seminormal if and only if each x in the field of fractions satisfying $x^n, x^{n+1} \in R$ for some positive integer n is actually an element of R.

2.30   Let k be a field, t a variable over k, and $R = k[t(t-1), t^2(t-1)]$. Prove that R is the set of all polynomials $f \in k[t]$ such that $f(0) = f(1)$. Prove that R is seminormal but not normal.

2.31   Let X, Y variables over a field k and $R = k[X^2, Y^2, XY, X^2Y, XY^2]$. Prove that R is seminormal but not normal.

2.32   Let R be a seminormal domain and S a domain containing R. Prove that $R :_R S$ is a radical ideal in S.

Seminormality was first defined by Traverso in [296] as follows: the seminormalization of a reduced Noetherian ring R with module–finite integral closure $\overline{R}$ is the set of all elements $s \in \overline{R}$ of R such that for every prime ideal P of R, $s \in A_P + J(\overline{R}_{R \backslash P})$, where $J(\_)$ denotes the intersection of all the maximal ideals of the ring. The ring R is said to be seminormal if it is equal to its seminormalization. The definition appearing in Exercise 2.29 is a characterization due to Hamann [103]. Hamann proved the equivalence for pseudo-geometric rings; in full generality the equivalence was proved by Gilmer and Heitmann in [92]. Swan [277] redefined seminormality and constructed seminormalizations in greater generality: R is seminormal if whenever $s, t \in R$ and $s^3 = t^2$, there exists $r \in R$ such that $s = r^2$ and $t = r^3$. Hamann's and Swan's formulations are equivalent when the total ring of fractions of R is a finite direct product of fields.

Here is another notion of integral dependence: Swan [277] defined a **subintegral** extension to be an integral extension $R \subset S$ of rings such that the contraction of ideals from S to R gives a one-to-one correspondence between prime ideals in S and prime ideals in R, and the induced maps of residue fields are all isomorphisms. The seminormalization of a reduced ring R turns out to be the largest subintegral extension of R in its total quotient ring. Roberts and Singh [246] gave element–wise definition for subintegrality over a $\mathbb{Q}$–algebra. Later, Reid, Roberts, and Singh [242] removed the $\mathbb{Q}$–algebra assumption, and introduced a more general definition, **quasisubintegrality** (later renamed **weak subintegrality**), which agrees with subintegrality for $\mathbb{Q}$–algebras. Vitulli and Leahy in [314] defined the notion of weak subintegrality for ideals.

3
Separability

Our purpose in this chapter is to develop basic results on separability and the tensor product of fields. We need to some of these results to prove the main theorems on the behavior of integral closure under flat homomorphisms in Chapter 19. A basic case to understand is the behavior of integral closures under field extensions, and that immediately leads to the notion of separability. One of the main results of this chapter, Theorem 3.1.3, relates to integral closure: it shows that the integral closure of integrally closed domains in separable field extensions of the field of fractions is always module–finite. More connections to integral and algebraic closure are in Section 3.3. The main section, Section 3.2, contains many formulations of separability.

3.1. Algebraic separability

An algebraic field extension $k \subseteq \ell$ is **separable** if for every $x \in \ell$, the minimal polynomial $f \in k[X]$ of x over k is relatively prime to its formal derivative f' in the polynomial ring $k[X]$. In other words, in an algebraic closure of k, all the roots of f are distinct. Some relevant properties of separable extensions are summarized without proof in the following (see for example [210, Corollaries 5.7 and 8.17]):

Theorem 3.1.1 *Let $k \subseteq \ell$ be a finite separable field extension. Then the following properties hold:*
(1) (Primitive Element Theorem) There exists $x \in \ell$ such that $\ell = k(x)$.
(2) The trace map $\mathrm{Tr} : \ell \to k$ is not identically zero.

Discussion 3.1.2 A consequence of Theorem 3.1.1 is that for any basis $\{x_1, \ldots, x_n\}$ of a separable field extension ℓ over k, the $n \times n$ matrix $[\mathrm{Tr}(x_i x_j)]$ is invertible: if not, there exist $c_1, \ldots, c_n \in k$, not all zero, such that for all $j = 1, \ldots, n$, $0 = \sum_{i=1}^n c_i \mathrm{Tr}(x_i x_j) = \mathrm{Tr}(\sum_{i=1}^n c_i x_i x_j)$. Set $c = \sum_{i=1}^n c_i x_i$. By the assumption on the c_i not being all zero, $c \neq 0$. Let $a \in \ell$ such that $\mathrm{Tr}(a) \neq 0$, and write $a/c = \sum_j a_j x_j$ for some $a_j \in k$. Then

$$0 = \sum_{j=1}^n a_j \mathrm{Tr}\left(\sum_{i=1}^n c_i x_i x_j\right) = \mathrm{Tr}\left(c \sum_{j=1}^n a_j x_j\right) = \mathrm{Tr}(a),$$

which is a contradiction. Thus in particular there exist $a_{ij} \in k$ such that $[a_{ij}][\mathrm{Tr}(x_i x_j)]$ is the identity matrix. Set $y_i = \sum_{i=1}^n a_{ij} x_i$. Then $\mathrm{Tr}(y_i x_j) = \delta_{ij}$, and one can easily verify that $\{y_1, \ldots, y_n\}$ is a basis of ℓ over k.

This is enough to show that integral closure behaves better on separable extensions:

Theorem 3.1.3 *Let R be a Noetherian integral domain that is integrally closed in its field of fractions K. Let L be a finite separable field extension of K. The integral closure of R in L is module–finite over R.*

Proof: Let S be the integral closure of R in L and let $\{x_1, \ldots, x_n\}$ be a vector space basis for L over K.

Each x_i is algebraic over K, so that for some non–zero $r_i \in R$, $r_i x_i$ is integral over R. Each $r_i x_i \in S$, and $\{r_1 x_1, \ldots, r_n x_n\}$ is still a vector space basis for L over K. Thus by replacing x_i by $r_i x_i$ we may assume that each x_i is in S.

As $K \subseteq L$ is finite and separable, as in Discussion 3.1.2, there exists a dual basis $\{y_1, \ldots, y_n\}$ of L over K such that for all i, j, $\mathrm{Tr}(x_i y_j) = \delta_{ij}$, where Tr stands for the trace of L over K and δ_{ij} is Kronecker's delta function. Let $s \in S$. Write $s = \sum_i k_i y_i$ for some $k_i \in K$. Then
$$\mathrm{Tr}(x_j s) = \sum_i \mathrm{Tr}(x_j k_i y_i) = \sum_i k_i \mathrm{Tr}(x_j y_i) = k_j.$$
But the trace of any element of S is in K and is still integral over R. Thus $\mathrm{Tr}(x_j s) = k_j$ is in R. It follows that $s \in \sum_{i=1}^n R y_i$, hence S is an R–submodule of $\sum_{i=1}^n R y_i$, and consequently S is module–finite over R. $\qquad\square$

3.2. General separability

In this section we give many criteria for a ring to be separable over a field. The basic definition is the following.

Definition 3.2.1 *Let k be a field, and R a k–algebra. We say that R is* **separable** *over k if for every field extension $k \subseteq \ell$, the ring $R \otimes_k \ell$ is reduced.*

It is not clear that this notion of separability generalizes the definition introduced in the first section, but it does. See Proposition 3.2.4 (5) below.

Proposition 3.2.2 *Let k be a field and R a k–algebra.*
(1) Suppose that $k \subseteq S \subseteq R$. If R is separable over k then S is separable over k.
(2) If R is separable over k and k is contained in a field ℓ, then $R \otimes_k \ell$ is separable over ℓ.
(3) A k–algebra R is separable over k if and only if for every subalgebra $S \subseteq R$ that is finitely generated as a k–algebra, S is separable over k.
(4) R is separable over k if and only if $R \otimes_k \ell$ is reduced for all finitely generated field extensions ℓ of k.

Proof: (1) Let ℓ be a field extension of k. Since ℓ is flat as a k–algebra, the injection of S to R induces an injection from $S \otimes_k \ell$ into $R \otimes_k \ell$. As $R \otimes_k \ell$ is reduced, so is $S \otimes_k \ell$, proving that S is separable over k.

(2) Let L be a field extension of ℓ. Then $(R \otimes_k \ell) \otimes_\ell L \cong R \otimes_k L$, so that $(R \otimes_k \ell) \otimes_\ell L$ is reduced, and hence $R \otimes_k \ell$ is separable over ℓ.

(3) One direction follows immediately from (1). If R is not separable over k then there is a field extension ℓ of k and a non–zero nilpotent element $y \in R \otimes_k \ell$. Suppose $y^n = 0$. The tensor product of R and ℓ over k is isomorphic to a free module F modulo the submodule N generated by certain universal relations that give the tensor product its universal property. To say $y^n = 0$ means that we can represent the element corresponding to y^n in this F as a finite sum of generators of N. We can collect the coefficients needed to define y as an element of the tensor product, and the coefficients appearing when writing y^n as a finite sum of elements of N. Let S be the subalgebra of R generated by all the coefficients of R appearing. Define $z \in S \otimes_k \ell$ using the same coefficients as for y. Then $z^n = 0$, and z is non–zero since it maps to y under the injection of $S \otimes_k \ell$ to $R \otimes_k \ell$ (using the flatness of ℓ over k).

Part (4) has a proof similar to that of (3) and we leave it as an exercise. $\square$

Discussion 3.2.3   If $k \subseteq \ell \subseteq R$ and R is separable over k, then R is not necessarily separable over ℓ. For example, let k be a field of positive characteristic p, and let t be a variable over k. Set $\ell = k(t^p)$ and $R = k(t)$. By Proposition 3.2.4 below, R is separable over k, but R is not separable over ℓ. Although this example is clear, the reader should consider the following "proof": suppose that $k \subseteq \ell \subseteq R$ and R is separable over k. We will "prove" that R is separable over ℓ. Let $\ell \subseteq L$ be an arbitrary field extension. First note that $R \otimes_k \ell$ is separable over ℓ by Proposition 3.2.2 (2). Hence $(R \otimes_k \ell) \otimes_\ell L$ is reduced. After tensoring with R over k, the inclusion of $k \subseteq \ell$ induces an inclusion $R \subseteq R \otimes_k \ell$. Tensor the latter inclusion with L over ℓ (which is flat and will preserve the injection), to obtain that $R \otimes_\ell L \subseteq (R \otimes_k \ell) \otimes_\ell L$. Since $(R \otimes_k \ell) \otimes_\ell L$ is reduced, so is $R \otimes_\ell L$, which implies that R is separable over ℓ. What is wrong with this "proof"?

Proposition 3.2.4   *Let k be a field and R a k–algebra.*

(1) *If $t_1, \ldots, t_n$ are algebraically independent over k, then $k(t_1, \ldots, t_n)$ is separable over k.*

(2) *R is separable over k if and only if for every reduced k–algebra S, the ring $R \otimes_k S$ is reduced.*

(3) *Suppose that $R = k[t]/(f)$. Then R is separable over k if and only if $(f, f') = 1$, where f' is the derivative of f.*

(4) *Suppose that $k \subseteq \ell \subseteq R$ and that ℓ is a field. If R is separable over ℓ and ℓ is separable over k, then R is separable over k.*

(5) *If $k \subseteq \ell$ is an algebraic field extension, then ℓ is separable over k if and only if for every $x \in \ell$, the minimal polynomial $f \in k[t]$ of x over k satisfies that $(f, f') = 1$.*

Proof: (1) Let ℓ be an extension field of k. Then $k(t_1, \ldots, t_n) \otimes_k \ell \cong k[t_1, \ldots, t_n]_W \otimes_k \ell$, where W is the multiplicatively closed set consisting of the non–zero elements of the polynomial ring $k[t_1, \ldots, t_n]$. But $k[t_1, \ldots, t_n]_W \otimes_k \ell$ is isomorphic to $(k[t_1, \ldots, t_n] \otimes_k \ell)_W \cong (\ell[t_1, \ldots, t_n])_W$, which is a domain.

(2) One direction is trivial. For the other, we first reduce to the case where S is Noetherian. If R is separable over k and S is a reduced k–algebra such that $R \otimes_k S$ is not reduced, then there exists a subalgebra $T \subseteq S$ that is finitely generated over k such that $R \otimes_k T$ is not reduced. One simply needs to use the argument in Proposition 3.2.2 (3) above. By replacing S by T we can assume that S is Noetherian. Let $P_1, \ldots, P_n$ be the minimal primes of S. Since S is reduced, S embeds in $T = S/P_1 \times \cdots \times S/P_n$. Let K_i be the field of fractions of S/P_i. As R is flat over k, $R \otimes_k S \subseteq \prod_i R \otimes_k K_i$. If R is separable over k, each $R \otimes_k K_i$ is reduced, and thus so is $R \otimes_k S$.

(3) Assume that R is separable over k. Let $\overline{k}$ be an algebraic closure of k. We have that $\overline{k}[t]/(f) \cong R \otimes_k \overline{k}$ is reduced, which implies that f has no multiple roots over $\overline{k}$. Hence $(f, f') = 1$. Conversely suppose that $(f, f') = 1$. Let ℓ be an extension field of k. If $R \otimes_k \ell$ is not reduced, then neither is $R \otimes_k \overline{\ell}$, where $\overline{\ell}$ is an algebraic closure of ℓ. Thus $f(t)$ has multiple roots over $\overline{\ell}$, which gives that f and f' are not relatively prime.

(4) Let L be an extension field of ℓ. As ℓ is separable over k, the algebra $S = \ell \otimes_k L$ is reduced. By (2), we then have that $R \otimes_\ell S$ is reduced, and this tensor product is isomorphic with $R \otimes_k L$, proving that R is separable over k.

(5) If ℓ is separable over k, then for every $x \in \ell$, by Proposition 3.2.2, $k(x)$ is separable over k. This is also an algebraic extension, so that $k(x) = k[t]/(f)$, where f is the minimal polynomial for x over k. By (3), $(f, f') = 1$. Conversely, assume that for every $x \in \ell$, the minimal polynomial $f \in k[t]$ of x over k satisfies that $(f, f') = 1$, and suppose that ℓ is not separable over k. Then there exists a field extension L of k such that $\ell \otimes_k L$ is not reduced. By collecting the coefficients as in the proof of Proposition 3.2.2 (3), there exist $x_1, \ldots, x_n \in \ell$ such that $k(x_1, \ldots, x_n) \otimes_k L$ is not reduced. Let $f_i(t)$ be the minimal polynomial for x_i over k. The criterion $(f_i, f_i')k[t] = 1$ implies that if g_i is a minimal polynomial for x_i over $k(x_1, \ldots, x_{i-1})$, then $(g_i, g_i')k(x_1, \ldots, x_{i-1})[t] = 1$. Hence by (3), $k \subseteq k(x_1) \subseteq k(x_1, x_2) \subseteq \cdots \subseteq k(x_1, \ldots, x_n)$ is a chain of separable extensions. Thus by (4), $k \subseteq k(x_1, \ldots, x_n)$ is separable, whence $k(x_1, \ldots, x_n) \otimes_k L$ is reduced. $\square$

Definition 3.2.5 *Let L and K be subfields of a common field E, both of them containing a subfield k. We say that L and K are* **linearly disjoint** *over k if the subalgebra LK of E generated by L and K is isomorphic to $L \otimes_k K$ via the map sending $\sum_i l_i \otimes k_i$ to $\sum_i l_i k_i$.*

Eventually we want to establish several equivalent criteria for an algebra to be separable over a field. First we treat the case in which the algebra is a field extension, and then handle the case of an arbitrary algebra. In the finitely generated case we need the concept of a separably generated field.

Definition 3.2.6 *Let L be a finitely generated field extension of a field k. We say that L is* **separably generated** *over k if there exists a transcendence basis $z_1, \ldots, z_d$ for L over k such that L is separable over $k(z_1, \ldots, z_d)$.*

Theorem 3.2.7 (MacLane's criterion [198]) *Let k be a field and let L be a field extension of k. The following are equivalent:*

(1) L is separable over k.

(2) $L \otimes_k \overline{k}$ is reduced, where $\overline{k}$ is an algebraic closure of k.

(3) $L \otimes_k k'$ is reduced for every purely inseparable field extension k' of k.

(4) Either k has characteristic 0, or k has positive characteristic p and $L \otimes_k k^{1/p}$ is reduced.

(5) Either k has characteristic 0, or k has positive characteristic p and L and $k^{1/p}$ are linearly disjoint over k.

Furthermore, in case that L is finitely generated over k, these conditions are equivalent to:

(6) If $L = k(z_1, \ldots, z_n)$, there exists a transcendence basis $z_{i_1}, \ldots, z_{i_d}$ for L over k such that L is separable over $k(z_{i_1}, \ldots, z_{i_d})$.

(7) L is separably generated over k.

Proof: Clearly (1) implies (2), (2) implies (3), and (3) implies (4).

We prove that (4) implies (5). There is nothing to prove unless k has positive characteristic p. Assume (4). Consider the natural map from $L \otimes_k k^{1/p} \to Lk^{1/p}$, where $Lk^{1/p}$ is the subalgebra of the field $L^{1/p}$ generated by the images of L and $k^{1/p}$. Let P be the kernel of this map. We claim in general that P is nilpotent; this will prove (5) since we are assuming that $L \otimes_k k^{1/p}$ is reduced. Let $\alpha = \sum_{i=1}^m l_i \otimes a_i^{1/p} \in P$, where $l_i \in L$ and $a_i \in k$. Then $\alpha^p = \sum_{i=1}^m l_i^p \otimes a_i = (\sum_{i=1}^m l_i^p a_i) \otimes 1$. However, $\sum_{i=1}^m l_i^p a_i = (\sum_{i=1}^m l_i a_i^{1/p})^p = 0$, proving that P is nilpotent.

We next prove that (5) implies (6) under the assumption that L is finitely generated over k. Say $L = k(z_1, \ldots, z_n)$. We may assume that the characteristic of k is a positive prime p, and that $z_1, \ldots, z_n$ are not algebraically independent. Thus there exists a non–zero polynomial $f \in k[T_1, \ldots, T_n]$ of minimal degree such that $f(z_1, \ldots, z_n) = 0$. Not every monomial in f is a pth power since L and $k^{1/p}$ are linearly disjoint over k. So by possibly reindexing we may assume that T_n appears in some monomial to the ith power, where i an integer that is not a multiple of p. Then by the minimal degree assumption on f the coefficient of T_n^i in $f(z_1, \ldots, z_{n-1}, T_n)$ is non–zero. Hence $k(z_1, \ldots, z_{n-1}) \subseteq k(z_1, \ldots, z_n) = L$ is separable algebraic. Now $k(z_1, \ldots, z_{n-1})$ and $k^{1/p}$ are still linearly disjoint over k, so by induction, after reindexing, $z_1, \ldots, z_d$ are algebraically independent over k and $k(z_1, \ldots, z_{n-1})$ is separable algebraic over $k(z_1, \ldots, z_d)$. Then by Proposition 3.2.4 (4), $L = k(z_1, \ldots, z_n)$ is separable algebraic over $k(z_1, \ldots, z_d)$. This proves (6).

Clearly (6) implies (7).

Assume (7). Then L is separable over $k(z_1, \ldots, z_d)$ for some transcendence basis $z_1, \ldots, z_d$ of L over k. By Proposition 3.2.4 we know that $k(z_1, \ldots, z_d)$ is separable over k, and an application of the same proposition then shows that L is separable over k.

It remains to prove that (5) implies (1) without necessarily assuming that L is finitely generated over k. By Proposition 3.2.2, to prove that L is separable over k it suffices to prove that every finitely generated subfield of L is separable over k. Since (5) clearly passes to subfields, it suffices to prove the case where L is finitely generated over k. But with this assumption, the equivalence of (1) through (7) has been proved. $\square$

Theorem 3.2.8 *Let k be a field and let R be a Noetherian k–algebra. The following are equivalent.*
(1) For every reduced k–algebra S, $R \otimes_k S$ is reduced.
(2) R is separable over k.
(3) There exists an algebraically closed field extension field ℓ of k such that $R \otimes_k \ell$ is reduced.
(4) For all purely inseparable field extensions k' of k, $R \otimes_k k'$ is reduced.
(5) R is reduced and for every minimal prime P of R, $\kappa(P)$ is separable over k.

Proof: Clearly (1) implies (2) and (2) implies (3). Assume (3). Every purely inseparable field extension k' of k can be embedded into ℓ. Then $R \otimes_k k' \subseteq R \otimes_k \ell$, proving that $R \otimes_k k'$ is reduced, which proves that (3) implies (4).

Assume (4). By assumption $R = R \otimes_k k$ is reduced. Let W be the multiplicatively closed set of R equal to $R \setminus P$. Let k' be a purely inseparable field extension of k. To prove that $\kappa(P)$ is separable over k, it suffices to prove that $\kappa(P) \otimes_k k'$ is reduced, by Theorem 3.2.7, part (3). But $\kappa(P) \otimes_k k' \cong (R \otimes_k k')_W$, which is reduced since $R \otimes_k k'$ is reduced. This proves (5).

Finally assume (5). Let $P_1, \dots, P_n$ be the minimal primes of R. Set $R_i = R/P_i$, and let $K_i = \kappa(P_i)$. To prove (1) we may assume S is a finitely generated k–algebra, hence Noetherian. Let $Q_1, \dots, Q_m$ be the minimal primes of S, and set $L_i = \kappa(Q_i)$. There are embeddings of R into $\prod_i K_i$ and S into $\prod_i L_i$ which induce an embedding of $R \otimes_k S$ into $\prod_{i,j} K_i \otimes_k L_j$. As K_i is separable over k for all i, the product is reduced, proving (1). $\square$

3.3. Relative algebraic closure

This section examines the behavior of algebraic closures of a subfield in a larger field under various extensions, such as tensoring and adjoining indeterminates.

Theorem 3.3.1 *Let $K = k(x_1, \dots, x_n)$ be a purely transcendental extension of a field k. Let L be an arbitrary extension of k such that k is algebraically closed in L. Then K is algebraically closed in $L(x_1, \dots, x_n)$.*

Proof: By induction on n it suffices to prove the theorem for $n = 1$. Suppose that $\eta \in L(x)$ is algebraic over $K = k(x)$. Let $f(T) = T^n + \alpha_1 T^{n-1} + \cdots + \alpha_n$ be the minimal polynomial for η over K. Write $\alpha_i = \frac{a_i}{b_i}$ with $a_i, b_i \in k[x]$. Let b be the least common multiple of the b_i. Then $b\eta$ is algebraic over K

with minimal polynomial $T^n + b\alpha_1 T^{n-1} + \cdots + b^n \alpha_n$ having coefficients in $k[x]$. Since $b\eta$ is integral over $k[x]$ it follows that $b\eta \in L[x]$. Write $g(x) = b\eta$. If $g(x) \in k[x]$ we are done since then $\eta \in k(x)$.

We have reduced the problem to proving that if k is algebraically closed in L, and $g(x) \in L[x]$ is integral over $k[x]$, then $g(x) \in k[x]$. By Theorem 2.3.2 the integral closure of $k[x]$ in $L[x]$ is graded. If $g(x) = c_n x^n + \cdots + c_0$ this fact implies that each $c_i x^i$ is integral over $k[x]$. Thus each c_i must be algebraic over k forcing $c_i \in k$ for all i.   □

Proposition 3.3.2 *Let $K \subseteq L$ be a finitely generated extension of fields. Let F be the algebraic closure of K in L. Then F is finite over K.*

Proof: Choose a transcendence basis $x_1, \ldots, x_d$ for L over K. If $d = 0$, then $L = F$ is finitely generated over K and algebraic over K, hence finite. Assume that $d > 0$. Let L_1 be the algebraic closure of $K(x_1)$ in L. By induction on d, L_1 is finite over $K(x_1)$, and is therefore finitely generated over K. Clearly F is the algebraic closure of K in L_1 as well. Thus we can replace L by L_1 and assume that $d = 1$. Let $x = x_1$.

If linearly independent elements of F over K remain linearly independent over $K(x)$, then $[F : K] \leq [L : K(x)]$. The latter is finite since L is finitely generated and algebraic over $K(x)$. If there are linearly independent elements of F that are not linearly independent over $K(x)$, then there is a non–zero polynomial $f(T) \in F[T]$ such that $f(x) = 0$. In this case x is algebraic over F, contradicting the choice of F as the algebraic closure of K in L.   □

Theorem 3.3.3 *Let F be a field, let $F \subseteq K$ be an arbitrary separable field extension of F, and let $F \subseteq L$ be a finitely generated field extension of F such that F is algebraically closed in L. Then K is integrally closed in $K \otimes_F L$.*

Proof: We may write K as a (directed) union of finitely generated extensions K_i of F, and by Proposition 3.2.2 (1) each of these is separable. If we prove that K_i is integrally closed in $K_i \otimes_F L$, then as $K \otimes_F L = \cup K_i \otimes_F L$, it will follow that K is integrally closed in $K \otimes_F L$. Thus we reduce to the case where K is finitely generated over F. Then by Theorem 3.2.7 this extension is separably generated, and so there exists a transcendence basis Y for K over F so that K is finite separable algebraic over $F(Y)$.

By Theorem 3.3.1, the field $F(Y)$ is algebraically closed in $L(Y)$. Note that $K \otimes_F L = (K \otimes_{F(Y)} F(Y)) \otimes_F L = K \otimes_{F(Y)} L(Y)$, so that by replacing F by $F(Y)$ and L by $L(Y)$ we have reduced to the case where K is finite separable algebraic over F.

There exists a primitive element $a \in K$ so that $K = F(a)$. By Lemma 3.3.4 below, $[L(a) : L] = [F(a) : F]$. We claim that $F(a)$ is algebraically closed in $L(a)$. Suppose that b is in $L(a)$ and is algebraic over $F(a)$ (and hence also algebraic over F). Since $L(a)/L$ is separable algebraic it follows that $F(b)$ is also a separable algebraic field extension of F. We claim that $b \in F(a)$. If not, then $F(a,b)$ is separable algebraic over F, and by applying Lemma 3.3.4

below we obtain that $[L(a) : L] = [L(a,b) : L] = [F(a,b) : F] > [F(a) : F] = [L(a) : L]$, which is a contradiction. Therefore $b \in F(a)$. Note that $K \otimes_F L = F(a) \otimes_F L = L(a)$, and this identification finishes the proof. $\square$

Lemma 3.3.4 *Let F be a field, let $F \subseteq K$ be a finite separable field extension of F, and let $F \subseteq L$ be a finitely generated field extension of F such that F is algebraically closed in L. For every subfield ℓ of K containing F, $[\ell : F] = [\ell L : L]$, where ℓL is the subfield of an algebraic closure of L generated by the images of elements of ℓ.*

Proof: Without loss of generality, $\ell = K$. As K is separable over F, there exists a primitive element $a \in K$ so that $K = F(a)$. Let $g(X) \in F[X]$ be the minimal polynomial for a. Since F is algebraically closed in L, $g(X)$ must be irreducible in $L[X]$ because the coefficients of every factor of $g(X)$ are sums of products of roots of $g(X)$, hence algebraic over K. Hence $[KL : L] = [L(a) : L] = [F(a) : F]$. $\square$

3.4. Exercises

3.1   Prove (4) of Proposition 3.2.2.

3.2   Determine exactly what is wrong with the "proof" given in Discussion 3.2.3.

3.3   Let E and L be extension fields of a field k. Find necessary and sufficient conditions for $E \otimes_k L$ to be local.

3.4   Let k be an algebraically closed field, and let E and L be arbitrary extension fields of k. Prove that $E \otimes_k L$ is a domain.

3.5   Let k be a field, and let E be a separable algebraic extension field of k and L be an arbitrary extension field of k. If k is separably algebraically closed in L, prove that $E \otimes_k L$ is a field.

3.6   Let k be an algebraically closed field, and let R be a k–algebra. Prove that k is algebraically closed in R if and only if R is reduced and has no non–trivial idempotents.

3.7*   (Sweedler [283]) Prove the Units Theorem: if k is an algebraically closed field and R and S are two k–algebras that are reduced and have no non–trivial idempotents, then every unit in $R \otimes_k S$ has the form $a \otimes b$, where a is a unit in R and b is a unit in S.

3.8   Let R and S be k–algebras. Let $r \in R$, $s \in S$ be transcendental over k. Prove that $r \otimes 1 + 1 \otimes s$ in $R \otimes_k S$ is not invertible. (Hint: Use Exercise 3.7.)

3.9*   (Sweedler [284]) Let R and S be k–algebras, k a field. Prove that the following are equivalent:
   (i) $R \otimes_k S$ is local.
   (ii) R is local with maximal ideal $\mathfrak{m}$, S is local with maximal ideal $\mathfrak{n}$, either R or S is algebraic over k, and $(R/\mathfrak{m}) \otimes_k (S/\mathfrak{n})$ is local.

3.10 Let k be a field, and let E and F be two finitely generated extension fields of k. Prove that every maximal ideal of the ring $R = E \otimes_k F$ has height $\min\{e, f\}$, where e is the transcendence degree of E over k and f is the transcendence degree of E over k.

3.11 Let F be a field, K a finite separable field extension of F and R an integrally closed F–domain. Prove that if $K \otimes_F R$ is a domain, it is integrally closed.

3.12 Let L and K be subfields of a field E, and let k be a subfield of $L \cap K$. Prove that L and K are linearly disjoint over k (with respect to E) if and only if elements of L that are linearly independent over k remain linearly independent over K.

3.13 Let k be a field and X a variable over k.

  (i) Prove that the field of fractions of $k[[X]]$ is uncountable.

  (ii) Prove that if k is countable, so is the algebraic closure of any finitely generated field extension of k.

  (iii) Prove that the transcendence degree of the field of fractions of $k[[X]]$ over $k(X)$ is infinity.

  (iv) Prove that $\sum_{i>0} X^{i!} \in k[[X]]$ is transcendental over $k(X)$.

4
Noetherian rings

This book is mostly about Noetherian rings. However, the integral closure of a Noetherian domain R need not be Noetherian (Exercise 4.9), so we need to develop some results for non–Noetherian rings as well. Nevertheless, integral closure of Noetherian rings is fairly well–behaved, and we concentrate on their good properties in this chapter. In Section 4.10 we present Krull domains and the Mori–Nagata Theorem that the integral closures of Noetherian domains are Krull domains. In Sections 4.3 and 4.6 we present different scenarios in which we can conclude that the integral closure of a given ring in an extension ring is necessarily module–finite over the base ring. In Section 4.9 we prove the Krull–Akizuki Theorem: the integral closure of a Noetherian domain of dimension one is Noetherian, though not necessarily module–finite. In Section 4.10 we prove that the integral closure of a two–dimensional Noetherian domain is Noetherian. The first two sections analyze principal ideals in integrally closed rings and normalization theorems. Section 4.4 covers Jacobian ideals, Section 4.5 covers Serre's criteria. Section 4.8 analyzes Lying–Over and preservation of heights under integral extensions.

4.1. Principal ideals

Proposition 1.5.2 shows that principal ideals in integrally closed reduced Noetherian rings that are generated by non–zerodivisors are themselves integrally closed. Such ideals also have good primary decompositions:

Proposition 4.1.1 *(Cf. Proposition 1.5.2.) Let R be a Noetherian ring that is integrally closed in its total ring of fractions. The set of associated primes of an arbitrary principal ideal generated by a non–zerodivisor x consists exactly of the set of minimal prime ideals over (x).*

Furthermore, all such associated prime ideals are locally principal.

Proof: All minimal prime ideals over (x) are associated to (x). Let P be a prime ideal associated to xR. By Prime Avoidance there exists a non–zerodivisor y in R such that $P = xR :_R y$. We may localize at P and assume without loss of generality that R is a local ring with maximal ideal P. By definition $\frac{y}{x}P \subseteq R$. If $\frac{y}{x}P \subseteq P$, then by Lemma 2.1.8, $\frac{y}{x} \in \overline{R} = R$, so that $y \in xR$ and $P = xR :_R y = R$, which is a contradiction. Thus necessarily $\frac{y}{x}P = R$. Hence there exists $z \in P$ such that $\frac{y}{x}z = 1$. Then y is a non–zerodivisor and $P = xR :_R y = yzR :_R y = zR$, so P is a prime ideal of height 1. Thus P is minimal over xR.

The last statement follows immediately. $\qquad\square$

The proposition implies that for any principal ideal generated by a non–zerodivisor in a reduced integrally closed Noetherian ring, the primary decomposition is particularly simple: $xR = \cap_P (xR_P \cap R)$, as P varies over the minimal prime ideals over xR.

A special case of integrally closed domains are one–dimensional Noetherian integrally closed domains, also known as **Dedekind domains**. The following is an easy consequence of Proposition 4.1.1, and we leave the details to the reader (cf. Proposition 6.4.4).

Proposition 4.1.2 *Let $(R, \mathfrak{m})$ be a Noetherian local ring. Then $\mathfrak{m}$ is principal generated by a non–zerodivisor if and only if R is a Dedekind domain, and that holds if and only if every non–zero ideal in R is principal and generated by a non–zerodivisor.* $\qquad\qquad\square$

The non–zerodivisor assumption is necessary. For example, let k be a field, X and Y variables over k, and R the reduced ring $k[X] \times k[Y]$. Then R is integrally closed, but the principal ideal generated by the zero–divisor X^2 has an embedded prime ideal.

4.2. Normalization theorems

We present several versions of the Noether Normalization Theorem.

Lemma 4.2.1 *Let R be a finitely generated algebra over a field k, $x_1, \ldots, x_n$ elements in R, $X_1, \ldots, X_n$ variables over R, and $f \in k[X_1, \ldots, X_n]$ a non–zero polynomial such that $f(x_1, \ldots, x_n) = 0$.*
(1) There exist infinitely many positive integers $e_1, \ldots, e_{n-1}$ such that x_n is integral over the ring $k[x_1 + x_n^{e_1}, \ldots, x_{n-1} + x_n^{e_{n-1}}]$.
(2) If R is a domain and $\frac{\partial f}{\partial X_n}(x_1, \ldots, x_n) \neq 0$, then there are infinitely many $e_1, \ldots, e_{n-1}$ so that x_n satisfies a monic polynomial $h(X_n)$ of integral dependence over $k[x_1 + x_n^{e_1}, \ldots, x_{n-1} + x_n^{e_{n-1}}]$ such that $h'(x_n) \neq 0$.
(3) If k is an infinite field, there exists a non–empty Zariski–open subset U of k^{n-1} such that whenever $(c_1, \ldots, c_{n-1}) \in U$, then x_n is integral over $k[x_1 - c_1 x_n, \ldots, x_{n-1} - c_{n-1} x_n]$.
(4) Suppose that k is infinite and that R is a domain. If $\frac{\partial f}{\partial X_n}(x_1, \ldots, x_n) \neq 0$, there exist infinitely many $c_1, \ldots, c_{n-1} \in k$ such that x_n satisfies a monic polynomial $h(X_n)$ of integral dependence over $k[x_1 - c_1 x_n, \ldots, x_{n-1} - c_{n-1} x_n]$ such that $h'(x_n) \neq 0$.

Proof: Let e be a positive integer strictly larger than the degree of f. Set $e_1 \geq e$, and for each $i \geq 2$, set $e_i \geq e \cdot e_{i-1}$. Then $g(X_1, \ldots, X_n) = f(X_1 - X_n^{e_1}, \ldots, X_{n-1} - X_n^{e_{n-1}}, X_n)$ is a non–zero polynomial. By the choice of $e_1, \ldots, e_{n-1}$, g is, up to a scalar multiple, monic in X_n. But $g(x_1 + x_n^{e_1}, \ldots, x_{n-1} + x_n^{e_{n-1}}, x_n) = f(x_1, \ldots, x_n) = 0$, so that g is an equation of integral dependence of x_n over $k[x_1 + x_n^{e_1}, \ldots, x_{n-1} + x_n^{e_{n-1}}]$. This finishes the proof of (1).

Suppose that $\frac{\partial f}{\partial X_n}(x_1, \ldots, x_{n-1}, x_n) \neq 0$. By the chain rule, $\frac{\partial g}{\partial X_n}$ equals $\sum_{i=1}^{n-1}(-e_i X_n^{e_i-1})\frac{\partial f}{\partial X_i} + \frac{\partial f}{\partial X_n}$ evaluated at $(X_1 - X_n^{e_1}, \ldots, X_{n-1} - X_n^{e_{n-1}}, X_n)$. After passing to $X_n \mapsto x_n$ and $X_i \mapsto x_i + x_n^{e_i}$ for $i < n$, by possibly increasing the previously determined e_i, this is not zero, which proves that $g(x_1 + x_n^{e_1}, \ldots, x_{n-1} + x_n^{e_{n-1}}, X_n)$ is separable in X_n, and hence it proves (2).

If k is infinite, let d be the degree of f, and let f_d the component of f of degree d. As f_d is a non–zero polynomial and k is infinite, there exist $a_1, \ldots, a_n \in k$ such that $f_d(a_1, \ldots, a_n) \neq 0$ and $a_n \neq 0$. The polynomial $g(X_1, \ldots, X_n) = f(X_1 + a_1 X_n, \ldots, X_{n-1} + a_{n-1} X_n, a_n X_n)$ has then degree at most d, and has coefficient of X_n^d equal to $f_d(a_1, \ldots, a_n)$, which is non–zero. Thus up to a scalar multiple g is a monic polynomial in X_n such that

$$g(x_1 - a_1 a_n^{-1} x_n, \ldots, x_{n-1} - a_{n-1} a_n^{-1} x_n, a_n^{-1} x_n) = f(x_1, \ldots, x_n) = 0,$$

whence x_n is integral over $k[x_1 - a_1 a_n^{-1} x_n, \ldots, x_{n-1} - a_{n-1} a_n^{-1} x_n]$. Setting $c_i = a_i a_n^{-1}$ for $i = 1, \ldots, n-1$ and $U = \{(c_1, \ldots, c_{n-1}) \mid f_d(c_1, \ldots, c_{n-1}, 1) \neq 0\}$ finishes the proof of (3).

Now suppose that k is infinite and in addition that $\frac{\partial f}{\partial X_n}(x_1, \ldots, x_n) \neq 0$. By the chain rule, $\frac{\partial g}{\partial X_n}(x_1 - a_1 a_n^{-1} x_n, \ldots, x_{n-1} - a_{n-1} a_n^{-1} x_n, X_n)$ equals $\sum_{i=1}^{n} a_i \frac{\partial f}{\partial X_i}(x_1 - a_1 a_n^{-1} x_n + a_1 X_n, \ldots, x_{n-1} - a_{n-1} a_n^{-1} x_n + a_{n-1} X_n, a_n X_n)$. This is not zero if $a_1 = \cdots = a_{n-1} = 0$ and $a_n = 1$. Since k is infinite, it is possible to find infinitely many $a_1, \ldots, a_n$ such that this is non–zero and such that $f_d(a_1, \ldots, a_n) a_n \neq 0$, which proves (4). $\qquad\square$

Theorem 4.2.2 (Noether normalization) *Let k be a field and R a finitely generated k–algebra. Then there exist elements $x_1, \ldots, x_m \in R$ such that $k[x_1, \ldots, x_m]$ is a transcendental extension of k (i.e., $k[x_1, \ldots, x_m]$ is isomorphic to a polynomial ring in m variables over k) and such that R is integral over $k[x_1, \ldots, x_m]$.*

If k is infinite, $x_1, \ldots, x_m$ may be taken to be k–linear combinations of elements of a given generating set of R.

In any case, if R is a domain and the field of fractions of R is separably generated over k, then $x_1, \ldots, x_m$ above can be chosen so that the field of fractions of R is separable over $k[x_1, \ldots, x_m]$.

Proof: Write $R = k[y_1, \ldots, y_n]$. We use induction on n to prove the first claim, including the assertion that if k is infinite, $x_1, \ldots, x_m$ may be taken to be the sufficiently general k–linear combinations of elements of the y_i. We may choose the y_i so that $k[y_1, \ldots, y_l]$ is transcendental over k and such that $k[y_1, \ldots, y_l] \subseteq R$ is algebraic. If $n = l$, set $x_i = y_i$ and $l = m$, and we are done. We thus assume that $n > l$. By Lemma 4.2.1, there exist $z_1, \ldots, z_{n-1}$ such that $k[z_1, \ldots, z_{n-1}, y_n] = k[y_1, \ldots, y_n]$ and such that y_n is integral over $k[z_1, \ldots, z_{n-1}]$. If k is infinite, $z_1, \ldots, z_{n-1}$ may be taken to be k–linear combinations of $y_1, \ldots, y_n$. By induction on n we may assume that there exist elements $x_1, \ldots, x_m \in k[z_1, \ldots, z_{n-1}]$ such that $k[z_1, \ldots, z_{n-1}]$ is integral over $k[x_1, \ldots, x_m]$ and such that $x_1, \ldots, x_m$ are transcendental over k. If k is infinite, $x_1, \ldots, x_m$ may be taken to be k–linear combinations of $z_1, \ldots, z_{n-1}$, and therefore k–linear combinations of $y_1, \ldots, y_{n-1}$. It follows that R is integral over $k[x_1, \ldots, x_m]$.

If $Q(R)$ is separable over k, by Theorem 3.2.7 (6), there exist algebraically independent $x_1, \ldots, x_m \in \{y_1, \ldots, y_n\}$ such that $k(x_1, \ldots, x_m) \subseteq Q(R)$ is separable algebraic. By the choice of the x_i, $k[x_1, \ldots, x_m] \subseteq R$. Suppose that $y_1, \ldots, y_r$ satisfy separable equations of integral dependence over $k[x_1, \ldots, x_m]$ but $y_{r+1}, \ldots, y_n$ do not. If $r = n$ we are done, so we suppose that $r < n$. By Lemma 4.2.1 (2) and (4), there are $x_1', \ldots, x_m' \in R$, with the differences $x_i - x_i'$ all powers of y_{r+1} or all scalar multiples of y_{r+1} in case k is infinite, such that y_{r+1} is separable and integral over $k[x_1', \ldots, x_m']$. Then $k[x_1', \ldots, x_m'] \subseteq k[x_1', \ldots, x_m', y_{r+1}] = k[x_1, \ldots, x_m, y_{r+1}] \subseteq k[x_1, \ldots, x_m, y_1, \ldots, y_{r+1}]$ are all separable integral extensions, and necessarily the extension $k(x_1', \ldots, x_m') \subseteq Q(R)$ is separable algebraic. By repeating this step, we eventually get that all y_i are separable and integral over an m–generated polynomial subring. $\square$

Observe that the proof above is quite explicit about how to choose a generating set for a Noether normalization. The same holds for the graded version below.

Theorem 4.2.3 (Graded Noether normalization) *Let k be a field and R a finitely generated $\mathbb{N}$–graded k–algebra. There exist algebraically independent $x_1, \ldots, x_m \in R$, homogeneous of the same degree, such that R is integral over $k[x_1, \ldots, x_m]$. If k is infinite and R is generated over k by elements of degree 1, then the x_i may be taken to be of degree 1.*

Proof: Write $R = k[y_1, \ldots, y_n]$, with $y_1, \ldots, y_n$ homogeneous elements. Let d be an integer multiple of all $\deg(y_i)$. Then R is module–finite over the k–subalgebra $\tilde{R}$ generated by all elements of degree d. Thus it suffices to replace R by $\tilde{R}$ and assume that all the y_i have the same degree d.

Suppose that k is infinite. By Theorem 4.2.2, there exist linear combinations $x_1, \ldots, x_m$ of the y_i that are algebraically independent over k such that R is integral over $k[x_1, \ldots, x_m]$. Note that each x_i is homogeneous of degree d.

Now assume that k is a finite field. Let p be the characteristic of k. Let k' be the purely inseparable closure of k (inside an algebraic closure of k). Then k' is an infinite field. Set $R' = R \otimes_k k'$. Then R' is generated over k' by homogeneous elements $y_1, \ldots, y_n$ of degree d. By the infinite field case, there exist k'–linear combinations $z_1, \ldots, z_m$ of $y_1, \ldots, y_n$ such that R' is integral over the polynomial ring $k'[z_1, \ldots, z_m]$. For each $i = 1, \ldots, n$, let $f_i(X)$ be a monic polynomial in X with coefficients in $k'[z_1, \ldots, z_m]$ such that $f_i(y_i) = 0$. Let L be the finite field extension of k generated by the finitely many elements of k' that appear as coefficients in $f_1, \ldots, f_n$, and as coefficients when the z_i are written as linear combinations of the y_j. Then $L[y_1, \ldots, y_n]$ is integral over $L[z_1, \ldots, z_m]$, and each z_i is an L–linear combination of $y_1, \ldots, y_n$. As L/k

is a finite purely inseparable extension, there exists a non–negative integer e such that for all $x \in L$, $x^{p^e} \in k$. For each $i = 1, \ldots, n$, set $F_i(X) = f_i(X)^{p^e}$. Then F_i is a monic polynomial in X with coefficients in $k[z_1^{p^e}, \ldots, z_m^{p^e}]$, each $z_i^{p^e}$ is a k–linear combination of $y_1^{p^e}, \ldots, y_n^{p^e}$, and $F_i(y_i) = 0$. Thus if we set $x_1 = z_1^{p^e}, \ldots, x_m = z_m^{p^e}$, $k[x_1, \ldots, x_m]$ is a transcendental extension of k and $R = k[y_1, \ldots, y_n]$ is integral over $k[x_1, \ldots, x_m]$.   □

Observe that in the theorems above, given any finite set of algebra generators of R over k, the x_i in the conclusion of the theorem may be obtained from the given set by successively applying exponentiation and addition.

Theorem 4.2.4 *Let R be a Noetherian ring and S a finitely generated R–algebra containing R that is a domain. Then there exist elements $x_1, \ldots, x_m \in S$ and $r \in R$ such that $R[x_1, \ldots, x_m]$ is a transcendental extension of R and such that S_r is integral over $R[x_1, \ldots, x_m]_r$ (localization at one element).*

Proof: Let $W = R \setminus \{0\}$, which is a multiplicatively closed subset of R. Then $W^{-1}R \subseteq W^{-1}S$ is a finitely generated algebra extension of the field $W^{-1}R$, so that by Theorem 4.2.2 and the observation above, there exist elements $x_1, \ldots, x_m \in S$ such that $W^{-1}R \subseteq W^{-1}R[x_1, \ldots, x_m]$ is transcendental and $W^{-1}R[x_1, \ldots, x_m] \subseteq W^{-1}S$ is integral. Thus clearly $R \subseteq R[x_1, \ldots, x_m]$ is transcendental. Let $\{s_1, \ldots, s_n\}$ be a generating set of S as an algebra over $R[x_1, \ldots, x_m]$. Each s_i is integral over $W^{-1}R[x_1, \ldots, x_m]$, so that there exists $u_i \in W$ such that s_i is integral over $R[x_1, \ldots, x_m]_{u_i}$. Let $r = u_1 \cdots u_n$. Then each algebra generator s_i of S_r over $R[x_1, \ldots, x_m]_r$ is integral over $R[x_1, \ldots, x_m]_r$, which proves the theorem.   □

4.3. Complete rings

In this section we present a first set of connections between integral closure and complete rings. Many more connections between integral closure and completion are in Sections 4.6, 4.8, and in Chapter 9 on analytically unramified rings.

The main result of this section, Theorem 4.3.4, is that the integral closure of a complete local Noetherian domain is a module–finite extension, hence Noetherian. We start with showing that module–finite extensions of complete local Noetherian rings are direct products of complete local rings.

Proposition 4.3.1 *Let R be a ring and I an ideal in R such that I is complete in the I–adic topology. Then any idempotent of R/I lifts to an idempotent in R.*

Proof: Let $e \in R$ such that $e + I$ is an idempotent in R/I, i.e., $e^2 - e \in I$. Fix $n \geq 1$. Write the expansion of $1 = (e + (1 - e))^{2n-1}$ as $1 = e^n p_n + (1 - e)^n q_n$, and set $e_n = e^n p_n$. Then e_n is a multiple of e^n, $1 - e_n$ is a multiple of $(1 - e)^n$, and $e_n(1 - e_n) \in (e^n(1 - e)^n) \subseteq I^n$. Thus $e_n + I^n$ is an idempotent in R/I^n. Furthermore, $e_n + I = \sum_{i=n}^{2n-1} \binom{2n-1}{i} e^i (1 - e)^{2n-1-i} + I = e^{2n-1} + I = e + I$.

The sequence $\{e_n\}_n$ of elements of R is a Cauchy sequence in the I–adic topology:

$$e_{n+1} - e_n + I^n = e_{n+1}^2 - e_n^2 + I^n = (e_{n+1} - e_n)(e_{n+1} + e_n) + I^n$$
$$= (1 - e_n - 1 + e_{n+1})(e_{n+1} + e_n) + I^n$$
$$= ((1 - e)^n q_n - (1 - e)^{n+1} q_{n+1}))(e^{n+1} p_{n+1} + e^n p_n) + I^n$$
$$= 0.$$

By assumption the limit of this sequence exists in R. Let f be the limit. Then $f(1 - f) = \lim e_n(1 - e_n) = 0$, so that f is an idempotent in R. But $e_n + I^n$ is a lift of $e + I$ and the e_n converge to f, so that $f + I = e + I$.   □

The lifting of idempotents has the following important consequence:

Proposition 4.3.2 *Let $(R, \mathfrak{m})$ be a complete local Noetherian ring, and let $R \subseteq S$ be a module–finite extension ring of R. Then S has finitely many maximal ideals, say $\{\mathfrak{m}_1, \ldots, \mathfrak{m}_t\}$, and $S \cong S_{\mathfrak{m}_1} \times \cdots \times S_{\mathfrak{m}_t}$.*

Proof: There are only finitely many prime ideals $\mathfrak{m}_1, \ldots, \mathfrak{m}_t$ in S that are minimal over $\mathfrak{m}S$, and these are exactly the maximal ideals of S. By the Chinese Remainder Theorem, $S/\mathfrak{m}S \cong (S/\mathfrak{m}S)_{\mathfrak{m}_1} \times \cdots \times (S/\mathfrak{m}S)_{\mathfrak{m}_t}$ as rings. In particular there are mutually orthogonal idempotents $e_1, \ldots, e_t \in S/\mathfrak{m}S$ that give this decomposition, so that $e_1 + \cdots + e_t = 1$. Note that S is complete in the $\mathfrak{m}$–adic topology and Noetherian. Idempotents can be lifted in complete rings by Proposition 4.3.1, so that each e_j lifts to $f_j \in S$ that are mutually orthogonal and such that $f_1 + \cdots + f_t = 1$. Hence $S \cong S_{\mathfrak{m}_1} \times \cdots \times S_{\mathfrak{m}_t}$.   □

Complete local domains have a normalization result reminiscent of normalization results in the previous section:

Theorem 4.3.3 (Cohen Structure Theorem) *Let $(R, \mathfrak{m})$ be a complete Noetherian local domain, and k a coefficient ring of R. In case k is a discrete valuation domain of rank one with maximal ideal generated by p, we assume that $p, x_1, \ldots, x_d$ is a system of parameters. If R contains a field, we assume that $x_1, \ldots, x_d$ is a system of parameters. Then the subring $k[[x_1, \ldots, x_d]]$ of R is a regular local ring and R is module–finite over it.*

We only comment on a proof. The most difficult part is establishing the existence of the coefficient ring k, and we do not provide a proof of that in this book. Once a coefficient ring and a system of parameters are chosen, the theorem is proved as follows. Let I be the ideal generated by the system of parameters. For notation we set x_0 to be either 0 if R contains a field and p otherwise. As the R–module R/I has finite length, there exists a finite set S of elements in R with the property that whenever $r \in R$, then there exist $s \in S$ and a unit $u \in k$ such that $r - su \in I$. Clearly $A = k[[x_1, \ldots, x_d]]$ of R is a subring of R. We claim that R is module–finite over A and is generated by the elements of S. If $r \in R$, choose $s_1 \in S$, a unit $u_1 \in k$ and $r_{10}, \ldots, r_{1d} \in R$ such that $r = s_1 u_1 + \sum_i r_{1i} x_i$. Repeat this for r_{1i} to obtain that for some $s_{2i} \in S$, for some units $u_{2i} \in k$ and some $r_{2ij} \in R$, $r_{1i} = s_{2i} u_{2i} + \sum_j r_{1ij} x_j$. Substitute this expression in r to obtain $r = s_1 u_1 + \sum_i r_{1ij} s_{2i} u_{2i} x_i + \sum_{ij} r_{1ij} x_i x_j$. We continue this process: $r = \sum_{s \in S} r_{ns} s + t_n$, with $t_n \in I^n$, r_{ns} an element of $k[x_1, \ldots, x_d]$, and $r_{ns} - r_{n-1,s} \in I^n$ for each n, s. Thus $\{r_{ns}\}$ is a Cauchy sequence in $k[x_1, \ldots, x_d]$, whence $r_s \in \lim_n r_{ns} \in k[[x_1, \ldots, x_d]] \subseteq R$ and $r = \sum_{s \in S} r_s s$. This proves that R is module–finite over A. By Theorem 2.2.5 $\dim R = \dim A$, so that there cannot be any algebraic relations among $x_1, \ldots, x_d$. Thus A is a regular local ring.

Theorem 4.3.4 *The integral closure of a complete local Noetherian domain R is module–finite over R. More generally, if L is a finite field extension of the field of fractions K of R, then the integral closure S of R in L is module–finite over R. Furthermore, S is a complete Noetherian local domain.*

Proof: Let $\mathfrak{m}$ be the maximal ideal and k a coefficient ring of R. If R contains a field, let $x_1, \ldots, x_d$ be a system of parameters in R. If R does not contain a field, then k is a complete local discrete valuation domain with maximal ideal (x_0). In this case we choose a system of parameters $x_0, x_1, \ldots, x_d$. By the Cohen Structure Theorem 4.3.3, $A = k[[x_1, \ldots, x_d]]$ is a regular local subring of R and R is module–finite over it. Then L is a finite algebraic extension of the field of fractions of A and the integral closure of A in L equals S. Thus by replacing R by A we may assume that R is a regular local ring.

As R is regular, it is integrally closed. If the characteristic of K is zero, $K \subseteq L$ is separable, and then the theorem holds by Theorem 3.1.3.

Thus it remains to consider the case when the characteristic of K is a positive prime integer p. In this case, k is a field of characteristic p. Then there exists a finite extension L' of L and an intermediate field K' between K and L' such that $K \subseteq K'$ is purely inseparable and $K' \subseteq L'$ is separable. Suppose that the integral closure R' of R in K' is module–finite over R. As the integral closure of R' in L' is the same as the integral closure of R in L', by Theorem 3.1.3, the integral closure of R in L' is module–finite over R. But S is contained in the integral closure of R in L', so that by the Noetherian assumption S is also module–finite over R. Thus it suffices to prove that the integral closure R' of R in K' is module–finite over R.

By switching notation we may assume that $K' = L$ is a purely inseparable extension of K. Let $q = [L : K]$. If $r \in S$, then $r^q \in K \cap S = R$, so that $r \in R^{1/q} = k^{1/q}[[x_1^{1/q}, \ldots, x_d^{1/q}]]$. Observe that $R^{1/q}$ is a regular local ring. For every non–zero $r \in S$ we define its lowest form to be the non–zero homogeneous component of r of smallest possible degree, when r is considered as an element of $R^{1/q}$.

Let $a_1, \ldots, a_s \in S$ and let f_i be the lowest form of a_i. If $f_1, \ldots, f_s$ are linearly independent over R, so are $a_1, \ldots, a_s$. As L is finite–dimensional over

K, the cardinality of sets of such linearly independent lowest forms must be finite. Let $\{f_1, \ldots, f_s\}$ be a maximal linearly independent set of such lowest forms. Let $c_1, \ldots, c_t \in k^{1/q}$ be all the coefficients appearing in the f_i. Note that $f_i \in k(c_1, \ldots, c_t)[[x_1^{1/q}, \ldots, x_d^{1/q}]]$.

For any $a \in S \subseteq R^{1/q}$, the lowest form f of a also lies in $R^{1/q}$. Furthermore, by the maximal linearly independent assumption, there exists a homogeneous $r \in R \setminus \{0\}$ such that $rf \in k(c_1, \ldots, c_t)[[x_1^{1/q}, \ldots, x_d^{1/q}]]$. This implies that all the coefficients of f lie in $k(c_1, \ldots, c_t)$, and hence that all the lowest forms of elements of S lie in $k(c_1, \ldots, c_t)[[x_1^{1/q}, \ldots, x_d^{1/q}]]$. Let T be the subring of $R^{1/q}$ generated over R by the lowest forms of elements of S. Then $T \subseteq k(c_1, \ldots, c_t)[[x_1^{1/q}, \ldots, x_d^{1/q}]]$, and as the latter is module–finite over R, so is T. Let $g_1, \ldots, g_l$ be the generators of T over R. For each i, let $b_i \in S$ be such that g_i is the lowest form of b_i. Set $R' = R[b_1, \ldots, b_l]$. Then R' is module–finite over R, hence complete.

Clearly $R' \subseteq S$, and we next prove that $S \subseteq R'$. Let $a \in S$. For each $i \in \mathbb{N}$ we define $a_i \in R'$ such that the lowest form of $a - a_i$ has degree at least i. Start with $a_0 = 0$. Suppose that $a_0, a_1, \ldots, a_n$ have been found with this property. Let f be the lowest form of $a - a_n$. Then we can write $f = \sum_j h_j g_j$ for some homogeneous forms $h_j \in R$. Set $a_{n+1} = a_n + \sum_j h_j b_j$. Then $a - a_{n+1} = (a - a_n) - \sum_j h_j b_j$ has the lowest form of degree strictly greater than the degree of the lowest form of $a - a_n$. This constructs a Cauchy sequence $\{a_n\}$ in R'. As R' is complete, this Cauchy sequence has a limit in R', but the limit equals a. This proves that $S = R'$, and so S is finitely generated over R.

In particular, S is Noetherian, and complete in the $\mathfrak{m}$–adic topology. By Proposition 4.3.2, then S is the direct product of the r rings S_i, where r is the number of maximal ideals in S. But S is an integral domain, so that $r = 1$, and S has only one maximal ideal.   □

It is also true that if R is an integral domain that is finitely generated over a complete Noetherian local ring, then $\overline{R}$ is module–finite over R. A proof is worked out in Exercise 9.7.

4.4. Jacobian ideals

In this section we define Jacobian ideals and prove basic properties about them. Jacobian ideals will be used in the subsequent section to prove a few criteria for when a ring is locally an integrally closed integral domain. Further uses of Jacobian ideals be found in Chapters 7, 12, 13, 15.

Definition 4.4.1 *Let A be a commutative ring and R a localization of a finitely generated A–algebra. Write R as $W^{-1}A[\underline{X}]/(f_1, \ldots, f_m)$, where $A[\underline{X}] = A[X_1, \ldots, X_n]$, $X_1, \ldots, X_n$ are variables over A, $f_i \in A[\underline{X}]$, and W is a multiplicatively closed subset of $A[\underline{X}]$. A **Jacobian matrix** of R over*

A is defined as the $m \times n$ matrix whose (i,j) entry is $\frac{\partial f_i}{\partial X_j}$. These partial derivatives are symbolic:

$$\frac{\partial (X_1^{a_1} \cdots X_n^{a_n})}{\partial X_j} = a_j X_1^{a_1} \cdots X_{j-1}^{a_{j-1}} X_j^{a_j - 1} X_{j+1}^{a_{j+1}} \cdots X_n^{a_n},$$

and extend A–linearly to all of $A[\underline{X}]$.

Assume furthermore that A is universally catenary, and that there exists a non–negative integer h such that for each prime ideal P in $A[\underline{X}]$ that is minimal over $(f_1, \ldots, f_m)$ and such that $P \cap W = \emptyset$, $A[\underline{X}]_P$ is equidimensional of dimension h. Observe that this set of prime ideals is in one–to–one correspondence with the minimal primes of R. Under these conditions, the **Jacobian ideal** *of R over A, denoted $J_{R/A}$, is the ideal in R generated by all the $h \times h$ minors of the Jacobian matrix of R over A.*

Example 4.4.2 Let k be a field, X, Y variables, and $R = k[X]/(X^3) \cong k[X,Y]/(XY, Y - X^2)$. Both presentations of R allow us to define the Jacobian ideal and both give $J_{R/k} = 3X^2 R$. The second presentation gives the Jacobian matrix

$$\begin{bmatrix} Y & X \\ -2X & 1 \end{bmatrix}.$$

Example 4.4.3 It can happen that $J_{R/A} = 0$. For instance, let $A = k$ be a field of characteristic 2, and let $R = k[X]/(X^2)$. Then $J_{R/A} = 0$. Even when R contains $\mathbb{Q}$, the Jacobian ideal can be zero. For example, let $R = \mathbb{Q}[X,Y]/(X^2, XY, Y^2)$. The Jacobian ideal $J_{R/\mathbb{Q}}$ is the image of (X^2, XY, Y^2) in R, which is 0.

Clearly the Jacobian matrix depends on the choice of the f_i and the X_j. However, the Jacobian ideal $J_{R/A}$ is independent:

Proposition 4.4.4 *The Jacobian ideal $J_{R/A}$ is well–defined, i.e., if R and R' are both localizations of finitely generated A–algebras and are A–isomorphic, then the isomorphism takes one Jacobian ideal to the other.*

Proof: The proof proceeds in several steps.

If $A[X_1, \ldots, X_n] = A[\underline{X}]$, $R = W^{-1} A[\underline{X}]/(f_1, \ldots, f_m)$, and we let W' be the saturation of W. In other words, W is the set of all polynomials in $A[\underline{X}]$ whose image is invertible in R. Then $R = (W')^{-1} A[\underline{X}]/(f_1, \ldots, f_m)$ and the Jacobian ideals computed either with W or with W' are equal. Thus we may always make the multiplicatively closed set W as large as possible.

Next, we prove that if we fix the images of $X_1, \ldots, X_n$ in R, then the Jacobian ideal is independent of the presenting ideal and its generators. Namely, let $X_1, \ldots, X_n$, $f_1, \ldots, f_m$, W, and h be as in the definition of the Jacobian ideal. Let x_i be the image of X_i in R.

Suppose that $(g_1, \ldots, g_s)$ is another ideal in $A[\underline{X}]$ such that via the A–algebra map $X_i \mapsto x_i$, $W^{-1}(A[\underline{X}]/(g_1, \ldots, g_s))$ is isomorphic to R, and such that for each prime ideal P in $A[\underline{X}]$ that is minimal over $(g_1, \ldots, g_s)$ and with

$P \cap W = \emptyset$, $A[\underline{X}]_P$ is equidimensional of dimension h. Then there exists an element $u \in W$ such that for all $i = 1, \ldots, s$, $ug_i \in (f_1, \ldots, f_m)$. Write $ug_i = \sum_k D_{ik} f_k$ for some $D_{ik} \in A[\underline{X}]$. Then

$$u \frac{\partial g_i}{\partial X_j} + \frac{\partial u}{\partial X_j} g_i = \sum_k \frac{\partial D_{ik}}{\partial X_j} f_k + \sum_k D_{ik} \frac{\partial f_k}{\partial X_j},$$

so that in R, $u \frac{\partial g_i}{\partial X_j} = \sum_k D_{ik} \frac{\partial f_k}{\partial X_j}$. As u is a unit in R, $I_h(\frac{\partial g_i}{\partial X_j}) \subseteq I_h(\frac{\partial f_i}{\partial X_j})R$. Hence by symmetry, $I_h(\frac{\partial g_i}{\partial X_j})R = I_h(\frac{\partial f_i}{\partial X_j})R$, which proves that $J_{R/A}$ is independent of the presentation ideal.

Thus we may assume that $(f_1, \ldots, f_m) = W^{-1}(f_1, \ldots, f_m) \cap A[\underline{X}]$. We need to check that this kernel satisfies the condition that $A[\underline{X}]_P$ is equidimensional of dimension h for every prime ideal P in $A[\underline{X}]$ that is minimal over $(f_1, \ldots, f_m)$ and satisfies $P \cap W = \emptyset$. But such primes correspond exactly to the minimal primes of R, and by our assumption they satisfy the condition.

Let $x_0 \in R$. Let X_0 a variable over $A[\underline{X}]$, and under the natural A–algebra homomorphism taking $X_i \mapsto x_i$, let K be the kernel and U the set of all elements in $A[X_0, \ldots, X_n]$ that map to a unit in R. Observe that $W \subseteq U$. As $R \cong W^{-1}(A[\underline{X}]/(f_1, \ldots, f_m))$, there exist $u \in W$ and $f \in A[\underline{X}]$ such that $uX_0 - f \in K$. Let $\pi : W^{-1}A[\underline{X}] \to R$ be the surjection as before with kernel $(f_1, \ldots, f_m)$. Let φ be the natural homomorphism from $U^{-1}A[X_0, \ldots, X_n]$ onto R extending π and sending X_0 to x_0. We claim that the kernel of φ is generated by $f_1, \ldots, f_m, uX_0 - f$. These elements are clearly in the kernel. Suppose that $\varphi(g) = 0$ for some $g \in U^{-1}A[X_0, \ldots, X_n]$. Without loss of generality we may even assume that $g \in A[X_0, \ldots, X_n]$. If the degree of g in X_0 is e, then we can write $u^e g - (uX_0 - f)g' \in A[\underline{X}]$ for some $g' \in A[X_0, \ldots, X_n]$, whence $\pi(u^e g - (uX_0 - f)g') = \varphi((u^e g - (uX_0 - f)g') = 0$. Since the kernel of π is generated by $f_1, \ldots, f_m$, we obtain that $u^e g \in U^{-1}(f_1, \ldots, f_m, uX_0 - f)$, and therefore $g \in U^{-1}(f_1, \ldots, f_m, uX_0 - f)$.

We have proved that $U^{-1}A[X_0, \ldots, X_n]/(f_1, \ldots, f_m, uX_0 - f) \cong R$. Let Q be a prime ideal in $A[X_0, \ldots, X_n]$ that is minimal over $(f_1, \ldots, f_m, uX_0 - f)$ and such that $Q \cap U = \emptyset$. We will prove that $A[X_0, \ldots, X_n]_Q$ is equidimensional of dimension $h + 1$, so that we can use this representation to compute the Jacobian ideal. Let q be an arbitrary minimal prime ideal in $A[X_0, \ldots, X_n]_Q$. We know that q is naturally an extension of a minimal prime ideal p in A. For any ideal I in $A[\underline{X}]$, let I^e denote $IA[X_0, \ldots, X_n]$. Let P be a prime ideal in $A[\underline{X}]$ that is contained in Q and is minimal over $(f_1, \ldots, f_m)$. Since $W \subseteq U$, $Q \cap U = \emptyset$ implies that $P \cap W = \emptyset$. We claim that $\mathrm{ht}(Q/P^e) = 1$. By Krull's Height Theorem, $\mathrm{ht}(Q/P^e) \leq 1$. If the height is 0, then $P^e = Q$, and it follows that $uX_0 - f \in P^e$, which implies that $u \in P$. But $P \cap W = \emptyset$, a contradiction. Thus $\mathrm{ht}(Q/P^e) = 1$.

Since $Q \cap A = P \cap A$, P contains p_0. By assumption $A[\underline{X}]_P$ is equidimensional of dimension h, hence $\mathrm{ht}(P/pA[\underline{X}]) = h$. The assumption that A is universally catenary gives that

$$\mathrm{ht}(Q/q) = \mathrm{ht}(Q/pA[X_0,\ldots,X_n])$$
$$= \mathrm{ht}(P^e/pA[X_0,\ldots,X_n]) + \mathrm{ht}(Q/P^e) = h+1.$$

The value of $\mathrm{ht}(Q/q)$ is independent of Q and q. It follows that the Jacobian ideal R over A can also be defined using the algebra generators $x_0,\ldots,x_n$. We claim that this is the same as the Jacobian ideal defined using the algebra generators $x_1,\ldots,x_n$. The Jacobian matrix for R over A with respect to the algebra generators $x_0,\ldots,x_n$ and the relations $f_1,\ldots,f_m, uX_0 - f$ is

$$C = \begin{bmatrix} 0 & \left(\frac{\partial f_i}{\partial X_j}\right) \\ u & * \end{bmatrix}.$$

Clearly $I_{h+1}(C)R = uI_h(\frac{\partial f_i}{\partial X_j})R = I_h(\frac{\partial f_i}{\partial X_j})R$.

Using a straightforward induction, it follows that whenever the Jacobian ideal R over A is defined using the algebra generators $x_1,\ldots,x_n$, then the same Jacobian ideal is obtained after adding further elements of R to the algebra generating list. Hence if R can also be written as a localization of $A[Y_1,\ldots,Y_r]/(g_1,\ldots,g_s)$ at a multiplicatively closed set U, and there is a fixed integer l such that for every minimal prime P over $(g_1,\ldots,g_s)$ with $P \cap U = \emptyset$, we have that $A[Y_1,\ldots,Y_r]_P$ is equidimensional of dimension l, then if y_i is the image in R of Y_i, by above the Jacobian ideal defined via the algebra generators $x_1,\ldots,x_n$ is the same as the Jacobian ideal defined via the algebra generators $x_1,\ldots,x_n,y_1,\ldots,y_r$, which in turn is the same as the Jacobian ideal defined via the algebra generators $y_1,\ldots,y_r$. $\square$

Corollary 4.4.5 *Let A be a universally catenary Noetherian ring, and let R be $W^{-1}A[\underline{X}]/(f_1,\ldots,f_m)$ for some variables $X_1,\ldots,X_n$ over A, some $f_i \in A[\underline{X}]$, and some multiplicatively closed subset W of $A[\underline{X}]$. Assume that there exists a non–negative integer h such that for each prime ideal P in $A[\underline{X}]$ that is minimal over $(f_1,\ldots,f_m)$ and such that $P \cap W = \emptyset$, $A[\underline{X}]_P$ is equidimensional of dimension h. Let U be an arbitrary multiplicatively closed subset of R. The Jacobian ideal of $U^{-1}R$ over A is defined as well, and*

$$U^{-1}(J_{R/A}) = J_{(U^{-1}R)/A}.$$

Proof: The fact that the Jacobian ideal is defined follows at once from the definition since the set of primes for which the equidimensionality condition needs to be checked can only get smaller after localizing at W. The equality follows since we may use the same elements $f_1,\ldots,f_m$ to compute the Jacobian ideal. $\square$

We defined Jacobian matrices in greater generality than Jacobian ideals. We show on an example that without the height restrictions on $(f_1,\ldots,f_m)$ the Jacobian ideal need not localize as expected (as in this corollary):

Example 4.4.6 Let k be a field of characteristic other than 2, let X,Y,Z be variables over k and $R = k[X,Y,Z]/(XY^2, XZ^2)$. The Jacobian matrix of

R over k is

$$\begin{bmatrix} Y^2 & 2XY & 0 \\ Z^2 & 0 & 2XZ \end{bmatrix},$$

whose ideal of 1×1 minors is $I_1 = (Y^2, 2XY, Z^2, 2XZ)R$ and whose ideal of 2×2 minors is $I_2 = 4X^2YZR$. The minimal prime ideals of R are $P = XR$ and $Q = (Y, Z)R$. Observe that R_P is a localization of $k[X, Y, Z]/(X)$, hence its Jacobian ideal is clearly R_P. If we want Corollary 4.4.5 to apply to this R, since $I_2 R_P \ne R_P$, necessarily I_2 is not the Jacobian ideal, so we would then require a possible Jacobian ideal of R/k to be I_1. But R_Q is a localization of $k[X, Y, Z]/(Y^2, Z^2)$, so its Jacobian ideal is $4YZR_Q$, which is not $I_1 R_Q$. Thus we cannot define a Jacobian ideal for this non–equidimensional R and still expect Corollary 4.4.5 to hold.

Discussion 4.4.7 Although we have chosen to give a "bare–hands" proof of the fact that the Jacobian ideal is well–defined (under our assumptions), a possibly better way of presenting this material is through the module of Kähler differentials. We highlight relevant statements in this discussion. We opted not to develop Jacobian ideals in this way simply because we did not want to include a detailed discussion of Kähler differentials, but wanted this book to be largely self–contained.

First, for simplicity, let $A = k$ be a field, and let $R = k[\underline{X}]/(f_1, \ldots, f_m)$ be a finitely generated k–algebra. We let J be the Jacobian matrix of R over k. By viewing J as a matrix with entries in R, J is actually a presentation matrix for the module of k–linear Kähler differentials $\Omega_{R/k}$. Thus the various ideals generated by the minors of J (the Fitting ideals of $\Omega_{R/k}$) are independent of the chosen generators of I or the presentation of R; they depend only on the k–isomorphism class of R. The formation of Fitting ideals commutes with localization and base change, e.g., if W is a multiplicatively closed subset of R, then $\Omega_{W^{-1}R/k} \cong W^{-1}R \otimes_R \Omega_{R/k}$ (cf. Corollary 4.4.5). The problem with defining *the* Jacobian ideal with this approach is that $\Omega_{R/k}$ may not have a rank, so that there is not a good unique choice of which Fitting ideal should be the Jacobian ideal. In other words, after localizing at a minimal prime ideal of R, $\Omega_{R/k}$ may not be free, or even if it is free, the rank may vary depending on which minimal prime is used. However, if R is an equidimensional finitely generated k–algebra of dimension d and has no embedded primes, then the ideal of $(n - d) \times (n - d)$ minors of J reduced modulo I can be taken to be the Jacobian ideal. Of course, when we replace k by an arbitrary ring A, one needs some additional equidimensionality assumptions, which accounts for the version we give above.

Lemma 4.4.8 *Let $R \subseteq S \subseteq T$ be regular Noetherian rings with S and T localizations of finitely generated rings over R. Assume that the total ring of fractions of T is algebraic over the total ring of fractions of R. Then*

$$J_{T/R} = J_{T/S} J_{S/R}.$$

Proof: We proved above that Jacobian ideals localize. The equality in the lemma holds if and only if it holds after localizing at all the prime ideals of T. Thus we may assume that T is local, and by also inverting the elements in R and S that are units in T we may assume in addition that R and S are local. As these are regular rings, they are domains.

We may write $S = (R[X_1, \ldots, X_n]/Q)_P$ for some variables $X_1, \ldots, X_n$ over R, and some prime ideals $Q \subseteq P$ in $R[X_1, \ldots, X_n]$. Necessarily there exist $f_1, \ldots, f_l \in R[X_1, \ldots, X_n]$ that after localization at P form part of a regular system of parameters and generate Q_P. By Prime Avoidance we may assume that $f_1, \ldots, f_l$ generate an ideal of height l in $R[X_1, \ldots, X_n]$, so that as R is regular, $(f_1, \ldots, f_l)$ has all minimal prime ideals of the same height l. Note that $S = (R[X_1, \ldots, X_n]/(f_1, \ldots, f_l))_P$. By the algebraic dependence assumption on the fields of fractions of S and R, $l \geq n$. But letting K be the field of fractions of R, $QK[X_1, \ldots, X_n]$ is a prime ideal of height l. Necessarily $l \leq n$, hence $n = l$.

Thus $R[X_1, \ldots, X_n]/(f_1, \ldots, f_n)$ is equidimensional and S is its localization. Similarly, for some variables $Y_1, \ldots, Y_m$ over S and some polynomials $g_1, \ldots, g_m \in S[Y_1, \ldots, Y_m]$, T is a localization of the equidimensional ring $S[Y_1, \ldots, Y_m]/(g_1, \ldots, g_m)$. Hence $J_{S/R}$ is generated by the determinant of the $m \times m$ matrix $(\frac{\partial f_i}{\partial X_j})$, $J_{T/S}$ by the determinant of the $n \times n$ matrix $(\frac{\partial g_i}{\partial Y_j})$, and $J_{T/R}$ by the determinant of the $(n+m) \times (n+m)$ matrix

$$\begin{pmatrix} \frac{\partial f_i}{\partial X_j} & 0 \\ \frac{\partial g_i}{\partial X_j} & \frac{\partial g_i}{\partial Y_j} \end{pmatrix},$$

whence the lemma follows. □

We prove in the next theorem that the Jacobian ideal determines the regularity and thus the normality of finitely generated algebras over a field. We also prove a partial converse, namely that regularity of the ring supplies some information on the Jacobian ideal, if in addition we assume a separable condition on residue fields. (For separability, see Chapter 3.)

Theorem 4.4.9 (Jacobian criterion) *Let k be a field and R an equidimensional finitely generated k–algebra. Let J be the Jacobian ideal $J_{R/k}$. Let P be a prime ideal in R. If J is not contained in P, then R_P is a regular ring.*

Conversely, if R_P is a regular ring and $\kappa(P)$ is separable over k (say if k is a perfect field, by Theorem 3.2.7 (4)), then J is not contained in P.

Proof: Let $S = k[X_1, \ldots, X_n]$, and $(f_1, \ldots, f_m)$ be an ideal in S such that $R = S/(f_1, \ldots, f_m)$. As R is equidimensional and polynomial rings over a field are catenary domains, all the prime ideals in S minimal over $(f_1, \ldots, f_m)$ have the same height h. Let Q be the preimage of P in S.

Suppose that J is not contained in P. By possibly reindexing, we may assume that the minor of the submatrix of the Jacobian matrix consisting of the first h columns and the first h rows is not in Q.

Claim: $f_1, \ldots, f_h$ are elements of Q whose images are linearly independent in the vector space QS_Q/Q^2S_Q. Indeed when $r_1, \ldots, r_h \in S$ such that $\sum_i r_i f_i \in Q^2 S_Q$, for some $s \in S \setminus Q$, $s \sum_i r_i f_i \in Q^2$. Then for all j, $\sum_i sr_i \frac{\partial f_i}{\partial X_j} + \sum_i \frac{\partial (sr_i)}{\partial X_j} f_i \in Q$. Hence in R_P, for all j, $\sum_i r_i \frac{\partial f_i}{\partial X_j} \in P$. This gives a linear dependence relation on the columns of the fixed $h \times h$ submatrix after passing to R_P/PR_P. By our assumption on P, this is the trivial relation, so that all r_i are in P. This proves the claim.

But S_Q is a regular local ring, so that by the claim, $f_1, \ldots, f_h$ is part of a regular system of parameters in S_Q. Thus $(f_1, \ldots, f_h)S_Q$ is a prime ideal of height h, and by equidimensionality $(f_1, \ldots, f_m)S_Q = (f_1, \ldots, f_h)S_Q$. Then $R_P = (S/(f_1, \ldots, f_h))_Q$ is a regular local ring.

For the converse, suppose that k is a perfect field and that R_P is regular. By Theorem 3.2.7 we may choose x_i in $\kappa(P)$ such that $k \subseteq k(x_1, \ldots, x_t)$ is transcendental and $k(x_1, \ldots, x_t) \subseteq \kappa(P)$ is separable algebraic. By the Primitive Element Theorem (Theorem 3.1.1), there exists $g(Y) \in k[x_1, \ldots, x_t][Y]$ that is separable in Y such that $\kappa(P) = k(x_1, \ldots, x_t)[Y]/(g(Y))$. Then $J_{\kappa(P)/k}$ is the ideal generated by the $\frac{\partial g}{\partial x_1}, \ldots, \frac{\partial g}{\partial x_t}, \frac{\partial g}{\partial Y}$. As $\frac{\partial g}{\partial Y}$ is non–zero, $J_{\kappa(P)/k}$ is non–zero.

Since R_P is regular, we may choose $f_1, \ldots, f_h \in S$ such that $(f_1, \ldots, f_h)$ has height h in S, such that $f_1, \ldots, f_h$ is part of a regular system of parameters in S_Q, and such that R_P as a localization of $k[X_1, \ldots, X_n]/(f_1, \ldots, f_h)$ at Q. As $\kappa(Q) = \kappa(P)$ is a regular quotient of R_P, there exist $k_1, \ldots, k_s \in S$ such that $(f_1, \ldots, f_h, k_1, \ldots, k_s)$ has height $h + s = \operatorname{ht} Q$, and such that after localization at Q, $S/(f_1, \ldots, f_h, k_1, \ldots, k_s)$ equals $\kappa(Q)$. By computing the Jacobian ideal of $\kappa(P)$ over k this way, we get that

$$J_{\kappa(P)/k} = I_{h+s} \begin{pmatrix} \frac{\partial f_i}{\partial X_j} \\ \frac{\partial k_i}{\partial X_j} \end{pmatrix} \kappa(P) \subseteq I_h \left(\frac{\partial f_i}{\partial X_j} \right) \kappa(P) = J_{R/k}\kappa(P).$$

As $J_{\kappa(P)/k} \neq 0$, it follows that $J_{R/k} \not\subseteq P$. $\qquad \square$

Example 4.4.10 Let k be a field of positive characteristic p. Suppose that $\alpha \in k$ is not a pth power. Set $R = k[T]/(T^p - \alpha)$. Obviously $J_{R/k} = 0$, but R is isomorphic to $k(\alpha^{1/p})$ and is a field. This explains why in the converse direction of the Jacobian criterion we need some condition on separability.

The assumption in Theorem 4.4.9 that k be a field cannot be relaxed to k being a regular ring. Namely, if $R \subseteq S$ are regular local rings with the same field of fractions, the Jacobian ideal $J_{S/R}$ need not be the unit ideal. For example, let $R = k[[X, Y]]$, where k is a field and X and Y variables over k. Let S be a localization of $R[\frac{Y}{X}] \cong \frac{R[T]}{(XT-Y)}$. Then both R and S are regular and $J_{S/R} = XS$, but $J_{S/R}$ is a unit ideal if and only if X is a unit in S.

If a ring is regular, it is normal, so the Jacobian criterion gives a sufficient condition of normality. More conditions for normality are in the next section.

4.5. Serre's conditions

Serre's conditions (R_k) and (S_k) are convenient conditions for normality, regularity, Cohen–Macaulayness and other properties of rings.

Definition 4.5.1 *Let R be a Noetherian ring and k a non–negative integer. R is said to satisfy* **Serre's condition (R_k)** *if for all prime ideals P in R of height at most k, R_P is a regular local ring.*

The ring R is said to satisfy **Serre's condition (S_k)** *if for all prime ideals P in R, the depth of R_P is at least $\min\{k, \operatorname{ht} P\}$.*

In other words, R satisfies Serre's condition (R_k) if and only if for all $P \in \operatorname{Spec} R$, $\mu(PR_P) = \operatorname{ht} P$, and R satisfies Serre's condition (S_k) if and only if for all $P \in \operatorname{Spec} R$, PR_P contains a regular sequence of length at least $\min\{k, \operatorname{ht} P\}$.

It is clear that every zero–dimensional Noetherian ring satisfies all (S_k), and more generally, that every Cohen–Macaulay ring satisfies all (S_k). Recall that R is Cohen–Macaulay if and only if for all $P \in \operatorname{Spec} R$, PR_P contains an R_P-regular sequence of length $\operatorname{ht} P$.

Theorem 4.5.2 *A Noetherian ring is reduced if and only if it satisfies Serre's conditions (R_0) and (S_1).*

Proof: Every reduced Noetherian ring clearly satisfies (S_1) and (R_0). Conversely, by the (S_1) condition, the only associated primes of the zero ideal are the minimal prime ideals. By the (R_0) condition, localization at each minimal prime ideal is a regular local ring, thus a domain. Thus the only primary components of the zero ideal are the minimal prime ideals, proving that R is reduced. $\qquad\square$

Serre's conditions characterize integrally closed rings:

Theorem 4.5.3 (Serre's conditions) *A Noetherian ring R is normal if and only if it satisfies Serre's conditions (R_1) and (S_2).*

Proof: First assume that R is normal. Let P be a prime ideal in R of height one. Then R_P is a one–dimensional integrally closed domain. By Proposition 4.1.1, the maximal ideal of R_P is principal, so that R_P is a regular local ring. Thus R satisfies (R_1). Now let P be a prime ideal of height at least 2. As R_P is an integrally closed domain of dimension at least 2, there exists $x \in P$ that is a non–zerodivisor in R_P. By Proposition 4.1.1, PR_P is not associated to xR_P, so that there exists an element $y \in P$ such that x, y is a regular sequence in R_P. Thus R satisfies (S_2).

Assume that R satisfies (R_1) and (S_2). By Theorem 4.5.2, R is reduced. We need to show that for every prime ideal P in R, R_P is an integrally closed domain. As R is reduced, so is R_P. We first show that R_P is integrally closed in its total ring of fractions. Let $x, y \in R_P$ be non–zero elements such that y is not in any minimal prime ideal of R_P and x/y is integral over R_P. Set

$I = yR_P :_{R_P} x$, an ideal of R_P. By assumption (S_2), all the associated primes of I have height one. Let Q be an associated prime ideal of I. As R_Q satisfies (R_1), $\frac{x}{y} \in R_Q$, so that $I_Q = R_Q$, and that contradicts the hypothesis that Q was associated to I. Thus I has no associated primes, so that $I = R_P$. This means that x is an R_P–multiple of y, so $\frac{x}{y} \in R_P$, and so R_P is integrally closed. Hence by Corollary 2.1.13, R_P is a domain. $\square$

In geometric terms, this says in particular that a normal variety of dimension d has singular locus of dimension at most $d - 2$ (and it is a closed subscheme, by Theorem 4.4.9).

Definition 4.5.4 *Let R be a ring. The* **singular locus** *of R the set of all $P \in \operatorname{Spec} R$ such that R_P is not regular.*

Let $(R, \mathfrak{m}) \to (S, \mathfrak{n})$ be a flat local homomorphism of Noetherian local rings. We assume familiarity with the following facts:
(1) If S is regular (respectively Cohen–Macaulay), so is R;
(2) If R and $S/\mathfrak{m}S$ are regular (respectively Cohen–Macaulay), then S is also regular (respectively Cohen–Macaulay).

Theorem 4.5.5 *Let $(R, \mathfrak{m}) \to (S, \mathfrak{n})$ be a flat local map of Noetherian local rings, and let k be in $\mathbb{N}$.*
(1) If S satisfies Serre's condition (R_k), respectively (S_k), so does R.
(2) If R and all $\kappa(P) \otimes_R S$ with $P \in \operatorname{Spec}(R)$ satisfy Serre's condition (R_k), respectively (S_k), so does S.

Proof: Let $P \in \operatorname{Spec} R$ be of height at most k. Let $Q \in \operatorname{Spec} S$ be minimal over PS. By Proposition B.2.3, $\operatorname{ht} Q = \operatorname{ht} P \leq k$. Thus S_Q satisfies Serre's condition (R_k), respectively (S_k). In either case, S_Q is Cohen–Macaulay. Then by the remark above, since $R_P \to S_Q$ is faithfully flat, R_P also satisfies Serre's condition (R_k), respectively (S_k).

Now let $Q \in \operatorname{Spec} S$ have height at most k. Let $P = Q \cap R$. Since $PS \subseteq Q$, by Proposition B.2.3, $\operatorname{ht} P \leq \operatorname{ht} Q \leq k$. By assumption R_P and $\kappa(P) \otimes_R S$ satisfy Serre's condition (R_k), respectively (S_k). Again, the conclusion follows from the remark above. $\square$

Corollary 4.5.6 *Let $(R, \mathfrak{m}) \to (S, \mathfrak{n})$ be a local flat homomorphism of Noetherian local rings. Then S is normal if and only if R and all the fibers $\kappa(P) \otimes_R S$ are normal, as P varies over the prime ideals of R.*

Proof: Apply the theorem above and Theorem 4.5.3. $\square$

Checking Serre's conditions on an arbitrary ring may be an impossible task, as a property has to be checked for each of the possibly infinitely many prime ideals. However, for localizations of finitely generated algebras over a field the task is reduced to a finite task by using the Jacobian ideals:

Theorem 4.5.7 (Serre's conditions) *(Matsumoto [199, Proposition 2]) Let R be a Noetherian ring and J an ideal in R such that $V(J)$ is exactly the* *singular (non–regular) locus of R. (For example, if k is a perfect field, R is a localization of an equidimensional finitely generated k–algebra, and J the Jacobian ideal of this extension.) Then for any integer r, J has grade at least r if and only if R satisfies (R_{r-1}) and (S_r).*

Proof: Suppose that J has grade at least r. If P is a prime ideal of height at most $r - 1$, then $J \not\subseteq P$, so that by assumption R_P is regular. Thus R satisfies (R_{r-1}). If a prime ideal P contains J, it has grade at least r, so depth $R_P \geq r$. If a prime ideal P does not contain J, then R_P is regular, so again depth $R_P = \text{ht } P$. Thus R satisfies (S_r).

Conversely, assume that R satisfies (R_{r-1}) and (S_r). By assumption that $V(J)$ is the singular locus and that the condition (R_{r-1}) holds, J has height at least r. Hence by assumption (S_r), J has grade at least r. □

An immediate corollary of this theorem and of Theorems 4.5.2 and 4.5.3 is as follows:

Corollary 4.5.8 *(Matsumoto [199, Corollary 3]) Let R be a Noetherian ring, K its total ring of fractions, and J an ideal in R such that $V(J)$ is the singular locus of R. Then*
(1) R is reduced if and only if the grade of J is at least one, and
(2) R is a direct product of normal domains if and only if J has grade at least two, which holds if and only if $0 :_R J = 0$ and $R :_K J = R$. □

A consequence is about the normal locus of a ring, which we define next.

Definition 4.5.9 *The set of all prime ideals P in a ring R for which R_P is normal is called the **normal locus** of R.*

Corollary 4.5.10 *Let R be a Noetherian domain, and S a module–finite extension domain of R. Assume that there exists a non–zero element $f \in R$ such that S_f is normal. Then the subset of $\text{Spec } R$ consisting of those prime ideals P for which S_P is normal is open in $\text{Spec } R$.*

Proof: Let $Q_1, \ldots, Q_r$ be the prime ideals in S minimal over fS such that S_{Q_i} does not satisfy (R_1). Set $P_i = Q_i \cap R$. Then S_{P_i} does not satisfy (R_1). Let T_1 be the closed subset of $\text{Spec } R$ consisting of the prime ideals that contain one of the P_i. If $P \in T_1$, as S_{P_i} does not satisfy (R_1), neither does S_P. If instead $P \in \text{Spec } R$ such that S_P does not satisfy (R_1), then for some prime ideal Q in S of height one such that $Q \cap (R \setminus P) = \emptyset$, S_Q does not satisfy (R_1). Thus S_Q is not integrally closed, so necessarily $f \in Q$, and Q is minimal over fS. Thus $Q = Q_i$ for some i, so Q contains P_i, and $P_i \subseteq P$. This proves that $P \in T_1$ if and only if S_P does not satisfy Serre's condition (R_1).

Let T_2 be the closed subset of $\text{Spec } R$ consisting of those prime ideals that contain one of the embedded prime ideals of the R–module S/fS. By assumption on normality of S_f, for any prime ideal P in R, $P \in T_2$ if and only if S_P does not satisfy Serre's condition (S_2).

Thus $T_1 \cup T_2$ is a closed subset of Spec R consisting of those prime ideals P for which S_P does not satisfy either (R_1) or (S_2), i.e., for which S_P is not normal. Thus Spec $R \setminus (T_1 \cup T_2)$ is open. $\square$

Corollary 4.5.11   *Let R be a Noetherian domain whose normal locus is non–empty and open. Then the integral closure of R is module–finite over R if and only if for each maximal ideal $\mathfrak{m}$ in R, the integral closure of $R_{\mathfrak{m}}$ is module–finite over $R_{\mathfrak{m}}$.*

Proof: By assumption there exists $f \in R$ such that R_f is normal. For each maximal ideal $\mathfrak{m}$ in R let $S(\mathfrak{m})$ be a finitely generated R–module contained in $\overline{R}$ such that $S(\mathfrak{m})_{\mathfrak{m}} = \overline{R}_{\mathfrak{m}}$. As $S(\mathfrak{m})_f = R_f$, by Corollary 4.5.10, the normal locus of $S(\mathfrak{m})$ is open. Set $T_{\mathfrak{m}} = \{P \in \operatorname{Spec} R \mid S(\mathfrak{m})_P \text{ is integrally closed}\}$. Then $\mathfrak{m} \in T_{\mathfrak{m}}$ and by Corollary 4.5.10, $T_{\mathfrak{m}}$ is open in Spec R. Thus $\cup_{\mathfrak{m}} T_{\mathfrak{m}}$ is an open subset of Spec R that contains every maximal ideal, whence $\cup_{\mathfrak{m}} T_{\mathfrak{m}} =$ Spec R. As Spec R is quasi–compact, there exist maximal ideals $\mathfrak{m}_1, \ldots, \mathfrak{m}_r$ such that $\cup_{i=1}^{r} T_{\mathfrak{m}_i} = \operatorname{Spec} R$.

Set $S = \sum_{i=1}^{r} S(\mathfrak{m}_i)$. Then S is a module–finite extension of R contained in $\overline{R}$. Let $P \in \operatorname{Spec} R$. Then $P \in T_{m_i}$ for some i, so that $S(\mathfrak{m}_i)_P$ is normal. Thus $S(\mathfrak{m}_i)_P = \overline{R}_P = S_P$. It follows that S is the integral closure of R. $\square$

4.6. Affine and $\mathbb{Z}$–algebras

In this section we develop some criteria for module–finiteness via completion, and prove that finitely generated algebras over fields and over $\mathbb{Z}$ that are domains have module–finite integral closures.

Lemma 4.6.1   *Let R be a semi–local Noetherian domain and let x be a non–zero element contained in every maximal ideal of R. Assume that for all $P \in \operatorname{Ass}(R/xR)$, R_P is a one–dimensional integrally closed ring and the completion of R/P with respect to its Jacobson radical is reduced. Then the completion of R is reduced.*

Proof: Let $P_1, \ldots, P_r$ be the prime ideals associated to R/xR. By Proposition 4.1.1, for each $i = 1, \ldots, r$, there exists $y_i \in P_i$ such that $y_i R_{P_i} = P_i R_{P_i}$. Let $Q_{i1}, \ldots, Q_{in_i}$ be the associated prime ideals of $\widehat{R}/P_i \widehat{R}$. By assumption, $\widehat{R}/P_i \widehat{R}$ is reduced, so the Q_{ij} are minimal over $P_i \widehat{R}$ and $P_i \widehat{R} = \cap_{j=1}^{n_i} Q_{ij}$. Thus $Q_{ij} \widehat{R}_{Q_{ij}} = P_i \widehat{R}_{Q_{ij}} = y_i \widehat{R}_{Q_{ij}}$, and y_i is a non–zerodivisor in $\widehat{R}$. Thus the maximal ideal in $R_{Q_{ij}}$ is principal generated by a non–zerodivisor, so that by Proposition 4.1.2, $\widehat{R}_{Q_{ij}}$ is a one–dimensional integrally closed domain.

Clearly, these Q_{ij} are all the minimal prime ideals over $x\widehat{R}$. Let $Q \in \operatorname{Ass}(\widehat{R}/x\widehat{R})$. As $R/xR \to \widehat{R}/x\widehat{R}$ is flat, non–zerodivisors on R/xR map to non–zerodivisors on $\widehat{R}/x\widehat{R}$, so that $Q \cap R$ is contained in some P_i. As P_i has height 1, necessarily $Q \cap R = P_i$. Then Q contains y_i. Since the depth of

$\widehat{R}_Q$ is one and y_i is a non–zerodivisor in Q it follows that Q is also associated to $y_i\widehat{R}$. Set W to be the multiplicatively closed set $R \setminus P_i$. Then $W^{-1}Q$ is associated to $y_iW^{-1}\widehat{R} = P_iW^{-1}\widehat{R}$, which is a radical ideal with minimal primes $Q_{i1}, \ldots, Q_{in_i}$. Thus Q is one of the minimal primes over $x\widehat{R}$, and $\mathrm{Ass}(\widehat{R}/x\widehat{R}) = \{Q_{ij} \,|\, i,j\}$. Write $x\widehat{R} = \cap q_{ij}$, where q_{ij} is Q_{ij}–primary. Let φ_{ij} be the natural map $\widehat{R} \to \widehat{R}_{Q_{ij}}$. Then $\ker \varphi_{ij} \subseteq q_{ij}$, so $\cap_{i,j} \ker \varphi_{ij} \subseteq x\widehat{R}$. For any $y_0 \in \cap_{i,j} \ker \varphi_{ij}$, write $y_0 = y_1 x$ for some $y_1 \in \widehat{R}$. As x is a non–zerodivisor in $\widehat{R}$, necessarily $y_1 \in \cap_{i,j} \ker \varphi_{ij}$. By repeating this argument, $y_0 \in \cap_n x^n \widehat{R} = 0$. Thus $\cap_{i,j} \ker \varphi_{ij} = 0$. If $r \in \widehat{R}$ and $r^n = 0$, then $0 = \varphi_{ij}(r^n) = \varphi_{ij}(r)^n$, and as $\widehat{R}_{Q_{ij}}$ is a domain, necessarily $\varphi_{ij}(r) = 0$. As this holds for all i, j, $r = 0$. Thus $\widehat{R}$ is reduced. $\qquad\square$

With this we can prove a criterion for module–finiteness:

Corollary 4.6.2 *Let R be a Noetherian semi–local domain, with Jacobson radical $\mathfrak{m}$. Let $\widehat{R}$ be the $\mathfrak{m}$–adic completion of R.*
(1) If $\widehat{R}$ is reduced, then $\overline{R}$ is a module–finite extension of R.
(2) If for every $P \in \mathrm{Spec}\, R$ the integral closure of R/P in any finite field extension of $\kappa(P)$ is a module–finite extension of R/P, then $\widehat{R}$ is reduced.

Proof: Assume that $\widehat{R}$ is reduced. Let $\mathrm{Min}(\widehat{R}) = \{P_1, \cdots, P_n\}$. As in Corollary 2.1.13, the integral closure $\widehat{\overline{R}}$ of $\widehat{R}$ is $\overline{\widehat{R}/P_1} \times \cdots \times \overline{\widehat{R}/P_n}$. By Theorem 4.3.4 the integral closure of a complete semi–local domain is a module–finite extension, so that $\widehat{\overline{R}}$ is module finite over $\widehat{R}/P_1 \times \cdots \times \widehat{R}/P_n$, and hence it is module–finite over $\widehat{R}$. Let K be the total ring of fractions of R. Then $R \subseteq \overline{R} \subseteq K$. Since $\widehat{R}$ is faithfully flat over R, $\widehat{R} \subseteq \overline{R} \otimes_R \widehat{R} \subseteq K \otimes_R \widehat{R}$. Every non-zerodivisor on R is a non–zerodivisor on $\widehat{R}$, so $K \otimes_R \widehat{R}$ is contained in the total ring of fractions of $\widehat{R}$. Therefore the elements of $\overline{R} \otimes_R \widehat{R}$ are in the total ring of fractions of $\widehat{R}$ and are integral over $\widehat{R}$, so they are contained in $\widehat{\overline{R}}$, which implies that $\overline{R} \otimes_R \widehat{R}$ is a finitely generated $\widehat{R}$–module. But $\widehat{R}$–module if faithfully flat over R, so necessarily $\overline{R}$ is module–finite over R.

Now assume that for all $P \in \mathrm{Spec}\, R$, the integral closure of R/P in any finite field extension of $\kappa(P)$ is module–finite over R. If the dimension of R is zero, there is nothing to prove: $\widehat{R} = R$ is a field. So assume that $\dim R > 0$. By assumption $\overline{R}$ is module–finite over R, so $\widehat{R} \subseteq \widehat{R} \otimes_R \overline{R}$, which is the completion of $\overline{R}$ in the topology defined by the Jacobson radical of $\overline{R}$.

For every prime ideal Q in $\overline{R}$, $\overline{R}/Q$ is module–finite over $R/(Q \cap R)$. Let L be a finite field extension of $\kappa(Q)$. Then L is a finite field extension of $\kappa(Q \cap R)$, so by assumption the integral closure of $R/(Q \cap R)$ and thus of $\overline{R}/Q$ in L is module–finite over $R/(Q \cap R)$. Thus $\overline{R}$ satisfies the same hypotheses on integral closures as does R.

Let x be a non–zero element contained in every maximal ideal of $\overline{R}$. Let $P \in \mathrm{Ass}(\overline{R}/x\overline{R})$. As $\overline{R}$ is integrally closed, by Proposition 4.1.1, P is minimal over $x\overline{R}$, so the height of P is one. Hence $\overline{R}_P$ is a one–dimensional integrally closed local domain. As $\dim(\overline{R}/P) < \dim R$, by induction on dimension the completion of $\overline{R}/P$ is reduced. By Lemma 4.6.1, $\widehat{\overline{R}}$ is reduced. As $\widehat{R} \subseteq \widehat{\overline{R}}$, it follows that $\widehat{R}$ is reduced. $\square$

An example for which the conclusion of (1) in Corollary 4.6.2 holds but the hypothesis does not is in Exercise 4.11.

Theorem 4.6.3 *Let R be a domain that is a finitely generated algebra over a field. Let K be the field of fractions of R and L a finite field extension of K. The integral closure of R in L is module–finite over R.*

Proof: By Theorem 4.2.2 (Noether normalization), R is module–finite over a polynomial subring $A = k[x_1, \ldots, x_m]$ in m variables over a subfield k. The integral closure $\overline{R}$ of R in L is the integral closure of A in L, A is integrally closed, and L is a finite extension over $k(x_1, \ldots, x_m)$. If the characteristic of k is zero, by Theorem 3.1.3, $\overline{R}$ is module–finite over A and hence over R.

Now suppose that the characteristic of k is a positive prime p. By standard field theory there exists a finite extension L' of L such that the extension $k(x_1, \ldots, x_m) \subseteq L'$ can be factored with $k(x_1, \ldots, x_m) \subseteq F \subseteq L'$, where $k(x_1, \ldots, x_m) \subseteq F$ is purely inseparable, and $F \subseteq L'$ is separable algebraic. By possibly enlarging F and L', we may assume that F is the field of fractions of $k'[x_1^{1/p^e}, \ldots, x_m^{1/p^e}]$ for some non–negative integer e and some purely inseparable finite field extension k' of k. If the integral closure of A in L' is module–finite, so is the integral closure of A in L. Thus it suffices to assume that $L = L'$. It also suffices to prove that the integral closure of $k'[x_1^{1/p^e}, \ldots, x_m^{1/p^e}]$ in L is module–finite. But as L is a separable extension of F, this follows by Theorem 3.1.3. $\square$

The next goal is to prove the analogous result for finitely generated $\mathbb{Z}$–algebras. First we need a lemma:

Theorem 4.6.4 *Let A be a Noetherian integrally closed domain satisfying the following properties:*
(1) The field of fractions has characteristic 0.
(2) For every non–zero prime ideal p in A, and every finitely generated (A/p)–algebra R that is a domain, the integral closure of R in a finite field extension L of the field of fractions of R is module–finite over R.

Let R be a domain that is finitely generated over A. Let K be the field of fractions of R and L a finite field extension of K. Then the integral closure of R in L is module–finite over R.

Proof: Let p be the kernel of the map $A \to R$. If p is not zero, the conclusion follows from assumption (2). Thus without loss of generality $A \subseteq R$. By Theorem 4.2.4, there exist a finitely generated polynomial ring B over A contained in R and an element $f \in B$ such that $B_f \subseteq R_f$ (localization at one element) is an integral extension. By Exercise 2.3, B is integrally closed. Let C be the integral closure of B in L. Then L is the field of fractions of C, and by the characteristic assumption L is a finite separable extension of $Q(R) = Q(R_f) \supseteq Q(B_f) = Q(B)$. By Theorem 3.1.3, C is module–finite over B. Then $R[C]$ is module–finite over R, and it suffices to prove that the integral closure of $R[C]$ is module–finite over $R[C]$.

Thus we may rename $R[C]$ to be R: this new R contains an integrally closed finitely generated A–algebra C, R is finitely generated over C, $R_f = C_f$, and the fields of fractions of R and C coincide. Under these assumptions we need to prove that the integral closure of R is module–finite over R. By induction on the number of generators of R over C without loss of generality $R = C[x]$.

As, $C_f = R_f$, by Corollary 4.5.10, the normal locus of R is open, and as R is a domain, the normal locus of R is non–empty. Thus by Corollary 4.5.11 it suffices to prove that for every maximal ideal Q in R, $\overline{R}_Q$ is module–finite over R_Q. Choose a maximal ideal Q in R.

First assume that $x \in Q$. We claim that, with X being a variable over C, the kernel J of the natural map $C[X] \to C[x] = R$ is generated by elements of the form $aX - b$, where $ax = b$: any element of the kernel can be written in the form $a_n X^n + \cdots + a_0$ for some $a_i \in C$. Then $a_n x$ is integral over C, so $b = a_n x \in C$. By subtracting $X^{n-1}(a_n X - b)$ from $a_n X^n + \cdots + a_0$ we get a polynomial in the kernel of degree strictly smaller than n, and an induction establishes our claim. Hence

$$\frac{R}{xR} \cong \frac{C[X]}{XC[X] + J} = \frac{C[X]}{XC[X] + I} \cong \frac{C}{I},$$

where

$$I = xC \cap C = \{b \in C \mid \text{there exists } a \in C \text{ such that } ax = b\}.$$

Let $P \in \mathrm{Ass}(R/xR)$, $P \subseteq Q$. Set $p = P \cap C$. As $R/xR \cong C/I$, then $p \in \mathrm{Ass}(C/I)$. If $x = \frac{a}{b}$ for elements $a, b \in C$, then as C–modules, $xC \cap C$ is isomorphic to $aC \cap bC$, so that as C is integrally closed, by Proposition 4.1.1, all the associated primes of $I = xC \cap C$ have height one. Thus p has height one, and as C is integrally closed, C_p is a one–dimensional regular ring by Serre's conditions. As $C_p \subseteq R_P$ have the same field of fractions, necessarily $C_p = R_P$. By induction on dimension, R/P has the property that for every quotient that is a domain, its integral closure is a module–finite extension. Thus by Corollary 4.6.2, the completion of $(R/P)_Q$ is reduced. By Lemma 4.6.1, the completion of R_Q is reduced, and then by Corollary 4.6.2, the integral closure of R_Q is module–finite over R_Q.

It remains to prove that $\overline{R}_Q$ is module–finite over R_Q whenever $x \notin Q$. As Q is a maximal ideal in $C[x]$, there exists $q \in Q$ and $a \in C[x]$ such that $ax + q = 1$. Write $a = \sum_{i=0}^{r} b_i x^i$ for some $b_i \in C$. Then $q = 1 - \sum_{i=0}^{r} b_i x^{i+1}$, and necessarily at least one b_i is not in Q, so that that there exists a monic polynomial $f(X) \in C_{Q \cap C}[X]$ such that $f(x) \in QR_{Q \cap C}$. Let L' be a finite field extension of the field of fractions of R that contains all the roots of $f(X)$. Let C' be the integral closure of $C_{Q \cap C}$ in L'. By Theorem 3.1.3, C' is module–finite over $C_{Q \cap C}$. Set $R' = C'[x] = R[C']$, and let M be a maximal ideal in R' containing Q. As $f(x) \in QR' \subseteq M$, there exists $a_i \in C'$ such that $x - a_i \in M$. Then $R' = C'[x - a_i]$, the fields of fractions of R' and C' are identical, C' is a localization of an integrally closed finitely generated A–algebra, and $C'_f = R'_f$. Thus by the previous case, as $x - a_i \in M$, the integral closure of R'_M is module–finite over R'_M. Thus by Corollaries 4.5.10 and 4.5.11, $\overline{R'}_Q$ is module–finite over R'_Q. But R' is module–finite over $R_{Q \cap C}$, and $\overline{R}$ is a submodule of $\overline{R'}$, so that $\overline{R}_Q$ is module–finite over R_Q. $\qquad\square$

Corollary 4.6.5 *Let R be a finitely generated $\mathbb{Z}$–algebra that is an integral domain. Let L be a finite field extension of the field of fractions of R. Then the integral closure of R in L is module–finite over R.*

Proof: By Theorem 4.6.3, the hypotheses of Theorem 4.6.4 are satisfied by $A = \mathbb{Z}$. Thus the conclusion of Theorem 4.6.4 holds. $\qquad\square$

For an alternate proof via Krull domains modify Exercise 9.7.

4.7. Absolute integral closure

There is the concept of the "largest" integral extension:

Definition 4.7.1 *Let R be a reduced ring with finitely many minimal prime ideals, let $K = K_1 \times \cdots \times K_s$ be the total ring of fractions of R, and for each $i = 1, \ldots, s$, let $\overline{K}_i$ be an algebraic closure of K_i. We define **the algebraic closure** of K to be $\overline{K} = \overline{K}_1 \times \cdots \times \overline{K}_s$, and we define **the absolute integral closure** of R to be the integral closure R^+ of R in $\overline{K}$.*

Clearly R^+ is the smallest ring contained in $\overline{K}$ that contains all the roots in $\overline{K}$ of all the monic polynomials in a variable X with coefficients in R. As $\overline{K}$ is unique up to isomorphism, similarly, R^+ is unique up to isomorphism. If S is any integral extension of R that is contained in $\overline{K}$, then S is a subring of R^+. If R is a domain, the field of fractions of R^+ is algebraically closed.

Discussion 4.7.2 Let R be an integrally closed domain with an algebraically closed field of fractions. For example, R could be the integral closure of a domain in an algebraic closure of its field of fractions. Let $\{P_i\}_{i \in I}$ be any family of prime ideals of R. Then $\sum_i P_i$ is either a prime ideal or it is equal to R. This remarkable property was first discovered by Michael Artin in [14], and a simpler proof, that we give below, is from Hochster and Huneke [125].

The proof reduces at once to a finite family of primes, and then by induction to the case of two primes, P and P'. Suppose that $xy \in P + P'$. Let $z = y - x$, so that $x^2 + zx = a + b$ with $a \in P$ and $b \in P'$. The equation $T^2 + zT = a$ has a solution $t \in R$, and since $t(t + z) \in P$, either $t \in P$ or $t + z \in P$. Now $x^2 + zx = t^2 + zt + b$, and so $(x - t)(x + t + z) = b \in P'$, so that either

$x - t \in P'$ or $x + t + z \in P'$. Since $x = (x - t) + t = (x + t + z) - (t + z)$ and $x + z = (x - t) + (t + z) = (x + t + z) - t$, we see that in all four cases, either $x \in P + P'$ or $y = x + z \in P + P'$, as required.   □

The absolute integral closure of a Noetherian local ring is contained in the absolute integral closure of the completion (modulo nilradicals). Part of the following is from Nagata [213, 33.10]:

Proposition 4.7.3 *Let $(R, \mathfrak{m})$ be a local reduced Noetherian ring, $\widehat{R}$ its $\mathfrak{m}$-adic completion, and $T = (\widehat{R})_{red}$. Let K be the total ring of fractions of R and L the total ring of fractions of T. Then $K \subseteq L$ and $\overline{T} \cap K = \overline{R}$.*

Write $L = L_1 \times \cdots \times L_s$, where each L_i is a field. Set $\overline{L} = \overline{L}_1 \times \cdots \times \overline{L}_s$, where $\overline{L}_i$ denotes an algebraic closure of L_i. Let $\overline{K}$ be the algebraic closure of K in $\overline{L}$. Let T^+ denote the integral closure of T in $\overline{L}$, and R^+ the integral closure of R in $\overline{K}$. Then $T^+ \cap \overline{K} = R^+$.

Proof: Let $\mathrm{Min}(\widehat{R}) = \{Q_1, \ldots, Q_s\}$. Then $T \subseteq \widehat{R}/Q_1 \times \cdots \times \widehat{R}/Q_s$, and the two rings have the same integral closure. As R is reduced, the composition $R \hookrightarrow \widehat{R} \twoheadrightarrow T$ is an inclusion. Every non–zerodivisor x of R is a non–zerodivisor in T, for otherwise x is contained in some Q_i, which contradicts the flatness of the map $R \to \widehat{R}$. Thus K is contained in L.

By possibly reindexing we have that L_i is the field of fractions of the complete Noetherian integral domain $\widehat{R}/Q_i$. Let $\overline{T}_i$ be the integral closure of $T\widehat{R}/Q_i$ in L_i, and T_i^+ the integral closure of $\overline{T}_i$ in L_i^+. Clearly $\overline{T} = \overline{T}_1 \times \cdots \times \overline{T}_s$, $T^+ = T_1^+ \times \cdots \times T_s^+$, and $\overline{R} \subseteq \overline{T} \cap K$ and $R^+ \subseteq T^+ \cap \overline{K}$.

If $\frac{x}{y} \in \overline{T} \cap K$, with x, y non–zero elements of R and y not contained in any minimal prime ideal of R, then $\frac{x}{y}$ satisfies an integral equation over T. As $T = (\widehat{R})_{red}$, $\frac{x}{y}$ also satisfies an equation of integral dependence over $\widehat{R}$: there exists a positive integer n such that $(\frac{x}{y})^n + r_1(\frac{x}{y})^{n-1} + \cdots + r_n = 0$ for some $r_i \in \widehat{R}$. Then $x^n \in y(x,y)^{n-1}\widehat{R} \cap R = y(x,y)^{n-1}R$, which can be converted to an integral equation of $\frac{x}{y}$ over R. Thus $\frac{x}{y} \in \overline{R}$, which proves that $\overline{R} = \overline{T} \cap K$.

Observe that $\overline{K}$ is the total ring of fractions of R^+. Thus if $\alpha \in T^+ \cap \overline{K}$, then $\alpha = \frac{x}{y}$ for some elements x, y of R^+, y not in any minimal prime ideal of R^+. By assumption, $\frac{x}{y}$ satisfies an integral equation over T, and hence over $\widehat{R}$: there exists a positive integer n such that $(\frac{x}{y})^n + r_1(\frac{x}{y})^{n-1} + \cdots + r_n = 0$ for some $r_i \in \widehat{R}$. Set $T' = R[x, y]$, so that T' is a module–finite extension of R. Then $x^n \in y(x,y)^{n-1}(T' \otimes_R \widehat{R}) \cap T'$, which by faithful flatness means that $x^n \in y(x,y)^{n-1}T'$. This can be converted to an integral equation of $\frac{x}{y}$ over T'. Thus $\frac{x}{y}$ is integral over T' and thus also over R. It follows that $R^+ = T^+ \cap \overline{K}$.   □

More on the absolute integral closure is in Section 16.7. See also Exercises 4.3 and 4.4.

4.8. Finite Lying–Over and height

If $R \subseteq S$ are rings and S is Noetherian, then certainly for every prime ideal P in R there exist at most finitely many prime ideals Q in S that are minimal over PS. In particular, if $R \subseteq S$ is in addition an integral extension so that the Incomparability property holds, then for every P, there exist only finitely many prime ideals Q in S that contract to P. In this section we prove that the finite Lying–Over property holds more generally, and that under some conditions the heights are preserved.

First note that arbitrary integral extension do not have this finite Lying–Over property: $\mathbb{C}[X] \subseteq S = \mathbb{C}[X^{1/n} \,|\, n = 1, 2, \ldots]$ is an integral extension. For each positive integer n, if $u_{n1}, \ldots, u_{nn}$ denote the nth roots of unity in $\mathbb{C}$, then each $X^{1/n} - u_{ni}$ is a factor of $X - 1$ in S. Thus $(X^{1/n} - u_{ni})S \cap \mathbb{C}[X] = (X - 1)\mathbb{C}[X]$, and there exists a prime ideal Q in S that contains $X^{1/n} - u_{ni}$ and contracts to $(X - 1)\mathbb{C}[X]$. For each n, a proper ideal in S can contain at most one such factor, for otherwise this proper ideal has to contain the non-zero complex number $u_{ni} - u_{nj} = (X^{1/n} - u_{nj}) - (X^{1/n} - u_{ni})$. Thus S must have at least n prime ideals contracting to $(X - 1)\mathbb{C}[X]$. But n was arbitrary, so that S contains infinitely many prime ideals that lie over $(X - 1)\mathbb{C}[X]$.

Finite Lying–Over holds over a complete Noetherian local domain:

Lemma 4.8.1 *If $(R, \mathfrak{m})$ is a complete Noetherian local domain, then R^+ (the integral closure of R in an algebraic closure of the field of fractions of R) has only one maximal ideal.*

Proof: Write $R^+ = \cup R_i$, where each R_i is module–finite over R. By the module–finite assumption, R_i is complete in the $\mathfrak{m}$–adic topology and Noetherian. By Proposition 4.3.2 each R_i is a direct product of rings corresponding to the maximal ideals of R_i. Since each R_i is a domain, there can be only a unique maximal ideal $\mathfrak{m}_i$ of R_i. Clearly then R^+ has a unique maximal ideal that is the union of all the $\mathfrak{m}_i$. $\square$

For non–complete rings there is a weaker version of the finite Lying–Over:

Proposition 4.8.2 *Let R be a reduced Noetherian ring, and $K = K_1 \times \cdots \times K_s$ its total ring of fractions, with each K_i a field. For each $i = 1, \ldots, s$, let L_i be a finite field extension of K_i. Let S be contained in the integral closure of R in $L = L_1 \times \cdots \times L_s$. Then for any prime ideal P of R, there are only finitely many prime ideals in S that contract to P.*

If R is an integral domain, the number of prime ideals in the integral closure of R that contract to P is at most the number of minimal prime ideals in the P–adic completion of R_P.

An integral domain R' is said to be **almost finite** over a subdomain R if R' is integral over R and the field of fractions of R' is finite algebraic over the field of fractions of R. In [110, Theorem 2], Heinzer proves that if R is a Noetherian domain and R' is an almost finite extension domain of R, then R'

has a Noetherian spectrum, meaning in particular that for every ideal I in R', the set of prime ideals in R' minimal over I is finite. This result of Heinzer's implies the first statement of the proposition above. The first statement also follows by using Exercises 2.6 and 2.8. We give a different proof below, which in addition gives the second statement.

Proof of Proposition 4.8.2: By the set–up, R has only finitely many minimal prime ideals. We label them as $P_1, \ldots, P_s$, with $P_i = \ker(R \to L_i)$. By Lying–Over, we may assume that S is the integral closure of R in L.

Let S_i be the integral closure of R/P_i in L_i. By Corollary 2.1.13,

$$R \subseteq (R/P_1) \times \cdots \times (R/P_s) \subseteq S_1 \times \cdots \times S_s = S$$

are all integral extensions. Clearly there are only finitely many prime ideals in $(R/P_1) \times \cdots \times (R/P_s)$ that contract to P. If we can prove that for each $i = 1, \ldots, s$, every prime ideal of R/P_i is the contraction of only finitely many prime ideals in S_i, then by the structure of prime ideals in direct sums, every prime ideal in R contracts from only finitely many prime ideals in S. Thus by working with each R/P_i instead of with R, without loss of generality we may assume that R is a Noetherian domain, with field of fractions K, and that L is a finite field extension of K.

Let $s_1, \ldots, s_n \in S$ such that $L = K(s_1, \ldots, s_n)$. Set $R' = R[s_1, \ldots, s_n]$. Then R' is module–finite over R, and the field of fractions of R' is L. By the Incomparability Theorem (Theorem 2.2.3), every prime ideal in R' that contracts to P in R has to be minimal over PR', and as R' is Noetherian, there are only finitely many such prime ideals in R'. Thus it suffices to prove that each prime ideal in R' is contracted from only finitely many prime ideals in S. This reduces the proof to showing that whenever R is a Noetherian domain with field of fractions K, if S is the integral closure of R, then a prime ideal P in R is contracted from at most finitely many prime ideals in S.

We may in addition assume that R is local. As $S_{R \setminus P}$ is the integral closure of R_P in L, and as the set of prime ideals in S contracting to P is, after localization at $R \setminus P$, the same as the set of prime ideals in $S_{R \setminus P}$ contracting to PR_P, without loss of generality we may replace R by the localization of R at P and assume that R is a Noetherian local domain with maximal ideal P.

Let $Q_1, \ldots, Q_r$ be the minimal prime ideals in the P–adic completion $\widehat{R}$ of R. Set $T_i = \widehat{R}/Q_i$, $T = T_1 \times \cdots \times T_r$. Then T_i is a complete local domain. By Theorem 4.3.4, $\overline{T_i}$ has only one maximal ideal, i.e., exactly one prime ideal contracting to PT_i. It follows that $\overline{T}$ has exactly r prime ideals contracting to P in R.

We claim that there are at most r prime ideals in $\overline{R}$ that contract to P. Suppose not. Let $P_1, \ldots, P_{r+1}$ be distinct prime ideals in $\overline{R}$ that contract to P. By the Incomparability Theorem (Theorem 2.2.3), there are no inclusion relations among the P_i. Thus for all $i, j \in \{1, \ldots, r+1\}$ with $i \neq j$, there exists $a_{ij} \in P_i \setminus P_j$. Let $B = R[a_{ij} \, | \, i, j]$. Then B is a module–finite ring extension of R contained in $\overline{R} \subseteq K$. Set $p_i = P_i \cap B$. By the choice of the a_{ij}, all the p_i are distinct, and they all contract to P in R. A prime ideal in B contracts to P if and only if it is a maximal ideal, and this holds if and only if it is minimal over PB. But B is Noetherian, so that there are only finitely many prime ideals in B contracting to P. Let n be the number of these prime ideals. We have proved that $n \geq r + 1$.

The completion $\widehat{B}$ of B in the P–adic topology is the direct sum of n rings. Also, $\widehat{R} \subseteq \widehat{B} \cong B \otimes_R \widehat{R} \subseteq K \otimes_R \widehat{R}$. Observe that $\widehat{R}/\sqrt{0} \subseteq K \otimes_R (\widehat{R}/\sqrt{0})$, as non–zero elements in R are non–zerodivisors on $\widehat{R}$. The image B' of $\widehat{B}$ in $K \otimes_R (\widehat{R}/\sqrt{0})$ is generated by the a_{ij} over $\widehat{R}/\sqrt{0}$, so it is integral over $\widehat{R}/\sqrt{0}$. As $K \otimes_R (\widehat{R}/\sqrt{0})$ lies in the total ring of fractions of $\widehat{R}/\sqrt{0}$, it follows that B' lies in $\overline{T}$. By Lying–Over for $B' \subseteq \overline{T}$, B' has at most r maximal ideals. But $B' = \widehat{B}/\sqrt{0}$, so that B itself has at most r maximal ideals. $\square$

Observe that the proof above shows even more:

Lemma 4.8.3 *Let R be a Noetherian local domain with maximal ideal P. Let Q be a prime ideal in $\overline{R}$ that contracts to P. Let $\overline{T}$ be the integral closure of $\widehat{R}/\sqrt{0}$. Then Q is the contraction of a maximal ideal in $\overline{T}$.*

Proof: We use notation as in the previous proof.

Since $B' \subseteq \overline{T}$ is an integral extension, every maximal ideal in B' is contracted from a maximal ideal in $\overline{T}$. The maximal ideals in B' correspond to maximal ideals in B. By Proposition 4.7.3, $B \subseteq \overline{T}$, so that every maximal ideal in B is contracted from a maximal ideal in $\overline{T}$. By construction of B, there is a one–to–one correspondence between maximal ideals in $\overline{R}$ and B (via contraction), so that as $\overline{R} \subseteq \overline{T}$, every maximal ideal in $\overline{R}$ is contracted from a maximal ideal in $\overline{T}$. $\square$

Lemma 4.8.4 *Let R be a Noetherian domain and Q a prime ideal in the integral closure $\overline{R}$ of R. Then there exist a finitely generated R–algebra extension $C \subseteq \overline{R}$ and a prime ideal P in C such that Q is the only prime in $\overline{R}$ contracting to P.*

In particular, $\operatorname{ht} P = \operatorname{ht} Q$ and $\overline{R}_Q$ is the integral closure of C_P.

Proof: Let $q = Q \cap R$. By Proposition 4.8.2 there are only finitely many prime ideals in $\overline{R}$ that contract to q. Let $Q = Q_1, \ldots, Q_r$ be all such prime ideals. Choose an element $x \in Q \setminus (Q_2 \cup \cdots \cup Q_r)$. Set $C = R[x]$ and $P = C \cap Q$. Then $R \subseteq C \subseteq \overline{R}$ and by construction Q is the only prime ideal in $\overline{R}$ that contracts to P. Thus $C_P \subseteq \overline{R}_{C \setminus P}$ is a local module–finite extension, so that $\operatorname{ht} P = \dim(C_P) = \dim(\overline{R}_{C \setminus P}) = \operatorname{ht} Q$. Also, as $\overline{R}$ is integrally closed, so is $\overline{R}_{C \setminus P}$, hence $\overline{R}_Q$ is the integral closure of C_P. $\square$

The following is adapted from Nagata [213, 33.10]:

Proposition 4.8.5 *Let R be a reduced Noetherian ring, K its total ring of fractions, and $\overline{K}$ the algebraic closure of K. Let L be a module–finite*

*extension of K contained in $\overline{K}$ and S a subring of the integral closure of R
in L. For any prime ideal P of R and any prime ideal Q in S lying over P,
$\kappa(P) \subseteq \kappa(Q)$ is a finite field extension.*

Proof: By the structure theorem of integral closures of reduced rings (Corollary 2.1.13), it is easy to reduce to the case where R is an integral domain, L is a field, and S is the integral closure of R in L. By localizing at $R \setminus P$, we may assume that R is local with maximal ideal P.

As L is finitely generated over the field of fractions of R, it is even generated by finitely many elements of S. Let R' be the subring of S generated over R as an algebra by these finitely many generators. Then R' is Noetherian, with the same field of fractions as S, and module–finite over R, so that $R/P = \kappa(P) \subseteq \kappa(Q \cap R') = R'/(Q \cap R')$ is a finite field extension. Thus it suffices to prove that $\kappa(Q \cap R') \subseteq \kappa(Q)$ is a finite field extension, and so we may replace R by R' and assume that S is the integral closure of R. Again, by localizing, we may assume that R is local with maximal ideal P.

By Lemma 4.8.3, if $\overline{T}$ is the integral closure of $\widehat{R}/\sqrt{0}$, Q is the contraction of a prime ideal $\mathfrak{M}$ in $\overline{T}$. Then $\kappa(P\widehat{R}) = \kappa(P) \subseteq \kappa(Q) \subseteq \kappa(\mathfrak{M})$, so that it suffices to prove that $\kappa(P\widehat{R}) \subseteq \kappa(\mathfrak{M})$ is a finite field extension. But this is clear as $\widehat{R}/\sqrt{0} \to \overline{T}$ is a module–finite extension by Corollary 2.1.13 and Theorem 4.3.4. □

We saw in Example 2.2.6 that even if $R \subseteq S$ is a module–finite extension of Noetherian rings, it can happen that for a prime ideal Q in S, $\operatorname{ht} Q \neq \operatorname{ht}(Q \cap R)$. The following proposition gives a condition for equality:

Proposition 4.8.6 *Let R be a Noetherian integral domain that satisfies the dimension formula. Let S be an integral extension of R contained in a finite field extension of the field of fractions of R. Then for any prime ideal Q in S, $\operatorname{ht}(Q) = \operatorname{ht}(Q \cap R)$.*

Proof: Set $P = Q \cap R$. By the Incomparability Theorem, $\operatorname{ht}(P) \geq \operatorname{ht}(Q)$. By Proposition 4.8.2, there exist only finitely many prime ideals $Q_1, \ldots, Q_s$ in S minimal over PS. By the Incomparability Theorem, there are no inclusion relations among the Q_i and after renumbering, $Q = Q_1$. For each $i \neq j$, let $r_{ij} \in Q_i \setminus Q_j$. Let $R' = R[r_{ij}]$. Then R' is module–finite over R and $Q_i' = Q_i \cap R'$ are incomparable prime ideals in R', all contracting to P in R. As R satisfies the dimension formula and R' is module–finite over R, $\operatorname{ht}(P) = \operatorname{ht}(Q_i')$ for all i. Thus it suffices to prove that $\operatorname{ht}(Q_1') = \operatorname{ht}(Q_1)$. Let $W = R' \setminus Q_1'$. Then W is a multiplicatively closed subset of R', $W^{-1}R' \subseteq W^{-1}S$ is an integral extension of rings, $W^{-1}R'$ is Noetherian local of dimension $\operatorname{ht}(P)$, and all the maximal ideals of $W^{-1}S$ contract to $Q_1'W^{-1}R'$ in R' and thus to P in R. As W contains an element $r \in Q_2' \cap \cdots \cap Q_s' \setminus Q_1'$, necessarily $W^{-1}S$ has only one maximal ideal, namely $QW^{-1}S$. Thus $\operatorname{ht}(Q) = \operatorname{ht}(QW^{-1}S) = \dim W^{-1}S = \dim W^{-1}R' = \operatorname{ht}(Q_1'W^{-1}R') = \operatorname{ht}(Q_1') = \operatorname{ht}(P)$. □

See Example 2.2.6 showing that the locally formally equidimensional assumption above is necessary.

4.9. Dimension one

The previous sections showed that many Noetherian domains have module–finite integral closures. However, not every Noetherian domain has this property. Already in dimension one the integral closure of a Noetherian integral domain need not be a module–finite extension, see Example 4.9.1 below. Nevertheless, the integral closure of a one–dimensional Noetherian domain and even of a two–dimensional Noetherian domain is always Noetherian. This was proved by Krull and Akizuki for dimension one and by Nagata for dimension two. In this section we prove the case of dimension one: the result is stronger for dimension one and the proof is more accessible. We postpone proving the case of dimension two until introducing Krull domains in Section 4.10.

We first give an example of a non–local one–dimensional Noetherian ring for which the integral closure is not a module–finite extension. A local example, due to Nagata, is in Exercise 4.8.

Example 4.9.1   Let k be a field and $X_1, X_2, \ldots$ variables over k. Let A be the subring $k[X_n^n, X_n^{n+1} \mid n = 1, 2, \ldots]$ of the polynomial ring $S = k[X_1, X_2, \ldots]$. The rings A and S are certainly not Noetherian. Let W be the subset of A consisting of all polynomials that, as elements of S, do not have any variable as a factor. Then W is a multiplicatively closed subset of A. Set $R = W^{-1}A$. It is straightforward to verify that the only prime ideals in R are 0 and $P_n = (X_n^n, X_n^{n+1})R$ as n varies over positive integers. All of these prime ideals are finitely generated, so that R is Noetherian and of dimension one. It is clear that the integral closure of R is $W^{-1}S$. This is not module–finite over R as for each n, after localizing at P_n, $(W^{-1}S)_{R \setminus P_n}$ is minimally generated by n elements over R_{P_n}. We leave verification to the reader.

Thus the integral closure of a one–dimensional Noetherian domain need not be a module–finite extension. However, the integral closure of a one–dimensional Noetherian domain is still Noetherian. This follows from the more general result of Krull and Akizuki:

Theorem 4.9.2   (Krull–Akizuki) *Let R be a one–dimensional Noetherian reduced ring with total ring of fractions K. If S is any ring between R and K, with $S \neq K$, then S is one–dimensional and Noetherian. In particular, the integral closure of R is one–dimensional and Noetherian.*

Proof: Let $P_1, \ldots, P_r$ be the minimal prime ideals of R. For each $i = 1, \ldots, r$, let K_i be the field of fractions of R/P_i. We know that $K = K_1 \times \cdots \times K_r$. Let J_i be the kernel of the composition map $S \to K \to K_i$, so that $R/P_i \subseteq S/J_i \subseteq K_i$. If the theorem holds for integral domains, then S/J_i is Noetherian for all i. Furthermore, the dimension of S/J_i is 1 if $S/J_i \neq K_i$ and is 0 otherwise.

As $J_1 \cap J_2 \cap \cdots \cap J_r = 0$, it follows that S is Noetherian. Clearly S has dimension at most 1 and has dimension 0 exactly when each $S/J_i = K_i$. Thus by assumption S has dimension 1.

Hence it suffices to prove the theorem under the assumption that R is an integral domain. We first prove that any non–zero ideal I in S is finitely generated. As I is a non–zero ideal, there exists a non–zero element $a \in I \cap R$. The images of the set of ideals $\{a^n S \cap R + aR\}_n$ form a descending chain of ideals in the Artinian ring R/aR, so that there exists an integer l such that for all $n \geq l$, $(a^n S \cap R) + aR = (a^{n+1} S \cap R) + aR$.

Claim: $a^l S \in a^{l+1} S + R$.

Note that it suffices to prove the claim after localization at each maximal ideal $\mathfrak{m}$ of R, so that without loss of generality we may assume that R is a local ring with maximal ideal $\mathfrak{m}$. If a is a unit, there is nothing to show, so we may assume that $a \in \mathfrak{m}$. Let x be a non–zero element in S. Write $x = \frac{b}{c}$ for some non–zero $b, c \in R$. Then there exists an integer $n \geq l$ such that $\mathfrak{m}^{n+1} \subseteq cR \subseteq \frac{1}{x}R$. In particular, $a^{n+1}x \in R$. Thus $a^{n+1}x \in a^{n+1} S \cap R \subseteq a^{n+2} S \cap R + aR$, so that $a^n x \in a^{n+1} S + R$. Now let n be the smallest integer greater than or equal to l such that $a^n x \in a^{n+1} S + R$. If $n > l$, then

$$a^n x \in (a^{n+1} S + R) \cap a^n S = a^{n+1} S + a^n S \cap R$$
$$\subseteq a^{n+1} S + a^{n+1} S \cap R + aR = a^{n+1} S + aR,$$

so that $a^{n-1}x \in a^n S + R$, contradicting the minimality of n. This proves that $n = l$. Thus

$$\frac{S}{aS} \cong \frac{a^l S}{a^{l+1} S} \subseteq \frac{a^{l+1} S + R}{a^{l+1} S} \cong \frac{R}{a^{l+1} S \cap R},$$

so that S/aS has finite length as an R–module. In particular, the image of I in S/aS is finitely generated, hence I is finitely generated. This proves that S is Noetherian. Furthermore, the display above shows that S/aS has dimension zero, so S has dimension one. $\square$

The Krull–Akizuki Theorem fails in dimension 2 or higher. For example, let k be a field, X and Y variables over k, and $R = k[X,Y]$. Then $S = k[X, \frac{Y}{X}, \frac{Y^2}{X^3}, \frac{Y^3}{X^7}, \ldots, \frac{Y^n}{X^{2n-1}}, \ldots]$ is a ring between R and $Q(R)$ that is not Noetherian (the maximal ideal $(X, \frac{Y^n}{X^{2n-1}} \mid n)$ is not finitely generated).

Nevertheless, the integral closure of a two–dimensional Noetherian domain is still Noetherian. We give a proof of this fact that uses Krull domains, so we only present it in the section on Krull domains. See Theorem 4.10.7.

Here is a strengthening of the Krull–Akizuki Theorem:

Proposition 4.9.3 *Let R be a reduced Noetherian ring, S its integral closure, and Q a prime ideal in S such that S/Q has dimension 1. Then S/Q is Noetherian.*

Proof: Let $P = Q \cap R$. Then $R/P \subseteq S/Q$ is an integral extension, so that $\dim(R/P) = \dim(S/Q) = 1$. By Proposition 4.8.5, $\kappa(P) \subseteq \kappa(Q)$ is a finite field extension. Thus there exist finitely many elements $s_1, \ldots, s_m \in S$ such that $\kappa(Q)$ is generated over $\kappa(P)$ by the images of these s_i. Let $R' = R[s_1, \ldots, s_m]$. Then R' is Noetherian and S is its integral closure. Set $P' = Q \cap R'$. Then $R'/P' \subseteq S/Q$ is an integral extension and $\kappa(P') = \kappa(Q)$. Thus by the Krull–Akizuki Theorem, S/Q is Noetherian. $\square$

Lemma 4.9.4 *Let R be a Noetherian domain, K its field of fractions, and Q a prime ideal of height one in the integral closure $\overline{R}$ of R. Then $\overline{R}_Q$ is a discrete valuation ring of rank one between R and K. Moreover, there exist a finitely generated R–algebra extension $C \subseteq \overline{R}$ and a prime ideal P of height one in C such that $\overline{R}_Q$ is the integral closure of C_P.*

Proof: By Lemma 4.8.4 there exist a finitely generated R–algebra extension $C \subseteq \overline{R}$ and a prime ideal P in C such that $\overline{R}_Q$ is the integral closure of C_P. Then $\operatorname{ht} P = \dim C_P = \dim(\overline{R}_Q) = \operatorname{ht} Q = 1$. $\square$

Lemma 4.9.5 *Let R be a Noetherian domain and let $\overline{R}$ be the integral closure of R. If Q is a height one prime ideal of $\overline{R}$ and $q = Q \cap R$, then $\operatorname{depth} R_q = 1$.*

Proof: By localization we may assume that q is the unique maximal ideal of R. By Lemma 4.9.4 there exists a module–finite R–subalgebra $C \subseteq \overline{R}$ such that if $P = Q \cap C$. Then Q is the unique prime lying over P and furthermore the height of P is 1. Since C is module–finite over R, $J = R :_R C \neq 0$. Choose a non–zero element $x \in R$. If $J \not\subseteq P$, then $C_P = R(= R_q)$ and so R has dimension one, whence $\operatorname{depth} R_q = 1$. Now assume that $J \subseteq P$. Suppose for contradiction that $\operatorname{depth} R > 1$. Choose $y \in q$ that is a non–zerodivisor on R/xR. Let $P = P_1, \ldots, P_k$ be the primes in C that are minimal over xC. Since $\sqrt{xC} = P_1 \cap \cdots \cap P_k$, we can choose $b \in P_2 \cap \cdots \cap P_k \setminus P_1$ such that $by^n \in xC$ for some n. Since $xC \subseteq R$, we find that $(xbC)y^n = x(by^n C) \subseteq x(xC) \subseteq xR$. Because y is a non–zerodivisor on R/xR, it follows that $xbC \subseteq xR$ and thus $bC \subseteq R$. Then b is in the conductor, a contradiction as $b \notin P$. $\square$

4.10. Krull domains

Integral closures of Noetherian domains need not be Noetherian. See Nagata's examples in the exercises. Nevertheless, integral closure of Noetherian domains have some good Noetherian–like properties, as will be proved below. In particular, in such rings there is a primary decomposition of principal ideals.

Definition 4.10.1 *An integral domain R is a **Krull domain** if*

(1) for every prime ideal P of R of height one, R_P is a Noetherian integrally closed domain,

(2) $R = \cap_{\operatorname{ht}(P)=1} R_P$, and

(3) every non–zero $x \in R$ lies in at most finitely many prime ideals of R of height one.

Krull domains need not be Noetherian. They are always integrally closed because for each prime ideal P of height one in R, R_P is integrally closed, hence their intersection R is integrally closed.

Proposition 4.10.2 *Let R be a Krull domain. If α is in the field of fractions of R, and I is an ideal in R of height at least two such that $\alpha I \subseteq R$, then $\alpha \in R$.*

Proof: For any prime ideal P in R of height 1, $\alpha \subseteq \alpha I R_P \subseteq R_P$. Thus by assumptions, $\alpha \in \cap_P R_P = R$. □

Primary decomposition in non–Noetherian rings is not easy, but principal ideals in Krull domains have a simple primary decomposition:

Proposition 4.10.3 *Let R be a Krull domain and x a non–zero element of R. Let $P_1, \ldots, P_s$ be all the prime ideals in R of height one containing x. Then $xR = \cap_i (xR_{P_i} \cap R)$ is a minimal primary decomposition of xR. In particular, principal ideals in Krull domains have no embedded primes.*

Proof: Let $P_1, \ldots, P_s$ be all the prime ideals of height one in R containing x. Let $I = \cap_i (xR_{P_i} \cap R)$ and $y \in I$. Then $\frac{y}{x} \in R_{P_i}$ for all $i = 1, \ldots, s$. Furthermore, if P is any other prime ideal in R of height one, then $\frac{y}{x} \in R_P$. Thus by the second property of Krull domains, $\frac{y}{x} \in R$, so that $y \in xR$. This proves that $I = xR$. Furthermore, as each P_i has height one, xR_{P_i} is primary, so that $xR = \cap_i (xR_{P_i} \cap R)$ is a minimal primary decomposition of xR. □

A partial converse holds:

Proposition 4.10.4 *Any integrally closed Noetherian domain is a Krull domain.*

Proof: Let R be an integrally closed Noetherian domain. Properties (1) and (3) of Krull domains are clear.

The inclusion $R \subseteq \cap_{\text{ht}(P)=1} R_P$ is straightforward. To prove the other inclusion let $s \in \cap_{\text{ht}(P)=1} R_P$. Write $s = \frac{x}{y}$ for some $x, y \in R$, with $y \neq 0$. Consider the ideal $J = yR :_R x$. For every prime ideal P of R of height 1, $\frac{x}{y} \in R_P$, so that $JR_P = R_P$. Thus J is not contained in any height one prime ideal of R. The associated prime ideals of J are contained in the set of associated primes of yR and by Proposition 4.1.1, all of these prime ideals are of height 1. This forces J to have no associated primes, i.e., J is forced to be the unit ideal. Thus $x \in yR$, so $s \in R$. □

A much stronger result is the following result of Mori and Nagata proving that the integral closure of a Noetherian domain is a Krull domain.

Theorem 4.10.5 (Mori–Nagata) *The integral closure $\overline{R}$ of a reduced Noetherian ring R in its total ring of fractions is a direct product of r Krull domains, where r is the number of minimal prime ideals of R.*

Proof: Let $P_1, \ldots, P_r$ be all the minimal prime ideals of R. Then the total ring of fractions of R is $K_1 \times \cdots \times K_r$, where K_i is the field of fractions of R/P_i. Let R_i be the integral closure of R/P_i in K_i. By Corollary 2.1.13, $R_1 \times \cdots \times R_r$ is the integral closure of R. Thus it suffices to prove that each R_i is a Krull domain. Thus we may assume that R is a Noetherian integral domain. By Lemma 4.9.4, $\overline{R}$ satisfies the first property of Krull domains.

Next we verify the third property of Krull domains for $\overline{R}$. For an arbitrary non–zero $x \in \overline{R}$, there exists a non–zero $y \in R$ such that $yx \in R$. If yx is contained in only finitely many prime ideals of $\overline{R}$ of height 1, then the same also holds for x. Thus by replacing x by yx without loss of generality we may assume that $x \in R$. Lemma 4.9.5 proves that the contractions of height one primes minimal over $x\overline{R}$ are among the associated primes of xR. In particular, there are only finitely many such contractions of minimal primes over $x\overline{R}$. By Proposition 4.8.2, there are then only finitely many prime ideals in $\overline{R}$ that lie over these contractions. This proves that $\overline{R}$ satisfies the third property of Krull domains.

It remains to prove that $\overline{R}$ satisfies the second property of Krull domains. Suppose that this is known when R is local. By Proposition 2.1.6, $\overline{R} = \bigcap_{\mathfrak{m}} \overline{R_{\mathfrak{m}}}$ as $\mathfrak{m}$ varies over all the maximal ideals of R. If P varies over all the height one prime ideals in $\overline{R_{\mathfrak{m}}}$, by assumption $\overline{R_{\mathfrak{m}}} = \bigcap_P (\overline{R})_P$. Thus $\bigcap_{\mathfrak{m}} \overline{R_{\mathfrak{m}}} = \bigcap_{\mathfrak{m}} \bigcap_P (\overline{R})_P, = \bigcap_P (\overline{R})_P$, where P varies over all the height one prime ideals in $\overline{R}$. Thus without loss of generality we have to prove the second property of Krull domains under the additional assumption that R is a local domain.

Let $x \in \bigcap_P (\overline{R})_P$, where P varies over the height one prime ideals of $\overline{R}$. Then x is contained in the field of fractions K of R. We need to prove that x is contained in $\overline{R}$. Let $Q_1, \ldots, Q_s$ be all the minimal prime ideals in the completion $\widehat{R}$ of R in the topology determined by its maximal ideal. We know that R embeds canonically in $T = \widehat{R}/Q_1 \times \cdots \times \widehat{R}/Q_s$, that K is contained in the total ring of fractions L of T, and that by Proposition 4.7.3, $\overline{R} = \overline{T} \cap K$, where $\overline{T}$ is the integral closure of T in L. To prove that $x \in \overline{R}$, it suffices to prove that $x \in \overline{T}$.

By Corollary 2.1.13 and by Theorem 4.3.4, $\overline{T}$ is a direct product of integrally closed complete local domains, and $\overline{T}$ is module–finite over T. In particular, $\overline{T}$ is a Noetherian direct product of integrally closed domains. By the previous proposition, $\overline{T}$ is a direct product of Noetherian Krull domains.

By Corollary B.3.7, each factor of $\overline{T}$ satisfies the dimension formula. Let Q be a prime ideal in $\overline{T}$ of height one. By the structure of prime ideals in a direct product of rings and by the dimension formula, $Q \cap T$ has height one, and hence the preimage of Q in $\widehat{R}$ has height one. As $R \to \widehat{R}$ is flat, by Proposition B.2.3, the height of $Q \cap R$ is either 0 or 1. By the Dimension Inequality B.2.5 for $R \subseteq \overline{R}$, $\mathrm{ht}(Q \cap \overline{R})$ is either 0 or 1. In any case, $x \in \overline{R}_{Q \cap \overline{R}} \subseteq \overline{T}_Q$. Since Q was an arbitrary prime ideal of height one and since $\overline{T}$ is a direct product of Krull domains, necessarily $x \in \overline{T}$, whence $x \in R$. □

Theorem 4.10.6 (Nagata) *The integral closure of a two–dimensional Noetherian domain is Noetherian.*

Proof: Let R be a two–dimensional Noetherian domain. By Theorem 4.10.5, $\overline{R}$ is a Krull domain. By Proposition 4.9.3, for every non–zero prime ideal P of $\overline{R}$, $\overline{R}/P$ is Noetherian. The theorem follows by applying the next theorem of Mori and Nishimura, Theorem 4.10.7. $\qquad\square$

Theorem 4.10.7 (Mori–Nishimura Theorem) *Let R be a Krull domain such that R/P is Noetherian for all prime ideals P of height one. Then R is Noetherian.*

Proof: Let P be a prime ideal of height one. Choose $a \in P \setminus R \cap P^2 R_P$. As R is a Krull domain, a is contained in only finitely many prime ideals of height one, say in $P, P_1, \ldots, P_m$. For each $i = 1, \ldots, m$, let $b_i \in P_i \setminus P$ such that $b_i R_{P_i} = a R_{P_i}$. Then $x = a/(b_1 \cdots b_m) \in R_P$ has the property that $x \in P R_P \setminus P^2 R_P$, and that x is not in any $Q R_Q$ as Q varies over other prime ideals of height one.

Let n be a positive integer. As $(x^n R[x] \cap R)_P = P^n R_P$, then $x^n R[x] \cap R \subseteq P^n R_P \cap R$. Let $y \in P^n R_P \cap R = x^n R_P \cap R$. Choose $s \in R \setminus P$ such that $sy \in x^n R$. Write $sy = r x^n$ for some $r \in R$. Then for any height one prime ideal $Q \neq P$, $r x^n = sy \in s R_Q \cap R$ implies that $r \in s R_Q \cap R$. As $r \in R_P = s R_P$ and as R is a Krull domain, then $r \in sR$. Hence $y \in x^n R \cap R \subseteq x^n R[x] \cap R$, which proves that $x^n R[x] \cap R = P^n R_P \cap R$.

Thus $P^{n+1} R_P \cap R$ is the kernel of the composition $P^n R_P \cap R \hookrightarrow x^n R[x] \to (x^n R[x])/(x^{n+1} R[x])$. It follows that $(P^n R_P \cap R)/(P^{n+1} R_P \cap R)$ is an R–submodule of $(x^n R[x])/(x^{n+1} R[x]) \cong R[x]/xR[x] \cong R/P$, whence it is a Noetherian R–module. Then the short exact sequences

$$0 \longrightarrow \frac{P^n R_P \cap R}{P^{n+1} R_P \cap R} \longrightarrow \frac{R}{P^{n+1} R_P \cap R} \longrightarrow \frac{R}{P^n R_P \cap R} \longrightarrow 0$$

together with induction on n prove that each $R/(P^n R_P \cap R)$ is Noetherian.

Let a be an arbitrary non–zero element of R. As R is a Krull domain, by Proposition 4.10.3, aR has a primary decomposition $\cap_i (a R_{P_i} \cap R)$ for some finitely many prime ideals P_i of height one. As R_{P_i} is a Dedekind domain, $a_i R_{P_i} = P_i^{n_i} R_{P_i}$ for some positive integer n_i. By above, each $R/(P_i^{n_i} R_{P_i} \cap R)$ is Noetherian, so that $R/(a)$ is Noetherian. As a was arbitrary, R is Noetherian. $\qquad\square$

Thus the integral closure of Noetherian domains of dimension at most two is still Noetherian. However, the integral closure of a three–dimensional Noetherian domain need not be Noetherian; see Nagata's example in Exercise 4.9.

4.11. Exercises

4.1   Let R be a Noetherian domain, S an integral extension contained in a finite field extension of the field of fractions of R, and Q a prime ideal in S. Prove that there exists a module–finite extension R' of R contained in S such that $R'_{Q \cap R'} \subseteq S_{R' \setminus (Q \cap R')} = S_Q$.

4.2   Let R be an integral domain that is not a field. Assume that the field of fractions of R is algebraically closed. Prove that R is not Noetherian.

4.3   Let R be an integrally closed domain with algebraically closed field of fractions. Suppose that q_i is primary to a prime P_i in R for every i. Suppose that $P = \sum_i P_i$ is a proper ideal (necessarily a prime ideal, by Discussion 4.7.2). Prove that $\sum_i q_i$ is primary to P.

4.4   Let R be a domain. Prove that for any multiplicatively closed subset W of R, $(W^{-1}R)^+ = W^{-1}(R^+)$, and that for any prime ideal P in R^+, $R^+/P = (R/(P \cap R))^+$.

4.5   (Generalized Jacobian criterion) Let A be a regular ring and R a quotient of a polynomial ring over A by a radical ideal L all of whose minimal prime ideals have the same height such that A is a subring of R. Let J be the Jacobian ideal of R over A. Prove that $V(J)$ contains the singular locus of R. (See for example [200, Theorem 30.4].)

4.6   (Abhyankar and Moh [7]) Let $K \subseteq L$ be fields of characteristic zero, X a variable over L, and $\sum_{i=0}^{\infty} a_i X^i \in L[[X]]$ integral over $K[[X]]$. Prove that $[K(a_i \,|\, i \geq 0) : K] < \infty$.

4.7   (Nagata [213, page 206]) Let k be a field of positive prime characteristic p. Let $X_1, \ldots, X_n$ be variables over k. Set R to be the subring of $k[[X_1, \ldots, X_n]]$ generated by $k^p[[X_1, \ldots, X_n]][k]$. Prove that R is a regular local ring whose completion equals $k[[X_1, \ldots, X_n]]$.

4.8   (Nagata) Let k be a field of positive prime characteristic p that is an infinite–dimensional vector space over the subfield $k^p = \{x^p \,|\, x \in k\}$. Let $b_1, b_2, \ldots$ be a countable subset of k of elements that are linearly independent over k^p. Let X be a variable over k and $c = \sum_n b_n X^n$. Set R to be the subring of $k[[X]]$ generated by $k^p[[X]][c]$ and k.
   (i) Prove that R is a one–dimensional Noetherian local domain.
   (ii) Prove that $\overline{R}$ is not finitely generated over R.
   (iii) Prove that $Q(R)$ is a finitely generated R–algebra.

4.9   ([213, Example 5, page 209]) Let k be a field of positive prime characteristic 2 that is an infinite–dimensional vector space over the subfield $k^2 = \{x^2 \,|\, x \in k\}$. Let $b_1, b_2, \ldots$ be a countable subset of k of elements such that for all n, $[k^2(b_1, \ldots, b_n) : k^2] = 2^n$. Let X, Y, Z be variables over k and $d = Y \sum_{n>0} b_{2n} X^n + Z \sum_{n \geq 0} b_{2n+1} X^n$. Set R to be the subring of $k[[X, Y, Z]]$ generated by $k^2[[X, Y, Z]][k][d]$.
   (i) Prove that R is a three–dimensional Noetherian local domain.
   (ii) Prove that the integral closure of R is not Noetherian.

4.10   Let R and X be as in the previous exercise. Let S be the finitely generated R–algebra $R[\frac{1}{X}]$. Prove that the integral closure of R in S is not Noetherian.

4.11   ([213, Appendix] or [212]) Let k_0 be a perfect field of characteristic 2, $X, Y, X_1, Y_1, X_2, Y_2, \ldots$ variables over k_0, $k = k_0(X_1, Y_1, X_2, Y_2, \ldots)$, set $A = k^2[[X, Y]][k]$, $f = \sum_{i=0}^{\infty}(X_i X^i + Y_i Y^i)$, and $R = A[f]$.

   (i) Prove that R is a normal Noetherian local ring whose completion contains non–zero nilpotent elements. (Cf. Corollary 4.6.2 (1).)

   (ii) Prove that R is a module–finite extension of A.

   (iii) Prove that $k^2[[X, Y, Z]][k]$ has module–finite integral closure, but that some quotient of it does not.

4.12   Let R be a Noetherian domain, with field of fractions K and integral closure $\overline{R}$. Prove that $\{P \in \operatorname{Spec} R \mid P = (\overline{R} :_R x)$ for some $x \in K\}$ equals $\{Q \cap R \mid Q \in \operatorname{Spec} \overline{R}, \operatorname{ht} Q = 1\}$.

4.13   Prove that a Krull domain is a Dedekind domain if and only if after localization at each maximal prime ideal it is a principal ideal domain.

4.14   Prove that a one–dimensional Noetherian domain is a Krull domain if and only if it is a Dedekind domain.

4.15   Let R be a Krull domain. Prove that the integral closure of R in a finite field extension of the field of fractions of R is a Krull domain.

4.16   Let R be a Krull domain, X a variable over R. Prove that $R[X]$ and $R[[X]]$ are Krull domains.

4.17   Let R be a reduced Noetherian ring and $\overline{R}$ its integral closure. Let I be an ideal in $\overline{R}$ of height at least 2 and α in the total ring of fractions of R such that $I\alpha \subseteq \overline{R}$. Prove that $\alpha \in \overline{R}$.

4.18   Let $(R, \mathfrak{m})$ be a one–dimensional reduced Noetherian local ring. Prove that $\overline{R}$ is module–finite over R if and only if $\widehat{R}$ is reduced.

4.19   Prove that any unique factorization domain is a Krull domain.

4.20   The purpose of this exercise: to construct a Noetherian domain R that is essentially of finite type over $\mathbb{C}$, with maximal ideal $\mathfrak{m}$ of positive height d, a prime ideal P strictly contained in $\mathfrak{m}$ such that the $\mathfrak{m}$–adic completion $\widehat{R}$ has two minimal primes p_1, p_2 with $\dim(\widehat{R}/p_1) = \dim(\widehat{R}/p_2) = d$ and $P\widehat{R} + p_1$ is $\mathfrak{m}\widehat{R}$–primary.

   Let $S = \mathbb{C}[x_1, \ldots, x_d]$, where $x_1, \ldots, x_d$ are variables over $\mathbb{C}$, let $\mathfrak{m}_1, \mathfrak{m}_2$ be distinct maximal ideals in S, and $J = \mathfrak{m}_1 \cap \mathfrak{m}_2$. Set $R = \mathbb{C} + J$.

   (i) Prove that R is a Noetherian domain, finitely generated over $\mathbb{C}$, and that $\mathfrak{m} = J$ is a maximal ideal of R.

   (ii) Prove that S is the integral closure of R and that $\mathfrak{m}_1$ and $\mathfrak{m}_2$ are the only prime ideals in S lying over $\mathfrak{m}$.

   (iii) Prove that the integral closure of $\widehat{R}$ equals $\widehat{S_{\mathfrak{m}_1}} \times \widehat{S_{\mathfrak{m}_2}}$. Let p_i be the kernel of the natural map $\widehat{R} \to \widehat{S_{\mathfrak{m}_i}}$. Prove that $\widehat{R}$ has two minimal primes, p_1 and p_2, both of dimension d.

    (iv) Prove that R has an isolated singularity at $\mathfrak{m}$, i.e., that for any prime ideal p strictly contained in $\mathfrak{m}$, R_p is regular.

    (v) Assume that $d \geq 2$. Show that there exists a prime ideal Q in S such that $\dim(S/Q) = 1$ and $Q \subseteq \mathfrak{m}_2$, $Q \not\subseteq \mathfrak{m}_1$. Set $P = Q \cap R$. Prove that P is a prime ideal strictly contained in $\mathfrak{m}$ and that $P\widehat{R} + p_1$ is $\mathfrak{m}\widehat{R}$–primary.

4.21   Let R be a Noetherian domain satisfying Serre's condition (R_1).

    (i) Let M be a finitely generated R–module that satisfies the condition that for each prime ideal P in R, the PR_P–grade on M_P is at least $\min\{2, \operatorname{ht} P\}$. Prove that $\operatorname{Hom}_R(\operatorname{Hom}_R(M, R), R)$ is isomorphic to M.

    (ii) Assume that S is a module–finite extension domain. Prove that its R–double dual $\operatorname{Hom}_R(\operatorname{Hom}_R(S, R), R)$ is a subring of $\overline{S}$ containing R.

    (iii) (Vasconcelos [304]) Let S be a module–finite extension of R that is a domain and that satisfies Serre's condition (R_1). Prove that $\operatorname{Hom}_R(\operatorname{Hom}_R(S, R), R)$ is $\overline{S}$.

4.22   Let R be a finitely generated algebra over a perfect field k. If R is a domain that satisfies (S_2), prove that $\operatorname{Hom}_R(J^{-1}, J^{-1})$ satisfies (S_2), where J is the Jacobian ideal of R over k.

4.23   Let R be a Noetherian domain and $\overline{R}$ the integral closure of R in its field of fractions. Let I be a non–zero ideal in R.

    (i) Prove that if I is a radical ideal contained in an associated prime of $\overline{R}/R$, then $\operatorname{Hom}_R(I, I)$ properly contains R.

    (ii) Prove that if I is an ideal not contained in any associated prime of $\overline{R}/R$, then $\operatorname{Hom}_R(I, I) = R$.

4.24*   (Ulrich [44]) (Ulrich's proof uses canonical modules) Let $(R, \mathfrak{m})$ be a Cohen–Macaulay local ring, I an $\mathfrak{m}$–primary ideal and J an ideal in I that is a complete intersection. Prove that $J : (J : I) \subseteq \overline{I}$.

4.25   (Tate) Let R be a Noetherian integrally closed domain and x a non–zero element of R. Assume that xR is a prime ideal, that R is complete and separated in the topology determined by xR, and that the integral closure of R/xR in any finite field extension of $\kappa(xR)$ is module–finite over R/xR. Prove that the integral closure of R in any finite field extension of $Q(R)$ is module–finite over R.

4.26*   (Zariski's Main Theorem) Let R be a ring, S a finitely generated R–algebra, $\overline{R}$ the integral closure of R in S. Let Q be a prime ideal in S, $P = Q \cap R$, and $\overline{P} = Q \cap \overline{R}$. Assume that S_Q/PS_Q is a finitely generated module over $\kappa(P)$.

    (i) Suppose that $S = R[x]$. Prove that x is integral over $\overline{R}_{\overline{P}}$.

    (ii) Prove that $\overline{R}_{\overline{P}} = S_{\overline{R} \setminus \overline{P}} = S_Q$. (Hint: Use induction on the number of algebra generators of S over R.)

    (iii) Prove that there exists $f \in \overline{R} \setminus Q$ such that $S_f = \overline{R}_f$.

4.27* (Zariski's Main Theorem, version by Evans [74]) Let R be a ring, S a finitely generated R–algebra and T an integral extension of S such that R is integrally closed in T. Let $Q \in \operatorname{Spec} T$ such that the image of Q in $\kappa(Q \cap R) \otimes_R T$ is a minimal and maximal prime ideal. Prove that there exists $s \in R \setminus (Q \cap R)$ such that $T_s = R_s$.

4.28* (Zariski's Main Theorem, version in [213, Theorem (37.4)]) Let $(R, \mathfrak{m})$ be a Noetherian normal local domain whose $\mathfrak{m}$–adic completion $\widehat{R}$ is a domain. Let $(R', \mathfrak{m}')$ be a Noetherian local ring in $Q(R)$ containing R such that
   (i) $\mathfrak{m}' \cap R = \mathfrak{m}$,
   (ii) $R'/\mathfrak{m}'R$ is finite over $R/\mathfrak{m}$,
   (iii) $\dim R' = \dim R$.
   Prove that $R = R'$.

4.29* (Abhyankar [8]) Let d be a positive integer, k a field, and $(R, \mathfrak{m})$ a d–dimensional domain that is essentially of finite type over k. Let K be the field of fractions of R and L a finite algebraic extension. Let Ω be a set of d–dimensional normal local domains $(S, \mathfrak{n})$ with field of fractions L such that $\mathfrak{n} \cap R = \mathfrak{m}$, $S/\mathfrak{n}$ is finite algebraic over $R/\mathfrak{m}$, and $\mathfrak{m}S$ is $\mathfrak{n}$–primary. (Ω could be the set of all localizations of the integral closure of R in L. Why?) Then $|\Omega| \le [L : K]$.

5
Rees algebras

5.1. Rees algebra constructions

Definition 5.1.1 *Let R be a ring, I an ideal and t a variable over R. The* **Rees algebra of** I *is the subring of $R[t]$ defined as*

$$R[It] = \left\{ \sum_{i=0}^{n} a_i t^i \,\middle|\, n \in \mathbb{N}; a_i \in I^i \right\} = \bigoplus_{n \geq 0} I^n t^n,$$

and the **extended Rees algebra of** I *is the subring of $R[t, t^{-1}]$ defined as*

$$R[It, t^{-1}] = \left\{ \sum_{i=-n}^{n} a_i t^i \,\middle|\, n \in \mathbb{N}; a_i \in I^i \right\} = \bigoplus_{n \in \mathbb{Z}} I^n t^n,$$

where, by convention, for any non–positive integer n, $I^n = R$.

For every ideal J of R,

$$J \subseteq JR[It] \cap R \subseteq JR[It, t^{-1}] \cap R \subseteq JR[t, t^{-1}] \cap R = J,$$

so that equality holds throughout. Thus every ideal of R is contracted from an ideal of $R[It]$ and $R[It, t^{-1}]$. Also,

$$\frac{R}{J} \subseteq \frac{R[It]}{JR[t, t^{-1}] \cap R[It]} \subseteq \frac{R[It, t^{-1}]}{JR[t, t^{-1}] \cap R[It, t^{-1}]} \subseteq \frac{R[t, t^{-1}]}{JR[t, t^{-1}]},$$

where the two rings in the middle are isomorphic to the Rees algebra, respectively the extended Rees algebra, of the image of I in R/J. In particular, if P is a minimal prime ideal of R, then $PR[t, t^{-1}] \cap R[It]$ and $PR[t, t^{-1}] \cap R[It, t^{-1}]$ are minimal prime ideals in their respective rings. Any nilpotent element of $R[It]$ or $R[It, t^{-1}]$ is also nilpotent in $R[t, t^{-1}]$, so it lies in $\cap_P PR[t, t^{-1}]$, as P varies over the minimal prime ideals of R. Thus all the minimal prime ideals of the two Rees algebras are contracted from the minimal prime ideals of $R[t, t^{-1}]$, each of which is of the form $PR[t, t^{-1}]$, for some minimal prime ideal P of R. Thus

$$\dim R[It] = \max\left\{ \dim\left(\frac{R}{P}\left[\frac{I+P}{P} t \right] \right) \,\middle|\, P \in \operatorname{Min} R \right\}, \tag{5.1.2}$$

$$\dim R[It, t^{-1}] = \max\left\{ \dim\left(\frac{R}{P}\left[\frac{I+P}{P} t, t^{-1} \right] \right) \,\middle|\, P \in \operatorname{Min} R \right\}. \tag{5.1.3}$$

Theorem 5.1.4 *Let R be a Noetherian ring and I an ideal of R. Then $\dim R$ is finite if and only if the dimension of either Rees algebra is finite. If $\dim R$ is finite, then*

$$(1) \quad \dim R[It] = \begin{cases} \dim R + 1, & \text{if } I \not\subseteq P \text{ for some prime ideal } P \\ & \text{with } \dim(R/P) = \dim R, \\ \dim R, & \text{otherwise.} \end{cases}$$

(2) $\dim R[It, t^{-1}] = \dim R + 1$.

(3) *If $\mathfrak{m}$ is the unique maximal ideal in R, and if $I \subseteq \mathfrak{m}$, then $\mathfrak{m}R[It, t^{-1}] + ItR[It, t^{-1}] + t^{-1}R[It, t^{-1}]$ is a maximal ideal in $R[It, t^{-1}]$ of height $\dim R + 1$.*

Proof: First we compute $\dim(R[It])$. By Equation (5.1.2), it suffices to prove that for an integral domain R, $\dim R[It] = \dim R$ if I is the zero ideal and is $\dim R + 1$ otherwise. Thus we may assume that R is a domain. Dimension Inequality, Theorem B.2.5, implies that for every prime ideal Q in $R[It]$, $\operatorname{ht} Q \leq \operatorname{ht}(Q \cap R) + 1 \leq \dim R + 1$, which proves that $\dim R[It] \leq \dim R + 1$. Clearly $\dim R[It] = \dim R$ if I is the zero ideal. So assume that I is non–zero. Let $P_0 = ItR[It]$. Then $P_0 \cap R = (0)$, $It \subseteq P_0$, $\operatorname{ht} P_0 > 0$, and $R[It]/P_0 \cong R$, proving that P_0 is prime, and that $\dim R[It] \geq \dim R + 1$. This proves (1).

Similarly, by Equation (5.1.3), to verify the equality for $\dim R[It, t^{-1}]$, it suffices to assume that R is a domain. By the Dimension Inequality (Theorem B.2.5), $\dim R[It, t^{-1}] \leq \dim R + 1$, and the other inequality follows from $\dim R[It, t^{-1}] \geq \dim R[It, t^{-1}]_{t^{-1}} = \dim R[t, t^{-1}] = \dim R + 1$.

Let $P_0 \subsetneq P_1 \subsetneq \cdots \subsetneq P_h = \mathfrak{m}$ be a saturated chain of prime ideals in R, with $h = \operatorname{ht} \mathfrak{m}$. Set $Q_i = P_i R[t, t^{-1}] \cap R[It, t^{-1}]$. As $Q_i \cap R = P_i$, $Q_0 \subseteq Q_1 \subseteq \cdots \subseteq Q_h$ is a chain of distinct prime ideals in R. The biggest one is $Q_h = \mathfrak{m}R[t, t^{-1}] \cap R[It, t^{-1}] = \mathfrak{m}R[It, t^{-1}] + ItR[It, t^{-1}]$, which is properly contained in the maximal ideal $Q_h + t^{-1}R[It, t^{-1}]$. This proves (3). $\square$

There are two other closely related rings:

Definition 5.1.5 *The **associated graded ring** of I is*

$$\operatorname{gr}_I(R) = \oplus_{n \geq 0}(I^n/I^{n+1}) = R[It]/IR[It] = R[It, t^{-1}]/t^{-1}R[It, t^{-1}].$$

*If R is Noetherian local with maximal ideal $\mathfrak{m}$, the **fiber cone** of I is the ring*

$$\mathcal{F}_I(R) = \frac{R[It]}{\mathfrak{m}R[It]} \cong \frac{R}{\mathfrak{m}} \oplus \frac{I}{\mathfrak{m}I} \oplus \frac{I^2}{\mathfrak{m}I^2} \oplus \frac{I^3}{\mathfrak{m}I^3} \oplus \cdots.$$

*We also denote it as F_I. The Krull dimension of $\mathcal{F}_I$ is also called the **analytic spread** of I and is denoted $\ell(I)$.*

Clearly $\operatorname{gr}_I(R) = R[It, t^{-1}]/t^{-1}R[It, t^{-1}]$ and $\mathcal{F}_I = \operatorname{gr}_I(R)/\mathfrak{m}\operatorname{gr}_I(R)$.

Proposition 5.1.6 *For any ideal I in $(R, \mathfrak{m})$,*

$$\ell(I) = \dim \mathcal{F}_I \leq \dim(\operatorname{gr}_I(R)) = \dim R.$$

Furthermore, if M is the maximal ideal in $\operatorname{gr}_I(R)$ consisting of all elements of positive degree, then $\dim(\operatorname{gr}_I(R)) = \operatorname{ht} M$.

Proof: As $\mathcal{F}_I$ is a quotient of $\operatorname{gr}_I(R)$, $\dim \mathcal{F}_I \leq \dim \operatorname{gr}_I(R)$. As t^{-1} is a non–zerodivisor in $R[It, t^{-1}]$ and $\operatorname{gr}_I(R) = R[It, t^{-1}]/t^{-1}R[It, t^{-1}]$, by Theorem 5.1.4 (2) we conclude that $\dim \operatorname{gr}_I(R) \leq \dim R$. Let $Q = \mathfrak{m}R[It, t^{-1}] + ItR[It, t^{-1}] + t^{-1}R[It, t^{-1}]$. By Theorem 5.1.4 (3), Certainly $\dim \operatorname{gr}_I(R) \geq \dim(\operatorname{gr}_I(R))_Q$, and by Theorem 5.1.4 (3), $\dim(\operatorname{gr}_I(R))_Q$ is $\operatorname{ht} Q - 1 = h = \dim R$. $\qquad\square$

Just as the integral closure of ideals reduces to the integral closure of ideals modulo the minimal prime ideals, the same goes for the analytic spread:

Proposition 5.1.7 *Let $(R, \mathfrak{m})$ be a Noetherian local ring and I an ideal in R. Then $\ell(I) = \max\{\ell(I(R/P)) \mid P \in \operatorname{Min}(R)\}$.*

Proof: By the one–to–one correspondence between minimal prime ideals in R and minimal prime ideals in $R[It]$,

$$
\begin{aligned}
\dim \mathcal{F}_I &= \dim R[It]/\mathfrak{m}R[It] \\
&= \max\{\dim(R/P)[((I+P)/P)t]/\mathfrak{m}(R/P)[((I+P)/P)t] \mid P \in \operatorname{Min}R\} \\
&= \max\{\dim \mathcal{F}_{I(R/P)} \mid P \in \operatorname{Min}R\}. \qquad\square
\end{aligned}
$$

The associated graded ring can be thought of as the special fiber of the extended Rees algebra (i.e., the ring one gets by setting $t^{-1} = 0$), while the "general fiber" (setting t^{-1} equal to a non–zero constant) is isomorphic to the base ring R. This has an important effect: for most ring–theoretic properties, if $\operatorname{gr}_I(R)$ has the property, so does R. For example, let I be an ideal in a Noetherian ring R such that $\cap_n I^n = 0$. If $\operatorname{gr}_I(R)$ is either a reduced ring, an integral domain, or an integrally closed domain, then R has the corresponding property. We leave the proofs to the exercises. See Exercise 5.9.

5.2. Integral closure of Rees algebras

We proved in Theorem 2.3.2 that if $A \subseteq B$ is an extension of $\mathbb{N}$–graded rings, then the integral closure of A in B is also $\mathbb{N}$–graded. Thus in particular, the integral closure of $R[It]$ in $R[t]$ is

$$
I_0 \oplus I_1 t \oplus I_2 t^2 \oplus I_3 t^3 \oplus \cdots,
$$

for some R–submodules I_i of R. These ideals I_i are integrally closed ideals:

Proposition 5.2.1 *Let R be a ring and t a variable over R. For any ideal I in R, the integral closure of $R[It]$ in $R[t]$ equals the graded ring*

$$
R \oplus \bar{I}t \oplus \overline{I^2}t^2 \oplus \overline{I^3}t^3 \oplus \cdots.
$$

Similarly, the integral closure of $R[It, t^{-1}]$ in $R[t, t^{-1}]$ equals the graded ring

$$
\cdots \oplus Rt^{-2} \oplus Rt^{-1} \oplus R \oplus \bar{I}t \oplus \overline{I^2}t^2 \oplus \overline{I^3}t^3 \oplus \cdots.
$$

Proof: Let S be the integral closure of $R[It]$ in $R[t]$. By Theorem 2.3.2, S is an $\mathbb{N}$–graded submodule of $R[t]$. Denote the graded piece of S of degree $k \in \mathbb{N}$ by S_k.

Let $s \in S_k$, $k \in \mathbb{N}$. Write $s = s_k t^k$ for some $s_k \in R$. As s is integral over $R[It]$, there exist a positive integer n and $a_i \in R[It]$, $i = 1, \ldots, n$, such that

$$s_k^n t^{kn} + a_1 s_k^{n-1} t^{k(n-1)} + a_2 s_k^{n-2} t^{k(n-2)} + \cdots + a_{n-1} s_k t^k + a_n = 0.$$

Expand each $a_i = \sum_{j=0}^{k_i} a_{i,j} t^j$, with $a_{i,j} \in I^j$. The homogeneous part of degree kn in the equation above is exactly

$$t^{kn}(s_k^n + a_{1,k} s_k^{n-1} + a_{2,2k} s_k^{n-2} + \cdots + a_{n-1,(n-1)k} s_k + a_{n,nk}) = 0.$$

As $a_{i,ik} \in I^{ik}$, this equation says that s_k is integral over the ideal I^k. Thus $S_k \subseteq \overline{I^k} t^k$. The other inclusion is easy to prove.   □

With this, as the integral closure of the ring in an overring is integrally closed in that overring, it follows easily that for every ideal I in R, $\overline{I}$ is an ideal and $\overline{\overline{I}} = \overline{I}$. (Compare with Corollary 1.3.1.)

An extension of rings is integral if and only if it is generated by elements that are integrally dependent on the subring (see Proposition 2.1.10). Similar result holds for ideals:

Corollary 5.2.2   *For any ideals $I \subseteq J$ in R, every element of J is integral over I if and only if each element in some generating set of J over I is integral over I.*   □

It follows that the integral closure of a homogeneous ideal is homogeneous:

Corollary 5.2.3   *Let I be a homogeneous ideal in a G–graded ring R, where G is $\mathbb{Z}^d \times \mathbb{N}^d$. Then $\overline{I}$ is also G–graded.*

Proof: When R is G–graded, then $R[It]$ is $G \oplus \mathbb{N}$–graded. By Theorem 2.3.2, the integral closure of $R[It]$ in $\overline{R}[t]$ is $G \oplus \mathbb{N}$–graded, so that as $\overline{R}[t]$ is integrally closed, the integral closure of $R[It]$ is $G \oplus \mathbb{N}$–graded. Thus by the structure of this integral closure, each $\overline{I^n}$ is G–graded.   □

In particular, as already proved in Proposition 1.4.2, the integral closure of a monomial ideal in a polynomial ring is again a monomial ideal.

One can also compute the absolute integral closure of Rees algebras:

Proposition 5.2.4   *Let R be a ring and $\overline{R}$ the integral closure of R in its total ring of fractions. The integral closure of $R[It]$ in its total ring of fractions equals*

$$\overline{R} \oplus \overline{I\overline{R}}t \oplus \overline{I^2\overline{R}}t^2 \oplus \overline{I^3\overline{R}}t^3 \oplus \cdots,$$

and the integral closure of $R[It, t^{-1}]$ in its total ring of fractions equals

$$\cdots \oplus \overline{R}t^{-1} \oplus \overline{R} \oplus \overline{I\overline{R}}t \oplus \overline{I^2\overline{R}}t^2 \oplus \overline{I^3\overline{R}}t^3 \oplus \cdots.$$

Proof: Observe that for all n, $\overline{I^n\overline{R}} = \overline{I^n}\,\overline{R}$. The integral closure of $R[It]$ clearly contains $\overline{R}[I\overline{R}t]$. By Proposition 5.2.1, the integral closure of the latter ring in $\overline{R}[t]$ is $\overline{R} \oplus \overline{I\overline{R}}t \oplus \overline{I^2\overline{R}}t^2 \oplus \overline{I^3\overline{R}}t^3 \oplus \cdots$. But $\overline{R}[t]$ is integrally closed (see Exercise 2.3), so that integral closure of $R[It]$ is as displayed above.

The proof of the second part is similar. $\square$

Observe that the integral closure of $R[It]$ equals the integral closure of $\overline{R}[I\overline{R}t]$ in $\overline{R}[t]$, and that the integral closure of $R[It, t^{-1}]$ equals the integral closure of $\overline{R}[I\overline{R}t, t^{-1}]$ in $\overline{R}[t, t^{-1}]$.

The following is a criterion for when the integral closure of a Rees algebra is Noetherian:

Proposition 5.2.5 *Let R be a Noetherian ring and I an ideal in R. Let S be the Rees algebra $R[It]$ of I, and T be an $\mathbb{N}$–graded ring containing S. Then T is module–finite over S if and only if there exists an integer k such that for all $n \geq k$, $T_n = I^{n-k}T_k$.*

In particular, if T is the integral closure of S in $R[t]$, then T is module–finite over S if and only if there exists an integer k such that for all $n \geq k$, $\overline{I^n} = I^{n-k}\overline{I^k}$.

Proof: First assume that T is module–finite over R. As T is $\mathbb{N}$–graded, there exist homogeneous generators of T over S of degrees at most k. Then for all $n \geq k$, $T_n = S_{n-k}T_k$.

Conversely, assume that there exists an integer k such that for all $n \geq k$, $T_n = S_{n-k}T_k$. Then $T = ST_0 + ST_1 + \cdots + ST_k$, so that T is a finitely generated module over S.

By Proposition 5.2.1, if T is the integral closure of S, then T is $\mathbb{N}$–graded and $T_n = \overline{I^n}$. The rest follows easily. $\square$

5.3. Integral closure of powers of an ideal

Extended Rees algebras help analyze powers of an ideal: since

$$t^{-n}R[It, t^{-1}] \cap R = I^n,$$

many properties of I^n descend from the corresponding properties of the powers of the principal ideal $t^{-1}R[It, t^{-1}]$ generated by a non–zerodivisor. Principal ideals generated by a non–zerodivisor are in general easier to handle.

Similarly, the integral closure of the Rees algebra is used to capture some properties of the integral closures of powers of an ideal. We present the general method of descent in this section, and apply it in Section 5.4 to study the associated primes of the integral closures of powers of an ideal. A lot of the work on the sets of associated primes of powers and of integral closures of powers of an ideal was done by Ratliff [227]; Brodmann [26]; McAdam and Eakin [204]; Katz [156]; McAdam and Ratliff [205]. Katz [156] characterized the sets for integral closures with asymptotic sequences, which in turn were first defined by Rees in [234]. A good overview of this area is [203]. Also see Exercises 5.17–5.19.

Proposition 5.3.1 *Let I be an ideal in a ring R, $S = R[It, t^{-1}]$ and $\bar{S}$ the integral closure of S in $R[t, t^{-1}]$. We grade S by giving t degree 1 (therefore $\bar{S}$ is also graded by Theorem 2.3.2).*

(1) The ideal $\overline{t^{-n}S}$ is $\mathbb{Z}$–graded, and the degree m component of $\overline{t^{-n}S}$ equals $(\overline{I^{m+n}} \cap I^m)t^m$.

(2) Also, $\overline{t^{-n}\bar{S}}$ is $\mathbb{Z}$–graded in $\bar{S}$, and its degree m component equals $\overline{I^{m+n}}t^m$.

Proof: The ideal $\overline{t^{-n}S}$ is $\mathbb{Z}$–graded by Corollary 5.2.3. Every element of degree m in S is of the form rt^m for some $r \in I^m$. If $rt^m \in \overline{t^{-n}S}$, then there is an equation of integral dependence $(rt^m)^k + a_1(rt^m)^{k-1} + \cdots + a_k = 0$, with $a_i \in t^{-ni}S$. Write $a_i = b_i t^{-ni}$, for some $b_i \in S$. By looking at terms of t–degree mk in the equation, without loss of generality $b_i = c_i t^{(m+n)i}$ for some $c_i \in I^{(m+n)i}$. Thus r is integral over I^{m+n}, and $rt^m \in S$ implies that $r \in (\overline{I^{m+n}} \cap I^m)t^m$.

Conversely, if $r \in \overline{I^{m+n}}$, write $r^k + c_1 r^{k-1} + \cdots + c_k = 0$ for some $c_i \in I^{(m+n)i}$. If, further, r is in I^m, then $(rt^m)^k + c_1 t^m (rt^m)^{k-1} + \cdots + c_k t^{mk} = 0$, and for each i, $c_i t^{mi} \in t^{-ni}S$. Thus $rt^m \in \overline{t^{-n}S}$. This proves the first part.

By Corollary 5.2.3, $\overline{t^{-n}\bar{S}}$ is $\mathbb{Z}$–graded. Let rt^m be a homogeneous element of $\bar{S}$ that is integral over $t^{-n}\bar{S}$. Write

$$(rt^m)^k + b_1 t^{-n}(rt^m)^{k-1} + b_2 t^{-2n}(rt^m)^{k-2} + \cdots + c_k t^{(m+n)k} t^{-nk} = 0,$$

for some $b_i \in \bar{S}$. By looking at the terms of degree nk, without loss of generality $b_i = c_i t^{i(m+n)} \in \bar{S}$ of t–degree $i(m + n)$. By Proposition 5.2.1, $c_i \in \overline{I^{i(m+n)}}$. Multiply the equation above through by t^{-mk}. We see that r is integral over $\overline{I^{m+n}}$, so that by Corollary 1.3.1, $r \in \overline{I^{m+n}}$. Now for any $r \in \overline{I^{m+n}}$, let $r^k + c_1 r^{k-1} + \cdots + c_k = 0$ be an equation of integral dependence of r over I^{m+n}. Multiply this equation through by t^{mk} to get an equation of integral dependence of $rt^m \in \bar{S}_m$ over $t^{-n}\bar{S}$. $\qquad\square$

We have already seen that each ideal of R contracts from an ideal in a Rees algebra. But some ideals in R contract from principal ideals in the extended Rees algebra:

Proposition 5.3.2 *Let I be an ideal in a ring R and t a variable over R. Let S be the extended Rees algebra $R[It, t^{-1}]$ of I. Then for any $n \in \mathbb{N}$,*

$$t^{-n}S \cap R = I^n \quad and \quad \overline{t^{-n}S} \cap R = \overline{I^n}.$$

If R is reduced and $\bar{S}$ the integral closure of S, then also $t^{-n}\bar{S} \cap S = \overline{t^{-n}S}$.

Proof: Clearly $I^n \subseteq t^{-n}S \cap R$ and $\overline{I^n} \subseteq \overline{t^{-n}S} \cap R$. If $r \in t^{-n}S \cap R$, then $r = t^{-n}(rt^n)$, with $rt^n \in S_n$, so that necessarily $r \in I^n$, which proves the first equality. The second equality follows from Proposition 5.3.1.

If R is reduced, so is S. As $t^{-n}\bar{S}$ is integrally closed by Proposition 1.5.2, it follows that $\overline{t^{-n}S} \subseteq t^{-n}\bar{S} \cap S$. The opposite inclusion holds by Proposition 1.6.1. $\qquad\square$

Here are some immediate applications to the associated primes of integral closures of powers of ideals.

Proposition 5.3.3 *Let I be an ideal in a ring R, and let $S = R[It, t^{-1}]$.*
(1) Let P be a prime ideal associated to some $\overline{I^n}$. There exists $Q \in \operatorname{Spec} S$ such that $Q \cap R = P$ and Q is associated to $\overline{t^{-n}S}$.
(2) For any integer n and $Q \in \operatorname{Ass}(S/\overline{t^{-n}S})$ there exists an integer $m \geq n$ such that $Q \cap R \in \operatorname{Ass}(R/\overline{I^m})$.
(3) Let $\overline{S}$ denote the integral closure of S in $R[t, t^{-1}]$. For every integer n and $Q \in \operatorname{Ass}(\overline{S}/t^{-n}\overline{S})$ there exists an integer $m \geq n$ such that $Q \cap R \in \operatorname{Ass}(R/\overline{I^m})$.

Proof: By Proposition 5.3.2, $\overline{I^n}$ contracts from $\overline{t^{-n}S}$, and so an irredundant primary decomposition of $\overline{t^{-n}S}$ contracts to a possibly redundant primary decomposition of $\overline{I^n}$. This proves (1).

To prove the rest of the proposition, by localizing at $Q \cap R$ without loss of generality we may assume that R is a ring with only one maximal ideal, that ideal being $P = Q \cap R$. First let $Q = \overline{t^{-n}S} : xt^r$ for some $x \in I^r$. By Proposition 5.3.1, the degree r component of $\overline{t^{-n}S}$ is $(\overline{I^{r+n}} \cap I^r)t^r$, so that $P = (\overline{I^{r+n}} \cap I^r) : x = \overline{I^{r+n}} : x$. This proves (2).

If instead $Q \in \operatorname{Ass}(\overline{S}/t^{-n}\overline{S})$, by Proposition 5.2.1, $Q = t^{-n}\overline{S} : xt^r$ for some $x \in \overline{I^r}$. Thus by Proposition 5.3.1, $P = \overline{I^{n+r}} : x$, so that $P \in \operatorname{Ass}(R/\overline{I^{n+r}})$. This proves (3). $\square$

For many of the rings analyzed in this section, the integral closure of Rees algebras behaves well:

Proposition 5.3.4 *Let R be one of the following:*
(1) $(R, \mathfrak{m})$ is complete local Noetherian,
(2) R is finitely generated over a field or over $\mathbb{Z}$,
(3) or, more generally, R is finitely generated over a Noetherian integrally closed domain satisfying the property that every finitely generated A-algebra has a module–finite integral closure.
Let I be an ideal in R, and S either the Rees algebra of I or the extended Rees algebra of I. Then the integral closure of S is a module–finite extension of S, and there exists an integer k such that for all $n \geq k$, $\overline{I^n} = I^{n-k}\overline{I^k}$.

Proof: By the forms of the integral closures given in Proposition 5.2.4, it suffices to prove that the integral closure of the Rees algebra $S = R[It]$ is module–finite over S.

Rings of form (2) are rings of form (3), by Theorem 4.6.3 and by Corollary 4.6.5. Thus if R is of the form (2) or (3), the integral closure $\overline{S}$ of S is module–finite over S, and the last statement follows by Proposition 5.2.5.

Now assume that R is of form (1), i.e., that $(R, \mathfrak{m})$ is a complete local ring. Let $I = (a_1, \ldots, a_r)$, and t a variable over R. Let T be the subring of the ring of formal power series $R[[t]]$ generated over R by power series in $a_1 t, \ldots, a_r t$.

This is a Noetherian ring, and since it is a subring of $R[[t]]$, it contains no non–zero nilpotent elements.

Let $\mathfrak{M}$ be the ideal in T generated by all power series whose constant term is in $\mathfrak{m}$. It is easy to see that $T/\mathfrak{M} = R/\mathfrak{m}$ and that the elements of T not in $\mathfrak{M}$ are units in T. Thus T is a Noetherian local ring with maximal ideal $\mathfrak{M}$.

Claim: T is complete in the $\mathfrak{M}$–adic topology.

Proof of claim: An element $\sum_i c_i t^i$ in T, with c_i in R, lies in $\mathfrak{M}^n$ if for all i, $c_i \in \mathfrak{m}^{n-i} I^i$. Let $\{x_j\}_j$ be a Cauchy sequence in T, with $x_j = \sum_i c_{j,i} t^i$, $c_{j,i} \in R$, and $x_j - x_{j+1} \in \mathfrak{M}^j$ for all j. Hence for all i, $c_{j,i} - c_{j+1,i} \in \mathfrak{m}^{j-i} I^i$. It follows that for each i, $c_{1,i}, c_{2,i}, c_{3,i}, \ldots$ is a Cauchy sequence in R, with limit c_i an element of I^i. Hence the limit $\sum_i c_i t^i$ of $\{x_j\}_j$ is an element of T. This proves the claim.

It follows by Theorem 4.3.4 that the integral closure $\overline{T}$ of T is a module–finite extension. Clearly $T \subseteq R[[t]]$. As in Proposition 5.2.4, $\overline{T} = \prod_{n \geq 0} \overline{I^n} t^n$. But finite generation of $\overline{T}$ over T implies that there exists an integer k such that for all $n \geq k$, $\overline{I^n} = I^{n-k} \overline{I^k}$, which is by Proposition 5.2.5 equivalent to saying that $\overline{S}$ is module–finite over S. $\qquad\square$

5.4. Powers and formal equidimensionality

In this section we prove Ratliff's Theorem that was stated in Chapter 1, and we also prove its converse. We then expand on the associated primes of powers of an ideal and on the associated primes of integral closures of powers of an ideal.

Theorem 5.4.1 (Ratliff [228]) *Let R be a locally formally equidimensional Noetherian ring, and let $(x_1, \ldots, x_n)$ be a parameter ideal, i.e., the height of $(x_1, \ldots, x_n)$ is at least n. For all $m \geq 1$,*

$$\overline{(x_1, \ldots, x_{n-1})^m} : x_n = \overline{(x_1, \ldots, x_{n-1})^m}.$$

Proof: Set $J = (x_1, \ldots, x_{n-1})$. We fix an element $r \in \overline{J^m} : x_n$. If $r \notin \overline{J^m}$, choose a prime ideal minimal over $\overline{J^m} : r$. We can localize at this prime ideal without changing the assumptions; note that localization commutes with integral closure. Thus without loss of generality R is a local ring with maximal ideal $\mathfrak{m}$, and some power of $\mathfrak{m}$ multiplies r into $\overline{J^m}$.

Next, pass to the completion to assume that R is a complete Noetherian local ring. Note that parameters are preserved under passage to completion, and Proposition 1.6.2 implies that $r \notin \overline{J^m \widehat{R}}$. Henceforth we assume that R is complete. We claim that the images of $x_1, \ldots, x_n$ stay parameters after reducing R modulo an arbitrary minimal prime of R. Let P be a minimal prime of R, and let Q be a prime ideal that is minimal over $P + (x_1, \ldots, x_n)$. Since Q contains $(x_1, \ldots, x_n)$, the height of Q is at least n, so there is some minimal prime $P' \subseteq Q$ such that $\operatorname{ht}(Q/P') \geq n$. By Lemma B.4.2, $\dim R = \dim(R/Q) + \operatorname{ht}(Q/P)$ and $\dim R = \dim(R/Q) + \operatorname{ht}(Q/P')$. Thus $\operatorname{ht}(Q/P) = \operatorname{ht}(Q/P') \geq n$, so that modulo each minimal prime ideal, $x_1, \ldots, x_n$ is still a system of parameters. If r is in the integral closure of the image of J^m modulo each minimal prime ideal of R, then by Proposition 1.1.5, $r \in \overline{J^m}$. Hence we may assume that R is a complete Noetherian local domain. Let $\overline{R}$ be the integral closure of R. By Theorem 4.3.4, $\overline{R}$ is a Noetherian local domain, module–finite over R. By Corollary B.3.7, R satisfies the dimension formula, so parameters of R stay parameters in $\overline{R}$. If r is in the integral closure of $J^m \overline{R}$, then by Proposition 1.6.1, $r \in \overline{J^m}$.

Thus by changing notation we may assume that $(R, \mathfrak{m})$ is a complete Noetherian local integrally closed domain, $r \in \overline{J^m} : x_n$, and that $\mathfrak{m}^l r \subseteq \overline{J^m}$ for some l. It remains to prove that $r \in \overline{J^m}$. If $n = 1$, this is trivial, so we may assume that $n > 1$.

Set $S = R[Jt]$. Let $\overline{S}$ denote the integral closure of S inside its field of fractions. By Proposition 5.3.4, $\overline{S}$ is module–finite over S. By Corollary B.3.7, R is universally catenary, so by Theorem 5.1.4, $\operatorname{ht}(\mathfrak{m}S) = \dim S - \dim(S/\mathfrak{m}S)$ as $\mathfrak{m}S$ is contained in the maximal ideal of S of maximal height. Since $S/\mathfrak{m}S$ is the homomorphic image of a polynomial ring over a field in $n - 1$ variables, it follows that $\dim S/\mathfrak{m}S \leq n - 1$. Hence $\operatorname{ht}(\mathfrak{m}S) \geq 2$. By Corollary B.3.7 and by Lemma B.3.4, S satisfies the dimension formula, so that $\operatorname{ht}(\mathfrak{m}\overline{S}) \geq 2$. As $\overline{S}$ satisfies Serre's condition (S_2), it follows that $\mathfrak{m}^l \overline{S}$ contains a regular sequence of length two. The element rt^m is in the field of fractions of S (and not necessarily in $\overline{S}$), and by Proposition 5.2.4, $(\mathfrak{m}^l \overline{S})(rt^m) = (\mathfrak{m}^l r)t^m \overline{S}$ is contained in $\overline{S}$. Thus by Exercise 4.17, $rt^m \in \overline{S}$, forcing $r \in \overline{(x_1, \ldots, x_{n-1})^m}$, a contradiction. □

Corollary 5.4.2 *Let R be a locally formally equidimensional Noetherian ring, and let $(x_1, \ldots, x_n)$ be a parameter ideal, i.e., an ideal with the property that the height of $(x_1, \ldots, x_n)$ is n. Every associated prime of $\overline{(x_1, \ldots, x_n)^m}$ has height n.*

Proof: By localizing at an associated prime ideal of $\overline{(x_1, \ldots, x_n)^m}$ we may assume that $(R, \mathfrak{m})$ is local and that $\mathfrak{m}$ is associated to $\overline{(x_1, \ldots, x_n)^m}$. If the dimension of R is n, there is nothing to prove. As $x_1, \ldots, x_n$ are parameters, the dimension of R is at least n. Henceforth we assume that $\dim R > n$ and we will reach a contradiction.

Choose $y, x_{n+1} \in R$ such that $\mathfrak{m} = \overline{(x_1, \ldots, x_n)^m} : y$ and such that $x_1, \ldots, x_n, x_{n+1}$ are parameters. We have that $y \in \overline{(x_1, \ldots, x_n)^m} : x_{n+1}$. By Theorem 5.4.1, we obtain that $y \in \overline{(x_1, \ldots, x_n)^m}$, a contradiction. □

The converse of Ratliff's Theorem holds as well, and we prove it later in this section (Theorem 5.4.5).

Corollary 5.4.3 (Ratliff and Rush [229]) *Let R be a locally formally equidimensional Noetherian ring, and $I = (x_1, \ldots, x_n)$ a parameter ideal of height*

n that contains a non–zerodivisor. For any integer m, define $I^{(m)}$ to be the intersection of the minimal primary components of I^m. Then for all m, i,

$$I^{(m+i)} : I^{(i)} \subseteq \overline{I^m}.$$

Proof: It is enough to check the inclusion after localizing at each associated prime ideal of $\overline{I^m}$. So let P be an associated prime of $\overline{I^m}$. By Theorem 5.4.1, P is minimal over I. Hence $(I^{(m+i)} : I^{(i)})_P = I^{m+i}R_P : I^i R_P$, which by Corollary 1.1.8 is contained in $\overline{I^m}R_P = \overline{I^m R_P}$. $\qquad\square$

We now return to the associated primes of powers and of integral closures of powers of arbitrary ideals, with the goal of proving the converse of Ratliff's theorem. We start with a lemma from Katz's thesis [155].

Lemma 5.4.4 *Let R be a Noetherian ring and I an ideal.*
(1) Let $P \in \mathrm{Ass}(R/\overline{I})$. Then there exists a minimal prime ideal $p \subseteq P$ such that P/p is associated to the integral closure of $I(R/p)$.
(2) Let p be a minimal prime ideal, and P a prime ideal minimal over $I + p$. Then for all sufficiently large integers n, for any ideal J such that $I^n \subseteq J \subseteq \overline{I^n}$, P is associated to J.

Proof: Without loss of generality P is the unique maximal ideal of R.

(1) Write $P = \overline{I} : r$ for some $r \in R$. So $r \notin \overline{I}$, and by Proposition 1.1.5 there exists a minimal prime ideal $p \subseteq P$ such that r is not in the integral closure of the image of I in R/p. Hence $P/p \subseteq \overline{I(R/p)} : r \neq R/p$, so that P/p is associated to the integral closure of $I(R/p)$.

(2) Choose k such that $P^k \subseteq I + p$ and p^k is contained in the p–primary component of 0. Let $x \in R \setminus p$ annihilate p^k. By Exercise 5.14 by possibly increasing k, $x \notin \overline{I^k}$. Choose $n \geq k$. Then $P^{2nk} \subseteq (I+p)^{2n} \subseteq I^n + p^n \subseteq J + p^n$. As $x \notin \overline{I^k}$, $x \notin J$. But also $xP^{2nk} \subseteq J$, so that P is associated to J. $\qquad\square$

The converse of Ratliff's Theorem 5.4.1 also holds:

Theorem 5.4.5 (Ratliff [228]) *Let R be a Noetherian ring with the property that for every parameter ideal I and every integer m, $\overline{I^m}$ has no embedded prime ideals. Then R is locally formally equidimensional.*

Proof: Without loss of generality R is a Noetherian local ring with maximal ideal $\mathfrak{m}$. Suppose that R is not formally equidimensional. Then there exists $q \in \mathrm{Min}\,\widehat{R}$ such that $j = \dim(\widehat{R}/q) < \dim R$. Necessarily $j \geq 1$. By the Prime Avoidance lemma, we may inductively choose elements $x_1, \ldots, x_j$ such that for each $k \in \{1 \ldots, j\}$, the height of $(x_1, \ldots, x_k)R$ is k and such that the images of $x_1, \ldots, x_k$ in $\widehat{R}/q$ also generate an ideal of height k. Set $I = (x_1, \ldots, x_j)$. Then $\mathfrak{m}\widehat{R}$ is minimal over $I\widehat{R} + q$. By Lemma 5.4.4, $\mathfrak{m}\widehat{R}$ is associated to $\overline{I^n\widehat{R}}$ for all sufficiently large n, as this ideal lies between $I^n\widehat{R}$ and $\overline{I^n}\widehat{R}$. Since $R \to \widehat{R}$ is faithfully flat, then $\mathfrak{m}$ is associated to $\overline{I^n}$. But then by assumption, $\mathfrak{m}$ must be minimal over I, whence $\mathrm{ht}(\mathfrak{m}) = j < \dim R$, which is a contradiction. $\qquad\square$

Thus a Noetherian local ring is formally equidimensional if and only if for every parameter ideal I, the associated primes of all $\overline{I^n}$ are all minimal over I.

Theorem 5.4.6 (McAdam [202]) *Let $(R, \mathfrak{m})$ be a formally equidimensional Noetherian local ring and I an ideal in R. Then $\mathfrak{m} \in \mathrm{Ass}(R/\overline{I^n})$ for some n if and only if $\ell(I) = \dim R$.*

Proof: Assume that $\mathfrak{m} \in \mathrm{Ass}(R/\overline{I^n})$ for some n. By Lemma 5.4.4, there exists a minimal prime ideal p in R such that if $R' = R/p$, then $\mathfrak{m}R'$ is associated to the integral closure of $I^n R'$. Let $S = R'[IR't, t^{-1}]$. By Proposition 5.3.3 (1), there exists $Q \in \mathrm{Spec}\, S$ such that $Q \cap R' = \mathfrak{m}R'$ and Q is associated to $\overline{t^{-n}S}$. As R is formally equidimensional, by Lemma B.4.2, R' is formally equidimensional, so by Theorem B.5.2, R' satisfies the dimension formula and S is locally formally equidimensional. Thus by Theorem 5.4.1, Q has height 1. By the Dimension Formula (Theorem B.5.1), $\mathrm{ht}(Q) + \mathrm{tr.deg}_{\kappa(\mathfrak{m}R')}\kappa(Q) = \mathrm{ht}(\mathfrak{m}R') + \mathrm{tr.deg}_{R'} S$, so that $\mathrm{tr.deg}_{\kappa(\mathfrak{m}R')}\kappa(Q) = \mathrm{ht}(\mathfrak{m}R') = \dim R$. But $\kappa(Q)$ is a localization of a homomorphic image of $\mathcal{F}_I(R)$, so that $\dim R = \mathrm{tr.deg}_{\kappa(\mathfrak{m}R')}\kappa(Q) \le \dim \mathcal{F}_I(R) \le \dim R$. Thus $\ell(I) = \dim \mathcal{F}_I(R) = \dim R$.

Conversely, assume that $\ell(I) = \mathrm{ht}(\mathfrak{m})$. Let $S = R[It, t^{-1}]$ and $Q \in \mathrm{Spec}\, S$ such that the image of Q in $\mathcal{F}_I(R)$ is minimal and $\dim(S/Q) = \dim(\mathcal{F}_I(R))$. As $\ell(I) = \dim(\mathcal{F}_I(R))$, it follows that $\dim(S/Q) = \mathrm{ht}(\mathfrak{m})$. By construction, $Q \cap R = \mathfrak{m}$ and Q contains t^{-1}. As S has dimension 1 more than R, necessarily Q has height 1. Thus Q is minimal over $t^{-1}S$ and thus associated to $\overline{t^{-n}S}$ for all n. Hence by Proposition 5.3.3 (2), $\mathfrak{m}$ is associated to $\overline{I^n}$ for all large n. $\square$

The proof of one direction, in the last paragraph, did not require formal equidimensionality, which immediately proves the following:

Proposition 5.4.7 (Burch [34]) *Let $(R, \mathfrak{m})$ be a Noetherian local ring and I an ideal in R. If $\ell(I) = \dim R$, then $\mathfrak{m} \in \mathrm{Ass}(R/\overline{I^n})$ for all large n.* $\square$

When $(R, \mathfrak{m})$ is a Noetherian formally equidimensional local ring, much more can be said about the minimal primes of $\mathrm{gr}_I(R)$, by taking advantage of the fact that the associated graded ring is the extended Rees algebras modulo a principal ideal.

Proposition 5.4.8 *Let $(R, \mathfrak{m})$ be a Noetherian formally equidimensional local ring, and let I be an ideal of R. For every minimal prime ideal P of $\mathrm{gr}_I(R)$, $\dim(\mathrm{gr}_I(R)/P) = \dim R$.*

Proof: We use the isomorphism $\mathrm{gr}_I(R) \cong R[It, t^{-1}]/t^{-1}R[It, t^{-1}]$. Since every minimal prime ideal of $R[It, t^{-1}]$ comes from a minimal prime ideal of R as in the introduction to this chapter, we can assume that R is a domain. As R is formally equidimensional, it is universally catenary by Theorem B.5.1. Hence $R[It, t^{-1}]$ is catenary, and therefore $\mathrm{ht}\, P + \dim(R[It, t^{-1}]/P)$ equals $\dim R[It, t^{-1}] = \dim R + 1$ (observe that P is contained in the maximal ideal of $R[It, t^{-1}]$ generated by $t^{-1}, \mathfrak{m}$, and It, which has height equal to the dimension of $R[It, t^{-1}]$). By Krull's Height Theorem, $\operatorname{ht} P = 1$, showing $\dim(R[It, t^{-1}]/P) = \dim R$. □

5.5. Defining equations of Rees algebras

In this section we prove some basic theorems, used several times in this book, concerning the equations defining the Rees algebra of an ideal. In particular, the results apply in the case the ideal is generated by a regular sequence. For more on Rees algebras and their defining equations, see [305] and [307].

We begin with some discussion. Let $I = (x_1, \ldots, x_n)$ be an ideal. The Rees algebra $R[It] = R[x_1 t, \ldots, x_n t]$ can be written as a homomorphic image of the polynomial ring $R[T_1, \ldots, T_n]$ by the map π sending T_i to $x_i t$. The kernel is an ideal $\mathcal{A} \subseteq R[T_1, \ldots, T_n]$.

Our goal in this section is to give cases where we can describe the generators of $\mathcal{A}$, which we refer to as the **defining equations** of the Rees algebra. We also will describe how the equations defining the extended Rees algebra or affine pieces of the blowup of I relate to the defining ideal of the Rees algebra; see Propositions 5.5.7 and 5.5.8.

The map π is graded of degree 0, so that the kernel $\mathcal{A}$ is generated by homogeneous polynomials $F(T_1, \ldots, T_n) \in R[T_1, \ldots, T_n]$ with the property that $F(x_1 t, \ldots, x_n t) = 0$. Since F is homogeneous, $F(x_1 t, \ldots, x_n t) = 0$ if and only if $F(x_1, \ldots, x_n) = 0$. Generators for the homogeneous polynomials of degree one in $\mathcal{A}$ can always be obtained from a presentation of I. Namely, let $R^m \to R^n \to I \to 0$ be a presentation of I, where we represent the map from $R^m \to R^n$ by an $n \times m$ matrix A. Let T denote the $1 \times n$ matrix of the variables $T_1, \ldots, T_n$, and let L be the ideal generated in $R[T_1, \ldots, T_n]$ by the entries of the matrix TA; the entries are linear polynomials in the T_i that vanish after the substitution $T_i \to x_i$. Hence $L \subseteq \mathcal{A}$, and L is exactly the subideal $\mathcal{A}_1$ of $\mathcal{A}$ generated by all linear polynomials in $\mathcal{A}$ (Exercise 5.22). The algebra $R[T_1, \ldots, T_n]/\mathcal{A}_1$ is isomorphic to the symmetric algebra of I as an R–module, and the Rees algebra is particularly easy to analyze in the case $\mathcal{A}_1 = \mathcal{A}$. Valla introduced the following definition:

Definition 5.5.1 I *is said to be of* **linear type** *if* $\mathcal{A}_1 = \mathcal{A}$.

Ideals of linear type are the simplest in terms of the defining equations of their Rees algebras. We shall see that every regular sequence is of linear type. More generally, every d-sequence is of linear type.

Definition 5.5.2 *Let R be a commutative ring. Set $x_0 = 0$. A sequence of elements $x_1, \ldots, x_n$ is said to be a* **d-sequence** *if one (and hence both) of the following equivalent conditions hold:*

(1) $(x_0, \ldots, x_i) : x_{i+1} x_j = (x_0, \ldots, x_i) : x_j$ for all $0 \le i \le n - 1$ and for all $j \ge i + 1$.

(2) $(x_0, \ldots, x_i) : x_{i+1} \cap (x_1, \ldots, x_n) = (x_1, \ldots, x_i)$ for all $0 \le i \le n - 1$.

The equivalence of these two conditions is left to the reader (Exercise 5.23). The first condition was introduced in Huneke [135], and the second was introduced in Fiorentini [79]. The second definition is often more useful in practice, but the first definition is usually easier to check.

Example 5.5.3 Clearly any regular sequence is a d-sequence as well. A single element x is a d-sequence if and only if $0 : x^2 = 0 : x$. Let $R = k[x, y, u, v]/(xu - yv)$ with k an infinite field. The elements x, y form a d-sequence, but not a regular sequence. In the polynomial ring $R = k[x, y, z]$, the ideal generated by xy, xz, yz is generated by a d-sequence, namely any set of three general generators, but no rearrangement of the original three generators form a d-sequence. See Exercise 5.24 for more examples.

The main theorem of this section states that if I is generated by a d-sequence, then I is of linear type. This theorem was proved independently by Huneke [134] and Valla [301]. A more general statement, due to Raghavan [225], is given in Theorem 5.5.4.

Let $F \in R[T_1, \ldots, T_n]$. We define the weight of F to be i if $F \in (T_1, \ldots, T_i)$ but $F \notin (T_1, \ldots, T_{i-1})$. We set the weight to be 0 if $F = 0$.

Theorem 5.5.4 Let R be a ring, $x_1, \ldots, x_n$ a d-sequence in R, and $I = (x_1, \ldots, x_n)$. Let $\mathcal{A} \subseteq S = R[T_1, \ldots, T_n]$ be the defining ideal of the Rees algebra of I, and let $\mathcal{A}_1 \subseteq \mathcal{A}$ be the ideal in $\mathcal{A}$ generated by all homogeneous polynomials in $\mathcal{A}$ having degree 1. If $F(T_1, \ldots, T_n)$ is a form in S of degree d such that $F(x_1, \ldots, x_n) \in (x_1, \ldots, x_j)$, then there exists a form $G(T_1, \ldots, T_n)$ of degree d and weight at most j such that $F - G \in \mathcal{A}_1$.

Proof: Use induction on d. Suppose that $d = 1$. Since $F(x_1, \ldots, x_n) \in (x_1, \ldots, x_j)$, we may write $F(x_1, \ldots, x_n) = \sum_{i=1}^{j} r_i x_i$. Set $G = \sum_{i=1}^{j} r_i T_i$. Clearly the weight of G is at most j. Since $(F - G)(x_1, \ldots, x_n) = 0$ and since the degree of F is one, $F - G \in \mathcal{A}_1$.

Let $d > 1$, and assume the theorem is true for smaller values of d. Now use induction on the weight of F. If the weight of F is at most j, set $F = G$. If not, write $F = T_k F_1 + F_2$, where the weight of F is k and the weight of F_2 is at most $k - 1$, and both F_1 and F_2 are homogeneous. Note that $\deg(F_1) = d - 1$. We have that $F(x_1, \ldots, x_n) = x_k F_1(x_1, \ldots, x_n) + F_2(x_1, \ldots, x_n) \in (x_1, \ldots, x_j)$, and $F_2(x_1, \ldots, x_n) \in (x_1, \ldots, x_{k-1})$. Hence

$$F_1(x_1, \ldots, x_n) \in ((x_1, \ldots, x_{k-1}) : x_k) \cap I = (x_1, \ldots, x_{k-1}).$$

Apply induction to F_1; we obtain that there exists a homogeneous polynomial G_1 of degree $d - 1$ such that the weight of G_1 is at most $k - 1$ and such that $F_1 - G_1 \in \mathcal{A}_1$.

Set $G' = T_k G_1 + F_2$. The weight of G' is at most $k - 1$. Moreover, $F - G' = T_k(F_1 - G_1) \in \mathcal{A}_1$. We apply the induction to G'. Notice that $G'(x_1, \ldots, x_n) = F(x_1, \ldots, x_n) \in (x_1, \ldots, x_j)$ and G' has weight at most $k - 1$ and has degree d. By induction there exists a homogeneous polynomial

G of degree d and weight at most j such that $G - G' \in \mathcal{A}_1$. It follows that $F - G = (F - G') + (G' - G) \in \mathcal{A}_1$, finishing the proof. $\qquad\square$

Corollary 5.5.5 *Let R be a ring. If $x_1, \ldots, x_n$ is a d-sequence in R, then the ideal $I = (x_1, \ldots, x_n)$ is of linear type.*

Proof: Let $\mathcal{A}$ and $\mathcal{A}_1$ be as in the theorem above. If $F(T_1, \ldots, T_n) \in \mathcal{A}$ and F is homogeneous of degree d, then since $F(x_1, \ldots, x_n) = 0$ we can apply Theorem 5.5.4 with $j = 0$ to conclude that $F \in \mathcal{A}_1$. $\qquad\square$

Corollary 5.5.6 *Let R be a ring and $x_1, \ldots, x_n$ be a regular sequence. The defining ideal of the Rees algebra of $(x_1, \ldots, x_n)$ is generated by the 2×2 minors of the matrix*

$$\begin{pmatrix} x_1 & x_2 & \cdots & x_n \\ T_1 & T_2 & \cdots & T_n \end{pmatrix}.$$

Proof: Since a regular sequence is a d-sequence, by Corollary 5.5.5 the defining equations of the Rees algebra are linear, coming from a presentation matrix of $(x_1, \ldots, x_n)$. But the relations on $x_1 \ldots, x_n$ are generated by the Koszul relations, $x_i x_j - x_j x_i = 0$, which translate into the equations $x_i T_j - x_j T_i$ in $R[T_1, \ldots, T_n]$; these are exactly the 2×2 minors of the given matrix. $\qquad\square$

We can pass from the equations of the Rees algebra to not only the equations defining the extended Rees algebra, but also the equations of the various affine pieces of the blowup of the ideal I. The next two propositions detail this process.

Proposition 5.5.7 *Let R be a Noetherian ring, $x_1, \ldots, x_n \in R$, $T_1, \ldots, T_n, y$ variables over R. Set $S = R[T_1, \ldots, T_n]$, and write $R[x_1 t, \ldots, x_n t] \cong S/\mathcal{A}$, where the isomorphism identifies T_i with $x_i t$. Consider the induced surjective map $\varphi : S[y] \to R[x_1 t, \ldots, x_n t, t^{-1}]$ sending y to t^{-1} and T_i to $x_i t$. Then the kernel of φ is equal to $\mathcal{A}S[y] + (yT_i - x_i)$.*

Proof: We can make φ a graded map of degree 0 between $\mathbb{Z}$–graded rings by giving y degree -1 and T_i degree 1 for all $1 \leq i \leq n$. Let $T = S[y]/(\mathcal{A}S[y] + (yT_i - x_i))$. T clearly surjects onto $R[x_1 t, \ldots, x_n t, t^{-1}]$. Notice that $T/(y) \cong S/(\mathcal{A} + (x_1, \ldots, x_n)S) \cong R[x_1 t, \ldots, x_n t, t^{-1}]/(t^{-1})$, as these algebras are all isomorphic to the associated graded algebra of I. Let K denote the kernel of the homomorphism of T onto $R[x_1 t, \ldots, x_n t, t^{-1}]$. Tensor the exact sequence

$$0 \to K \to T \to R[x_1 t, \ldots, x_n t, t^{-1}] \to 0$$

of T–modules with T/yT. Since the image of y in $R[x_1 t, \ldots, x_n t, t^{-1}]$ is a regular element (namely t^{-1}), it follows that the sequence

$$0 \to K/yK \to T/yT \to R[x_1 t, \ldots, x_n t, t^{-1}]/(t^{-1}) \to 0$$

is exact. The right–hand terms are isomorphic, which proves that $K = yK$. We claim that $K = 0$. It suffices to prove that $K_M = 0$ for every maximal ideal M of T. If $y \in M$, then Nakayama's Lemma proves that $K_M = 0$. If $y \notin M$, it suffices to prove that $T_M \cong R[x_1t, \ldots, x_nt, t^{-1}]_M$. We prove the stronger statement that $T[y^{-1}] \cong R[x_1t, \ldots, x_nt, t^{-1}, t] = R[t, t^{-1}]$ via the map determined by φ. We first claim that $\mathcal{A}S[y, y^{-1}] \subseteq (yT_i - x_i)S[y, y^{-1}]$. This follows since if $F(T_1, \ldots, T_n)$ is homogeneous of degree d and $F \in \mathcal{A}$, then modulo the ideal $(yT_i - x_i)S[y, y^{-1}] = (T_i - y^{-1}x_i)S[y, y^{-1}]$, F is congruent to $F(y^{-1}x_1, \ldots, y^{-1}x_n) = y^{-d}F(x_1, \ldots, x_n) = 0$. Thus $T[y^{-1}] \cong S[y, y^{-1}]/(T_i - y^{-1}x_i) \cong R[y, y^{-1}] \cong R[t^{-1}, t]$. Hence $K = 0$. $\square$

Proposition 5.5.8 *Let R be a Noetherian ring, $x_1, \ldots, x_n \in R$, $T_1, \ldots, T_n$ variables over R. Write $R[x_1t, \ldots, x_nt] \cong R[T_1, \ldots, T_n]/\mathcal{A}$, where the isomorphism sends T_i to x_it. Then there is an isomorphism*

$$R\left[\frac{x_2}{x_1}, \ldots, \frac{x_n}{x_1}\right][x_1t, (x_1t)^{-1}] \cong (R[T_1, \ldots, T_n]/\mathcal{A})[T_1^{-1}],$$

which identifies $R[\frac{x_2}{x_1}, \ldots, \frac{x_n}{x_1}]$ as the component of $R[x_1t, \ldots, x_nt, (x_1t)^{-1}]$ of degree zero, where by $R[\frac{x_2}{x_1}, \ldots, \frac{x_n}{x_1}]$ we denote the natural image of R together with $\frac{x_i}{x_1}$ inside R_{x_1}. Moreover, writing $R[Y_2, \ldots, Y_n]/J \cong R[\frac{x_2}{x_1}, \ldots, \frac{x_n}{x_1}]$, where the isomorphism sends the variable Y_i to $\frac{x_i}{x_1}$, generators for J can be determined as follows: let $F_1(T_1, \ldots, T_n), \ldots, F_m(T_1, \ldots, T_n)$ be a homogeneous generating set for $\mathcal{A}$. Set $d_i = \deg F_i$, $Y_i = T_iT_1^{-1}$, and $f_i(Y_2, \ldots, Y_n) = T_1^{-d_i}F_i$. Then $J = (f_1, \ldots, f_m)$.

Proof: Note that $(R[T_1, \ldots, T_n]/\mathcal{A})[T_1^{-1}] \cong R[x_1t, \ldots, x_nt, (x_1t)^{-1}]$. This latter ring is equal to $R[\frac{x_2}{x_1}, \ldots, \frac{x_n}{x_1}][x_1t, (x_1t)^{-1}]$. The subring $R[\frac{x_2}{x_1}, \ldots, \frac{x_n}{x_1}]$ is exactly the subring of elements of degree 0 in $R[\frac{x_2}{x_1}, \ldots, \frac{x_n}{x_1}][x_1t, (x_1t)^{-1}]$.

We first prove that $f_i \in J$. We need to prove that $f_i(\frac{x_2}{x_1}, \ldots, \frac{x_n}{x_1}) = 0$. It suffices to prove that $x_1^{d_i}f_i(\frac{x_2}{x_1}, \ldots, \frac{x_n}{x_1}) = 0$. However, $x_1^{d_i}f_i(\frac{x_2}{x_1}, \ldots, \frac{x_n}{x_1}) = F_i(x_1, \ldots, x_n) = 0$.

Now let $f \in J$. Set $d = \deg f$. Define $F(T_1, \ldots, T_n) = T_1^d f(\frac{T_2}{T_1}, \ldots, \frac{T_n}{T_1})$. Then $F(x_1, \ldots, x_n) = x_1^d f(\frac{x_2}{x_1}, \ldots, \frac{x_n}{x_1}) = 0$, so that F is a homogeneous polynomial in $\mathcal{A}$. By assumption there are homogeneous polynomials $G_1, \ldots, G_m$ in $R[T_1, \ldots, T_n]$ such that $F = \sum_{i=1}^m G_iF_i$. By counting degrees, we see that $\deg G_i = d - d_i$. Hence

$$f(Y_2, \ldots, Y_n) = T_1^{-d}F(T_1, \ldots, T_n) = \sum_{i=1}^m T_1^{d_i-d}G_i(T_1, \ldots, T_n)f_i(Y_2, \ldots, Y_n)$$

and then letting $g_i(Y_2, \ldots, Y_n) = T_1^{d_i-d}G_i(T_1, \ldots, T_n)$ we have that $f = \sum_{i=1}^m g_if_i$. $\square$

Finally we apply these propositions to the case of a regular sequence:

Corollary 5.5.9 *Let R be a Noetherian ring and let $x_1, \ldots, x_n$ be a regular sequence in R. There are isomorphisms*

$$R[x_1t, \ldots, x_nt, t^{-1}] \cong R[T_1, \ldots, T_n, y]/(yT_i - x_i),$$

which sends T_i to $x_i t$ and y to t^{-1}, and

$$R[y_2, \ldots, y_n]/(x_1 y_j - x_j) \cong R\Big[\frac{x_2}{x_1}, \ldots, \frac{x_n}{x_1}\Big],$$

sending y_j to $\frac{x_j}{x_1}$. In particular, $R[\frac{x_2}{x_1}, \ldots, \frac{x_n}{x_1}]/(x_1)$ is isomorphic to the polynomial ring $(R/(x_1, \ldots, x_n))[y_2, \ldots, y_n]$ over $R/(x_1, \ldots, x_n)$. In addition, the associated graded ring of $J = (x_1, \ldots, x_n)$ is a polynomial ring in n variables over the ring R/J.

Proof: The proof is immediate from Propositions 5.5.7 and 5.5.8 and from Corollary 5.5.6. □

5.6. Blowing up

In this section we summarize some of the main results concerning blowing up ideals. We refer to [119] for a fuller treatment. There is no doubt that blowing up is one of the key operations in birational algebraic geometry. While it often can be replaced by considerations involving Rees algebras, this can lead to awkward and difficult issues. On the other hand, without knowledge of resolution of singularities, blowing up loses much of its power. We have chosen to keep this book largely self–contained, and will not use resolution of singularities as a tool. But certainly deeper studies of integral closure often rely both on blowing up and on resolution of singularities.

Definition 5.6.1 *Let $R = \oplus_{i \geq 0} R_i$ be a graded ring, and set $R_+ = \oplus_{i > 0} R_i$, the so–called* **irrelevant ideal**. *We define*

$$\mathrm{Proj}(R) = \{homogeneous\ primes\ P\,|\ R_+ \not\subseteq P\}.$$

We give $\mathrm{Proj}(R)$ a topological structure by choosing a basis of open sets to be $D_+(f) = \{P \in \mathrm{Proj}(R)\,|\ f \notin P\}$, where f is a homogeneous element of positive degree. We define a scheme structure on $X = \mathrm{Proj}(R)$ by letting $\mathfrak{O}_X(D_+(f)) = [R_f]_0$, the degree zero part of the localization of R at f, where f is homogeneous of positive degree.

We refer the reader to standard books such as [109] or [67] for information about schemes. Suffice it to say here that the scheme structure on $\mathrm{Proj}(R)$ consists of a collection of rings that paste together in a natural way, e.g., if R is generated by R_1 over R_0 and $a_1, \ldots, a_n$ generate R_1, then $[R_{a_i}]_0 = R_0[\frac{a_1}{a_i}, \ldots, \frac{a_n}{a_i}]$ and $\mathrm{Proj}(R)$ consists of these rings pasted along their natural overlaps.

Definition 5.6.2 *Let R be a ring and I an ideal. We give the natural grading to the Rees algebra, $R[It]$, by setting $\deg(t) = 1$. The* **blowup** *of I is by definition $\mathrm{Proj}(R[It])$, which we denote as $\mathrm{Bl}_I(R)$.*

When $I = (a_1, \ldots, a_n)$, $\mathrm{Bl}_I(R)$ is covered by $\mathrm{Spec}(R[\frac{I}{a_i}])$, $i = 1, \ldots, n$, a so–called affine covering of $\mathrm{Bl}_I(R)$.

We can use this covering to compute the Čech cohomology of $\mathrm{Bl}_I(R)$, which in turn is the sheaf cohomology. Specifically, set $X = \text{Bl}_I(R)$. The ith cohomology $H^i(X, \mathcal{O}_X)$ is the ith cohomology of the complex

$$0 \to \oplus_i R\Big[\frac{I}{a_i}\Big] \to \oplus_{i<j} R\Big[\frac{I^2}{a_i a_j}\Big] \to \cdots \to R\Big[\frac{I^n}{a_1 \cdots a_n}\Big] \to 0,$$

where the maps are induced from the natural maps coming from degree 0 parts of localizations at products of the elements a_i with signs induced from the Koszul cohomology. The first module $\oplus_i R[\frac{I}{a_i}]$ is the 0th component in the complex.

There is an important exact sequence, discovered by Sancho de Salas [257], which relates various cohomologies of the blowup with local cohomology. This sequence was developed further in [188], and used, for example, in [144] and [143]. Let $R = \oplus_{i \geq 0} R_i$ be a Noetherian graded ring, let I be an ideal of R_0, and let $M = \oplus_{i \in \mathbb{Z}} M_i$ be a graded R–module. Set $X = \text{Proj}(R)$, $E = X \times_{\text{Spec}(R)} \text{Spec}(R/I)$, $\mathfrak{m} = IR + R_{>0}$, and let $\mathcal{M}_n$ denote the quasi–coherent sheaf on X associated to the graded R–module $M(n)$ (where $M(n)_m = M_{m+n}$). The Sancho de Salas sequence is:

$$\cdots \to H^i_{\mathfrak{m}}(M) \to \oplus_{n \in \mathbb{Z}} H^i_I(M_n) \to \oplus_{n \in \mathbb{Z}} H^i_E(X, \mathcal{M}_n) \to H^{i+1}_{\mathfrak{m}}(M) \to \cdots.$$

This sequence has been especially useful to study the Cohen–Macaulay property of the Rees algebra of an ideal.

5.7. Exercises

5.1 Let I and J be ideals in a Noetherian ring R. Prove that $JR[t, t^{-1}] \cap R[It, t^{-1}] = \oplus_{i \in \mathbb{Z}} (J \cap I^i) t^i$ and $JR[t, t^{-1}] \cap R[It] = \oplus_{i \in \mathbb{N}} (J \cap I^i) t^i$.

5.2 Prove or disprove:
   (i) For ideals I and J in a Noetherian ring R, the Rees algebra of the image of I in R/J is $R[It]/JR[It]$.
   (ii) For ideals I and J in a Noetherian ring R, the extended Rees algebra of the image of I in R/J is $R[It, t^{-1}]/JR[It, t^{-1}]$.

5.3 Let R be a Noetherian ring and I, J ideals in R.
   (i) Prove that $JR[t, t^{-1}] \cap R[It, t^{-1}] + t^{-1} R[It, t^{-1}]$ is an ideal in $R[It, t^{-1}]$ of height at least $\text{ht } J + 1$.
   (ii) Prove that if R is local and equidimensional, then $JR[t, t^{-1}] \cap R[It, t^{-1}] + t^{-1} R[It, t^{-1}]$ is an ideal in $R[It, t^{-1}]$ of height exactly $\text{ht } J + 1$.

5.4 Let R be a Noetherian ring, I, J ideals in R. Prove that $JR[t, t^{-1}] \cap R[It, t^{-1}]$ and $JR[t, t^{-1}] \cap R[It]$ are ideals of height at least $\text{ht } J$. Prove that if J is a prime ideal, then the two ideals have height $\text{ht } J$ if and only if $I \not\subseteq J$. Formulate a similar statement for heights of $JR[t, t^{-1}] \cap S$, where S is either the integral closure of $R[It]$ or of $R[It, t^{-1}]$ in $R[t, t^{-1}]$.

5.5 Prove that for any ideal I in a Noetherian ring R, $\dim \mathrm{gr}_I(R) = \sup\{\mathrm{ht}\, P \mid P \in \mathrm{Spec}\, R \text{ such that } I \subseteq P\}$.

5.6 Let I be an ideal such that the Rees algebra $R[It]$ is integrally closed in $R[t]$. Prove that for any $n \in \mathbb{N}$, $R[I^n t]$ is integrally closed in $R[t]$.

5.7 Let I be an ideal in a Noetherian ring R such that $\mathrm{gr}_I(R)$ is a reduced ring. Prove that for all n, I^n is integrally closed. In other words, prove that I is normal.

5.8 Let I be an ideal in a Noetherian ring R such that $\mathrm{gr}_I(R)$ is a reduced ring. Prove that $R[It]$ is integrally closed in $R[t]$ and that $R[It, t^{-1}]$ is integrally closed in $R[t, t^{-1}]$.

5.9 Let I be an ideal in a Noetherian ring R satisfying $\cap_n I^n = 0$ such that $\mathrm{gr}_I(R)$ is reduced, a domain, or respectively an integrally closed domain. Prove that R has the corresponding property.

5.10 (The blowup of a blowup is a blowup) Let R be a Noetherian domain, let I be an ideal in R and x a non–zero element in I. Let $S = R[\frac{I}{x}]$. Let J be an ideal in S and y a non–zero element in J. Prove that there exist an ideal K in R and a non–zero element $z \in K$ such that $R[\frac{K}{z}] = R[\frac{I}{x}][\frac{J}{y}]$.

5.11 Let $I = (a_1, \ldots, a_n)$ be an ideal in R, $X_1, \ldots, X_n$ variables over R, and $\varphi : R[X_1, \ldots, X_n] \to R[It]$ the R–algebra map with $\varphi(X_i) = a_i t$ (as in Section 5.5). Clearly φ is graded and surjective. For $n \geq 0$, set $\mathcal{A}_n$ to be the ideal generated by $\{r \in \ker(\varphi) \mid \deg r \leq n\}$, where the degree is the $(X_1, \ldots, X_n)$–degree.

   (i) Prove that $\mathcal{A}_0 \subseteq \mathcal{A}_1 \subseteq \mathcal{A}_2 \subseteq \cdots$, and that $\cup_n \mathcal{A}_n = \ker(\varphi)$. The **relation type** of I is the least integer n such that $\mathcal{A}_n = \ker(\varphi)$.

   (ii) The ideal I is said to be of **quadratic type** if $\mathcal{A}_2 = \ker(\varphi)$. Let $I = (X^2, XY, Y^2) \subseteq k[X, Y]$ (polynomial ring in variables X and Y over a field k). Prove that I is of quadratic type and not of linear type. (In fact, if I is an ideal in a Noetherian local ring, then for all large n, the relation type of I^n is two. See [317].)

5.12 Let R be a Noetherian domain and I a non–zero ideal in R. Prove that the following are equivalent:

   (i) The symmetric algebra of I is isomorphic to $R[It]$, i.e., I is of linear type.

   (ii) The symmetric algebra of I is an integral domain.

   (iii) The symmetric algebra of I has no R–torsion.

5.13 (Equation of integral dependence) Let R be a Noetherian ring, let I be an ideal in R, and let x be an element of R. Prove that $x \in \bar{I}$ if and only if there exist an integer n and elements $r_i \in \overline{I^i}$ such that $x^n + r_1 x^{n-1} + \cdots + r_n = 0$.

5.14 Let R be a Noetherian ring and I an ideal. Prove that $\cap_n \overline{I^n} = \cap_P P$, where P varies over those minimal prime ideals of R for which $I + P \neq R$.

5.15   Let R be a Noetherian ring and I an ideal in R containing a non–zerodivisor such that $I^n = \overline{I^n}$ for infinitely many n. Prove that $I^n = \overline{I^n}$ for all large n.

5.16   (Katz [157]; Schenzel [260]; Verma [308], [310]) Let R be a Noetherian ring. For any ideal I and any positive integer n define $I^{(n)}$ to be the intersection of the minimal components of I^n. Prove that the following are equivalent:

(i) R is locally formally equidimensional.

(ii) For every ideal I in R satisfying $\mu(I) = \operatorname{ht}(I)$, there exists an integer k such that for all $n \in \mathbb{N}$, $I^{(n+k)} \subseteq I^n$.

(iii) For every ideal I in R satisfying that $\mu(I) = \operatorname{ht}(I)$ and for every $n \in \mathbb{N}_{>0}$ there exists $k \in \mathbb{N}_{>0}$ such that $I^{(k)} \subseteq I^n$.

(iv) For any prime ideal P that is associated to all high powers of an ideal I but is not minimal over I, and for any prime ideal $q \in \operatorname{Ass}(\widehat{R_P})$, $\ell(I(\widehat{R_P})/q) < \dim((\widehat{R_P})/q)$.

(v) Let S be the set of prime ideals P containing I that satisfy the property that for some prime ideal q that is associated to $(\widehat{R_P})$, the quotient $I(\widehat{R_P})/q$ is primary to $P(\widehat{R_P})/q$. (Such primes are called **essential divisors** of I.) Then S is the set of minimal prime ideals over I.

5.17   (See [203].) Let $(R, \mathfrak{m})$ be a Noetherian ring. A sequence $x_1, \ldots, x_s$ in R is said to be an **asymptotic sequence** if $(x_1, \ldots, x_s) \neq R$ and if for all $i = 1, \ldots, s$, x_i is not in any associated prime of $\overline{(x_1, \ldots, x_i)^n}$, where n is very large.

(i) Prove that this definition does not depend on n.

(ii) Prove that every regular sequence is an asymptotic sequence.

(iii) Prove that $x_1, \ldots, x_s$ is an asymptotic sequence in R if and only if it is an asymptotic sequence in the $\mathfrak{m}$–adic completion of R.

(iv) Prove that the asymptotic sequence $x_1, \ldots, x_s$ cannot be prolonged to a strictly longer asymptotic sequence if and only if $\mathfrak{m} \in \operatorname{Ass}(R/\overline{(x_1, \ldots, x_s)})$.

(v) Prove that $x_1, \ldots, x_s$ is an asymptotic sequence in R if and only if for every minimal prime ideal p in R, $x_1 + p, \ldots, x_s + p$ is an asymptotic sequence in R/p.

5.18   Let $(R, \mathfrak{m})$ be a complete local Noetherian domain, and $x_1, \ldots, x_s \in \mathfrak{m}$. Prove that $x_1, \ldots, x_s$ is an asymptotic sequence if and only if for all $i = 1, \ldots, s$, $\operatorname{ht}(x_1, \ldots, x_i) = i$.

5.19   Prove that a Noetherian local ring is formally equidimensional if and only if every system of parameters is an asymptotic sequence.

5.20   Let $(R, \mathfrak{m})$ be a Noetherian local domain of positive dimension. Assume that there exists an integer n such that whenever a non–zero ideal I in R is contained in $\overline{\mathfrak{m}^n}$, then $\mathfrak{m}$ is associated to I. Prove that the integral closure of R contains a maximal ideal of height 1.

5.21   Let R be a Noetherian domain and I an ideal. Prove that if $P \in$ Ass$(R/\overline{I^n})$ for some large n, then there exist $a \in I$ and a prime ideal Q in $S = R[\frac{I}{a}]$ such that Q is associated to $\overline{a^n S}$ for all large n and such that $Q \cap R = P$.

5.22   Let R be a Noetherian ring, $I = (x_1, \ldots, x_n)$ an ideal, and $T_1, \ldots, T_n$ variables over R. Define $\pi : R[T_1, \ldots, T_n] \to R[It]$ to be the R–algebra homomorphism with $\pi(T_i) = x_i t$. Let $\mathcal{A}_1$ be the ideal in $R[T_1, \ldots, T_n]$ generated by all homogeneous elements of total $(T_1, \ldots, T_n)$–degree 1 that are in the kernel of π. Let $R^m \xrightarrow{A} R^n \to I \to 0$ be a presentation of I, i.e., an exact complex, where A is an $n \times m$ matrix. If T is the $1 \times n$ matrix $[T_1, \ldots, T_n]$, let L be the ideal in $R[T_1, \ldots, T_n]$ generated by the entries of the matrix TA. Prove that $\mathcal{A}_1 = L$.

5.23   Prove that the two conditions of Definition 5.5.2 are equivalent.

5.24   Let R be a Noetherian ring, and let P be a prime in R of height g. Assume that there is a regular sequence of length g in P, that R_P is regular, and that P has $g + 1$ generators. Prove that P is generated by a d-sequence.

5.25   Let R be a ring and $x_1, \ldots, x_n$ a d-sequence. Prove that the elements $x_1 t, \ldots, x_n t$ in the Rees algebra $R[x_1 t, \ldots, x_n t]$ form a d-sequence.

5.26   Prove that the ideal generated by the 2×2 minors of a generic $n \times m$ matrix over $\mathbb{Z}$ is a prime ideal by using Exercises 5.22, 5.25 and Theorem 5.5.4 repeatedly.

5.27   (Huckaba and Huneke [128, Theorem 3.11]) Let k be a field of characteristic not three, let $R = k[x, y, z]$, and let $I = (x^4, x(y^3 + z^3), y(y^3 + z^3), z(y^3 + z^3)) + (x, y, z)^5$. Prove that I is a height three normal ideal, that $\text{gr}_{I^n}(R)$ is not Cohen–Macaulay for any $n \geq 1$, and if X is the blowup of I, that then X is normal but $H^2(X, \mathcal{O}_X) \neq 0$. (Hint: use Exercise 1.17.)

6
Valuations

6.1. Valuations

This chapter presents the theory of valuations, which besides being a major topic in itself, is also an important tool in the study of the integral closure of ideals. A classic reference for this material is Chapter VI of Zariski and Samuel's book [322]. We owe much to their presentation and make no pretense of covering this material as thoroughly or as well in this book. Zariski's interest in valuation theory was motivated by resolution of singularities. In [321] he gave a classification of valuations in a two–dimensional algebraic function field. He used this to prove local uniformization for surface singularities. See Cutkosky's book [53] for a modern treatment. For much more on classical valuations a reader may consult [73], [245] or [197]. For a history of valuations from its beginning with the 1912 paper of Josef Kürschák see [249]. For valuation theory from a constructive point of view see for example [206].

In most of this book, the emphasis is on Noetherian rings, but we treat more general rings in this chapter. A reader may want to read this chapter by concentrating only on the Noetherian valuation rings, that is, on the valuation rings with value group $\mathbb{Z}$.

Definition 6.1.1 *Let K be a field. A **valuation** on K (or a K–**valuation**) is a group homomorphism v from the multiplicative group $K^* = K \setminus \{0\}$ to a totally ordered abelian group G (written additively) such that for all x and y in K,*

$$v(x + y) \geq \min\{v(x), v(y)\}.$$

It follows immediately from the properties of group homomorphisms that $v(1) = 0$, and that for all $x \in K \setminus \{0\}$, $v(x^{-1}) = -v(x)$.

When R is a domain with field of fractions K and G is a totally ordered abelian group, then a function $v : R \setminus \{0\} \to G$ satisfying the properties

$$v(xy) = v(x) + v(y), \qquad v(x + y) \geq \min\{v(x), v(y)\}$$

for all $x, y \in R$ can be extended uniquely to a valuation $v : K \setminus \{0\} \to G$ by setting $v(\frac{x}{y}) = v(x) - v(y)$ for any non–zero $x, y \in R$. It is easy to verify that this is well–defined and yields a valuation on K. For this reason we also sometimes call such a "partial" function $v : R \setminus \{0\} \to G$ a **valuation**.

Sometimes we write $v : K \to G \cup \{\infty\}$ by assigning $v(0) = \infty$, where $G \cup \{\infty\}$ is totally ordered and extends the structure of G via the relations $\infty + g = g + \infty = \infty + \infty = \infty$, and $g < \infty$ for all $g \in G$.

Remark 6.1.2   Let K be a field, $x_1, \ldots, x_n \in K$, and v a K–valuation. Then $v(\sum_{i=1}^{n} x_i) \geq \min\{v(x_1), \ldots, v(x_n)\}$. If $v(x_i)$ are all distinct, then $v(\sum_{i=1}^{n} x_i) = \min\{v(x_i)\}$.

The first statement follows from the definition by an easy induction. The second statement we prove by induction on n. It suffices to prove the equality in the case $n = 2$. In this case we relabel $x = x_1, y = x_2$. Without loss of generality we may assume that $v(x) > v(y)$. Suppose that $v(x + y) > v(y)$ as well. Then $v(y) = v(-y) = v(x - (x + y)) \geq \min\{v(x), v(x + y)\}$, which is a contradiction.

Remark 6.1.3   (This is an example of a **Gauss extension**.) Let K be a field and X a variable over K. Any K–valuation v can be extended to $K(X)$ as follows: whenever $a_0, \ldots, a_n \in K$, define $w'(a_0 + a_1 X + \cdots + a_n X^n) = \min\{v(a_0), \ldots, v(a_n)\}$. This gives a function w' on $K[X]$ that satisfies the properties $w'(fg) = w'(f) + w'(g)$ and $w'(f + g) \geq \min\{w'(f), w'(g)\}$. Hence w' extends to a valuation w on $K(X)$. In case R is a domain with field of fractions K such that v is non–negative on R, then w is non–negative on $R[X]$.

The simplest valuations to define and work with are the so–called monomial valuations:

Definition 6.1.4 *A valuation v on the field of fractions of the polynomial ring $k[X_1, \ldots, X_d]$ over a field k is said to be* **monomial** *with respect to $X_1, \ldots, X_d$ if for any polynomial f, $v(f)$ equals the minimum of all $v(\underline{X}^\nu)$ as $\underline{X}^\nu$ varies over all the monomials appearing in f with a non–zero coefficient.*

By uniqueness of the representation of polynomials, a monomial valuation v is well–defined on the polynomial ring by specifying the v–values of the variables, and it extends uniquely to the field of fractions. Note that a monomial valuation v is determined uniquely by $v(X_1), \ldots, v(X_d)$.

Some particular examples of such valuations are :

Example 6.1.5   Let k be a field, X and Y variables over k, and $K = k(X, Y)$. Let v be a monomial valuation on K defined by $v(X) = 1$, $v(Y) = 1/2$. Observe that every element of the subfield $k(\frac{X}{Y^2})$ has value 0.

Example 6.1.6   Let k be a field, X and Y variables over k, and $K = k(X, Y)$. Let v be a monomial valuation on K defined by $v(X) = \sqrt{2}$, $v(Y) = 1$. Note that a monomial $X^i Y^j$, $i, j \in \mathbb{Z}$ has value at least n if and only if $i\sqrt{2} + j \geq n$. The elements of value 0 are exactly the non–zero elements of k.

Non–monomial valuations exist:

Example 6.1.7   Let $R = k[X, Y]$, where k is a field and X, Y variables over k. Let $e(X)$ be an element of $Xk[[X]]$ that is transcendental over $k[X]$. (Note by Exercise 3.13, $k[[X]]$ has infinite transcendence degree over $k[X]$.) Write $e = \sum_{i \geq 1} e_i X^i$. Define $v : R \setminus \{0\} \to \mathbb{Z}$ by

$$f(X, Y) \mapsto \max\{n \,|\, f(X, e(X)) \in X^n k[[X]]\}.$$

Then v is a valuation on R, so that its extension to the field of fractions of R (also denoted v) is a valuation. However, v is not a monomial valuation with respect to X, Y. To prove this, let n be the smallest positive integer such that $e_n \neq 0$. Then $v(Y) = n$, $v(e_n X^n) = n$, and $v(Y - e_n X^n) > n$. Thus this v is not a monomial valuation.

More generally, without the restriction that v be a monomial valuation, there may be infinitely many valuations on the function field $k(X_1, \ldots, X_n)$ with given values for the variables (see Exercise 6.12).

There is a natural way to identify some valuations:

Definition 6.1.8 *Let K be a field. We say that valuations $v : K^* \to G_v$ and $w : K^* \to G_w$ are* **equivalent** *if there exists an order–preserving isomorphism $\varphi : image(v) \to image(w)$ such that for all $\alpha \in K^*$, $\varphi(v(\alpha)) = w(\alpha)$.*

Observe that a given valuation $v : K^* \to G$ is trivially equivalent to the natural valuation $v : K^* \to image(v)$. Also, any valuation $v : K^* \to \mathbb{Z}$ is equivalent to $2v : K^* \to \mathbb{Z}$.

6.2. Value groups and valuation rings

Going hand–in–hand with valuations are valuation rings and value groups:

Definition 6.2.1 *Let K be a field and v a K–valuation. Then the image $\Gamma_v = v(K^*)$ of v is a totally ordered abelian group, called the* **value group** *of v.*

Definition 6.2.2 *Let K be a field. A* **K-valuation ring**, *or simply a* **valuation ring** *or a* **valuation domain**, *is an integral domain V whose field of fractions is K that satisfies the property that for every non–zero element $x \in K$, either $x \in V$ or $x^{-1} \in V$.*

The set of ideals in a valuation domain V is totally ordered by inclusion. Namely, if I and J are ideals in V and $x \in I \setminus J$, then for each non–zero $y \in J$, either $xy^{-1} \in V$ or $yx^{-1} \in V$. If $xy^{-1} \in V$ then $x = (xy^{-1})y \in J$, which is a contradiction. Thus for all $y \in J$, $yx^{-1} \in V$, so that $y = (yx^{-1})x \in I$ and hence $J \subseteq I$.

It follows that a valuation ring V has a unique maximal ideal, which is the ideal of all non–units: $\{x \in V \mid x = 0 \text{ or } x^{-1} \notin V\}$. The maximal ideal is usually denoted as $\mathfrak{m}_V$.

One can construct a valuation domain from a valuation: given a valuation $v : K^* \to G$, define

$$R_v = \{r \in K^* \mid v(r) \geq 0\} \cup \{0\}.$$

It is easy to see that R_v is a subring of K with a unique maximal ideal $\mathfrak{m}_v = \{r \in K^* \mid v(r) > 0\}$ and is a valuation domain. We call R_v the valuation ring corresponding to the valuation v. We denote the **residue field** of R_v by $\kappa(v)$ or by $\kappa(V)$ with $V = R_v$.

If v and w are equivalent valuations, then clearly $R_v = R_w$.

Proposition 6.2.3 *Let V be a valuation domain V with field of fractions K. Let $\Gamma_V = K^*/V^*$, where $V^* \subseteq K^*$ are the multiplicative groups of units, and let $v : K^* \to \Gamma_V$ be the natural group homomorphism. By convention the operation on Γ_V is written as $+$. Then Γ_V is a totally ordered abelian group, v is a K–valuation, and Γ_V is the value group of v.*

Proof: As K^* is abelian under multiplication, so is Γ_V. We order Γ_V as follows. Let $x, y \in K$ such that the image of x in Γ_V is α, and the image of y in Γ_V is β. Then define $\alpha \leq \beta$ if $yx^{-1} \in V$. We leave it to the reader to prove that this order makes Γ_V a totally ordered abelian group whose group structure is compatible with the order.

Now we prove that v is a K–valuation. Certainly $v(xy)$ equals the image of x times the image of y, which in additive notation of Γ_V is written as $v(x) + v(y)$. To prove the second property of valuations, observe that either $xy^{-1} \in V$ or $x^{-1}y \in V$. Assume that xy^{-1} is in V. Then $(x + y)/y \in V$ so that $v(x + y) \geq v(y) \geq \min\{v(x), v(y)\}$. The case where $x^{-1}y \in V$ is handled similarly.

As v is surjective, the value group of v is exactly Γ_V. $\qquad\square$

In particular, tracing the argument above shows that if K is a field and v is a K–valuation, the valuation obtained as in the proposition above from the valuation ring R_v of v is equivalent to v. Furthermore, if V is a K–valuation ring and v is the valuation obtained from V as in the proposition above, then the valuation ring of v is exactly V.

Thus K–valuation rings and equivalence classes of K–valuations are in natural one–to–one correspondence. In the sequel, if the valuation v corresponds to a valuation ring V, then the value group of V is also called the **value group** of v, and is denoted Γ_V or by Γ_v.

Of special interest are the real–valued valuations. These are characterized by having the Archimedean property.

Definition 6.2.4 *Let Γ be a totally ordered abelian group. We say that Γ is* **Archimedean** *if for any elements $g, h \in \Gamma$ such that $g > 0$, there exists a positive integer n such that $ng > h$.*

Theorem 6.2.5 (Hölder) *Let Γ be a totally ordered abelian group that is Archimedean. Then Γ is isomorphic to a subgroup of $\mathbb{R}$.*

Proof: For $r = \frac{n}{m} \in \mathbb{Q}$, we write $ra < b$ if $na < mb$. This does not depend on the representation of r as a quotient of integers.

Let a be a fixed positive element in Γ (i.e., $a > 0$). For $b \in \Gamma$ positive, set $S_b = \{r \in \mathbb{Q} \,|\, ra \leq b\}$. Since there is $n \in \mathbb{N}_{>0}$ with $b < na$, n is an upper bound for S_b. Define $\varphi : \Gamma \to \mathbb{R}$ by $\varphi(0) = 0$, for $b > 0$, $\varphi(b) = \sup(S_b)$, and $\varphi(-b) = -\varphi(b)$. To show that φ is a homomorphism, it is enough to prove that $\varphi(b + c) = \varphi(b) + \varphi(c)$ for $b, c > 0$. Set $x = \varphi(b)$ and $y = \varphi(c)$, and $z = \varphi(b + c)$. Then $x + y \leq z$, since if $r, s \in \mathbb{Q}$ with $r \leq x$ and $s \leq y$, then

$ra \leq b$ and $sa \leq c$, so $(r+s)a \leq b+c$, yielding $r+s \leq z$. Next, suppose that $x + y < z$. Then there exist rational numbers r and s such that $x < r$, $y < s$ and $r + s < z$. From these inequalities we see that $(r + s)a \leq b + c$ but that $ra > b$ and $sa > c$. This is impossible. Therefore, $x + y = z$, proving that φ is a homomorphism.

We next show that φ preserves inequalities, and so it is one–to–one. It is enough to prove that if $b > 0$, then $\varphi(b) > 0$. To see this, there is $m \in \mathbb{N}_{>0}$ with $a < mb$, so $\frac{1}{m} \in S_b$. Therefore, $\frac{1}{m} \leq \varphi(b)$, and hence $\varphi(b) > 0$, as desired.

It is worth remarking that in this embedding, $\varphi(a) = 1$. The element a can be chosen to be an arbitrary positive element of Γ. □

A special subset of the real–valued valuations are the integer–valued valuations, and they play a major role in the theory of integral closure of ideals in Noetherian rings (see subsequent sections, especially 6.8, and Chapter 10).

6.3. Existence of valuation rings

In this section we prove the existence of many valuation rings. In particular, we prove that every prime ideal in an integral domain is a contraction of the maximal ideal of some valuation overring. When the domain is Noetherian, the valuation overring may be taken to be Noetherian as well.

The following lemma first appeared in Cohen and Seidenberg [41].

Lemma 6.3.1   *Let R be a domain with field of fractions K. Let $\mathfrak{m}$ be a prime ideal of R. For all $x \in K$, either $\mathfrak{m}R[x] \neq R[x]$ or $\mathfrak{m}R[x^{-1}] \neq R[x^{-1}]$.*

Proof: The hypotheses and conclusion are not affected if we first localize at the multiplicatively closed set $R \setminus \mathfrak{m}$. Thus we may assume that $\mathfrak{m}$ is the unique maximal ideal of R. Suppose that $\mathfrak{m}R[x^{-1}] = R[x^{-1}]$. Then $1 = a_0 + a_1 x^{-1} + a_2 x^{-2} + \cdots + a_n x^{-n}$ for some $a_i \in \mathfrak{m}$, hence $(1 - a_0)x^n = a_1 x^{n-1} + a_2 x^{n-2} + \cdots + a_n$. As $a_0 \in \mathfrak{m}$, $1 - a_0$ is a unit in R, so that x is integral over R. Thus $R[x]$ is an integral extension of R, and by Lying–Over (Theorem 2.2.2), $\mathfrak{m}R[x] \neq R[x]$. □

In the proof above, one can avoid the usage of the Lying–Over Theorem by instead assuming for contradiction that $\mathfrak{m}R[x] = R[x]$. Then for some $c_1, \ldots, c_m \in \mathfrak{m}R$, $1 = c_m x^m + c_{m-1}x^{m-1} + \cdots + c_0$. We may assume that both m and n are chosen to be minimal, and that by possibly switching x and x^{-1}, $n \leq m$. But then substituting $x^n = (1 - a_0)^{-1}(a_1 x^{n-1} + \cdots + a_n)$ into this equation produces a strictly smaller m, which gives the desired contradiction.

Theorem 6.3.2   (Existence of valuation domains) *Let R be an integral domain, not necessarily Noetherian, and let P be a non–zero prime ideal in R. Then there exists a valuation domain V between R and the field of fractions K of R such that $\mathfrak{m}_V \cap R = P$.*

Proof: By localizing at P we may assume that R is local with maximal ideal $\mathfrak{m}$.

Let Σ be the collection of all local rings $(S, \mathfrak{m}_S)$ such that $R \subseteq S$, $\mathfrak{m}S \subseteq \mathfrak{m}_S$, and $S \subseteq K$. The set Σ is not empty as it contains $(R, \mathfrak{m})$. We put a partial order $\leq$ on Σ: $(S, \mathfrak{m}_S) \leq (S', \mathfrak{m}_{S'})$ if $S \subseteq S'$ and $\mathfrak{m}_S S' \subseteq \mathfrak{m}_{S'}$. It is easy to prove that every ascending chain in Σ has an upper bound, so that by Zorn's lemma, Σ has a maximal element $(V, \mathfrak{m}_V)$.

By construction, $\mathfrak{m}_V \cap R = \mathfrak{m}$. We claim that V is a valuation domain.

Let $x \in K$. By Lemma 6.3.1, $\mathfrak{m}_V$ stays a proper ideal either in $V[x]$ or in $V[x^{-1}]$. Say $\mathfrak{m}_V V[x] \neq V[x]$. Let M be a maximal ideal in $V[x]$ containing $\mathfrak{m}_V V[x]$. Set $S = V[x]_M$ and $\mathfrak{m}_S = MS$. Then $(S, \mathfrak{m}_S)$ is an element of Σ, containing V, so by the choice of V it has to equal V. Hence $x \in S = V$. This proves that V is a valuation domain whose maximal ideal contracts to $\mathfrak{m}$ in R. $\square$

Moreover, when the starting ring is Noetherian, the valuation may be chosen to be Noetherian:

Theorem 6.3.3 (Existence of Noetherian valuation domains) *Let R be a Noetherian integral domain and let P be a non–zero prime ideal in R. Then there exists a Noetherian valuation domain V between R and the field of fractions K of R such that if $\mathfrak{m}_V$ is the maximal ideal of V, then $\mathfrak{m}_V \cap R = P$.*

Proof: By localization we may assume that P is the unique maximal ideal of R. Let $G = \mathrm{gr}_P(R)$. If every element of $P/P^2 \subseteq G$ is nilpotent, then clearly the dimension of G is zero. Hence by Proposition 5.1.6, $\dim R = 0$. As R is a domain, this implies that R is a field, contradicting the assumption that R has a non–zero prime ideal. Thus not every element of $P/P^2 \subseteq G$ is nilpotent. Let $x \in P \setminus P^2$ be such that its image in P/P^2 is not nilpotent in G.

Set $S = R[\frac{P}{x}]$. Since S is a finitely generated R–algebra and R is Noetherian, S is Noetherian. If $xS = S$, write

$$1 = x \sum_{i=0}^{n} \frac{a_i}{x^i} = \frac{a}{x^{n-1}}$$

for some $a_i \in P^i$, $a \in P^n$. Then $x^{n-1} \in P^n$, contradicting the choice of x. Thus $xS = PS$ is a proper ideal in S.

Let Q be a prime ideal in S minimal over xS. By Krull's Height Theorem (Theorem B.2.1), $\dim S_Q = 1$. Let T be the integral closure of S_Q. By Lying–Over (Theorem 2.2.2), there exists a maximal ideal M in T containing QT. By the Krull–Akizuki Theorem (Theorem 4.9.2), T is one–dimensional, Noetherian, and integrally closed, hence T_M is a Noetherian valuation domain. Set $V = T_M$ so that $\mathfrak{m}_V = MT_M$. It follows that $Q \subseteq \mathfrak{m}_V$, so $PS = xS \subseteq \mathfrak{m}_V$ and finally $P \subseteq \mathfrak{m}_V$, hence $\mathfrak{m}_V \cap R = P$. $\square$

Observe that the proof of the existence of valuation overrings over arbitrary domains and the proof of the existence of Noetherian valuation overrings over Noetherian domains are quite different.

6.4. More properties of valuation rings

In this section we prove some basic properties of valuation rings and passages between various valuation rings. For example, we prove in Proposition 6.4.1 that valuation domains are integrally closed. We show that intersecting a valuation ring with a subfield of its field of fractions gives a valuation ring (Proposition 6.4.7).

Proposition 6.4.1 *A valuation domain V is integrally closed.*

Proof: Let x be in the field of fractions of V such that $x^n + r_1 x^{n-1} + \cdots + r_n = 0$ for some r_i in V. If $x \notin V$, then $x^{-1} \in V$, so that $1 + r_1 x^{-1} + \cdots + r_n x^{-n} = 0$, whence 1 is in the ideal $x^{-1} V$. Hence x^{-1} is a unit in V, contradicting the assumption that $x \notin V$. $\qquad\square$

From the definition it follows that every ring between a K–valuation domain and K is also a valuation domain. Thus overrings of valuation domains are special. But also ideals are special:

Lemma 6.4.2 *Let V be a valuation domain.*
(1) Let I be an ideal of V and let G be a finite generating set of I. Then there exists $z \in G$ such that $zV = I$. In other words, if v is the valuation corresponding to V, then $\{v(i) \,|\, i \in I \setminus 0\}$ achieves a minimum on G.
(2) If for some x, y in V, $(x, y)V \neq yV$, then for all $r \in V$, $(x - ry)V = (x, y)V$.

Proof: Proof of (1): By induction it suffices to prove that every two–generated ideal is principal. So let $x, y \in V$ be non–zero elements. Then either xy^{-1} or yx^{-1} is in V, which says that either $x \in yV$ or $y \in xV$.

We already proved (2) for valuations in Remark 6.1.2, and (2) is a restatement for valuation rings. $\qquad\square$

Thus every finitely generated ideal of a valuation ring is principal. Even more can be said:

Lemma 6.4.3 *Let R be a ring, $V_1, \ldots, V_n$ valuation domains that are R–algebras. Assume that for each $j = 1, \ldots, n$, IV_j is a principal ideal. (This holds, for example, if I is a finitely generated ideal in R or if all the V_i are Noetherian.)*
(1) There exists an integer m and $x \in I^m$ such that for all i, $xV_i = I^m V_i$.
(2) Let $\mathfrak{m}_i$ be the maximal ideal of V_i. Assume that R contains elements $u_1, \ldots, u_{n-1}$ with the property that modulo each $\mathfrak{m}_i \cap R$, all u_j are distinct and non–zero. Then there exists an element $x \in I$ such that for all $i = 1, \ldots, n$, $xV_i = IV_i$.

Proof: The case $n = 1$ is trivial. So assume that $n > 1$.

Proof of (1): for all $i = 1, \ldots, n$, by induction on n there exist an integer m_i and an element $x_i \in I^{m_i}$ such that for all $j \neq i$, $x_i V_j = I^{m_i} V_j$. If for some i, $x_i V_i = I^{m_i} V_i$, we are done, so assume the contrary. Set $m = \prod m_i$, $r_i = m/m_i$, $x = \sum_j x_1^{r_1} \cdots \widehat{x_j^{r_j}} \cdots x_n^{r_n}$. Then $x \in I^{m(n-1)}$ and by Lemma 6.4.2, $xV_i = I^{m(n-1)}V_i$ for all i.

Proof of (2): By induction we may assume that there exist $x, y \in I$ such that for all $i < n$ and all $j > 1$, $xV_i = IV_i$, $yV_j = IV_j$. If $xV_n = IV_n$, we are done, so without loss of generality $xV_n \neq IV_n$. Similarly, $yV_1 \neq IV_1$. By Lemma 6.4.2, for any unit u in R, $(x - uy)V_1 = IV_1$ and $(x - uy)V_n = IV_n$.

It remains to find a unit u such that for all $i = 2, \ldots, n-1$, $(x-uy)V_i = IV_i$. Let $\mathfrak{m}_i$ be the maximal ideal of V_i. If $(x - uy)V_i \neq IV_i$, then $x - uy \in \mathfrak{m}_i I$. So now using our units, if $x - u_j y, x - u_k y \in \mathfrak{m}_i I$, then $(u_j - u_k)y \in \mathfrak{m}_i I$. By assumption, if $j \neq k$, $u_j - u_k$ is a unit in V_i, so that $y \in \mathfrak{m}_i I$. But $IV_i = yV_i$, so that for some $r \in \mathfrak{m}_i$, $y = ry$, contradicting that V_i is a domain. Thus for each i, there is at most one u_j in R for which $x - u_j y \in \mathfrak{m}_i I$. As there are $n - 1$ of the u_j and only $n - 2$ valuations $V_2, \ldots, V_{n-1}$ to consider, for at least one u_j, $(x - u_j y)V_i = IV_i$ for $i = 2, \ldots, n - 1$, and hence also for all $i = 1, \ldots, n$. $\qquad\square$

It follows that if a valuation domain is Noetherian, it is a local principal ideal domain, and hence has dimension one or zero. Here are some equivalent formulations of Noetherian valuation rings (cf. Proposition 4.1.1):

Proposition 6.4.4 *Let $(R, \mathfrak{m})$ be a local domain, K its field of fractions, and $R \neq K$. Then the following are equivalent:*
(1) R is a Noetherian valuation domain.
(2) R is a principal ideal domain.
(3) R is Noetherian and the maximal ideal $\mathfrak{m}$ is principal.
(4) R is Noetherian, and there is no ring properly between R and K.
(5) R is Noetherian, one–dimensional, and integrally closed.
(6) $\cap_n \mathfrak{m}^n = 0$ and $\mathfrak{m}$ is principal.
(7) R is a valuation domain with value group isomorphic to $\mathbb{Z}$.

Proof: (1) implies (2) by Lemma 6.4.2, and the other equivalences of (1) through (4) are easy. (1) implies (5) has been established in this section. The implication (5) implies (6) follows by Proposition 4.1.1 and Krull's Intersection Theorem. Assume (6). Let $x \in R$ such that $\mathfrak{m} = xR$. Let I be an arbitrary non–zero proper ideal in R and let y be a non–zero element in I. By properness of I there exists $r_1 \in R$ such that $y = r_1 x$. From r_1 we construct by induction finitely many $r_n \in R$ satisfying $y = r_n x^n$ as follows. If $r_{n-1} \in \mathfrak{m}$, there exists $r_n \in R$ such that $y = r_n x^n$. As $\cap_i \mathfrak{m}^i = 0$ and y is non–zero, necessarily for some n the corresponding r_n is not in $\mathfrak{m}$. Thus $yR = x^n R$. Let y be an element in I for which such n is least possible. Then $I = x^n R = yR$ is principal, so (6) implies (2). (2) clearly implies (7). Now assume (7), namely that R is a valuation domain with value group isomorphic to $\mathbb{Z}$. Let v be the composition of the maps $K^* \to K^*/R^* \cong \mathbb{Z}$. Then there exists $x \in K$ such that its image in $\mathbb{Z}$ under v is 1, and so $x \in R$ is not a unit. Let I be an arbitrary non–zero ideal in R. There exists $y \in I$ such that $v(y) = \min\{v(i) \mid i \in I\} = n \in \mathbb{N}$. Then $v(yx^{-n}) = 0$ and $v(zx^{-n}) \geq 0$ for all $z \in I$. Thus yx^{-n} is a unit in R and $I = x^n R = yR$, so R is a principal ideal domain, and (2) holds.   □

One can immediately deduce the following:

Corollary 6.4.5 *Let R be a valuation domain. Then R is Noetherian if and only if Γ_V is $\mathbb{Z}$ or 0.*

There exist one–dimensional valuation domains that are not Noetherian and thus their value groups are not $\mathbb{Z}$. However, their valuations are still real–valued:

Lemma 6.4.6 *The value group of a one–dimensional valuation ring V is isomorphic to a subgroup of $\mathbb{R}$.*

Proof: Let Γ be the value group of V. We will prove that Γ is Archimedean, and then apply Theorem 6.2.5. Suppose that Γ contains g, h such that $g > 0$ and such that for all positive integers n, $ng < h$. Let $x \in V$ have value g and $y \in V$ have value h. Then yV is a non–zero ideal in V whose radical does not contain x. Hence there exists a prime ideal P containing y but not x, which contradicts the one–dimensionality of V.   □

A method of generating valuations, and Noetherian valuations as well, is via intersections with a subfield of its field of fractions:

Proposition 6.4.7 *Let V be a K–valuation domain and F a subfield of K.*
(1) The intersection $V \cap F$ is an F–valuation domain.
(2) If V is Noetherian, then so is $V \cap F$.
(3) If $F \subseteq K$ is an algebraic extension, then $\Gamma_V \otimes_{\mathbb{Z}} \mathbb{Q} = \Gamma_{V \cap F} \otimes_{\mathbb{Z}} \mathbb{Q}$.

Proof: Let $x \in F^*$. Then $x \in K^*$ so that either $x \in V$ or $x^{-1} \in V$. Thus either $x \in V \cap F$ or $x^{-1} \in V \cap F$, so that $V \cap F$ is an F–valuation domain.
Certainly $\Gamma_{V \cap F} \subseteq \Gamma_V$.
If V is Noetherian, then by Corollary 6.4.5 Γ_V is a subgroup of $\mathbb{Z}$, hence $\Gamma_{V \cap F}$ is a subgroup of $\mathbb{Z}$, whence $V \cap F$ is Noetherian by the same corollary.
Assume that K/F is algebraic, and let $x \in K$. Let $x^n + a_1 x^{n-1} + \cdots + a_n = 0$ be an equation of algebraic dependence of x over F. Set $a_0 = 1$. Then for some i, j with $0 \leq i < j \leq n$, $v(a_i x^{n-i}) = v(a_j x^{n-j})$. Hence $(i - j)v(x) = v(a_i) - v(a_j) \in \Gamma_{V \cap F}$.   □

6.5. Valuation rings and completion

Proposition 6.5.1 *Let V be a valuation ring with maximal ideal $\mathfrak{m}$ and W the $\mathfrak{m}$–adic completion of V. Then W is a valuation ring.*

Proof: Let $\{a_n\}_{n \geq 0}$ and $\{b_n\}_{n \geq 0}$ be two Cauchy sequences (in the $\mathfrak{m}$–adic topology) of elements in V whose product $\{a_n b_n\}_{n \geq 0}$ is zero in W. Then for all positive N there exists a real number M such that for all integers $n \geq M$, $a_n b_n \in \mathfrak{m}^{2N}$. Thus there exists a finitely generated ideal I_n in $\mathfrak{m}$ such that $a_n b_n \in I_n^{2N}$. As every finitely generated ideal is principal, there exists $c_n \in I_n$ such that $a_n b_n \in c_n^{2N} V$. Then necessarily for each n, either $a_n \in c_n^N V$ or $b_n \in c_n^N V$. As $\{a_n\}_{n \geq 0}$ and $\{b_n\}_{n \geq 0}$ are Cauchy sequences, there exists M' such that for all integers $n \geq M'$, $a_n - a_{n+1} \in \mathfrak{m}^N$ and $b_n - b_{n+1} \in \mathfrak{m}^N$. Hence if for some $n_0 \geq M, M'$, $a_{n_0} \in c_{n_0}^N V \subseteq \mathfrak{m}^N$, it follows that for all $n \geq n_0$, $a_n \in \mathfrak{m}^N$. It follows that $\{a_n\}_{n \geq 0}$ is zero in W. Similarly, if for some $n_0 \geq M, M'$, $b_{n_0} \in c_{n_0}^N V \subseteq \mathfrak{m}^N$, then $\{b_n\}_{n \geq 0}$ is zero in W. Thus W is a domain.

Let x be in the field of fractions of W. Write $x = \{a_n\}_{n \geq 0} / \{b_n\}_{n \geq 0}$, where $\{a_n\}_{n \geq 0}$ and $\{b_n\}_{n \geq 0}$ are two Cauchy sequences in V. For each n, either $v(a_n) \geq v(b_n)$ or $v(a_n) \leq v(b_n)$. Thus after choosing subsequences, either for all n, $v(a_n) \geq v(b_n)$, or for all n, $v(a_n) \leq v(b_n)$. Thus either $x \in W$ or $1/x \in W$. $\qquad\square$

In our applications, we will be considering valuation domains V containing a (Noetherian) domain R with the same field of fractions. In that case, we call the prime ideal $\mathfrak{m}_V \cap R$ of R **the center** of V on R.

The following result of Abhyankar (Abhyankar [5, pages 513–514]), here slightly generalized, enables the passage between a Noetherian local domain and its completion while preserving some basic properties of valuation overrings (a weaker version of this already appeared in Abhyankar and Zariski [9]):

Proposition 6.5.2   *Let $(R, \mathfrak{m})$ be a Noetherian local domain with field of fractions K. Let $\widehat{R}$ be the $\mathfrak{m}$–adic completion of R, and Q a minimal prime ideal in $\widehat{R}$ (so that $Q \cap R = 0$). Let L be the field of fractions of $\widehat{R}/Q$. For any real–valued L–valuation centered on $\mathfrak{m}(\widehat{R}/Q)$ the contraction to K gives a K–valuation centered on $\mathfrak{m}$ that preserves the value group and the residue field. No two such L–valuations contract to the same K–valuation. Thus the contraction of real–valued L–valuations centered on $\mathfrak{m}(\widehat{R}/Q)$ to K–valuations is a one–to–one map.*

Proof: Let w be a real–valued L–valuation, and v its contraction to K^*. Let V be the K–valuation ring corresponding to v and W the L–valuation ring corresponding to w. By abuse of notation we write the image of an element x of $\widehat{R}$ in $\widehat{R}/Q$ also as x. Let x be any element of $\widehat{R} \setminus Q$. Set $t = w(x)$ and $s = w(\mathfrak{m}(\widehat{R}/Q))$. For any integer $u > t/s$ there exists $x' \in R$ such that $x - x' \in \mathfrak{m}^u \widehat{R}$. Then $w(x) = w(x' + (x - x')) = w(x') = v(x')$. Thus w and v have the same value groups.

Now let $x, y \in \widehat{R} \setminus Q$ with $x/y \in W$. As above, there exist non–zero $x', y' \in R$ such that $w(x - x') > w(x) = w(x') = v(x')$, $w(y - y') > w(y) = w(y') = v(y')$. and

$$\frac{x}{y} - \frac{x'}{y'} = \frac{x - x'}{x} \cdot \frac{x}{y} + \frac{y' - y}{y} \cdot \frac{x'}{y'} \in \mathfrak{m}_W.$$

Since $x'/y' \in V$, the residue fields of V and W are the same. Clearly v is uniquely obtained from w, and if a real–valued L–valuation w' contracts to v, then $w' = w$. $\qquad\square$

With notation as in the proposition, it is not true that every real–valued K–valuation centered on $\mathfrak{m}$ extends to a real–valued L–valuation:

Example 6.5.3   Consider Example 6.1.7, this time with $R = k[X, Y]_{(X,Y)}$ (localization). Let w be an extension of v to $k[[X, Y]]$. Then for all positive integers n,

$$w(Y - \sum_{i \geq 1} e_i X^i) \geq \min\{w(Y - \sum_{i \geq 1} e_i X^i + \sum_{i \geq n} e_i X^i), w(- \sum_{i \geq n} e_i X^i)\}$$

$$\geq \min\{w(Y - \sum_{i < n} e_i X^i), w(X^n)\}$$

$$= \min\{v(Y - \sum_{i < n} e_i X^i), v(X^n)\} \geq n,$$

so that $w(Y - \sum_{i \geq 1} e_i X^i)$ cannot be a real number.

However, for Noetherian local rings whose completions are domains, every valuation centered on the maximal ideal does extend to some valuation on its completion, it just need not be real–valued:

Proposition 6.5.4   *Let R be a Noetherian domain and S a faithfully flat extension that is a domain. Let K be the field of fractions of R and L the field of fractions of S. Then any K–valuation that is non–negative on R extends to an L–valuation that is non–negative on S. In particular, if R is local with completion a domain (i.e., R is analytically irreducible), then for any valuation domain V between R and K there exists a valuation domain W between $\widehat{R}$ and L such that W contracts to V.*

Proof: Let V be a K–valuation ring with center on a prime ideal $\mathfrak{m}$. Let T be the smallest subring of L containing V and S. Suppose that $\mathfrak{m}_V T = T$. Then we can write $1 = \sum_{i=1}^n a_i s_i$ for some $a_i \in \mathfrak{m}_V$ and $s_i \in S$. Write $a_i = b_i/c$ for some $b_1, \ldots, b_n, c \in R$. Then $c = \sum_{i=1}^n b_i s_i \in (b_1, \ldots, b_n)S \cap R = (b_1, \ldots, b_n)R$, the latter equality by the faithful flatness of S over R. Then $1 \in (a_1, \ldots, a_n)V \subseteq \mathfrak{m}_V$, which is a contradiction. Thus necessarily $\mathfrak{m}_V T$ is a proper ideal in T. By Theorem 6.3.2, there exists a valuation ring W between T and L such that $\mathfrak{m}_V W \neq W$. Hence $W \cap K \subseteq V$, and as $V \subseteq T \subseteq W$, then also $W \cap K = V$. $\qquad\square$

Discussion 6.5.5   Let $(R, \mathfrak{m})$ be an analytically irreducible local ring with field of fractions K and completion $\widehat{R}$. Let V be a valuation domain such that $R \subseteq V \subseteq K$. The question of when there exists a unique extension of V to the field of fractions of $\widehat{R}$ is interesting and delicate. In [115], Heinzer and Sally studied this problem when $\widehat{R}$ is assumed to be integrally closed. Among their results is the statement that if W is a valuation ring containing

$\widehat{R}$ and contained in the field of fractions of $\widehat{R}$ such that every non–zero prime of W lies over either $\mathfrak{m}\widehat{R}$ or a height one prime of $\widehat{R}$, then W is the unique extension of $V = W \cap K$ to a valuation domain birationally dominating $\widehat{R}$. In [54], Cutkosky and Ghezzi study when there is an *immediate* extension of V to a valuation ring W birationally dominating $\widehat{R}$ in the sense that the value groups and residue fields of V and W are the same. For more on valuations on an analytically irreducible ring $(R, \mathfrak{m})$ and on its $\mathfrak{m}$–adic completion, see Proposition 9.3.5.

The following shows that some extensions of valuations to completions can be obtained without the assumption that the completion of R be an integral domain. We leave the straightforward proof to the reader.

Lemma 6.5.6 *Let V be a valuation ring and v the corresponding valuation. Assume that v is real–valued.*

(1) Let W be the set of all sequences $\{a_n\}_{n \geq 0}$ of elements of V satisfying the property that for each positive number C there exists a positive integer M such that for all $n \geq M$, $v(a_{n+1} - a_n) \geq C$. We define addition and multiplication on W to be componentwise. Then W is a commutative ring with 1.

(2) Define two elements $\{a_n\}_n$ and $\{b_n\}_n$ of W to be equivalent if for all positive numbers C there exists a positive integer M such that for all $n \geq M$, $v(a_n - b_n) \geq C$. Then the set of all equivalence classes $\tilde{V}$ is a valuation domain containing V whose value group contains Γ_v and is a subgroup of $\mathbb{R}$.

(3) Let $(R, \mathfrak{m})$ be a Noetherian local domain with field of fractions K such that $R \subseteq V$ and $\mathfrak{m}_V \cap R = \mathfrak{m}$. Then there is a natural map $\widehat{R} \to \tilde{V}$, and the maximal ideal of $\tilde{V}$ contracts to the maximal ideal of $\widehat{R}$.

6.6. Some invariants

There are several invariants one can define for a valuation that give insight into the structure of the valuation.

Definition 6.6.1 *Let $V = R_v$ be a valuation domain corresponding to the valuation v. The **rank** of v is defined to be the Krull dimension of V.*

*The **rational rank** of v, denoted $\mathrm{rat.rk}\, v$, is the rank of the value group Γ_v of V over $\mathbb{Q}$, i.e., $\mathrm{rat.rk}\, v = \dim_{\mathbb{Q}}(\Gamma_v \otimes_{\mathbb{Z}} \mathbb{Q})$.*

*Suppose that k is a subfield of the residue field $\kappa(v)$. We define the **transcendence degree of v over k**, $\mathrm{tr.deg}_k(v)$, to be the transcendence degree of $\kappa(v)$ over k.*

*The valuation and the valuation ring are said to be **(generalized) discrete** if the value group is isomorphic to $\mathbb{Z}^n$ with the lexicographic ordering. Recall that this ordering declares $(a_1, \ldots, a_n) \in \mathbb{Z}^n$ greater than or equal to $(b_1, \ldots, b_n) \in \mathbb{Z}^n$ if and only if the first non–zero entry of $(a_1 - b_1, \ldots, a_n - b_n)$ is positive.*

Proposition 6.6.2 *A valuation domain that is not a field is Noetherian if and only if it is a rank one discrete valuation domain.*

Proof: This is rephrasing of Proposition 6.4.4 by using the new notation. □

In the literature Noetherian valuation domains are often called **DVRs**, or **discrete valuation rings**, or **classical discrete valuation rings**. We allow discrete valuation rings that are not Noetherian.

The rank of the valuation is defined above as the Krull dimension of V. But the rank of the valuation can also be determined from its value group. First we need a definition:

Definition 6.6.3 *Let Γ be a totally ordered abelian group. A non–empty subset S of Γ is called a* **segment** *if for any $s \in S$, the set $\{g \in \Gamma \mid -s \le g \le s\}$ is contained in S.*

A saturated filtration of segments of the value group of v is the same as the rank of v:

Proposition 6.6.4 *Let Γ be the value group of a valuation v, and let V be the corresponding valuation ring. Then the following hold:*

(1) For any subgroup G of Γ that is a segment (i.e., an **isolated, or convex subgroup**), *$P_G = \{r \in V \mid$ for all $g \in G, v(r) > g\}$ is a prime ideal in V.*

(2) For any prime ideal P in V, $G_P = \{\pm v(s) \mid s \in V \setminus P\}$ is a subgroup of Γ that is a segment.

(3) For any prime ideal P in V, $P_{(G_P)} = P$, and for any subgroup G of Γ that is a segment, $G_{(P_G)} = G$.

(4) The set of all subgroups of Γ that are segments is totally ordered by inclusion.

(5) $\dim V (= \operatorname{rk} v)$ is the supremum of all integers n for which Γ has n distinct subgroups that are segments.

(6) If a subgroup G of Γ is a segment, then Γ/G is a totally ordered abelian group that is isomorphic to the value group of the valuation ring V_{P_G}.

Proof: Let $x, y \in V \setminus P_G$. Choose $g, h \in G$ with $v(x) \le g$ and $v(y) \le h$. Then $v(xy) \le g + h$, so that $xy \notin P_G$. This proves (1).

Let $g \in \Gamma$ such that for some $s \in V \setminus P$, $-v(s) \le g \le v(s)$. By possibly replacing g by $-g$, without loss of generality $g \ge 0$. Let x be in the field of fractions of V such that $g = v(x)$. Then $v(x), v(sx^{-1}) \ge 0$, so that $x, sx^{-1} \in V$. Hence $s \in xV$, whence $x \in V \setminus P$. This proves (2).

Observe that $P_{(G_P)} = \{r \in V \mid v(r) > v(s)$ for all $s \in V \setminus P\}$. Thus if $x \in P_{(G_P)}$, then $v(x) > v(s)$ for all $s \in V \setminus P$, whence x is not in $V \setminus P$, which means that $x \in P$. Conversely, should $x \in P$ such that $v(x) \le v(s)$ for some $s \in V \setminus P$, then $s \in xV \subseteq P$, which is a contradiction. Thus $P \subseteq P_{(G_P)}$. Similarly, $G_{(P_G)} = \langle v(s) \mid s \in V \setminus P_G \rangle = \langle v(s) \mid s \in V, v(s) \le g$ for some $g \in G \rangle$, so clearly $G \subseteq G_{(P_G)}$. If $g \in G_{(P_G)}$, then without loss of generality $g = v(s)$ for some $s \in V$ and $v(s) \le h$ for some $h \in G$. As G is a segment, $g = v(s) \in G$. This proves (3). Since the ideals in a valuation domain are totally ordered, so are the isolated segments of its value group, so (4) and (5) follow as well.

To prove (6), we first prove that Γ/G inherits a total order from Γ. Let $h, h' \in \Gamma$ represent distinct cosets in Γ/G, and suppose that $h < h'$. We claim that for all $g \in G$, $h + g < h'$. If not, there exists $g \in G$ such that $h + g > h'$. Hence $g > h' - h > -g$. Since G is a segment, it follows that $h' - h \in G$, a contradiction. Thus Γ/G inherits a total order from Γ. Observe that the localization V_{P_G} is a valuation ring with the same field of fractions, K, as V. We can identify $\Gamma \cong K^*/V^*$. Under this identification, we claim that $\Gamma/G \cong K^*/(V_{P_G})^*$, which will finish the proof of (6). Clearly $\Gamma \cong K^*/V^*$ surjects onto $K^*/(V_{P_G})^*$, so it suffices to prove that the kernel of this surjection is exactly G. Let $\alpha \in (V_{P_G})^*$, and write $\alpha = \frac{r}{s}$, where $r, s \in V \setminus P_G$. Then $v(s)$ is a non–negative element of Γ and there exists an element $g \in G$ such that $v(s) < g$. Hence $v(s) \in G$ since G is a segment. Similarly, $v(y) \in G$, and since G is a group, $v(\alpha) \in G$. Hence the inverse image is contained in G. Conversely, let $x \in K^*$ such that $v(x) \in G$. Since $v(\frac{1}{x}) \in G$ as well, without loss of generality we may assume that $x \in V$. Then the definition of P_G shows that $x \notin P_G$, which means that $\frac{1}{x} \in V_{P_G}$. Therefore $x \in (V_{P_G})^*$. $\square$

The following is now immediate:

Proposition 6.6.5 *A totally ordered non–trivial Archimedean abelian group has rank one.*

Now that both the rank and the rational rank of a valuation are expressed in terms of its value group, we can also compare them:

Proposition 6.6.6 *For any valuation v, $\operatorname{rk} v \leq \operatorname{rat.rk} v$.*

Proof: We prove more generally that if Γ is a totally ordered abelian group, then its rank, defined as the length of a saturated chain of subgroups that are segments, is at most its rational rank, defined as $\dim_{\mathbb{Q}}(\Gamma \otimes_{\mathbb{Z}} \mathbb{Q})$. By Proposition 6.6.4, this finishes the proof by setting $\Gamma = \Gamma_v$.

Let $n = \operatorname{rk} \Gamma$, and let $0 = G_0 \subsetneq G_1 \subsetneq \cdots \subsetneq G_n = \Gamma$ be a saturated chain of subgroups of Γ that are segments. If n is 0 or 1, then clearly $\operatorname{rat.rk} \Gamma \geq n$. If $n > 1$, by induction, the rank $n - 1$ of G_{n-1} is at most $\dim_{\mathbb{Q}}(G_{n-1} \otimes_{\mathbb{Z}} \mathbb{Q})$. The group G_n/G_{n-1} is a totally ordered abelian group that is not zero, so its rational rank is at least one. Thus $\operatorname{rat.rk} \Gamma = \operatorname{rat.rk}(G_{n-1} \oplus (G_n/G_{n-1})) \geq n - 1 + 1 = n = \operatorname{rk} \Gamma$. $\square$

In this book we use primarily rank one Noetherian valuations, however, the study of arbitrary valuations is intricate and beautiful. We present a noted result of Abhyankar on (rational) ranks of valuations:

Theorem 6.6.7 (Abhyankar [6]) *Let $(R, \mathfrak{m})$ be a Noetherian local integral domain with field of fractions K and residue field k. Let V be a K–valuation* domain such that $R \subseteq V$, $\mathfrak{m}_V \cap R = \mathfrak{m}$, and let v be the corresponding valuation. Set $n = \dim(R)$. Then

(1) rat.rk $v + \text{tr.deg}_k v \le n$.

(2) If rat.rk $v + \text{tr.deg}_k v = n$, then $\Gamma_v \cong \mathbb{Z}^n$ and $\kappa(v)$ is finitely generated over k.

(3) If rk$(v) + \text{tr.deg}_k v = n$, then Γ_v is discrete.

First we prove a lemma:

Lemma 6.6.8 (Abhyankar [6]) *Let $F \subseteq K$ be an inclusion of fields, let w be a K–valuation, and let v be the restriction of w to F. Then*

$$\text{rat.rk}\, w \le \text{rat.rk}\, v + \text{tr.deg}_F K.$$

Proof: There is nothing to show if the transcendence degree of K over F is infinite. So assume that $\text{tr.deg}_F K < \infty$. We use induction on $\text{tr.deg}_F K$. If $\text{tr.deg}_F K = 0$, then by Proposition 6.4.7 (3), rat.rk $w = $ rat.rk v. Now assume that $\text{tr.deg}_F K > 0$. If for all elements $x \in K$ that are transcendental over F, $w(x)$ is up to a positive integer multiple in Γ_v, then rat.rk $w = $ rat.rk v, and we are done. So we may assume that for some $x \in K$ that is transcendental over F, no positive integer multiple of $w(x)$ is in Γ_v. If the lemma holds for the restrictions of w to the field extensions $F \subseteq F(x)$ and $F(x) \subseteq K$, then

$$\begin{aligned}
\text{rat.rk}\, w &\le \text{rat.rk}(w|_{F(x)}) + \text{tr.deg}_{F(x)} K \\
&\le \text{rat.rk}\, v + \text{tr.deg}_F F(x) + \text{tr.deg}_{F(x)} K \\
&= \text{rat.rk}\, v + \text{tr.deg}_F K,
\end{aligned}$$

which proves the lemma. Thus by induction it suffices to prove the case $K = F(x)$. As $w(x)$ is not rationally dependent on Γ_v, for every $f = a_0 + a_1 x + \cdots + a_n x^n$ with $a_i \in F$, $w(f) = \min\{v(a_i) + iw(x) \,|\, i = 0, \ldots, n\}$ by Remark 6.1.2. It follows that $\Gamma_w \otimes_{\mathbb{Z}} \mathbb{Q}$ is generated by $w(x)$ and Γ_v, so that rat.rk $w = $ rat.rk $v + 1 = $ rat.rk $v + \text{tr.deg}_F K$. $\square$

With this, we can prove Theorem 6.6.7:

Proof of 6.6.7: We use induction on the dimension n of R. If $n = 0$, then V is trivial, and (1) follows easily.

If $n = 1$, then by the Krull–Akizuki Theorem 4.9.2, the integral closure $\bar{R}$ of R is Noetherian. As $\bar{R} \subseteq V$, if $\mathfrak{M} = \mathfrak{m}_V \cap \bar{R}$, then $\bar{R}_{\mathfrak{M}} \subseteq V$. But $\bar{R}_{\mathfrak{M}}$ is a Noetherian one–dimensional integrally closed domain, hence a principal ideal domain, so that by Proposition 6.4.4, $V = \bar{R}_{\mathfrak{M}}$ or $V = K$. The latter case implies that $\mathfrak{m} = \mathfrak{m}_V \cap R = 0$, so that $R = K$, which is impossible. So necessarily $V = \bar{R}_{\mathfrak{M}}$, hence $\Gamma_v = \mathbb{Z}$, so that rat.rk $v = 1$, $\text{tr.deg}_k v = 0$, so again (1) follows.

Now let $\dim R = n > 1$. First assume that $\text{tr.deg}_k v > 1$. Let $x \in V$ such that its image in $V/\mathfrak{m}_V$ is transcendental over k. Set $S = R[x]_{\mathfrak{m}_V \cap R[x]}$. Then S is Noetherian local with maximal ideal $\mathfrak{m}_S$, $\text{tr.deg}_{\kappa(\mathfrak{m}_S)} v = \text{tr.deg}_k v - 1$, and by the Dimension Inequality Theorem B.2.5, $\dim S \le \dim R - 1$. Thus it suffices to prove (1) with S in place of R, and hence it suffices to prove (1) in case $\mathrm{tr.deg}_k v = 0$. More generally, it suffices to prove that $\mathrm{rat.rk}\, v \le n$.

Suppose that v is real–valued. By Lemma 6.5.6, there exists a valuation $\tilde{v}$ whose value group contains Γ_v and is contained in $\mathbb{R}$, and furthermore $\tilde{v}$ is non–negative on $\widehat{R}$, and is positive on the maximal ideal of $\widehat{R}$. Let P be the set of elements in $\widehat{R}$ on which $\tilde{v}$ is infinite. If we can show that $\mathrm{rat.rk}(\tilde{v}|_{Q(\widehat{R}/P)}) \le \dim(\widehat{R}/P)$, then it follows that $\mathrm{rat.rk}\, v \le \dim(\widehat{R}/P) \le \dim R$, which would prove (1). Thus it suffices to prove (1) when v is real–valued, in the case that R is a complete local domain. By the Cohen Structure Theorem, there exists a subring S of R that is a power series ring in finitely many variables over a field or over a complete local principal ideal domain and for which $S \subseteq R$ is module–finite. The rational ranks of V and of $V \cap Q(S)$ are the same by Proposition 6.4.7, and as $\dim S = \dim R$, without loss of generality by replacing R with S we may assume that R is a complete regular local ring. By Proposition 6.5.2, we may replace R by $A[X_1, \ldots, X_n]_{(X_0, X_1, \ldots, X_n)}$, where A is either a field or a local principal ideal domain, $X_1, \ldots, X_n$ are variables over A, and X_0 generates the maximal ideal of A. If A is a principal ideal domain, we assume that $X_n = 0$ (recall that the dimension of R is n). Let F be the field of fractions of A. By Lemma 6.6.8, $\mathrm{rat.rk}\, v \le \mathrm{rat.rk}(v|_F) + d$, where d is the transcendence degree of the field of fractions of R over F. Since $\dim A + d = \dim R$, and $\mathrm{rat.rk}(v|_F) \le \dim A$ as $\dim A \le 1$, we have proved (1) in the case v is real–valued.

Now we prove (1) in full generality. By the reductions above, without loss of generality $\mathrm{tr.deg}_k v = 0$, $\dim R > 1$, and v is not real–valued. By Lemma 6.4.6, v is not Archimedean. In other words, there exist non–zero $x, y \in \mathfrak{m}$ such that for all positive integers n, $nv(x) < v(y)$. No power of x can be in yV, so that there exists a prime ideal Q in V containing y but not x. Thus $0 \ne Q \subsetneq \mathfrak{m}_V$. Let $P = Q \cap R$. As $x \notin Q$, $P \subsetneq \mathfrak{m}$, and as $R \subseteq V$ have the same field of fractions, $P \ne 0$. Observe that $R_P \subseteq V_Q$ and that V_Q is a valuation domain. Let v_Q be the corresponding valuation. Since $\dim R_P < \dim R$, induction gives that $\mathrm{rat.rk}(v_Q) + \mathrm{tr.deg}_{\kappa(P)} \kappa(Q) \le \dim R_P$. Also, we have inclusion $R/P \subseteq V/Q$, and V/Q is a valuation ring. Let v^* be its valuation. By Lemma 6.6.8,

$$\mathrm{rat.rk}(v^*) \le \mathrm{rat.rk}(v^*|_{Q(R/P)}) + \mathrm{tr.deg}_{\kappa(P)} \kappa(Q),$$

and by induction on dimension, $\mathrm{rat.rk}(v^*|_{Q(R/P)}) \le \dim(R/P)$. Thus

$$\mathrm{rat.rk}(v_Q) + \mathrm{rat.rk}(v^*) \le \dim R_P + \mathrm{rat.rk}(v^*|_{Q(R/P)})$$
$$\le \dim R_P + \dim(R/P) \le \dim R.$$

As in Proposition 6.6.4, Q corresponds to the subgroup G_Q of Γ_v, and G_Q is the value group of the valuation ring V_Q. The quotient group Γ_v/G_Q is the value group of the valuation ring V/Q. Thus $\mathrm{rat.rk}\, v = \mathrm{rat.rk}(v_Q) + \mathrm{rat.rk}(v^*)$, which finishes the proof of (1).

Now assume that rat.rk v + tr.deg$_k v$ = dim R = n (respectively, that rk v + tr.deg$_k v$ = dim R = n). We want to show that $\kappa(v)$ is finitely generated over k, and that $\Gamma_v \equiv \mathbb{Z}^n$ (respectively, that Γ_v is discrete). The same reduction as for (1) shows that without loss of generality we may assume that tr.deg$_k v$ = 0. In case v is real–valued, we may similarly assume that $R = A[X_1, \ldots, X_d]_{(X_0, \ldots, X_d)}$, where A is a field or a discrete valuation ring of rank one, X_0 generates the maximal ideal of A, and $X_1, \ldots, X_d$ are variables over A. With F the field of fractions of A, then as before, rat.rk$(v|_F)$ = dim A and rat.rk v = dim $A + d$. Thus for some $y_1, \ldots, y_d \in k[X_1, \ldots, X_d]$, $\Gamma_v \otimes_{\mathbb{Z}} \mathbb{Q}$ is generated by the images of $v(y_1), \ldots, v(y_d)$, and also by $v(X_0)$ if A is not a field. Without loss of generality $y_1, \ldots, y_d \in (\underline{X}) \setminus (X_0, \underline{X})^2$ and thus without loss of generality $y_1 = X_1, \ldots, y_d = X_d$. As Γ_v is generated by the values of the polynomials in $X_0, \ldots, X_d$, and as all these values are rationally independent, Γ_v is generated by $v(X_0), \ldots, v(X_d)$ and $\Gamma_v \cong \mathbb{Z}^n$. In this case, $k_v = k$. This proves (2) when v is real–valued. If also rk v = dim R, then necessarily after possibly reordering the X_i, for all $i = 0, \ldots, d - 1$ and all positive integers m, $mv(X_i) > v(X_{i+1})$. This proves that V is discrete so that (3) holds for real–valued valuations.

Now assume the hypothesis of (2) for arbitrary v. With reductions and notation as in the last paragraph of the proof of (1), rat.rk(v_Q) + tr.deg$_{\kappa(P)} \kappa(Q)$ = dim R_P, rat.rk(v^*) = dim(R/P) + tr.deg$_{\kappa(P)} \kappa(Q)$, and tr.deg$_k v$ = 0. By induction, the value group Γ_Q of v_Q is isomorphic to $\mathbb{Z}^{\text{ht } P}$ and its residue field $\kappa(Q)$ is finitely generated over $\kappa(P)$. Thus there exist $x_1, \ldots, x_s \in \mathfrak{m}_V/Q$ that are transcendental over $\kappa(P)$ and such that $\kappa(Q)$ is algebraic over $\kappa(P)(x_1, \ldots, x_s)$. Let R' be the ring $(R/P)[x_1, \ldots, x_s]$ localized at the contraction of $\mathfrak{m}_V/Q$. Then R' is a Noetherian local ring of dimension dim$(R/P) + s$ such that $\kappa(Q)$ is finite over $Q(R')$. There exist finitely many elements $y_1, \ldots, y_r \in \kappa(Q)$ that are integral over R' and such that $\kappa(Q) = Q(R')(y_1, \ldots, y_r)$. Let R'' be the localization of $R'[y_1, \ldots, y_r]$ at the contraction of $\mathfrak{m}_V/Q$. Then R'' is a Noetherian local domain of dimension at most dim R' = dim$(R/P) + s'$ whose field of fractions is $\kappa(Q)$. By (1), rat.rk$(v^*) \leq$ dim$(R'') \leq$ dim$(R/P) + s$ = dim(R/P) + tr.deg$_{\kappa(P)} \kappa(Q)$ = rat.rk(v^*), so that equality holds throughout. Thus by induction on dimension, the value group of $\overline{v}^*$ is isomorphic to $\mathbb{Z}^{\dim(R'')}$, and the residue field of v^* is finitely generated over the residue field of R''. As the latter is finitely generated over k, then the residue field of v^* is finitely generated over k. But the residue field of v^* is k_v. Thus, $\Gamma_v \cong \Gamma_{v_Q} \oplus \Gamma_{v^*}$, which is isomorphic to the direct sum of ht P + dim(R/P) + tr.deg$_{\kappa(P)} \kappa(Q)$ copies of $\mathbb{Z}$. But the rational rank of v is dim R (under the assumption tr.deg$_k v$ = 0), so that Γ_v is isomorphic to a direct sum of dim R copies of $\mathbb{Z}$. This proves (2). A similar argument proves that rk v + tr.deg$_k v$ = dim R implies that Γ_v is discrete. $\square$

6.7. Examples of valuations

We analyze basic examples of valuations in light of the definitions from the previous section.

Example 6.7.1 (Cf. Example 6.1.5.) Let k be a field, X and Y variables over k, and $K = k(X, Y)$. Let v be a monomial valuation on K defined by $v(X) = 1$, $v(Y) = 1/2$. The valuation ring R_v is the localization of $k[Y, \frac{X}{Y^2}]$ at the prime ideal generated by Y. The residue field of the valuation ring of v is the field $k(\frac{X}{Y^2})$; hence tr.deg$_k v = 1$. The maximal ideal of the valuation ring is generated by Y. For each positive integer n, the set of all elements r of $R = k[X, Y]$ satisfying $v(r) \geq n$ is the ideal $(v) = (X, Y^2)^n$. This example is a discrete rank one valuation, with a Noetherian valuation ring. Of course, the value group is isomorphic to $\mathbb{Z}$, but is literally $\mathbb{Z} \cdot (1/2)$.

Notice that we could this valuation by letting $v(X) = 2$, $v(Y) = 1$. In this case the valuation ring does not change, nor does any essential property of the valuation, although the value group is now $\mathbb{Z}$. In fact, the two valuations (the original and the scaled one) are equivalent (see Definition 6.1.8).

Example 6.7.2 (Cf. Example 6.1.6.) Let k be a field, let X and Y be variables over k, and let $K = k(X, Y)$. Let v be the monomial valuation on K defined by $v(X) = \sqrt{2}$, $v(Y) = 1$. In this case the value group is $\Gamma = \{a + b\sqrt{2} \mid a, b \in \mathbb{Z}\}$, so that the rational rank of v is 2. However, the associated valuation ring is not discrete. Namely, the value group, though isomorphic to the abelian group $\mathbb{Z}^2$, is not isomorphic to the totally ordered abelian group $\mathbb{Z}^2$ under the lexicographic ordering. By Proposition 6.6.5, the rank of Γ is one. Thus by Proposition 6.6.4, the maximal ideal is the only non–zero prime ideal in the valuation ring V of v. Here is also an easy direct proof: let P be a non–zero prime ideal in V. Let z be an arbitrary non–unit in V. Write $v(z) = a_0 + b_0\sqrt{2}$ for some non–zero $(a_0, b_0) \in \mathbb{Z}^2$. Let $r \in P \setminus \{0\}$. Write $v(r) = a + b\sqrt{2}$ for some non–zero $(a, b) \in \mathbb{Z}^2$. Let n be a sufficiently large integer such that $nv(z) - v(r) \geq 0$. Let $s \in V$ such that $v(s) = nv(z) - v(r) \geq 0$. Then rs has value $nv(z) = v(z^n)$, which forces rs to be a unit multiple of z^n. As P is a prime ideal and $rs \in P$, then $z \in P$. This proves that there is only one non–zero prime ideal in V. Thus V has rank one. The ideal of all elements r of $R = k[X, Y]$ satisfying $v(r) \geq n$ is generated by all monomials $X^i Y^j$, $i, j \in \mathbb{Z}$ such that $i\sqrt{2} + j \geq n$.

Example 6.7.3 (Cf. Example 6.1.7.) Let $R = k[X, Y]$, where k is a field and X, Y variables over k. Let e be an element of $Xk[[X]]$ that is transcendental over $k[X]$ (it exists by Exercise 3.13). Write $e(X) = \sum_{i \geq 1} e_i X^i$. Define $v : R \setminus \{0\} \to \mathbb{Z}$ by

$$f(X, Y) \mapsto \max\{n \mid f(X, e(X)) \in X^n k[[X]]\}.$$

In other words, we embed $k[X, Y]$ in $k[[X]]$, which is a valuation ring, and v is the restriction of the valuation of $k[[X]]$ to R. Thus v is a valuation.

The value group of this valuation is $\mathbb{Z}$, so that its associated valuation ring is Noetherian. As the value group is $\mathbb{Z}$, clearly the rational rank of the valuation is 1, and by Proposition 6.4.4, the rank is 1. From $k = \kappa((X,Y)R) \subseteq \kappa(v) \subseteq \kappa(Xk[[X]]) = k$ we deduce that $\operatorname{tr.deg}_k(v) = 0$.

Example 6.7.4   Let $R = k[X, Y, Z]$, where k is a field and X, Y, Z are variables over k. Let $e(X)$ be an element of $Xk[[X]]$ that is transcendental over $k[X]$ (it exists by Exercise 3.13). Every non–zero $g \in R$ can be written as $g = Z^r f(X, Y, Z)$ for some $f(X, Y, 0) \neq 0$. We define $v : R \mapsto \mathbb{Z}^2$ by $v(g) = (r, \sup\{n \mid f(X, e(X), 0) \in X^n k[[X]]\})$. Under the lexicographic order on $\mathbb{Z}^2$, this extends to a valuation on $k(X, Y, Z)$. The value group is $\mathbb{Z}^2$, and v is not a monomial valuation.

Example 6.7.5   Let k be a field and t a variable over k. Let V be the set of all generalized power series of the form $\sum_{n=0}^{\infty} a_n t^{e_n}$, where all a_n are in k and where $\{e_n\}_n$ is a strictly increasing sequence of rational numbers. The elements of V can be added and multiplied in a natural way, which makes V into a commutative domain with identity. By using the identity $\frac{1}{1-x} = \sum_{n=0}^{\infty} x^n$ it is easy to prove that an element of V is a unit if and only if it is of the form $\sum_{n=0}^{\infty} a_n t^{e_n}$ with $a_0 \neq 0$. Every element of V can then be written uniquely as some rational power of t times a unit, which gives that V is a valuation domain, with valuation v reading off the exponent of t.

Example 6.7.6   Let k be a field and let X and Y be variables over k. If V and v are as in the previous example, map $k[X, Y] \to V$ by sending X to t and Y to $\sum_{i=2}^{\infty} t^{n + \frac{1}{n}}$. Then $v(X) = 1$ and $v(Y) = 2 + \frac{1}{2}$. It follows that $v(Y/X^2) = \frac{1}{2}$. Set $f_1 = X$, $f_2 = Y/X^2$. Then $Y - X^2 f_2$ has value $3 + \frac{1}{3}$, so that $f_3 = (Y - X^2 f_2)/X^3$ has v–value $\frac{1}{3}$. Similarly, $f_n = (Y - \sum_{i=2}^{n-1} f_i)/X^n$ has v–value $\frac{1}{n}$. Consequently the value group is $\mathbb{Q}$. (A more general example giving real valuations with preassigned value groups can be found in [322], Chapter VI, Section 15, Example 3.)

Valuations also arise from the order function whose special cases were already used in this chapter, say in Examples 6.1.7 and 6.5.3:

Definition 6.7.7   *Let I be an ideal in a ring R. The function $\operatorname{ord}_I : R \to \mathbb{Z}_{\geq 0} \cup \{\infty\}$ defined by $\operatorname{ord}_I(r) = \sup\{m \mid r \in I^m\}$ is called the* **order** *of I.*

We next prove that under some conditions on I, ord_I is a valuation. It is then called the I–**adic valuation**.

Theorem 6.7.8   *Let R be a Noetherian ring with an ideal I such that $\cap_{n \geq 0} I^n = 0$. Then the associated graded ring $\operatorname{gr}_I(R)$ is an integral domain if and only if the order function ord_I yields a discrete valuation of rank one.*

Proof: First assume that $\operatorname{gr}_I(R)$ is an integral domain. It is easy to verify that R is an integral domain, see Exercise 5.9. It suffices to prove that ord_I satisfies the properties $\operatorname{ord}_I(xy) = \operatorname{ord}_I(x) + \operatorname{ord}_I(y)$ and $\operatorname{ord}_I(x + y) \geq$

$\min\{\mathrm{ord}_I(x), \mathrm{ord}_I(y)\}$ for all $x, y \in R$. Let $\mathrm{ord}_I(x) = m$ and $\mathrm{ord}_I(y) = n$. Then $x \in I^m$ and $y \in I^n$, so that $x + y \in I^{\min\{m,n\}}$, which proves the second property. Certainly $xy \in I^{m+n}$. The associated element x^* of x in $\mathrm{gr}_I(R)$ lies in the component of degree m (namely is the element $x + I^{m+1} \in I^m/I^{m+1}$) and the associated element y^* of y in $\mathrm{gr}_I(R)$ lies in the component of degree n. As $\mathrm{gr}_I(R)$ is an integral domain, $x^*y^* = xy + I^{m+n+1} \in I^{m+n}/I^{m+n+1}$ is a non–zero element in $\mathrm{gr}_I(R)$ of degree $m + n$. Thus xy is not in I^{m+n+1}. This proves the first property.

Now assume that ord_I is a discrete valuation of rank one. To prove that $\mathrm{gr}_I(R)$ is an integral domain it is enough to prove that the product of non–zero homogeneous elements is non–zero. Let $x \in I^m \setminus I^{m+1}$, $y \in I^n \setminus I^{n+1}$. Then $\mathrm{ord}_I(xy) = \mathrm{ord}_I(x) + \mathrm{ord}_I(y) = m + n$, so that $xy \in I^{m+n} \setminus I^{m+n+1}$. This proves that $\mathrm{gr}_I(R)$ is an integral domain. $\square$

Theorem 6.7.9 *Let R be a regular ring and $\mathfrak{m}$ a maximal ideal. Assume that $\mathfrak{m}$ is not zero. Then the order function relative to $\mathfrak{m}$ is a discrete valuation of rank one and the residue field of the corresponding valuation ring is purely transcendental over $R/\mathfrak{m}$ of transcendence degree $\dim R - 1$. Explicitly, the $\mathfrak{m}$–adic valuation ring equals $(R[\frac{\mathfrak{m}}{x}])_{(x)}$ for any $x \in \mathfrak{m} \setminus \mathfrak{m}^2$.*

Proof: We may localize at $\mathfrak{m}$ to assume that R is local with maximal ideal $\mathfrak{m}$. The associated graded ring of a maximal ideal in a regular ring, being a polynomial ring, is an integrally closed Noetherian domain. By Theorem 6.7.8 the order function induces a rank one discrete valuation. Let $d = \dim R$ and $\mathfrak{m} = (x_1, \ldots, x_d)$. Set $S = R[\frac{x_i}{x_1} \,|\, i = 2, \ldots, d]$. By Corollary 5.5.9 $S \cong R[Y_2, \ldots, Y_d]/(x_1 Y_2 - x_2, \ldots, x_1 Y_d - x_d)$. After inverting x_1, S is clearly regular, and if a prime ideal P in S contains x_1, then S_P is regular. Thus S is a regular ring, whence integrally closed. The ideal $Q = x_1 S$ is a prime ideal, and gives a valuation ring $V = S_Q$. Let the corresponding valuation be v. For any $f \in R$, let $r = \mathrm{ord}_{\mathfrak{m}}(f)$. Then $f \in \mathfrak{m}^r \subseteq \mathfrak{m}^r S = x_1^r S$, so that $v(f) \geq r$. If $v(f) \geq r + 1$, then $f \in x_1^{r+1} S_Q \cap R = x_1^{r+1} S \cap R = \mathfrak{m}^{r+1} S \cap R$. It is straightforward to compute that this intersection is $\mathfrak{m}^{r+1}$, which is a contradiction. Thus v agrees with the order valuation of $\mathfrak{m}$. The residue field of v is $k(Y_2, \ldots, Y_d)$, proving the rest of the claims. $\square$

The theorem above provides a rich source of valuations. For example let R be a Noetherian integral domain with field of fractions K. For every regular local ring S such that $R \subseteq S \subseteq K$, we get a discrete rank one valuation that is non–negative on R by taking the order valuation associated to S. The set of all such S is plentiful; for example in general we can take any finitely generated R–subalgebra of K, take its integral closure, and take the associated localizations at height one primes of the integral closure. We explore this topic in more detail in later chapters.

It is worth noting that whenever the order function of a localization R_P gives a valuation v, then the set of all elements in the ring whose v–value is at least n equals $P^{(n)}$, the nth symbolic power of P. This observation allows one to study the growth of symbolic powers in the context of valuations. This has been done for example by Spivakovsky [275]; Cutkosky [52]; and Ein, Lazarsfeld and Smith [65].

6.8. Valuations and the integral closure of ideals

We prove in this section that valuations determine the integral closures of ideals and integral domains. We also prove that the integral closure of ideals in Noetherian rings is determined by the Noetherian valuation domains.

We need some preliminary results, such as that every ideal in a valuation domain is integrally closed. If the ideal is principal, this follows easily from Propositions 1.5.2 and 6.4.1, but here is a more general result:

Proposition 6.8.1 *Let R be an integral domain with field of fractions K. Let I be an ideal in R and let V be a valuation ring between R and K. Then $IV = \bar{I}V = \overline{IV}$.*

Proof: As $I \subseteq \bar{I}$, it follows that $IV \subseteq \bar{I}V$, and by persistence of integral closure, $\bar{I}V \subseteq \overline{IV}$. Now let $r \in \overline{IV}$. Let $r^n + a_1 r^{n-1} + \cdots + a_{n-1} r + a_n = 0$ be an equation of integral dependence of r over IV, with each $a_i \in I^i V$. There is a finitely generated ideal J contained in I such that $a_i \in J^i V$, $i = 1, \ldots, n$. Thus by Lemma 6.4.2 there exists $j \in J$ such that $JV = jV$, and so r satisfies an equation of integral dependence of degree n over jV. By Proposition 1.5.2, $r \in jV = JV \subseteq IV$, which proves that $\overline{IV} \subseteq IV$. □

Furthermore, valuations determine the integral closure of ideals:

Proposition 6.8.2 *Let R be an integral domain, not necessarily Noetherian, and let I be an ideal in R. Then*
$$\bar{I} = \bigcap_V IV \cap R,$$
where V varies over all valuation domains of the field of fractions of R that contain R. When R is Noetherian, in addition V may be taken to vary only over all discrete valuation domains of rank one.

Proof: By Proposition 6.8.1, $\bar{I} \subseteq \bigcap_V \bar{I}V \cap R = \bigcap_V IV \cap R$. To prove the other inclusion, let r be a non–zero element of $\bigcap_V IV \cap R$. Let S be the ring $R[\frac{I}{r}]$, i.e., the ring generated over R by the elements $\frac{x}{r}$, $x \in I$. Note that R and S have the same field of fractions. Thus by the choice of r, for all valuation rings V between S and the field of fractions, $r \in IV$. Hence for each such V, the ideal $\frac{I}{r}S$ of S extends to the unit ideal in V.

By Theorems 6.3.2 and 6.3.3, whenever J is a proper ideal of an integral domain S, there exists a valuation ring V between S and its field of fractions K such that $JV \neq V$. Furthermore, if S is Noetherian, V may be taken to be Noetherian as well.

It follows that $\frac{I}{r}S = S$. Thus we can write $1 = \sum_{i=1}^{n} \frac{a_i}{r^i}$ for some a_i in I^i. Multiplying this equation through by r^n yields an equation of integral dependence of r over I of degree n, so that r is integral over I. $\qquad\square$

Combining the result above and Proposition 1.1.5 yields:

Theorem 6.8.3 (Valuative criterion) *Let R be a ring, I be an ideal in R, and $r \in R$. The following are equivalent:*
(1) $r \in \bar{I}$,
(2) for all $P \in \mathrm{Min}(R)$ and for all valuation rings V between R/P and its field of fractions $\kappa(P)$, $r \in IV$.
In case R is Noetherian, the conditions above are equivalent to:
(3) for all $P \in \mathrm{Min}(R)$ and for all rank one discrete valuation rings V between R/P and $\kappa(P)$, $r \in IV$.

In the Noetherian case one can be even more selective with the discrete valuations that determine the integral closures of ideals:

Proposition 6.8.4 *Let R be a Noetherian domain and I an ideal in R. Then the integral closure $\bar{I}$ of I equals $\cap_V IV \cap R$, where V varies over those discrete valuation rings of rank one between R and its field of fractions for which the maximal ideal of V contracts to a maximal ideal of R.*

Proof: By Proposition 1.1.4, for all maximal ideals $\mathfrak{m}$ of R, $\overline{IR_\mathfrak{m}} = \bar{I}R_\mathfrak{m}$. As $\bar{I} = \cap_\mathfrak{m} \bar{I}R_\mathfrak{m} \cap R$, where $\mathfrak{m}$ varies over the maximal ideals of R, it suffices to prove that the proposition holds for the ideal $IR_\mathfrak{m}$ in $R_\mathfrak{m}$. Thus without loss of generality R is a Noetherian local ring with maximal ideal $\mathfrak{m}$.

Let r be a non–zero element of the intersection $\cap_V IV \cap R$. Set $S = R[\frac{I}{r}]$. If $Q = \mathfrak{m}S + \frac{I}{r}S$ is a proper ideal in S, by Theorem 6.3.3 there exists a discrete valuation ring V of rank one between S and the field of fractions K whose maximal ideal $\mathfrak{m}_V$ contains Q. Thus $\mathfrak{m}_V \cap R = \mathfrak{m}$, so by assumption, $r \in IV$. But also $IS \subseteq rS$, so that $rV = IV$ and $\frac{I}{r}V = V$, contradicting the properness of Q. Necessarily $\mathfrak{m}S + \frac{I}{r}S = S$. We can write $1 = \sum_{i=0}^{n} \frac{a_i}{r^i}$, with $a_0 \in \mathfrak{m}$, and for $i = 1, \ldots, n$, $a_i \in I^i$. Hence $r^n = \sum_{i=0}^{n} a_i r^{n-i}$, and thus $r^n(1 - a_0) = \sum_{i=1}^{n} a_i r^{n-i}$. As $1 - a_0$ is a unit, this equation can be rewritten as an equation of integral dependence of r over I, so that $r \in \bar{I}$. $\qquad\square$

Valuations enable easy proofs of some ideal inclusions. We give three examples below.

Corollary 6.8.5 *Let $(R, \mathfrak{m})$ be a Noetherian local ring and I an ideal in R. Then $\bar{I} = \cap_n \overline{I + \mathfrak{m}^n}$.*

Proof: Clearly $\bar{I} \subseteq \cap_n \overline{I + \mathfrak{m}^n}$.

By Theorem 6.8.3 we may assume that R is an integral domain. By Proposition 6.8.4, $\bar{I} = \cap_V IV \cap R$, where V varies over all discrete valuations V of rank one between R and its field of fractions whose maximal ideals contain $\mathfrak{m}$. Let $r \in R \setminus \bar{I}$. There exists a valuation V as above such that r is not in

IV. As $\mathfrak{m}$ is contained in the maximal ideal of V, there exists an integer n such that $r \notin (I + \mathfrak{m}^n)V$ since $IV = \cap_n (I + \mathfrak{m}^n)V$. Thus by Proposition 6.8.4, $r \notin \overline{I + \mathfrak{m}^n}$. $\qquad\square$

Corollary 6.8.6 *Let R be a Noetherian ring and let I and J be two ideals of R. Then $\overline{I} \cdot \overline{J} \subseteq \overline{IJ}$.*

Proof: Let $r \in \overline{I}, s \in \overline{J}$. For any $P \in \operatorname{Min} R$ and any $\kappa(P)$ valuation ring V containing R/P, by Theorem 6.8.3, $r \in IV$ and $s \in JV$. Hence $rs \in IJV$, whence since P and V were arbitrary, by Theorem 6.8.3 again, $rs \in \overline{IJ}$. $\qquad\square$

Corollary 6.8.7 *Let R be a Noetherian integral domain, and I and J ideals in R with $I = (a_1, \ldots, a_d)$. Then for any positive integer n, $\overline{JI^n} : I^n = \cap_i(\overline{JI^n} : a_i^n) = \overline{J}$.*

Proof: Certainly $\overline{JI^n} : I^n \subseteq \cap_i(\overline{JI^n} : a_i^n)$. Let $r \in \cap_i(\overline{JI^n} : a_i^n)$. Let V be a discrete valuation ring of rank one between R and its field of fractions. There exists i such that $IV = a_i V$. By the assumption on r, $a_i^n rV \subseteq \overline{JI^n}V$, so that $a_i^n rV \subseteq JI^n V = Ja_i^n V$, whence $r \in JV$. By Proposition 6.8.2, $r \in \overline{J}$. The other inclusion is by Corollary 6.8.6. $\qquad\square$

A consequence of this corollary is the following important proposition.

Proposition 6.8.8 *Let I be an ideal in a Noetherian ring R. Then for all $n \geq 1$, $\operatorname{Ass}(R/\overline{I^n}) \subseteq \operatorname{Ass}(R/\overline{I^{n+1}})$.*

Proof: Let $P \in \operatorname{Ass}(R/\overline{I^n})$. To prove that $P \in \operatorname{Ass}(R/\overline{I^{n+1}})$ we may localize at P and thus assume that P is the unique maximal ideal of R. If the height of P is zero, the conclusion follows at once, so we may assume that the height of P is positive. Write $P = \overline{I^n} : x$ for some $x \in R$. Then $P \subseteq \overline{I^{n+1}} : Ix$. If Ix is not contained in $\overline{I^{n+1}}$, then P is associated to $\overline{I^{n+1}}$. Assume that $Ix \subseteq \overline{I^{n+1}}$. We will reach a contradiction by proving that $x \in \overline{I^n}$. Using Proposition 1.1.5 it suffices to prove that for every minimal prime ideal Q of R, $x' \in \overline{(I')^n}$, where by I' and x' we denote the images of I and x in R/Q. Fix a minimal prime ideal Q. If $I \subseteq Q$, then since P has positive height and $Px \subseteq \overline{I^n}$, it follows that $x \in Q$ as well, proving that $x' \in \overline{(I')^n} = 0$. If I is not contained in Q, then after reducing modulo Q and using Corollary 6.8.7, $x' \in \overline{(I')^{n+1}} : I' \subseteq \overline{(I')^n}$. $\qquad\square$

Expressions about the integral closure of ideals involving valuation rings can also be translated into expressions involving valuations.

Definition 6.8.9 *Let R be an integral domain, V a valuation ring between R and its field of fractions. For every non–empty subset S of non–zero elements of R, $v(s) \geq 0$ for all $s \in S$. Whenever the set $\{v(x) \mid x \in S\}$ has a minimum, we define $v(S) = \min\{v(x) \mid x \in S\}$. Also, we define $v(0) = \infty$, and if I is a non–zero ideal in R, we define*

$$v(I) = \min\{v(x) \mid x \in I \setminus \{0\}\}.$$

In particular, when I is an ideal of R, $v(I)$ is defined whenever I is finitely generated, or when every subset of Γ consisting of elements greater than 0 has a minimum element. In particular, $v(I)$ is defined if v is a discrete valuation of rank one.

It is straightforward to prove that whenever $v(I)$ is defined, so is $v(I^n)$ for every positive integer n, and furthermore $v(I^n) = nv(I)$.

The following is a valuation analog of Proposition 6.8.1.

Proposition 6.8.10 *Let R be an integral domain with field of fractions K, v a valuation on K that is non–negative on R, and I an ideal in R such that $v(I)$ is defined (see definition above). Then $v(I) = v(\bar{I})$.*

Proof: As $I \subseteq \bar{I}$, $v(I) \geq v(\bar{I})$. For any $r \in \bar{I}$, $r^n + a_1 r^{n-1} + \cdots + a_{n-1} r + a_n = 0$ for some integer n and some $a_i \in I^i$. Then

$$nv(r) = v(r^n) \geq \min\{v(a_i r^{n-i}) \mid i = 1, \ldots, n\}$$
$$\geq \min\{iv(I) + (n-i)v(r) \mid i = 1, \ldots, n\}.$$

By cancelling, for some $i > 0$, $iv(r) \geq iv(I)$. Hence $v(r) \geq v(I)$, and thus $v(\bar{I})$ is defined and $v(\bar{I}) \geq v(I)$. $\qquad\square$

Corollary 6.8.11 *Assume that R is Noetherian, I an ideal in R and $r \in R$. Then $r \in \bar{I}$ if and only if there exists an integer n such that for all integers $m > n$, $r^m \in I^{m-n}$.*

Proof: Suppose that $r \in \bar{I}$. Then $I \subseteq I + (r)$ is a reduction, so that for some integer n, $I(I + (r))^n = (I + (r))^{n+1}$. Thus for all $m > n$, $(I + (r))^m = I^{m-n}(I + (r))^n \subseteq I^{m-n}$, whence $r^m \in I^{m-n}$.

Now assume that there exists n such that for all $m > n$, $r^m \in I^{m-n}$. Then for all $P \in \operatorname{Min} R$, $r^m \in I^{m-n}(R/P)$. Let v be any $\kappa(P)$–valuation that is non–negative on R/P, and is discrete and Noetherian. Then $v(r^m) \geq v(I^{m-n})$, or $v(r) \geq \frac{m-n}{m}v(I)$ for all $m > n$. As v is $\mathbb{Z}$–valued, it follows that $v(r) \geq v(I)$. As P and v were arbitrary, by Theorem 6.8.3, $r \in \bar{I}$. $\qquad\square$

Compare the following corollary with Exercise 1.5. The next corollary is inspired by the definition of tight closure and is quite important. Recall that R^o is the set of elements in R that are not in any minimal prime of R.

Corollary 6.8.12 *Let R be a Noetherian ring and I an ideal in R. An element r is in $\bar{I}$ if and only if there is an element $c \in R^o$ such that for infinitely many integers $m \geq 0$, $cr^m \in \overline{I^m}$. When this occurs, there exists an element $c' \in R^o$ such that $c'r^n \in I^n$ for all large n.*

Proof: First assume that $r \in \bar{I}$. By the previous corollary, $r \in \bar{I}$ if and only if there exists an integer n such that for all $m \geq n$, $r^m \in I^{m-n}$. Let $P_1, \ldots, P_l$ be the minimal primes of R. By relabeling them if necessary we may assume that $I \subseteq P_1 \cap \ldots \cap P_k$, and I is not in P_i for $k+1 \leq i \leq l$. Choose $d \in I^n \setminus (P_{k+1} \cup \cdots \cup P_l)$, and choose $e \in \cap_{k+1 \leq i \leq l} P_i$ with $e \notin \cup_{1 \leq i \leq k} P_i$ such that for some fixed N, $e(P_1 \cap \ldots \cap P_k)^N = 0$. Set $c' = d + e$. We claim that for all $m \geq N + n$, $c' r^m \in I^m$. Since $d \in I^n$, $d r^m \in I^n I^{m-n} \subseteq I^m$. Moreover, $e r^m \in e I^{n-m} \subseteq e I^N \subseteq e(P_1 \cap \ldots \cap P_k)^N = 0$. This proves our claim. Finally note that $c' \in R^o$ by the choice of d and e.

Conversely, assume that $c r^m \in \overline{I^m}$ for infinitely many $m \geq 0$. Let P be an arbitrary minimal prime of R, and let V be a Noetherian valuation domain of rank one lying between R/P and its field of fractions. For infinitely many m, $c r^m \in I^m V$ and $cV \neq 0$. Letting v be the corresponding valuation, this says that $v(c) \geq m(v(I) - v(r))$ for infinitely many m, so necessarily $v(I) \leq v(r)$. In other words, $r \in IV$. As this holds for all such V, by Theorem 6.8.3 it follows that $r \in \overline{I}$. □

With $I = (x_1, \ldots, x_n)$, it need not be the case that for every i, $\overline{I^n} : x_i^m = \overline{I^{n-m}}$. However, if the x_i form a system of parameters in a locally formally equidimensional ring, this does hold:

Corollary 6.8.13 *Let R be a locally formally equidimensional Noetherian ring, and let $(x_1, \ldots, x_n)$ be a parameter ideal. For all $m \geq 1$ and $i = 1, \ldots, n$,*

$$\overline{(x_1, \ldots, x_i)^m} : x_i = \overline{(x_1, \ldots, x_i)^{m-1}}.$$

Proof: Certainly $(x_1, \ldots, x_i)^{m-1} \subseteq \overline{(x_1, \ldots, x_i)^m} : x_i$. We proved on page 7 that $\overline{(x_1, \ldots, x_i)^m} : x_i$ is integrally closed, so it also contains $\overline{(x_1, \ldots, x_i)^{m-1}}$. This proves one inclusion. Now let $r \in \overline{(x_1, \ldots, x_i)^m} : x_i$. By Corollary 6.8.12, there exists an element $c \in R^o$ such that for all large k,

$$c(r x_i)^k \in (x_1, \ldots, x_i)^{mk} \subseteq x_i^k (x_1, \ldots, x_i)^{mk-k} + (x_1, \ldots, x_{i-1})^{mk-k}.$$

Let $u \in (x_1, \ldots, x_i)^{mk-k}$ and $v \in (x_1, \ldots, x_{i-1})^{mk-k}$ such that $c(r x_i)^k = x_i^k u + v$. In this case, $x_i^k (c r^k - u) = v$, so that $c r^k - u \in (x_1, \ldots, x_{i-1})^{mk-k} : x_i^k \subseteq \overline{(x_1, \ldots, x_{i-1})^{mk-k}}$ by Theorem 5.4.1. It follows that for all large k, $c r^k \in \overline{(x_1, \ldots, x_i)^{mk-k}}$. An application of Corollary 6.8.12 then gives that $r \in \overline{(x_1, \ldots, x_i)^{m-1}}$. □

Just as the integral closure of an ideal is determined by passage to valuation domains, so is the integral closure of an integral domain determined by intersecting valuation domains:

Proposition 6.8.14 *Let R be an integral domain. Then the integral closure of the ring R equals $\cap_V V$, where V varies over all the valuation domains between R and its field of fractions. If R is Noetherian, all the V may be taken to be Noetherian.*

Proof: Certainly the integral closure of R is contained in each V. Now let x be a non–zero element in $\cap_V V$. Write $x = a/b$ for some $a, b \in R$. If a is not integral over (b), then by the Proposition 6.8.4, there exists a valuation domain (Noetherian if R is Noetherian) such that $a \notin bV$. Hence $x = a/b \notin V$, contradicting the assumption. So $a \in \overline{(b)}$. Thus there exists an equation of integral dependence $a^n + r_1 ba^{n-1} + \cdots + r_n b^n = 0$ for some $r_i \in R$. Division through by b^n produces an equation of integral dependence of $x = a/b$ over R. So $\cap_V V$ equals the integral closure of the ring R. $\qquad\square$

6.9. The asymptotic Samuel function

Another numerical characterization of integral dependence is via the order function, using the valuative criterion:

Corollary 6.9.1 *Let R be a Noetherian ring, I an ideal in R, $r \in R \setminus \{0\}$, $c \in \mathbb{N}$. Then $r \in \overline{I^c}$ if and only if $\limsup_{m \to \infty} \frac{\mathrm{ord}_I(r^m)}{m} \geq c$.*

Proof: Without loss of generality $c > 0$. First assume that $r \in \overline{I^c}$. By Remark 1.2.3 there exists an integer n such that for all $m \geq n$, $r^m \in (I^c)^{m-n+1}$. Thus $\mathrm{ord}_I(r^m) \geq c(m-n+1)$ and

$$\limsup \frac{\mathrm{ord}_I(r^m)}{m} \geq \limsup \frac{c(m-n+1)}{m} = c.$$

Conversely, assume that $\limsup \frac{\mathrm{ord}_I(r^m)}{m} \geq c$. For arbitrary positive k, this means that for infinitely many m, $\mathrm{ord}_I(r^m) \geq cm - \frac{m}{k}$. Let P be a minimal prime ideal in R and let V be any rank one discrete valuation ring between R/P and $\kappa(P)$. Let v be its corresponding valuation. Then for infinitely many m, $r^m \in I^{\lfloor cm - \frac{m}{k} \rfloor}$, whence $mv(r) \geq \lfloor cm - \frac{m}{k} \rfloor v(I) > (cm - \frac{m}{k} - 1)\rfloor v(I)$, so that $v(r) > (c - \frac{1}{k} - \frac{1}{m})v(I)$. Since this holds for infinitely many positive integers m for each positive k, it follows that $v(r) \geq cv(I)$ for all v. By the Valuative criterion (Theorem 6.8.3), $r \in \overline{I^c}$. $\qquad\square$

In the corollary above, $\limsup$ can be replaced by $\lim$, by the following:

Lemma 6.9.2 (Rees [231]) *Let I be an ideal in a Noetherian ring R. For any $x \in R$,*

$$\lim_{n \to \infty} \frac{\mathrm{ord}_I(x^n)}{n}$$

exists.

Proof: Let $u = \limsup_{n \to \infty} \frac{\mathrm{ord}_I(x^n)}{n}$ (possibly ∞). Let N be an arbitrary number strictly smaller than u. Choose $n_0 \in \mathbb{N}_{>0}$ such that $\frac{\mathrm{ord}_I(x^{n_0})}{n_0} > N$. Let n be an arbitrary positive integer. Write $n = qn_0 + r$ for some $q, r \in \mathbb{N}$ with $r < n_0$. Since clearly for all $i, j \in \mathbb{N}_{>0}$, $\mathrm{ord}_I(x^{i+j}) \geq \mathrm{ord}_I(x^i) + \mathrm{ord}_I(x^j)$, it follows that

$$\frac{\mathrm{ord}_I(x^n)}{n} = \frac{\mathrm{ord}_I(x^{qn_0+r})}{qn_0 + r} \geq q\frac{\mathrm{ord}_I(x^{n_0})}{qn_0 + r} + \frac{\mathrm{ord}_I(x^r)}{qn_0 + r}$$

$$\geq \frac{qn_0}{qn_0 + r}\frac{\mathrm{ord}_I(x^{n_0})}{n_0} \geq \frac{qn_0}{qn_0 + r}N \geq \frac{qn_0}{n_0(q+1)}N.$$

Thus $u \geq N$. Since this holds for all N, the limit exists. $\qquad\square$

Definition 6.9.3 *For an ideal I in a ring R, the function $\overline{v}_I : R \to \mathbb{R}_{\geq 0} \cup \{\infty\}$ defined by $\overline{v}_I(x) = \lim_{n \to \infty} \frac{\mathrm{ord}_I(x^n)}{n}$, is called the* **asymptotic Samuel function**.

We prove with Rees valuations in Chapter 10 that the range of $\overline{v}_I$ is a subset of $\mathbb{Q}_{\geq 0} \cup \{\infty\}$.

6.10. Exercises

6.1 Prove that for a field k and a totally ordered abelian group Γ, the function $k[\Gamma] \setminus \{0\} \to \Gamma$ defined as $\sum_g k_g g \mapsto \min\{g : k_g \neq 0\}$ is a (partial) valuation whose value group is Γ and whose residue field is k.

6.2 Suppose that the value group Γ of a valuation v is finitely generated. Prove that $\mathrm{rk}\, v = \mathrm{rat.rk}\, v$ if and only if the value group of v is isomorphic to $\mathbb{Z}^{\mathrm{rk}\, v}$ ordered lexicographically.

6.3 Prove that any ring between a valuation domain V and its field of fractions is a localization of V.

6.4 Prove that the radical of a proper ideal in a valuation domain is a prime ideal.

6.5 Let $(R, \mathfrak{m})$ be a local domain that is not a field. Prove that R is a Noetherian valuation domain if and only if R is a discrete valuation domain of rank 1.

6.6 Let K be a field, V a K–valuation ring, and S the set of all K–valuation rings that contain V. Prove that S is totally ordered (by inclusion). Is the set of all K–valuation rings totally ordered?

6.7 Let R be an integral domain and v a valuation on the field of fractions of R. Let Γ be the value group of v, and $\gamma \in \Gamma$, and assume that for all $r \in R \setminus \{0\}$, $v(r) \geq 0$. Let $I_\gamma = \{r \in R \,|\, v(r) \geq \gamma\}$. Prove that I_γ is integrally closed in R.

6.8 Let R be a local integral domain with field of fractions K and infinite residue field. Let $v_1, \ldots, v_n$ be discrete K–valuations of rank one such that $v_i(r) \geq 0$ for all $r \in R \setminus \{0\}$. Let I be an ideal in R. Prove that there exists $x \in I$ such that for all $i = 1, \ldots, n$, $v_i(x) = v_i(I)$.

6.9 Let R be a regular ring. Prove that for any non–minimal prime ideal P in R there exists a natural discrete valuation v_P of rank one satisfying the property that for any $x \in R$, $v_P(x) = n$ if and only if $x \in P^n R_P \setminus P^{n+1} R_P$.

6.10 Let Γ be a subgroup of $\mathbb{Q}$ and let R be a polynomial ring in two variables over a field. Prove that Γ is the value group of a valuation on R that is non–negative on R.

6.11 Let k be a field, $X_1, \ldots, X_d$ variables over k, and $R = k[[X_1, \ldots, X_d]]$. Prove that the valuation ring of the $(X_1, \ldots, X_d)$–adic valuation on R is $k(\frac{X_2}{X_1}, \ldots, \frac{X_d}{X_1})[[X_1]]$.

6.12   Let $R = k[X, Y, Z]$, polynomial ring in variables X, Y, and Z over a field k. The ring $V = k(\frac{X+Y}{X^n}, \frac{Z}{X})[X]_{(X)}$ is a discrete valuation domain between R and its field of fractions. Let v be the corresponding valuation. Prove that $v(X) = v(Y) = v(Z) = 1$ and $v(X + Y) = n$. (Hint: cf. Example 6.7.6.)

6.13   Let k be a field, $X_1, \dots, X_d$ variables over k and R the polynomial ring $k[X_1, \dots, X_d]$. Assume that $d \geq 3$. Let n be a positive integer. Prove that there exists a discrete valuation v of rank one on the field of fractions of R that is non–negative on R and such that $v(X_1) = \cdots = v(X_d) = 1$ and $v(X_1 + X_2) = v(X_2 + X_3) = n$.

6.14   Let R be a Noetherian ring and I an ideal in R.
   (i) Prove that $\bar{I}$ has a primary decomposition all of whose primary components are integrally closed.
   (ii) Find an ideal I such that $I = \cap_{i=1}^n q_i$ is a primary decomposition, but $\bar{I} \neq \cap_{i=1}^n \bar{q_i}$.

6.15   Let V be a valuation domain containing a field, and t a variable over V. Prove that $V[[t]]$ is integrally closed if and only if V is of rank one. (Cf. Exercise 2.24.)

6.16   Let R be a principal ideal domain and K its field of fractions. Prove that the only valuation rings in K containing R are the rings of the form $R_{(p)}$, where p is a prime element in R.

6.17   Let R be an integral domain with field of fractions K. Prove that the following are equivalent:
   (i) R is integrally closed.
   (ii) R is an intersection of K–valuation domains.
   (iii) R is an intersection of K–valuation domains V such that $\mathfrak{m}_V \cap R$ is a maximal ideal in R and such that $V/\mathfrak{m}_V$ is algebraic over $R/(\mathfrak{m}_V \cap R)$.

6.18   Let $(R, \mathfrak{m})$ be a local integrally closed domain with field of fractions K, $x \in K^*$ such that x^{-1} is not in R. Assume that x satisfies a polynomial with coefficients in R and that one of these coefficients is a unit. Prove that $x \in R$.

6.19   (Nagata [213, (11.10), (11.11)]) Let K be a field and let $V_1, \dots, V_n$ be K–valuation rings such that for all $i \neq j$, $V_i \not\subseteq V_j$. Set R to be the ring $\cap_i V_i$.
   (i) Prove that for any $x \in K$ there exists an integer m such that $(1 + x + x^2 + \cdots + x^m)^{-1}$ and $x(1 + x + x^2 + \cdots + x^m)^{-1}$ both belong to R.
   (ii) Prove that for each $i = 1, \dots, n$, the localization of R at $\mathfrak{m}_{V_i}$ equals V_i.
   (iii) Prove that $\{\mathfrak{m}_{V_1} \cap R, \dots, \mathfrak{m}_{V_n} \cap R\}$ is the set of all the maximal ideals of R.
   (iv) If $V_1, \dots, V_n$ are all Noetherian, prove that R is a principal ideal domain.

6.20   Let K be a field. Two K–valuation domains V and W are said to be **independent** if the smallest subring of K containing V and W is K.

(i) Prove that distinct discrete K–valuation rings are independent.

(ii) (Approximation of valuations) Let $v_1, \ldots, v_n$ be K–valuations with respective value groups $\Gamma_1, \ldots, \Gamma_n$. Assume that the valuation domains of the v_i are pairwise independent. Let $\gamma_i \in \Gamma_i$ and $x_i \in K$, $i = 1, \ldots, n$. Then there exists $x \in K$ such that for all $i = 1, \ldots, n$, $v_i(x - x_i) = \gamma_i$.

(iii) Let $v_1, \ldots, v_n$ distinct discrete valuations of rank one defined on K. Let $k_1, \ldots, k_n$ be arbitrary integers and $x_1, \ldots, x_n$ arbitrary non–zero elements of K. Prove that there exists $x \in K$ such that for all $i = 1, \ldots, n$, $v_i(x_i - x) = k_i$.

6.21   (Nagata [213, (11.4)]) Let K be a field, V a K–valuation domain, and W a $(V/\mathfrak{m}_V)$–valuation domain. Prove that $U = \{x \in V \mid x + \mathfrak{m}_V \in W\}$ is a K–valuation domain such that $U_{\mathfrak{m}_V} = V$ and $U/\mathfrak{m}_V = W$. Moreover, prove that if U is a discrete valuation ring of rank n and W is a discrete valuation ring of rank m, then V is a discrete valuation ring of rank $m + n$.

The construction in the previous exercise is related to the "D + M" construction in Gilmer [91, Appendix 2], in which it is assumed that $V/\mathfrak{m}_V$ is a contained in V. More on this construction is in the exercise below.

6.22   (Gilmer [91, Appendix 2, page 560]) Let V be a valuation ring with field of fractions K, $V \neq K$. Suppose that V is of the form $L + M$, where L is a field and M is the maximal ideal of V. Let D be a proper subring of L, and let $R = D + M$.

(i) Prove that if X is a variable over L, then $L[[X]]$ and $L[X]_{(X)}$ are valuation domains of the form $L + M$.

(ii) Prove that M is the conductor of R in V.

(iii) Prove that V is the complete integral closure of R.

(iv) Prove that if $\overline{D}$ is the integral closure of D in L, then $\overline{D} + M$ is the integral closure of R in K.

(v) Prove that every ideal in R containing M is of the form $I + M$, where I is an ideal of D. Prove that I is maximal/prime/primary if and only if $I + M$ is maximal/prime/primary. A generating set of I in D is also a generating set of $I + M$ in R.

(vi) Prove that $\dim R = \dim D + \dim V$.

(vii) Prove that R is a valuation ring if and only if D is a valuation ring.

(viii) Prove that R is Noetherian if and only if V is Noetherian, D is a field, and $[L : D] < \infty$.

6.23   (Nagata [213, (11.9)]) Let R be an integral domain with field of fractions K. Let $P_0 \subseteq P_1 \subseteq \cdots \subseteq P_n$ be a chain of prime ideals in R. Prove that there exists a K–valuation domain V with prime ideals

$Q_0 \subseteq Q_1 \subseteq \cdots \subseteq Q_n$ such that for each $i = 1, \ldots, n$, $Q_i \cap R = P_i$. (Hint: Use Theorem 6.3.2, induction, and Exercise 6.21.)

6.24 (Abhyankar [6, Lemma 7]) Let K be a field and $(R_1, \mathfrak{m}_1) \subseteq (R_2, \mathfrak{m}_2) \subseteq \cdots$ a sequence of integrally closed domains with field of fractions K, such that for all $n = 2, 3, \ldots$, $\mathfrak{m}_n \cap R_{n-1} = \mathfrak{m}_{n-1}$. Set $R = \cup_n R_n$.

  (i) Prove that R is an integrally closed domain with field of fractions K, maximal ideal $\mathfrak{m} = \cup_n \mathfrak{m}_n$ and residue field $\cup_n (R_n / \mathfrak{m}_n)$.

  (ii) Assume that R is not a valuation ring. Prove that there exist infinitely many K–valuations v such that for each n, v has center $\mathfrak{m}_n$ on R_n and the residue field of v has positive transcendence degree over $R_n / \mathfrak{m}_n$. (Hint: use Exercise 6.18.)

6.25 Let $(R, \mathfrak{m})$ be a Noetherian regular local ring and I an integrally closed $\mathfrak{m}$–primary ideal. Prove that there exists an integrally closed $\mathfrak{m}$–primary ideal J contained in I such that $I/J \cong R/\mathfrak{m}$. (This was first proved by Lipman [191] and Noh [214] in dimension two. Watanabe [319] proved that if $(R, \mathfrak{m})$ is an excellent normal ring with $R/\mathfrak{m}$ algebraically closed, then every integrally closed $\mathfrak{m}$–primary ideal I has an adjacent integrally closed ideal J as above. A monomial ideal version is proved in Crispin Quiñonez [48].)

6.26 Let k be a field, X_{ij} for $i = 1, \ldots, m$ and $j = 1, \ldots, n$ variables over k. Set $R = k[X_{ij} \,|\, i, j]$. Let M be the $m \times n$ matrix whose entry (i, j) is X_{ij}. For any $r \leq \min\{m, n\}$ let $I_r(M)$ be the ideal in R generated by the $r \times r$ minors of M. It is well known that $I_r(M)$ is a prime ideal.

  (i) Let v_r be the valuation on the field of fractions of R associated to the prime ideal $I_r(M)$ as in Exercise 6.9. Let x be any $s \times s$ minor of M. Prove that $v_r(x) = \max\{s - r + 1, 0\}$.

  (ii) Let $u_1, \ldots, u_k$ and $v_1, \ldots, v_l$ be two non–increasing sequences of positive integers, with all $u_i, v_j \leq \min\{m, n\}$. For any integer u let $\widehat{u}$ denote an arbitrary $u \times u$ minor of M. Prove that for all r, $v_r(\widehat{u_1} \cdots \widehat{u_k}) \leq v_r(\widehat{v_1} \cdots \widehat{v_l})$ if and only if for all $i = 1, \ldots, r$, $\sum_{j=1}^i u_j \leq \sum_{j=1}^i v_j$.

  (iii) Prove that every $I_r(M)$ is a contraction of an ideal in a valuation overring. Show by example that not every power of $I_r(M)$ is a contraction of an ideal in a valuation overring.

6.27 Construct a non–Noetherian valuation domain.

6.28 Let R be a Cohen–Macaulay integral domain, $x, y \in R$ such that $\operatorname{ht}(x, y) = 2$. Prove that there exists a discrete valuation domain V of rank one between R and its field of fractions such that $xV = yV \neq V$.

7
Derivations

7.1. Analytic approach

Let $\mathcal{Y}$ be a subset of $\mathbb{C}^n$, typically open and connected. Recall that a function $f : \mathcal{Y} \to \mathbb{C}$ is called **analytic, (complex-)differentiable**, or **holomorphic** (any and all of the three names) if f is complex–differentiable at every point of $\mathcal{Y}$.

The set of all holomorphic functions on $\mathcal{Y}$ forms a commutative associative ring $\mathcal{O}$ with identity under pointwise addition and multiplication.

First we consider the case $\mathcal{Y} = \mathbb{C}^n$. Let $\mathcal{O}_{\mathbb{C}^n}$ be the ring of holomorphic functions on $\mathbb{C}^n$. Every holomorphic function on $\mathbb{C}^n$ can be written locally as a convergent power series in n variables $X_1, \ldots, X_n$ with coefficients in $\mathbb{C}$.

What maximal ideals does $\mathcal{O}_{\mathbb{C}^n}$ have? For any point $(\alpha_1, \ldots, \alpha_n) \in \mathbb{C}^n$, $(X_1 - \alpha_1, \ldots, X_n - \alpha_n)\mathcal{O}_{\mathbb{C}^n}$ is the ideal in $\mathcal{O}_{\mathbb{C}^n}$ of all holomorphic functions vanishing at $(\alpha_1, \ldots, \alpha_n)$. Clearly it is a maximal ideal in $\mathcal{O}_{\mathbb{C}^n}$. However, $\mathcal{O}_{\mathbb{C}^n}$ contains also other maximal ideals that are not of this form, and $\mathcal{O}_{\mathbb{C}^n}$ is not Noetherian. For example, with $n = 1$ and $X = X_1$, the ideal I generated by all $\cos(2^{-n}X)$, $n = 1, 2, \ldots$, is not finitely generated. Furthermore, I is not contained in any prime ideal of the form $X - \alpha$, with α a complex number.

At localizations at maximal ideals corresponding to points, $\mathcal{O}_{\mathbb{C}^n}$ is Noetherian:

Lemma 7.1.1 *Locally at each maximal ideal of holomorphic functions vanishing at a point, $\mathcal{O}_{\mathbb{C}^n}$ is Noetherian.*

Proof: Let R be the ring of holomorphic functions on $\mathbb{C}^n$. Every element of R can be written locally as a convergent power series in n variables $X_1, \ldots, X_n$ with coefficients in $\mathbb{C}$. It suffices to prove that the R_M is Noetherian, where $M = (X_1, \ldots, X_n)$.

Let I be a non–zero ideal in R_M. We will prove that I is finitely generated. It suffices to prove that I is finitely generated after applying a ring automorphism. Clearly after a homogeneous linear change of variables I contains a power series f such that the lowest degree term appearing in f has degree m and that X_n^m appears in f with a non–zero complex coefficient. By the Weierstrass Preparation Theorem, I contains an element of the form $f = X_n^m + f_1 X_n^{m-1} + \cdots + f_m$, with $f_1, \ldots, f_m$ holomorphic functions in variables $X_1, \ldots, X_{n-1}$. Hence an argument similar to the proof of the Hilbert's Basis Theorem shows that I is finitely generated. Namely, any element g in $I \cap R$ can be written as $g = rf + h$, where $r \in R$ and h is a polynomial in X_n

of degree at most $m - 1$ and with coefficients being holomorphic functions in $X_1, \ldots, X_{n-1}$. Let S_k be the set of elements in $I \cap R$ of X_n–degree at most k, and let J_k be the ideal generated by the coefficients of X_n^k in elements of S_k. By induction on n, for each $k = 1, \ldots, m$, J_k is finitely generated in the localized $\mathcal{O}_{\mathbb{C}^{n-1}}$. Let $I_k = (h_{k1}, \ldots, h_{kl_k})R$, where $h_{kj} \in S_k$, such that J_k is generated by the coefficients of X_n^k in $h_{k1}, \ldots, h_{kl_k}R$. The claim is that $I = (f, h_{kj} \,|\, k, j)R_M$. Certainly I contains f and all the h_{kj}. For the other inclusion it suffices to prove that $I \cap R \subseteq (f, h_{kj} \,|\, k, j)R_M$. Let $g \in I \cap R$. Write $g = rf + h$ as above. If $h = 0$, we are done. Otherwise let $k = \deg_{X_n}(h)$, so that $h \in S_k \setminus S_{k-1}$. Then the leading coefficient of h is in the ideal J_k, so after subtracting an appropriate linear combination of the h_{kj}, h is either 0 or an element of S_{k-1}. By repeating this, we see that g is a linear combination of f and the h_{kj}. $\qquad\square$

This ring structure makes the pair $(\mathbb{C}^n, \mathcal{O}_{\mathbb{C}^n})$ into a **locally ringed space**. A subset D of $\mathbb{C}^n$ is **closed** if and only if D is the common zero set of a subset $\mathcal{I}$ of all holomorphic functions on $\mathcal{O}_{\mathbb{C}^n}$.

Similarly, if $U \subseteq \mathbb{C}^n$ is the polydisc $\{|z_i| < 1 \,|\, i = 1, \ldots, n\}$, then the ring $\mathcal{O}_U$ of all holomorphic functions on U is a locally ringed space.

The following is a more general result:

Definition 7.1.2 ([109, page 438]) *A **complex analytic space** is a topological space $\mathcal{Y}$ together with a sheaf of rings $\mathcal{O}_{\mathcal{Y}}$, that can be covered by open sets U_i, such that for each i, there exist an integer n and holomorphic functions $f_1, \ldots, f_q$ on $U = \{|z_i| < 1 \,|\, i = 1, \ldots, n\} \subseteq \mathbb{C}^n$, such that $\mathcal{O}_{U_i}$ is isomorphic, as a locally ringed space, to the sheaf $\mathcal{O}_U/(f_1, \ldots, f_q)$.*

When the underlying space $\mathcal{Y}$ is the unit disc in $\mathbb{C}$, the structure sheaf is locally a convergent power series ring in one variable, which is a discrete valuation ring of rank one. Furthermore, every discrete valuation ring that arises as a localization of a complex analytic variety (the set of common solutions of several equations involving analytic functions) is isomorphic to a ringed space of the unit disc. The general valuative criterion of integrality as in Theorem 6.8.3 or Proposition 6.8.4 has an analog for complex analytic spaces (see also Lejeune-Jalabert and Teissier [182]):

Theorem 7.1.3 (Valuative criterion for complex analytic spaces) *Let $\mathcal{I}$ be a coherent sheaf of ideals on a complex space $\mathcal{Y}$, and $h \in \Gamma(\mathcal{Y}, \mathcal{O}_{\mathcal{Y}})$. Let D be the unit disc in $\mathbb{C}$. Then $h \in \Gamma(\mathcal{Y}, \bar{\mathcal{I}})$ if and only if for every morphism $\varphi : D \to \mathcal{Y}$, $h \circ \varphi \in \Gamma(D, \varphi^{-1}\mathcal{I})$.*

A consequence of this analytic criterion is the following:

Corollary 7.1.4 *Let R be the convergent power series $\mathbb{C}\{X_1, \ldots, X_n\}$ in n variables $X_1, \ldots, X_n$ over $\mathbb{C}$, and $f \in R$ such that $f(\underline{0}) = 0$. Then*

$$f \in \overline{\left(X_1 \frac{\partial f}{\partial X_1}, \ldots, X_n \frac{\partial f}{\partial X_n} \right)}.$$

In Lemma 7.1.1 we proved that $\mathbb{C}\{X_1, \ldots, X_n\}$ is a local ring with maximal ideal $(X_1, \ldots, X_n)$. It is clearly of dimension n, and thus a regular local ring.

Proof: Let D be the unit disc in $\mathbb{C}$. Let $\varphi : D \to \mathbb{C}^n$ be a morphism (of locally ringed spaces). Assume that the origin 0 in $\mathbb{C}^n$ is $\varphi(d)$ for some $d \in D$. There is a corresponding map of rings $\varphi^{\#} : \mathcal{O}_{\mathbb{C}^n} \to \mathcal{O}_D$, with the induced map of local rings $(\mathcal{O}_{\mathbb{C}^n})_0 = R \to (\mathcal{O}_D)_d$. But $(\mathcal{O}_D)_d \cong \mathbb{C}\{t\}$, a convergent power series ring in one variable t over $\mathbb{C}$. Thus by the valuative criterion it suffices to prove that for any local map $\psi : R \to \mathbb{C}\{t\}$,

$$\psi(f) \in \psi\left(X_1 \frac{\partial f}{\partial X_1}, \ldots, X_n \frac{\partial f}{\partial X_n} \right) \mathbb{C}\{t\}.$$

Note that for any element $g \in R$ with $g(\underline{0}) = 0$, $\psi(g)\mathbb{C}\{t\} = t\frac{d\psi(g)}{dt}\mathbb{C}\{t\}$. This clearly holds if $g = 0$, so we assume that g is non–zero. Then $\psi(g) = t^m u$ for some unit $u \in \mathbb{C}\{t\}$. By assumption that $g(\underline{0}) = 0$, $m > 0$. Then

$$\psi(g)\mathbb{C}\{t\} = t \cdot t^{m-1}\mathbb{C}\{t\} = t \cdot t^{m-1}\left(mu + t\frac{du}{dt} \right)\mathbb{C}\{t\}$$

$$= t \cdot \frac{d(t^m u)}{dt}\mathbb{C}\{t\} = t \cdot \frac{d\psi(g)}{dt}\mathbb{C}\{t\}.$$

By the chain rule,

$$\psi(f) \in t\frac{d\psi(f)}{dt}\mathbb{C}\{t\} = t\left(\sum_{i=1}^n \frac{\partial f}{\partial X_i}(\psi(X_1), \ldots, \psi(X_n))\frac{d\psi(X_i)}{dt} \right)\mathbb{C}\{t\}$$

$$\subseteq \sum_{i=1}^n \left(\psi\left(\frac{\partial f}{\partial X_i} \right) t\frac{d\psi(X_i)}{dt}\mathbb{C}\{t\} \right)$$

$$= \sum_{i=1}^n \psi\left(\frac{\partial f}{\partial X_i} \right) \psi(X_i)\mathbb{C}\{t\} = \sum_{i=1}^n \psi\left(X_i\frac{\partial f}{\partial X_i} \right)\mathbb{C}\{t\},$$

which proves the corollary. $\square$

An analogous result with essentially the same proof holds for power series rings as well, but for this we have to use some results from future sections and chapters. We state and prove it here for completeness.

Theorem 7.1.5 *Let k be a field of characteristic zero, $X_1, \ldots, X_n$ variables over k, and $R = k[[X_1, \ldots, X_n]]$, the power series ring. Let $f \in R$ be a non–unit (its constant coefficient is zero). With $\frac{\partial f}{\partial X_i}$ being defined formally,*

$$f \in \overline{\left(X_1\frac{\partial f}{\partial X_1}, \ldots, X_n\frac{\partial f}{\partial X_n} \right)}.$$

Proof: We use the valuative criterion from Proposition 6.8.4. It suffices to prove that for any Noetherian valuation ring V containing R and with the same field of fractions, $f \in (X_1 \frac{\partial f}{\partial X_1}, \ldots, X_n \frac{\partial f}{\partial X_n})V$. By Proposition 1.6.2, it suffices to show that $f \in (X_1 \frac{\partial f}{\partial X_1}, \ldots, X_n \frac{\partial f}{\partial X_n})\widehat{V}$, where $\widehat{V}$ is the completion of V in the topology defined by the maximal ideal. By Cohen's Structure Theorem, $\widehat{V} = L[[t]]$ for some field L of characteristic zero and some indeterminate t over L. Under the inclusion map $\varphi : R \to \widehat{V}$, $X_1, \ldots, X_n$ map to non–units. In other words, for each $i = 1, \ldots, n$, there exist a positive integer n_i and a unit u_i in $\widehat{V}$ such that $\varphi(X_i) = u_i t^{n_i}$. By the chain rule,

$$\frac{d\varphi(f)}{dt} = \sum_{i=1}^{n} \varphi\left(\frac{\partial f}{\partial X_i}\right) \cdot \left(\frac{d\varphi(X_i)}{dt}\right).$$

As $\widehat{V}$ is a valuation domain, there exists i such that $\frac{d\varphi(f)}{dt}$ is a multiple of $\varphi(\frac{\partial f}{\partial X_i}) \cdot (\frac{d\varphi(X_i)}{dt})$. But up to multiplication by a unit in $\widehat{V}$, $\varphi(f)$ equals $t\frac{d\varphi(f)}{dt}$, so that $\varphi(f)$ is a $\widehat{V}$–multiple of $t \cdot \varphi(\frac{\partial f}{\partial X_i}) \cdot (\frac{d\varphi(X_i)}{dt})$, whence a $\widehat{V}$–multiple of $\varphi(\frac{\partial f}{\partial X_i})\varphi(X_i)$. This proves the theorem. □

Note that Corollary 7.1.4 and Theorem 7.1.5 prove that in a convergent power series ring over $\mathbb{C}$ and in a power series ring over an arbitrary field of characteristic zero,

$$f \in (X_1, \ldots, X_n) \overline{\left(\frac{\partial f}{\partial X_1}, \ldots, \frac{\partial f}{\partial X_n}\right)}.$$

However, it is an open question whether

$$f \in \overline{(X_1, \ldots, X_n)\left(\frac{\partial f}{\partial X_1}, \ldots, \frac{\partial f}{\partial X_n}\right)}.$$

A special case is when f is homogeneous in $X_1, \ldots, X_n$ of degree d. It is easy to verify in this case that

$$d \cdot f = \sum_{i=1}^{n} X_i \frac{\partial f}{\partial X_i}. \tag{7.1.6}$$

This is called **Euler's formula**. In particular, $f \in (\frac{\partial f}{\partial X_1}, \ldots, \frac{\partial f}{\partial X_n})$ whenever f is homogeneous.

A similar formula holds if f is **quasi–homogeneous**, that is, if it is possible to assign degrees to the variables to make f homogeneous. Remarkably, Saito [253] proved that whenever $(0, \ldots, 0)$ is an isolated critical point of f, $f \in (\frac{\partial f}{\partial X_1}, \ldots, \frac{\partial f}{\partial X_n})$ if and only if after a biholomorphic change of coordinates, f is quasi–homogeneous.

We quote another criterion for integral dependence for complex analytic spaces (see Lejeune-Jalabert and Teissier [182] or Lipman and Teissier [193]):

Theorem 7.1.7 *Assume that the subspace of $\mathcal{X}$ defined by $\mathcal{I}$ is nowhere dense in a neighborhood of some point $x \in \mathcal{X}$. Set $R = \mathcal{O}_{\mathcal{X},x}$ and $I = \mathcal{I}_x \subseteq R$. Let*

$(f_{1,x}, \ldots, f_{m,x})$ be generators for I, where the f_i generate $\Gamma(U, \mathfrak{I})$ for some open neighborhood U of x. Then given $h \in \Gamma(U, \mathcal{O}_X)$, $h_x \in \bar{I}$ if and only if there exists a neighborhood $U' \subseteq U$ of x and a real constant $C > 0$ such that for every $y \in U'$,

$$|h(y)| \le C \cdot \sup\{|f_i(y)| \mid i = 1, \ldots, m\}.$$

7.2. Derivations and differentials

In this section we prove a result due to Hübl [131] that shows the close connection between the module of differentials and the integral closure of ideals.

Definition 7.2.1 *Let k be a commutative ring and let R be a k–algebra and M an R–module. A **k–derivation** $D : R \to M$ is a k–linear map from R to M satisfying $D(ab) = aD(b) + bD(a)$ for all $a, b \in R$. The set of all k–derivations from R to M is denoted $\mathrm{Der}_k(R, M)$.*

The fact that D is k–linear forces $D(a) = 0$ if $a \in R$ is the image of an element of k.

Definition 7.2.2 *Let k be a commutative ring and let R be a k–algebra. A* **universally finite module of differentials with a universally finite derivation** *is a finite R–module $\widetilde{\Omega^1_{R/k}}$ and a k–derivation $d = d_{R/k} : R \to \widetilde{\Omega^1_{R/k}}$ with the following universal property: if M is a finite R–module and $D : R \to M$ is a k–derivation, then there exists a unique R–homomorphism $f : \widetilde{\Omega^1_{R/k}} \to M$ such that $D = f \circ d$.*

See [178] for a detailed treatment of this definition and for the following remarks.

Many rings do not have a universally finite derivation. But the following rings do:
(1) R is essentially of finite type over k.
(2) k is a field with a valuation and R is an analytic k–algebra, i.e., R is module–finite over a ring of convergent power series over k.
(3) k is a field, R is complete, and the residue field of R is a finitely generated field extension of k.

Assume that $(R, \mathfrak{m})$ is a local domain with a universally finite derivation. Let K be the field of fractions of R. By $(\Omega^1_{R/k})^*$ we denote $\widetilde{\Omega^1_{R/k}}$ modulo its torsion submodule, and by d^* the composite map obtained by composing $d_{R/k}$ with the natural surjection of $\widetilde{\Omega^1_{R/k}}$ onto $(\Omega^1_{R/k})^*$. The following proposition is proved in [131]:

Proposition 7.2.3 *Let $(R, \mathfrak{m})$ be a local domain with a universally finite derivation. Let K be the field of fractions of R. Then $(\Omega^1_{R/k})^*$ is an $\mathfrak{m}$–adically separated and torsion–free R–module and d^* is a k–derivation. Furthermore, if M is an $\mathfrak{m}$–adically separated and torsion–free R–module and $D : R \to M$ is*

a k–derivation, then there exists a unique R–homomorphism $f : (\Omega^1_{R/k})^* \to M$
such that D is the composite of f and d^*. *There is a canonical isomorphism*

$$\mathrm{Hom}_R((\Omega^1_{R/k})^*, M) \to \mathrm{Der}_k(R, M)$$

taking f to $f \circ d^*$.

Theorem 7.2.4 *Let k be a field of characteristic* 0 *and let* $(R, \mathfrak{m})$ *be a local domain that is a k–algebra such that the derivation* $d^* : R \to (\Omega^1_{R/k})^*$ *exists. Let I and J be two proper ideals of R. Assume that*

$$d^*(I) \subseteq Rd^*(J) + I(\Omega^1_{R/k})^*.$$

Then $I \subseteq \bar{J}$.

Proof: By Proposition 6.8.4 it suffices to prove that if K is the quotient field of R and V is a rank one discrete valuation ring of K containing R such that the maximal ideal of V contracts to the maximal ideal of R, then $IV \subseteq JV$. It suffices to prove this containment after completing V. By the Cohen–Structure Theorem, the completion of V is isomorphic to $L[[t]]$ for some field L containing k. We define a derivation $\delta : R \to L[[t]]$ by first taking the injection of R into V and V into $L[[t]]$, then taking partial derivatives with respect to t. As $L[[t]]$ is $\mathfrak{m}$–adically separated and a torsion–free R–module, there exists an R–homomorphism $f : (\Omega^1_{R/k})^* \to L[[t]]$ such that $\delta = f \circ d^*$. Hence

$$\delta(I) = f(d^*(I)) \subseteq f(Rd^*(J) + I(\Omega^1_{R/k})^*) \subseteq \delta(J)L[[t]] + IL[[t]].$$

Since $IL[[t]]$ is a proper ideal, we have that $\delta(I)L[[t]]$ is not contained in $IL[[t]]$. Hence we must have that $\delta(I)L[[t]] \subseteq \delta(J)L[[t]]$, which implies that $IL[[t]] \subseteq JL[[t]]$ as needed. $\square$

Corollary 7.2.5 *Let k be a field of characteristic* 0 *and let* $(R, \mathfrak{m})$ *be a local domain that is a k–algebra such that the derivation defined above,* $d^* : R \to (\Omega^1_{R/k})^*$, *exists. Let* $r \in \mathfrak{m}$ *such that* $d^*(r) \in J(\Omega^1_{R/k})^*$. *Then* $r \in \bar{J}$.

Proof: Apply the above theorem to the ideal $I = (J, r)$. We have that $d^*(I) = d^*(J + Rr) \subseteq Rd^*(J) + Rd^*(r) + r(\Omega^1_{R/k})^* \subseteq Rd^*(J) + I(\Omega^1_{R/k})^*.$ $\square$

The next corollary is one of the most important and non–trivial ways in which integral closure arises.

Corollary 7.2.6 *Let k be a field of characteristic* 0, *and let* $R = k[[t_1, \ldots, t_n]]$ *be a formal power series over k. If* $f \in R$ *is not a unit, then f is integral over the ideal generated by its partial derivatives (partial derivatives are defined formally).*

Proof: Under the conditions of this corollary, the universally finite module of differentials is a free R–module on generators $dt_1, \ldots, dt_n$, and this is torsion free. The universally finite derivation $d = d^* : R \longrightarrow \widetilde{\Omega^1_{R/k}} = (\Omega^1_{R/k})^*$ is given by $d^*(f) = \sum_i \frac{\partial f}{\partial t_i} dt_i$. The corollary above applies directly with J the ideal generated by the partial derivatives of f. $\square$

In positive characteristic this corollary does not hold. For example, in characteristic p, let $f = x^p$. The partial derivative of f with respect to x is 0, and f is not integral over 0.

7.3. Exercises

7.1   Set $R = \mathbb{C}[X_1, \ldots, X_n]$, and let $f \in R$ such that $f(\underline{0}) = 0$. Let $I = (\frac{\partial f}{\partial X_1}, \ldots, \frac{\partial f}{\partial X_n})R$. Recall from Corollary 7.1.4, that f is contained in the integral closure of $I\mathbb{C}[[X_1, \ldots, X_n]]$. Is $f \in \overline{IR}$? Try $f = X_1^4 - X_1^2 X_2^3 - X_1^2 X_2^5 + X_2^8$.

7.2   Let $R = k[[x_1, \ldots, x_n]]$ be a power series ring over a field k. Let P be a prime ideal, and let $f \in P^{(m)}$. Prove that for all i, $\frac{\partial f}{\partial X_i} \in P^{(m-1)}$.

7.3   Let $R = k[[x_1, \ldots, x_n]]$ be a power series ring over a field k of characteristic 0. Assume that for each $f \in \mathfrak{m} = (x_1, \ldots, x_n)$, f is in $\mathfrak{m}$ times the integral closure of the Jacobian ideal $(\frac{\partial f}{\partial X_1}, \ldots, \frac{\partial f}{\partial X_n})$. Prove that for every prime ideal P in R, $P^{(2)} \subseteq \mathfrak{m}P$. (Without the assumption on the Jacobian ideals the conclusion of this exercise is an open problem, first raised by Eisenbud and Mazur in [70]. For other work, see [130].)

8
Reductions

The study of reductions started with the influential paper [215] of Northcott and Rees, published in 1954. Northcott and Rees defined minimal reductions, analytic spread, analytically independent elements, proved existence theorem, and connected these ideas with multiplicity.

Recall from Definition 1.2.1 that for an ideal I, a subideal J of I is said to be a **reduction** of I if there exists a non–negative integer n such that $JI^n = I^{n+1}$. Reductions always exist as every ideal is its own reduction. A connection between reductions and the integral closure of ideals was proved in Proposition 1.2.5: if I is finitely generated, then J is a reduction of I if and only if $I \subseteq \bar{J}$. In this chapter we explore further the theory of reductions.

Of particular interest are minimal reductions, that is, reductions that are minimal with respect to inclusion. Their existence and properties are proved in Section 8.3. When the residue field is infinite, the existence of minimal reductions of an ideal I is intricately connected to graded Noether normalizations of the fiber cone of I, which we discuss in Sections 8.2 and 8.3. In Section 8.4 we explain the standard procedure of reducing to the case of infinite residue field and show some applications. In Section 8.5 we develop the theory of superficial elements, which can be used to prove the existence of (minimal) reductions. Most of the results of this chapter are over Noetherian local rings, but in Section 8.7 we present some results for non–local rings as well. Section 8.8 presents a theorem of J. Sally regarding the behavior of analytic spread under certain maps between regular local rings.

8.1. Basic properties and examples

Proposition 8.1.1 *Let R be a ring, $J \subseteq I$ ideals. Consider the conditions:*
(1) J is a reduction of I.
(2) $W^{-1}J$ is a reduction of $W^{-1}I$ for every multiplicatively closed subset W of R.
(3) J_P is a reduction of I_P for every prime ideal P of R.
(4) J_M is a reduction of I_M for every maximal ideal M of R.
Then $(1) \Rightarrow (2) \Rightarrow (3) \Rightarrow (4)$. If R is Noetherian, in addition (4) implies (1).

Proof: (1) implies (2) because the condition $JI^n = I^{n+1}$ localizes. (2) trivially implies (3) and (3) trivially implies (4). Now assume (4) and that R is Noetherian. Observe that $(J : I) \subseteq (JI : I^2) \subseteq (JI^2 : I^3) \subseteq (JI^3 : I^4) \subseteq \cdots$. As R is Noetherian, this chain stabilizes, i.e., there exists an integer l such that for all $n \geq l$, $JI^n : I^{n+1} = JI^l : I^{l+1}$. Equality is preserved after localization at every maximal ideal. By assumption (4), for each maximal ideal M, for all large n, $JI^n : I^{n+1}$ is not contained in M, so that $JI^l : I^{l+1}$ is not contained in M. Hence $JI^l : I^{l+1} = R$, so that $JI^l = I^{l+1}$ and J is a reduction of I. $\square$

Without the Noetherian assumption (4) need not imply (1):

Example 8.1.2 Let k be a field, $X_1, Y_1, Z_1, X_2, Y_2, Z_2, \ldots$ variables over k, and $S = k[X_1, Y_1, Z_1, X_2, Y_2, Z_2, \ldots]$. Set

$$J = (X_1 \cdots X_i Y_i^i, X_1 \cdots X_i Z_i^i \,|\, i \geq 1)S,$$
$$I = J + (X_1 \cdots X_{i+1} Y_i Z_i^{i-1} \,|\, i \geq 1)S,$$

and for $i \geq 1$, set $M_j = (Y_1, \ldots, Y_{i-1}, Z_1, \ldots, Z_{i-1}, X_i)S$. Let W be the multiplicatively closed subset of S consisting of elements that are not in any M_j. It is straightforward to see that a prime ideal P containing J either contains some variable X_i, in which case it contains some M_j, or it contains $M = (\underline{Y}, \underline{Z})$. If P is in addition disjoint from W, then necessarily P equals some M_j or M. Set

$$K = \sum_{i=1}^{\infty} X_1 \cdots X_{i+2} (Y_i^i - Y_{i+1}^{i+1}, Z_i^i - Z_{i+1}^{i+1}, Y_i Z_i^{i-1} - Y_i^i).$$

Then $K \subseteq M$, $K \subseteq M_j$ for all i, and

$$J_{M_j} = (X_i, Y_1, Y_2^2, \ldots, Y_{i-1}^{i-1}, Z_1, Z_2^2, \ldots, Z_{i-1}^{i-1})_{M_j},$$
$$K_{M_j} = (X_i, Y_1 - Y_2^2, Y_2^2 - Y_3^3, \ldots, Y_{i-3}^{i-3} - Y_{i-2}^{i-2},$$
$$Z_1 - Z_2^2, Z_2^2 - Z_3^3, \ldots, Z_{i-3}^{i-3} - Z_{i-2}^{i-2},$$
$$Y_1 - Y_1, Y_2 Z_2 - Y_2^2, Y_3 Z_3^2 - Y_3^3, \ldots, Y_{i-3} Z_{i-3}^{i-4} - Y_{i-3}^{i-3}).$$

With $R = W^{-1}(S/K)$, $JR_{M_j} = (Y_1, Z_1, Y_{i-1}^{i-1}, Z_{i-1}^{i-1})R_{M_j}$,

$$IR_{M_j} = (Y_1, Z_1, Y_{i-1}^{i-1}, Z_{i-1}^{i-1}, Y_{i-1} Z_{i-1}^{i-2})R_{M_j},$$

and $JI^l R_{M_j} = I^{l+1} R_{M_j}$ with $l \geq i-2$. Similarly, $JR_M = IR_M = (Y_1, Z_1)R_M$. Thus $JR \subseteq IR$ is locally a reduction. However, $J \subseteq I$ is not a reduction, for otherwise if $JI^l = I^{l+1}$, the same still holds after localization at M_{l+3}, which is false. (Compare with Exercise 1.8.)

Reductions remain reductions under ring homomorphisms:

Lemma 8.1.3 Let $R \to S$ be a ring homomorphism and $J \subseteq I$ ideals in R.
(1) If J is a reduction of I, then JS is a reduction of IS.
(2) If S is faithfully flat over R and JS is a reduction of IS, then J is a reduction of I.

Proof: The first part is clear. If JS is a reduction of IS, then there exists an integer n such that $(JS)(I^n S) = I^{n+1}S$. By faithful flatness, $I^{n+1} = I^{n+1}S \cap R = JI^n S \cap R = JI^n$, which proves the second part. $\square$

In particular, reductions are preserved under localization, as already proved in Proposition 8.1.1. Also, reductions are preserved under passage to quotient rings, with the following partial converse which follows easily from Proposition 1.1.5:

Lemma 8.1.4 *Let R be a Noetherian ring, $J \subseteq I$ ideals in R. Then $J \subseteq I$ is a reduction if and only if for every minimal prime ideal P of R, $J(R/P) \subseteq I(R/P)$ is a reduction.*

The Noetherian assumption above is non–trivial: if k is a field, $X_1, X_2, \ldots$ are variables over k and $R = k[\underline{X}]/(X_1, X_2^2, X_3^3, \ldots)$, then $0 \subseteq (\underline{X})$ is not a reduction even though it is a reduction modulo the one minimal prime ideal.

Proposition 8.1.5 *Let $J = (a_1, \ldots, a_k) \subseteq I$ be ideals in a ring R.*
(1) If J is a reduction of I, then for any positive integer m, $(a_1^m, \ldots, a_k^m)$ and J^m are reductions of I^m.
(2) If for some positive integer m, $(a_1^m, \ldots, a_k^m)$ or J^m is a reduction of I^m, then J is a reduction of I.

Proof: (1) Choose n such that $JI^n = I^{n+1}$. Then for all $m \geq 1$, $J^m I^n = I^{n+m}$, and multiplying the last equation through by I^{mn-n} gives $J^m (I^m)^n = (I^m)^{n+1}$. Thus J^m is a reduction of I^m.

We claim that for every positive integer m,
$$(a_1^m, \ldots, a_k^m)(a_1, \ldots, a_k)^{(k-1)(m-1)} = (a_1, \ldots, a_k)^{(m-1)k+1}. \qquad (8.1.6)$$
It suffices to prove $J^{(m-1)k+1} \subseteq (a_1^m, \ldots, a_k^m)(a_1, \ldots, a_k)^{(k-1)(m-1)}$. Observe that $J^{(m-1)k+1}$ is generated by elements of the form $a_1^{n_1} \cdots a_k^{n_k}$, with each n_i a non–negative integer and $\sum n_i = (m-1)k + 1$. If $n_i < m$ for all i, then $(m-1)k + 1 = \sum n_i \leq (m-1)k$, which is a contradiction. Thus necessarily $n_i \geq m$ for at least one i, which proves the claim.

The claim implies that $(a_1^m, \ldots, a_k^m)$ is a reduction of J^m. By transitivity (Proposition 1.2.4) then also $(a_1^m, \ldots, a_k^m)$ is a reduction of I^m.

(2) If either $(a_1^m, \ldots, a_k^m)$ or J^m is a reduction of I^m, then by Proposition 1.2.4, J^m is a reduction of I^m. Thus there exists an integer n such that $J^m(I^m)^n = (I^m)^{n+1}$. Hence $I^{mn+m} \subseteq JI^{mn+m-1} \subseteq I^{mn+m}$, and equality holds throughout. Thus J is a reduction of I. $\square$

Examples of reductions can also be built via sums and products of ideals:

Proposition 8.1.7 *Let R be a ring, J_1, J_2, I_1, I_2 ideals in R, such that J_1 is a reduction of I_1 and J_2 is a reduction of I_2. Then*
(1) $J_1 + J_2$ is a reduction of $I_1 + I_2$, and
(2) $J_1 \cdot J_2$ is a reduction of $I_1 \cdot I_2$.

Proof: We may choose n such that $J_1 I_1^n = I_1^{n+1}$ and $J_2 I_2^n = I_2^{n+1}$. Then
$$\begin{aligned}
(I_1 + I_2)^{2n+1} &\subseteq I_1^{n+1}(I_1 + I_2)^n + I_2^{n+1}(I_1 + I_2)^n \\
&= J_1 I_1^n (I_1 + I_2)^n + J_2 I_2^n (I_1 + I_2)^n \\
&\subseteq (J_1 + J_2)(I_1 + I_2)^{2n} \subseteq (I_1 + I_2)^{2n+1},
\end{aligned}$$

so that equality holds throughout and $J_1 + J_2$ is a reduction of $I_1 + I_2$. Part (2) follows even more easily: by assumption $(I_1 I_2)^{n+1} = J_1 J_2 I_1^n I_2^n$, which immediately implies that $J_1 J_2$ is a reduction of $I_1 I_2$.   □

There are more ways to generate reductions via the Jacobson radical:

Lemma 8.1.8 *Let R be a Noetherian ring, $\mathfrak{m}$ be its Jacobson radical (i.e., the intersection of all the maximal ideals), $J, J' \subseteq I$ ideals, and L any ideal contained in $\mathfrak{m}I$. If $J + L = J' + L$, then J is a reduction of I if and only if J' is a reduction of I.*

Proof: Suppose that J is a reduction of I. Then there exists an integer n such that $JI^n = I^{n+1}$. Thus $I^{n+1} = JI^n \subseteq (J + L)I^n \subseteq (J' + \mathfrak{m}I)I^n$, so that by Nakayama's Lemma $I^{n+1} \subseteq J'I^n$, whence $I^{n+1} = J'I^n$, and J' is a reduction of I. The rest is easy.   □

The integral closure of a homogeneous ideal is homogeneous, but not every reduction of a homogeneous ideal is homogeneous:

Example 8.1.9 Consider the ideal $I = (X^3, XY, Y^4)$ in the polynomial ring $k[X, Y]$. Then $(XY, X^3 + Y^4)$ is easily seen to be a reduction of I and is not homogeneous under the usual grading. Furthermore, no 2–generated reduction of I is homogeneous: in any reduction J, if $JI^n = I^{n+1}$, by degree considerations XY appears as a term in one of the generators of J. If J is to be homogeneous, necessarily XY is one of the two generators of J up to scalar multiple. Without loss of generality the other generator is a linear combination of X^3 and Y^4 with coefficients in the ring. Again by degree count, X^3 appears with a unit coefficient in the generator, so that by homogeneity the second generator should be X^3 (up to scalar multiple). But (XY, X^3) is not a reduction of I as for all $n \geq 0$, $(Y^4)^{n+1} \notin (XY, X^3)I^n$.

The set of minimal prime ideals of a reduction ideal is independent of the reduction:

Lemma 8.1.10 *Let $J \subseteq I$ be a reduction. Then $\sqrt{J} = \sqrt{I}$, $\mathrm{Min}(R/I) = \mathrm{Min}(R/J)$, and $\mathrm{ht}\, J = \mathrm{ht}\, I$.*

Proof: This follows easily from $J \subseteq I$ and $JI^n = I^{n+1}$ for some n.   □

However, $\mathrm{Ass}(R/I)$ need not equal $\mathrm{Ass}(R/J)$, see Exercises 1.21, 1.22, or the following:

Example 8.1.11 Let R be the polynomial ring $k[X, Y, Z]$ in three variables X, Y and Z over a field k. Let $J = (X^3, Y^3, XY^2Z) \subseteq I = (X^3, Y^3, XY^2, X^2Y(Z - 1))$. Then both JI^2 and I^3 equal

$$(X^9, X^8Y(Z - 1), X^7Y^2, X^6Y^3, X^5Y^4, X^4Y^5, X^3Y^6, X^2Y^7, XY^8, Y^9),$$

so that J is a reduction of I. Note that $\mathrm{Ass}(R/J) = \{(X, Y), (X, Y, Z)\}$ and $\mathrm{Ass}(R/I) = \{(X, Y), (X, Y, Z - 1)\}$, so that no inclusion relations hold among the two sets.

8.2. Connections with Rees algebras

Theorem 8.2.1 *Let $J \subseteq I$ be ideals in a Noetherian ring R. Then J is a reduction of I if and only if $R[It]$ is module–finite over $R[Jt]$.*

Proof: Assume that J is a reduction of I. There exists an integer n such that $JI^n = I^{n+1}$, and for all $k \geq 1$, $J^k I^n = I^{n+k}$. It follows that $(R[It])_{k+n} = I^n t^n (R[Jt])_k$. For $i = 0, \ldots, n$, let $s_{i1}, \ldots, s_{ik_i}$ be the generators of the R–module I^i. Then $R[It] = \sum s_{ij} t^i R[Jt]$, so that $R[It]$ is a finitely generated module over $R[Jt]$.

Conversely, assume that $R[It]$ is a module–finite over $R[Jt]$. Both of these rings are $\mathbb{N}$–graded, so there exist finitely many homogeneous elements that generate $R[It]$ as an $R[Jt]$–module. Let n be the largest degree of one of these generators. Then $I^{n+1}t^{n+1} = (R[It])_{n+1} = \sum_{i=1}^{n+1}(J^i t^i)(I^{n+1-i}t^{n+1-i})$ $= JI^n t^{n+1}$, so that $I^{n+1} = JI^n$, and J is a reduction of I. $\qquad\square$

The proof above shows the following:

Corollary 8.2.2 *The minimum integer n such that $JI^n = I^{n+1}$ is the largest degree of an element in a homogeneous minimal generating set of the ring $R[It]$ over the subring $R[Jt]$.*

Definition 8.2.3 *Let J be a reduction of I. The **reduction number** of I with respect to J is the minimum integer n such that $JI^n = I^{n+1}$. It is denoted by $r_J(I)$. The (absolute) **reduction number** of I equals*

$$\min\{r_J(I) \,|\, J \text{ a minimal reduction of } I\}.$$

The reason that $J \subseteq I$ from Example 8.1.9 is not globally a reduction even if it is a reduction locally is that there is no bound on the local reduction numbers.

We show next that over a Noetherian local ring $(R, \mathfrak{m})$, the reduction number of I can also be determined via the fiber cone

$$\mathcal{F}_I(R) = \frac{R[It]}{\mathfrak{m}R[It]} \cong \frac{R}{\mathfrak{m}} \oplus \frac{I}{\mathfrak{m}I} \oplus \frac{I^2}{\mathfrak{m}I^2} \oplus \frac{I^3}{\mathfrak{m}I^3} \oplus \cdots.$$

Proposition 8.2.4 *Let n be a positive integer, $(R, \mathfrak{m})$ a Noetherian local ring, J, I ideals in R, $J \subseteq I^n$, and B the subalgebra of $\mathcal{F}_{I^n}(R)$ generated over $R/\mathfrak{m}$ by $(J + \mathfrak{m}I^n)/\mathfrak{m}I^n$. Then $J \subseteq I^n$ is a reduction if and only if $B \subseteq \mathcal{F}_I(R)$ is module–finite. (Yes, $\mathcal{F}_I(R)$, not $\mathcal{F}_{I^n}(R)$.)*

If either condition holds, the reduction number of I^n with respect to J is the largest degree of an element in a homogeneous minimal generating set of $\mathcal{F}_{I^n}$ over B.

Proof: The last part follows from Corollary 8.2.2 by using Nakayama's Lemma.

It is easy to show that $\mathcal{F}_I$ is module–finite over $\mathcal{F}_{I^n}$. Thus it suffices to prove the case $n = 1$.

First assume that $J \subseteq I$ is a reduction. By Theorem 8.2.1, $R[Jt] \subseteq R[It]$ is a module–finite extension. Hence $R[Jt]/(\mathfrak{m}R[It] \cap R[Jt]) \subseteq N_I$ is a module–finite extension. But $R[Jt]/(\mathfrak{m}R[It] \cap R[Jt])$ is canonically isomorphic to B, which proves that $B \subseteq \mathcal{F}_I$ is module–finite.

Now assume that $B \subseteq \mathcal{F}_I$ is module–finite. Let the homogeneous generators of $\mathcal{F}_I$ as a B–module be in degrees d or smaller. Then $I^{d+1}/\mathfrak{m}I^{d+1} \subseteq ((J + \mathfrak{m}I)/\mathfrak{m}I)(I^d/\mathfrak{m}I^d)$. Thus $I^{d+1} \subseteq JI^d + \mathfrak{m}I^{d+1}$, so that by Nakayama's Lemma J is a reduction of I. $\qquad\qquad\square$

As the analytic spread $\ell(I)$ of I is the dimension of $\mathcal{F}_I$, this proves:

Corollary 8.2.5 *Let R be a Noetherian local ring and $J \subseteq I$ a reduction. Then the minimal number $\mu(J)$ of generators of J is at most $\ell(I)$.*

If R is a Noetherian local ring and I an ideal, the graded ring $\mathrm{gr}_I(R)$ contains the information on the (minimal) reductions of I, their reduction number, and on the analytic spread of I. Ooishi in [220] generalized this to other graded rings and ideals to obtain a theory of reductions of graded rings. He also used this theory to study the properties of graded modules, in particular pseudo–flat modules.

8.3. Minimal reductions

In Noetherian rings, in general there is no descending chain condition and thus there may not be a reduction of an ideal that is minimal with respect to inclusion (cf. Exercise 8.10). However, in Noetherian local rings, minimal reductions do exist, see Theorem 8.3.6 and Proposition 8.3.7.

Definition 8.3.1 *A reduction J of I is called **minimal** if no ideal strictly contained in J is a reduction of I. An ideal that has no reduction other than itself is called **basic**.*

Here is an illustrative example:

Example 8.3.2 Let $R = \mathbb{Z}/2\mathbb{Z}[[X, Y]]/(XY(X+Y))$, where X, Y are variables over $\mathbb{Z}/2\mathbb{Z}$. Let $I = \mathfrak{m} = (X, Y)R$. We claim that I is basic. Suppose that this is not the case. Then there exists a reduction J properly contained in I. By Lemma 8.1.8, there exists an ideal J' generated by linear forms such that $J + I^2 = J' + I^2$. As J is properly contained in I, so is J', so that necessarily J' is generated by at most one linear form. This ring has only three linear forms: X, Y, and $X + Y$. By change of variables (and symmetry) we may assume that $J' = (X)R$. However, XR is not a reduction of I as $Y^{n+1} \notin XI^n$ for all n (in other words, at least if $n \geq 3$, $Y^{n+1} \notin (X^{n+1}, XY^n, Y^{n+1}, XY(X+Y))(\mathbb{Z}/2\mathbb{Z})[[X, Y]])$. Thus I cannot have any proper reductions, so it is basic.

If in the example above we enlarge $\mathbb{Z}/2\mathbb{Z}$ to a field k containing a unit u other than 1, then $J = (X + uY)$ is a minimal reduction of I:
$$JI^2 = (X + uY)(X, Y)^2 + (XY(X + Y))$$
$$= (X^3 + uX^2Y, X^2Y + uXY^2, XY^2 + uY^3) + (XY(X + Y))$$

$$= (X^3 + u^2 XY^2, XY^2(1+u), XY^2 + uY^3) + (XY(X+Y))$$
$$= (X^3, XY^2, Y^3) + (XY(X+Y)) = I^3.$$

In particular, when k has many units u, the ideal I has many minimal (1-generated) reductions.

Proposition 8.3.3   *Let $(R, \mathfrak{m})$ be a Noetherian local ring and J a minimal reduction of I. Then the following hold.*
(1) $J \cap \mathfrak{m}I = \mathfrak{m}J$.
(2) *For any ideal K such that $J \subseteq K \subseteq I$, every minimal generating set of J can be extended to a minimal generating set of K.*

Proof: Set $L = J \cap \mathfrak{m}I$. Let t be an integer such that $J/L \cong (R/\mathfrak{m})^t$, so that $J = (x_1, \ldots, x_t) + L$ for some $x_1, \ldots, x_t \in J$. By Lemma 8.1.8, $(x_1, \ldots, x_t)$ is also a reduction of I, which by the minimality of J implies that $J = (x_1, \ldots, x_t)$. Thus t is the minimal number of generators of J, implying that $L \subseteq \mathfrak{m}J$, which proves (1).

In particular, $J \cap \mathfrak{m}K = \mathfrak{m}J$. Thus $\{x_1, \ldots, x_t\}$ form part of a minimal generating set of K, which proves (2). □

The assumption about minimality in the last proposition is necessary, as can be seen by Example 8.1.11 after passing to the localization $k[X, Y, Z]_{(X,Y,Z)}$, or by the following:

Example 8.3.4   Let k be a field, X a variable, and $R = k[X^2, X^3]/(X^5, X^6)$. Set $I = \mathfrak{m} = (X^2, X^3)R$ and $J = (X^3, X^4)R$. Then $J \subseteq I$ and J is a reduction of I as $JI^2 = I^3 = 0$. However, $\mathfrak{m}I = (X^4, X^5)R = X^4 R$, $J \cap \mathfrak{m}I = X^4 R$, and $\mathfrak{m}J = 0$.

By Corollary 8.2.5, every reduction of I, and so every minimal reduction of I, has at least $\ell(I)$ generators. There are minimal reductions with strictly more generators, as in Example 8.3.2, but the reductions with exactly $\ell(I)$ generators are all minimal with further good properties:

Corollary 8.3.5   *Let $(R, \mathfrak{m})$ be a Noetherian local ring and $J \subseteq I$ a reduction such that $\mu(J) = \ell(I)$. Then*
(1) J *is a minimal reduction of I.*
(2) $\mathcal{F}_J$ *is canonically isomorphic to the subalgebra of $\mathcal{F}_I$ generated over $R/\mathfrak{m}$ by $(J + \mathfrak{m}I)/\mathfrak{m}I$, and is isomorphic to a polynomial ring in $\ell(I)$ variables over $R/\mathfrak{m}$.*
(3) *For all positive integers k, $J^k \cap \mathfrak{m}I^k = \mathfrak{m}J^k$.*

Proof: Let B be the subalgebra of $\mathcal{F}_I$ generated over $R/\mathfrak{m}$ by $(J + \mathfrak{m}I)/\mathfrak{m}I$. By Proposition 8.2.4, $\mathcal{F}_I$ is module–finite over B so that $\dim B = \dim \mathcal{F}_I = \ell(I)$. As J is generated by $\ell(I)$ elements, necessarily B is isomorphic to a polynomial ring over $R/\mathfrak{m}$. There is a natural surjective graded map $\mathcal{F}_J \to B$. As $\mathcal{F}_J$ is generated over $R/\mathfrak{m}$ by $\ell(J)$ elements, this surjective map onto a polynomial ring has to be an isomorphism. In particular for each k, the kernel of $J^k/\mathfrak{m}J^k \to (J^k + \mathfrak{m}I^k)/\mathfrak{m}I^k$ is zero, or in other words, $J^k \cap \mathfrak{m}I^k = \mathfrak{m}J^k$. If $K \subseteq J$ and K is a minimal reduction of I, then by Proposition 8.3.3, a minimal generating set of K can be extended to a minimal generating set of J. But by Corollary 8.2.5, $\mu(K) \geq \ell(I) = \mu(J)$, so that $K = J$, whence J is a minimal reduction of I. $\square$

In general there is no unique minimal reduction of an ideal (see second part of Example 8.3.2), but minimal reductions exist in Noetherian local rings:

Theorem 8.3.6 *Let $(R, \mathfrak{m})$ be a Noetherian local ring. If J is a reduction of I, then there exists at least one ideal K in J such that K is a minimal reduction of I.*

Proof: Let Σ be the set of all reductions K of I contained in J. Since $J \in \Sigma$, Σ is not empty. Since R is Noetherian, $\frac{I}{\mathfrak{m}I}$ is a finite dimensional $(R/\mathfrak{m})$–vector space, so there exists $K \in \Sigma$ such that $\frac{K+\mathfrak{m}I}{\mathfrak{m}I}$ is smallest under inclusion. Let the vector space dimension of this be n, and let $k_1, \ldots, k_n \in K$ be the preimages of a basis of $\frac{K+\mathfrak{m}I}{\mathfrak{m}I}$ in K. Set $K_0 = (k_1, \ldots, k_n)$. By Lemma 8.1.8, K_0 is a reduction of I. By renaming, without loss of generality $K = K_0$.

Both $\frac{K}{\mathfrak{m}K}$ and $\frac{K+\mathfrak{m}I}{\mathfrak{m}I}$ are n–dimensional $(R/\mathfrak{m})$–vector spaces, so that the canonical surjection $\frac{K}{\mathfrak{m}K} \to \frac{K}{K\cap\mathfrak{m}I} \cong \frac{K+\mathfrak{m}I}{\mathfrak{m}I}$ is an isomorphism. This implies that $K \cap \mathfrak{m}I = \mathfrak{m}K$.

We claim that K is a minimal reduction of I contained in J. It remains to prove minimality. If $L \subseteq K$ is a reduction of I, then by the minimality of K in Σ, $K + \mathfrak{m}I = L + \mathfrak{m}I$. Thus $K \subseteq (L + \mathfrak{m}I) \cap K = L + (\mathfrak{m}I \cap K)$, which by the previous paragraph means that $K \subseteq L + \mathfrak{m}K$. Hence by Nakayama's Lemma, $K = L$. Thus K is a minimal reduction of I. $\square$

The theorem gives that whenever $K_1 \supseteq K_2 \supseteq K_3 \cdots$ are all reductions of I, then $\bigcap_n K_n$ is also a reduction of I. However, without the local assumption this property may fail, see Exercise 8.10.

Example 8.3.2 shows that the number of generators of a minimal reduction depends on the cardinality of the residue field. That example also indicates that when the cardinality of the set of units is large enough, ideals generally have more reductions. In fact, when the residue field is infinite, there exist minimal reductions whose minimal generating sets have cardinality equal to the analytic spread of the ideal:

Proposition 8.3.7 *Let $(R, \mathfrak{m})$ be a Noetherian local ring with infinite residue field, I an ideal, and $l = \ell(I)$, the analytic spread of I. If $R/\mathfrak{m}$ is infinite, then every minimal reduction of I is minimally generated by exactly l elements. In particular, every reduction of I contains a reduction generated by l elements.*

Proof: By assumption, $l = \dim(\mathcal{F}_I)$. Let J be a reduction of I (possibly $J = I$). Let B be the subalgebra of $\mathcal{F}_I$ generated over $R/\mathfrak{m}$ by $(J + \mathfrak{m}I)/\mathfrak{m}I$. Proposition 8.2.4, $B \subseteq \mathcal{F}_I$ is a module–finite extension. By the Noether

Normalization Theorem (Theorem 4.2.3), there exist $\overline{a}_1, \ldots, \overline{a}_l \in B_1 = (J + \mathfrak{m}I)/\mathfrak{m}I$ such that $A = k[\overline{a}_1, \ldots, \overline{a}_l]$ is a polynomial subring of B and such that B is a module–finite extension of A. Hence $\mathcal{F}_I$ is module–finite over A. Let $a_i \in J$ be such that its image in $(J + \mathfrak{m}I)/\mathfrak{m}I$ is $\overline{a}_i$. Set $K = (a_1, \ldots, a_l)R$. Then $K \subseteq J$. By Proposition 8.2.4, K is a reduction of I, and by Corollary 8.3.5, K is a minimal reduction of I. Thus every minimal reduction is generated by exactly l elements. $\qquad\square$

This shows that the fiber cone of I and its dimension are useful for finding (minimal) reductions of I. But even without the assumption on the infinite cardinality of the residue field, the number $\ell(I)$ plays a role in reductions:

Proposition 8.3.8 *Let $(R, \mathfrak{m})$ be a Noetherian local ring and I an ideal. Then there exists an integer n such that I^n has a minimal reduction generated by $\ell(I)$ elements.*

Proof: By the graded version of the Noether Normalization Theorem (see Theorem 4.2.3), there exist an integer n and elements $\overline{a}_1, \ldots, \overline{a}_l \in (\mathcal{F}_I)_n = I^n/\mathfrak{m}I^n$ such that $A = k[\overline{a}_1, \ldots, \overline{a}_l]$ is a polynomial subring of $\mathcal{F}_I$ and such that $\mathcal{F}_I$ is a module–finite extension of A. Let $a_i \in I^n$ be such that its image in $I^n/\mathfrak{m}I^n$ is $\overline{a}_i$. Set $J = (a_1, \ldots, a_l)R$. Using Proposition 8.2.4, J is a reduction of I^n and by Corollary 8.3.5, J is a minimal reduction of I^n. $\qquad\square$

We revisit Example 8.3.2, with $R = k[[X, Y]]/(XY(X + Y))$, $I = (X, Y)R$. The fiber cone of I is $k[X, Y]/(XY(X + Y))$, which has Krull dimension 1. Under the assumption that k has more than one unit, we found a 1–generated minimal reduction of I, and under the assumption that k has exactly one unit, we proved that I has no 1–generated minimal reductions. By the proposition above, some power of I has a one–generated reduction. In fact, one can easily verify that $X^2 + XY + Y^2$ generates a minimal reduction of I^2.

Corollary 8.3.9 *Let $(R, \mathfrak{m})$ be a Noetherian local ring and I an ideal. Then $\dim R \geq \ell(I) \geq \operatorname{ht} I$.*

Proof: The first inequality holds by Proposition 5.1.6. By Proposition 8.3.8 there exists a positive integer n such that I^n has a minimal reduction J generated by $\ell(I)$ elements. Then $\operatorname{ht} J \leq \ell(I)$. But by Lemma 8.1.10, $\operatorname{ht} I = \operatorname{ht} I^n = \operatorname{ht} J$. $\qquad\square$

Even though monomial ideals need not have minimal monomial reductions (see Example 8.1.9), still, the geometry of monomial ideals carries some information. For example, Bivià-Ausina [16] proved that the analytic spread of a monomial ideal (over the complex numbers, and of more general **Newton non–degenerate ideals**) can be computed from its Newton polyhedron. Explicitly, the minimal number of generators of a minimal reduction of a monomial ideal is one more than the maximum dimension of a compact face of its Newton polytope. Crispin Quiñonez [48, Corollary 6.2.9] proved that a monomial ideal in a power series ring in two variables over a field has a minimal reduction with the following pattern: arrange the monomial generators in the order of increasing exponents in one of the variables (and decreasing in the other), let a be the sum of every other monomial in this arrangement, and let b be the sum of the remaining monomials. Then (a, b) is a minimal reduction of the ideal.

8.4. Reducing to infinite residue fields

In the previous section we proved stronger existence results for reductions under the extra condition that the residue field of the ambient Noetherian local ring is infinite. Many good properties of reductions still hold without the infinite residue field assumption, but often proofs reduce to the infinite residue case.

More examples of the passage to infinite residue fields are for example in Chapter 11 on multiplicities, in Remark 16.4.8, in Theorem 18.5.2, and elsewhere in the book.

Here is a general construction for this purpose: for a Noetherian local ring $(R, \mathfrak{m})$, let X be a variable over R. Set $S = R[X]_{\mathfrak{m}R[X]}$. Then $R \subseteq S$ is a faithfully flat extension of Noetherian local rings of the same Krull dimension. The residue field $S/\mathfrak{m}S$ of S contains the residue field $R/\mathfrak{m}$ of R. In fact, $\frac{S}{\mathfrak{m}S} \cong \left(\frac{R[X]}{\mathfrak{m}R[X]}\right)_{\mathfrak{m}R[X]}$, which is the field of fractions of $(R/\mathfrak{m})[X]$ and thus an infinite field.

Definition 8.4.1 Let $R(X)$ denote $R[X]_{\mathfrak{m}R[X]}$, where X is a variable over R.

As $R(X)$ is faithfully flat over R, the following are easy to prove:

Lemma 8.4.2 Let J, I be ideals in a Noetherian local ring $(R, \mathfrak{m})$. Then
(1) $J \subseteq I$ if and only if $JR(X) \subseteq IR(X)$.
(2) $\mathrm{ht}(I) = \mathrm{ht}(IR(X))$. In particular, $\dim R = \dim R(X)$, and I is $\mathfrak{m}$–primary if and only if $IR(X)$ is $\mathfrak{m}R(X)$–primary.
(3) $J \subseteq I$ is a reduction if and only if $JR(X) \subseteq IR(X)$ is a reduction.
(4) $\mu(I) = \mu(IR(X))$, $\ell(I) = \ell(IR(X))$.
(5) R is regular (resp. Cohen–Macaulay) if and only if $R(X)$ is regular (resp. Cohen–Macaulay).
(6) If I is $\mathfrak{m}$–primary, $\lambda(R/I) = \lambda(R(X)/IR(X))$. (Thus the Hilbert–Samuel functions of I and $IR(X)$ are the same; see Chapter 11.)
(7) I is generated by a regular sequence if and only if $IR(X)$ is generated by a regular sequence.
(8) If $I = q_1 \cap \cdots \cap q_k$ is a (minimal) primary decomposition, then $IR(X) = q_1 R(X) \cap \cdots \cap q_k R(X)$ is a (minimal) primary decomposition.
(9) $\overline{I}R[X] = \overline{IR[X]}$ and thus $\overline{I}R(X) = \overline{IR(X)}$. In particular, I is integrally closed if and only if $IR[X]$ is integrally closed, which holds if and only if $IR(X)$ is integrally closed.

Proof: We only prove (9). By faithful flatness, $I = IR[X] \cap R$, so I is integrally closed if $IR[X]$ is. Similarly, $IR[X]$ is integrally closed if $IR(X)$ is. To prove the rest we can replace I by $\bar{I}$, and assume that I is integrally closed. It suffices to prove that $IR[X]$ is integrally closed. Let $s \in \overline{IR[X]}$. Under the $\mathbb{N}$–grading $\deg(X) = 1$ and $\deg(R) = 0$, $IR[X]$ is a homogeneous ideal, so by Corollary 5.2.3, $\overline{IR[X]}$ is also homogeneous. Thus each graded component of s is in $\overline{IR[X]}$. Thus it suffices to prove that if $r \in R$, $m \in \mathbb{N}$ and $rX^m \in \overline{IR[X]}$, then $rX^m \in IR[X]$. Write $(rX^m)^n + a_1(rX^m)^{n-1} + \cdots + a_n = 0$ for some $a_i \in I^i R[X]$. Let $b_i X^{im}$ be the part of a_i of degree im. Then $b_i \in I^i$ and $(rX^m)^n + b_1 X^m (rX^m)^{n-1} + \cdots + b_n X^{mn} = 0$ yields $r^n + b_1 r^{n-1} + \cdots + b_n = 0$, whence $r \in \bar{I} = I$, and $rX^m \in IR[X]$. $\qquad\square$

We reprove Corollary 8.3.9 by using $R(X)$:

Corollary 8.4.3 *Let $(R, \mathfrak{m})$ be a Noetherian local ring. Then for every ideal I of R, $\mathrm{ht}(I) \le \ell(I) \le \dim(R)$. Also, $\ell(I)$ is bounded above by $\mu(I)$, the minimal cardinality of a generating set of I.*

Proof: The invariants ht, ℓ, $\dim$, and μ remain unchanged under passage from R to $R(X)$ and I to $IR(X)$, so without loss of generality we may assume that the residue field of R is infinite. By Proposition 5.1.6, $\ell(I) \le \dim(R)$. As $\mathcal{F}_I$ is generated over the field $R/\mathfrak{m}$ by $I/\mathfrak{m}I$, it follows that the dimension $\ell(I)$ of $\mathcal{F}_I$ is at most the number $\mu(I)$ of generators of $I/\mathfrak{m}I$. It remains to prove that $\mathrm{ht}(I) \le \ell(I)$. By Proposition 8.3.7 there exists a minimal reduction J of I generated by exactly $\ell(I)$ elements. Hence $\mathrm{ht}(I) = \mathrm{ht}(J) \le \mu(J) = \ell(I)$. $\square$

Proposition 8.4.4 *Let $(R, \mathfrak{m})$ be a Noetherian local ring, and I and J ideals in R. If $\ell(I) + \ell(J) > 0$, then $\ell(IJ) < \ell(I) + \ell(J)$.*

Proof: Without loss of generality we may pass to a Noetherian local ring with infinite residue field. We may also replace I and J by their respective minimal reductions. Thus we may assume that I and J are basic ideals. Now the proposition follows immediately from Corollary 1.7.6. $\qquad\square$

8.5. Superficial elements

Definition 8.5.1 *Let R be a ring, I an ideal, and M an R–module. We say that $x \in I$ is a **superficial element of I with respect to M** if there exists $c \in \mathbb{N}$ such that for all $n \ge c$, $(I^{n+1}M :_M x) \cap I^c M = I^n M$. If x is superficial with respect to R, we simply say that x is a superficial element of I.*

For any $x \in I$ and $n \ge c$, $I^n M$ is contained in $(I^{n+1}M :_M x) \cap I^c M$. It is the other inclusion that makes superficial elements special.

Remark 8.5.2 Let R be a ring, I an ideal, $x \in I$, and M an R–module. Assume that x is a superficial element of I with respect to M. Then for all $m \ge 1$, x^m is a superficial element of I^m with respect to M.

Proof: There exists an integer c such that for all $n \geq c$, $(I^{n+1}M :_M x) \cap I^c M = I^n M$. We claim that for all $n \geq c$, $(I^{m(n+1)}M :_M x^m) \cap I^{cm}M = I^{mn}M$, proving the remark. Let $r \in (I^{m(n+1)}M :_M x^m) \cap I^{cm}M$. A straightforward induction on $m - i$ proves that for all $0 \leq i \leq m$, $rx^i \in I^{m(n+1)-m+i}M$. When $i = 0$, we have established the claim. $\qquad\square$

Clearly the superficial property of an element is preserved under localization. Superficial elements that are non–zerodivisors behave better:

Lemma 8.5.3 *A element $x \in I$ that is a non–zerodivisor on M is a superficial element of I with respect to M if and only if for all sufficiently large integers n, $I^n M :_M x = I^{n-1}M$.*

Proof: The condition $I^n M :_M x = I^{n-1}M$ for all sufficiently large n clearly implies that x is superficial, say by taking $c = 0$. Now assume that x is superficial. By the Artin–Rees Lemma there exists an integer k such that for all $n \geq k$, $I^n M \cap xM = I^{n-k}(I^k M \cap xM) \subseteq xI^{n-k}M$. As $I^n M \cap xM = x(I^n M :_M x)$ and as x is a non–zerodivisor on M, it follows that $I^n M :_M x \subseteq I^{n-k}M$. Let $n \geq k + c$. Then $I^n M :_M x = (I^n M :_M x) \cap I^c M = I^{n-1}M$ by the assumption on superficiality, which proves the lemma. $\qquad\square$

The following is a partial converse:

Lemma 8.5.4 *Let R be a Noetherian ring, I an ideal in R, and M a finitely generated R–module. Assume that $\cap_n (I^n M) = 0$ and that I contains an element that is a non–zerodivisor on M. Then every superficial element of I with respect to M is a non–zerodivisor on M.*

Proof: Let x be a superficial element of I. Let c be an integer such that for all integers $n \geq c$, $(I^{n+1}M :_M x) \cap I^c M = I^n M$. Then $(0 :_M x)I^c \subseteq I^n M$ for all n. Thus $(0 :_M x)I^c = 0$, so that as I contains a non–zerodivisor on M, $0 :_M x = 0$. $\qquad\square$

Lemma 8.5.5 *Let R be a Noetherian ring, M a finitely generated R–module, I an ideal in R, and $x \in R$. Assume that either x is a non–zerodivisor on M or that for some ideal J, $I \subseteq \sqrt{J}$, $\cap_n J^n M = 0$, and that $x \in R$ is a superficial element for J with respect to M. Then there exists $e \in \mathbb{N}$ such that for all $n \geq e$,*

$$I^n M :_M x \subseteq (0 :_M x) + I^{n-e}M, \quad \text{and} \quad (0 :_M x) \cap I^e M = 0.$$

If x is superficial for I, then for all sufficiently large n, $I^n M :_M x = (0 :_M x) + I^{n-1}M$.

Proof: By the Artin–Rees Lemma, there exists $k \geq 0$ such that for all $n \geq k$, $I^n M \cap xM \subseteq xI^{n-k}M$. In particular, $I^n M :_M x \subseteq (0 :_M x) + I^{n-k}M$.

In case x is a non–zerodivisor on M, $(0 :_M x) = 0$, which shows that $I^n M :_M x \subseteq I^{n-k}M$, thus proving the displayed inclusions in case x is a non–zerodivisor on M.

If instead x is superficial for J with respect to M and if $I^m \subseteq J$, let c be such that for all $n \geq c$, $(J^n M :_M x) \cap J^c M = J^{n-1} M$. Set $e = cm + k$. From the first paragraph we conclude that for all $n \geq e$, $I^n M :_M x \subseteq (0 :_M x) + I^{n-e} M$. Also,

$$(0 :_M x) \cap I^e M \subseteq \bigcap_n (J^n M :_M x) \cap J^c M \subseteq \bigcap_n J^{n-1} M = 0,$$

which finishes the proof of the displayed inclusions. If in addition $I = J$, i.e., if x is superficial for I, then for $n \geq e + c$,

$$\begin{aligned}
I^n M :_M x &= ((0 :_M x) + I^c M) \cap (I^n M :_M x) \\
&= (0 :_M x) + I^c M \cap (I^n M :_M x) \\
&= (0 :_M x) + I^{n-1} M. \qquad \square
\end{aligned}$$

Superficial elements do not always exist:

Example 8.5.6 (Cf. Example 8.3.2.) Let $R = (\mathbb{Z}/2\mathbb{Z})[[X, Y]]/(XY(X+Y))$, where X, Y are variables over $\mathbb{Z}/2\mathbb{Z}$. Let $I = (X, Y)R$. Then I has no superficial element. For otherwise let r be a superficial element. As I is not nilpotent, by Exercise 8.3, necessarily $r \notin (X, Y)^2 R$, so that the degree 1 part of r is non-zero. By symmetry and a possible linear change of variables, we may assume that the degree one part of r is X. We may write $r = aX + bY^2$ for some $a, b \in \mathbb{Z}/2\mathbb{Z}[X, Y]$. For all $n > \max\{2, c\}$, $Y^{n-2}(X + Y)r \in I^{n+1}$, $Y^{n-2}(X + Y) \in I^c \setminus I^n$, contradicting the superficiality assumption. Thus I has no superficial element.

As is the case for reductions, stronger existence results hold when R is a local ring with infinite residue field:

Proposition 8.5.7 *Let R be a Noetherian ring, I an ideal, and M a finitely generated R–module. Then*

(1) There exists an integer m such that I^m has a superficial element x with respect to M, and even better, there exists an integer c such that for all $n \geq m, c$, $(I^n M :_M x) \cap I^c M = I^{n-m} M$.

(2) If R has infinite residue fields, then m above can be taken to be 1.

(3) If $(R, \mathfrak{m})$ is a Noetherian local ring with infinite residue field, then every ideal I has a superficial element with respect to M. Furthermore in this case, there exists a non–empty Zariski–open subset U of $I/\mathfrak{m}I$ such that whenever $r \in I$ with image in $I/\mathfrak{m}I$ in U, then r is superficial for I with respect to M.

Proof: The module $\mathrm{gr}_I(M) = (M/IM) \oplus (IM/I^2 M) \oplus (I^2 M/I^3 M) \oplus \cdots$ over the Noetherian ring $\mathrm{gr}_I(R)$ is finitely generated. Let $0 = N_1 \cap \cdots \cap N_r$ be a primary decomposition of the zero submodule in $\mathrm{gr}_I(M)$. For $i = 1, \ldots, r$, set P_i to be the radical of $(N_i :_{\mathrm{gr}_I(R)} \mathrm{gr}_I(M))$. Each P_i is a prime ideal. Without loss of generality $P_1, \ldots, P_s$ contain all elements of $\mathrm{gr}_I(R)$ of positive degree, and $P_{s+1}, \ldots, P_r$ do not. There exists an integer c such that $N_1, \ldots, N_s$ all contain $I^c M/I^{c+1} M$.

By Prime Avoidance there exists a homogeneous element h of positive degree in $\text{gr}_I(R)$ that is not contained in any P_i for $i > s$. Say $h = x + I^{m+1}$ for some $x \in I^m$. If R has infinite residue fields, then by Theorem A.1.2, m may be taken to be 1.

By way of contradiction suppose that $n \geq c$ and $y \in (I^n M :_M x) \cap I^c M \setminus I^{n-m}M$. Let k be the largest integer such that $y \in I^k M$. Then $c \leq k < n$. In $\text{gr}_I(M)$, $(x + I^{m+1}) \cdot (y + I^{k+1}M) = 0$. Thus by the choice of $x + I^{m+1}$, $y + I^{k+1}M \in N_{s+1} \cap \cdots \cap N_r$. By the choice of c, $y + I^{k+1}M \in N_1 \cap \cdots \cap N_s$, so that $y + I^{k+1}M = 0$, which contradicts the choice of k. This proves (1) and (2).

Now assume that in addition R is Noetherian local with maximal ideal $\mathfrak{m}$ and infinite residue field $R/\mathfrak{m}$. By Nakayama's Lemma, the images of $P_{s+1}, \ldots, P_r$ in $I/\mathfrak{m}I$ are proper $(R/\mathfrak{m})$–vector subspaces of $I/\mathfrak{m}I$. As $R/\mathfrak{m}$ is infinite, the Zariski–open subset of $I/\mathfrak{m}I$ avoiding all these proper subspaces is not empty. $\qquad\square$

An easy corollary is the following:

Lemma 8.5.8 *Let I be an ideal in a Noetherian ring R such that the associated graded ring $\text{gr}_I(R)$ has a non–zerodivisor of positive degree. Then for all large integers n, $I^{n+1} : I = I^n$.*

Proof: Let x and m be as in part (1) of Proposition 8.5.7. As in that proof, we may further assume that $h = x + I^{m+1}$ avoids all the associated primes of $\text{gr}_I(R)$ and that $c = 0$. Certainly $I^n \subseteq I^{n+1} : I$, and if $y \in I^{n+1} : I$, then $yx \in yI^m = yII^{m-1} \subseteq I^{n+m}$, so that $y \in (I^{n+m} : x) = I^n$. $\qquad\square$

Superficial elements of an ideal I exist even if we require further properties:

Corollary 8.5.9 *Let $(R, \mathfrak{m})$ be a Noetherian local ring with infinite residue field. Let I be an ideal of R and $P_1, \ldots, P_r$ ideals in R not containing I. Then for any finitely generated R–module M there exists a superficial element for I with respect to M that is not contained in any P_i.*

In particular, if I contains a non–zerodivisor on M, then there exists a superficial element of I with respect to M that is a non–zerodivisor on M.

Proof: By Proposition 8.5.7, there exists a non–empty Zariski–open subset U of $I/\mathfrak{m}I$ such that whenever $x \in I$ and $x + \mathfrak{m}I \in U$, then x is superficial for I with respect to M.

Let $W_i = ((P_i \cap I) + \mathfrak{m}I)/\mathfrak{m}I$, a vector subspace of $I/\mathfrak{m}I$. By assumption P_i does not contain I, so that by Nakayama's Lemma $(P_i \cap I) + \mathfrak{m}I$ does not contain I, whence W_i is a proper vector subspace of $I/\mathfrak{m}I$.

Let $U' = U \cap (I/\mathfrak{m}I \setminus (W_1 \cup \cdots \cup W_r))$. Then U' is a non–empty Zariski–open subset of $I/\mathfrak{m}I$. As it is a subset of U, it follows that whenever $x \in I$ with $x + \mathfrak{m}I \in U'$, then x is superficial for I. Furthermore, by construction of U', whenever $x + \mathfrak{m}I \in U'$, then x is not in P_i for all i.

The last statement is immediate from Lemma 8.5.4. $\qquad\square$

Definition 8.5.10 *A sequence $x_1, \ldots, x_s \in I$ is said to be a* **superficial sequence** *for I with respect to M if for all $i = 1, \ldots, s$, the image of x_i in $I/(x_1, \ldots, x_{i-1})$ is a superficial element of $I/(x_1, \ldots, x_{i-1})$ with respect to $M/(x_1, \ldots, x_{i-1})M$.*

We proved that every ideal in a Noetherian local ring with an infinite residue field contains a superficial element. Thus every superficial sequence can be continued to a longer superficial sequence.

Lemma 8.5.11 *Let $x_1, \ldots, x_s$ be a superficial sequence for I with respect to M. Then for all n sufficiently large,*

$$I^n M \cap (x_1, \ldots, x_s)M = (x_1, \ldots, x_s)I^{n-1}M.$$

Proof: We proceed by induction on s. First assume that $s = 1$. By assumption there exists an integer c such that for all $n \geq c$, $(I^{n+1}M :_M x_1) \cap I^c M = I^n M$. By the Artin–Rees lemma there exists an integer l such that for all $n \geq l$, $I^n M \cap x_1 M \subseteq x_1 I^{n-l} M$. Let $n > c + l$, and $y \in I^n M \cap x_1 M$. Write $y = x_1 a = x_1 b$ for some $a \in I^n M :_M x_1$ and some $b \in I^{n-l} M \subseteq I^c M$. Then $a - b \in (0 :_M x_1) \subseteq (I^n M :_M x_1)$, so that $b = a - (a - b) \in I^n M :_M x_1$. This proves that $I^n M \cap x_1 M \subseteq x_1 ((I^n M :_M x_1) \cap I^c M)$, whence by the superficiality of x_1, $I^n M \cap x_1 M \subseteq x_1 I^{n-1} M$. Thus $I^n M \cap x_1 M = x_1 I^{n-1} M$.

Now assume that $s > 1$. By induction there exists an integer k such that for all $n \geq k$,

$$I^n M \cap (x_1, \ldots, x_{s-1})M = (x_1, \ldots, x_{s-1})I^{n-1}M.$$

By the case $s = 1$, there exists an integer c such that for all $n \geq c$,

$$(I^n M + (x_1, \ldots, x_{s-1})M) \cap (x_1, \ldots, x_s)M = (x_1, \ldots, x_{s-1})M + x_s I^{n-1} M.$$

Hence for all $n \geq k + c$,

$$
\begin{aligned}
I^n M \cap (x_1, \ldots, x_s)M &\subseteq I^n M \cap \big((x_1, \ldots, x_{s-1})M + x_s I^{n-1} M\big) \\
&= I^n M \cap (x_1, \ldots, x_{s-1})M + x_s I^{n-1} M \\
&= (x_1, \ldots, x_s)I^{n-1}M. \qquad \square
\end{aligned}
$$

The power of superficial sequences lies in the power of non–empty Zariski–open sets: the intersection of any two such sets is not empty and still Zariski–open, which means that further conditions of Zariski–open nature can be imposed freely. Here is one such example:

Lemma 8.5.12 *Let k be an infinite field and $m, n \in \mathbb{N}$. Let V be a non–empty Zariski–open subset of k^m, U a non–empty Zariski–open subset of k^{m+n}, and $a \in k^n$ and $b \in k^m$ such that $(b, a) \in U$. Then $\{u \in V \mid (u, a) \in U\}$ is a non–empty Zariski–open subset of k^m.*

Proof: Let $I \subseteq k[X_1, \ldots, X_m]$ be an ideal such that $v \in V$ if and only if there exists $F \in I$ such that $F(v) \neq 0$. Similarly, let $J \subseteq k[X_1, \ldots, X_{m+n}]$

be such that $u \in U$ if and only if there exists $F \in J$ such that $F(u) \neq 0$. Set $K = (F(X_1, \ldots, X_m, a) \mid F \in J)$. As there exists $F \in J$ such that $F(b, a) \neq 0$, it follows that K is a non–zero ideal in $k[X_1, \ldots, X_m]$. But

$$\{u \in V \mid (u, a) \in U\} = \{u \in k^m \mid \text{ there exists } G \in IK \text{ such that } G(u) \neq 0\}.$$

As IK is non–zero, this Zariski–open set is not empty. $\qquad\square$

8.6. Superficial sequences and reductions

We have seen that whenever $J \subseteq I$ is a reduction, then modulo any ideal K, $J(R/K) \subseteq I(R/K)$ is a reduction, but the converse fails in general. However, if $K \subseteq J$ is generated by a superficial sequence for I, the converse holds:

Proposition 8.6.1 *Let R be a Noetherian ring, $J \subseteq I$ ideals in R, and let $x_1, \ldots, x_s$ be a superficial sequence for I contained in J. Set $K = (x_1, \ldots, x_s)$. If $J(R/K) \subseteq I(R/K)$ is a reduction, so is $J \subseteq I$.*

Proof: Choose a positive integer n such that $I^{n+1}(R/K) = JI^n(R/K)$. Then $I^{n+1} \subseteq (JI^n + K) \cap I^{n+1} = JI^n + (K \cap I^{n+1})$, so by the lemma above, for all n sufficiently large, $JI^n + (K \cap I^{n+1}) \subseteq JI^n + KI^n = JI^n$. Thus for all n sufficiently large, $I^{n+1} = JI^n$. $\qquad\square$

Lemma 8.5.11 has a somewhat remarkable corollary when applied to the associated graded module of I and M.

Corollary 8.6.2 *Let R be a Noetherian ring, I an ideal, and M a finitely generated R–module. Assume that $x \in I$ is superficial for I with respect to M. Set $\mathrm{gr}_I(M) = \oplus_{i \geq 0}(I^i M/I^{i+1}M)$. For all n sufficiently large,*

$$[gr_I(M)/x^* gr_I(M)]_n = [gr_{\overline{I}}(\overline{M})]_n,$$

where $\overline{I} = I/(x)$, $\overline{M} = M/xM$, and x^* is the leading form of x in $\mathrm{gr}_I(R)$.

Proof: The degree n part of $\mathrm{gr}_I(M)/x^* gr_I(M)$ is $I^n M/(xI^{n-1}M + I^{n+1}M)$, which clearly surjects onto the degree n piece of $\mathrm{gr}_{\overline{I}}(\overline{M})$, which is isomorphic to $I^n M/(I^{n+1}M + (xM \cap I^n M))$. The kernel of this surjection is isomorphic in degree n to $(I^{n+1}M + (xM \cap I^n M))/(xI^{n-1}M + I^{n+1}M)$. For large n, this kernel is 0 by Lemma 8.5.11. $\qquad\square$

This corollary is called a "miracle" in [21], the point being that for an element r to be superficial the multiplication map by r^* on $\mathrm{gr}_I(M)$ needs to be injective in high degrees, while the corollary says that this injectivity is enough to give the "correct" cokernel in high degrees.

There is a natural upper bound on the length of the shortest superficial sequence for I that generates a reduction of I:

Theorem 8.6.3 *Let $(R, \mathfrak{m})$ be a Noetherian local ring with infinite residue field and I an ideal of R. There exists a superficial sequence $(x_1, \ldots, x_l)$ for I of length $l = \ell(I)$ such that $(x_1, \ldots, x_l)$ is a minimal reduction of I.*

Proof: If $\ell(I) = 0$, then I is nilpotent, so the empty superficial sequence generates the zero ideal, which is the minimal reduction of I.

Now assume that $\ell(I) > 0$. By Proposition 8.5.7 there exists a non–empty Zariski–open subset U of $I/\mathfrak{m}I$ such that whenever $x \in I$ with $x + \mathfrak{m}I \in U$, then x is superficial for I. As $\mathcal{F}_I$ is generated over $R/\mathfrak{m}$ by elements of $I/\mathfrak{m}I$, and as $\mathcal{F}_I$ has positive dimension, it follows that no minimal prime ideal of $\mathcal{F}_I$ contains all of $I/\mathfrak{m}I$. Thus as $R/\mathfrak{m}$ is infinite, there exists an element $x_1 + \mathfrak{m}I$ in U that avoids all the minimal prime ideals of $\mathcal{F}_I$.

Set $J = I/(x_1)$ in $R/(x_1)$. Since $\mathcal{F}_J$ is a homomorphic image of $\mathcal{F}_I/x_1\mathcal{F}_1$, so that $\ell(J) = \dim \mathcal{F}_J \le \dim(\mathcal{F}_I/x_1\mathcal{F}_1) = \dim \mathcal{F}_I - 1 = \ell(I) - 1$. It follows by induction that there exists a superficial sequence $x_2 + (x_1), \ldots, x_l + (x_1)$ of J with $l = \ell(J)$ that generates a reduction of J. For all n sufficiently large, $I^{n+1} \subseteq (x_2, \ldots, x_l)I^n + x_1R$, so that $I^{n+1} \subseteq x_1R \cap I^{n+1} + (x_2, \ldots, x_l)I^n$. Hence by Lemma 8.5.11, for all n sufficiently large, $I^{n+1} \subseteq (x_1, x_2, \ldots, x_l)I^n$. Thus $(x_1, \ldots, x_l)$ is a reduction of I and by construction $l \le \ell(I)$, whence by Proposition 8.3.7, $l = \ell(I)$. $\qquad\square$

Discussion 8.6.4   The theorem above together with Lemma 8.5.11 have an important consequence, used by J. Sally in [255]. We leave the proof of the following theorem as an exercise (Exercise 8.25):

Theorem 8.6.5 (Sally's machine) *Let $(R, \mathfrak{m})$ be a Noetherian local ring, and let I be an ideal of R. Let $(x_1, \ldots, x_n) \subseteq I$ be a minimal reduction of I generated by a superficial sequence of length n. Fix $r \le n$, and set $J = (x_1, \ldots, x_r)$. Then*

$$\operatorname{depth}(\operatorname{gr}_I(R)) \ge r + 1 \text{ if and only if } \operatorname{depth}(\operatorname{gr}_{I/J}(R/J)) \ge 1.$$

This "machine" has been used with great effectiveness to study Hilbert co-efficients and the depth of Rees algebras by a number of researchers, especially by the Genova school. See [102], [72], [129], [149], [152], [250], [251], [252], [316], and their references.

Theorem 8.6.3, together with Proposition 8.5.7, shows that one can *successively* find sufficiently generic elements $x_1, \ldots, x_{\ell(I)}$ in I, that is, elements in some Zariski–open subsets of $I/\mathfrak{m}I$, such that $(x_1, \ldots, x_{\ell(I)})$ is a reduction of I. But one can also find such sufficiently general elements *all at once*:

Theorem 8.6.6   (Northcott and Rees [215], Trung [298]) *Let $(R, \mathfrak{m})$ be a Noetherian local ring with infinite residue field and I an ideal of analytic spread at most l. There exists a non–empty Zariski–open subset U of $(I/\mathfrak{m}I)^l$ such that whenever $x_1, \ldots, x_l \in I$ with $(x_1 + \mathfrak{m}I, \ldots, x_l + \mathfrak{m}I) \in U$, then $(x_1, \ldots, x_l)$ is a reduction of I.*

Furthermore, if there exists a reduction of I with reduction number n, then there exists a non–empty Zariski–open subset U of $(I/\mathfrak{m}I)^l$ such that whenever $x_1, \ldots, x_l \in I$ with $(x_1 + \mathfrak{m}I, \ldots, x_l + \mathfrak{m}I) \in U$, then $(x_1, \ldots, x_l)$ is a reduction of I with reduction number at most n.

Proof: Let $I = (a_1, \ldots, a_m)$. Let $X_1, \ldots, X_m$ be variables over $R/\mathfrak{m}$ let $A = (R/\mathfrak{m})[X_1, \ldots, X_m]$, and let $\varphi : A \to \mathcal{F}_I$ be the ring homomorphism sending X_i to $a_i + \mathfrak{m}I$. Then φ is a graded homomorphism and its kernel is a homogeneous ideal.

For any $(u_{ij}) \in R^{ml}$, set $x_i = \sum_{j=1}^m u_{ij}a_j$, $J = (x_1, \ldots, x_l)R$, and set J_X to be the ideal $(\sum_{j=1}^m \overline{u}_{1j}X_j, \ldots, \sum_{j=1}^m \overline{u}_{lj}X_j)A$, where $\overline{u}_{ij}$ is the image of u_{ij} in $R/\mathfrak{m}$. Let J_a be the image of J_X in $\mathcal{F}_I$.

Claim: J is a reduction of I if and only if $(X_1, \ldots, X_m)A = \sqrt{\ker(\varphi) + J_X}$. Namely, if J is a reduction of I, then for a large integer N, $I^{N+1} = JI^N$, so that in $\mathcal{F}_I$, $(\mathcal{F}_I)_+ = \sqrt{J_a}$. The preimages in A show that $(X_1, \ldots, X_m)A = \sqrt{\ker(\varphi) + J_X}$. Now assume that $(X_1, \ldots, X_m)A = \sqrt{\ker(\varphi) + J_X}$. Let $N \in \mathbb{N}$ such that $(X_1, \ldots, X_m)^{N+1} \subseteq \ker(\varphi) + J_X$. Then by homogeneity, $(X_1, \ldots, X_m)^{N+1} \subseteq \ker(\varphi) + J_X(X_1, \ldots, X_m)^N$. By passing to the homomorphic image of this inclusion under φ, and then by lifting to R, we get that $I^{N+1} \subseteq JI^N + \mathfrak{m}I^{N+1}$, so that by Nakayama's Lemma, J is a reduction of I. This proves the claim.

For each $n \geq 1$, let V_n be the $(R/\mathfrak{m})$–vector space in A whose basis consists of all monomials of degree n and let B_n be the vector subspace generated by $(\ker(\varphi) + J_X) \cap A_n$. Let $v_n = \dim V_n$, $b_n = \dim B_n$. Fix a generating set of B_n of b_n elements. Write each of the generators of B_n as a linear combination in the monomials. The coefficients are polynomials in the $\overline{u}_{ij}$ of degree at most one. Let K_n be the ideal in A generated by all the $v_n \times v_n$ minors of the $b_n \times v_n$ matrix obtained from the coefficients. If $b_n < v_n$, then K_n is automatically zero. However, by the existence of minimal reductions in local rings over an infinite residue field there exists (u_{ij}) such that the corresponding ideal J is a reduction of I. This means that for some n, K_n is not a trivially zero ideal. Set $U = \{(\overline{u}_{ij}) \in V_n \mid K_n(\overline{u}_{ij}) \neq 0\}$. Then U is a Zariski–open subset of $(I/\mathfrak{m}I)^l$ and by construction for every $(\overline{u}_{ij}) \in U$, the corresponding J is a reduction of I with reduction number at most n. □

This proves that any $\ell(I)$ "sufficiently generic" elements of I form a minimal reduction of I.

The proof above shows even more:

Corollary 8.6.7 *Let $(R, \mathfrak{m})$ be a Noetherian local ring. Then for every ideal I in R there exists an integer N such that the reduction number of any reduction of I is at most N.*

Proof: By the methods as in Section 8.4, without loss of generality we may assume that R has an infinite residue field. Let J be an arbitrary reduction of I. Let $K \subseteq J$ be a minimal reduction of I. Then $\mu(K) = \ell(I)$ by Proposition 8.3.7. Clearly the reduction number of K is an upper bound on the reduction number of J, so it suffices to prove that there is an upper bound on the reduction numbers of minimal reductions. Let $n = r_K(I)$.

The proof of Theorem 8.6.6 shows that there exists a non–empty Zariski–open subset U_n of $(I/\mathfrak{m}I)^l$ such that whenever $(x_1, \ldots, x_l) \in I^l$ such that $(x_1 + \mathfrak{m}I, \ldots, x_l + \mathfrak{m}I) \in U_n$, then $(x_1, \ldots, x_l)$ is a reduction of I with reduction number at most n. Then $\bigcup_i U_i$ gives an open cover of the subset of $(I/\mathfrak{m}I)^l$ that yields all the reductions of I. But $(I/\mathfrak{m}I)^l$ is a finite–dimensional vector space, so there exists $N \in \mathbb{N}$ such that U_N contains all the U_i. Hence N is an upper bound on all the reduction numbers. $\qquad\qquad\qquad\square$

An upper bound on $\ell(I)$ and also on reduction numbers can be obtained via the following:

Theorem 8.6.8   (Eakin and Sathaye [64]) *Let $(R, \mathfrak{m})$ be a Noetherian local ring with infinite residue field, and let I be an ideal in R. Suppose that there exist integers n and r such that I^n can be generated by strictly fewer than $\binom{n+r}{r}$ elements. Then there exist $y_1, \ldots, y_r \in I$ such that $(y_1, \ldots, y_r)I^{n-1} = I^n$. In particular, $(y_1, \ldots, y_r)$ is a reduction of I. In fact, any "sufficiently generic" $y_1, \ldots, y_r \in I$ form a reduction of I.*

Proof: The last statement follows from Theorem 8.6.6.

The interpretation for $r = 0$ is trivial: by assumption then $I^n = 0$, so indeed I has a reduction generated by zero elements. The result for $n = 1$ is immediate. Thus we may assume that $r > 0$ and that $n > 1$. If there exists a counterexample to the theorem, we may choose a counterexample in which r is minimal and n is minimal for this r. Thus any generating set of I^{n-1} has at least $\binom{n-1+r}{r}$ elements.

Suppose that there exists $y \in I \setminus \mathfrak{m}I$ such that $(yI^{n-1} + \mathfrak{m}I^n)/\mathfrak{m}I^n$ is an $(R/\mathfrak{m})$–vector space of rank at least $\binom{n-1+r}{r}$. If $r = 1$, then by assumption I^n is generated by at most n elements, so that by Nakayama's Lemma, $yI^{n-1} = I^n$, and we are done. Thus in this case we may assume that $r > 1$. Now pass to $R' = R/(yI^{n-1} + \mathfrak{m}I^n)$. Then as $\binom{n+r-1}{r-1} + \binom{n+r-1}{r} = \binom{n+r}{r}$, it follows that $I^n R'$ is generated by strictly fewer than $\binom{n+r-1}{r-1}$ elements. By induction on r, there exist $y_2, \ldots, y_r \in I$ such that $(y_2, \ldots, y_r)I^{n-1}R' = I^n R'$. Hence $I^n \subseteq (y_2, \ldots, y_r)I^{n-1} + yI^{n-1} + \mathfrak{m}I^n$, which after applying Nakayama's Lemma finishes the proof of this case.

Now assume that for all $y \in I \setminus \mathfrak{m}I$, $(yI^{n-1} + \mathfrak{m}I^n)/\mathfrak{m}I^n$ is an $(R/\mathfrak{m})$–vector space of rank less than $\binom{n-1+r}{r}$. Set $R' = R/(\mathfrak{m}I^n : y)$. Then $I^{n-1}R'$ is generated by fewer than $\binom{n-1+r}{r}$ elements. By induction on n there exist $y_1, \ldots, y_r \in I$ such that $(y_1, \ldots, y_r)I^{n-2}R' = I^{n-1}R'$. By Theorem 8.6.6, there exists a non–empty Zariski–open subset U of $(I/\mathfrak{m}I)^r$ such that whenever $(y_1 + \mathfrak{m}I, \ldots, y_r + \mathfrak{m}I) \in U$, then $(y_1, \ldots, y_r)R'$ is a reduction of R' with reduction number at most $n - 2$. Let $I = (x_1, \ldots, x_s)$, with each $x_i \in I \setminus \mathfrak{m}I$. Then by construction for each x_i there exists a non–empty Zariski–open set U_i in $(I/\mathfrak{m}I)^r$ such that whenever $(y_1 + \mathfrak{m}I, \ldots, y_r + \mathfrak{m}I) \in U_i$, then $(y_1, \ldots, y_r)R/(\mathfrak{m}I^n : x_i)$ is a reduction of $I(R/(\mathfrak{m}I^n : x_i))$ with reduction number at most $n - 2$. Let $U = \cap_i U_i$. Choose $y_1, \ldots, y_r \in I$

such that $(y_1 + \mathfrak{m}I, \ldots, y_r + \mathfrak{m}I) \in U$. Then for all $i = 1, \ldots, s$, $I^{n-1} \subseteq$ $(y_1, \ldots, y_r)I^{n-2} + (\mathfrak{m}I^n : x_i)$, so that $x_i I^{n-1} \subseteq (y_1, \ldots, y_r)I^{n-1} + \mathfrak{m}I^n$. It follows that $I^n \subseteq (y_1, \ldots, y_r)I^{n-1} + \mathfrak{m}I^n$, which by Nakayama's Lemma proves the theorem. □

Caviglia [36] gave an alternate proof that is perhaps more natural, but requires different background. Namely, Caviglia's proof is based on Green's hyperplane restriction theorem, which we state here without proof:

Theorem 8.6.9 (Green's hyperplane restriction theorem [98]) *Let R be a standard graded algebra over an infinite field K and let L be a generic linear form in R. Set $S = R/(L)$. Then for all positive integers d,*

$$\dim_K S_d \leq (\dim R_d)_{\langle d \rangle}.$$

(This is the Macaulay representation of numbers, Definition A.5.2.)

Using Green's hyperplane restriction theorem, an alternate proof of Theorem 8.6.8 is as follows:

Alternate proof of Theorem 8.6.8: (Caviglia [36]) The fiber cone $\mathcal{F}_I$ is a standard graded $(R/\mathfrak{m})$–algebra such that $\dim_{R/\mathfrak{m}}(\mathcal{F}_I)_n \leq \binom{n+r}{r} - 1 = \binom{n+r-1}{n} + \binom{n+r-2}{n-1} + \cdots + \binom{r}{1}$. Let $h_1, \ldots, h_r$ be generic elements in $(\mathcal{F}_I)_1$. By Green's hyperplane restriction theorem (Theorem 8.6.9) and by induction,

$$\dim_{R/\mathfrak{m}}\left(\frac{\mathcal{F}_I}{(h_1, \ldots, h_r)}\right)_n \leq \binom{n-1}{n} + \binom{n-2}{n-1} + \cdots + \binom{0}{1} = 0.$$

Hence $I^n \subseteq (h_1, \ldots, h_r)I^{n-1} + \mathfrak{m}I^n$, so that by Nakayama's Lemma, $I^n \subseteq (h_1, \ldots, h_r)I^{n-1}$. □

8.7. Non–local rings

In the previous sections we mainly used reductions in local rings. But there are also nice results in non–local rings. Lyubeznik [196] proved that every ideal in a polynomial ring in n variables over an infinite field has an n–generated reduction. This result is stronger than saying that every ideal in a polynomial ring in n variables is up to radical generated by at most n elements. Over Noetherian local rings, the result is known by Proposition 8.3.8 and Corollary 8.3.9. The method of proof is very different for polynomial rings, however. Katz generalized Lyubeznik's existence results and we present Katz's version. For both Lyubeznik's and Katz's result one needs a result of Mohan Kumar [176], which we use without proof.

First we need a proposition:

Proposition 8.7.1 *Let R be a d–dimensional Noetherian ring and I an ideal contained in the Jacobson radical $\mathfrak{m}$ of R. Then*
(1) Some power of I has a d–generated reduction.
(2) If R has infinite residue fields, then I has a d–generated reduction.

Proof: If $d = 0$, then I is nilpotent, so 0 is a reduction of I (and of any power of I). Now assume that $d > 0$. If I consists of zerodivisors, let $L = \cup_n (0 : I^n)$. Say $L = 0 : I^l$. Then $I(R/L)$ contains non–zerodivisors. If the proposition holds for $I(R/L)$, then for some positive integer m, there exists a d–generated ideal J in R contained in I^m such that $J(R/L) \subseteq I^m(R/L)$ is a reduction. Thus for some integer n, $I^{m(n+1)} \subseteq JI^{mn} + L$, so that $I^{m(n+1)+l} \subseteq JI^{mn+l}$. Hence we may assume that I contains non–zerodivisors. As I is contained in the Jacobson radical of R, necessarily $\cap_n I^n = 0$. Thus by Lemma 8.5.4, every superficial element of I is a non–zerodivisor.

By Proposition 8.5.7, some power I^m of I contains a superficial element x, and if R has infinite residue fields, then m can be taken to be 1. By induction on d, some power of $I^m(R/(x))$ has a $(d-1)$ generated reduction, say $I^{ml}R/(x)$ has a $(d-1)$ generated reduction, with $ml = 1$ if the residue field is infinite. It follows that $I^{ml}R/(x^l)$ also has a $(d-1)$ generated reduction. By Remark 8.5.2, x^l is superficial for I^{ml} since x is superficial for I^m. By Proposition 8.6.1, I^{ml} has a d–generated reduction. If the residue field is infinite, $ml = 1$. $\qquad\square$

Here are some facts that we will use below. Part (3) is a difficult result due to Mohan Kumar, and we provide no proof here.

Lemma 8.7.2 *Let R be a Noetherian ring R. Let I be an ideal, and W the multiplicatively closed set $\{1 - i \,|\, i \in I\}$. Then*
(1) $\mu(W^{-1}I) = \mu(I/I^2) \geq \mu(I) - 1$. In fact, if $I = (x_1, \ldots, x_d) + I^2$, then there exists $x \in I$ such that for all n, $I = (x_1, \ldots, x_d, x^n)$.
(2) If $\mu(I/I^2) = d > \dim R$, then $\mu(I) = d$.
(3) (Mohan Kumar [176, proof of Theorem 2]; we provide no proof) If $R = A[X]$ is a reduced polynomial ring of dimension d, I an ideal in R of positive height such that I/I^2 is generated by at most d elements, then I is generated by the same number of elements.

Proof: The first equality in (1) follows by the Nakayama's Lemma: $W^{-1}I$ is in the Jacobson radical of $W^{-1}R$. Let $d = \mu(I/I^2)$. Let $x_1, \ldots, x_d \in I$ such that $W^{-1}I = W^{-1}(x_1, \ldots, x_d)$. Let $x \in I$ such that $(1 - x)I \subseteq (x_1, \ldots, x_d)$. Then $I = (x_1, \ldots, x_d, x)$, so that I has at most $d + 1$ generators. Furthermore, as $(1 - x^{n-1})x \in (x_1, \ldots, x_d)$, it follows that $x \in (x_1, \ldots, x_d, x^n)$, whence $I = (x_1, \ldots, x_d, x^n)$. This proves (1).

If in addition $d > \dim R$, then by prime avoidance $x_1, \ldots, x_d$ may be chosen such that the radical of $(x_1, \ldots, x_d)$ equals $\sqrt{I}$. Thus for some large n, $x^n \in (x_1, \ldots, x_d)$, whence by (1), $I = (x_1, \ldots, x_d)$. $\qquad\square$

Katz generalized Lyubeznik's result as follows:

Theorem 8.7.3 (Katz [161]) *Let R be a Noetherian ring and I an ideal in R. Let d be the maximum of the heights of maximal ideals containing I and suppose that $d < \infty$. Then*

(1) Some power of I has a $(d+1)$–generated reduction.

(2) If R has infinite residue fields, then I has a $(d+1)$–generated reduction.

(3) If $R = A[X]$ is a polynomial ring and $d = \dim R$, then d may be taken in place of $d+1$ in (1) and (2).

Proof: Without loss of generality I is a proper ideal. Let $W = \{1 - i \mid i \in I\}$. Then W is a multiplicatively closed subset of R that does not contain zero, $W^{-1}R$ is a d–dimensional Noetherian ring, and $W^{-1}I$ is in the Jacobson radical of $W^{-1}R$. By Proposition 8.7.1 there exists a d–generated ideal J contained in a power I^m of I ($m = 1$ if R has infinite residue fields) such that $W^{-1}J$ is a reduction of $W^{-1}I^m$. Let $K = W^{-1}J \cap R$. For some $i \in I$, $(1 - i)K \subseteq J \subseteq I^m$. It follows that $(1 - i^m)K \subseteq I^m$, so that $K \subseteq I^m$. Let n be such that $W^{-1}I^{m(n+1)} = W^{-1}JI^{mn} = W^{-1}KI^{mn}$. Then for some $s = 1 - i' \in W$, $sI^{m(n+1)} \subseteq KI^{mn} \subseteq K$, so that $I^{m(n+1)} \subseteq K$. Since $(1-i')I^{m(n+1)} \subseteq KI^{mn}$, we obtain that for all N, $(1 - (i')^N)I^{m(n+1)} \subseteq KI^{mn}$, so that $I^{m(n+1)} \subseteq KI^{mn} + I^N$ for all N. Since $I^N \subseteq KI^{mn}$ for $N \gg 0$, we see that K is a reduction of I^m. As $\sqrt{K} = \sqrt{I}$, then $W^{-1}R = T^{-1}R$, where $T = \{1 - k \mid k \in K\}$. The number of generators of $T^{-1}K$ is at most d, and by Lemma 8.7.2, K has at most $d+1$ generators. This proves (1) and (2).

Now let $R = A[X]$ be a polynomial ring of dimension d. By Lemma 8.7.2, the assumption $\mu(T^{-1}K) \leq d$ implies that $\mu(K/K^2) \leq d$. If $\mu(K/K^2) < d$, then again by Lemma 8.7.2, $\mu(K) \leq d$. So we may assume that $\mu(K/K^2) = d$. If K is nilpotent, so is I, and every power of I has a 0–generated reduction.

It remains to prove (3) in the case $\mu(K/K^2) = d$ and K is not nilpotent. Let L be the intersection of the height zero prime ideals that do not contain K. Clearly $\mu((K + L)/(K^2 + L)) \leq d$. If $\mu((K + L)/(K^2 + L)) < d$, then by Lemma 8.7.2 (1), $\mu((K + L)/L) \leq d$, and if $\mu((K + L)/(K^2 + L)) = d$, then by Lemma 8.7.2 (2) and (3), $\mu((K + L)/L) \leq d$. Thus in any case, $\mu((K + L)/L) \leq d$. It follows that there exists a d–generated ideal $K' \subseteq K$ such that $K' + L = K + L$. Thus $K = K' + L \cap K$. But $L \cap K$ is nilpotent, so that K' is a reduction of K, which is a reduction of I^m. $\qquad\square$

8.8. Sally's theorem on extensions

In Chapter 14 on two–dimensional regular local rings we prove that blowups along a divisorial valuation (a "good" Noetherian valuation) eventually produce a two–dimensional regular local ring whose order valuation is the original valuation.

A related statement (in arbitrary dimension) proved by Judith Sally in [254] is the topic of this section.

Theorem 8.8.1 *Let $(R, \mathfrak{m}) \subsetneq (S, \mathfrak{n})$ be an extension of d–dimensional regular local rings with the same field of fractions. Then the analytic spread $\ell(\mathfrak{m}S)$ of $\mathfrak{m}S$ is strictly smaller than d.*

As in Theorem 6.7.9, to a regular local ring $(R, \mathfrak{m})$ we associate its order valuation. We need a general lemma for the proof.

Lemma 8.8.2 *Let* $(R, \mathfrak{m}) \subseteq (S, \mathfrak{n})$ *be* d*–dimensional local rings with the same field of fractions* K. *Assume that* $(R, \mathfrak{m})$ *is regular and* $\ell(\mathfrak{m}S) = d$. *Let* $(V, \mathfrak{m}_V)$ *be the valuation ring of the order valuation of* R. *Then:*
(1) $(S, \mathfrak{n}) \subseteq (V, \mathfrak{m}_V)$,
(2) $S/\mathfrak{n} = R/\mathfrak{m}$, *and*
(3) $\mathfrak{n}^k \cap R = \mathfrak{m}^k$ *for all* k.

Proof: We begin by proving that (1) implies (2). By the Dimension Inequality Theorem B.2.5 (note: S need not be finitely generated over R), $\dim R + \mathrm{tr.deg}_K K \geq \dim S + \mathrm{tr.deg}_{R/\mathfrak{m}}(S/\mathfrak{n})$. Hence $S/\mathfrak{n}$ is algebraic over $R/\mathfrak{m}$. Then $R/\mathfrak{m} \subseteq S/\mathfrak{n} \subseteq V/\mathfrak{m}_V = \kappa(V)$, and by Theorem 6.7.9, $\kappa(V)$ is purely transcendental over $R/\mathfrak{m}$, proving that $R/\mathfrak{m} = S/\mathfrak{n}$.

Next observe that (1) implies (3), as $\mathfrak{m}^k \subseteq \mathfrak{n}^k \cap R \subseteq \mathfrak{m}_V^k \cap R = \mathfrak{m}^k$, the latter equality holding since V is the order valuation of R.

It remains to prove (1). Write $\mathfrak{m} = (x_1, \ldots, x_d)$. By definition, the analytic spread of $\mathfrak{m}S$ is the dimension of the fiber ring $\frac{S[x_1 t, \ldots, x_d t]}{\mathfrak{n} \cdot S[x_1 t, \ldots, x_d t]}$. By assumption this dimension is d, forcing this quotient ring to be isomorphic to a polynomial ring over $S/\mathfrak{n}$ in d variables. In particular, $\mathfrak{n}S[\mathfrak{m}t]$ is prime ideal. Since $\mathrm{ht}(\mathfrak{n}S[\mathfrak{m}t]) + \dim(\frac{S[x_1 t, \ldots, x_d t]}{\mathfrak{n} \cdot S[x_1 t, \ldots, x_d t]}) \leq \dim S[\mathfrak{m}t] = d + 1$, it follows that $\mathfrak{n}S[\mathfrak{m}t]$ is prime ideal of height 1.

Recall that by Theorem 6.7.9 $(V, \mathfrak{m}_V) = R[\frac{\mathfrak{m}}{x_1}]_{(x_1)}$. As $R[\frac{\mathfrak{m}}{x_1}] \subseteq S[\frac{\mathfrak{m}}{x_1}]$, and as $\mathfrak{n}S[\mathfrak{m}t]$ is a height one prime ideal, we obtain that $\mathfrak{n}S[\mathfrak{m}t]_{x_1 t}$ is a height one prime ideal. Then $\mathfrak{n}S[\frac{\mathfrak{m}}{x_1}]$ is also a height one prime ideal, and $V = R[\frac{\mathfrak{m}}{x_1}]_{\mathfrak{m}R[\frac{\mathfrak{m}}{x_1}]} \subseteq S[\frac{\mathfrak{m}}{x_1}]_{\mathfrak{n}S[\frac{\mathfrak{m}}{x_1}]}$, implying that V is this latter ring. Clearly then $\mathfrak{n}$ is contracted from the maximal ideal of V, which proves (1). $\square$

Proof of Theorem 8.8.1: For contradiction assume that $\ell(\mathfrak{m}S) = d$. Write $\mathfrak{m} = (x_1, \ldots, x_d)$. We claim that $\mathfrak{m}S = \mathfrak{n}$. It suffices to prove that $\dim_{S/\mathfrak{n}}((\mathfrak{m} + \mathfrak{n}^2)/\mathfrak{n}^2) = d$. Suppose that $\dim_{S/\mathfrak{n}}(\mathfrak{m} + \mathfrak{n}^2)/\mathfrak{n}^2 < d$. Then there is a relation, $\sum_{i=1}^d \lambda_i x_i \in \mathfrak{n}^2$, where λ_i are either units in S or are zero. By Lemma 8.8.2, $R/\mathfrak{m} = S/\mathfrak{n}$, so we can write $\lambda_i = \alpha_i + s_i$ where α_i are units (or zero) in R, and $s_i \in \mathfrak{n}$. Then $\sum_{i=1}^d \alpha_i x_i \in R \cap \mathfrak{n}^2 = \mathfrak{m}^2$, using Lemma 8.8.2, which is a contradiction. Thus $\mathfrak{m}S = \mathfrak{n}$.

Next we prove that $S \subseteq \widehat{R}$. Let $s \in S$ with $s \in \mathfrak{n}^e \setminus \mathfrak{n}^{e+1}$. Write $s = f(x_1, \ldots, x_d)$ for some homogeneous polynomial $f(T_1, \ldots, T_d)$ in $S[T_1, \ldots, T_d]$ of degree e with at least one coefficient a unit in S. Write $f = \sum_\nu s_\nu \underline{x}^\nu$, where $\nu = (\nu_1, \ldots, \nu_d)$ is a multi–index with $\nu_1 + \cdots + \nu_d = e$ and all $\nu_i \geq 0$. Define a polynomial $g(T_1, \ldots, T_d) \in R[T_1, \ldots, T_d]$ as follows: $g(T_1, \ldots, T_d) = \sum_\nu r_\nu \underline{x}^\nu$, where $r_\nu = 0$ if $s_\nu \in \mathfrak{n}$ and r_ν is a unit in R congruent to s_ν modulo $\mathfrak{n}$ if $s_\nu \notin \mathfrak{n}$. Set $r_e = g(x_1, \ldots, x_d)$. Then $s - r_e \in \mathfrak{n}^{e+1}$. We replace s by $s - r_e$ and continue. This gives a Cauchy sequence of elements of R converging to s and gives the embedding of S into $\widehat{R}$. Then $S \subseteq \widehat{R} \cap Q(R) = R$ (see Exercise 8.1), which proves the theorem. $\square$

8.9. Exercises

8.1   Let R be a local Noetherian domain. Prove that $\widehat{R} \cap Q(R) = R$.

8.2   Let $(R, \mathfrak{m})$ be a Noetherian local ring, I an ideal and X an indeterminate over R. Prove that the reduction number of I equals the reduction number of $IR[X]_{\mathfrak{m}R[X]}$.

8.3   Assume that I is a non–nilpotent ideal. Prove that if r is superficial for I, then $r \notin I^2$.

8.4   Prove that any element of a nilpotent ideal I is superficial for I.

8.5   Let r be a superficial element for I. Prove that for any integer l, the image of r in $R/(0 : I^l)$ is superficial for the image of I in $R/(0 : I^l)$.

8.6   Give an example of a Noetherian local ring $(R, \mathfrak{m})$, an ideal I and an element $r \in I$ such that r is superficial for I and $r \in \mathfrak{m}I$.

8.7   Let $(R, \mathfrak{m})$ be a Noetherian local ring with infinite residue field and I an ideal in R. Prove that there exist a non–empty Zariski–open subset U of $I/\mathfrak{m}I$ and an integer l such that whenever $r \in I$ with image in $I/\mathfrak{m}I$ in U, and $i \geq 1$, then r^i is superficial for I^i with $(I^n : r^i) \cap I^l = I^{n-i}$ for all $n \geq i + l$.

8.8   Let R be a Noetherian ring and I an ideal. Prove that there exists an integer l such that for all $n \geq l$, I^n has a superficial element.

8.9   Let R be an $\mathbb{N}$–graded integral domain and $J \subseteq I$ homogeneous ideals. If J is a reduction of I, then the smallest $n \in \mathbb{N}$ such that I contains a non–zero element of degree n equals the smallest $n \in \mathbb{N}$ such that J contains a non–zero element of degree n.

8.10 Let X, Y, Z be variables over a field k and let $R = k[X, Y, Z]$. Let I be the ideal $(X^5Y, Y^5(Z - 1), X^3Y^2Z, X^2Y^3(Z - 1))$ in R. For each $l \in \mathbb{N}_{>0}$, set $J_l = (X^5Z - Y^5(Z - 1), X^3Y^2Z^l, X^2Y^3(Z - 1)^l)$. Prove that each J_l is a reduction of I, but that $\cap_l J_l$ is not a reduction of I.

8.11 Let $(R, \mathfrak{m})$ be a Noetherian local ring and I an $\mathfrak{m}$–primary ideal.
  (i) Prove that there exists an integer n such that I^n has a reduction generated by a system of parameters.
  (ii) Prove that if $R/\mathfrak{m}$ is infinite, then I has a reduction generated by a system of parameters.

8.12 Give an example of ideals I and J in a Noetherian ring R showing that $\sqrt{J} = \sqrt{I}$ and $J \subseteq I$ do not imply that J is a reduction of I. Prove that if I is a nilpotent ideal, then any ideal contained in I is its reduction.

8.13 Let R be a Noetherian local ring and I an ideal with $\operatorname{ht} I = \mu(I)$. Prove that I is basic.

8.14 Let J_1, J_2, I_1 and I_2 be ideals in a Noetherian ring R. Prove or disprove: if J_1 is a reduction of I_1 and J_2 is a reduction of I_2, then $J_1 \cap J_2$ is a reduction of $I_1 \cap I_2$.

8.15 Prove that if $J \subseteq I$ is a reduction in a Noetherian local ring, then $\ell(J) = \ell(I)$.

8.16 Give examples of ideals $J \subseteq I$ in a Noetherian local ring R such that
(i) $\ell(J) > \ell(I)$,
(ii) $\ell(J) < \ell(I)$.

8.17 Let $(R, \mathfrak{m})$ be a Noetherian local domain and I a non–zero ideal. Let P be a prime ideal of R. Prove that $\ell(I) \geq \ell(IR_P)$.

8.18 Let $(R, \mathfrak{m})$ be a Noetherian local domain with infinite residue field and I a non–zero ideal. Let $\overline{R}$ be the integral closure of R. Prove that $\ell(I) = 1$ if and only if $I\overline{R}$ is principal.

8.19 (Cf. Lemma 8.5.4.) Let $R = k[X, Y]/(X - X^2Y^2)$, where k is a field and X and Y variables over k. (Or replace R by its localization at the complement of the union of the prime ideals (X, Y) and $(1 - XY^2)$.) Let $I = (Y)$.
(i) Prove that $X \in I^n$ for all n, so that $\cap_n I^n$ is non–zero.
(ii) Let $r = (1 - XY^2)Y$. Prove that $r \in I$ is a zerodivisor in R that is superficial for I.

8.20 Let $R = (\mathbb{Z}/2\mathbb{Z})[[X, Y, Z]]/(Z^5, Z^4 + XY(X + Y))$, where X, Y, Z are variables over $\mathbb{Z}/2\mathbb{Z}$. Let $J = (X, Y)R$, $I = \mathfrak{m} = (X, Y, Z)R$.
(i) Prove that J is a minimal reduction of I.
(ii) Prove that $\mu(J) \neq \ell(I)$.
(iii) Prove that $J^3 \cap \mathfrak{m}I^3 \neq \mathfrak{m}J^3$. (Cf. Corollary 8.3.5 (3).)

8.21 Let k be a field, and $X_1, \ldots, X_n$ variables over k. Set R to be either $k[[X_1, \ldots, X_n]]$ or $k[X_1, \ldots, X_n]_{(X_1, \ldots, X_n)}$. Then R is a Noetherian local ring with maximal ideal $\mathfrak{m} = (X_1, \ldots, X_n)$. Let $a_1, \ldots, a_s$ be homogeneous elements in $k[X_1, \ldots, X_n]$ all of the same degree, and I the ideal in R generated by $a_1, \ldots, a_s$.
(i) Prove that $\mathcal{F}_I(R) = \oplus_i(I^i/\mathfrak{m}I^i)$ is isomorphic to $k[a_1, \ldots, a_s]$, the subring of $k[X_1, \ldots, X_n]$ generated by $a_1, \ldots, a_s$.
(ii) Assume that $a_1, \ldots, a_s$ are monomials of degree d, Write $a_i = X_1^{c_{i1}} \cdots X_n^{c_{in}}$ for some non–negative integers c_{ij}. Prove that $\ell(I)$ equals the rank of the matrix (c_{ij}).
See [16] for more on the analytic spread of monomial ideals.

8.22 Suppose that R is a Noetherian ring and I is an ideal of linear type. Prove for all prime ideals P containing I, the analytic spread of I_P is equal to the minimal number of generators of I_P, i.e., that I_P is its own minimal reduction.

8.23 Let k be a field, m, n positive integers, and X_{ij}, $i = 1, \ldots, m$, $j = 1, \ldots, n$, variables over k. Set $R = k[X_{ij}]_{(X_{ij})}$. Let $t < \min\{m, n\}$ and I the ideal of all $t \times t$ minors of the $m \times n$ matrix (X_{ij}). Prove that $\ell(I) = mn$.

8.24   Let k be a field, $t, m, n \in \mathbb{N}$, $2 \leq t < \min\{m, n\}$, and X_{ij} variables over k, where $i = 1, \ldots, m$, $j = 1, \ldots, n$. Set $R = K[X_{ij}]$, and P the prime ideal generated by the $t \times t$ minors of (X_{ij}). Prove that $\mathrm{gr}_P(R)$ is not an integral domain. (Hint: use Exercise 8.23.)

8.25   Let $(R, \mathfrak{m})$ be a Noetherian local ring and let I be an ideal in R. Let $(x_1, \ldots, x_n) \subseteq I$ be a minimal reduction of I generated by a superficial sequence of length n. Set $J = (x_1, \ldots, x_r)$. Prove that $\mathrm{depth}(\mathrm{gr}_I(R)) \geq r + 1$ if and only if $\mathrm{depth}(\mathrm{gr}_{I/J}(R/J)) \geq 1$.

8.26   Let $(R, \mathfrak{m})$ be a Noetherian local ring and I an ideal in R. Elements $a_1, \ldots, a_l$ are said to be **analytically independent** in I if for any polynomial F in variables $X_1, \ldots, X_l$ over R that is homogeneous of degree d, the condition $F(a_1, \ldots, a_l) \in \mathfrak{m}I^d$ implies that all the coefficients of F are in $\mathfrak{m}$. Assume that $a_1, \ldots, a_l$ are analytically independent in I.

   (i) Prove that if $J = (a_1, \ldots, a_l)$, then $\mathcal{F}_J(R)$ is isomorphic to a polynomial ring in l variables over $R/\mathfrak{m}$.
   (ii) Prove that $a_1, \ldots, a_l$ minimally generate $(a_1, \ldots, a_l)$.
   (iii) Assume that $(b_1, \ldots, b_l) = (a_1, \ldots, a_l)$. Prove that $b_1, \ldots, b_l$ are analytically independent.

8.27   Let $(R, \mathfrak{m})$ be a Noetherian local ring and I an ideal in R.

   (i) Prove that if $a_1, \ldots, a_l$ are analytically independent in I, then $l \leq \ell(I)$.
   (ii) Prove that if $R/\mathfrak{m}$ is infinite, there exist $\ell(I)$ analytically independent elements in I that generate a reduction of I.
   (iii) Prove that if $R/\mathfrak{m}$ is infinite, every minimal reduction of I is generated by analytically independent elements.

8.28   Let R be a local Noetherian ring with a fixed prime ideal q. Prove that there exists a local homomorphism $R \to T$, with kernel q, such that T is a one–dimensional Noetherian local domain with field of fractions $\kappa(q)$, essentially of finite type over R whose residue field L is a purely transcendental extension of the residue field k of R. (Hint: Choose elements $y_1, \ldots, y_d$ in R/q that form a system of parameters. Set $T = (R/q)[\frac{y_2}{y_1}, \ldots \frac{y_d}{y_1}]_{(\mathfrak{m}/q)}$. Since $y_1, \ldots, y_d$ are a system of parameters, they are analytically independent and $\mathfrak{m}/q$ generates a prime ideal.)

8.29   (Swanson [281]) Let $(R, \mathfrak{m})$ be a Noetherian local ring with infinite residue field and let I be an ideal. Define $S(I)$ to be the set of all $s \in \mathbb{N}$ for which there exists a non-empty Zariski–open subset $U \subseteq (I/\mathfrak{m}I)^s$ such that whenever $x_1, \ldots, x_s \in I$ and $(\overline{x}_1, \ldots, \overline{x}_s) \in U$, then $I \subseteq \sqrt{(x_1, \ldots, x_s)}$. Prove that $\ell(I) = \min(S(I))$.

8.30*   Let $(R, \mathfrak{m})$ be a Noetherian local ring and I an ideal. Is the number of generators of minimal reductions of I independent of the reduction? The answer is yes if $R/\mathfrak{m}$ is infinite (Proposition 8.3.7), but not known in general.

8.31 (Compare with the proof of Theorem 8.6.6.) Let $(R, \mathfrak{m})$ be a Noetherian local ring with infinite residue field and I an ideal of analytic spread at most l. For each $n \geq 1$, define U_n to be the set of all $(\overline{u}_{ij}) \in (I/\mathfrak{m}I)^l$ such that $(\sum_{j=1}^m u_{1j}a_j, \ldots, \sum_{j=1}^m u_{lj}a_j)$ is a reduction of I of reduction number at most n.

(i) Prove that $\min\{n \,|\, U_n \neq \emptyset\}$ is the (absolute) reduction number of I.

(ii) Prove that "almost all" minimal reductions of I have the same reduction number. In precise terms, if $s = \dim(I/\mathfrak{m}I)$, prove that there exists a Zariski–open subset U of $(R/\mathfrak{m})^{sl} \cong (I/\mathfrak{m}I)^l$ such that whenever $(u_{ij}) \in R^{sl}$ with image $(\overline{u}_{ij}) \in (I/\mathfrak{m}I)^l$ being in U, then $(\sum_{j=1}^m u_{1j}a_j, \ldots, \sum_{j=1}^m u_{lj}a_j)$ is a reduction of I with minimal possible reduction number.

8.32 Let $(R, \mathfrak{m})$ be a Noetherian local ring. Let $a_1, \ldots, a_d$ be a system of parameters. Set $S = R[\frac{a_2}{a_1}, \ldots, \frac{a_d}{a_1}]$, and $Q = \mathfrak{m}S$.

(i) Prove that $\dim S = d$.

(ii) Prove that Q is a prime ideal of height one and that $Q = \sqrt{a_1 S}$.

(iii) Prove that the elements $\frac{a_2}{a_1}, \ldots, \frac{a_d}{a_1}$ are algebraically independent over $R/\mathfrak{m}$.

(iv) Prove that $\operatorname{Ass} S = \{pR[a_1^{-1}] \cap S \,|\, p \in \operatorname{Ass} R, a_1 \notin p\}$.

8.33 (Burch [34]) Let $(R, \mathfrak{m})$ be a Noetherian local ring and I a basic ideal.

(i) Let $f_1, \ldots, f_r \in \mathfrak{m}$, and $C_1, C_2, C_3, \ldots$ ideals of R such that for each n, $I^{n+1} \subseteq C_{n+1} \subseteq C_n \subseteq \overline{I^n}$ and $(f_1, \ldots, f_r)(R/C_n)$ is generated by a regular sequence. Prove that $\ell(I + (f_1, \ldots, f_r)R) = \ell(I) + \ell((f_1, \ldots, f_r)R)$.

(ii) Let $C_1, C_2, C_3, \ldots$ ideals of R such that for each n, $I^{n+1} \subseteq C_{n+1} \subseteq C_n \subseteq \overline{I^n}$ and $\operatorname{depth}(R/C_n) \geq k$. Prove that $\ell(I) \leq \dim R - k$.

8.34 ([34]) Let $(R, \mathfrak{m})$ be a Noetherian local ring, I an ideal in R. Prove

(i) $\ell(I) \leq \dim R - \min\{\operatorname{depth}(R/I^n) \,|\, n \in \mathbb{N}\}$.

(ii) $\ell(I) \leq \dim R - \min\{\operatorname{depth}(R/\overline{I^n}) \,|\, n \in \mathbb{N}\}$.

8.35 Let k be a field, X, Y variables over k, $R = k[[X^2, X^3, Y]]$, $I = (X^2, X^3Y)R$. Prove that the bound for $\ell(I)$ in Exercise 8.34 (ii) is sharper than the bound for $\ell(I)$ in Exercise 8.34 (i).

9
Analytically unramified rings

It is not surprising that the behavior of integral closure of ideals under extension of rings should relate to the behavior of nilpotents under extensions, since the nilradical is always inside the integral closure of every ideal. It is also not surprising that the behavior of the integral closure of ideals under extension rings should directly relate to the module–finite property of the integral closures of finitely generated extension rings of a given ring, since the integral closures of Rees algebras and affine pieces of the blowup of an ideal relate directly to the integral closure of the ideal. In this chapter we study these problems and show that they all relate to a basic idea that the completions of the localizations of a ring R should not have "new" nilpotents. We formalize this in the following definition:

Definition 9.0.1 *A local Noetherian ring* $(R, \mathfrak{m})$ *is* **analytically unramified** *if its* $\mathfrak{m}$*–adic completion is reduced.*

If $(R, \mathfrak{m})$ is analytically unramified, then R must itself be reduced as R embeds in its completion. However, not every reduced Noetherian ring is analytically unramified (see Exercises 4.9 and 4.11).

The main results of this chapter, Theorems 9.1.2 and 9.2.2, are two classic theorems of David Rees characterizing the locally analytically unramified rings. The first theorem characterizes analytically unramified rings in terms of integral closures of powers of ideals, and the second theorem has to do with module–finiteness of integral closures of analytically unramified rings.

In the last section, we use the work on analytically unramified rings to study an important set of rank one discrete valuations, called divisorial valuations.

We give here an incomplete list of locally analytically unramified rings.

(1) Every reduced ring that is a localization of finitely generated algebra over a field; Theorems 4.6.3, 9.2.2.

(2) Every reduced ring that is a localization of finitely generated algebra over $\mathbb{Z}$; Corollary 4.6.5 and Theorem 9.2.2.

(3) Every complete local domain (by definition).

(4) Every pseudo–geometric Noetherian reduced ring; see Exercise 9.6.

(5) Every reduced ring that is finitely generated over a complete local ring; see Exercise 9.7.

(6) The ring of convergent power series in n variables over $\mathbb{Q}$, $\mathbb{R}$, or $\mathbb{C}$ (completion is the power series ring, which is a domain).

(7) Every excellent reduced ring. (We do not define excellent rings in this book; the analytically unramified property follows quickly from the definition.)

9.1. Rees's characterization

The main result of this section is Rees's characterization of analytically un-ramified local rings in terms of integral closures of powers of ideals. We also discuss constructions of analytically unramified rings.

We first prove a lemma: even though a necessary condition that integral closure commute with passage to completion for all ideals is that the ring be analytically unramified, there is a class of ideals for which the integrally closed property is preserved for all Noetherian local rings:

Lemma 9.1.1 *Let* $(R, \mathfrak{m})$ *be a Noetherian local ring, and let* I *be an* $\mathfrak{m}$–*primary ideal. Then* $\overline{I}\widehat{R} = \overline{I\widehat{R}}$.

Proof: Obviously $\overline{I}\widehat{R} \subseteq \overline{I\widehat{R}}$. Let $s \in \overline{I\widehat{R}}$ and consider an integral equation

$$s^n + a_1 s^{n-1} + \cdots + a_n = 0,$$

where $a_j \in I^j \widehat{R}$. Choose ℓ such that $\mathfrak{m}^\ell \subseteq I$. Choose $s' \in R$ and $a'_i \in I^i$ such that $s - s', a_i - a'_i \in \mathfrak{m}^{n\ell}\widehat{R}$. Then

$$(s')^n + a'_1 (s')^{n-1} + \cdots + a'_n \in \mathfrak{m}^{n\ell}\widehat{R} \cap R = \mathfrak{m}^{n\ell}.$$

By the choice of ℓ, $\mathfrak{m}^{n\ell} \subseteq I^n$, so that we can modify this equation to give an integral equation for s' over I. Since $s - s' \in \mathfrak{m}^{n\ell}\widehat{R} \subseteq I^n\widehat{R} \subseteq \overline{I\widehat{R}}$, it follows that $s \in \overline{I}\widehat{R}$. $\qquad\square$

With this we can prove that the analytically unramified property is closely related to the integral closures of powers of ideals:

Theorem 9.1.2 (Rees [233]) *Let* $(R, \mathfrak{m})$ *be a Noetherian local ring. The following are equivalent:*

(1) R is analytically unramified.

(2) For all $I \subseteq R$ there is an integer k such that for all $n \geq 0$, $\overline{I^{n+k}} \subseteq I^n$.

(3) There exist an $\mathfrak{m}$–primary ideal J and an integer k such that $\overline{J^{n+k}} \subseteq J^n$ for all $n \geq 0$.

(4) There exist an $\mathfrak{m}$–primary ideal J and a function $f : \mathbb{N} \to \mathbb{N}$ such that $\lim_{n\to\infty} f(n) = \infty$ and such that $\overline{J^n} \subseteq J^{f(n)}$ for all n.

Proof: Obviously (2) implies (3) and (3) implies (4).

We assume (4) and prove (1): let N denote the nilradical of $\widehat{R}$. Then $N \subseteq \cap_{n\geq 1} \overline{J^n \widehat{R}}$, so that by Lemma 9.1.1, $N \subseteq \cap_{n\geq 1} \overline{J^n}\widehat{R}$. By assumption this intersection is contained in $\cap_n J^{f(n)}\widehat{R} = 0$. Hence R is analytically unramified.

Finally we assume (1) and prove (2). Fix any $I \subseteq R$. It is enough to prove that for any ideal $I \subseteq R$ there exists an integer k such that for all $n \geq 0$

$$\overline{I^{n+k}\widehat{R}} \subseteq I^n \widehat{R},$$

since contracting to R gives (2) in this case. Thus we may assume that R is complete and reduced.

We next reduce to the case in which R is a domain. Let the minimal primes of R be $\{P_1, \cdots, P_t\}$. By $(\_)_i$ we denote images in R/P_i. Suppose that there exist integers k_i such that

$$\overline{I_i^{n+k_i}} \subseteq I_i^n$$

for all $n \geq 0$ and for $1 \leq i \leq t$. Choose $c_i \in P_1 \cap \cdots \cap \widehat{P_i} \cap \cdots \cap P_t \setminus P_i$. Set $c = c_1 + \cdots + c_t$, so that in particular $c \notin \bigcup_{i=1}^t P_i$. Let k be the maximum of the k_i. Since $\overline{I^{n+k}} R_i \subseteq \overline{I_i^{n+k}} \subseteq I_i^n$, we obtain that $\overline{I^{n+k}} \subseteq I^n + P_i$ for all $1 \leq i \leq t$. Multiply by c_i to obtain that

$$c_i \overline{I^{n+k}} \subseteq I^n + c_i P_i = I^n$$

for all $1 \leq i \leq t$. Hence $c\overline{I^{n+k}} \subseteq I^n$ for all $n \geq 0$ and so $c\overline{I^{n+k+\ell}} \subseteq I^{n+\ell} \cap (c) \subseteq cI^n$ for some ℓ and for all $n \geq 0$, the latter containment following by the Artin–Rees Lemma. As R is reduced, c is a non–zerodivisor, and hence

$$\overline{I^{n+k+\ell}} \subseteq I^n.$$

Thus we have reduced the implication $(1) \Longrightarrow (2)$ to the case in which R is a complete local domain. Now the theorem follows from Proposition 5.3.4. $\square$

Theorem 13.4.8 in Chapter 13 shows that under some extra conditions, k can be taken to be independent of I.

This enables us to build further analytically unramified rings from the known ones.

Proposition 9.1.3 *Let $(R, \mathfrak{m})$ be a Noetherian local ring R. If R is analytically unramified, then for every minimal prime ideal P in R, R/P is also analytically unramified.*

If R/P is analytically unramified for all $P \in \mathrm{Min}(R)$ and if R is reduced, then R is analytically unramified.

Proof: First assume that R is analytically unramified. To prove that R/P is analytically unramified, we use Theorem 9.1.2. It suffices to prove that for all ideals $I \subseteq R$ there exists an integer k such that $\overline{I^{n+k}(R/P)} \subseteq I^n(R/P)$ for all $n \geq 0$. Fix I.

By Theorem 9.1.2 there exists an integer k_1 such that for all $n \geq 0$, $\overline{I^{n+k_1}} \subseteq I^n$. Next, choose $c \in R$ such that $P = (0 : c)$. By the Artin–Rees Lemma, we may choose k_2 such that

$$(c) \cap I^{n+k_2} \subseteq c \cdot I^n \text{ for all } n \geq 0.$$

We claim that $\overline{I^{n+k_1+k_2}(R/P)} \subseteq I^n(R/P)$ for $n \geq 0$. Let $u \in R$ such that $u + P \in \overline{I^{n+k_1+k_2}(R/P)}$. Then there exists an equation:

$$u^l + a_1 u^{l-1} + \cdots + a_l \in P, \quad a_j \in I^{j(n+k_1+k_2)}.$$

Multiply by c^l:

$$(cu)^l + a_1 c(cu)^{l-1} + \cdots + a_l c^l = 0,$$

which implies that $cu \in \overline{I^{n+k_1+k_2}} \subseteq I^{n+k_2}$, $n \geq 0$. Hence $cu \in I^{n+k_2} \cap (c) \subseteq cI^n$, which implies that there exists $v \in I^n$ such that $u - v \in (0 : c) = P$. Then $u + P = v + P \in I^n(R/P)$.

Now assume that R/P is analytically unramified for all minimal prime ideals P of R, and that R is reduced. Then $\cap_{P \in \mathrm{Min}(R)} P = 0$, and so $\cap_P (P\widehat{R}) = 0$. By assumption, each $P\widehat{R}$ is a radical ideal in $\widehat{R}$, so that $\widehat{R}$ is reduced. Thus R is analytically unramified. $\qquad\square$

Here is another example of construction of analytically unramified rings:

Proposition 9.1.4 *An analytically unramified Noetherian local ring is locally analytically unramified.*

Proof: Let $Q \in \mathrm{Spec}(R)$. For an arbitrary ideal J in R_Q, let I be an ideal in R such that $IR_Q = J$. As R is analytically unramified, there exists an integer k such that for all $n \geq k$, $\overline{I^{n+k}} \subseteq I^n$. Thus $\overline{J^{n+k}} \subseteq J^n$. As J was an arbitrary ideal in R_Q, it follows by Theorem 9.1.2 that R_Q is analytically unramified. $\qquad\square$

It is not true that module–finite extensions of analytically unramified rings are also analytically unramified, or that an arbitrary quotient domain of an analytically unramified ring is analytically unramified; see Exercise 4.11.

9.2. Module–finite integral closures

Chapter 4 is about Noetherian rings whose integral closures are module–finite extensions. With the newly introduced analytically unramified rings, we can now prove even more cases of Noetherian rings whose integral closures are module–finite extensions.

Corollary 9.2.1 *Let $(R, \mathfrak{m})$ be an analytically unramified Noetherian local ring and I an ideal in R.*
(1) *Let t be a variable over R, and set $S = R[It]$, the Rees algebra of I. Then the integral closure of S in $R[t]$ is module–finite over S.*
(2) *There exists an integer ℓ such that for all $n \geq 1$,*
$$\overline{I^{n+\ell}} = I^n \overline{I^\ell} \quad and \quad (\overline{I^\ell})^n = \overline{I^{\ell n}}.$$

Proof: By Proposition 5.2.1, the integral closure of S in $R[t]$ is $\oplus_{n \geq 0} \overline{I^n} t^n$. By Theorem 9.1.2 there exists an integer k such that $\overline{I^n} \subseteq I^{n-k}$ for all $n \geq k$. Hence $\oplus_{n \geq k} \overline{I^n} t^n \subseteq R[It] t^k$ and then $\oplus_{n \geq 0} \overline{I^n} t^n$ is a finitely generated $R[It]$–module. Therefore by Proposition 5.2.5, there exists an integer ℓ such that $\overline{I^n} = I^{n-\ell} \cdot \overline{I^\ell}$ for all $n \geq \ell$. Hence
$$(\overline{I^\ell})^n \subseteq \overline{I^{\ell n}} = I^{\ell n - \ell} \cdot \overline{I^\ell} \subseteq (\overline{I^\ell})^n,$$
which implies that $(\overline{I^\ell})^n = \overline{I^{\ell n}}$ for all $n \geq 1$. $\qquad\square$

The following theorem due to Rees proves that whenever R is an analytically unramified domain and S is a finitely generated R–algebra with the same field of fractions as R, then S is also analytically unramified:

Theorem 9.2.2   (Rees [233]) *Let R be a Noetherian local domain. The following are equivalent:*

(1) R is analytically unramified.

(2) For every finitely generated R–algebra S such that S is between R and its field of fractions, the integral closure of S is a finitely generated S–module.

Proof: We begin by proving that (1) implies (2). By using Corollary 4.6.2, the integral closure $\overline{R}$ of R is a finitely generated R–module. Let r be a non–zero element of R such that $r\overline{R} \subseteq R$. Write $S = R[\frac{x_1}{x}, \ldots, \frac{x_m}{x}]$ and set $I = (x, x_1, \ldots, x_m)$. Choose an integer k as in Theorem 9.1.2 such that for all $n \geq 0$, $\overline{I^{n+k}} \subseteq I^n$. Note that $S = R[\frac{I}{x}] = \cup_{l \geq 0} \frac{I^l}{x^l}$. Let $\overline{S}$ be the integral closure of S and let $z \in \overline{S}$. There is an equation $z^n + s_1 z^{n-1} + \cdots + s_n = 0$ for some $s_i \in S$. Write $s_i = \frac{a_i}{x^l}$, where we may increase l if necessary to assume that $l \geq k$ and where $a_i \in I^l \subseteq R$, $1 \leq i \leq n$. Multiply by x^{nl}:

$$(x^l z)^n + a_1 (x^l z)^{n-1} + \cdots + a_j (x^l)^{j-1} (x^l z)^{n-j} + \cdots + a_n (x^l)^{n-1} = 0.$$

Note that all coefficients $a_j (x^l)^{j-1} \in R$. This proves that $x^l z$ is integral over R and hence $r x^l z \in R$. As $a_j (x^l)^{j-1} \in I^{jl}$, $r x^l z \in \overline{I^l \overline{R}} \cap R = \overline{I^l}$. Hence $r x^l z \in I^{l-k}$. Then $r x^k z = \frac{I^{l-k}}{x^{l-k}} \subseteq S$ and so $z \in S \cdot \frac{1}{r x^k}$. Therefore $\overline{S} \subseteq S \cdot \frac{1}{r x^k}$, so that it is a finitely generated S–module.

Assume condition (2). To prove (1), it suffices by Theorem 9.1.2 to prove that there exist an $\mathfrak{m}$–primary ideal J and an integer k such that $\overline{J^n} \subseteq J^{n-k}$ for all $n \geq k$. By Proposition 8.5.7 there exists an integer l such that $J = \mathfrak{m}^l$ has a non–zero superficial element x. By Lemma 8.5.3 this means that there exists an integer q such that for all $n \geq q$, $J^n : x = J^{n-1}$. Set $S = R[\frac{J}{x}]$. The integral closure $\overline{S}$ of S is finite over S, and hence $\overline{S} \cap R_x$ is a finitely generated S–module, so there exists p such that $\overline{S} \cap R_x \subseteq S \cdot \frac{1}{x^p}$. Let $u \in \overline{J^n}$ and assume that $n \geq p + q$. Then $\frac{u}{x^n} \in \overline{S}$ and $\frac{u}{x^n} \in R_x$, so that $\frac{u}{x^n} \in S \cdot \frac{1}{x^p}$. Now, $S = \bigcup_{m \geq 0} \frac{J^m}{x^m}$, so there exist a $m \in \mathbb{N}$ and $s \in J^m$ such that $\frac{u}{x^n} = \frac{s}{x^{m+p}}$. Hence $u x^{m+p} = s x^n \in J^{m+n}$ and so $u \in J^{m+n} : x^{m+p}$ We can cancel one x and one exponent of J at a time provided that the power of J never goes below q. The final cancellation gives $u \in J^{(m+n)-(m+p)} = J^{n-p}$ for $n - p \geq q$, i.e., $u \in J^{n-p}$ whenever $n \geq p + q$. Hence $\overline{J^n} \subseteq J^{n-p-q}$ for all $n \geq p + q$. Setting $k = p + q$ proves (1). $\square$

Here is a special case of this:

Theorem 9.2.3   *Let R be a regular domain, K its field of fractions and L a finite separable field extension of K. Let S be a finitely generated R–algebra contained in L. Then the integral closure $\overline{S}$ of S is module–finite over S.*

Proof: Let R' be the integral closure of R in L. By Theorem 3.1.3, R' is module–finite over R. The integral closure $\overline{S}$ of S contains R'. Let T be the algebra generated over R by the generators of R' and S. Then T is finitely generated over R, module–finite over S, the field of fractions of T is L, and the integral closure of T equals $\overline{S}$. Thus it suffices to prove that the integral closure of T is module–finite over T. By changing notation we may thus assume that L is the field of fractions of S, and that S contains the integral closure R' of R in L.

By Theorem 4.2.4, there exists a non–zero $r \in R$ such that S_r is integral over R_r. Thus $S_r = R'_r$ is integrally closed. By Corollary 4.5.10, the normal locus of S is open. It is certainly not empty as S is a domain. Corollary 4.5.11 shows that to prove the theorem it suffices to prove that for each maximal ideal $\mathfrak{m}$ in S, $\overline{S}_\mathfrak{m}$ is module–finite over $S_\mathfrak{m}$. Let $\mathfrak{m}$ be a prime ideal in S, and set $P = \mathfrak{m} \cap R$. We may now change notation: we replace R by the regular local ring R_P and S by the finitely generated domain extension $S_{R \setminus P}$ (S need not be local). The integral closure R' of R in L is still contained in S.

As R is local and R' is module–finite over it, R' is Noetherian and semi–local. Let $\widehat{R}$ be the $\mathfrak{m}$–adic completion of R. As R is a regular local ring, $\widehat{R}$ is a domain. Let $\widehat{K}$ be the field of fractions of $\widehat{R}$. Then

$$R' \otimes_R \widehat{R} \subseteq L \otimes_R \widehat{R} \cong (L \otimes_K K) \otimes_R \widehat{R} \cong L \otimes_K (K \otimes_R \widehat{R}) \subseteq L \otimes_K \widehat{K}.$$

By Proposition 3.2.4 (2), $L \otimes_K \widehat{K}$ is reduced, so that $R' \otimes \widehat{R}$ is reduced. But $R' \otimes \widehat{R}$ is a finite direct product of completions of $(R')_Q$, where Q varies over the maximal ideals of R', so that for each Q, the completion of $(R')_Q$ is reduced, so that each $(R')_Q$ is analytically unramified. Set $Q = \mathfrak{m} \cap R'$. As $Q \cap R = \mathfrak{m} \cap R' = P$, Q is a maximal ideal in R'. Thus $(R')_Q$ is analytically unramified, so that by Theorem 9.2.2, $\overline{S}_{R' \setminus Q}$ is module–finite over $S_{R' \setminus Q}$, whence $\overline{S}_\mathfrak{m}$ is module–finite over $S_\mathfrak{m}$. $\qquad\square$

In Chapter 19 we prove that under some quite general conditions on $(R, \mathfrak{m})$, for any ideal I in R, $\overline{I}\widehat{R}$ is integrally closed and equals the integral closure of $I\widehat{R}$. In Section 19.2, we prove the easier special case in which R is analytically unramified. The reader may proceed to that section right now (via the introductory section Section 19.1 on normal maps).

9.3. Divisorial valuations

Let R be a Noetherian integral domain and let S be a domain containing R. Let Q be a prime ideal in S and $P = Q \cap R$. By the Dimension Inequality (Theorem B.2.5),

$$\operatorname{ht} Q + \operatorname{tr.deg}_{\kappa(P)} \kappa(Q) \leq \operatorname{ht} P + \operatorname{tr.deg}_R S.$$

If S happens to be a Noetherian valuation ring V properly contained in the field of fractions K of R, then with $Q = \mathfrak{m}_V$, the formula above says that

$\text{tr.deg}_{\kappa(P)}\kappa(v) \leq \text{ht}\,P - 1$. The same inequality also follows from Theorem 6.6.7 (1). This motivates the following definition:

Definition 9.3.1 *Let R be a Noetherian integral domain with field of fractions K, and let $(V, \mathfrak{m}_V)$ be a valuation ring of K such that $R \subseteq V$. Let $P = \mathfrak{m}_V \cap R$. If $\text{tr.deg}_{\kappa(P)}\kappa(v) = \text{ht}\,P - 1$ then the corresponding valuation v of V is said to be a* **divisorial valuation** *with respect to R, and V is said to be a* **divisorial valuation ring** *of K.*

The set of all divisorial valuation rings of K that are non–negative on R is denoted $D(R)$.

The Noetherian property is not part of the definition of divisorial valuations, but in fact this property is forced. See Theorem 9.3.2 below.

An example of a non–divisorial valuation is in Example 6.7.3. Divisorial valuations are classically called **prime divisors**. Prime divisors are further divided into the first and second kinds. V is of the **first kind** if the center of V on R is a height one prime ideal. This forces V to be the localization at a height one prime of the integral closure of R. All the others are said to be of the **second kind** (with respect to R).

Our use of the word "divisorial" is not completely standard. Some authors refer to a divisorial valuation of R to mean only prime divisors of the first kind.

Divisorial valuations are critically important in the theory of the integral closure of ideals. Namely, as we prove in Chapter 10, the integral closures of the powers of a given ideal in a Noetherian domain are determined by finitely many discrete divisorial valuations. These special valuations are called the Rees valuations of the ideal. Also, in Chapter 18 the divisorial valuations are used to define the adjoints of ideals (multiplier ideals).

Divisorial valuations tend to be essentially of finite type:

Theorem 9.3.2 *Let R be a Noetherian integral domain. Every divisorial valuation domain V with respect to R is Noetherian. If moreover R is locally analytically unramified, then every divisorial valuation domain V with respect to R is essentially of finite type over R.*

Proof: Assume that V is a divisorial valuation domain with respect to R with associated valuation v. Necessarily R has positive dimension. Let $\mathfrak{m}_V$ be the maximal ideal of V, and $P = \mathfrak{m}_V \cap R$. The hypotheses and the conclusion do not change if we pass to the localization at P, so that we may assume that P is the unique maximal ideal of R. Let d be the height of P. Note that $\text{tr.deg}_{\kappa(P)}\kappa(v) = d - 1$.

Let $x, x_2, \ldots, x_d$ be elements in P such that $\frac{x_2}{x}, \ldots, \frac{x_d}{x}$ are units in V and such that their images in $\kappa(v)$ are algebraically independent over $\kappa(P)$. Let I be the ideal $(x, x_2, \ldots, x_d)$ in R. If R is assumed to be analytically unramified, Theorem 9.2.2 shows that the integral closure S of $R[\frac{I}{x}]$ is a finitely generated R–algebra contained in V. Let $Q = \mathfrak{m}_V \cap S$. Then $Q \cap R = P$. By the

Dimension Inequality (Theorem B.2.5):

$$\text{ht}\,Q + \text{tr.deg}_{\kappa(P)}\kappa(Q) \le \text{ht}\,P + \text{tr.deg}_R S = d,$$

so that $\text{ht}\,Q \le d - \text{tr.deg}_{\kappa(P)}\kappa(Q)$. Any algebraic relation among the $\frac{x_i}{x}$ in $\kappa(Q)$ with coefficients in $\kappa(P)$ remains an algebraic relation in $\kappa(v)$, so that $\text{tr.deg}_{\kappa(P)}\kappa(Q) = d - 1$. Hence $\text{ht}\,Q \le 1$. But Q contracts to the non–zero prime ideal P, so that $\text{ht}\,Q = 1$. Thus as S is a Krull domain, S_Q is a Noetherian valuation domain, so it necessarily equals V. If R is analytically unramified, then since S is a finitely generated R–algebra, V is essentially of finite type over R. □

Small modifications of the proof above show relations between $D(R)$ and $D(S)$ for various extensions S of R. These facts will be needed in Chapter 18:

Lemma 9.3.3 *Let R be a Noetherian locally analytically unramified integral domain with field of fractions K. Let V be a discrete valuation domain of rank one between R and K. Let S be a finitely generated R–algebra between R and V. If $V \in D(R)$ then $V \in D(S)$. If the dimension dormula holds between $R \subseteq S$, then $V \in D(R)$ if and only if $V \in D(S)$, i.e., the valuation domain V is divisorial over R if and only if it is divisorial over S.*

Proof: Let $Q = \mathfrak{m}_V \cap S$ and $P = \mathfrak{m}_V \cap R$. By the Dimension Inequality for $R \subseteq S$,

$$\begin{aligned}
\text{ht}\,P - 1 - \text{tr.deg}_{\kappa(P)}\kappa(v) &= \text{ht}\,P - 1 - \text{tr.deg}_{\kappa(P)}\kappa(Q) - \text{tr.deg}_{\kappa(Q)}\kappa(v) \\
&\ge \text{ht}\,Q - 1 - \text{tr.deg}_{\kappa(Q)}\kappa(v).
\end{aligned}$$

Equality holds above if $R \subseteq S$ satisfies the dimension formula. Suppose that V is divisorial over R. By the definition, $\text{ht}\,P - 1 - \text{tr.deg}_{\kappa(P)}\kappa(v) = 0$, and by Theorem 9.3.2, V is essentially of finite type over R and hence over S. Thus $0 \ge \text{ht}\,Q - 1 - \text{tr.deg}_{\kappa(Q)}\kappa(v) \ge 0$, so that V is divisorial over S. If equality holds in the display above, then the implication also goes the other way. □

Here is another relation of $D(R)$ and divisorial valuations on an extension of R:

Lemma 9.3.4 *Let R be a Noetherian locally analytically unramified universally catenary integral domain with field of fractions K. Let t be a variable over K. Let V be a discrete valuation domain of rank one between $R[t]_t$ (inverting t) and $K(t)$. If $V \in D(R[t]_t)$ then either $V \cap K = K$ or $V \cap K \in D(R)$.*

Proof: We may assume that $V \cap K \ne K$. By Proposition 6.4.7, $V \cap K$ is a valuation domain containing R with field of fractions K, it is Noetherian and thus a discrete valuation domain of rank one. Let $\mathfrak{n}$ be the maximal ideal of $V \cap K$. By Exercise 9.5 and Theorem 9.3.2, V is essentially of finite type over $R[t]$ and thus also essentially of finite type over R, and therefore it is essentially of finite type over $V \cap K$. By the Dimension Inequality $\text{tr.deg}_{\kappa(\mathfrak{n})}\kappa(V) \le 1$. But as t is a unit in V, $\text{tr.deg}_{\kappa(\mathfrak{n})}\kappa(V) \ge 1$, so that $\text{tr.deg}_{\kappa(\mathfrak{n})}\kappa(V) = 1$.

Let $Q = \mathfrak{m}_V \cap R[t]_t$ and $P = \mathfrak{m}_V \cap R$. By the additivity of transcendence degrees and by the Dimension Formula (Theorem B.3.2),

$$
\begin{aligned}
\operatorname{ht} P - 1 - \operatorname{tr.deg}_{\kappa(P)} \kappa(\mathfrak{n}) &= \operatorname{ht} P - 1 - \operatorname{tr.deg}_{\kappa(P)} \kappa(V) + \operatorname{tr.deg}_{\kappa(\mathfrak{n})} \kappa(V) \\
&= \operatorname{ht} P - 1 - \operatorname{tr.deg}_{\kappa(P)} \kappa(Q) - \operatorname{tr.deg}_{\kappa(Q)} \kappa(V) + 1 \\
&= \operatorname{ht} Q - \operatorname{tr.deg}_R R[t]_t - \operatorname{tr.deg}_{\kappa(Q)} \kappa(V) \\
&= \operatorname{ht} Q - 1 - \operatorname{tr.deg}_{\kappa(Q)} \kappa(V),
\end{aligned}
$$

which is zero as $V \in D(R[t]_t)$. Thus $V \cap K \in D(R)$. $\qquad\square$

Proposition 9.3.5 *Let $(R, \mathfrak{m})$ be a Noetherian local domain whose $\mathfrak{m}$–adic completion $\widehat{R}$ is an integral domain (R is analytically irreducible). Let K be the field of fractions of R and L the field of fractions of $\widehat{R}$. There is a one–to–one correspondence between elements $w \in D(\widehat{R})$ with center on $\mathfrak{m}\widehat{R}$ and elements of $v \in D(R)$ with center on $\mathfrak{m}$, given by restriction of the L–valuations to K. This correspondence preserves value groups and residue fields.*

Proof: By Proposition 6.5.2, real–valued L–valuations with center on $\mathfrak{m}$ contract to K–valuations with center on $\mathfrak{m}$ that preserve value groups and residue fields. Thus

$$
\operatorname{tr.deg}_{\kappa(\mathfrak{m})} \kappa(v) = \operatorname{tr.deg}_{\kappa(\mathfrak{m}\widehat{R})} \kappa(w),
$$

and $\operatorname{ht}(\mathfrak{m}\widehat{R}) - 1 = \operatorname{ht}(\mathfrak{m}) - 1$. Thus $W \in D(\widehat{R})$ if and only if $V \in D(R)$. It remains to prove that for any $V \in D(R)$ there exists $W \in D(R)$ such that $W \cap K = V$, but this follows from Proposition 6.5.4. $\qquad\square$

9.4. Exercises

9.1 Let R be a local analytically unramified Noetherian domain. Prove that the integral closure of R is also a locally analytically unramified Noetherian domain.

9.2 Let R be an integrally closed domain that is locally analytically irreducible. Let L be a finitely separable field extension of the field of fractions of R and let S be a finitely generated R-algebra contained in L. Prove that the integral closure of S is module–finite over S.

9.3 (Rees [233]) Let $(R, \mathfrak{m})$ be a Noetherian local ring. Prove that R is analytically unramified if and only if for any ideal I in R there exists a constant $k \in \mathbb{N}$ such that for all non–zero $x \in R$,

$$
0 \le \overline{v}_I(x) - \operatorname{ord}_I(x) = \lim_{n \to \infty} \frac{\operatorname{ord}_I(x^n)}{n} - \operatorname{ord}_I(x) \le k.
$$

(Cf. Theorem 13.4.8, where k can be taken to be independent of I.)

9.4 Let $(R, \mathfrak{m})$ be an analytically unramified Noetherian local domain. Prove that for any $x_1, \ldots, x_n$ in the field of fractions of R and for all $P \in \operatorname{Spec} R[x_1, \ldots, x_n]$, $(R[x_1, \ldots, x_n])_P$ is analytically unramified.

9.5   Let $(R, \mathfrak{m})$ be an analytically unramified Noetherian local ring, and X a variable over R. Prove that $R[X]$ is locally analytically unramified, i.e., that for arbitrary $Q \in \operatorname{Spec}(R[X])$, $R[X]_Q$ is analytically unramified.

9.6   A ring R is said to be **pseudo–geometric** if for every prime ideal P in R, the integral closure of R/P in any finite field extension of $\kappa(P)$ is module–finite over R/P. Pseudo–geometric is Nagata's terminology, Matsumura's terminology is **Nagata**, and EGA's is **universally Japanese**.

   (i) Prove that a pseudo–geometric Noetherian domain is locally analytically unramified.

   (ii) Prove that every finitely generated algebra over a field or $\mathbb{Z}$ is pseudo–geometric.

   (iii) Prove that if R is pseudo–geometric, then every finitely generated R–algebra is pseudo–geometric.

9.7   Let R be a finitely generated algebra over a complete local ring A. Prove that R is pseudo–geometric.

9.8   (Abhyankar [6, Proposition 8]) Let d be a positive integer and $(R_0, \mathfrak{m}_0)$ a d–dimensional regular local ring with field of fractions K. Let V be a Noetherian K–valuation ring containing R and centered on $\mathfrak{m}$. Let v be the corresponding valuation. For every integer $n \geq 0$, we build $(R_{n+1}, \mathfrak{m}_{n+1})$ if $\dim R_n > 1$ as follows: let $x_n \in \mathfrak{m}_n$ be such that $x_n V = \mathfrak{m}_n V$, $S_{n+1} = R_n[\mathfrak{m}_n/x_n]$, and R_{n+1} the localization of S_{n+1} at $\mathfrak{m}_V \cap S_{n+1}$.

   (i) Prove that R_{n+1} is a regular local ring with field of fractions K such that the center of V on R_{n+1} is $\mathfrak{m}_{n+1}$.

   (ii) Let $x \in K$ such that $v(x) \geq 0$. For any n, write $x = a_n/b_n$ for some $a_n, b_n \in R_n$. Prove that if $v(a_n) > 0$ and $v(b_n) > 0$, then we may choose a_{n+1} and b_{n+1} in R_{n+1} such that $v(a_{n+1}) < v(a_n)$ and $v(b_{n+1}) < v(b_n)$.

   (iii) Prove that $\cup_n R_n = V$.

9.9   With the set–up as in Exercise 9.8, assume in addition that v is a divisorial valuation with respect to R.

   (i) Prove that v is a divisorial valuation with respect to each R_n.

   (ii) Prove that for all sufficiently large $n \geq 0$, $R_n/\mathfrak{m}_n \subseteq R_{n+1}/\mathfrak{m}_{n+1}$ is an algebraic extension.

   (iii) Let N be the largest integer such that $R_N/\mathfrak{m}_N \subseteq R_{N+1}/\mathfrak{m}_{N+1}$ is transcendental. Prove that $\cup_n R_n = R_{N+1} = V$.

   (iv) Prove that $V/\mathfrak{m}$ is purely transcendental over $R_N/\mathfrak{m}_N$.

9.10   (Chevalley's Theorem) Let R be a complete semi–local ring, $\mathfrak{m}$ the intersection of the maximal ideals of R and $\{I_n\}$ a descending sequence of ideals such that $\cap_n I_n = (0)$. Prove that there exists a function $f : \mathbb{N} \to \mathbb{N}$ that tends to infinity such that $I_n \subseteq \mathfrak{m}^{f(n)}$.

10
Rees valuations

The valuative criterion for integral closure, Theorem 6.8.3, shows that the integral closure of an arbitrary ideal I in a ring R is the intersection of contractions of extensions of I to possibly infinitely many valuation rings. Thus whenever R/I satisfies the descending chain condition, the integral closure of I is the intersection of contractions of extensions of I to finitely many valuation rings. We prove in this chapter more generally that whenever R is Noetherian, for an arbitrary ideal I there exist finitely many valuation rings that determine not just the integral closure of I but also the integral closures of all the powers of I.

10.1. Uniqueness of Rees valuations

Definition 10.1.1 *Let R be a ring and I an ideal in R. Suppose that there exist finitely many discrete valuation rings $V_1, \ldots, V_r$ of rank one satisfying the following properties:*
(1) For each $i = 1, \ldots, r$, there exists a minimal prime ideal P of R such that $R/P \subseteq V_i \subseteq \kappa(P)$. Let $\varphi_i : R \to V_i$ be the natural ring homomorphism.
(2) For all $n \in \mathbb{N}$, $\overline{I^n} = \cap_{i=1}^r \varphi_i^{-1}(\varphi_i(I^n)V_i)$.
(3) The set $\{V_1, \ldots, V_r\}$ satisfying (2) is minimal possible.
*Then $V_1, \ldots, V_r$ are called the **Rees valuation rings** of I, and the corresponding valuations $v_1, \ldots, v_r$ are called the **Rees valuations** of I. If $v_1, \ldots, v_r$ are Rees valuations of I as in the definition above, then the set $\{v_1, \ldots, v_r\}$ is denoted as $\mathcal{RV}(I)$. We also sometimes by abuse of notation write $\mathcal{RV}(I) = \{V_1, \ldots, V_r\}$ to be the set of valuation rings themselves. It should be clear from the context which set we mean.*

Note that if I is the zero ideal in an integral domain, any valuation ring V between R and its field of fractions is by definition a Rees valuation ring of I. Thus in this case the set of Rees valuations is not unique. But this is an exceptional case. We will prove below that the set of Rees valuations $\mathcal{RV}(I)$ of I is unique in non–exceptional cases.

Example 10.1.2 The simplest and in some ways the *only* example of Rees valuations arises from the case in which R is an integrally closed Noetherian domain, and $I = (x)$ is a principal ideal. We claim that the rings $\{R_P | P \in \mathrm{Min}(R/xR)\}$ are Rees valuations rings of (x). To see this, we need to prove first that $(x^n) = \overline{(x^n)} = \bigcap_{P \in \mathrm{Min}(x)} (x^n R_P \cap R)$. The first equality holds since principal ideals in integrally closed domains are integrally closed by Proposition 1.5.2. Since R is integrally closed, every associated prime of (x^n)

is minimal (Proposition 4.1.1), and then the desired formula follows at once from the primary decomposition of the ideal (x^n). By the uniqueness theorem of primary decomposition, no R_P can be deleted in the formula

$$(x^n) = \overline{(x^n)} = \bigcap_{P \in \mathrm{Min}(x)} (x^n R_P \cap R).$$

Discussion 10.1.3 As the example above indicates, it is helpful to think of what the existence of a set of Rees valuations means for the primary decompositions of the ideals $\overline{I^n}$. Suppose in general that $R \subseteq S$ are Noetherian rings and that Q is a P–primary ideal in S, where P is a prime in S. Set $q = Q \cap R$ and $p = P \cap R$. Then p is a prime ideal in R, and q is p–primary. This follows immediately from the definition of a primary ideal. Further assume that R is a domain, and that $S = V$ is a Noetherian valuation domain. Let I be a non–zero ideal in R. Automatically, IV is $\mathfrak{m}_V$–primary, and hence $IV \cap R$ is p–primary in R, where p is the center of the valuation ring V on R, i.e., $p = \mathfrak{m}_V \cap R$.

Suppose that $\{V_1, \ldots, V_r\}$ are the Rees valuations for I, where R is a domain and I is a non–zero ideal of R. Condition (2) in the definition says that

$$\overline{I^n} = \bigcap_{i=1}^{r} I^n V_i \cap R.$$

Since $I^n V_i$ is primary in V_i, then $I^n V_i \cap R$ is primary in R, so that $\overline{I^n} = \cap_{i=1}^r I^n V_i \cap R$ is a primary decomposition of $\overline{I^n}$. In particular, the associated primes of $\overline{I^n}$ are among the centers of the Rees valuations, $P_1 = \mathfrak{m}_{V_1} \cap R, \ldots, P_r = \mathfrak{m}_{V_r} \cap R$.

Condition (3) means in particular that every one of $P_1, \ldots, P_r$ is associated to $\overline{I^n}$ for some n. To see this, let I be an ideal with a primary decomposition $I = q_1 \cap \cdots \cap q_r$, and suppose that q_1 cannot be removed. Let $p_1 = \sqrt{q_1}$. We prove that p_1 is associated to I. If $I : (q_2 \cap \cdots \cap q_r)$ is not contained in p_1, then choose an element $s \in I : (q_2 \cap \cdots \cap q_r)$ with $s \notin p_1$. Since $s(q_2 \cap \cdots \cap q_r) \subseteq I \subseteq q_1$ and $s \notin p_1$ it would follow that $q_2 \cap \cdots \cap q_r \subseteq q_1$, contradicting the assumption that q_1 cannot be removed. Hence $I : (q_2 \cap \cdots \cap q_r) \subseteq p_1$. This means that after localizing at p_1, $(q_1)_{p_1}$ is still needed in the primary decomposition of I_{p_1} given by localizing the given primary decomposition at p_1. Hence we may assume that p_1 is the unique maximal ideal of R. If p_1 is not associated to I, we can choose an element $z \in q_1$ that is not a zerodivisor on R/I. Thus $z(q_2 \cap \cdots \cap q_r) \subseteq q_1 \cap \cdots \cap q_r = I$, which forces $I = q_2 \cap \cdots \cap q_r$, a contradiction.

We conclude that the centers of the Rees valuations are exactly the associated primes of $\overline{I^n}$ as n varies. In particular, for any ideal I in a Noetherian ring R, $\cup_n \mathrm{Ass}(R/\overline{I^n})$ is finite. We record this important fact in Corollary 10.2.4, after proving the existence of Rees valuations.

Though the centers of the Rees valuations will be the associated primes of the integral closures of large powers of an ideal, the correspondence is not one–to–one, i.e., it is possible that $P_i = P_j$ for $i \neq j$. For example, if $(R, \mathfrak{m})$ is local and I is $\mathfrak{m}$–primary then $\mathfrak{m}$ is the center of every Rees valuation of I.

A valuation v corresponding to a valuation ring V as in Definition 10.1.1 is only defined on the field of fractions of R/P for some minimal prime ideal P in R. With abuse of notation (as in Chapter 6 on valuations), it will be convenient in what follows to think of v to be also defined on all of R as follows:

$$v(r) = \begin{cases} \infty & \text{if } r \in P, \\ v((r + P)/P) & \text{otherwise.} \end{cases}$$

We adopt the following natural order on $\mathbb{R} \cup \{\infty\}$:

$$\infty/\infty = \infty, \infty + \infty = \infty, u/\infty = 0, u + \infty = \infty.$$

Remark 10.1.4 With the convention above, it is important to observe that (2) of Definition 10.1.1 is equivalent to saying that $r \in \overline{I^n}$ if and only if $v(r) \geq nv(I)$ for every Rees valuation v. This equivalent condition is often easier to work with and think about.

Lemma 10.1.5 *Let I be an ideal contained in no minimal prime ideal of R. Let $w : R \to \mathbb{Q}_{\geq 0} \cup \{\infty\}$ be a function such that for all $n \geq 1$, $\overline{I^n} = \{x \in R \mid w(x) \geq n\}$, and such that $w(x^n) = nw(x)$ for all $x \in R$ and $n \geq 1$. Let $\mathcal{RV}(I) = \{v_1, \ldots, v_r\}$ be a set of Rees valuations. Then*

$$w(x) = \min\{v_i(x)/v_i(I) \mid i = 1, \ldots, r\}.$$

Proof: We define $w' : R \to \mathbb{Q}_{\geq 0} \cup \{\infty\}$ to be

$$w'(x) = \min\{v_i(x)/v_i(I) \mid i = 1, \ldots, r\}.$$

With this notation, for any $n \geq 1$, $\overline{I^n} = \{x \in R \mid w'(x) \geq n\}$. Furthermore, for all $x \in R$ and all $n \geq 1$, $w'(x^n) = nw'(x)$.

If $w \neq w'$, then there exists a non–zero element x in R, such that $w'(x) \neq w(x)$. Assume that $w'(x) < w(x)$. For some large integer n, $w'(x^n) \leq w(x^n) - 1$. In case $w(x) = \infty$, set k to be an arbitrarily large integer, and otherwise set $k = \lfloor w(x^n) \rfloor$, the largest integer less than or equal to $w(x^n)$. By assumption $x^n \in \overline{I^k}$, and x^n is not in $\overline{I^k}$ by the definition of w', which is a contradiction. Thus necessarily $w'(x) \geq w(x)$. A symmetric argument proves that $w(x) = w'(x)$. $\square$

Thus by Corollary 6.9.1, assuming the existence of Rees valuations,

$$\overline{v}_I(x) = \lim_{m \to \infty} \frac{\mathrm{ord}_I(x^m)}{m} = \min\left\{ \frac{v(x)}{v(I)} \,\middle|\, v \in \mathcal{RV}(I) \right\}.$$

We next prove that for an ideal I not contained in any minimal prime ideal of R, a minimal set of valuations determining the function w is uniquely determined from w. This then proves that the set of Rees valuations of I, when it exists, is unique.

Theorem 10.1.6 (Uniqueness of Rees valuations) *Let R be a Noetherian ring and let I be an ideal contained in no minimal prime ideal of R. Let $v_1, \ldots, v_r$ be discrete valuations of rank 1 that are non–negative on R and each infinite exactly on one minimal prime ideal of R. Let $w : R \to \mathbb{Q}_{\geq 0} \cup \{\infty\}$ be defined by*

$$w(x) = \min\{v_i(x)/v_i(I) \,|\, i = 1, \ldots, r\}.$$

If no v_i can be omitted in this expression then the v_i are determined by the function w up to equivalence of valuations.

Therefore the set of Rees valuations of I, when it exists, is uniquely determined, up to equivalence of valuations.

Proof: The last statement follows from the first statement by using the previous lemma. We therefore only need to prove the first statement.

The conclusion is clear if $r = 1$. So suppose that $r > 1$.

We call a subset S of R to be w–consistent if for any $m \in \mathbb{N}$ and any $x_1, \ldots, x_m \in S$, $w(x_1 \cdots x_m) = \sum_{i=1}^m w(x_i)$. Note that for any element x of R, the set $\{x^m \,|\, m \in \mathbb{Z}\}$ is w–consistent. Thus the set of all non–empty w–consistent sets is not empty. Under the natural partial ordering by inclusion, each chain of w–consistent sets has its union as an upper bound. Thus by Zorn's lemma there exist maximal w–consistent sets.

For each $i = 1, \ldots, r$, define $S_i = \{x \in R \,|\, w(x) = \frac{v_i(x)}{v_i(I)}\}$. This set is w–consistent and non–empty as it contains 1. We claim that the maximal w–consistent sets are exactly the sets S_i. Since every S_i is contained in some maximal w–consistent set, it is enough to prove that every maximal w–consistent set S is equal to some S_i. If not, then for some maximal w–consistent set S and for each $i = 1, \ldots, r$, there exists $y_i \in S \setminus S_i$. As $y_i \notin S_i$, $w(y_i) < \infty$. Set $y = y_1 \cdots y_r$. As S is w–consistent, for each $j = 1, \ldots, r$,

$$w(y) = \sum_{i=1}^r w(y_i) < \sum_{i=1}^r \frac{v_j(y_i)}{v_j(I)} = \frac{v_j(y)}{v_j(I)}.$$

Hence $w(y) < \min\{\frac{v_j(y)}{v_j(I)} \,|\, j = 1, \ldots, r\}$, which contradicts the definition of w. This proves that every maximal w–consistent set S is equal to some S_i, and hence the S_i are exactly the maximal w–consistent sets. Since the v_i are irredundant, the S_i are all distinct. Notice that the the number of the v_i is uniquely determined by w as the number of maximal w–consistent sets.

We need to recover the valuation v_i from the sets S_i. Let c be an element of R such that $v_i(c) < \infty$. By the irredundancy of the v_i there exists $x_i \in S_i \setminus \cup_{j \neq i} S_j$. By the choice of x_i, for all $j \neq i$, $\frac{v_j(x_i)}{v_j(I)} > \frac{v_i(x_i)}{v_i(I)}$, so that for all sufficiently large positive integers d, whenever $j \neq i$,

$$\left(\frac{v_j(x_i)}{v_j(I)} - \frac{v_i(x_i)}{v_i(I)}\right) d > \frac{v_i(c)}{v_i(I)} - \frac{v_j(c)}{v_j(I)}.$$

This means that $\frac{v_j(cx_i^d)}{v_j(I)} > \frac{v_i(cx_i^d)}{v_i(I)}$. Thus $w(cx_i^d) = \frac{v_i(cx_i^d)}{v_i(I)}$, so that $cx_i^d \in S_i$.

Now let K be the set of all fractions $\frac{a}{b}$, with a, b elements of R, and b not in any minimal prime ideal of R. For each $i = 1, \ldots, r$, define a function $u_i : K \to \mathbb{Q} \cup \{\infty\}$ as follows. Let $\frac{a}{b} \in K$, with $a, b \in R$, b not in any minimal prime ideal of R. By the previous paragraph, for all sufficiently large integers d, $x_i^d b \in S_i$. If for some large positive integer d, $x_i^d a \in S_i$ and $x_i^d b \in S_i$, define $u_i(a/b) = w(x_i^d a) - w(x_i^d b)$. Clearly u_i does not depend on d. If instead for all positive integers d, $x_i^d a \notin S_i$, then define $u_i(a/b) = \infty$. Note that $x_i^d a \notin S_i$ for all d means that $w(a) < \infty = v_i(a)$. Now we show that u_i does not depend on a and b. So suppose that $a/b = a'/b'$ in K^*, i.e., that there exists $c \in R$ and not in any minimal prime ideal in R such that $c(ab' - a'b) = 0$. If we can show that $u_i(a/b) = u_i((ca')/(cb'))$, then similarly $u_i(a'/b') = u_i((ca')/(cb'))$. Thus without loss of generality we may assume that $c = 1$. First suppose that for all d, $x_i^d a \notin S_i$. Then $v_i(a) = \infty$, so $v_i(a') = \infty$, and $u_i(a'/b') = \infty = u_i(a/b)$. Equality also holds if for all e, $x_i^e a' \notin S_i$. Thus we may suppose that for all sufficiently large d and e, $x_i^d a, x_i^e a' \in S_i$. By the previous paragraph, for all sufficiently large integers e, d, $x_i^d b, x_i^e b' \in S_i$. Then

$$
\begin{aligned}
w(x_i^e a') + w(x_i^d b) &= \frac{v_i(x_i^e a')}{v_i(I)} + \frac{v_i(x_i^d b)}{v_i(I)} = \frac{v_i(x_i^{d+e} a' b)}{v_i(I)} \\
&= \frac{v_i(x_i^{d+e} ab')}{v_i(I)} = \frac{v_i(x_i^d a)}{v_i(I)} + \frac{v_i(x_i^e b')}{v_i(I)} = w(x_i^d a) + w(x_i^e b'),
\end{aligned}
$$

so that $u_i(a'/b') = u_i(a/b)$, and u_i depends only on w. Furthermore, it is straightforward to verify that u_i satisfies the two properties of valuations:

$$
u_i(\alpha\beta) = u_i(\alpha) + u_i(\beta), \quad u_i(\alpha + \beta) \geq \min\{u_i(\alpha), u_i(\beta)\}.
$$

Set P_i to be the prime ideal $\{r \in R \mid u_i(r) = \infty\}$ in R. Note that u_i is naturally a $\kappa(P_i)$–valuation. Set

$$
V_i = \{s \in \kappa(P_i)^* \mid s \text{ is the image of some } r \in K \text{ such that } u_i(r) \geq 0\}.
$$

As u_i depends only on w, the same also holds for V_i. As u_i and $\frac{v_i}{v_i(I)}$ agree on S_i and thus on all of $\kappa(P_i)^*$, V_i is the unique valuation ring of v_i. Hence u_i is equivalent to v_i, and since u_i only depends upon w, this completes the proof of the theorem. $\qquad\square$

10.2. A construction of Rees valuations

In this section we construct Rees valuations. In the case of an ideal I in a Noetherian domain our construction gives that all the localizations of all the normalizations of all the affine pieces of the blowup of I at height one prime ideals containing I are exactly the Rees valuations.

The following lemma will help prove irredundancy of the construction. It also highlights the critical point in understanding which affine pieces of the blowup of I contribute to the set of Rees valuations.

Lemma 10.2.1 *Let R be an integral domain, I a non–zero ideal in R, a, b non–zero elements of I, and V a discrete valuation ring of rank one that is the localization of the integral closure of $R[\frac{I}{b}]$ at a prime ideal containing b. Assume that $aV = IV$. Then V is the localization of the integral closure of $R[\frac{I}{a}]$ at a height one prime ideal containing a.*

Proof: Note that $bR[\frac{I}{b}] = IR[\frac{I}{b}]$, so that $bV = IV = aV$. Hence $\frac{a}{b}$ is a unit in V, and V is the localization of the integral closure of $R[\frac{I}{b}]_{\frac{I}{b}} = R[\frac{I}{a}]_{\frac{b}{a}}$ at a height one prime, so V is the localization of the integral closure of $R[\frac{I}{a}]$. Necessarily the localization is at a height one prime ideal of the integral closure of $R[\frac{I}{a}]$ containing a.   □

Theorem 10.2.2 (Existence of Rees valuations, Rees [230])
(1) Every ideal in a Noetherian ring has a set of Rees valuations.
(2) $\mathcal{RV}(I) \subseteq \cup \mathcal{RV}(I(R/P))$, as P varies over the minimal prime ideals of R.
(3) In case R is an integral domain, let $I = (a_1, \ldots, a_d)$, and for each $i = 1, \ldots, d$, set $S_i = R[\frac{I}{a_i}]$ and $\overline{S}_i$ to be the integral closure of S_i. Let T be the set of all discrete valuation domains $(\overline{S}_i)_p$, where p varies over the prime ideals in $\overline{S}_i$ minimal over $a_i \overline{S}_i$, and i varies from 1 to d. Then T is the set of Rees valuation rings of I. In particular, for all n,

$$\overline{I^n} = \bigcap_{i=1}^{d} \overline{I^n \overline{S}_i} \cap R = \bigcap_{i=1}^{d} \overline{a_i^n \overline{S}_i} \cap R = \bigcap_{i=1}^{d} (\bigcap_p (a_i^n (\overline{S}_i)_p) \cap R).$$

Proof: Proposition 1.1.5 proves (1) and (2) provided that Rees valuations exist for $I(R/P)$ as P varies over the minimal primes of R. Hence both (1) and (2) reduce to proving Rees valuations exist for ideals in Noetherian domains.

Thus we may assume that R is an integral domain. As any discrete valuation ring of rank one between R and its field of fractions will do for the zero ideal, we may assume that I is a non–zero ideal.

Clearly for all $i = 1, \ldots, d$, $IS_i = a_i S_i$. Note that by Proposition 5.5.8, S_i equals the homogeneous part of degree zero of the ring $R[It]_{a_i t}$, where t is a variable of degree 1 over the zero–degree ring R, and that the integral closure $\overline{S}_i$ of S_i equals the homogeneous part of degree zero of the ring $\overline{R[It]}_{a_i t}$. In particular, $\overline{S}_i = \cup_{m \geq 0} \frac{\overline{I^m \overline{R}}}{a_i^m}$. We claim that $\overline{I^n} = \cap_i \overline{a_i^n \overline{S}_i} \cap R$. Certainly $\overline{I^n}$ is contained in the intersection. Let $r \in \cap_i a_i^n \overline{S}_i \cap R$. There exists an integer $m \geq n$ such that for each i, $r = \frac{b_i a_i^n}{a_i^m}$ for some $b_i \in \overline{I^m \overline{R}}$. Thus for each i, $r a_i^{m-n} \in \overline{I^m \overline{R}}$, whence $r \in \overline{I^n \overline{R}}$ by Corollary 6.8.7. Hence $r \in \overline{I^n \overline{R}} \cap R = \overline{I^n}$ (by Proposition 1.6.1). Thus for all n, $\overline{I^n} = \cap_i a_i^n \overline{S}_i \cap R$.

As S_i is Noetherian, by the Mori–Nagata Theorem 4.10.5, $\overline{S}_i$ is Krull domain, and so every prime ideals p in $\overline{S}_i$ minimal over a_i has height one and $(\overline{S}_i)_p$ is an integrally closed domain. Thus by Proposition 6.4.4, $(\overline{S}_i)_p$ is a discrete valuation ring of rank one. Let T_i be the set of such valuation rings. This is a finite set (as $\bar{S}_i$ is a Krull domain). Hence by Proposition 4.10.3, for all $n \in \mathbb{N}$,

$$I^n \bar{S}_i = a_i^n \bar{S}_i = \bigcap_{V \in T_i} a_i^n V \cap \bar{S}_i$$

yields a minimal primary decomposition of $a_i^n \bar{S}_i$. Thus $\overline{I^n} = \bigcap_i \bigcap_{V \in T_i} I^n V \cap R$, and the set $T = \cup_i T_i$ satisfies the first two properties of Rees valuations.

It remains to prove the third condition, namely that none of the valuation rings in T is redundant. For this let V_0 be one of the valuation rings in T. Then $V_0 \in T_i$ for some i between 1 and d. By relabeling, say $i = 1$. By the primary decomposition of $a_1 \bar{S}_1$, there exists r in $\cap_{V \in T_1 \setminus \{V_0\}} a_1 V \cap \bar{S}_1$ such that $rV_0 = V_0$. Write $r = s/a_1^m$ for some integer m and some $s \in \overline{I^m R}$. Let V be any valuation ring in T that does not contain $\bar{S}_1$. Then $a_1 V$ is properly contained in IV. Hence as T is a finite set, there exists an integer n such that $a_1^n V \subseteq I^{n+1} V$ for all such V. Thus

$$s^n V_0 = r^n a_1^{mn} V_0 = r^n I^{mn} V_0 \nsubseteq I^{mn+1} V_0,$$
$$s^n V = r^n a_1^{mn} V \subseteq I^{mn+1} V, \quad \text{for all } V \in T \setminus \{V_0\},$$

so that s^n of $\bar{R}$ lies in $\bigcap_{V \in T \setminus \{V_0\}} I^{mn+1} V \cap \bar{R}$ but not in $\bigcap_{V \in T} I^{mn+1} V \cap \bar{R}$.

This proves that V_0 is not redundant in the set T as a Rees valuation ring of $I\bar{R}$. It remains to prove that V_0 is not redundant in the set T as a Rees valuation ring of I, which is done in the following proposition:

Proposition 10.2.3 *Let R be a Noetherian integral domain with field of fractions K. Let I be a non–zero ideal. Let T be a set of Noetherian K–valuation rings satisfying:*

(1) For all $n \geq 1$, $\overline{I^n \bar{R}} = \cap_{V \in T} I^n V \cap \bar{R}$.

(2) If $T' \subsetneq T$, there exists an integer n such that $\overline{I^n \bar{R}} \neq \cap_{V \in T'} I^n V \cap \bar{R}$.

Then T satisfies the same properties also in R, not just in $\bar{R}$. Explicitly:

(1) For all $n \geq 1$, $\overline{I^n} = \cap_{V \in T} I^n V \cap R$.

(2) If $T' \subsetneq T$, there exists an integer n such that $\overline{I^n} \neq \cap_{V \in T'} I^n V \cap R$.

Proof: By Proposition 1.6.1, $\overline{I^n} = \cap_V I^n V \cap R$, so we only have to prove (2).

By the assumption on the irredundancy of T on $\bar{R}$, for each $V_0 \in T$ there exist $n \geq 1$ and $r \in \bar{R}$ such that $r \in \bigcap_{V \neq V_0} I^n V \cap \bar{R}$ and $r \notin I^n V_0 \cap \bar{R}$. As $R[r]$ is a module–finite extension of R in $Q(R)$, there exists a non–zero element $c \in R$ such that $cR[r] \subseteq R$.

Choose m such that $c \notin I^m V_0$. As $r \notin I^n V_0$, $r^m \notin I^{m(n-1)+1} V_0$, so that $cr^m \notin I^{mn} V_0$. Thus $cr^m \in \bigcap_{V \neq V_0} I^{mn} V \cap R$, so that

$$\overline{I^n} = \bigcap_{V \in T}^r I^{mn} V \cap R \subsetneq \bigcap_{V \neq V_0} I^{mn} V \cap R.$$

Hence V_0 is not redundant. $\qquad\qquad\square$

This also finishes the proof of Theorem 10.2.2.

Corollary 10.2.4 (Ratliff [227]) *For an ideal I in a Noetherian ring R,
$\mathrm{Ass}(R/\bar{I}) \subseteq \mathrm{Ass}(R/\overline{I^2}) \subseteq \mathrm{Ass}(R/\overline{I^3}) \subseteq \cdots$. Furthermore, $\cup_n \mathrm{Ass}(R/\overline{I^n})$
equals the set of centers of the Rees valuations of I and is therefore a finite
set.*

Proof: Discussion 10.1.3 shows that $\cup_n \mathrm{Ass}(R/\overline{I^n})$ is exactly the set of centers
of the Rees valuations of I, so that it is a finite set. The rest follows from
Proposition 6.8.8.   □

It is not clear, but nonetheless is true, that $\cup_n \mathrm{Ass}(R/I^n)$ is also finite and
independent for large n (see Brodmann [26]).

The given construction of Rees valuations shows that the Rees valuations
of an ideal I in R correspond to certain valuations on the affine pieces of
$\mathrm{Proj}(R[It])$. But the finiteness of the set guarantees that one needs to consider
only one affine piece when there are sufficiently many units in the ring:

Proposition 10.2.5 *Let R be a Noetherian integral domain. Assume that R
contains an infinite field, or more generally, assume that for any prime ideals
$P_1, \ldots, P_r$ in R there exist elements $u_1, \ldots, u_{r-1} \in R$ that are distinct and
non–zero modulo each P_i. Let I be a non–zero proper ideal in R. Then there
exists an element $a \in I$ such that $\mathcal{RV}(I)$ is the set of all valuation domains
that are localizations of the integral closure $\bar{S}$ of $S = R[\frac{I}{a}]$ at prime ideals
minimal over $a\bar{S}$.*

*In fact, if $V_1, \ldots, V_r$ are the Rees valuation rings of I and there exists $a \in I$
such that for all i, $aV_i = IV_i$, then $V_1, \ldots, V_r$ are all the valuation domains
that are the localizations of $\bar{S}$ at prime ideals minimal over $a\bar{S}$.*

Proof: By Lemma 6.4.3, there exists an element $a \in I$ such that for all
$i = 1, \ldots, r$, $aV_i = IV_i$. Set $S = R[\frac{I}{a}]$. By the choice of a, S and its integral
closure $\bar{S}$ are both contained in each V_i.

We claim that for all n, $a^n \bar{S} \cap R = \overline{I^n}$. Certainly $\overline{I^n}$ is contained in
$a^n \bar{S} \cap R$. For the other inclusion, if $x \in a^n \bar{S} \cap R$, then for all $i = 1, \ldots, r$,
$xV_i \subseteq a^n V_i \subseteq I^n V_i$, so by the assumption on the Rees valuations, $x \in \overline{I^n}$.
This proves the claim.

Thus $\mathcal{RV}(I) \subseteq \mathcal{RV}(aS)$. But by Theorem 10.2.2, every valuation arising as
the localization of the integral closure of $S = R[\frac{I}{a}]$ at a prime ideal minimal
over a is a Rees valuation of I.   □

Observe that a as in the proposition is a "sufficiently general" element of
I. Existence of sufficiently general elements requires sufficiently many units
to exist in the appropriate rings. But even without this assumption on the
units, one can still find all Rees valuations of I to arise as localizations of the
normalization of one ring, see Exercise 10.3. However, a priori one does not
know which are these special rings without further work. Thus to find all the

Rees valuations of I, it may be more efficient to consider more than one affine piece of $\operatorname{Proj} R[It]$.

There are cases when the one needed affine piece is known in advance: A key point in recognizing this is that if all elements of a given set of minimal generators of I have minimal value when evaluated at each Rees valuation, then all the Rees valuations come from any of the affine pieces determined by those generators. This point is illustrated by the next two propositions.

Proposition 10.2.6 *Let $(R, \mathfrak{m})$ be a Noetherian formally equidimensional local ring of dimension d. Assume that I is an ideal generated by elements $x_1, \ldots, x_d$ that are analytically independent after going modulo each minimal prime of R. Let V be a Rees valuation ring of I with center on $\mathfrak{m}$.*
(1) For every $i = 1, \ldots, d$, $x_i V = IV$.
(2) Let P be the minimal prime ideal of R such that $R/P \subseteq V$. Write $(\_)'$ to denote images in R/P. For every $i = 1, \ldots, d$, V is the localization of the integral closure $\overline{S}$ of $S = R'[\frac{I'}{x_i'}]$ at a height one prime ideal minimal over $x_i' \overline{S}$.

Proof: To prove both (1) and (2) we may assume without loss of generality that R is a domain and that $P = 0$. Henceforth we omit the $(\_)'$ notation. There exists an integer $j \in \{1, \ldots, d\}$ such that V is the localization of the normalization of $R[\frac{I}{x_j}]$ at a height one prime ideal minimal over x_j. By Corollary 8.3.5, $\mathcal{F}_I$ is a polynomial ring over $R/\mathfrak{m}$ of dimension d, so that $\mathfrak{m} R[It]$ is a prime ideal in $R[It]$. By formal equidimensionality, $\mathfrak{m} R[It]$ has height 1. We claim that $x_i t \notin \mathfrak{m} R[It]$ for all $1 \le i \le d$. If not, then $x_i \in \mathfrak{m} I$, contradicting the fact that I must be minimally generated by $x_1, \ldots, x_d$. Thus $\mathfrak{m} R[It]_{x_j t}$ is a prime ideal of height 1, and $x_i/x_j = x_i t/x_j t$ is a unit in $R[It]_{\mathfrak{m} R[It]}$. As in Proposition 5.5.8, the degree zero component of $R[It]_{x_j t}$ is $R[\frac{I}{x_j}]$, so that $\mathfrak{m} R[\frac{I}{x_j}]$ is a prime ideal of height 1 and $\frac{x_i}{x_j}$ is a unit in $R[\frac{I}{x_j}]_{\mathfrak{m} R[\frac{I}{x_j}]}$. The maximal ideal $\mathfrak{m}_V$ of V contracts to $\mathfrak{m}$ in R, so that $\frac{x_i}{x_j}$ is a unit in V. In particular, $v(x_i) = v(x_j) = v(I)$, proving (1), and (2) follows by applying Lemma 10.2.1. $\square$

Corollary 10.2.7 (Sally [256, page 438]) *Let $(R, \mathfrak{m})$ be a Noetherian formally equidimensional local domain of dimension $d > 0$, and I an $\mathfrak{m}$–primary ideal satisfying $\mu(I) = d$. Let $I = (a_1, \ldots, a_d)$. Then for every Rees valuation ring V of I and every $i = 1, \ldots, d$, V is the localization of the normalization of $R[\frac{I}{a_i}]$ at a height one prime ideal minimal over a_i.*

Proof: Observe that all Rees valuations of I have center on $\mathfrak{m}$. The proof is immediate from the above proposition since I. $\square$

There are other cases in which we can reach the same conclusion, even when the ring is not equidimensional:

Proposition 10.2.8 *Let R be a Noetherian ring, $x_1, \ldots, x_r$ a regular sequence in R, and $I = (x_1, \ldots, x_r)$. Then:*
(1) for every Rees valuation v of I and every $i = 1, \ldots, r$, $v(x_i) = v(I)$.
(2) Let V be a Rees valuation ring of I, taking infinite value on a minimal prime P of R. Write $(\_)'$ to denote images in R/P. For every $i = 1, \ldots, r$, V is the localization of the integral closure $\overline{S}$ of $S = R'[\frac{I'}{x_i'}]$ at a height one prime ideal minimal over $x_i' \overline{S}$.

Proof: Let n be a positive integer. We first prove that $\overline{I^n} : x_i = \overline{I^{n-1}}$. Let $s \in \overline{I^n} : x_i$. By Remark 1.2.3, there exists a positive integer k such that for all $l \in \mathbb{N}$, $(sx_i)^{k+l} \in I^{ln}$. Hence $s^{k+l} \in I^{ln-l-k}$ for all positive integers l. Passing to any valuation domain V that is an R–algebra, since k is fixed and l varies over all positive integers, necessarily $sV \subseteq I^{n-1}V$. Thus by Theorem 6.8.3, $s \in \overline{I^{n-1}}$. This proves that $\overline{I^n} : x_i = \overline{I^{n-1}}$ for all n.

Now let v be a Rees valuation of I. By definition, $v(x_i) \geq v(I)$. Suppose that $v(x_i) > v(I)$. As the set of Rees valuations is irredundant, there exist an integer n and an element $y \notin \overline{I^n}$ such that for all Rees valuations w different from v, $w(y) \geq nw(I)$. Necessarily $v(y) < nv(I)$. If $\mathcal{RV}(I) = \{v\}$, set $y = 1$, $n = 1$. Choose an integer $l \geq \frac{nv(I)-v(y)}{v(x_i)-v(I)}$. This choice gives that $v(x_i^l y) \geq v(I^{l+n})$, and for all w as above, $w(x_i^l y) = lw(x_i) + w(y) \geq lw(I) + nw(I)$, so that by the definition of Rees valuations, $x_i^l y \in \overline{I^{n+l}}$. Thus by the first paragraph, $y \in \overline{I^n}$, which contradicts the choice of y. So necessarily $v(x_i) = v(I)$. This proves (1).

Let V be a Rees valuation as in (2). We may pass to R/P and assume that R is a domain. We still denote the image of x_j in R/P by x_j. Note that these elements need not form a regular sequence. We know that V is the localization of the normalization of some $R[\frac{I}{x_j}]$ at a height one prime ideal containing x_j. As $v(x_j) = v(I) = v(x_i)$ by (1), we have that $x_i V = IV$. We apply Lemma 10.2.1 to conclude that V is also the localization of the normalization of $R[\frac{I}{x_i}]$ at a height one prime ideal containing x_i. $\square$

Alternative constructions of Rees valuations are outlined in Exercises 10.5 and 10.6.

10.3. Examples

One of the main examples of Rees valuations comes from the order valuation of a regular local ring:

Example 10.3.1 Let $(R, \mathfrak{m})$ be a regular local ring. Then the $\mathfrak{m}$–adic valuation (the valuation $\text{ord}_\mathfrak{m}$) is a divisorial valuation that is the only Rees valuation of $\mathfrak{m}$. The residue field of the corresponding valuation ring is purely transcendental over $R/\mathfrak{m}$. Explicitly, the $\mathfrak{m}$–adic valuation ring equals $R[\frac{\mathfrak{m}}{x}]_{(x)}$ for any $x \in \mathfrak{m} \setminus \mathfrak{m}^2$.

In Theorem 6.7.9 it was proved that the order function is a valuation, the residue field is purely transcendental, and that this $\mathfrak{m}$–adic valuation ring equals $R[\frac{\mathfrak{m}}{x}]_{(x)}$ for any $x \in \mathfrak{m} \setminus \mathfrak{m}^2$. Since $xR[\frac{\mathfrak{m}}{x}]$ is prime, Proposition 10.2.6 gives that $R[\frac{\mathfrak{m}}{x}]_{(x)}$ is the unique Rees valuation.

Furthermore, this is a divisorial valuation: clearly the center of v is $\mathfrak{m}$, and $\kappa(xR[\frac{\mathfrak{m}}{x}]_{(x)})$ is generated over $\kappa(\mathfrak{m})$ by the images of $\frac{x_i}{x}$, $i = 2, \ldots, d$, where $\mathfrak{m} = (x, x_2, \ldots, x_d)$ with no relations among them, so that the transcendence degree of $\kappa(xR[\frac{\mathfrak{m}}{x}]_{(x)})$ over $\kappa(\mathfrak{m})$ is $d - 1 = \mathrm{ht}\,\mathfrak{m} - 1$.

Example 10.3.2   Let R be a one–dimensional Noetherian semi–local integral domain. Then $\{\overline{R}_P \mid P \in \mathrm{Max}\,\overline{R}\} = \cup_I \mathcal{RV}(I)$ is finite, as I varies over all the ideals of R. Namely, by Theorem 4.9.2 (the Krull–Akizuki Theorem), $\overline{R}$ is one–dimensional Noetherian. Thus for each maximal ideal $\mathfrak{m}$ in R, there are only finitely many prime ideals in $\overline{R}$ minimal over $\mathfrak{m}\overline{R}$. But every maximal ideal in $\overline{R}$ contracts to a maximal ideal in R (by Going–Up, Theorem 2.2.4), and by Incomparability Theorem, Theorem 2.2.3, every maximal ideal in $\overline{R}$ is minimal over the extension to $\overline{R}$ of some maximal prime ideal in R. Thus $\overline{R}$ has only finitely many maximal ideals. For each maximal prime ideal P of $\overline{R}$, $\overline{R}_P$ is a discrete valuation ring of rank one. By the previous example, $\overline{R}_P$ is a Rees valuation ring of any $x\overline{R}$, where $x \in P \cap R \setminus \{0\}$. Hence by Proposition 10.2.3, $\overline{R}_P$ is a Rees valuation ring of xR. (It is also a Rees valuation ring of $P \cap R$, see Exercise 10.8.) If V is a Rees valuation ring of some non–zero ideal of R, as V is integrally closed, V contains $\overline{R}$. Set $P = \mathfrak{m}_V \cap \overline{R}$. Then $P \in \mathrm{Max}\,\overline{R}$, $\overline{R}_P \subseteq V$, and as both $\overline{R}_P$ and V are discrete valuation rings of rank one with the same field of fractions, $\overline{R}_P = V$. This proves the example.

Example 10.3.3 Let $R = k[X, Y]$ be the polynomial ring in two variables X and Y over a field k. Let I be the ideal $I = (XY, X^3, Y^3)R$. We calculate its Rees valuations via the original construction.

For this let $S_1 = R[\frac{I}{X^3}]$, $S_2 = R[\frac{I}{XY}]$, and $S_3 = R[\frac{I}{Y^3}]$. Then

$$S_1 = R\left[\frac{I}{X^3}\right] = k\left[X, Y, \frac{XY}{X^3}, \frac{Y^3}{X^3}\right] \cong \frac{k[X, Y, Z, W]}{(X^2 Z - Y, W - X^3 Z^3)},$$

which is isomorphic to $k[X, Z]$, and hence integrally closed. We find all the minimal prime ideals over $X^3 S_1$: if a prime ideal P minimally contains X^3, it contains X, and hence also Y and $W = \frac{Y^3}{X^3}$. But $(X, Y, \frac{Y^3}{X^3})$ is a prime ideal in S_1, so that it equals P. The localization $(S_1)_P$ is a one–dimensional integrally closed domain. As $\frac{XY}{X^3}$ is a unit of $(S_1)_P$, $Y \in X^2 (S_1)_P \setminus X^3 (S_1)_P$. Also, $\frac{Y^3}{X^3} = X^3 (\frac{XY}{X^3})^3 \in X^3 (S_1)_P$. Thus $P(S_1)_P = X(S_1)_P$. Thus for the corresponding valuation v_1, $v_1(X) = 1$, $v_1(Y) = 2$, and v_1 is monomial (see Proposition 10.3.4 for a general result). In particular, $v_1(I) = 3$.

By symmetry, S_3 yields one monomial valuation $v_2(X) = 2$, $v_2(Y) = 1$.

It is straightforward to see that $S_2 = R[\frac{I}{XY}] = k[X, Y, \frac{X^3}{XY}, \frac{Y^3}{XY}]$ is isomorphic to $k[X, Y, Z, W]/(ZV - XY, X^2 - YZ, Y^2 - XW)$, which is an integrally closed ring by the Jacobian criterion (Theorem 4.4.9). A prime ideal in S_2 minimal over XY contains either X or Y. If it contains X, it also contains Y^2 so that it contains Y. Similarly, a prime ideal minimal over Y also contains X. Any prime ideal containing X and Y also contains $XY = \frac{X^3}{XY} \cdot \frac{Y^3}{XY}$, so that we get two prime ideals minimal over XYS_2:

$$P_1 = \left(X, Y, \frac{X^3}{XY}\right)S_2, \, P_2 = \left(X, Y, \frac{Y^3}{XY}\right)S_2.$$

It can be easily checked that these two prime ideals give the valuations v_1 and v_2 that were also obtained on S_1 and S_3. The geometric picture that goes with this example is given below:

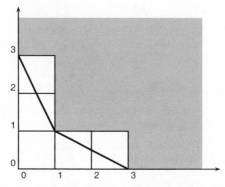

The three lattice points on the outside corners of the gray shaded area correspond to the generators X^3, XY, Y^3. All the integral lattice points to the right and up of the three points correspond to monomials in I. The two bold lines connecting the three generator points bound the Newton polytope. The equations of the two lines are

$$x + 2y = 3 \quad \text{and} \quad 2x + y = 3.$$

From the equation for the first line we can read off the valuation $v_1(X^a Y^b) = a + 2b$, $v_1(I) = 3$, and from the second one the valuation $v_2(X^a Y^b) = 2a + b$, $v_2(I) = 3$. Furthermore, the piece $R[\frac{I}{X^3}]$ gives only the first valuation as only the first line passes through the point corresponding to X^3, and the piece $R[\frac{I}{XY}]$ gives both valuations as both lines pass through the point corresponding to XY.

Notice in this example that the Rees valuations are monomial, and can be read off from the Newton polyhedron of I. We prove that this is true in general for monomial ideals. First, we prove that all Rees valuations are monomial.

Proposition 10.3.4 *Let I be a monomial ideal in a polynomial ring. Then the Rees valuations of I are monomial.*

Proof: By Theorem 10.1.6 it suffices to prove that we can construct a set of Rees valuations that are all monomial. Let $k[X_1, \ldots, X_d]$ be the polynomial ring.

Let $a_1, \ldots, a_m$ be the monomials generating I. For each $i = 1, \ldots, m$, $R[\frac{I}{a_i}]$ is a $\mathbb{Z}^d$–graded algebra under the grading $\deg X_i = (0, \ldots, 0, 1, 0, \ldots, 0)$, with 1 in the ith entry and 0 everywhere else. By Theorem 2.3.2, the integral closure S of $R[\frac{I}{a_i}]$ in $k[X_1^{\pm 1}, \ldots, X_d^{\pm 1}]$ is a $\mathbb{Z}^d$–graded algebra. As $k[X_1^{\pm 1}, \ldots, X_d^{\pm 1}]$ is integrally closed, S is the integral closure of $R[\frac{I}{a_i}]$, and is $\mathbb{Z}^d$–graded. As a_i is homogeneous, all the minimal prime ideals over $a_i S$ are homogeneous. Let P be such a minimal prime ideal. By Theorem 10.2.2, S_P is a Rees valuation ring of I. By Lemma 6.4.2 there exist integers $e_1, \ldots, e_d \in \mathbb{Z}$ such that $r = X_1^{e_1} \cdots X_d^{e_d}$ is in S and such that $PS_P = rS_P$. Let $\underline{e} = (e_1, \ldots, e_d)$.

It suffices to prove that the valuation v corresponding to the valuation ring S_P is monomial. For let $f = \sum_\nu a_\nu \underline{X}^\nu$ be a polynomial, with $a_\nu \in k$. For each ν, let $n_\nu = v(\underline{X}^\nu)$ and set $n = \min\{n_\nu \mid a_\nu \neq 0\}$. We need to show that $v(f) = n$. By subtracting homogeneous summands of f, without loss of generality we may assume that for all ν with $a_\nu \neq 0$, $\underline{X}^\nu$ is not in the homogeneous ideal $P^{(n+1)}$ (symbolic power). We may write $f = r^n \sum_\nu a_\nu \underline{X}^{\nu - n\underline{e}}$, where $\underline{X}^{\nu - n\underline{e}}$ are elements of S_P. Suppose that $\sum_\nu a_\nu \underline{X}^{\nu - n\underline{e}} \in PS_P$. Then $f \in P^{(n+1)}$, and there exists $s \in S \setminus P$ such that $sf \in P^{n+1}$. Without loss of generality no summand of s is in P. Both s and f are finite linear combinations of monomials with integer exponents. Under the lexicographic ordering of monomials, write $s = s_0 + s'$, $f = f_0 + f'$, where s_0 (respectively f_0) is a non–zero homogeneous summand of s (respectively f) of highest degree in the ordering. Then $sf = s_0 f_0 +$ lower terms, so that necessarily $s_0 f_0 \in P^{n+1}$. As $s_0 \notin P$, necessarily $f_0 \in P^{(n+1)}$, which contradicts the assumption. So necessarily $\sum_\nu a_\nu \underline{X}^{\nu - n\underline{e}} \notin PS_P$, so that $v(f) = v(r^n) = n$. This proves that v is a monomial valuation. $\qquad\square$

We now prove that the Rees valuations of an arbitrary monomial ideal I can be read off from the Newton polyhedron of I. Namely, the convex hull of the Newton polyhedron of a monomial ideal in $k[X_1, \ldots, X_d]$ is defined by finitely many non–redundant hyperplanes, and they are of the form $a_1 X_1 + \cdots + a_d X_d = a$ for some non–negative integers a_i, a. Furthermore, for an arbitrary positive integer n, the hyperplanes bounding the Newton polyhedron of I^n are simply the (translated) hyperplanes of the form $a_1 X_1 + \cdots + a_d X_d = na$, where $a_1 X_1 + \cdots + a_d X_d = a$ is a hyperplane bounding the Newton polyhedron of I. (The vector $(m_1, \ldots, m_d) \in \mathbb{Q}^d$ is in the Newton polyhedron of I^n if and only if $(m_1/n, \ldots, m_d/n)$ is in the Newton polyhedron of I.) From each such hyperplane we can read off the monomial valuation $v(X_1^{b_1} \cdots X_d^{b_d}) = a_1 b_1 + \cdots + a_d b_d$, $v(I) = a$. A monomial $X_1^{b_1} \cdots X_d^{b_d}$ is in the integral closure of I^n if and only if $(b_1, \ldots, b_d)$ is in the Newton polyhedron of I^n, and that holds if and only if $(b_1, \ldots, b_d)$ lies on the correct side of every boundary hyperplane of the Newton polyhedron of I^n. In other words, $X_1^{b_1} \cdots X_d^{b_d}$ is in the integral closure of I^n if and only if for all valuations v obtained from the hyperplanes as above, $v(X_1^{b_1} \cdots X_d^{b_d}) = a_1 b_1 + \cdots + a_d b_d \geq na = v(I^n)$. Thus the corresponding monomial valuations $v(X_1^{b_1} \cdots X_d^{b_d}) = a_1 b_1 + \cdots + a_d b_d$, $v(I) = a$, determine the integral closures of all the powers of I. Thus by the non–redundancies among the hyperplanes, we just proved:

Theorem 10.3.5 *Let I be a monomial ideal in a polynomial ring R over a field. Then the set of Rees valuations of I is the set of monomial valuations obtained from the bounding hyperplanes of the Newton polyhedron of I.*

In general, the set of Rees valuations of I may be strictly larger than a set of valuations determining the integral closure of I. For example, the integrally closed monomial ideal $I = (X^3, XY, Y^2)$ has two Rees valuations, namely the monomial valuations $v_1(X^a Y^b) = a + b$ and $v_2(X^a Y^b) = a + 2b$, which can be read off from the Newton polytope. The integral closure of I is determined by v_2 alone: $\bar{I} = \{r \in k[X, Y] \mid v_2(r) \geq v_2(I)\} = I$, but the integral closure of I^2 needs both valuations.

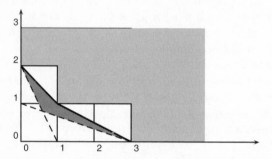

We now present a false attempt at finding the Rees valuations of I. Note that $I = (X^3, Y) \cap (X, Y^2)$. The Newton polytopes of (X^3, Y) and (X, Y^2) are bounded by $x + 3y = 3$ and $2x + y = 2$, respectively (see figure above). Do the two corresponding monomial valuations $v'(X^a Y^b) = a + 3b$ and $v''(X^a Y^b) = 2a + b$ constitute the set of Rees valuations of I? The answer is no. Here is the reason. Note that the Newton polytope of I is touching or above the two lines $x + 3y = 3$ and $2x + y = 2$. The gap between the area below the convex hull of the Newton polytope and the latter two lines contains the point $(1, 2/3)$, which translates to the "monomial" $XY^{2/3}$. The third power of this element, $X^3 Y^2$, is a legitimate monomial, and it is not in $\bar{I}^3$. However, $v'(X^3 Y^2) = 9 \geq 3 \cdot 3 = 3 \cdot v'(I)$, and $v''(X^3 Y^2) = 8 \geq 3 \cdot 2 = 3 \cdot v''(I)$. Thus $\{v', v''\}$ could not be the set of Rees valuations of I.

This example in particular shows that the Rees valuations of monomial ideals correspond to the non–coordinate boundary hyperplanes (so–called faces) and not to arbitrary hyperplanes bounding from below the integer lattice part of the polytope.

10.4. Properties of Rees valuations

Proposition 10.4.1 *Let R be a Noetherian ring, I an ideal of R, and W is any multiplicatively closed set in R. Then $\mathcal{RV}(W^{-1}I) = \{V \in \mathcal{RV}(I) \,|\, \mathfrak{m}_V \cap W = \emptyset\}$ (where $\mathfrak{m}_V$ is the unique maximal ideal of V).*

Proof: By definition, for all positive integers n, $\overline{I^n} = \cap_{V \in \mathcal{RV}(I)} I^n V \cap R$, so that $\overline{W^{-1}I^n} = W^{-1}\overline{I^n} = \cap_{V \in \mathcal{RV}(I), \mathfrak{m}_V \cap W = \emptyset} I^n V \cap R$. By irredundancy of elements of $\mathcal{RV}(I)$, the set $\{V \in \mathcal{RV}(I) \,|\, \mathfrak{m}_V \cap W = \emptyset\}$ is irredundant in determining the integral closures of $W^{-1}I^n$. $\qquad\square$

Centers of Rees valuations can be determined without explicit construction:

Theorem 10.4.2 *Let R be a Noetherian ring, I an ideal in R and P a prime ideal in R. If $\ell(IR_P) = \dim(R_P)$, then P is the center of a Rees valuation of I. Conversely, if R is locally formally equidimensional and P is the center of a Rees valuation of I, then $\ell(IR_P) = \dim(R_P)$.*

Proof: By Proposition 10.4.1, without loss of generality we may assume that P is the unique maximal ideal of R. If P is not the center of any Rees valuation of I, then by Discussion 10.1.3 P is not associated to any $\overline{I^n}$. But by Proposition 5.4.7, $\ell(I) = \dim(R)$ implies that P is associated to $\overline{I^n}$ for all large n, which forces P to be the center of some Rees valuation of I.

Conversely, suppose that P is the center of a Rees valuation of I. We may again assume that P is the unique maximal ideal of R. By Discussion 10.1.3 we know that P is associated to $\overline{I^n}$ for all large n. We can apply Theorem 5.4.6 to conclude that $\ell(IR_P) = \dim(R_P)$. $\qquad\square$

Every Rees valuation on a restricted class of rings is a divisorial valuation:

Proposition 10.4.3 *Let R be a Noetherian ring, I an ideal, and V a Rees valuation ring of I. Set $P = \mathfrak{m}_V \cap R$. There exist a minimal prime ideal q in the P-adic completion $\widehat{R}$ of R_P and a valuation ring W in the field of fractions of $\widehat{R}/q$ such that W is a Rees valuation of $I\widehat{R}$, V is a valuation ring on the field of fractions of $R/(q \cap R)$, and such that V is the contraction of W. For any such q,*

$$\mathrm{tr.deg}_{\kappa(P)}\kappa(\mathfrak{m}_V) = \dim(\widehat{R}/q) - 1.$$

In particular, if R_P is formally equidimensional, then $\mathrm{tr.deg}_{\kappa(P)}\kappa(\mathfrak{m}_V) = \mathrm{ht}\, P - 1$, implying that V is a divisorial valuation ring (see Definition 9.3.1).

Proof: Without loss of generality R is local with maximal ideal P, and R is an integral domain with field of fractions K. Let $\widehat{R}$ be the P-adic completion of R. Let $\{w_1, \ldots, w_r\}$ be a set of Rees valuations of $I\widehat{R}$. Let $v_i = w_{i|K}$. By Proposition 6.4.7, v_i is a Noetherian K valuation. Proposition 1.6.2 says that for all n, $\overline{I^n\widehat{R}} \cap R = \overline{I^n}$, so that $\overline{I^n} = \{r \in R \,|\, w_i(r) \geq nw_i(I)\} = \{r \in R \,|\, v_i(r) \geq nw_i(I)\}$. By uniqueness of Rees valuations, V must equal the valuation ring corresponding to some v_i. Let W be the valuation ring corresponding to w_i. Then $V = W \cap K$. Note that $\mathfrak{m}_W \cap R = P$ and $\mathfrak{m}_W \cap \widehat{R} = P\widehat{R}$. Let q be the minimal prime ideal in $\widehat{R}$ such that W is a valuation ring on $\kappa(q)$. As $\widehat{R}/q$ is analytically unramified, by the construction of Rees algebras and by Theorem 9.2.2, W is a localization of a finitely generated $\widehat{R}$–algebra. As $\widehat{R}/q$ is universally catenary, the Dimension Formula (Theorem B.5.1) applies:

$$\text{tr.deg}_{\kappa(P\widehat{R})}\kappa(\mathfrak{m}_W) = \dim(\widehat{R}/q) + \text{tr.deg}_{\widehat{R}/q}W - \text{ht}\,\mathfrak{m}_W.$$

As W is Noetherian, $\text{ht}\,\mathfrak{m}_W = 1$, and by the definition of Rees valuations, $\text{tr.deg}_{\widehat{R}/q}W = 0$. By Proposition 6.5.2, $\kappa(\mathfrak{m}_V) = \kappa(\mathfrak{m}_W)$, so that

$$\text{tr.deg}_{\kappa(P)}\kappa(\mathfrak{m}_V) = \dim(\widehat{R}/q) - 1 = \dim(\widehat{R}/q) - 1.$$

The rest follows trivially. □

Proposition 10.4.4 *Let R be a Noetherian local integral domain. Let V be a divisorial valuation ring between R and its field of fractions. There exists a non–zero ideal I in R such that V is one of its Rees valuations. Furthermore, we may assume that I is primary to $P = \mathfrak{m}_V \cap R$, where $\mathfrak{m}_V$ is the maximal ideal of V.*

We give essentially the same proof as for Theorem 9.3.2. Note that in general V will not be the only Rees valuation ring of I. In fact, there exists a two–dimensional complete normal local domain in which every ideal has at least two Rees valuations (see Cutkosky [52]).

Proof: By Proposition 10.4.1, without loss of generality we may localize at P and thus assume that P is the unique maximal ideal of R. Let $d = \text{ht}\,P$. By assumption on V, $\text{tr.deg}_{\kappa(P)}\kappa(\mathfrak{m}_V) = d - 1$. Thus there exist non–zero elements $x_1, \ldots, x_d \in P$ such that $\frac{x_2}{x_1}, \ldots, \frac{x_d}{x_1}$ are in V and their images in $V/\mathfrak{m}_V = \kappa(\mathfrak{m}_V)$ form a transcendence basis of $\kappa(\mathfrak{m}_V)$ over $\kappa(P)$.

Let n be a positive integer such that $P^n V \subseteq x_1 V$. Let I be any ideal between $(x_1, x_2, \ldots, x_d)$ and $(x_1, x_2, \ldots, x_d) + P^n$. Thus I can be taken to be P–primary.

Set $S = R[\frac{I}{x_1}]$, a finitely generated R–algebra. Let $\overline{S}$ be its integral closure. By the choice of I, $S \subseteq \overline{S} \subseteq V$. Set $Q = \mathfrak{m}_V \cap \overline{S}$. The images of the $\frac{x_i}{x_1}$ in $\kappa(Q \cap S) \subseteq \kappa(Q)$ form a transcendence basis of $\kappa(Q \cap S)$ over $\kappa(P)$, so that by the Dimension Inequality (Theorem B.2.5), $\text{ht}(Q \cap S) \leq 1$. Hence $\text{ht}\,Q = \dim(\overline{S})_Q \leq \dim S_{Q \cap S} \leq 1$, so that as $x_1 \in Q$, $\text{ht}\,Q = 1$, and Q is a prime ideal minimal over $x_1\overline{S}$. Thus by the construction of Rees valuations $\overline{S}_Q$ is one of the Rees valuation rings of I. But $\overline{S}_Q \subseteq V$ are both discrete valuation rings of rank 1 with the same field of fractions, so that they must be equal. Thus V is a Rees valuation ring of I. □

With the following proposition one can find some Rees valuations of products of ideals:

Proposition 10.4.5 *Let R be a Noetherian integral domain. Then for any non-zero ideals I and J in R, $\mathcal{RV}(I) \cup \mathcal{RV}(J) \subseteq \mathcal{RV}(IJ)$.*

Proof: We will prove that $\mathcal{RV}(I) \subseteq \mathcal{RV}(IJ)$. Let $V \in \mathcal{RV}(I)$. By Theorem 10.2.2 we may choose an element $a \in I$ such that if $\overline{S}$ is the integral closure of $S = R[\frac{I}{a}]$, and $P = \mathfrak{m}_V \cap \overline{S}$, then $(\overline{S})_P = V$. Choose $b \in J$ such that $JV = bV$. Then

$$S = R\left[\frac{I}{a}\right] \subseteq T = R\left[\frac{I}{a}\right]\left[\frac{J}{b}\right] = R\left[\frac{IJ}{ab}\right] \subseteq V.$$

Let $\overline{T}$ be the integral closure of T, and $Q = \mathfrak{m}_V \cap \overline{T}$. Then $V = (\overline{S})_P \subseteq (\overline{T})_Q \subseteq V$, so that $V = (\overline{T})_Q$. Thus $V \in \mathcal{RV}(IJ)$. $\square$

In general, the other inclusion need not hold:

Example 10.4.6 Let $R = k[X, Y, Z]$ be a polynomial ring over a field k. Let $I = (X, Y)$ and $J = (X, Z)$. We claim that $\mathcal{RV}(IJ) \neq \mathcal{RV}(I) \cup \mathcal{RV}(J)$. We have that $\mathcal{RV}(I) = \{v_1\}$, where v_1 is the monomial valuation that takes value 1 on X and Y and value 0 on Z. Similarly, $\mathcal{RV}(J) = \{v_2\}$, where v_2 is a monomial valuation taking value 1 on X and Z and value 0 on Y. Suppose that $\mathcal{RV}(IJ) = \{v_1, v_2\}$. Note that $v_1(IJ) = 1$ since $v_1(XZ) = 1$, and similarly $v_2(IJ) = 1$ since $v_2(XY) = 1$. However, $v_1(X) = v_2(X) = 1$ as well. If $\mathcal{RV}(IJ) = \{v_1, v_2\}$, then this would imply that $X \in \overline{IJ}$, which is clearly false.

This can also be seen from the equations of the Rees algebra of IJ. We can map $R[A, B, C, D]$ onto the Rees algebra of IJ by mapping $A \to X^2$, $B \to XY$, $C \to XZ$ and $D \to YZ$. With this map,

$$k[X, Y, Z, A, B, C, D]/P \cong R[IJt]$$

where $P = (XD - YC, YC - ZB, BC - AD, XC - ZA, XB - YA)$. The Rees algebra is Gorenstein and integrally closed, and the height of P is 3, as can be easily seen by using a computer algebra program or by hand. Since the Rees algebra is integrally closed, Remark 10.1.4 shows that the Rees valuations correspond to the minimal primes over $IJR[IJt]$, and there are exactly three minimal primes corresponding to the images of the prime ideals $P_1 = (X, Y, A, B)$ (which gives v_1), $P_2 = (X, Z, A, C)$ (which gives v_2) and a third prime $P_3 = (X, Y, Z, BC - AD)$, which accounts for an additional valuation v_3 in $\mathcal{RV}(IJ)$. The valuation ring of v_3 is $(R[IJt]_{P_3})$, and v_3 is the order valuation of the maximal ideal (X, Y, Z).

Below we prove two cases where equality holds in Proposition 10.4.5.

Proposition 10.4.7 *Let R be a Noetherian domain, I and J non-zero ideals in R such that I is locally principal. Then $\mathcal{RV}(I) \cup \mathcal{RV}(J) = \mathcal{RV}(IJ)$.*

Proof: By Proposition 10.4.5 it suffices to prove that $\mathcal{RV}(IJ) \subseteq \mathcal{RV}(I) \cup \mathcal{RV}(J)$, and it suffices to prove this locally. So without loss of generality R is local, and $I = (x)$ for some non–zero $x \in R$. Let V be a Rees valuation ring of IJ. Then there exists $b \in J$ such that $T = R\left[\frac{xJ}{xb}\right] \subseteq V$, and there exists a prime ideal Q in the integral closure $\overline{T}$ of T such that Q is minimal over xb and $(\overline{T})_Q = V$. If $b \in Q$, then as $T = R\left[\frac{J}{b}\right]$, it follows by construction of Rees valuations that V is a Rees valuation ring of J. Now suppose that $b \notin Q$. Then necessarily $x \in Q$, and as b is a unit in $(\overline{T})_Q$, V is a localization of the integral closure of R at a prime ideal necessarily minimal over xR, so that V is a Rees valuation ring of I. $\qquad\square$

Proposition 10.4.8 *Let R be a Noetherian locally formally equidimensional domain of dimension at most 2. Let I and J be non–zero ideals in R. Then $\mathcal{RV}(I) \cup \mathcal{RV}(J) = \mathcal{RV}(IJ)$.*

Proof: By Proposition 10.4.5 it suffices to prove that $\mathcal{RV}(IJ) \subseteq \mathcal{RV}(I) \cup \mathcal{RV}(J)$. Let V be a Rees valuation of IJ. By the construction (proof of Theorem 10.2.2) we know that there exist $a \in I$ and $b \in J$ such that $S = R\left[\frac{IJ}{ab}\right] \subseteq V$, and there exists a height one prime ideal Q in the integral closure $\overline{S}$ of S containing ab such that $(\overline{S})_Q = V$.

If $a \notin Q$, then $\overline{S}_Q$ is the same as the integral closure of $R\left[\frac{J}{b}\right]$ localized at (the image of) Q, so that then clearly $V \in \mathcal{RV}(J)$. So we may assume that $a \in Q$, and similarly that $b \in Q$.

Set $p = Q(\overline{S})_Q \cap R = \mathfrak{m}_V \cap R$. As $R \subseteq V$ is contained in the field of fractions of R, p is a non–zero prime ideal, so its height is either 1 or 2. By Proposition 10.4.3, $\mathrm{tr.deg}_{\kappa(p)}\kappa(\mathfrak{m}_V) = \mathrm{tr.deg}_{\kappa(p)}\kappa(Q) = \mathrm{ht}\, p - 1$.

If $\mathrm{tr.deg}_{\kappa(p)}\kappa(Q)$ is 0, then the height of p is 1, so that the height of IJ is one, and p is minimal either over I or over J. Say p is minimal over I. The normalization of R_p is a one–dimensional semi–local Noetherian domain contained in V, so for some maximal ideal $\tilde{p}$ in $\overline{R}_p$, $(\overline{R})_{\tilde{p}}$ equals V. Then by Example 10.3.2, V is a Rees valuation ring of I.

Now assume that $\mathrm{tr.deg}_{\kappa(p)}\kappa(Q)$ is 1. Then necessarily $\mathrm{ht}\, p = \dim R = 2$. Let $S_a = \overline{R[\frac{I}{a}]}$, $S_b = \overline{R[\frac{I}{b}]}$, $q_a = Q(\overline{S})_Q \cap S_a$, and $q_b = Q(\overline{S})_Q \cap S_b$. Both S_a and S_b are subrings of $\overline{S}$, and both S_a/q_a and S_b/q_b are subrings of $\overline{S}/Q$. There is a natural ring surjection

$$\left(\frac{S_a}{q_a}\right)\left(\frac{S_b}{q_b}\right) \to \frac{S_a S_b}{Q \cap S_a S_b}.$$

The composition of this map with inclusion into $\overline{S}/Q$ is an injection, so that $\left(\frac{S_a}{q_a}\right)\left(\frac{S_b}{q_b}\right) = \frac{S_a S_b}{Q \cap S_a S_b}$. Let W be the multiplicatively closed set $(S_a \setminus q_a)(S_b \setminus q_b)$ in $S_a S_b \subseteq \overline{S}$. The localization at W gives

$$\kappa(q_a)\kappa(q_b) = W^{-1}\left(\frac{S_a}{q_a}\right)\left(\frac{S_b}{q_b}\right) = W^{-1}\frac{S_a S_b}{Q \cap S_a S_b} \subseteq W^{-1}\left(\frac{\overline{S}}{Q}\right).$$

The last inclusion is integral as $\overline{S}$ is the integral closure of $S_a S_b$. Hence the transcendence degree of $\kappa(Q)$ over $\kappa(q_a)\kappa(q_b)$ is 0. Thus the transcendence degree of $\kappa(q_a)\kappa(q_b)$ over $\kappa(p)$ is 1, whence either tr.deg$_{\kappa(p)}\kappa(q_a) = 1$ or tr.deg$_{\kappa(p)}\kappa(q_b) = 1$. Suppose that tr.deg$_{\kappa(p)}\kappa(q_a) = 1$. By the Dimension Inequality B.2.5, ht$(q_a) \leq 1$, so that as q_a is non–zero, necessarily ht$(q_a) = 1$. Hence by Proposition 6.4.4, $(S_a)_{q_a}$ is a discrete valuation domain of rank one contained in V, and necessarily equal to V. Furthermore, by construction V is a Rees valuation ring of I. Similarly, if tr.deg$_{\kappa(p)}\kappa(q_b) = 1$, it follows that V is a Rees valuation of J. $\qquad\square$

Proposition 10.4.9 *Let R be a Noetherian domain and I an ideal in R with $\mathcal{RV}(I) = \{v_1, \ldots, v_r\}$. Let X be a variable over R, and for each $i = 1, \ldots, r$, let w_i be the Gauss extension of v_i to $Q(R)(X)$ as in Remark 6.1.3: w_i agrees with v_i on $Q(R)$ and takes X to 0. Then $\mathcal{RV}(IR[X]) = \{w_1, \ldots, w_r\}$.*

Proof: By definition, w_i is a Gauss extension of v_i. For $n \geq 1$, $\{u \in R[X] \,|\, w_i(u) \geq w_i(I^n), i = 1, \ldots, r\}$ is the ideal in $R[X]$ generated by elements in $u \in R$ with $v_i(u) \geq v_i(I^n)$, for $i = 1, \ldots, r$. This means that $\{u \in R[X] \,|\, w_i(u) \geq w_i(IR[X]), i = 1, \ldots, r\}$ equals $\overline{I^n}R[X]$, which is the integral closure of $I^n R[X]$ by Lemma 8.4.2. Thus $w_1, \ldots, w_r$ are contained in the set of Rees valuations of $IR[X]$. If one of them, say w_i, is redundant, then it is easy to show that v_i would be redundant as well, proving that $\mathcal{RV}(IR[X]) = \{w_1, \ldots, w_r\}$. $\qquad\square$

10.5. Rational powers of ideals

In this section we associate to an ideal its "rational powers", in the following sense:*

Definition 10.5.1 *Let R be a Noetherian ring, and let I be an ideal. Fix a rational number $\alpha = \frac{p}{q}$ with $p, q \in \mathbb{N}$, $q \neq 0$. We define $I_\alpha = \{x \in R \,|\, x^q \in \overline{I^p}\}$.*

The next proposition summarizes the basic properties of I_α, including the fact it is a well–defined ideal and is integrally closed. It would be appropriate to call this ideal the "αth" power of I except for the fact that the definition brings in the integral closure of I^p. If we simply took the set of elements x such that $x^q \in I^p$, this set would not necessarily be an ideal. Taking all of this into account, it is natural to think of I_α as the integral closure of the "αth" power of I.

Proposition 10.5.2 *Let R be a Noetherian ring, I an ideal in R, and $\alpha, \beta \in \mathbb{Q}$.*
(1) I_α is well–defined, i.e., does not depend on the representation of α as a quotient of two integers.
(2) If $\alpha < \beta$, then $I_\beta \subseteq I_\alpha$.
(3) $I_\alpha I_\beta \subseteq I_{\alpha+\beta}$.

---

*   We thank Mark Johnson for sharing notes on this topic. We also refer to [182] for more information.

(4) I_α is an integrally closed ideal.

(5) For $n \in \mathbb{N}$, $I_n = \overline{I^n}$.

(6) $x \in I_\alpha$ if and only if for all minimal primes P of R and for all rank one discrete $\kappa(P)$–valuations v that are non–negative on R/P, $v(x) \geq \alpha \cdot v(I)$.

(7) An element $x \in I_\alpha$ if and only if $v(x) \geq \alpha \cdot v(I)$ for all the Rees valuations v of I.

Proof: First of all, I_α is well–defined. Namely, if $\alpha = \frac{tm}{tn}$ for some integer t, certainly $r^n \in \overline{I^m}$ implies that $r^{tn} \in \overline{I^{tm}}$, and if $r^{tn} \in \overline{I^{tm}}$, then an equation of integral dependence of r^{tn} on I^{mn} is also an equation of integral dependence of r^n on I^m. Thus I_α is well–defined. It is clear that if α, β are two non–negative rational numbers, then $I_\alpha I_\beta \subseteq I_{\alpha+\beta}$, and if $\alpha < \beta$, then $I_\beta \subseteq I_\alpha$.

Let r be integral over I_α. Write $\alpha = \frac{m}{n}$ for some non–negative integers m, n. As r^n is integral over I_α^n and I_α^n is contained in I_m, it follows that r^n is integral over I^m, whence $r \in I_\alpha$. This proves that I_α is integrally closed.

Part (5) is clear from the definition.

Suppose that $x \in I_\alpha$ and v is a rank one discrete $\kappa(P)$–valuation (for some P a minimal prime of R). Write $\alpha = \frac{p}{q}$ with $p, q \in \mathbb{N}$. Apply v to the equation $x^q \in \overline{I^p}$. One obtains that $qv(x) \geq pv(I)$ or equivalently $v(x) \geq \alpha \cdot v(I)$. Conversely, suppose that for all rank one discrete $\kappa(P)$–valuations v (for P a minimal prime of R) $v(x) \geq \alpha \cdot v(I)$. Write $\alpha = \frac{p}{q}$ with $p, q \in \mathbb{N}$. We obtain that for all such valuations v, $v(x^q) = qv(x) \geq q\alpha \cdot v(I) = pv(I) = v(I^p)$. The valuative criterion for integral closure Theorem 6.8.3 gives that $x^q \in \overline{I^p}$.

Suppose that $x \in I_\alpha$ and $v = v_i$ is a Rees valuation of I. Write $\alpha = \frac{p}{q}$ with $p, q \in \mathbb{N}$. Apply v to the equation $x^q \in \overline{I^p}$. One obtains that $qv(x) \geq pv(I)$ or equivalently $v(x) \geq \alpha \cdot v(I)$. Conversely, suppose that for all Rees valuations v of I we know that $v(x) \geq \alpha \cdot v(I)$. Write $\alpha = \frac{p}{q}$ with $p, q \in \mathbb{N}$. We obtain that for all Rees valuations v of I, $v(x^q) = qv(x) \geq q\alpha \cdot v(I) = pv(I) = v(I^p)$. It follows that $x^q \in \overline{I^p}$.

Definition 10.5.3 *Let R be a Noetherian ring, I an ideal, and $\alpha \in \mathbb{Q}$. Define $I_{>\alpha} = \cup_{\beta > \alpha} I_\beta$.*

By Proposition 10.5.2, this union is actually an ascending union of ideals, hence it is also an ideal. Moreover, as R is Noetherian, the chain $\{I_\beta\}$ for $\beta > \alpha$ of ideals stabilizes, and the stable value is therefore $I_{>\alpha}$. In particular, all of the ideals $I_{\alpha+\epsilon}$ for small rational ϵ are the same, namely $I_{>\alpha}$.

Associated to the ideals I_α is the graded ring $G = \oplus_{\alpha \in \mathbb{Q}} I_\alpha / I_{>\alpha}$, which is a graded ring over the non–negative rational numbers by Proposition 10.5.2.

As we shall see, however, this graded ring is actually $\mathbb{N}$–graded. Its degree 0 piece is exactly $R/\sqrt{I}$ by the calculation in the example below.

Example 10.5.4 $I_{>0} = \sqrt{I}$. If $x \in \sqrt{I}$, then $x^n \in I$ for all large n, so that $x \in I_{\frac{1}{n}}$ for all large n. Thus $\sqrt{I} \subseteq I_{>0}$. Conversely, if $x \in I_{>0}$, then for some n, $x \in I_{\frac{1}{n}}$, which implies that $x^n \in \overline{I} \subseteq \sqrt{I}$, and $x \in \sqrt{I}$.

Proposition 10.5.5 *Let R be a Noetherian ring and I an ideal of R of positive height. Let $\{v_1, ..., v_m\}$ be the Rees valuations of I. Set e to be the least common multiple of $v_1(I), \ldots, v_m(I)$. Every ideal I_α with $\alpha \in \mathbb{Q}$ is equal to $I_{\frac{n}{e}}$ for some $n \in \mathbb{N}$.*

Proof: Set $e_i = v_i(I)$ and write $e = e_i d_i$. Given $\alpha \in \mathbb{Q}$, set $n = \lceil e\alpha \rceil$. We claim that $I_\alpha = I_{\frac{n}{e}}$. Since $\frac{n}{e} \geq \alpha$, we know that $I_{\frac{n}{e}} \subseteq I_\alpha$. To prove the opposite containment, let $x \in I_\alpha$. Write $\alpha = \frac{p}{q}$. By Proposition 10.5.2, $v_i(x) \geq \alpha e_i$ for all $i = 1, \ldots, m$. Hence $\min\{\frac{v_i(x)d_i}{e} \mid i = 1, \ldots, m\} \geq \alpha$. By the choice of n it follows that $n \leq \min\{v_i(x)d_i \mid i\}$, and therefore that $v_i(x) \geq \frac{n}{e} e_i$ for all $i = 1, \ldots, m$. Hence $v_i(x^e) \geq v_i(I^n)$ for all i, which implies that $x \in I_{\frac{n}{e}}$. $\qquad\square$

The main information provided by these rational "powers" of I is provided by the following theorem:

Theorem 10.5.6 *Let R be a Noetherian ring, let I be an ideal of positive height, and let u be a variable. Let $\{v_1, ..., v_n\}$ be the Rees valuations of I, and e the least common multiple of $v_1(I), \ldots, v_n(I)$. Put $t = u^e$. Let T be the integral closure of the extended Rees algebra $S = R[It, t^{-1}]$ in $R[u, u^{-1}]$.*

(1) T is a $\mathbb{Z}$–graded ring of the form $\oplus_{i \in \mathbb{Z}} J_i u^i$, where J_i are ideals of R and $J_i = R$ for $i \leq 0$.

(2) For $i > 0$, $J_i = I_{\frac{i}{e}}$.

(3) $T/u^{-1}T \cong \oplus_{\alpha \in \mathbb{Q}_{\geq 0}} I_\alpha/I_{>\alpha}$ is an $\mathbb{N}$–graded algebra with degree zero piece equal to $R/\sqrt{I}$.

(4) $u^{-1}T$ is radical, i.e., $\sqrt{u^{-1}T} = u^{-1}T$.

Proof: By Theorem 2.3.2, T is a $\mathbb{Z}$–graded ring, and clearly J_i is an ideal in R. Moreover, u^{-1} is integral over S, which implies that $J_i = R$ for $i < 0$. This proves (1).

Let $i > 0$, and let $z \in I_{\frac{i}{e}}$. Then $z^e \in \overline{I^i}$. Then $(zu^i)^e$ is integral over S. Hence $z \in J_i$. Conversely, suppose that $z \in J_i$. Consider $z^e \in J_{ei}$. Then $(zu^i)^e = z^e t^i \in T$ since it is integral over S and is inside $R[t, t^{-1}]$. Thus $z^e \in \overline{I^i}$, so by definition $z \in I_{\frac{i}{e}}$. This proves (2).

To prove (3), first observe that $\oplus_{\alpha \in \mathbb{Q}_{\geq 0}} I_\alpha/I_{>\alpha} = \oplus_{i \geq 0} J_i/J_{i+1}$. This follows since $I_\alpha = I_{>\alpha}$ if α is not of the form $\frac{i}{e}$ for some non–negative integer i. Moreover, if $\alpha = \frac{i}{e}$, then $I_{>\alpha} = I_{\frac{i+1}{e}} = J_{i+1}$. The isomorphism $\oplus_{i \geq 0} J_i/J_{i+1} \cong T/u^{-1}T$ is clear. The last statement follows from Example 10.5.4, which gives that $I_{>0} = \sqrt{I}$.

Let z be a nilpotent element in $T/u^{-1}T$ of degree α. Let l be a positive integer such that $z^l = 0$. Let $r \in I_\alpha$ be a preimage of z in R. Then $r^l \in I_{>l\alpha}$, so that there exists a rational number $f > l\alpha$ such that $r^l \in I_f$. Then $r \in I_{\frac{f}{l}} \subseteq I_{>\alpha}$, so that $z = 0$. Thus $T/u^{-1}T$ is reduced, which proves (4). $\qquad\square$

We remark that the eth Veronese subring of T, $\oplus_{ke} T_{ke}$, is isomorphic to $\overline{S}$, as the above proof shows. This is useful for calculations of cohomology. For more geometric understanding of the topic in this section, see [169] and [170].

10.6. Exercises

10.1   Let R be a Noetherian domain.
   (i) Prove that for any ideal I and any positive integer n, $\mathcal{RV}(I^n) = \mathcal{RV}(I)$.
   (ii) Let $I_1, \ldots, I_k$ be ideals of R. Prove that $\cup_{n_1,\ldots,n_k} \mathcal{RV}(I_1^{n_1} \cdots I_k^{n_k})$ is finite.

10.2   Let R be a Noetherian domain and I and J ideals in R. If for some $m, n \in \mathbb{N}_{>0}$, $\overline{I^m} = \overline{J^n}$, then $\mathcal{RV}(I) = \mathcal{RV}(J)$. Prove that the converse fails.

10.3   Let R be a Noetherian ring and I an ideal in R. Prove that there exist $m \in \mathbb{N}_{>0}$ and $a \in I^m$ such that each Rees valuation of I is a localization of the integral closure of $R[\frac{I^m}{a}]$ at a prime ideal minimal over a.

10.4   Let I be an ideal in a Noetherian ring R and let S be the integral closure of $R[It, t^{-1}]$. Let P be a prime ideal in S minimal over $t^{-1}S$. Prove that for each $n \in \mathbb{N}$, $t^{-n} \overline{S}_P \cap R = I^n \overline{S}_P \cap R$.

10.5   (Alternative construction of Rees valuations) Let R be a Noetherian integral domain with field of fractions K, I an ideal in R, and S the integral closure of $R[It, t^{-1}]$. Prove that the set of Rees valuations of I equals the set of all $S_Q \cap K$, as Q varies over prime ideals in S that are minimal over $t^{-1}S$.

10.6   (Alternative construction of Rees valuations) Let R be a Noetherian integral domain with field of fractions K, I an ideal in R, and S the integral closure of $R[It]$. Prove that the set of Rees valuations of I equals the set of all $S_Q \cap K$, as Q varies over prime ideals in S that are minimal over IS.

10.7   Let R be a Noetherian local ring with infinite residue field. Prove that for any ideal I in R there exists $x \in I$ such that $v(x) = v(I)$ for all $v \in \mathcal{RV}(I)$.

10.8   Let R be a Noetherian local domain of dimension 1. Let S be the integral closure of R and P a maximal ideal in S. Prove that S_P is a Rees valuation of $P \cap R$.

10.9   Let k be a field, $X_1, \ldots, X_d$ variables over k, and I a monomial ideal in $R = k[X_1, \ldots, X_d]$. Let m be a monomial generator of I, and let U be the set of all valuations that are localizations of the integral closure of $R[\frac{I}{m}]$ at height one prime ideals containing m. Prove that there is a natural one–to–one correspondence between U and the set of valuations obtained from those hyperplanes bounding the Newton polyhedron of I that pass through the point corresponding to m.

10.10 Let R be a Noetherian domain and I an ideal. If $P \in \mathrm{Ass}(R/\overline{I^n})$ for some large n, then there exists $a \in I$ and a prime ideal Q in $S = R[\frac{I}{a}]$ such that Q is associated to $\overline{a^n S}$ for all large n and such that $Q \cap R = P$.

10.11 (J. Watanabe [318]) Let $(R, \mathfrak{m})$ be a Noetherian local ring with infinite residue field. Let I be an integrally closed ideal in R. Prove that I is **m–full**, i.e., that there exists $x \in \mathfrak{m}$ such that $\mathfrak{m}I : x = I$.

10.12 ([318]) Let $(R, \mathfrak{m})$ be an arbitrary Noetherian local ring and let I be an an $\mathfrak{m}$–primary ideal that is **m–full** (meaning that there exists an element $x \in R$ such that $\mathfrak{m}I : x = I$). Prove that for all ideals J containing I, $\mu(J) \le \mu(I)$.

10.13 ([318]; Goto [94]) Let $(R, \mathfrak{m})$ be a Noetherian local ring and I an $\mathfrak{m}$–full ideal. Let J be any ideal in R containing I such that J/I has finite length. Prove that $\mu(J) \le \mu(I)$.

10.14 (Goto [94]) Let R be a Noetherian ring and I an integrally closed ideal in R whose minimal number of generators r equals its height. Let P be a prime ideal in R associated to I.
   (i) Prove that P is minimal over I.
   (ii) Prove that $\lambda_{R_P}((I+P^2)R_P/P^2 R_P) \ge r-1$ and that $\mu(PR_P) \le \mu(IR_P) = r$.
   (iii) Prove that R_P is regular local ring.
   (iv) Prove that the Rees algebra $R_P[IR_P t]$ is Cohen–Macaulay and satisfies Serre's condition (R_1).
   (v) Prove that for all positive integers n, $I^n = \overline{I^n}$, i.e., that I is a normal ideal.

10.15 Prove that for ideals $I \subseteq J$ in a Noetherian ring, $\overline{I} = \overline{J}$ if and only if for every Rees valuation ring V of I, $IV = JV$.

10.16 Let X, Y, Z be variables over $\mathbb{C}$, and $R = \mathbb{C}[X, Y, Z]/(X^2 + Y^3 + Z^5)$. Prove that R is a normal domain, that $\mathfrak{m} = (X, Y, Z)R$ has only one Rees valuation, but that $\mathrm{gr}_{\mathfrak{m}}(R)$ is not an integral domain.

10.17 Let $R = \mathbb{C}[X, Y, Z]$ be a polynomial ring in variables X, Y, Z. Let n be a positive integer, and A the integral domain $R/(X^2 + Y^2 + Z^n)$.
   (i) Prove that for $i < n/2$, the ideal $(X, Y, Z^i)A$ has more than one Rees valuation.
   (ii) Prove that if n is an even integer, then $(X, Y, Z^{n/2})$ has only one Rees valuation.
   (iii) Impose on A the grading $w(X) = w(Y) = n/\gcd(2, n)$, $w(Z) = 2/\gcd(2, n)$. Prove that w is a valuation function and that for any positive integer $i \le n/2$, $(X, Y, Z^i)A = \{r \in A \,|\, w(r) \ge i\}$.

10.18 Let R be the polynomial ring $k[X_1, \ldots, X_n]$, where k is a field.
   (i) Let I be an ideal of the form $(X_1^{a_1}, \ldots, X_n^{a_n})$, where $a_1, \ldots, a_m \in \mathbb{N}$ are not all zero. Prove that I has only one Rees valuation, namely the monomial valuation with $a_i v(x_j) = a_j v(x_i)$ for all i, j.

(ii) Give an example of ideals $I_1, \ldots, I_l$ in $k[X_1, \ldots, X_n]$ that are primary to $(X_1, \ldots, X_n)$ and of the form as in (i) such that with $\overline{I_1 \cap \cdots \cap I_l} \neq \overline{I_1} \cap \cdots \cap \overline{I_l}$.

10.19 Let R be a Noetherian local ring and I an ideal with only one Rees valuation. Prove that $\sqrt{I}$ is a prime ideal.

10.20 Let I be an ideal in an integrally closed Noetherian local domain R.

(i) Prove that $\mathrm{gr}_I(R)$ is reduced if and only if I is a normal ideal and if for each (normalized integer–valued) Rees valuation v of I, $v(I) = 1$.

(ii) Let $S = R/\bar{I} \oplus \bar{I}/\overline{I^2} \oplus \overline{I^2}/\overline{I^3} \oplus \cdots$. Prove that S is a reduced ring if and only if for each (integer–valued) Rees valuation v of I, $v(I) = 1$.

10.21 (Reid, Roberts, Vitulli [243]) Let $n > 2$, $k[X_1, \ldots, X_n]$ a polynomial ring over a field, and $a_1, \ldots, a_n$ positive integers whose greatest common divisor is strictly bigger than $n - 2$. Let I be the integral closure of $(X_1^{a_1}, \ldots, X_n^{a_n})$. Prove that I is normal.

10.22 (Katz [158]) Let $(R, \mathfrak{m})$ be a formally equidimensional Noetherian local domain. Let $x_1, \ldots, x_d$ be a system of parameters. Set $I = (x_1, \ldots, x_d)$ and

$$S = R\Big[\frac{x_2}{x_1}, \ldots, \frac{x_d}{x_1}\Big]_{\mathfrak{m} R\left[\frac{x_2}{x_1}, \ldots, \frac{x_d}{x_1}\right]}.$$

Let $\bar{S}$ be the integral closure of S.

(i) Prove that for all $n \geq 1$, $\overline{I^n} = I^n \bar{S} \cap R$.

(ii) Prove that the number of minimal prime ideals in the $\mathfrak{m}$–adic completion of R is at most the number of maximal ideals in $\bar{S}$.

(iii) Prove that there is a one–to–one correspondence between maximal ideals of $\bar{S}$ and the prime ideals in the integral closure of $R[It, t^{-1}]$ that are minimal over t^{-1}.

10.23 (Sally [256]) Let $(R, \mathfrak{m})$ be an analytically unramified Noetherian local ring. Assume that there exists an $\mathfrak{m}$–primary ideal I in R with only one Rees valuation. Prove that the $\mathfrak{m}$–adic completion of R is an integral domain, i.e., that R is analytically irreducible.

10.24 (Lipman [187, page 144]) Let $(R, \mathfrak{m})$ be a formally equidimensional Noetherian local ring, and I and J ideals in R satisfying $\mathrm{ht}(I + J) = \ell(I) + \ell(J)$. Prove that $\overline{IJ} = \bar{I} \cap \bar{J}$.

10.25 Let R be a Noetherian ring, I and J ideals of R.

(i) Assume that $\bar{I} = \bar{J}$. Prove that $\overline{v}_I(J) = \overline{v}_J(I) = 1$.

(ii) Assume that $\overline{v}_I(J) = \overline{v}_J(I) = 1$. Prove that $\bar{I} = \bar{J}$.

10.26 Let R be a Noetherian ring. Ideals I and J in R are called **projectively equivalent** if there exists a positive real number α such that $\overline{v}_I = \alpha \, \overline{v}_J$.

(i) Prove that α is a rational number.

(ii) Prove that I and J are projectively equivalent if and only if there exist positive integers m and n such that $\overline{I^m} = \overline{J^n}$.

10.27 Let R be a Noetherian domain, and I an ideal. Let T be the set of all rational numbers m/n for some positive integers m, n such that there exists an ideal J in R with $\overline{I^m} = \overline{J^n}$. Prove that T is a discrete set with no limit points.

10.28 Let $(R, \mathfrak{m})$ be a Noetherian local ring whose $\mathfrak{m}$–adic completion $\hat{R}$ is an integral domain. Let S be the integral closure of $\hat{R}$. Prove that R has an $\mathfrak{m}$–primary ideal with only one Rees valuation if and only if S has a zero–dimensional ideal with only one Rees valuation.

10.29 (Criterion of analytic irreducibility, Hübl and Swanson [132]) Let $(R, \mathfrak{m})$ be a Noetherian local domain.

  (i) Prove that R is analytically irreducible if and only if there exist positive integers a and b such that for all positive integers n and all $x, y \in R$, $xy \in \mathfrak{m}^{an+b}$ implies that either x or y is in $\mathfrak{m}^n$.

  (ii) Let $(R, \mathfrak{m})$ be a complete local domain of dimension 1, and let $\overline{R}$ be its integral closure. Let $\mathfrak{n}$ be the maximal ideal of $\overline{R}$, and let $\mathfrak{n}^a$ be the conductor ideal of R (why is there only one maximal ideal, and why is the conductor of this form?). Let I be a non–zero ideal in R, and i such that $I\overline{R} = \mathfrak{n}^i$. Prove that whenever $x, y \in R$ are such that $xy \in I^{2n + \lfloor 2a/i \rfloor}$, then either $x \in I^n$ or $y \in I^n$.

10.30 ([132]) Let $(R, \mathfrak{m})$ be a Noetherian local analytically irreducible domain, and I an $\mathfrak{m}$–primary ideal. Prove that I has only one Rees valuation if and only if there exists an integer b such that for all positive integers n and all $x, y \in R$, $xy \in I^{2n+b}$ implies that either x or y lies in I^n.

10.31 (Muhly and Sakuma [211, Lemma 4.1]) Let $I_1, \ldots, I_r$ be ideals in a two–dimensional universally catenary Noetherian integral domain R. Suppose that for $j = 1, \ldots, r$, $\mathcal{RV}(I_j) = \{v_j\}$, and that $v_1, \ldots, v_r$ are pairwise not equivalent. Prove that $\det(v_i(I_j)_{i,j}) \neq 0$.

10.32 Let $R = \mathbb{Q}[X, Y]$, and $I = (X^2Y + XY^2, Y^4, XY^3, X^4)$.

  (i) Prove that for all n, I^n is integrally closed.

  (ii) Prove that $J = (X^2Y + XY^2, Y^4 + XY^3 + X^4)$ is a minimal reduction of I.

  (iii) Let $S = R[\frac{I}{(X^2Y + XY^2)}]$. Prove that each Rees valuation of I equals S_Q for some prime ideal Q in S minimal over $X^2Y + XY^2$.

  (iv) Find all the Rees valuations of I.

  (v) For each Rees valuation v of I, compute $v(I)$.

10.33* Find the Rees valuations of generic determinantal ideals.

11
Multiplicity and integral closure

This chapter is devoted to the theory of multiplicity and its relationship to integral closure. We begin by developing the theory of Hilbert–Samuel polynomials. The basic result states that if $(R, \mathfrak{m})$ is a Noetherian local ring, I an $\mathfrak{m}$–primary ideal, and M a finitely generated R–module, then there exists a polynomial $P(n)$ with rational coefficients, such that for all sufficiently large n,

$$P(n) = \lambda_R(M/I^n M).$$

The leading coefficient of $P(n)$, suitably normalized, is an important invariant of I and M, called the multiplicity of I with respect to M. This invariant gives information concerning how (the integral closures of) the powers of I grow with respect to M. The main result in this chapter, Theorem 11.3.1, is a famous theorem due to David Rees, which states that if $(R, \mathfrak{m})$ is a formally equidimensional Noetherian local ring and $J \subseteq I$ are $\mathfrak{m}$–primary ideals, then $\overline{J} = \overline{I}$ if and only if the multiplicities of J and I with respect to R are the same. The general proof we give is due to Schaub [258]. For the case when R contains a field, we give another and easier proof, due to Scheja [259]. This proof introduces an important technique. The final section presents an algebraic approach, due to Lipman [187], related to Teissier's "principle of specialization of integral dependence".

11.1. Hilbert–Samuel polynomials

Throughout this section we let $(R, \mathfrak{m})$ be a Noetherian local ring of dimension d, I an $\mathfrak{m}$–primary ideal, and M a finitely generated R–module. We begin with two well–known lemmas concerning polynomials, which we will use in an inductive step.

Define a polynomial $\binom{t+i}{i} \in \mathbb{Q}[t]$ as $\binom{t+i}{i} = \frac{(t+i)(t+i-1)\cdots(t+1)}{i!}$. This is a polynomial of degree i in t, with leading coefficient $1/i!$. For any overfield K of $\mathbb{Q}$, a polynomial $P : \mathbb{N} \to K$ of degree d can be written uniquely as $P(n) = \sum_{i=0}^{d} k_i \binom{n+i}{i}$, with $k_i \in K$, and $k_d \neq 0$.

Lemma 11.1.1 *Let $P(t) \in \mathbb{Q}[t]$ be a polynomial of degree $d \geq 0$. (We use the convention that the degree of the constant zero polynomial is $-\infty$.) The following are equivalent:*
(1) $P(n) \in \mathbb{N}$ for all $n \gg 0$.
(2) There exist unique integers $a_1, \ldots, a_d$ such that $a_d > 0$ and $P(t) = \sum_{i=0}^{d} a_i \binom{t+i}{i}$.

Proof: It is easy to see that (2) implies (1). For the converse, observe that the set of polynomials $\binom{t+i}{i}$ forms a $\mathbb{Q}$–basis for $\mathbb{Q}[t]$, so that $P(t) = \sum_{i=0}^{d} a_i \binom{t+i}{i}$ for unique $a_i \in \mathbb{Q}$. We leave it as an exercise that (1) then forces $a_i \in \mathbb{Z}$ (Exercise 11.1). The leading coefficient must be positive. □

Lemma 11.1.2 *Let R be a domain containing $\mathbb{Q}$, and let $Q(n)$ be a polynomial in n of degree $d \geq 0$ with coefficients in R. Fix an integer j, and set $P(n) = \sum_{i=j}^{n} Q(i)$. Then $P(n)$ is a polynomial in n of degree equal to $\deg Q + 1$ and with coefficients in R. If the leading coefficient of Q is c, then the leading coefficient of P is $c/(\deg(Q) + 1)$.*

Proof: By assumption $Q(n)$ is a non–zero polynomial. By shifting the index and changing the sum by a constant, it is enough to prove that $\sum_{i=0}^{n} Q(i)$ is a polynomial in n of degree one greater than $Q(n)$ with the specified leading coefficient. Write $Q(n) = \sum_{j=0}^{d} a_j \binom{n+j}{j}$, for some $a_j \in \mathbb{Q}$, $a_d \neq 0$. Then

$$P(n) = \sum_{i=0}^{n} Q(i) = \sum_{j=0}^{d} a_j \sum_{i=0}^{n} \binom{i+j}{j} = \sum_{j=0}^{d} a_j \binom{n+j+1}{j+1},$$

so that $P(n)$ is a polynomial of degree $\deg Q + 1$, with coefficients in R. Clearly $c = a_d/d!$, and the leading coefficient of P is $a_d/(d+1)!$. □

Theorem 11.1.3 (The Hilbert–Samuel Polynomial) *Let $(R, \mathfrak{m})$ be a Noetherian local ring, let I be an $\mathfrak{m}$–primary ideal, and let M be a non–zero finitely generated R–module. Set $d = \dim(R)$. There exists a polynomial $P(n)$, with rational coefficients, such that for all $n \gg 0$,*

$$P(n) = \lambda_R(M/I^n M).$$

Furthermore, the degree in n of $P(n)$ is $\dim M$, which is at most d.

We call $\lambda(M/I^n M)$ the **Hilbert function** and $P(n)$ the **Hilbert–Samuel** polynomial of I with respect to M. If the dependence of $P(n)$ on I and M needs to be specified, we write $P_{I,M}(n)$.

Proof: Using Lemma 8.4.2 we may assume that the residue field k of R is infinite. We prove the theorem by induction on $d = \dim(M)$.

If $\dim M = 0$, then for all large n, $I^n M = 0$, and $P(n) = \lambda(M)$, the constant polynomial. Clearly in this case the degree of the polynomial is 0, equal to the dimension of M.

Assume that $\dim M > 0$. By Proposition 8.5.7, there exists a superficial element $x \in I$ with respect to M, and not contained in any prime ideal that is minimal over $\operatorname{ann} M$. For $n > 0$, the exact sequence

$$0 \to \frac{I^n M :_M x}{I^{n-1}M} \to \frac{M}{I^{n-1}M} \xrightarrow{x} \frac{M}{I^n M} \to \frac{M}{xM + I^n M} \to 0$$

of finite–length modules says that

$$\lambda\left(\frac{M}{I^n M}\right) - \lambda\left(\frac{M}{I^{n-1}M}\right) = \lambda\left(\frac{M}{xM + I^n M}\right) - \lambda\left(\frac{I^n M :_M x}{I^{n-1}M}\right).$$

By Lemma 8.5.5, for large n, say for $n \geq c$, $I^n M :_M x = 0 :_M x + I^{n-1}M$ and $(0 :_M x) \cap I^{n-1}M = 0$, so that $\frac{I^n M :_M x}{I^{n-1}M} \cong 0 :_M x$. It follows that

$$\lambda\left(\frac{M}{I^n M}\right) = \lambda\left(\frac{M}{I^c M}\right) + \sum_{i=c+1}^{n} \left(\lambda\left(\frac{M}{xM + I^i M}\right) - \lambda(0 :_M x)\right). \quad (11.1.4)$$

Since x is not in any minimal prime ideal over ann M, $\dim(M/xM) = \dim M - 1$, so that by induction on dimension, for all large n, $\lambda(M/(xM + I^n M))$ equals a polynomial $Q(n)$ of degree $\dim M - 1$, with rational coefficients. If $Q(n) = \lambda(0 :_M x)$ for all large n, then by this formula, $\lambda(M/I^n M)$ is a constant function for large n. It follows that $\lambda(I^n M/I^{n+1}M) = 0$ for large n, so that by Nakayama's Lemma, $I^n M = 0$, whence $\dim M = 0$, contradicting the assumption that the dimension of M was positive. Hence necessarily $Q(n) \neq \lambda(0 :_M x)$ for infinitely many n. But $Q(n)$ is a non–decreasing function, so that $Q(n) \neq \lambda(0 :_M x)$ for all large n. Thus by the displayed formula and by Lemma 11.1.2, for all large n, $\lambda(M/I^n M)$ equals a polynomial of degree $\dim M$ with rational coefficients. $\qquad \square$

Observe that for all n, $\lambda_R(M/I^n M) = \lambda_{R/\mathrm{ann}(M)}(M/I^n M)$, so that the Hilbert–Samuel polynomials of I and of $I(R/\mathrm{ann}(M))$ with respect to M are identical.

Definition 11.1.5 *Let $(R, \mathfrak{m})$ be a Noetherian local ring of dimension d, let I be an $\mathfrak{m}$–primary ideal, and let M be a finitely generated R–module. We define $e_R(I; M)$, the* **multiplicity of I on M**, *to be $d!$ times the coefficient of $P_{I,M}(n)$ of degree d. If $M = R$, we simply write $e(I)$ for $e_R(I; R)$, while if $M = R$ and $I = \mathfrak{m}$, we write $e(R) = e(\mathfrak{m})$, the* **multiplicity of R**.

In other words, $e(I; M) = \lim_{n \to \infty} \lambda(M/I^n M)d!/n^d$, where $d = \dim R$. By Lemma 11.1.1, $P_{I,M}(n)$ can be written in the following standard form:

$$P_{I,M}(n) = \sum_{i=0}^{\dim M} (-1)^{\dim M - i} \cdot e_{\dim M - i} \cdot \binom{n+i}{i},$$

where $e_0, \ldots, e_{\dim M} \in \mathbb{Z}$, and $e_0 > 0$. We conclude that $e_R(I; M) = 0$ if $\dim M < \dim R$, and $e_R(I; M) = e_0$ if $\dim M = \dim R$.

In the definition of multiplicity, we could have replaced d, the dimension of R, with the dimension of M. However, for purposes in this book, it is convenient to have the multiplicity be 0 when the dimension of M is less than the dimension of R.

Discussion 11.1.6 Let $(R, \mathfrak{m})$ be a Noetherian local ring of dimension d, M a finitely generated R–module of dimension d, and let I be an $\mathfrak{m}$–primary ideal. In computing $e(I; M)$ we can either use the polynomial $P(n)$ such that $P(n) = \lambda(M/I^n M)$ for all large n, whose leading coefficient is $\frac{e(I;M)}{d!}$, or use the function $\lambda(I^n M/I^{n+1}M)$, which is also a polynomial for large n of degree $d - 1$ whose leading coefficient is $\frac{e(I;M)}{(d-1)!}$. The function $\lambda(I^n M/I^{n+1}M)$ agrees with a polynomial for large n because $\lambda(I^n M/I^{n+1}M) = P(n+1) - P(n)$, and its leading coefficient is $\frac{e(I;M)}{(d-1)!}$ by Lemma 11.1.2.

The proof of Theorem 11.1.3 can be extended to show more:

Lemma 11.1.7 Let $(R, \mathfrak{m})$ be a d–dimensional Noetherian local ring, I an $\mathfrak{m}$–primary ideal, $x \in I$, and M a finitely generated R–module of dimension $d = \dim R$. Set $M' = M/xM$, $R' = R/xR$ or $R' = R/\mathrm{ann}(M')$, and $I' = IR'$. Then the following hold.

(1) If $\dim M' = d - 1$, then the leading coefficient of $(d-1)!P_{I,M'}$ is bigger than or equal to the leading coefficient of $d!P_{I,M}$, and for $n \gg 0$, $\lambda((I^n M :_M x)/I^{n-1}M)$ is a polynomial in n of degree at most $d - 1$ with rational coefficients.

(2) If $\dim R' = d - 1$, then $e_{R'}(I'; M') \geq e_R(I; M)$, and equality holds if and only if $\lambda((I^n M :_M x)/I^{n-1}M)$ is a polynomial in n of degree at most $d - 2$ for all large n.

Proof: If $\dim R' = d - 1$, necessarily $\dim M' = d - 1$. (The other implication need not hold.)

By Theorem 11.1.3, the polynomials $P_{I,M'}$ and $P_{I,M}$ have degrees $d-1$ and d, respectively. Let $g(n) = \lambda((I^n M :_M x)/I^{n-1}M)$. The short exact sequence

$$0 \to \frac{I^n M :_M x}{I^{n-1}M} \to \frac{M}{I^{n-1}M} \xrightarrow{x} \frac{M}{I^n M} \to \frac{M}{xM + I^n M} \to 0 \qquad (11.1.8)$$

gives that $g(n) = \lambda(M/(xM + I^n M)) + \lambda(M/I^{n-1}M) - \lambda(M/I^n M)$. Hence for all large n, $g(n) = P_{I,M'}(n) - P_{I,M}(n) + P_{I,M}(n-1)$ is a polynomial of degree at most $d - 1$, as the degree d–parts cancel. Furthermore, the degree $(d-1)$ part says that the difference of the leading coefficient of $(d-1)!P_{I,M'}$ minus the leading coefficient of $d!P_{I,M}$ is non–negative, and is zero exactly when $g(n)$ is a polynomial of degree at most $d - 2$. Part (2) is simply a translation of (1) into multiplicities. $\qquad \square$

Here are some cases where Lemma 11.1.7 applies:

Proposition 11.1.9 Let $(R, \mathfrak{m})$ be a Noetherian local ring, M a finitely generated R–module of dimension $d = \dim R > 0$, and I an $\mathfrak{m}$–primary ideal. Let $x \in I$, set $M' = M/xM$, and $R' = R/xR$ or $R' = R/\mathrm{ann}(M')$, and $I' = IR'$. Suppose that one of the following conditions hold:

(1) $x \in I$ is superficial with respect to M and not contained in any minimal prime ideal.

(2) The dimension of R' is $d - 1$, I has d generators, $x \in I \setminus \mathfrak{m}I$, and x is a non–zerodivisor on M or x is a superficial element for some $\mathfrak{m}$–primary ideal with respect to M.

Then

$$e_R(I; M) = \begin{cases} e_{R'}(I'; M') & \text{if } d > 1; \\ \lambda(M') - \lambda(0 :_M x) & \text{if } d = 1. \end{cases}$$

Proof: In case x is superficial for I with respect to M and not contained in any minimal prime ideal, it follows from Equation (11.1.4) and Lemma 11.1.2 that $e_R(I; M) = e_{R'}(I'; M')$ if $d > 1$, and $e_R(I; M) = e_{R'}(I'; M') - \lambda(0 :_M x)$ $= \lambda(M') - \lambda(0 :_M x)$ if $d = 1$. This proves the case (1), and also (2) if $d = 1$.

Now let $d > 1$ with set–up as in (2). Write $I = (x, x_2, \ldots, x_d)$, $J = (x_2, \ldots, x_d)$. By Lemma 8.5.5, there exists $e \in \mathbb{N}$ such that $(0 :_M x) \cap I^e M = 0$. As $I = (x) + J$, it follows that $I^n M :_M x = (xI^{n-1}M + J^n M) :_M x = I^{n-1}M + (J^n M :_M x)$. By the Artin–Rees Lemma, there exists $c \in \mathbb{N}$ such that for all $n \geq c$, $J^n M \cap xM \subseteq xJ^{n-c}M$. Thus for all $n \geq c$, $J^n M :_M x \subseteq J^{n-c}M + (0 :_M x)$, whence $I^n M :_M x \subseteq I^{n-1}M + J^{n-c}M + (0 :_M x)$, and

$$\frac{I^n M :_M x}{I^{n-1}M} \subseteq \frac{I^{n-1}M + J^{n-c}M}{I^{n-1}M} + \frac{I^{n-1}M + (0 :_M x)}{I^{n-1}M}.$$

For large n, by Lemma 8.5.5, the second module on the right is isomorphic to $(0 :_M x)$. The first module on the right is a module over R/I^c and its length is bounded above by $\mu(J^{n-c}M)\lambda(R/(I^c + \operatorname{ann}(M)))$, which in turn is bounded above by $\mu(M)\binom{n-c+d-2}{d-2}\lambda(R/(I^c + \operatorname{ann}(M)))$, which is a polynomial of degree at most $d - 2$. Thus for $d > 1$, the length of $(I^n M :_M x)/I^{n-1}M$ is for large n a polynomial of degree at most $d - 2$, so that by Lemma 11.1.7, $e_R(I; M) = e_{R'}(I'; M')$. $\qquad \square$

Proposition 11.1.10 *Let $(R, \mathfrak{m})$ be a Noetherian local ring of dimension d, $I = (x_1, \ldots, x_d)$ an $\mathfrak{m}$–primary ideal.*
(1) If M is a finitely generated R–module, then for all n, $\lambda(M/I^n M) \leq \lambda(M/IM)\binom{n+d-1}{d}$, and $e(I; M) \leq \lambda(M/IM)$.
(2) If R is Cohen–Macaulay, then for all n, $\lambda(R/I^n R) = \lambda(R/IR)\binom{n+d-1}{d}$ and $e(I; R) = \lambda(R/I)$.

Proof: Clearly $\lambda(M/I^n M) = \sum_{i=0}^{n-1} \lambda(I^i M/I^{i+1}M) \leq \sum_{i=0}^{n-1} \lambda(M/IM)\mu(I^i)$ $\leq \sum_{i=0}^{n-1} \lambda(M/IM)\binom{d+i-1}{d-1} = \lambda(M/IM)\binom{n+d-1}{d}$. If R is Cohen–Macaulay, then I is generated by a regular sequence, so the associated graded ring $R/I \oplus I/I^2 \oplus \cdots$ is isomorphic to a polynomial ring in d variables over R/I (see Corollary 5.5.9). In particular, I^i/I^{i+1} is a free (R/I)–module whose rank over R/I is exactly the number of monomials in d–variables of degree i. Hence the inequalities above, repeated with $M = R$, are all equalities. The rest follows easily. $\qquad \square$

An important theorem is that the converse statement in this proposition also is true. This is due to Serre [265, Appendice II]; see Exercise 11.7.

Example 11.1.11 *If $(R, \mathfrak{m})$ is a regular local ring, then $e(R) = 1$.*

Proof: Since $\mathfrak{m}$ is generated by a regular sequence we can apply Proposition 11.1.10 to see that $e(R) = \lambda(R/\mathfrak{m}) = 1$. $\qquad \square$

The converse of the statement in this example also holds if R is formally equidimensional. This is due to Nagata [213, Theorem 40.6]. For a proof, see

Exercise 11.8. If R is not equidimensional, then the multiplicity can be 1 even if R is not regular. A reader may verify this on $R = k[[X, Y, Z]]/(XY, XZ)$.

Remark 11.1.12   We defined multiplicities in Noetherian local rings. One gets a similar theory also on standard graded rings as follows. Let $A = \oplus_{n \geq 0} A_n$ be an $\mathbb{N}$–graded Noetherian ring, generated over the Artinian local ring A_0 by elements of A_1. For each $n \geq 0$, A_n is a finitely generated A_0–module, so it has finite A_0–length. We show that there exists a polynomial $Q(n)$ with rational coefficients such that for all large n, $Q(n) = \lambda_{A_0}(A_n)$. As $\lambda(A_n) = \lambda(\oplus_{i=0}^n A_i) - \lambda(\oplus_{i=0}^{n-1} A_i)$, it suffices to prove that $\lambda(\oplus_{i=0}^n A_i)$ equals a polynomial in n for large n. If $\mathfrak{m}$ is the maximal ideal of A_0, then $\mathfrak{M} = \mathfrak{m}A + A_1 A$ is the unique homogeneous maximal ideal of A, and $(A/\mathfrak{M}^{n+1})_{\mathfrak{M}} = A/\mathfrak{M}^{n+1} = \oplus_{i=0}^n A_i$. Thus by Theorem 11.1.3, $\lambda(\oplus_{i=0}^n A_i)$ equals a polynomial in n for large n, with rational coefficients, of degree $d = \dim(A_{\mathfrak{M}})$, and with leading coefficient $e(\mathfrak{M}; A)/d!$. It follows that $Q(n)$ has degree $d - 1 = \operatorname{ht} \mathfrak{M} - 1$, and that the leading coefficient is $e(\mathfrak{M}; A)/(d-1)!$.

11.2. Multiplicity

In this section we introduce several methods for manipulating and computing multiplicity. We prove that the multiplicity of a hypersurface variety is given by the order of the defining polynomial (Example 11.2.8) and that the multiplicity of ideals in Cohen–Macaulay rings are the co–lengths of certain parameter ideals (Propositions 11.2.1 and 11.2.2). The end of the section gives a geometric interpretation.

Proposition 11.2.1   *Let $(R, \mathfrak{m})$ be a Noetherian local ring, and assume that J and I are $\mathfrak{m}$–primary ideals having the same integral closure. Let M be a finitely generated R–module. Then $e(I; M) = e(J; M)$.*

Proof: It suffices to prove the proposition for the case $I = \overline{J}$. Then J is a reduction of I, and there exists an integer k such that $I^{k+1} = JI^k$. Hence for all $n \geq k+1$, $I^n = J^{n-k} \cdot I^k$, and $\lambda(M/J^n M) \geq \lambda(M/I^n M) \geq \lambda(M/J^{n-k} M)$ for all $n \geq k$. We know that for large n, $\lambda(M/J^n M) = \frac{e(J;M)}{d!} n^d + O(n^{d-1})$, where $d = \dim R$. Similarly, $\lambda(M/I^n M) = \frac{e(I;M)}{d!} n^d + O(n^{d-1})$. Note that $\lambda(M/J^{n-k} M) = \frac{e(J;M)}{d!} n^d + O(n^{d-1})$ as well, since $(n - k)^d = n^d + O(n^{d-1})$. After dividing by n^d and taking limits as $n \to \infty$, the proposition follows. $\square$

Proposition 11.2.2   *Let $(R, \mathfrak{m})$ be a Cohen–Macaulay Noetherian local ring with infinite residue field, and let I be an $\mathfrak{m}$–primary ideal. If J is an arbitrary minimal reduction of I, then $e(I; R) = \lambda(R/J)$.*

Proof: Since the residue field of R is infinite and R is Cohen–Macaulay, J is generated by a regular sequence. By Proposition 11.2.1, $e(I) = e(J)$, and by Proposition 11.1.10, $e(J) = \lambda(R/J)$. $\square$

Theorem 11.2.3 *Let $(R, \mathfrak{m})$ be a Noetherian local ring, and let I be an $\mathfrak{m}$–primary ideal. If $0 \to K \to M \to N \to 0$ is a short exact sequence of finitely generated R–modules, then $e(I; M) = e(I; K) + e(I; N)$.*

Proof: By tensoring the short exact sequence in the statement of the theorem with R/I^n, we get a right exact sequence $K/I^n K \to M/I^n M \to N/I^n N \to 0$, which proves that $\lambda(M/I^n M) \leq \lambda(N/I^n N) + \lambda(K/I^n K)$.

On the other hand, by the Artin–Rees Lemma, there exists an integer q such that for all $n \geq q$, $I^n M \cap K \subseteq I^{n-q} K$, and in particular, $\lambda(K/(I^n M \cap K)) \geq \lambda(K/I^{n-q} K)$. The sequence $0 \to K/(I^n M \cap K) \to M/I^n M \to N/I^n N \to 0$ is exact. Combining the above remarks, we see that

$$\lambda(N/I^n N) + \lambda(K/I^{n-q} K) \leq \lambda(M/I^n M) \leq \lambda(N/I^n N) + \lambda(K/I^n K).$$

Since $\lim_{n \to \infty} \lambda(K/I^n K) d!/n^d = \lim_{n \to \infty} \lambda(K/I^{n-q} K) d!/n^d$, we obtain that $e(I; M) = e(I; N) + e(I; K)$. $\square$

Theorem 11.2.4 (Associativity Formula) *Let $(R, \mathfrak{m})$ be a local Noetherian ring, let I be an $\mathfrak{m}$–primary ideal, and let M be a finitely generated R–module. Let Λ be the set of minimal prime ideals P of R such that $\dim(R/P) = \dim(R)$. Then*

$$e(I; M) = \sum_{P \in \Lambda} e(I; R/P) \lambda(M_P).$$

Proof: By Theorem 11.2.3, multiplicity is additive on short exact sequences. Fix a prime filtration of M, say $0 = M_0 \subseteq M_1 \subseteq M_2 \subseteq \cdots \subseteq M_n = M$, where $M_{i+1}/M_i \cong R/P_i$ (P_i a prime) for all $0 \leq i \leq n - 1$. As $e(I; R/Q) = 0$ if $\dim(R/Q) < \dim(R)$, the additivity of multiplicity applied to this filtration shows that $e(I; M)$ is a sum of the $e(I; R/P)$ for $P \in \Lambda$, counted as many times as R/P appears as some M_{i+1}/M_i. We count this number by localizing at P. In this case, we have a filtration of M_P, where all terms collapse except for those in which $(M_{i+1}/M_i)_P \cong (R/P)_P$, and the number of such copies is exactly the length of M_P. $\square$

Lemma 11.2.5 *Let $(R/\mathfrak{m})$ be a Noetherian local ring, and let M and N be finitely generated R–modules. Assume that I is an $\mathfrak{m}$–primary ideal. Let W be the complement of the set of minimal prime ideals of maximal dimension. If $W^{-1} M \cong W^{-1} N$, then $e(I; M) = e(I; N)$.*

Proof: Assume that $M_W \cong N_W$. There is a homomorphism from M to N such that the cokernel C is annihilated by some element of W. Tensoring the exact sequence $M \to N \to C \to 0$ with R/I^n yields an exact sequence that shows that the length of $N/I^n N$ is at most the sum of the lengths of $M/I^n M$ and $C/I^n C$. Since $\dim C < \dim R$, we obtain that $e_R(I; N) \leq e_R(I; M)$. Reversing the roles of M and N finishes the proof. $\square$

Corollary 11.2.6 *Let $(R, \mathfrak{m})$ be a Noetherian local domain, I an $\mathfrak{m}$–primary ideal, and M a finitely generated R–module. Then $e(I; M) = e(I; R) \operatorname{rk}_R M$.*

Proof: Let $W = R \setminus 0$ and $r = \mathrm{rk}_R M$. Then $r = \dim_{W^{-1}R}(W^{-1}RM)$ and by Lemma 11.2.5, $W^{-1}M \cong K^r \cong W^{-1}R^r$, and the corollary follows. $\qquad\square$

Theorem 11.2.7 *Let $(R, \mathfrak{m}, k)$ be a d–dimensional local Noetherian domain with field of fractions K, let I be an $\mathfrak{m}$–primary ideal, and suppose that S is a module–finite extension domain of R with field of fractions L. Then*

$$e_R(I) = \sum_{Q \in \mathrm{Max}(S), \dim S_Q = d} \frac{e_{S_Q}(IS_Q)[S/Q : k]}{[L : K]}.$$

Proof: Let $W = R \backslash 0$. Since $W^{-1}S \cong W^{-1}R^{[L:K]}$, we can apply Lemma 11.2.5 to conclude that $e_R(I; S) = e_R(I)[L : K]$. On the other hand,

$$e_R(I; S) = \lim_{n \to \infty} \frac{\lambda_R(S/I^n S)d!}{n^d}.$$

As every maximal ideal Q of S contains $\mathfrak{m}S$, the Chinese Remainder Theorem implies that $S/I^n S \cong \prod_{Q \in \mathrm{Max}(S)} S_Q/I^n S_Q$. In particular, $\lambda_R(S/I^n S) = \sum_{Q \in \mathrm{Max}(S)} \lambda_R(S_Q/I^n S_Q) = \sum_{Q \in \mathrm{Max}(S)} \lambda_{S_Q}(S_Q/I^n S_Q)[S/Q : k]$. Therefore $e_R(I; S)$ equals

$$\lim_{n \to \infty} \sum_{Q \in \mathrm{Max}(S)} \frac{\lambda_{S_Q}(S_Q/I^n S_Q)[S/Q : k]d!}{n^d}$$

$$= \lim_{n \to \infty} \sum_{\dim S_Q = d} \frac{\lambda_{S_Q}(S_Q/I^n S_Q)[S/Q : k]d!}{n^d}.$$

Hence

$$e_R(I) = \sum_{Q \in \mathrm{Max}(S), \dim S_Q = d} \frac{e_{S_Q}(IS_Q)[S/Q : k]}{[L : K]}. \qquad\square$$

Example 11.2.8   Let $(R, \mathfrak{m})$ be a regular local ring of dimension d, and let $f \in \mathfrak{m}$. Then $e(R/(f)) = \mathrm{ord}(f)$.

Proof: We will use that $\lambda(\mathfrak{m}^n/\mathfrak{m}^{n+1}) = \binom{n+d-1}{d-1}$, a polynomial of degree $d-1$ in n. This follows since the associated graded ring of R is a polynomial ring in d variables.

By Discussion 11.1.6, it suffices to calculate the polynomial giving the length of $(\mathfrak{m}^n + (f))/(\mathfrak{m}^{n+1} + (f))$. This module is isomorphic to $\mathfrak{m}^n/(\mathfrak{m}^{n+1} + ((f) \cap \mathfrak{m}^n)) = \mathfrak{m}^n/(\mathfrak{m}^{n+1} + f\mathfrak{m}^{n-t})$, where $t = \mathrm{ord}(f)$. The equality $(f) \cap \mathfrak{m}^n = f\mathfrak{m}^{n-t}$ follows from the fact that the leading form of f in the associated graded ring of R is a non–zerodivisor since this associated graded ring is a polynomial ring, hence a domain. There is a short exact sequence,

$$0 \to \mathfrak{m}^{n-t}/\mathfrak{m}^{n-t+1} \xrightarrow{f} \mathfrak{m}^n/\mathfrak{m}^{n+1} \to \mathfrak{m}^n/(\mathfrak{m}^{n+1} + f\mathfrak{m}^{n-t}) \to 0.$$

Hence $\lambda(\mathfrak{m}^n/(\mathfrak{m}^{n+1} + f\mathfrak{m}^{n-t})) = \lambda(\mathfrak{m}^n/\mathfrak{m}^{n+1}) - \lambda(\mathfrak{m}^{n-t}/\mathfrak{m}^{n-t+1})$, which equals $\binom{n+d-1}{d-1} - \binom{n+d-t-1}{d-1}$, a polynomial of degree $d - 2$ with leading coefficient $\frac{t}{(d-2)!}$. $\qquad\square$

Proposition 11.2.9 *Let $(R, \mathfrak{m})$ be a Noetherian local ring of dimension d, M a finitely generated R–module, and I an $\mathfrak{m}$–primary ideal.*
(1) If l is a positive integer, then $e(I^l; M) = l^d e(I; M)$.
(2) If $I = (x_1, \ldots, x_d)$, then for any $l_1, \ldots, l_d \in \mathbb{N}_{>0}$, $e((x_1^{l_1}, \ldots, x_d^{l_2}); M) = l_1 \cdots l_d\, e((x_1, \ldots, x_d); M)$.

Proof: If $\dim M < d$, all multiplicities are 0. So we may assume that $\dim M = d$. Part (1) follows from

$$e(I^l; M) = \lim_{n \to \infty} \frac{d!}{n^d} \lambda(M/I^{ln} M) = l^d \lim_{n \to \infty} \frac{d!}{(ln)^d} \lambda(M/I^{ln} M) = l^d e(I; M).$$

Part (2) follows immediately if $d = 1$. Now let $d > 1$. If (2) holds for $M = R/P$, where P is a prime ideal, then by the Associativity Formula (Theorem 11.2.4) it also holds for M. Thus it suffices to prove (2) in case $M = R$ is an integral domain. It also suffices to prove the case $l_2 = \cdots = l_d = 1$. Set $l = l_1$. Let $I = (x_1, \ldots, x_d)$ and $J = (x_1^l, x_2, \ldots, x_d)$. Let $(\_)'$ denote images in $R/x_d R$. By Proposition 11.1.9, $e_R(J; R) = e_{R'}(J'; R')$, and $e_R(I; R) = e_{R'}(I'; R')$. By induction on d, $e_{R'}(J'; R') = l\, e_{R'}(I'; R')$, whence (2) holds. $\square$

Theorem 11.2.10 (Lech's Formula [180]) *Let $(R, \mathfrak{m})$ be a d–dimensional Noetherian local ring, M a finitely generated R–module, $I = (x_1, \ldots, x_d)$ an $\mathfrak{m}$–primary ideal. Then*

$$\lim_{n_1, \ldots, n_d \to \infty} \frac{\lambda\left(\frac{M}{(x_1^{n_1}, \ldots, x_d^{n_d})M}\right)}{n_1 \cdots n_d} = e((x_1, \ldots, x_d); M).$$

(We mean limit as all n_i go to infinity, along any path.)

Proof: We use induction on the dimension d. The case $d = 1$ follows from the definition. So we may assume that $d > 1$.

By Proposition 11.1.10, for all $n_i > 0$,

$$e((x_1^{n_1}, \ldots, x_d^{n_d}); M) \le \lambda(M/(x_1^{n_1}, \ldots, x_d^{n_d})M).$$

Thus by Proposition 11.2.9,

$$e((x_1, \ldots, x_d); M) = \lim_{n_1, \ldots, n_d \to \infty} \frac{e((x_1^{n_1}, \ldots, x_d^{n_d}); M)}{n_1 \cdots n_d}$$

$$\le \lim_{n_1, \ldots, n_d \to \infty} \frac{\lambda(M/(x_1^{n_1}, \ldots, x_d^{n_d})M)}{n_1 \cdots n_d}.$$

We switch to the shorter notation $J = (x_1^{n_1}, \ldots, x_d^{n_d})$. If $0 \to K \to M \to N \to 0$ is a short exact sequence of finitely generated R–modules, then $K/JK \to M/JM \to N/JN \to 0$ is exact, so that $\lambda(M/JM) \le \lambda(N/JN) + \lambda(K/JK)$. Thus if $0 = M_0 \subseteq M_1 \subseteq M_2 \subseteq \cdots \subseteq M_n = M$ is a prime filtration of M, then $\lambda(M/JM) \le \sum_{i=1}^n \lambda(M_i/(JM_i + M_{i-1}))$. If we know the theorem for cyclic modules R/P where P is prime, then

$$\lim_{n_1, \ldots, n_d \to \infty} \frac{\lambda(M/JM)}{n_1 \cdots n_d} \le \sum_{i=1}^n e((x_1, \ldots, x_d); M_i/M_{i-1}) = e((x_1, \ldots, x_d); M)$$

by additivity of multiplicities, which proves the theorem for M. Thus it suffices to prove the theorem in case $M = R/P$ is a domain. If the dimension of R/P is strictly less than d, then both sides of the formula are 0. Hence we may assume that R/P has dimension d. We can then change R to R/P and assume that R is an integral domain of dimension d. In particular, each x_i is a non–zerodivisor.

Set $R' = R/(x_1)$. By induction on d,
$$\lim_{n_2,\ldots,n_d \to \infty} \frac{\lambda(R'/(x_2^{n_2},\ldots,x_d^{n_d})R')}{n_2\cdots n_d} = e((x_2,\ldots,x_d)R';R').$$
By Proposition 11.1.9, $e((x_2,\ldots,x_d)R';R') = e((x_1,x_2,\ldots,x_d);R)$. By the exact sequence
$$\frac{R}{(x_1,x_2^{n_2},\ldots,x_d^{n_d})} \xrightarrow{x_1^{n_1-1}} \frac{R}{(x_1^{n_1},x_2^{n_2},\ldots,x_d^{n_d})} \to \frac{R}{(x_1^{n_1-1},x_2^{n_2},\ldots,x_d^{n_d})} \to 0,$$
the length of $R/(x_1^{n_1},\ldots,x_d^{n_d})$ is at most $n_1\lambda(R/(x_1,x_2^{n_2},\ldots,x_d^{n_d}))$. Hence
$$\lim_{n_1,\ldots,n_d \to \infty} \frac{\lambda\left(\frac{R}{(x_1^{n_1},\ldots,x_d^{n_d})R}\right)}{n_1\cdots n_d} \leq \lim_{n_1,\ldots,n_d \to \infty} \frac{n_1\lambda\left(\frac{R}{(x_1,x_2^{n_2},\ldots,x_d^{n_d})R}\right)}{n_1\cdots n_d}$$
$$= \lim_{n_1,\ldots,n_d \to \infty} \frac{\lambda\left(\frac{R}{(x_1^n,x_2^{n_2},\ldots,x_d^{n_d})R}\right)}{n_2\cdots n_d}$$
$$= e((x_1,\ldots,x_d);R). \qquad \square$$

We end this section with a geometric interpretation of multiplicity. Suppose that $(R,\mathfrak{m})$ is a Noetherian local domain of dimension d, containing an infinite field. Suppose that for every system of parameters $x_1,\ldots,x_d$ in R, there is a Noether normalization $A = k[x_1,\ldots,x_d]$ or a Cohen subring $A = k[[x_1,\ldots,x_d]]$ of R such that $A \subseteq R$ is a module–finite extension. (Noether normalization exists if R is an affine k–algebra, by Theorem 4.2.2, and a Cohen subring exists if R is complete local, by the Cohen Structure Theorem 4.3.3.) Set $I = (x_1,\ldots,x_d)A$. By Corollary 11.2.6, $e(IR) = \mathrm{rk}_A(R)$. By Proposition 11.1.9, by induction on d, if $x_1,\ldots,x_d$ are a superficial sequence in $\mathfrak{m}$, then $e(IR) = e(\mathfrak{m})$. In general, for arbitrary $x_1,\ldots,x_d$, $e(IR) \leq e(\mathfrak{m})$ (as $\lambda(R/I^n) \leq \lambda(R/\mathfrak{m}^n)$). Superficial sequences arise from Zariski–open sets, so that for all sufficiently general choices of $x_1,\ldots,x_d$, the rank of R over A is finite, and equal to $e(\mathfrak{m})$. Lipman summarized this point of view for complex analytic rings in [187] as follows (we do not explain all the terms): "For a (reduced) d–dimensional complex germ $(V,0)$ in $\mathbb{C}^m$, and most linear projections $\pi\colon(\mathbb{C}^m,0) \to (\mathbb{C}^d,0)$, with restriction $\varphi\colon(V,0) \to (\mathbb{C}^d,0)$, almost all fibers of φ near the origin have the same finite cardinality — the *degree* of φ — which is greater than or equal to the multiplicity of $(V,0)$, with equality if and only if the linear space $\pi^{-1}(0)$ has no point in common with the Zariski tangent cone of V at the origin." (The avoidance of the Zariski tangent cone in our notation in the previous paragraph is guaranteed by the construction of superficial elements.)

11.3. Rees's theorem

The goal in this section is to prove a famous result of David Rees concerning the relationship between integral closures of ideals and multiplicity. A necessary assumption in the statement of his theorem is that the ring be formally equidimensional (cf. Exercise 11.5). One way to think about formally equidimensional rings is that in such rings parameters are 'true' parameters, i.e., their images stay parameters after completing and going modulo the minimal prime ideals. This is particularly relevant in work concerning integral closures since the integral closure of an ideal is determined by the integral closure of the image of the ideal in the completion of the ring modulo the minimal prime ideals. Moreover, if one deals with $\mathfrak{m}$–primary ideals in local rings with infinite residue field, then they are integral over ideals generated by parameters, and it is important that the parameters be true parameters in the sense above.

Theorem 11.3.1 (Rees [232]) *Let $(R, \mathfrak{m})$ be a formally equidimensional Noetherian local ring and let $I \subseteq J$ be two $\mathfrak{m}$–primary ideals. Then $J \subseteq \bar{I}$ if and only if $e(I) = e(J)$.*

Proof: We present a proof due to Schaub [258]. If $J \subseteq \bar{I}$, then $\bar{J} = \bar{I}$, and by Proposition 11.2.1, $e(J) = e(I)$.

Conversely, assume that $e(I) = e(J)$. We use induction on the dimension of R to prove the theorem. If the dimension is 0, then every proper ideal is integral over every other proper ideal as all proper ideals are nilpotent. Henceforth we assume that $\dim(R) > 0$.

We reduce to the case in which R is a complete local domain with infinite residue field and all powers of I and J are integrally closed. We first apply reductions as in Section 8.4: by passing to $S = R[X]_{\mathfrak{m}R[X]}$, $e(IS) = e(I) = e(J) = e(JS)$, and by Proposition 1.6.2, if JS and IS have the same integral closure, so do I and J. Thus by possibly renaming R, we may assume that R has an infinite residue field. Observe that we may also complete R: the multiplicities do not change and since $\widehat{I\widehat{R}} \cap R = \bar{I}$ by Proposition 1.6.2, we may descend the conclusion. Next we claim it is enough to prove the theorem going modulo the minimal prime ideals of R (which we now assume is complete). Let $P_1, \ldots, P_n$ be the minimal prime ideals of R. The Associativity Formula for multiplicity, Theorem 11.2.4, shows that $e(I) = \sum_{i=1}^{n} e(I_i)\lambda(R_{P_i})$, where I_i is the image of I in $R_i = R/P_i$. In general, the sum is taken over all minimal prime ideals of maximal dimension, which under our assumptions are all of the minimal prime ideals. Similarly, $e(J) = \sum_{i=1}^{n} e(J_i)\lambda(R_{P_i})$. As $e(J_i) \leq e(I_i)$ and $e(J) = e(I)$, we must have that $e(J_i) = e(I_i)$ for each i. If we have proved the theorem for complete domains, then we can conclude that $J_i \subseteq \bar{I_i}$ for each i. Now Proposition 1.1.5 shows that $J \subseteq \bar{I}$. Since R is a complete domain, it is analytically unramified, and Corollary 9.2.1 shows that both the rings $\oplus_n \bar{I^n}$ and $\oplus_n \bar{J^n}$ are finitely generated over R, and there exists an integer k (respectively ℓ) such that for all $n \geq 1$, $(\overline{I^k})^n = \overline{I^{kn}}$ (respectively $(\overline{J^\ell})^n = \overline{J^{\ell n}}$). By replacing I by the integral closure of $I^{k\ell}$ and J by the integral closure of $J^{k\ell}$, we still will have that I is contained in J and their multiplicities are the same. If we prove that $\overline{J^{k\ell}} \subseteq \overline{I^{k\ell}}$, then it follows that J is in the integral closure of I. Hence we may assume that both I and J are normal ideals, i.e., all powers of I and J are integrally closed.

Choose a superficial element x of I. By Lemma 8.5.3, there exists a positive integer c such that for all $n \geq c$, $I^n : x = I^{n-1}$. Thus for all $n \geq c$, $I^{cn} : x^c = I^{c(n-1)}$. By replacing I with I^c, and correspondingly x with x^c, and J with J^c, we may assume that for all $n \geq 1$, $I^n : x = I^{n-1}$. By $(\_)'$ we denote images in R/Rx. Lemma 11.1.7 implies that $e(I') = e(I)$. Then we have that $e(I') \geq e(J') \geq e(J) = e(I)$, where the first inequality follows since $I' \subseteq J'$ and the second inequality follows from Lemma 11.1.7. This proves that $e(J') = e(J)$.

We claim that the image x^* of x in $\mathrm{gr}_J(R)$ is a non–zerodivisor in $\mathrm{gr}_J(R)$ of degree one. Obviously $x \in J$. Suppose that $x \in J^2$. Then $\lambda((J^n : x)/J^{n-1}) \geq \lambda(J^{n-2}/J^{n-1})$, and the latter length is given by a polynomial in n of degree $d - 1$ for large n. Using Lemma 11.1.7 this contradicts that $e(J') = e(J)$. Hence $x \notin J^2$, proving that x^* has degree one in $\mathrm{gr}_J(R)$. Consider $0 : x^* \subseteq \mathrm{gr}_J(R)$. The nth graded piece of this annihilator is $[((J^{n+1} : x) \cap J^{n-1})/J^n]$. Since $e(J') = e(J)$, Lemma 11.1.7 implies that this length is bounded by a polynomial of degree at most $d - 2$ in n. In particular, $0 : x^*$ has dimension at most $d - 1$ as a $\mathrm{gr}_J(R)$–module. This means the dimension of $\mathrm{gr}_J(R)/(0 : (0 : x^*))$ is at most $d - 1$. For every minimal prime ideal Q of $\mathrm{gr}_J(R)$, $\dim \mathrm{gr}_J(R)/Q = d$ by Proposition 5.4.8. It follows that $(0 : (0 : x^*))$ is not contained in any minimal prime of $\mathrm{gr}_J(R)$. But $\mathrm{gr}_J(R)$ has no embedded prime ideals. This follows from the fact that $\mathrm{gr}_J(R) \cong R[Jt, t^{-1}]/(t^{-1})$: if $\mathrm{gr}_J(R)$ had an embedded prime, we could lift this prime ideal to an embedded prime of (t^{-1}) in the extended Rees algebra $R[Jt, t^{-1}]$. However, our reductions were made so that $R[Jt, t^{-1}]$ is an integrally closed domain, and hence principal ideals have no embedded prime ideals (Proposition 4.10.3, as $R[Jt, t^{-1}]$ is integrally closed, hence a Krull domain). Thus $0 : x^* = 0$, so that x^* is a non–zerodivisor.

By above, $e(J') = e(I')$. Using our induction we can conclude that $J' \subseteq \overline{I'}$. Then there exists an integer n such that $(J')^n = I'(J')^{n-1}$. Lifting this equality back to R, we obtain that $J^n \subseteq IJ^{n-1} + Rx \cap J^n$. Since x^* is a non–zerodivisor of degree one in $\mathrm{gr}_J(R)$, $Rx \cap J^n = xJ^{n-1}$. Hence $J^n = IJ^{n-1} + xJ^{n-1} = IJ^{n-1}$, proving the theorem. □

We can immediately extend Rees's Theorem to ideals used in Theorem 5.4.1:

Corollary 11.3.2 (Böger [19]) *Let $(R, \mathfrak{m})$ be a formally equidimensional local ring, and let $I \subseteq J$ be two ideals such that $\ell(I) = \mathrm{ht}(I)$. Then $J \subseteq \overline{I}$ if and only if $e(I_P) = e(J_P)$ for every prime P minimal over I.*

Proof: By Ratliff's Theorem 5.4.1, every associated prime ideal of $\bar{I}$ has height at most the height of I; hence every such prime is minimal over I. One direction of the corollary is clear: if $J \subseteq \bar{I}$ then $e(I_P) = e(J_P)$ for all minimal prime ideals P over I by a direct application of the easy part of Theorem 11.3.1. Assume that $e(I_P) = e(J_P)$ for all minimal prime ideals P over I. To prove that $J \subseteq \bar{I}$, it is enough to prove that $J_P \subseteq (\bar{I})_P$ for all associated prime ideals P of $\bar{I}$, and all these are minimal by above. Another application of Theorem 11.3.1 finishes the proof. $\square$

Discussion 11.3.3   We present Scheja's treatment [259] of Rees's Theorem in the case of the ring containing a field. The assumption on the existence of a subfield avoids the technicalities in the general proof, and makes Scheja's proof more accessible.

As in the proof of Rees's Theorem 11.3.1, we reduce to the case in which R is a complete domain with infinite residue field. It is this reduction that uses the hypothesis that R be formally equidimensional. We may also replace I by a minimal reduction of itself and J by $I + Ry$ for some $y \in J$. Write $I = (x_1, \ldots, x_d)$, where the dimension of R is d. Rees's theorem comes down to proving that y is integral over I if and only the multiplicities of I and $I + Ry$ are the same. When R contains a field, this question can be reduced to the hypersurface case as follows.

Choose a coefficient field $k \subseteq R$, and set $A = k[[x_1, \ldots, x_d]]$ and $B = A[[y]] = A[y]$. These are the complete subrings of R generated by the specified elements. Of course, A is isomorphic to a power series ring in d–variables over k, and R is module–finite over A. We let $\mathfrak{n}$ be the maximal ideal of B and let $K = (x_1, \ldots, x_d)B$. The residue fields of R and B are the same. If r is the rank of R as a B–module, then by Theorem 11.2.7, $r \cdot e_B(K) = e_R(KB) = e_R(I) = e_R(J) = e_R(\mathfrak{n}R) = r \cdot e_B(\mathfrak{n})$, proving that $e_B(K) = e_B(\mathfrak{n})$.

If we prove the theorem for the ideals $K \subseteq \mathfrak{n}$ in the ring B, it follows that y is integral over K, and hence over $I = KR$. Write $B \cong A[T]/(f(T))$, where $f(T) = T^s + a_1 T^{s-1} + \cdots + a_s$ is a monic polynomial with $a_i \in A$ and $f(y) = 0$. To prove that y is integral over K, it is enough to prove that $a_i \in K^i$. Note that $e(\mathfrak{n}) = e(B)$ and $e(B)$ is exactly the order of $f(T)$ thought of in the power series ring $k[[x_1, \ldots, x_d, T]]$, by Example 11.2.8. It is enough to prove that this order is at least s, since this forces a_i to be in $(x_1, \ldots, x_d)^i A$. The multiplicity $e(K)$ is $\lambda(B/K)$ as B is Cohen–Macaulay, and this is exactly s. Since our assumption is that $e(\mathfrak{n}) = e(K) = s$, we have proved the result. $\square$

Discussion 11.3.4   For some time it was not known if there was a generalization of the results of Rees and Böger to ideals I for which $\mathrm{ht}(I) \neq \ell(I)$. A generalization was proved by Gaffney and Gassler in 1999 [87, Corollary 4.9, p. 718] using Segre numbers. Another approach was taken in 2001 by Flenner and Manaresi [75] using j-multiplicities. We give a brief review of j-multiplicities from [10]. For additional information see [76]. We also benefited from the presentation provided in Ciupercă [38] and [39].

Let $(R, \mathfrak{m})$ be a Noetherian local ring and I an ideal of R. Let G be the associated graded ring $\mathrm{gr}_I(R)$ of I, and set $A = R/I$, $\mathfrak{m}_A = \mathfrak{m}/I$. Let Γ be the submodule of G consisting of all elements that are killed by some power of $\mathfrak{m}_A$. Since G is Noetherian, there exists an integer k such that Γ is a $(G/\mathfrak{m}_A^k G)$–module. If $\dim \Gamma < \dim R$, we define $j(I) = 0$. If $\dim \Gamma = \dim R$, we define $j(I) = e(\mathfrak{M}; \Gamma)$, where $\mathfrak{M}$ is the unique homogeneous maximal ideal of $G/\mathfrak{m}_A^k G$.

Since $\dim \Gamma \leq \dim G/\mathfrak{m}^k G = \dim G/\mathfrak{m}G = \ell(I)$, it follows that if $\ell(I) < \dim R$ then $j(I) = 0$. Conversely, if $\ell(I) = \dim R$, then one can prove that $j(I) \neq 0$. It is not difficult to prove that if I is $\mathfrak{m}$–primary, $j(I) = e(I)$.

The theorem of Flenner and Manaresi [75] states:

Theorem 11.3.5 (Flenner and Manaresi [75]) *Let $I \subseteq J$ be ideals in a formally equidimensional Noetherian local ring $(R, \mathfrak{m})$. Then I is a reduction of J if and only if $j(I_P) = j(J_P)$ for all prime ideals P.*

A proof of this theorem is beyond the scope of this book. More on j-multiplicities is in the book [76] by Flenner, O'Carroll, and Vogel.

Observe that Theorem 11.3.5 does recover Böger's theorem (and therefore also Rees's multiplicity theorem) since if $\lambda(I) = \mathrm{ht}(I)$, then $j(I_P) = 0$ for all prime ideals P not minimal over I, while if $P \in \mathrm{Min}(R/I)$, $j(I_P) = e(I_P)$.

Discussion 11.3.6 Rees's theorem gives algebraic and geometric meaning to the zeroth coefficient, namely the multiplicity, in the Hilbert–Samuel polynomial of an $\mathfrak{m}$–primary ideal I in a Noetherian local ring $(R, \mathfrak{m})$. Let $d = \dim R$. We can write $P_I(n) = \sum_{i=0}^d (-1)^{d-i} e_{d-i} \binom{n+i}{i}$ in standard form and speculate what meaning the other coefficients, $e_1, \ldots, e_d$, might have. Such speculation led Shah [266] to prove the existence of unique largest ideals I_i such that the first $i+1$ Hilbert coefficients of I and i agree, namely such that $e_j(I_i) = e_j(I)$ for $0 \leq j \leq i$. These ideals $I_0, \ldots, I_d$ are called the **coefficient ideals** of I. Rees's Theorem proves that $I_0 = \bar{I}$. The last coefficient ideal of I, I_d, is also known as the Ratliff–Rush closure of I (see [229] and [112]); it is the largest ideal such that for all large N, $I^N = (I_d)^N$, often denoted $\tilde{I}$. It plays an important role in the study of the Hilbert function of ideals. Liu [195] extended the theory to modules. The coefficient ideal I_1 is closely related to the S_2–ification of the Rees algebra of I; see [38] and [39]. One can also consider the Hilbert–Samuel polynomials associated with the integral closures of powers. See for example [209], [149], or [235].

11.4. Equimultiple families of ideals

In this section we give an algebraic overview of the "principle of specialization of integral dependence", introduced by Teissier in [292] and expanded upon in [295, Appendice I]. Our treatment follows that of Lipman in [187]. We are indebted to T. Gaffney for information about this topic (see Discussion 11.4.7).

Throughout this section $\varphi : (R, \mathfrak{m}, k) \to (S, \mathfrak{n}, \ell)$ is a local homomorphism of Noetherian local rings. We assume that S is formally equidimensional and that $\dim S = \dim R + \dim(S/\mathfrak{m}S)$. The second assumption is automatically satisfied if, for example, S is flat over R (Theorem B.2.2).

We fix an ideal I in S of height c such that $\varphi^{-1}(I) = 0$ and such that S/I is a finitely generated R–module.

For all $q \in \text{Spec}(R)$, define $S(q) = S \otimes_R \kappa(q)$, and $I(q) = IS(q)$. Then $S(q)/I(q) \cong (S/I) \otimes_R \kappa(q)$. Since S/I is a finitely generated R–module, $S(q)/I(q)$ is a finitely generated $\kappa(q)$–module, hence is Artinian as a ring. We may therefore think of I as a family of finite co–length ideals parametrized by $\text{Spec}(R)$. The main point of this section is to determine conditions that guarantee and are forced by the property that this family of ideals generated by ideal I in R is **equimultiple**, i.e., that $e_q(I) = e_{\mathfrak{m}}(I)$ for every prime ideal q in R. The first lemma gives us the ability to talk about the multiplicities of this family of ideals.

Lemma 11.4.1 *Use the notation as above.*
(1) $S(q)/I(q)^n$ *is module–finite over* $\kappa(q)$.
(2) *Let* $q \in \text{Spec}(R)$, *and let* Q *be a prime ideal in* S *containing* I *and contracting to* q. *(Such prime ideals always exist since* $R \hookrightarrow S/I$ *is a finite map.) Then* $c = \dim S - \dim R = \dim(S_Q/qS_Q)$.
(3) *There exists an integer* $e_q(I)$ *such that*

$$\dim_{\kappa(q)}(S(q)/I(q)^n) = \frac{e_q(I)}{c!}n^c + O(n^{c-1}).$$

(4) *Explicitly,*

$$e_q(I) = \sum_{I \subseteq Q, \varphi^{-1}(Q)=q} [\kappa(Q) : \kappa(q)]\, e(I(S_Q/qS_Q)).$$

(5) *(Semicontinuity) If* $q_1 \subseteq q_2 \in \text{Spec}(R)$, *then* $e_{q_2}(I) \geq e_{q_1}(I)$.
(6) *If* J *is a reduction of* I, *then for all prime ideals* q *in* R, $e_q(J) = e_q(I)$.

Proof: Part (1) is clear using induction on n and the exact sequences,

$$0 \to I(q)^{n-1}/I(q)^n \to S(q)/I(q)^n \to S(q)/I(q)^{n-1} \to 0.$$

To prove (2), note that $\dim R = \dim(S/I) = \dim S - \text{ht}(I)$. The first equality holds since S/I is module–finite over R and the second holds since S is formally equidimensional (Lemma B.4.2). Hence $c = \dim S - \dim R$.

By Theorem B.2.2, $\dim(S_Q/qS_Q) \geq \dim S_Q - \dim R_q \geq \dim S_Q + \dim(R/q)$ $- \dim R = \dim S_Q + \dim(S/Q) - \dim R$ (since S/Q is a finitely generated R/q–module), and this latter sum is exactly $\dim S - \dim R$ by assumption. Therefore

$$\dim(S_Q/qS_Q) \geq \dim S - \dim R.$$

We prove the opposite inequality. Notice that $\dim(S_Q/qS_Q)$ is exactly the degree of the polynomial that gives the length (over S_Q) of $S_Q/(I^n S_Q + qS_Q)$ for large n. This length is less than or equal to the dimension of $S(q)/I(q)^n$ over $\kappa(q)$, which by Nakayama's Lemma is the number of generators of $(S/I^n)_{R\setminus q}$ as an R_q–module, and this number is at most the number of generators of S/I^n as an R–module. In turn, this number is equal to the dimension over k of $S/(I^n + \mathfrak{m}S)$, which is exactly $[\ell : k]$ times the length of $S/(I^n + \mathfrak{m}S)$ as an S–module. Finally, this length is given by a polynomial of degree $\dim(S/\mathfrak{m}S)$ for large n. Comparing the degrees of the first polynomial, which gives the length of $S_Q/(I^n + qS_Q)$, and the last polynomial, which gives the k–dimension of $S/(I^n + \mathfrak{m}S)$, yields the inequality

$$\dim(S_Q/qS_Q) \le \dim(S/\mathfrak{m}S) = \dim S - \dim R.$$

Hence we obtain equality: $\dim(S_Q/qS_Q) = \dim S - \dim R = c$.

We prove (3) and (4) simultaneously. Note that $(S/I^n)_{R\setminus q}$ has finitely many maximal ideals, all contracting to q, since $(S/I^n)_{R\setminus q}$ is finite over R_q. The Chinese Remainder Theorem shows that $(S(q)/I(q)^n) \cong \prod_{I \subseteq Q, \varphi^{-1}(Q)=q}(S/I^n + qS)_Q$. Hence the dimension of $S(q)/I(q)^n$ over $\kappa(q)$ is the sum of the dimensions (over $\kappa(q)$) of $(S/I^n + qS)_Q$ as Q ranges over those prime ideals such that $I \subseteq Q$ and $\varphi^{-1}(Q) = q$. The length of $(S/I^n + qS)_Q$ as an S_Q–module is given by a polynomial of the form $\frac{e(I(S/qS)_Q)}{c(Q)!}n^{c(Q)} + O(n^{c(Q)-1})$, where $c(Q) = \dim(S/qS)_Q$. By (2), $c(Q) = c$. Hence the dimension of $(S/I^n + qS)_Q$ over $\kappa(q)$ is given by a polynomial, $\frac{[\kappa(Q):\kappa(q)]\cdot e(I(S/qS)_Q)}{c!}n^c + O(n^{c-1})$. Taking the sum of these quantities proves both (3) and (4).

By Nakayama's Lemma, the dimension of $S(q)/I(q)$ over $\kappa(q)$ is exactly the number of generators of $(S/I^n)_{R\setminus q}$ as an R_q–module. This clearly only increases as we change q to a larger prime. The stated claim in (5) follows because this number of generators is given by a polynomial of degree c for all q, whose leading coefficient is $\frac{e_q(I)}{c!}$.

Part (6) follows immediately from (4) by applying Proposition 11.2.1 together with the fact that the set of prime ideals containing I and the set containing J are the same.   □

Example 11.4.2   Consider the simplest non–trivial case, where R a one–dimensional local domain. There are two multiplicities to consider, namely $e_{(0)}(I)$ and $e_{\mathfrak{m}}(I)$. The proof above shows that

$$\mu_R(S/I^n) = \frac{e_{\mathfrak{m}}(I)}{c!}n^c + O(n^{c-1}) \ \text{ and } \ \mathrm{rk}_R(S/I^n) = \frac{e_{(0)}(I)}{c!}n^c + O(n^{c-1}).$$

Furthermore, $e_{\mathfrak{m}}(I) = [\ell : k]e(I; S/\mathfrak{m}S)$ by Lemma 11.4.1 (4).

We are interested in conditions that imply and are implied by $e_{\mathfrak{m}}(I) = e_{(0)}(I)$. From the description above, equality holds if and only if

$$\mu_R(S/I^n) - \mathrm{rk}_R(S/I^n) = O(n^{c-1}).$$

Though we will not use this formulation exactly in what follows, it gives a good idea of what issues one must expect.

The first theorem below shows that if the family of ideals generated by ideal I is equimultiple, then the analytic spread of I must be the height. We later prove a partial converse.

Theorem 11.4.3 *Let* $\varphi : (R, \mathfrak{m}) \to (S, \mathfrak{n})$ *be a local homomorphism of Noetherian local rings and let* I *be a proper ideal of* S *such that* S/I *is a finitely generated* R*–module and such that* $\varphi^{-1}(I) = 0$. *Assume that* S *is formally equidimensional and further that* $\dim(S) = \dim(R) + \dim(S/\mathfrak{m}S)$. *If for every prime ideal* q *of* R, $e_q(I) = e_{\mathfrak{m}}(I)$, *then* $\ell(I) = \operatorname{ht}(I)$.

Proof: We may assume that the residue field of S is infinite. We reduce to the case in which R is complete. The map φ induces a map from the completion $\widehat{R}$ of R with respect to the maximal ideal $\mathfrak{m}$ to the completion $\widehat{S}$ of S with respect to $\mathfrak{n}$. Let $\widehat{I} = I\widehat{S}$. Then $\widehat{S}/\widehat{I} = (S/I) \otimes_R \widehat{R}$ is a finitely generated $\widehat{R}$–module, and $\widehat{R} \to \widehat{S}/\widehat{I}$ is injective, since $R \to S/I$ is injective and $\widehat{R}$ is flat over R. As S is formally equidimensional, so is $\widehat{S}$. Moreover, since $\dim \widehat{R} = \dim R, \dim \widehat{S} = \dim S$, and $\dim S/\mathfrak{m}S = \dim \widehat{S}/\widehat{\mathfrak{m}}\widehat{S}$, we still have that

$$\dim \widehat{S} = \dim \widehat{R} + \dim \widehat{S}/\widehat{\mathfrak{m}}\widehat{S}.$$

Finally, if Q is a prime ideal in $\widehat{R}$, and $q = Q \cap R$, then for every integer n, we have that $\widehat{S}/\widehat{I^n} \otimes_{\widehat{R}} \kappa(Q) = ((S/I^n) \otimes_R \widehat{R}) \otimes_{\widehat{R}} \kappa(Q) = ((S/I^n) \otimes_R \kappa(q)) \otimes_{\kappa(q)} \kappa(Q)$, which implies that

$$e_Q(\widehat{I}) = e_q(I) = e_{\mathfrak{m}}(I) = e_{\widehat{\mathfrak{m}}}(\widehat{I}).$$

Here we used that S/I^n is a finitely generated R–module. Thus all our assumptions pass to the completions of R and S. If we prove the theorem in this case, then we can conclude that $\ell(\widehat{I}) = \operatorname{ht}(\widehat{I})$. Since the analytic spread of an ideal does not change under completion and neither does its height, we reach the conclusion that $\ell(I) = \operatorname{ht}(I)$.

Henceforth we assume that R and S are complete and $S/\mathfrak{n}$ is infinite. The ideal $I + \mathfrak{m}S$ is $\mathfrak{n}$–primary as S/I is finite over R. Set $c = \dim(S/\mathfrak{m}S)$. We can choose elements $z_1, \ldots, z_c \in I$ whose images in $S/\mathfrak{m}S$ generate a minimal reduction of the image of I in this ring. Set $J = (z_1, \ldots, z_c)$. We claim that the height of J is c. To prove this, set $d = \dim R$, and choose a system of parameters $x_1, \ldots, x_d$ of R. Then the ideal $(x_1, \ldots, x_d, z_1, \ldots, z_c)S$ is $\mathfrak{n}$–primary, and the dimension of S is $d + c$ by assumption. Since S is catenary and equidimensional it follows that the height of $(z_1, \ldots, z_c)$ is c as claimed.

To finish the proof it is enough to prove that J is a reduction of I. We first observe that S/J is a finite R–module. This follows at once from [66, Exercise 7.4, p. 203] or from [322, Corollary 2, p. 259], since R is complete and $S/(J + \mathfrak{m}S)$ is a finite $(R/\mathfrak{m})$–module. By Lemma 11.4.1 (5), we know that for every prime ideal q of R, $e_q(I) \leq e_q(J) \leq e_{\mathfrak{m}}(J) = e_{\mathfrak{m}}(I)$, which forces $e_q(I) = e_q(J)$ for all prime ideals q of R.

Since J is generated by c elements and has height c, and S is formally equidimensional, the only associated prime ideals of $\overline{J}$ all have height c and are minimal over J. (See Theorem 5.4.1.) Therefore to prove that J is a reduction of I, it suffices to prove that $I_Q \subseteq \bar{J}_Q$ for all prime ideals Q that are minimal over J. Set $q = \varphi^{-1}(Q)$. Since $\dim S/Q = \dim S - c = \dim R$, and since $\dim S/Q = \dim R/q$ (as S/Q is finite over R/q), it follows that $\dim R = \dim R/q$. Thus R_q is Artinian, and therefore qS_Q is a nilpotent ideal. Hence to prove that $I_Q \subseteq \bar{J}_Q$, it suffices to prove the same containment after passing to S_Q/qS_Q. Fix such a q. Since $J \subseteq I$, we know that every prime Q contracting to q that contains I also contains J and furthermore $e(J(S_Q/qS_Q)) \geq e(I(S_Q/qS_Q))$. Using Lemma 11.4.1 (4) we have that

$$e_q(I) = \sum_{I \subseteq Q, \varphi^{-1}(Q)=q} [\kappa(Q) : \kappa(q)] e(I(S_Q/qS_Q))$$

$$\leq \sum_{J \subseteq Q, \varphi^{-1}(Q)=q} [\kappa(Q) : \kappa(q)] e(J(S_Q/qS_Q)) = e_q(J).$$

By assumption $e_q(I) = e_q(J)$. This equality forces the set of primes Q in each sum above to be the same and furthermore $e(J(S_Q/qS_Q)) = e(I(S_Q/qS_Q))$. Because qS_Q is nilpotent, S_Q/qS_Q is formally equidimensional if and only if S_Q is formally equidimensional. But a localization of a formally equidimensional ring is also formally equidimensional by Theorem B.5.2. An application of Theorem 11.3.1 shows that $I(S_Q/qS_Q)$ and $J(S_Q/qS_Q)$ must have the same integral closure. $\square$

Remark 11.4.4 Theorem 11.4.3 shows that if I is an equimultiple family, then the analytic spread of I equals the height of I. By an abuse of language, any ideal I in a local ring such that $\ell(I) = \mathrm{ht}(I)$ has come to be called an **equimultiple ideal**.

Theorem 11.4.5 Let $\varphi : (R, \mathfrak{m}) \to (S, \mathfrak{n})$ be a flat local homomorphism of Noetherian local rings and let I be a proper ideal of S such that S/I is a finitely generated R–module and such that $\varphi^{-1}(I) = 0$. Assume that S is formally equidimensional. If $\ell(I) = \mathrm{ht}(I)$, then for every prime ideal q of R we have $e_q(I) = e_\mathfrak{m}(I)$.

Proof: We first prove the case in which R is a one–dimensional domain, and then reduce the general case to this one. We need to prove that $e_{(0)}(I) = e_\mathfrak{m}(I)$ assuming $\ell(I) = \mathrm{ht}(I)$.

Lemma 11.4.6 Let the notation and assumptions be as in Theorem 11.4.5 above. Further assume that R is a one–dimensional Noetherian local domain, the residue field of S is infinite, and choose a non–zero element $x \in \mathfrak{m}$. Then $\lambda((I^n :_S x)/I^n) = O(n^{c-1})$.

Proof: The length of $(I^n :_S x)/I^n$ is finite since $S/(I + \mathfrak{m}S)$ has finite length and the radical of xS contains $\mathfrak{m}S$. By the Artin–Rees Lemma there exists an integer k such that for all $n \geq k$,

$$x(I^n :_S x) = (x) \cap I^n = I^{n-k}((x) \cap I^k) = I^{n-k}x(I^k :_S x).$$

Since R is a domain, x is a non–zerodivisor in R and therefore is also a non–zerodivisor in S, as S is flat over R. Hence $(I^n :_S x) = I^{n-k}(I^k :_S x)$. Choose a minimal reduction J of I. There will exist an integer k' such that for all $n \geq k'$, $(I^n :_S x) = J^{n-k'}I^{k'-k}(I^k :_S x)$. Choose a generating set $y_1, \ldots, y_N$ of $J^{n-k'}$. Set $M = (I^{k'-k}(I^k :_S x))/I^{k'}$. There is a surjective homomorphism

$$M^{\oplus N} \to (I^n :_S x)/I^n \to 0$$

sending an element $\alpha = (a_1, \ldots, a_N) \in M^{\oplus N}$ to the image of $a_1 y_1 + \cdots + a_N y_N$ in $(I^n :_S x)/I^n$. In particular, the length of $(I^n :_S x)/I^n$ is bounded by $\lambda(M) \cdot N$. Since $\mu(J) = c$, N is at most $\binom{n-k'+c-1}{c-1} = O(n^{c-1})$.   □

We continue with the proof of Theorem 11.4.5 in case R is one–dimensional. As in the lemma, choose $x \in \mathfrak{m} \setminus \{0\}$. We calculate $e_S((x); S/I^n)$ in two ways. First, using Proposition 11.1.9, $e_S((x); S/I^n) = \lambda_S(S/(I^n, x)) - \lambda_S((I^n : x)/I^n)$, and hence by the lemma, $e_S((x); S/I^n) = \lambda_S(S/(I^n, x)) + O(n^{c-1})$. However, $\lambda_S(S/(I^n, x)) = \frac{e_S(I; S/xS)}{c!} n^c + O(n^{c-1})$, and by the Associativity Formula, Theorem 11.2.4,

$$e_S(I; S/xS) = \sum_{Q \in \mathrm{Min}(S/xS)} e_S(I; S/Q) \lambda_{S_Q}(S_Q/xS_Q).$$

We use that S is formally equidimensional to see that $\dim S/Q = \dim S/xS$ for every prime ideal Q in $\mathrm{Min}(S/xS)$. On the other hand, since $R \to S_Q$ is flat, we have that $\lambda_{S_Q}(S_Q/xS_Q) = \lambda_R(R/xR)\lambda_{S_Q}(S_Q/\mathfrak{m}S_Q)$. We conclude that

$$e_S((x); S/I^n) = \lambda_R(R/xR) \cdot \sum_{Q \in \mathrm{Min}(S/xS)} e_S(I; S/Q) \lambda_{S_Q}(S_Q/\mathfrak{m}S_Q)$$

$$= \lambda_R(R/xR) e_S(I; S/\mathfrak{m}S),$$

the latter equality following by Theorem 11.2.4 and the fact that $\sqrt{xS} = \sqrt{\mathfrak{m}S}$.

On the other hand, Corollary 11.2.6 implies that $e_S((x); S/I^n)$ equals

$$[\ell : k] e_R((x); S/I^n) = [\ell : k] e_R((x); R) \mathrm{rk}_R(S/I^n) = \lambda_R(R/xR)e_{(0)}(I).$$

The equality of $e_R((x); R)$ and $\lambda_R(R/xR)$ holds since R is Cohen–Macaulay, being a one–dimensional domain (see Proposition 11.1.10). We have proved that $[\ell : k] e_S(I; S/\mathfrak{m}S) = e_{(0)}(I)$. By Lemma 11.4.1 (4), $[\ell : k] e(I; S/\mathfrak{m}S) = e_{\mathfrak{m}}(I)$, proving $e_{(0)}(I) = e_{\mathfrak{m}}(I)$, and proving the one–dimensional domain case.

Using Exercise 8.28, there exists a local homomorphism $R \to T$, with kernel q, such that T is a one–dimensional Noetherian local domain with field of fractions $\kappa(q)$, essentially of finite type over R, whose residue field L is a purely transcendental extension of the residue field k of R.

We claim that there is a unique maximal ideal N in $S \otimes_R T = S_T$ that contains IS_T. Since $S_T/IS_T \cong (S/I) \otimes_R T$ is module–finite over T, every maximal ideal of S_T must contain the maximal ideal $\mathfrak{m}_T$ of T. Hence the maximal ideals correspond to maximal ideals in $(S/(I + \mathfrak{m}S)) \otimes_k (T/\mathfrak{m}_T)$. As $S/(I + \mathfrak{m}S)$ is Artinian local and $T/\mathfrak{m}_T$ is purely transcendental over k, the claim follows.

Set $S^* = (S_T)_N$. There is a natural map $\varphi^* : T \to S^*$. Note that S_T is flat over T by base change, and since S^* is a localization of S_T, the map $T \to S^*$ is a flat local homomorphism of Noetherian rings. Set $I^* = IS^*$. We have $S^*/I^* = S_T/IS_T$ (since S_T/IS_T is already local), and hence $S^*/I^* = S/I \otimes_R T$ is finite over T. Moreover, $(\varphi^*)^{-1}(I^*) = 0$. To prove this, choose a prime ideal $I \subseteq Q$ in S with $\varphi^{-1}(Q) = q$. We then get a natural homomorphism from $S/I \otimes_R T \to \kappa(q)$ whose kernel is a prime ideal contracting to q in R. It follows that I^* contracts to q in R and hence to 0 in T.

Recall that the field of fractions of T is $\kappa(q)$. Hence for every integer n,

$$(S^*/(I^*)^n) \otimes_T \kappa(q) = ((S/I^n) \otimes_R T) \otimes_T \kappa(q) = (S/I^n) \otimes_R \kappa(q)$$

and

$$
\begin{aligned}
(S^*/(I^*)^n) \otimes_T T/\mathfrak{m}_T &= ((S/I^n) \otimes_R T) \otimes_T T/\mathfrak{m}_T \\
&= (S/I^n) \otimes_R T/\mathfrak{m}_T \\
&= ((S/I^n) \otimes_R k) \otimes_k T/\mathfrak{m}_T.
\end{aligned}
$$

It follows that $e_{(0)}(I^*) = e_q(R)$ and $e_{\mathfrak{m}_T}(I^*) = e_\mathfrak{m}(I)$. But from the one–dimensional case done previously, we know that $e_{\mathfrak{m}_T}(I^*) = e_{(0)}(I^*)$, provided we prove that $\ell(I^*) = \mathrm{ht}(I^*)$ and that the map from T to S^* satisfies our other hypotheses. Since S^*/I^* is finite over its subring T, we know that $\dim S^*/I^* = 1$. Let Q be any non–maximal prime ideal of S^* containing I^*. Such Q correspond precisely to the maximal ideals in $S^* \otimes_T \kappa(q)$ that contain the image of I^*, hence to maximal ideals in $S(q) = S \otimes_R \kappa(q)$ that contain $I(q)$. But then $\dim S_Q^* = \mathrm{ht}(I) = c$. On the other hand, by flatness, $\dim S^* = \dim T + \dim S^*/\mathfrak{m}_T S^* = 1 + \dim S/\mathfrak{m}S = 1 + \dim S - \dim R = 1 + c$. Hence $\dim S_Q^* = c = \dim S^* - 1$. Furthermore, $\mathrm{ht}\, I^* = \dim S^* - 1 = c$. But $\ell(I^*) \le \ell(I) = c = \mathrm{ht}(I^*) \le \ell(I^*)$ so that $\ell(I^*) = \mathrm{ht}(I^*)$ as needed. $\square$

Discussion 11.4.7 Terence Gaffney communicated to us the following background on the Principle of Specialization of Integral Dependence:

The principle of specialization of integral dependence was first stated and proved by Teissier in [292]. In Proposition 3.1 of [292], Teissier considers the case in which X is a flat family of analytic spaces over the open unit disk centered at the origin in C via a map G, $I\mathcal{O}_X$ is a coherent ideal sheaf on X, the cosupport of I finite over C, σ a section of G, $\sigma(C) = $ cosupport of I and the multiplicity of $I_t = I\mathcal{O}_{X_t}$ independent of t. He proved that in this case the exceptional divisor of the blowup of X by I was equidimensional over D. In Corollary 3.2 of [292], he deduced under these hypotheses that if $f \in \mathcal{O}_X$, and $f_t \in \bar{I}_t$, $t \ne 0$, then $f \in \bar{I}$. This became known as the **principle of specialization of integral dependence**.

In the appendix to [295] Teissier improved the theorem to:

Consider a reduced equidimensional family $X \to Y$ of analytic spaces, and an ideal sheaf I on X with finite co–support over Y. Suppose h is a section of $\mathcal{O}_X$ so that for all t in a Zariski–open dense subset of Y the induced section of $\mathcal{O}_{X(t)}$ on the fiber over t is integrally dependent on the induced ideal sheaf $I\mathcal{O}_{X(t)}$. If the multiplicity $e(I\mathcal{O}_{X(t)})$ is independent of t in Y, then h is integrally dependent on I.

In [187], Lipman proved algebraic theorems (his Theorems 4a and 4b) related to Teissier's results, which we have given in this section with some modifications. Theorem 4a in [187] is reproduced in Theorem 11.4.5 and Theorem 4b of [187] is Theorem 11.4.3. In particular, Theorem 11.4.3 gives Teissier's principle of specialization of integral dependence result as a corollary. However, this implication is not obvious; the reader should consult [187] and [292] for details.

In [87], the principle of specialization of integral dependence is extended to modules of finite co–length ideals using the Segre numbers, while in [88] the principle is extended to modules using Buchsbaum–Rim multiplicity. See [85] and [86] for further work.

11.5. Exercises

11.1 Let $P(t) = \sum_{i=0}^{d} a_i \binom{t+i}{i}$, where $a_i \in \mathbb{Q}$ for $i = 0, \ldots, d$. Assume that $P(n) \in \mathbb{Z}$ for all integers $n \gg 0$. Prove that $a_i \in \mathbb{Z}$ for all i.

11.2 Let $R = k[[XU, XV, YU, YV]]$, where k is a field and X, Y, U, V are variables. Find the Hilbert–Samuel polynomial of the maximal ideal of R, and compute the multiplicity of R.

11.3 Let $R = k[[X_1, \ldots, X_n]]$, where k is a field. Set $f_i = X_1^{a_{i1}} + \cdots + X_n^{a_{in}}$ for $i = 1, \ldots, n$, and $a_{ij} > 0$ integers. Assume that the ideal generated by $f_1, \ldots, f_n$ is primary for the maximal ideal. Prove or disprove the following: $\lambda(R/(f_1, \ldots, f_n)) = \min_{\sigma \in S_n} \{a_{\sigma(1)1} \cdots a_{\sigma(n)n}\}$.

11.4 Let $(R, \mathfrak{m})$ be a regular local ring and let $f_1, \ldots, f_d$ be a system of parameters of R. Prove or disprove: $\lambda(R/(f_1, \ldots, f_d)) \geq \prod_{i=1}^{d} \mathrm{ord}(f_i)$.

11.5 Let $(R, \mathfrak{m})$ be a Noetherian local ring that is not formally equidimensional. Prove that there exist $\mathfrak{m}$–primary ideals $I \subseteq J$ such that $e(I) = e(J)$ but that I is not a reduction of J.

11.6 Let $(R, \mathfrak{m})$ be a complete Cohen–Macaulay local ring with $e(R) = 2$. Prove that $R \cong S/(f)$, where S is a regular local ring and $f \in S$ with $\mathrm{ord}(f) = 2$.

11.7 (Serre [265, Appendice II]) Let $(R, \mathfrak{m})$ be a Noetherian local ring of dimension d, $J = (x_1, \ldots, x_d)$ an $\mathfrak{m}$–primary ideal, and $X_1, \ldots, X_d$ variables over $A = R/J$. Set $B = A[X_1, \ldots, X_d]$, and let $\varphi : B \to \mathrm{gr}_J(R)$ be the natural surjective map, sending X_i to the image of x_i in $J/J^2 \subseteq \mathrm{gr}_J(R)$. Let $K = \ker \varphi$. We know that K is $\mathbb{N}$–graded.

Suppose that $\lambda(R/J) = e(J; R)$.

(i) Prove that there exists a polynomial $P(n)$ with rational co-efficients of degree at most $d - 2$ such that for all $n \gg 0$, $\lambda(K_n) = P(n)$.

(ii) Let F be a non–zero element of some component K_l of K. Let S be the set of all monomials of B of degree n. Prove that $\{F\alpha \,|\, \alpha \in S\}$ is a minimal generating set for $F \cdot [B]_n$ over A. Conclude that $\lambda(F \cdot [B]_n)$ is a polynomial of degree at least $d-1$, which contradicts the inequality $\lambda(F \cdot [B]_n) \leq \lambda(K_{n+l})$.

(iii) Prove that R is Cohen–Macaulay.

11.8 (Nagata [213, Theorem 40.6]) Let $(R, \mathfrak{m})$ be a Noetherian formally equidimensional local ring. Assume that $e(R) = 1$. Prove that R is regular. (Hint: reduce to the case that R has an infinite residue field, is complete, and a domain; use Proposition 11.1.9.)

11.9 Let $R = k[[t^{n_1}, \ldots, t^{n_l}]]$, where k is field and $1 \leq n_1 \leq n_2 \leq \cdots \leq n_l$ are integers. Prove that $e(R) = n_1$.

11.10 Let $(R, \mathfrak{m})$ be a Cohen–Macaulay local ring. Prove that $\mu(\mathfrak{m}) \leq e(R) + \dim R - 1$.

11.11 (Lipman [187, page 121], where he credits Dade's thesis [55]) Let $(R, \mathfrak{m})$ be a Noetherian local ring, and P a prime ideal in R such that R/P is regular.

(i) Prove that if $\ell(P) = \mathrm{ht}(P)$, then R and R_P have the same multiplicity.

(ii) Assume that R is formally equidimensional and that R and R_P have the same multiplicity. Prove that $\ell(P) = \mathrm{ht}(P)$.

11.12 Let $(R, \mathfrak{m})$ be a Noetherian local ring, and let I be an $\mathfrak{m}$–primary ideal. Prove that the j-multiplicity $j(I)$ equals the usual multiplicity $e(I)$.

12
The conductor

The conductor of a ring plays a crucial role in integral closure of rings and ideals. It was already introduced in Exercises 2.11 and 2.12, and we here recall the definition.

Definition 12.0.1 *Let R be a reduced ring with total ring of fractions K and integral closure $\overline{R}$. The **conductor** of R, denoted C_R, is the set of all elements $z \in K$ that satisfy the property $z\overline{R} \subseteq R$. In other words, $C_R = (R :_K \overline{R})$.*

Clearly $C_R \subseteq R$ and in fact C_R is the largest common ideal of R and $\overline{R}$ (Exercise 2.11).

Any $x \in K$ can be written as a fraction of two elements of R, the denominator being a non–zerodivisor $r \in R$, so that $rx \in R$. Thus any finitely generated R-module contained in K is multiplied by a non–zerodivisor $r \in R$ into R. In particular, if $\overline{R}$ is a finitely generated R–module, this gives an element in C_R that is a non–zerodivisor. Conversely, if C_R contains a non–zerodivisor z, then $\overline{R} \subseteq R \cdot \frac{1}{z}$, and if R is Noetherian, this implies that $\overline{R}$ is a finitely generated R–module. Because of this, generally when we speak of conductors we assume that the integral closure of R is a finitely generated R–module.

By Lemma 2.4.2, the conductor $C_R = \{x \in K \mid x\overline{R} \subseteq R\}$ can be naturally identified with $\operatorname{Hom}_R(\overline{R}, R)$.

We will prove several classic formulas concerning elements in the conductor. In the first section we prove a formula going back at least to Dedekind that gives explicit elements in the conductor. Namely, let R be an integrally closed domain and let $R[z]$ be a separable extension of R, i.e., $R[z]$ is an integral domain integral over R with field of fractions separable over the field of fractions of R. Let $f(X)$ be the minimal polynomial of z over R. Then the conductor of $R[z]$ always contains $f'(z)$.

In the Section 12.2 we present the case of one–dimensional rings and relate various length formulas involving the conductor to the property of being Gorenstein. In that section we use some background on the canonical module and on the Ext functor. A good reference is [28].

In Section 12.3 we prove a far–reaching generalization due to Lipman and Sathaye of the classical result of Dedekind that has played an important role in many aspects of the study of integral closure and of tight closure. More on the tight closure aspect is in Chapter 13.

12.1. A classical formula

The following theorem is well–known and goes back at least to the work of Dedekind.

Theorem 12.1.1 *Let R be an integrally closed domain with field of fractions K. Let $K \subseteq L$ be a separable algebraic field extension, and let $z \in L$ be integral over R. Set $S = R[z]$, and set $f(X)$ equal to the minimal polynomial of z over R. Then $f'(z) \in C_S$.*

Proof: By Theorem 2.1.17, f is the minimal polynomial of z over K. Label the K–homomorphisms of $K(z)$ into an algebraic closure of L as $\sigma_1, \ldots, \sigma_d$ (one of these is induced by the inclusion $K(z) \to L$). Set $z_i = \sigma_i(z)$. The roots of $f(X)$ are exactly these conjugates. We let $\mathrm{Tr} : K(z) \to K$ be the trace. Let $\overline{S}$ be the integral closure of S in its field of fractions. Observe that $\mathrm{Tr}(\overline{S}) \subseteq R$, since R is integrally closed and the elements of $\mathrm{Tr}(\overline{S})$ are integral over R and are in K.

Fix $u \in \overline{S}$. Note that

$$h(X) = \mathrm{Tr}(u(f(X)/(X - z))) = \sum_{1 \leq i \leq d} \sigma_i(u)(f(X)/(X - z_i)),$$

so that setting $X = z$ we obtain that $h(z) = u(z - z_2)(z - z_3) \cdots (z - z_d) = uf'(z)$. Since the coefficients of $h(X)$ are traces of elements in $\overline{S}$, $h(X) \in R[X]$. It follows that $uf'(z) \in R[z]$. Since u is arbitrary in $\overline{S}$, this proves that $f'(z) \in C_S$. $\qquad\square$

See Section 12.3 for generalizations due to Lipman and Sathaye.

12.2. One–dimensional rings

In this section we discuss the conductor of a one–dimensional Noetherian analytically unramified local ring $(R, \mathfrak{m})$ with total ring of fractions K. In this case we have shown in Corollary 4.6.2 that the integral closure of R is a finitely generated R–module and in particular that the conductor C_R contains a non–zerodivisor in R. Since R is one–dimensional, it follows that C_R is an $\mathfrak{m}$–primary ideal.

We will use facts about the Ext functor and the canonical module in this section. We refer the reader to Bruns and Herzog [28] for background.

Example 12.2.1 Let k be a field, t a variable over k, and let $R = k[[t^a, t^b]] \subseteq k[[t]]$, where a and b are positive relatively prime integers. The integral closure of R is $\overline{R} = k[[t]]$ and the conductor of R is the ideal of R generated by all powers of t past some exponent c, and c is the smallest possible such value, i.e., $C_R = (t^i)_{i \geq c} R$ and $t^{c-1} \notin R$. We claim that $c = (a - 1)(b - 1)$.

Proof: Write $1 = na + mb$ and without loss of generality assume that $n > 0$. Then $m < 0$. We choose such a representation with m as large as possible, but still negative. Since $1 = (n-a)b+(m+a)b$, we must have that $m+a \geq 0$, and then as a and b are relatively prime that $1 - a \leq m$.

Consider $(a-1)(b-1)+i$ for $i \geq 0$. As above we may write $i + 1 = ea + fb$ and assume without loss of generality that $f < 0$ and is as big as possible staying negative, so that $1-a \leq f$. Then $(a-1)(b-1)+i = ab-a-b+(i+1) = b(a - 1) - a + ea + fb = b(a - 1 + f) + a(e - 1)$, and both $a - 1 + f \geq 0$ and $e - 1 \geq 0$, proving that $t^{(a-1)(b-1)+i} \in R$.

We claim that $t^{(a-1)(b-1)-1} \notin R$. For suppose that $(a-1)(b-1)-1 = ca + db$ with both $c, d \geq 0$. We may assume that $d \leq a - 1$. Hence $b(a - 1 - d) + a(-1 - c) = 0$, which forces a to divide $a - 1 - d$. Since $0 \leq d \leq a - 1$ this implies that $a - 1 = d$, but then $c = -1$, a contradiction. $\square$

Theorem 12.2.2 *Let $(R, \mathfrak{m})$ be a one–dimensional local analytically unramified Gorenstein ring. Let K be the total ring of fractions of R and let $\overline{R}$ be the integral closure of R in K with conductor C_R. Then*

$$2 \cdot \lambda_R(R/C_R) = \lambda_R(\overline{R}/C_R).$$

Proof: The discussion at the beginning of this section shows that $\overline{R}$ is a finitely generated R–module and that C_R is an $\mathfrak{m}$–primary ideal of R.

Since $\overline{R}$ is a one–dimensional Cohen–Macaulay module and R is Gorenstein, $\text{Ext}^1_R(\overline{R}, R) = 0$ (e.g., see Theorem 3.3.10 in [28]). As $\overline{R}/R$ has finite length, $\text{Hom}_R(\overline{R}/R, R) = 0$. This means that the long exact sequence induced from applying $\text{Hom}_R(\_, R)$ to the short exact sequence $0 \to R \to \overline{R} \to \overline{R}/R \to 0$ gives the short exact sequence

$$0 \to \text{Hom}_R(\overline{R}, R) \to \text{Hom}_R(R, R) \to \text{Ext}^1_R(\overline{R}/R, R) \to 0.$$

By Lemma 2.4.2, we can identify C_R with $\text{Hom}_R(\overline{R}, R)$ via the identification of R with $\text{Hom}_R(R, R)$. Hence we can identify $R/C_R \cong \text{Ext}^1_R(\overline{R}/R, R)$. As $\overline{R}/R$ has finite length and R is Gorenstein, $\lambda(\overline{R}/R) = \lambda(\text{Ext}^1_R(\overline{R}/R, R))$ (Corollary 3.5.9 in [28] and duality). The theorem follows from the short exact sequence of finite length modules

$$0 \to R/C_R \to \overline{R}/C_R \to \overline{R}/R \to 0.   \square$$

We will prove the converse of this theorem in Corollary 12.2.4.

The assumptions in Theorem 12.2.2 imply that R has a canonical module ω_R that can be embedded into R (see Bruns, Herzog [28] for information about ω_R).

Theorem 12.2.3 *Let $(R, \mathfrak{m})$ be a one–dimensional local analytically unramified ring with infinite residue field and total ring of fractions K, conductor C_R and canonical module ω_R. Then*

$$\lambda(\overline{R}/C_R) \geq 2 \cdot \lambda(R/C_R) + \mu(\omega_R) - 1.$$

Proof: Without loss of generality $\omega_R \subseteq R$. We need to relate C_R and ω_R. We use that

$$C_R \cong \text{Hom}_R(\overline{R}, R) = \text{Hom}_R(\overline{R}, \text{Hom}_R(\omega_R, \omega_R)) \cong \text{Hom}_R(\overline{R} \otimes_R \omega_R, \omega_R),$$

where the first isomorphism is from Lemma 2.4.2 and the last isomorphism is from Hom–tensor adjointness. Since ω_R is torsion–free, all the torsion in $\overline{R} \otimes_R \omega_R$ is annihilated by any homomorphism to ω_R, which means that we can identify $C_R \cong \text{Hom}_R(\omega_R \overline{R}, \omega_R)$. As R has an infinite residue field, there exists $x \in \omega_R$ such that x generates a minimal reduction of ω_R as an ideal in R. Since $\overline{R}$ is a principal ideal ring (see Exercise 12.11), there exists a non–zerodivisor $x \in \omega_R$ such that $\omega_R \overline{R} = x\overline{R}$. The inclusions $xC_R \subseteq xR \subseteq \omega_R \subseteq x\overline{R} = \omega_R \overline{R}$ give the following length equalities:

$$\lambda(\overline{R}/C_R) = \lambda(x\overline{R}/xC_R) = \lambda(x\overline{R}/\omega_R) + \lambda(\omega_R/Rx) + \lambda(Rx/xC_R).$$

But $\lambda(Rx/xC_R) = \lambda(R/C_R)$ and by Proposition 8.3.3, $\lambda(\omega_R/Rx) \geq \mu(\omega_R)-1$. Apply $\text{Hom}_R(\_, \omega_R)$ to the short exact sequence

$$0 \to C_R \to R \to R/C_R \to 0$$

to obtain a short exact sequence

$$0 \to \text{Hom}_R(R, \omega_R) = \omega_R \to \text{Hom}_R(C_R, \omega_R) = x\overline{R} \to \text{Ext}^1_R(R/C_R, \omega_R) \to 0.$$

Since $\lambda(\text{Ext}^1_R(R/C_R, \omega_R)) = \lambda(R/C_R)$ (by local duality, see Corollary 3.5.9 in [28]), it follows that $\lambda(x\overline{R}/\omega_R) = \lambda(R/C_R)$, which finishes the proof. $\square$

Corollary 12.2.4 *Let $(R, \mathfrak{m})$ be a one–dimensional local analytically unramified ring with infinite residue field and total ring of fractions K, conductor C_R, and canonical module $\omega_R \subseteq R$. Then R is Gorenstein if and only if*

$$2 \cdot \lambda(R/C_R) = \lambda(\overline{R}/C_R).$$

Proof: If R is Gorenstein, then the length formula holds by Theorem 12.2.2. Conversely, if the length equality holds, then the inequality of Theorem 12.2.3 proves that $\mu(\omega_R) - 1 = 0$, i.e., that ω_R is principal. This means that R is Gorenstein. $\square$

12.3. The Lipman–Sathaye theorem

Theorem 12.1.1 was vastly generalized to finitely generated extensions of Cohen–Macaulay rings by Lipman and Sathaye [192]. This generalization is important because of its applications to tight closure theory and results concerning the so–called Briançon–Skoda Theorem. We begin by establishing some notation.

Remark 12.3.1 Throughout this section R denotes a Cohen–Macaulay Noetherian domain of positive dimension. We let $X_1, \ldots, X_n$ be variables over R and we set $T = R[X_1, \ldots, X_n]$. We assume that $R \subseteq S = T/P$ for some prime ideal P in T with $P \cap R = 0$. We let K be the field of fractions of R and we assume that the field of fractions L of S is a finite and separable field extension of K. If $g_1, \ldots, g_n \in P$, we let g_X denote the determinant of the Jacobian matrix whose (i, j)th entry is $\frac{\partial g_i}{\partial X_j}$. By Definition 4.4.1, the Jacobian ideal of S/R is $J = J_{S/R} = \{g_X \mid g_1, \ldots, g_n \in P\}S$.

Proposition 12.3.2 *Adopt the notation of Remark 12.3.1.*

(1) $L = K[X_1, \ldots, X_n]/P_K$, *where* P_K *is the image of* P *in* $K[X_1, \ldots, X_n]$. *In particular,* $T_P = K[X_1, \ldots, X_n]_{P_K}$ *is a regular local ring of dimension* n *and the height of* P *is* n, *the number of variables.*

(2) *Let* $Q \in \operatorname{Spec} T$ *with* $P \subseteq Q$. *If* $g_1, \ldots, g_n \in P$ *and* $(g_1, \ldots, g_n) : P \not\subseteq Q$, *then* $J_Q = g_X S_Q$.

(3) *If* $g_1, \ldots, g_n \in P$, *then* $g_X \notin P$ *if and only if* $(g_1, \ldots, g_n) : P \not\subseteq P$, *i.e., if and only if* $g_1, \ldots, g_n$ *generate* P *generically.*

(4) *The Jacobian ideal* J *is generated by elements* g_X *such that* $g_1, \ldots, g_n$ *is a regular sequence in* P *and such that* $(g_1, \ldots, g_n) : P \not\subseteq P$.

Proof: After inverting the non–zero elements of R, we have that $K \subseteq S \otimes_R K \subseteq L$. As L is a finite field extension of K, it follows that $S \otimes_R K$ is a domain integral over a field, hence is itself a field. As the field of fractions of S is L, we must have that $S \otimes_R K = L$. As $P \cap R = 0$ and $K[X_1, \ldots, X_n]/P_K$ is a field, we see that $T_P = K[X_1, \ldots, X_n]_{P_K}$ is a regular local ring of dimension n, and the height of P is n, the number of variables.

To prove (2), note that $(J_{S/R})_Q = J_{S_Q/R}$ by Corollary 4.4.5. Since $P_Q = (g_1, \ldots, g_n)_Q$ we can use $g_1, \ldots, g_n$ to compute the Jacobian ideal to obtain that $J_Q = g_X S_Q$.

First suppose that $(g_1, \ldots, g_n) : P \not\subseteq P$. By (2), $J_P = g_X S_P$, so it suffices to prove that $J_P \neq 0$. This follows immediately from Theorem 4.4.9 since by assumption L is separable over K.

Conversely, if $g_X \notin P$, then Theorem 4.4.9 shows that $(T/(g_1, \ldots, g_n))_P$ is regular. Moreover the height of $(g_1, \ldots, g_n)$ must be n, so that P is minimal over the ideal $(g_1, \ldots, g_n)$. It follows that $(g_1, \ldots, g_n)_P = P_P$ and hence that $(g_1, \ldots, g_n) : P \not\subseteq P$. This proves (3).

To prove (4), it is enough to prove that given any $g_1, \ldots, g_n \in P$ there exist $h_1, \ldots, h_n \in P$ such that $h_1, \ldots, h_n$ form a regular sequence and such that $g_X \equiv h_X$ modulo P. As the height of P is n and T is Cohen–Macaulay, by using Prime Avoidance we can find $h_i \equiv g_i$ modulo P^2 such that $h_1, \ldots, h_n$ form a regular sequence. It is clear that with such a choice $g_X \equiv h_X$ modulo P. The condition that $(g_1, \ldots, g_n) : P \not\subseteq P$ is automatically satisfied by any set of g_X that generate J, since by (3), $(g_1, \ldots, g_n) : P \not\subseteq P$ if and only if $g_X \notin P$, i.e., if and only if $g_X S \neq 0$. However, since $(g_1, \ldots, g_n) + P^2 = (h_1, \ldots, h_n) + P^2$, Nakayama's Lemma shows that $(g_1, \ldots, g_n) : P \not\subseteq P$ if and only if $(h_1, \ldots, h_n) : P \not\subseteq P$. This proves (4). □

Definition 12.3.3 *We say that* $\underline{g} = g_1, \ldots, g_n \in P$ *are* **acceptable** *if* $g_X \notin P$ *and* $g_1, \ldots, g_n$ *form a regular sequence, or equivalently, if* $g_1, \ldots, g_n$ *form a regular sequence and* $(\underline{g}) : P \not\subseteq P$.

Remark 12.3.4   Assume that $g_1, \ldots, g_n \in P$ are acceptable. We define an S–homomorphism $\varphi_{\underline{g}} : ((g_1, \ldots, g_n) : P)/(g_1, \ldots, g_n) \to L$ as follows: if $u \in (g_1, \ldots, g_n) : P$ represents a class $\overline{u} \in ((g_1, \ldots, g_n) : P)/(g_1, \ldots, g_n)$, define $\varphi_{\underline{g}}(\overline{u}) = \frac{u'}{g'_X}$, where we use $(\_)'$ to denote images in L. We set $M_{\underline{g}}$ to be the image of $\varphi_{\underline{g}}$.

A key point in the theorem of Lipman and Sathaye is the following:

Proposition 12.3.5   *Adopt the notation and assumptions of Remarks 12.3.1 and 12.3.4. Let $\underline{g}$ and $\underline{h}$ be acceptable sequences. Then $M_{\underline{g}} = M_{\underline{h}}$.*

Proof: For two acceptable sequences $\underline{g}$ and $\underline{h}$ in $T = R[X_1, \ldots, X_n]$ we define a distance $\rho(\underline{g}, \underline{h})$ between them as the minimum integer s such that for some invertible matrices E_1 and E_2 with coefficients in T and for some $w_{s+1}, \ldots, w_n$ in T, $\underline{g} \, E_1 = (g'_1, \ldots, g'_s, w_{s+1}, \ldots, w_n)$ and $\underline{h} E_2 = (h'_1, \ldots, h'_s, w_{s+1}, \ldots, w_n)$.

We prove the proposition by induction upon $\rho(\underline{g}, \underline{h})$. If $\rho(\underline{g}, \underline{h}) = 0$, then it is easy to check that $M_{\underline{g}} = M_{\underline{h}}$ as there is an invertible matrix E such that $\underline{g} = \underline{h}E$.

We need to do the case $\rho(\underline{g}, \underline{h}) = 1$ separately. By possibly multiplying by invertible matrices, we may assume without loss of generality that $\underline{g} = (y_1, \ldots, y_{n-1}, g)$ and $\underline{h} = (y_1, \ldots, y_{n-1}, h)$. Let $u \in (\underline{g}) : P$, and write

$$uh = \sum_{i=1}^{n-1} r_i y_i + vg.$$

It is straightforward to check that $\frac{u'}{g'_X} = \frac{v'}{h'_X}$.

We claim that $v \in (\underline{h}) : P$. Multiplying the displayed equation by an arbitrary element $z \in P$ yields $zuh = \sum_{i=1}^{n-1} r_i y_i z + vgz$. As $z \in P$ and $u \in (\underline{g}) : P$, there is an equation $zu = \sum_{i=1}^{n-1} s_i y_i + sg$, and upon substitution in the preceding equation one obtains that

$$g(sh - vz) \in (y_1, \ldots, y_{n-1}).$$

Since $\mathrm{ht}(y_1, \ldots, y_{n-1}, g) = n$, the fact that R is Cohen–Macaulay guarantees that $(sh - vz) \in (y_1, \ldots, y_{n-1})$. It follows that $vz \in (\underline{h})$, and hence that $v \in (\underline{h}) : P$. Note that $M_{\underline{g}}$ is generated by elements $\frac{u'}{g'_X}$ as u ranges over generators of $(\underline{g}) : P$. Since $\frac{u'}{g'_X} = \frac{v'}{h'_X}$, and $v \in (\underline{h}) : P$, this proves that $M_{\underline{g}} \subseteq M_{\underline{h}}$. By symmetry we obtain that $M_{\underline{h}} = M_{\underline{g}}$. This finishes the case $\rho(\underline{g}, \underline{h}) = 1$.

Suppose that $\rho(\underline{g}, \underline{h}) = m > 1$. We may assume that $g_i = h_i$ for $i > m$. Using the Prime Avoidance Theorem we may choose an element $h'_1 = h_1 + \sum_{i=2}^{n} \mu_i h_i$ such that all of the sequences $g_1, \ldots, g_{i-1}, h'_1, g_{i+1}, \ldots, g_n$ are regular sequences for $1 \leq i \leq m$. As $\underline{g}$ is acceptable, $(\underline{g}) : P \not\subseteq P$ and there is an element $u \in (\underline{g}) : P$ such that $u \notin P$. Write $uh'_1 = \sum_{i=1}^{n} \lambda_i g_i$. Some $\lambda_i \notin P$ for $i \leq m$; otherwise we would have that $uh'_1 \in P^2 + (g_{m+1}, \ldots, g_n) =$

$P^2 + (h_{m+1}, \ldots, h_n)T$, which contradicts the fact that $\underline{h}$ is acceptable and hence that $h'_1, h_2, \ldots, h_n$ form a minimal generating set of PT_P. Fix i between 1 and m such that $\lambda_i \notin P$, and let $\underline{h}' = (g_1, \ldots, g_{i-1}, h'_1, g_{i+1}, \ldots, g_n)$. Since $\lambda_i \notin P$, it follows that $(\underline{h}') : P \not\subseteq P$, and by choice, these elements form a regular sequence. Hence $\underline{h}'$ is acceptable. By construction, $\rho(\underline{g}, \underline{h}') \leq 1$. By the case in which the sequences are at most distance 1, we know that $M_{\underline{g}} = M_{\underline{h}'}$. In addition $\rho(\underline{h}', \underline{h}) < m$. Hence there exists an integer $l \leq \rho(\underline{h}', \underline{h})$ such that $M_{\underline{h}'} = M_{\underline{h}}$. The proposition follows with $k = l + 1$. $\qquad\square$

Definition 12.3.6 *If $\underline{g}$ is acceptable, we set $K_{S/R} = M_{\underline{g}}$.*

Proposition 12.3.5 proves that $K_{S/R}$ does not depend upon the choice of acceptable sequence $\underline{g}$. The notation $K_{S/R}$ is meant to suggest a relative canonical module, which is the role this module plays in the proofs.

Proposition 12.3.7 *Adopt the notation of 12.3.1. Let $\underline{g}$ be an acceptable sequence. Then*

(1) *The map $\varphi_{\underline{g}} : (\underline{g} :_T P)/(\underline{g}) \to L$ is injective. In particular, $(\underline{g} :_T P)/(\underline{g}) \cong K_{S/R}$.*

(2) *$K_{S/R} \subseteq (S :_L J_{S/R})$.*

(3) *Assume further that S is normal. Then $K_{S/R}$ is a reflexive S–module.*

(4) *Assume that S is normal and that for every height one prime Q of S, $R_{Q\cap R}$ is regular. Then $K_{S/R} = S :_L J_{S/R}$.*

Proof: Let $u \in (\underline{g} :_T P)$ represent the class of an element $\overline{u}$ in $(\underline{g} :_T P)/(\underline{g})$. If $\varphi_{\underline{g}}(\overline{u}) = 0$, then $u' = 0$ and hence $u \in P$. Thus the injectivity of $\varphi_{\underline{g}}$ is equivalent to the statement that $(\underline{g} :_T P) \cap P = (\underline{g})$. The assumption that $\underline{g}$ be acceptable means that $(\underline{g} :_T P) \not\subseteq P$. As $((\underline{g} :_T P) + P)((\underline{g} :_T P) \cap P) \subseteq (\underline{g})$, it is enough to prove that $(\underline{g} :_T P) + P$ is not contained in any associated prime of $\underline{g}$. However, T is Cohen–Macaulay as R is Cohen–Macaulay and $(\underline{g} :_T P) + P$ has height strictly greater than $n = \mathrm{ht}(\underline{g})$ since $(\underline{g} :_T P) \not\subseteq P$. The last statement of the first part follows by the definition that $K_{S/R} = M_{\underline{g}}$, which is the image of $\varphi_{\underline{g}}$.

Let $g_X \in J_{S/R}$, where $\underline{g}$ is acceptable (recall that $J_{S/R}$ is generated by g_X as $\underline{g}$ runs over acceptable sequences by Proposition 12.3.2). It suffices to prove that $g_X K_{S/R} \subseteq S$. Choose any element $\alpha \in K_{S/R}$. The definition of $M_{\underline{g}}$ then yields the desired containment of (2).

Assume that S is normal. To prove $K_{S/R}$ is reflexive, it suffices to prove that for an arbitrary acceptable sequence $\underline{g}$, the module $(\underline{g} :_T P)/(\underline{g})$ is reflexive. As S is normal, it suffices to prove that this module satisfies Serre's condition (S_2). Let Q be a prime in T containing P. If $\mathrm{ht}(Q/P) \geq 2$, then $\mathrm{depth}(T/(\underline{g} :_T P))_Q \geq 1$ since $\underline{g}$ is generated by a regular sequence and T is Cohen–Macaulay. Likewise, $\mathrm{depth}(T/\underline{g})_Q \geq 2$. The exact sequence

$$0 \to ((\underline{g} :_T P)/(\underline{g}))_Q \to (T/\underline{g})_Q \to (T/(\underline{g} :_T P))_Q \to 0$$

then gives $\mathrm{depth}(\underline{g} :_T P)/(\underline{g}))_Q \geq 2$. If $\mathrm{ht}(Q/P) \leq 1$, then the same exact sequence proves that $(g :_T P)/(\underline{g}))_Q$ is a maximal Cohen–Macaulay $(T/P)_Q$–module. Hence $(g :_T \underline{P})/(\underline{g})$ is reflexive.

By (2), $K_{S/R} \subseteq (S :_L J_{S/R})$, and by (3), $K_{S/R}$ is a reflexive S–module. To prove equality, it suffices to prove equality after localizing at an arbitrary height one prime of S. Let Q be such a prime and set $q = Q \cap R$. By assumption, R_q is regular, and as S is normal, S_Q is also regular. Lift Q to a prime in $T_q = R_q[X_1, \ldots, X_n]$, which we denote by Q'. Since $T_{Q'}$ and $S_Q = T_{Q'}/P_{Q'}$ are regular, $P_{Q'}$ is generated by a regular sequence, say $g_1, \ldots, g_n$, which we may assume are in P. Then $(g_1, \ldots, g_n) : P \not\subseteq Q'$, and Proposition 12.3.2 (2) shows that $(J_{S/R})_Q = g_X S_Q$. Hence $(S :_L J_{S/R})_Q = (S_Q :_L (J_{S/R})_Q) = (\frac{1}{g_X'})S_Q$. Choose $u \in ((g_1, \ldots, g_n) : P) \setminus Q'$. Then $\varphi_{\underline{g}}(u) = \frac{u'}{g_X'}$ is in the image of the map from K into $S :_L J_{S/R}$. This proves the needed equality at the localization of every height one prime ideal of S and hence proves (4).   □

We next compare $K_{S/R}$ and $K_{B/R}$ when B is a finite extension of S with the same field of fractions.

Remark 12.3.8 Adopt the notation of 12.3.1. Let $B = S[y] \subseteq L$ be an integral extension of S, where we assume that $y \notin S$. Extend the map of T onto S to an epimorphism $\varphi : T[Y] \to B$. Let Q be the kernel of φ. Clearly $Q \cap T = P$. Since the field of fractions are the same, Q contains an element $aY - b$ $(a, b \in T, a \notin P)$ and also Q contains a monic polynomial $h(Y)$ of degree m with coefficients in T.

Suppose that $\underline{g}$ is an acceptable sequence for P. The sequence $\underline{g}^* = g_1, \ldots, g_n, h$ is clearly a sequence in Q having height $n + 1$ and the Jacobian $g^*_{X,Y} = g_X \frac{\partial h}{\partial Y}$. If $\frac{\partial h}{\partial Y} \notin Q$, then $\underline{g}^*$ is an acceptable sequence for Q. Since $a \notin P = Q \cap T$, if $\frac{\partial h}{\partial Y} \in Q$, then $\frac{\partial(h + aY - b)}{\partial Y} = \frac{\partial h}{\partial Y} + a \notin Q$. Since $m \geq 2$, this change does not affect the fact that h is monic. Henceforth we assume that $\frac{\partial h}{\partial Y} \notin Q$.

We now have the map from $((\underline{g}^*) : Q)/(\underline{g}^*) \to L$ given by $v \in ((\underline{g}^*) : Q)$ goes to $\frac{v'}{(g^*_{X,Y})'}$. We denote the image as above by $M_{\underline{g}^*}$.

Lemma 12.3.9 *Let the notation be as above. Then $M_{\underline{g}^*} \subseteq M_{\underline{g}}$. Precisely, for every $v \in T[Y]$ with $vQ \subseteq (g_1, \ldots, g_n, h)T[Y]$, there is an element $u \in T$ with $uP \subseteq (g_1, \ldots, g_n)T$ such that*

$$v \equiv u \frac{\partial h}{\partial Y} \bmod Q.$$

Proof: We can replace T by $T/((g_1, \ldots, g_n)T$ to assume that $(g_1, \ldots, g_n) = 0$. Write $v = hw + a_1 Y^{m-1} + a_2 Y^{m-2} + \cdots + a_m$, where $w \in T[Y]$ and $a_i \in T$. Since $v(Q \cap T) \subseteq vQ \subseteq hT[Y]$, this forces $(Q \cap T)(a_1 Y^{m-1} + a_2 Y^{m-2} + \cdots + a_m) \subseteq hT[Y]$, and so $a_i(Q \cap T) = 0$ for $1 \leq i \leq m$.

We must also have that $v(aY - b) \in hT[Y]$ and so $(v - hw)(aY - b)$ is divisible by h. This expression is a polynomial in Y of degree m with leading coefficient aa_1, and hence $(v - hw)(aY - b) = aa_1 h$. Differentiating with respect to Y gives that $(v - hw)a \equiv a_1 a \frac{\partial h}{\partial Y}$ modulo Q. As $a \notin Q$, it follows that $v - hw \equiv a_1 \frac{\partial h}{\partial Y}$ modulo Q, and as $h \in Q$, that $v \equiv a_1 \frac{\partial h}{\partial Y}$ modulo Q. Setting $a_1 = u$ gives the conclusion. $\qquad\square$

Theorem 12.3.10 (Lipman–Sathaye Theorem) *Let R be a Cohen–Macaulay Noetherian domain with field of fractions K. Let S be a domain that is a finitely generated R–algebra. Assume that the field of fractions of S is separable and finite over K and that the integral closure $\overline{S}$ of S is a finitely generated S–module. Furthermore, assume that for all prime ideals Q in S of height one, $R_{Q \cap R}$ is a regular local ring. Then*

$$(\overline{S} :_L J_{\overline{S}/R}) \subseteq S :_L J_{S/R}.$$

In particular, $J_{S/R}\overline{S} \subseteq S$.

Proof: Write $S = R[X_1, \ldots, X_n]/P$ for some variables $X_1, \ldots, X_n$ over R and some prime ideal P in $T = R[X_1, \ldots, X_n]$. By Proposition 12.3.2, $J_{S/R}$ is generated by elements g_X, where $\underline{g}$ is acceptable for S. Fix one such $\underline{g}$. We write $\overline{S} = S[y_1, \ldots, y_l]$ and use Lemma 12.3.9 repeatedly. This lemma shows that there is an acceptable sequence $\underline{g}^*$ for $\overline{S}$ such that $M_{\underline{g}^*} \subseteq M_{\underline{g}}$. By Proposition 12.3.7 we have an equality $\overline{S} :_L J_{\overline{S}/R} = K_{\overline{S}/R} = M_{\underline{g}^*}$, and by definition $K_{S/R} = M_{\underline{g}}$. Hence $M_{\underline{g}^*} \subseteq M_{\underline{g}}$ gives that

$$\overline{S} :_L J_{\overline{S}/R} \subseteq K_{S/R} \subseteq S :_L J,$$

where the last containment follows from Proposition 12.3.7 (2). $\qquad\square$

See [122] and [139] for further generalizations of this theorem.

Remark 12.3.11 The assumption in Theorem 12.3.10 that $\overline{S}$ be module–finite over S is often satisfied. For example, it is satisfied if R is regular (Theorem 9.2.3), if R is finitely generated over a field (Theorem 4.6.3), if R is finitely generated over the integers (Corollary 4.6.5), or more generally, if R is excellent.

12.4. Exercises

12.1   Let k be a field, t a variable over k and $R = k[t^5, t^7, t^{11}]$. Prove that the integral closure of R is $k[t]$ and find the conductor of R.

12.2   Let $(R, \mathfrak{m})$ be an one–dimensional analytically unramified domain. Prove that every ideal in $\overline{R}$ that is contained in R is integrally closed.

12.3   Suppose that R is an analytically unramified local domain of dimension at least two. Prove that R is integrally closed if and only if there exists a regular sequence of length two in the conductor.

12.4   Let R be an analytically unramified local domain of dimension one. Prove that the conductor is never contained in a proper principal ideal.

12.5   Let R be an analytically unramified local Cohen–Macaulay domain. Prove or give a counterexample to following open question: can the conductor be contained in an ideal generated by a system of parameters?

12.6   Let R be an analytically unramified Cohen–Macaulay local domain that is not integrally closed. Prove that the conductor has height one and that all prime ideals in R minimal over it have the same height.

12.7   Let R be an analytically unramified local domain of dimension one. Prove that the conductor is an integrally closed ideal.

12.8   (Delfino [60]) Let R be an analytically unramified local domain of dimension one with conductor C and canonical module $\omega_R \subseteq R$. Prove that equality occurs in Theorem 12.2.3 if and only if $\omega_R/C\omega_R$ is a free (R/C)–module.

12.9   Let R be a Noetherian domain with finitely generated integral closure S. Let D be a derivation from R to R. (D induces a derivation, which we still call D, from the field of fractions of R back into itself.) If R contains $\mathbb{Q}$, prove that D takes S back into S.

12.10   Let R be a Noetherian domain with module–finite integral closure S. Let D be a derivation from R to R. Let C be the conductor of S to R. Prove that $D(C) \subseteq C$.

12.11   Let R be a Noetherian local reduced one–dimensional ring. Prove that the integral closure of R is a principal ideal ring.

12.12   Let R be a finitely generated domain over a field k. The goal of this exercise is to prove that the conductor of R is computable (even if $\overline{R}$ is not known) if the field of fractions L of R is separable over k. There exists a computable subset $x_1, \ldots, x_n \in R$ such that $A = k[x_1, \ldots, x_n]$ is a polynomial ring and such that R is module–finite over A (this is the (computable) Noether normalization; use Exercise 15.6). Let K be the field of fractions of A.

    (i) Prove that for any basis of L over K, its dual basis (with respect to trace) is computable.

    (ii) Prove that a basis of L over K is computable, and moreover, that there exists a computable basis S consisting of elements of R.

    (iii) With S as above, let T be its dual basis. Prove that there exists a non–zero computable element $c \in R$ such that for all $t \in T$, $ct \in R$.

    (iv) Prove that c is an element of the conductor. (Cf. Theorem 3.1.3.)

12.13   (Corso, Huneke, Katz, Vasconcelos [44]) Let R be a semi–local one–dimensional Noetherian domain such that $\overline{R}$ is module–finite over R.

    (i) Let C be the conductor of $R \subseteq \overline{R}$. Prove that $\mathrm{Hom}_R(C, R) = \overline{R}$. (Hint: use that $\overline{R}$ is a principal ideal domain.)

    (ii) Prove that for any ideal I in R, $(I^{-1})^{-1} \subseteq \overline{I}$.

13

The Briançon–Skoda Theorem

In this chapter we prove several basic theorems regarding the comparison of powers of an ideal with the integral closure of their powers. Results of this type are called the Briançon–Skoda Theorems, and we present several versions: one via tight closure, one due to Lipman and Sathaye, and a more general "uniform" version due to Huneke for broad classes of Noetherian rings. The Briançon–Skoda Theorem has played an important role in the development of many techniques in commutative algebra. These developments range from the theorem of Lipman and Sathaye, Theorem 13.3.3, to contributing to the development of tight closure, as well as Lipman's development of adjoint ideals. We also prove a theorem due to Shiroh Itoh concerning the intersection of integral closures of powers of ideals with minimal reductions (Theorem 13.2.4).

For additional Briançon–Skoda type theorems, see the joint reduction version 17.8.7 of Rees and Sally, Lipman's adjoint version 18.2.3, or papers Hyry and Villamayor [146]; Aberbach and Huneke [2] and [3]; and references therein.

Let $O_d = \mathbb{C}\{z_1, \ldots, z_d\}$ be the ring of convergent power series in d variables. Let $f \in O_d$ be a non–unit (i.e., f vanishes at the origin). The Jacobian ideal of f is $J(f) = (\partial f/\partial z_1, \ldots, \partial f/\partial z_d)O_d$ (see Definition 4.4.1). Since $f \in \overline{J(f)}$ by Corollary 7.1.4, by Corollary 1.2.2, $J(f) \subseteq J(f) + (f)$ is a reduction, so that there is an integer k such that $f^k \in J(f)$. John Mather first raised the following question:

Question 13.0.1 *Is there an exponent k that works for all non–units f?*

Briançon and Skoda [25] answered this question affirmatively by proving that the dth power of f always lies in $J(f)$, where d is the dimension of the convergent power series ring. To prove this they proved a stronger result concerning the integral closures of ideals which was later generalized to the case of regular local rings by Lipman and Sathaye [192] as follows:

Theorem 13.3.3 *Let R be a regular ring. Let I be any ideal of R that is generated by l elements. Then for any $n \geq 0$,*
$$\overline{I^{n+l}} \subseteq I^{n+1}.$$

Although it is not immediately apparent why this theorem answers the question of Mather, it does indeed provide an answer. This is because O_d is a (regular) local ring of dimension d with an infinite residue field, so that by Proposition 8.3.7 and Corollary 8.3.9 any ideal I in O_d has a reduction J generated by d elements, whence for all $n \geq 0$, $\overline{I^{n+d}} = \overline{J^{n+d}} \subseteq J^{n+1} \subseteq I^{n+1}$. Thus the uniform value of k sought by Mather can be taken to be $k = d$.

Lipman and Bernard Teissier proved in [193] that the theorem as above holds for pseudo–rational rings, but only in the case where I is generated by a regular sequence. As a consequence they obtained that for any ideal I in a pseudo–rational local ring of dimension d, $\overline{I^{d+n}} \subseteq I^{n+1}$. The full strength of Theorem 13.3.3 was later recovered for pseudo–rational rings by Aberbach and Huneke in [4].

The main results are in Sections 13.2, 13.3 and 13.4. Since Section 13.2 relies on tight closure, we have to develop the basics of tight closure in positive prime characteristic. This is done in Section 13.1. There is another reason for presenting tight closure: tight closure is basic for understanding integral closure. To present the most general Briançon–Skoda–type results of Section 13.4 we also need to develop the basics of test elements and F-finiteness, both parts of tight closure. We do this in the latter part of Section 13.1.

13.1. Tight closure

In this section we review the definition of tight closure in positive prime characteristic p. We also define test elements and F-finite rings and prove that under some conditions test elements exist in reduced F-finite rings.

The **Frobenius map** is the endomorphism $F : R \to R$ defined by $F(r) = r^p$. For an ideal I in a ring of characteristic p, if $q = p^e$, the ideal $I^{[q]}$ is generated in R by the eth Frobenius power of I, namely by the qth powers of elements of I.

Definition 13.1.1 *Let R be a Noetherian ring of prime characteristic $p > 0$ and let I be an ideal of R. An element $x \in R$ is said to be in the* **tight closure** *I^* of I if there is $c \in R^o$ such that for all large $q = p^e$, $cx^q \in I^{[q]}$.*

The definition of tight closure should be compared to the characterization of integral closure in Corollary 6.8.12, which states that an element $x \in \overline{I}$ if and only if there is an element $c \in R^o$ such that for all large n, $cx^n \in I^n$.

Some basic properties of tight closure are given in the following theorem.

Theorem 13.1.2 *Let R be a Noetherian ring of characteristic p and let I be an ideal.*

(1) $(I^*)^* = I^*$. If $I_1 \subseteq I_2 \subseteq R$, then $I_1^* \subseteq I_2^*$.

(2) If R is reduced or if I has positive height, then $x \in R$ is in I^* if and only if there exists $c \in R^o$ such that $cx^q \in I^{[q]}$ for all $q = p^e$.

(3) An element $x \in R$ is in I^* if and only if the image of x in R/P is in the tight closure of $(I + P)/P$ for every minimal prime P of R.

(4) $I^* \subseteq \overline{I}$.

(5) If I is tightly closed, then $I : J$ is tightly closed for every ideal J.

(6) If R is a regular local ring then $I^* = I$ for every ideal $I \subseteq R$.

Proofs of the first five parts are a straightforward application of the definitions and we leave them to the reader. To prove (6) we need a lemma:

Lemma 13.1.3 (Kunz [177]) *If R is a regular local ring of positive and prime characteristic, then the Frobenius map is flat.*

Proof: It suffices to prove that $\text{Tor}_i^R(M, S) = 0$ for any $i > 0$, any finitely generated R–module M, where S equals R but its R–module structure is via the Frobenius map. As R is a regular local ring, M has a finite free resolution. Applying the Frobenius homomorphism (that is, tensoring with S over R) simply raises the entries of (any) matrix in the resolution, which gives a map between consecutive free modules to the pth power. In particular, applying Frobenius homomorphism changes neither the ranks of the maps nor the depths of the ideals of minors since it only changes the minors up to radical. The Buchsbaum–Eisenbud Criterion ([31]) then proves that the ensuing complex is exact. Hence $\text{Tor}_i^R(M, S) = 0$ for $i > 0$ and it follows that the Frobenius homomorphism is flat. □

Proof of part (6) of Theorem 13.1.2: An important consequence of the flatness of the Frobenius homomorphism is that for all ideals I and elements x of the ring, $I^{[q]} : x^q = (I : x)^{[q]}$. More generally, if $R \to S$ is a flat homomorphism and if $I \subseteq R$ and $x \in R$, then $(I :_R x)S = (IS :_S x)$. This follows by tensoring the exact sequence $0 \to R/(I : x) \xrightarrow{x} R/I \to R/(I, x) \to 0$ with S. Applying this with the eth iteration of the Frobenius map proves $I^{[q]} : x^q = (I : x)^{[q]}$.

Let $(R, \mathfrak{m})$ be a regular local ring and suppose that $x \in I^*$ for some ideal I of R. There exists a non–zero element c such that $cx^q \in I^{[q]}$ for all large $q = p^e$. Hence $c \in \cap_q(I^{[q]} : x^q)$. The flatness of the Frobenius homomorphism gives that $I^{[q]} : x^q = (I : x)^{[q]}$. If $(I : x) \neq R$, then $c \in \cap_q \mathfrak{m}^q = 0$, a contradiction. Hence $x \in I$. □

This is enough tight closure background for Section 13.2.

In Section 13.4 we will need additional results about tight closure. We develop those below.

Definition 13.1.4 *Let R be a Noetherian ring of positive prime characteristic. A **test element** is an element $c \in R^o$ such that for every ideal I of R and every $x \in I^*$ (the tight closure of I), $cx \in I$.*

Observe that if $x \in I^*$, then $x^q \in (I^{[q]})^*$ for all $q = p^e$, so that if c is a test element, $cx^q \in I^{[q]}$ for all q. In other words, c can be used to test whether x is in the tight closure of I.

Test elements play a central role in the theory of tight closure. A key point is that they exist in abundance. Heuristically, one should think as follows: If $c \in R$ and R_c is regular, then every ideal is tightly closed in R_c, which implies that for every ideal I some power of c multiplies I^* into I. The next theorem shows that this power can be chosen uniformly, giving us a test element. See [126] for a more general result.

Definition 13.1.5 *A ring of characteristic p such that $R^{1/p}$ is finite as an R–module is said to be **F-finite**.*

Theorem 13.1.6 ([123, Theorem 3.4]) *Let R be an F-finite reduced ring of prime characteristic p and let c be any non–zero element of R such that R_c is regular. Then a power of c is a test element.*

We need a lemma first.

Lemma 13.1.7 *Let R be an F-finite regular Noetherian ring of characteristic p. Let $d \in R$. For all large $q = p^e$ there exists an R–linear homomorphism $f : R^{1/q} \to R$ with $f(d^{1/q}) = 1$.*

Proof: We first do the case in which R is local: Let $(R, \mathfrak{m})$ be an F-finite regular local ring of characteristic p. Since R is regular, the Frobenius map is flat, and hence $R^{1/q}$ is a flat R–module. As $R^{1/q}$ is finitely presented and R is local, it is actually free. Let $d \in R$ be non–zero. For sufficiently large $q = p^e$, $\mathfrak{m}^{[q]}$ does not contain d. Taking qth roots yields that $d^{1/q} \notin \mathfrak{m}R^{1/q}$. Since $R^{1/q}$ is free over R it follows that one may use $d^{1/q}$ as part of a free basis of $R^{1/q}$. In particular, there is an R–linear homomorphism $\varphi : R^{1/q} \to R$ that sends $d^{1/q}$ to 1.

Now let R be a regular (hence reduced) ring that is not necessarily local. Taking pth roots commutes with localization, so that $R^{1/q}$ will be projective over R for all $q = p^e$. Fix a maximal ideal $\mathfrak{m}$ of R. There will be a power of p, say $q = q(\mathfrak{m})$, depending upon $\mathfrak{m}$, such that $d \notin \mathfrak{m}^{[q]}$, and so $d^{1/q} \notin \mathfrak{m}R^{1/q}$. By the local case done above, there is a homomorphism from $R_\mathfrak{m}^{1/q}$ to $R_\mathfrak{m}$ sending $d^{1/q}$ to 1. Clearing denominators one sees that there is an element $r_\mathfrak{m} \notin \mathfrak{m}$ such that there is an R–linear map $\varphi_\mathfrak{m} : R_{r_\mathfrak{m}}^{1/q} \to R_{r_\mathfrak{m}}$ sending $d^{1/q}$ to 1. The ideal generated by all such $r_\mathfrak{m}$ is not contained in any maximal ideal so that there are finitely many of them, say $r_1 = r_{\mathfrak{m}_1}, \ldots, r_k = r_{\mathfrak{m}_k}$, that generate the unit ideal. Set $q = \max\{q(\mathfrak{m}_i)\}$. Let $\mathfrak{m}$ be an arbitrary maximal ideal of R. Some r_i is not contained in $\mathfrak{m}$, say $r = r_1$. As there is an R_r–linear map from $R_r^{1/q_1} \to R_r$ sending d^{1/q_1} to 1, there is such a map from $R_\mathfrak{m}^{1/q_1} \to R_\mathfrak{m}$. In particular, $d \notin \mathfrak{m}^{[q]}$, which proves the existence of a uniform q.

The existence of such a q proves that for each maximal ideal $\mathfrak{m}$ of R there are an element $r = r_\mathfrak{m}$ and an R_r–linear map from $R_r^{1/q} \to R_r$ sending $d^{1/q}$ to 1. For such r, there is $N_r \subseteq \mathbb{N}$ for which there is an R–linear map φ_r from $R^{1/q} \to R$ sending $d^{1/q}$ to r^{N_r}. There exists a finite number of such r generating the unit ideal. If N is an integer larger than the corresponding N_r, we may express $1 = \sum s_i r_i^N$ for some $s_i \in R$. Then $\varphi = \sum s_i \varphi_{r_i}$ is an R–linear map taking $d^{1/q}$ to 1. $\qquad\square$

Proof of Theorem 13.1.6: The ring R_c is regular. Lemma 13.1.7 proves that for every non–zero element $d \in R$ there is a sufficiently high power of p, say Q, such that there exists an R_c–linear map from $R_c^{1/Q}$ to R_c sending $d^{1/Q}$ to 1. Lifting back to R one obtains an R–linear map from $R^{1/Q}$ to R sending $d^{1/Q}$ to a power of c. Taking $d = 1$ yields an R–linear map from $R^{1/Q}$ to R sending 1 to c^N for some N. The embedding of $R^{1/p}$ into $R^{1/Q}$ composed with this R linear map yields an R–linear map φ from $R^{1/p}$ to R sending 1 to c^N. Relabel this power of c as c. Then there is an R–linear map φ from $R^{1/p}$ to R sending 1 to c.

We claim that for any such c, c^3 is a test element. Let I be an arbitrary ideal of R and let $z \in I^*$. There is an element $d \in R^o$, such that for all q, $dz^q \in I^{[q]}$. From the results of the paragraphs above, there are a power q' of p and an R–linear map f from $R^{1/q'} \to R$ sending $d^{1/q'}$ to c^N for some N. In this case, $c^N z^q \in I^{[q]}$ for all q. Simply take (q')th roots of the equation $dz^{qq'} \in I^{[qq']}$ to obtain that $d^{1/q'} z^q \in I^{[q]} R^{1/q'}$. Applying f yields that $c^N z^q \in I^{[q]}$ for all q. We need to prove that this power N can be chosen independently of the element z and the ideal I. Choose N least with the property that $c^N z^q \in I^{[q]}$ for all q. Write $N = p(\lfloor N/p \rfloor) + i$. Taking pth roots yields that $c^{\lfloor N/p \rfloor + i/p} z^q \in I^{[q]} R^{1/p}$ for all q. Hence $c^{\lfloor N/p \rfloor + 1} z^q \in I^{[q]} R^{1/p}$ for all q. Applying φ we obtain that $c^{\lfloor N/p \rfloor + 2} z^q \in I^{[q]}$ for all q. As N was chosen least, necessarily $\lfloor N/p \rfloor + 2 \geq N$. It easily follows that $N \leq 3$. $\square$

13.2. Briançon–Skoda via tight closure

In this section we prove a version of the Briançon–Skoda Theorem for rings in positive prime characteristic via tight closure. The standard method of reducing to characteristic p also gives the same version for rings containing a field. However, the process of reduction to positive characteristic is beyond the scope of this book. We refer the reader to [124] for a systematic development of this technique.

A recurrent theme in studying powers of integrally closed ideals is that stronger theorems with easier proofs can often be given in positive characteristic by making use of the Frobenius homomorphism.

As an application we prove a characteristic p version of the Huneke–Itoh Theorem connecting powers of a reduction generated by a regular sequence and integral closure of powers. See Theorem 13.2.4. We also prove a corollary that in a two–dimensional regular local ring an ideal with a two–generated reduction has reduction number two. By Proposition 8.3.7 and Corollary 8.3.9 this means that whenever the residue field of a two–dimensional regular local ring is infinite, then every ideal has reduction number two.

Theorem 13.2.1 *Let R be a ring of characteristic p. Let I be any ideal generated by l elements. Then for all $n \geq 0$,*

$$\overline{I^{n+l}} \subseteq (I^{n+1})^*.$$

In particular, if R is regular or if every ideal of R is tightly closed, then $\overline{I^{n+l}} \subseteq I^{n+1}$ for all $n \geq 0$.

Proof: Write $I = (a_1, \ldots, a_l)$. By Equation (8.1.6) in Proposition 8.1.5, for all $h \geq 0$, $I^{lh+nh} \subseteq (a_1^h, \ldots, a_l^h)^{n+1} I^{h(l-1)}$.

Let $z \in \overline{I^{l+n}}$. There exists an element $c \in R^o$ such that $cz^N \in I^{(l+n)N}$ for all $N \gg 1$, by Corollary 6.8.12. By the above remarks, this latter ideal is contained in $(a_1^N, \ldots, a_l^N)^{n+1} I^{N(l-1)}$. Put $N = q = p^e$. We obtain that $cz^q \in (I^{n+1})^{[q]}$ by ignoring the term $I^{N(l-1)}$ in the containment of the line above. Hence $z \in (I^{n+1})^*$ as claimed. The last statement follows by applying Theorem 13.1.2 (6). □

This proof is amazingly simple. As mentioned above, one can use reduction to characteristic p to assert the validity of this theorem for regular local rings containing a field. A consequence of this theorem is that for regular local rings there is a fixed choice of l that works for all ideals:

Corollary 13.2.2 *If R is a Noetherian local ring with an infinite residue field and of dimension d, then for all $n \geq 0$ and for all ideals I,*

$$\overline{I^{d+n}} \subseteq (I^{n+1})^*.$$

If R is regular, then $\overline{I^{d+n}} \subseteq I^{n+1}$.

Proof: Each ideal I has a d–generated reduction J by Proposition 8.3.7 and Corollary 8.3.9. By Theorem 13.2.1, $\overline{I^{n+d}} = \overline{J^{n+d}} \subseteq J^{n+1} \subseteq I^{n+1}$. □

We end this section with a theorem of S. Itoh [148]:

Theorem 13.2.3 (Itoh [148]) *Let $(R, \mathfrak{m})$ be a Noetherian local ring and let $x_1, \ldots, x_r$ be a regular sequence in R. Set $I = (x_1, \ldots, x_r)$. Then for all $n \geq 1$,*

$$(x_1, \ldots, x_r)^n \cap \overline{I^{n+1}} = (x_1, \ldots, x_r)^n \overline{I}.$$

This theorem was proved at the same time independently by Huneke in [137] for rings that contain a field. Huneke's proof uses characteristic p methods and by the standard reduction to characteristic p gives the theorem for rings containing fields, whereas Itoh's proof is harder but works in full generality. Itoh also proved a similar theorem for systems of parameters in formally equidimensional local rings [147]. The two proofs illustrate a theme from this chapter: often results have easy proofs in positive characteristic, and then by reduction to positive characteristic can be claimed for rings containing fields. However, full generality usually requires other methods. We only give the characteristic p proof below, but of a more general statement:

Theorem 13.2.4 *Let R be a Noetherian ring of positive and prime characteristic p and let $x_1, \ldots, x_r$ be a regular sequence in R. Set $I = (x_1, \ldots, x_r)$. Then for all $n \geq 1$ and all $j = 1, \ldots, r$,*

$$(x_1, \ldots, x_j)^n \cap \overline{I^{n+1}} = (x_1, \ldots, x_j)^n \overline{I}.$$

Proof: It is clear that $(x_1, \ldots, x_j)^n \overline{I} \subseteq (x_1, \ldots, x_j)^n \cap \overline{I^{n+1}}$. We need to prove the reverse containment. It suffices to prove the opposite containment after localizing at an arbitrary associated prime of $(x_1, \ldots, x_j)^n \overline{I}$. Since finite intersections, products and integral closures commute with localization, we may assume R is local.

If $A = (a_1, \ldots, a_j)$ with $a_i \geq 0$, we will write x^A to denote $x_1^{a_1} x_2^{a_2} \cdots x_j^{a_j}$. We let $|A| = a_1 + \cdots + a_j$. If q is an integer we write $qA = (qa_1, \ldots, qa_j)$. Let $z \in (x_1, \ldots, x_j)^n \cap \overline{I^{n+1}}$ and write $z = \sum_{|A|=n} r_A x^A$.

Choose $c \in R^\circ$ such that for all large N, $cz^N \in (x_1, \ldots, x_r)^{N(n+1)}$, which is possible by Corollary 6.8.12. Apply this last equation with $N = q$ for large $q = p^e$. We obtain that

$$cz^q = \sum_{|A|=n} cr_A^q x^{qA} \in (x_1, \ldots, x_r)^{q(n+1)}.$$

Hence $cr_A^q \in ((x_1, \ldots, x_r)^{q(n+1)} + (x^{qB} \mid n = |B|, B \neq A)) : x^{qA}$, which is contained in $(x_1, \ldots, x_r)^q$ since $x_1, \ldots, x_r$ form a regular sequence (Exercise 13.1). It follows that $r_A \in \bar{I}$ by Corollary 6.8.12. $\qquad\square$

The reason why characteristic p is so useful in considering powers of ideals is apparent in the proof above. If we were not in positive characteristic, then when computing z^q in the proof above we would get terms involving all monomials in $x_1, \ldots, x_r$ of degree nq instead of those involving qth powers. Proceeding as in the proof we could then only conclude that $cr_A^q \in (x_1, \ldots, x_r)$, which does not come close to what we need.

By combining Itoh's theorem with the Briançon–Skoda Theorem of Lipman and Sathaye (see Section 13.3), the following corollary is almost immediate.

Corollary 13.2.5 *Let $(R, \mathfrak{m})$ be a two–dimensional regular local ring, and let x, y be a system of parameters of R. Set $I = \overline{(x, y)}$. Then $\overline{I^2} = (x, y)I$.*

Proof: By Theorem 13.3.3, $\overline{I^2} \subseteq (x, y)$, so that $\overline{I^2} = \overline{I^2} \cap (x, y) = (x, y)I$, the last equality coming from applying Theorem 13.2.3. $\qquad\square$

13.3. The Lipman–Sathaye version

For general rings, i.e., rings that do not necessarily contain a field, proofs of the Briançon–Skoda Theorem require different techniques. In this section we prove the general version due to Lipman and Sathaye. We first observe:

Lemma 13.3.1 *Let R be a Noetherian domain and let $a_i, t_i \in R$ for $1 \leq i \leq n$. Set $S = R[\frac{a_1}{t_1}, \ldots, \frac{a_n}{t_n}]$. Then $t_1 \cdots t_n$ is contained in the Jacobian ideal $J_{S/R}$.*

Proof: Let $f : R[X_1, \ldots, X_n] \to S$ be the R–homomorphism determined by sending X_i to $\frac{a_i}{t_i}$. Let P be the kernel of f. Inside P are the n elements $g_i = t_i X_i - a_i$. Since the height of P is n and P is prime, the element $t_1 \cdots t_n$ is in $J_{S/R}$. $\qquad\square$

This lemma almost suffices to prove the Briançon–Skoda Theorem of Lipman and Sathaye when applied to the extended Rees ring of an ideal I, using

Theorem 13.3.3. However, the powers obtained are off by one, so we need a different and more difficult statement.

Lemma 13.3.2 *Let R be a regular Noetherian domain with field of fractions K. Let L be a finite separable field extension of K and S a finitely generated R–algebra in L with integral closure T. Let $0 \neq t \in R$ be such that R/tR is regular. If $tS \cap R \neq tR$, then $J_{T/R} \subseteq tT$.*

Proof: First observe that T is a finite S–module by Theorem 9.2.3; in particular T is a finitely generated R–module so that the Jacobian ideal makes sense. We can then drop S from our discussion, as both the conclusion and hypothesis deal with T.

It suffices to prove that $(J_{T/R})_p \subseteq tT_p$ for every associated prime p of tT. Thus we may replace T by $V = T_p$, since $(J_{T/R})_p = J_{T_p/R}$, where p is an associated prime of tT. We can further replace R by $R_{p \cap R}$, which we relabel as R. In other words it suffices to prove that $J_{V/R} \subseteq tV$. Note that V is no longer finitely generated over R, but it is a localization of a finitely generated R–algebra. Since R is universally catenary, the Dimension Formula (Theorem B.5.1) applies, and hence V is a rank one discrete valuation ring of residual transcendence degree $d - 1$ over R, where $d = \dim(R)$. Set $R' = V \cap K$, a rank one discrete valuation with the same field of fractions as R. By Lemma 4.4.8, $J_{V/R} = J_{V/R'} J_{R'/R}$, so that it suffices to prove that $J_{R'/R} \subseteq tR'$. Note that $tS \cap R \subseteq tR' \cap R$, so that $tR' \cap R \neq tR$. Thus without loss of generality, we can replace V by R' and assume that V is a rank one discrete valuation ring with the same field of fractions as R. Using Exercise 9.9 there is a finite sequence of regular local rings, $R = R_0 \subseteq R_1 \subseteq R_2 \subseteq \cdots \subseteq R_n = V$, where each R_i is the local ring of a closed point on the blowup of the maximal ideal of R_{i-1}. Necessarily $n \geq 1$ since $tV \cap R \neq tR$. Set $J_i = J_{R_i/R_{i-1}}$ for $1 \leq i \leq n$. By Lemma 4.4.8, $J_{V/R} = J_n J_{n-1} \cdots J_1$.

We need to describe the ideals J_i. Set $d_i = \dim(R_i)$, and let $\mathfrak{m}_i$ be the center of V on R_i. Fix a regular system of parameters $x_{1i}, \ldots, x_{d_i i}$ for $\mathfrak{m}_i$ such that $x_{i1} V = \mathfrak{m}_i V$. Since these regular systems of parameters form regular sequences, R_{i+1} is a localization of the ring $R_i[\frac{x_{2i}}{x_{1i}}, \ldots, \frac{x_{d_i i}}{x_{1i}}] \cong R_i[X_{2i}, \ldots, X_{d_i i}]/(x_{1i} X_{2i} - x_{2i}, \ldots, x_{1i} X_{d_i i} - x_{d_i i})$, the isomorphism following from Corollary 5.5.9. It follows that J_{i+1} is generated by the single element $x_{1i}^{d_i - 1}$, and hence

$$J_{V/R} = J_n J_{n-1} \cdots J_1 = \prod_{i=1}^{n} x_{1i}^{d_i - 1} V = \prod_{i=1}^{n} (\mathfrak{m}_i)^{d_i - 1} V.$$

We inductively define a sequence of ideals as follows: $I_0 = tR$, and $I_{i+1} = R_{i+1}$ if $I_i = R_i$; if not $I_{i+1} = I_i R_{i+1} (\mathfrak{m}_i)^{-1} R_{i+1}$, which is an ideal in R_{i+1}. By construction, $tV = I_n \prod (\mathfrak{m}_i)^{\epsilon_i} V$ where ϵ_i is 1 if $I_i \neq R_i$, and is 0 otherwise. Since $(\mathfrak{m}_i)^{d_i - 1} V \subseteq (\mathfrak{m}_i)^{\epsilon_i}$ in any case, to finish the proof of the lemma it is enough to prove that $I_n = R_n$.

Let v_i be the order valuation of the ring R_i. We need only to prove that $v_{n-1}(I_{n-1}) \leq 1$ since then either $I_{n-1} = R_{n-1}$ or else $I_{n-1} = \mathfrak{m}_{n-1}$. By induction on j we prove more generally that $v_j(I_j) \leq 1$ for all $j < n$. For $j = 0$ this holds since R/tR is regular, and hence $v_0(t) = 1$. If $v_j(I_j) = 0$ for some j, then this is true for all larger values, so assume that $v_j(I_j) = 1$. If $I_j = x_{1j}R_j$, then $I_{j+1} = R_{j+1}$. If not, after relabeling we may assume that $I_j = x_{2j}R_j$. It suffices to prove that $R_{j+1}/(\frac{x_{2j}}{x_{1j}})R_{j+1}$ is regular. This follows at once from the isomorphism $R_{j+1} \cong R_j[X_{2j}, \ldots, X_{d_ij}]/(x_{1j}X_{2j} - x_{2j}, \ldots, x_{1j}X_{d_jj} - x_{d_jj})$ by noting that $R_{j+1}/(x_{1j}, \frac{x_{2j}}{x_{1j}})R_{j+1}$ is isomorphic to a polynomial ring in $d_j - 2$ variables over the residue field of R_j. $\square$

We are ready to prove:

Theorem 13.3.3   (Lipman and Sathaye [192]) *Let R be a regular ring. Let I be any ideal of R that is generated by l elements. Then for any $n \geq 0$,*
$$\overline{I^{n+l}} \subseteq I^{n+1}.$$

Proof: We may localize and assume that $(R, \mathfrak{m})$ is a regular local ring. Write $I = (a_1, \ldots, a_l)$ and set $S = R[t^{-1}, a_1 t, \ldots, a_l t]$, the extended Rees algebra of I. Set $B = R[t^{-1}]$ and note that B is a regular Noetherian domain and S is a finitely generated B–algebra with the same field of fractions. Letting $u = t^{-1}$ we also observe that B/uB is regular. We can apply Lemma 13.3.2 to get that $J_{\overline{S}/B} \subseteq u\overline{S}$. Note that $S = R[u, \frac{a_1}{u}, \ldots, \frac{a_l}{u}] = B[\frac{a_1}{u}, \ldots, \frac{a_l}{u}]$. Thus by Lemma 13.3.1, $u^l \in J_{S/B}$.

Since $J_{\overline{S}/B} \subseteq u\overline{S}$, it follows that $t\overline{S} = u^{-1}\overline{S} \subseteq \overline{S} : J_{\overline{S}/B} \subseteq S : J_{S/R}$, where the last containment is the content of Theorem 12.3.10, using Remark 12.3.11. Hence $J_{S/R}t\overline{S} \subseteq S$. Therefore $u^{l-1}\overline{S} \subseteq S$. By Proposition 5.2.4, the $(n+l)$th graded piece of $\overline{S}$ is $\overline{I^{n+l}}t^{n+l}$, so that $u^{l-1}\overline{I^{n+l}}t^{n+l} = \overline{I^{n+l}}t^{n+1} \subseteq S$. It follows that $\overline{I^{n+l}} \subseteq I^{n+1}$. $\square$

Corollary 13.3.4   *Let R be a regular ring of dimension d. Let I be any ideal of R. Then for any $n \geq 0$,*
$$\overline{I^{d+n}} \subseteq I^{n+1}.$$

Proof: We may localize R to prove the statement and assume that R is local. By Lemma 8.4.2, replacing R by $R(t)$ does not change the dimension of R nor the fact that it is regular. Moreover, $R(t)$ has an infinite residue field. If we prove the result for $R(t)$, then
$$\overline{I^{d+n}} \subseteq \overline{I^{d+n}R(t)} \cap R \subseteq I^{n+1}R(t) \cap R = I^{n+1},$$
the last equality holding since $R(t)$ is faithfully flat over R. Hence we may replace R by $R(t)$ and assume the residue field is infinite.

We may replace I by a minimal reduction J of I; in this case $\overline{I^{d+n}} = \overline{J^{d+n}}$, while $J^{n+1} \subseteq I^{n+1}$. But a minimal reduction is generated by at most d elements by Proposition 8.3.7 and Corollary 8.3.9, finishing the proof. $\square$

13.4. General version

In Corollary 13.3.4, we proved that if R is regular of dimension d, then for *every* ideal $I \subseteq R$, $\overline{I^{d+n}} \subseteq I^{n+1}$ for all $n \geq 0$. Can we remove the assumption that R is regular? The answer is certainly "no" if we insist on keeping the dimension d in the equation. For example, in the hypersurface $z^2 - x^5 - y^7 = 0$, z is integral over $(x, y)^2$, and the dimension of this hypersurface is 2. The conclusion of Corollary 13.3.4 or even the conclusion of Theorem 13.3.3 with $n = 0$ would give that $z \in (x, y)$, which is false.

Remarkably, however, for "good" rings there is always a uniform bound k such that for all ideals I, $\overline{I^{k+n}} \subseteq I^{n+1}$ for all $n \geq 0$. This result is due to Huneke [138]. In this section we prove it assuming Huneke's result concerning the uniform Artin–Rees Theorem, a theorem that is beyond the scope of this book. We write its statement as we use it below in Theorem 13.4.2. For additional results on this topic, see [63], [217], [218], and [248].

The main result in this section is that for all ideals I of R and for all $n \geq 1$, $\overline{I^{n+k}} \subseteq I^n$ if R is a reduced Noetherian ring that is either local containing the rational numbers or F-finite of positive prime characteristic p (recall that F-finite means that $R^{1/p}$ is finite as an R–module). A more general result is in [138].

Definition 13.4.1 *Whenever a ring R has the property that given two finitely generated R–modules $N \subseteq M$ there exists a positive integer k such that for all $n \geq 1$ and all ideals I, $I^{n+k} M \cap N \subseteq I^n N$, we say that R has the* **uniform Artin–Rees property**.

We now quote without proof a uniform Artin–Rees Theorem from [138]:

Theorem 13.4.2 (Huneke [138]) *Let R be a Noetherian ring. Let $N \subseteq M$ be two finitely generated R–modules. If R satisfies any of the conditions below, then there exists an integer k such that for all ideals I of R and for all $n \geq 1$*

$$I^{n+k} M \cap N \subseteq I^n N.$$

(1) R is essentially of finite type over a Noetherian local ring.
(2) R is a reduced ring of characteristic p and R is F-finite.
(3) R is essentially of finite type over $\mathbb{Z}$.

In order to use the uniform Artin–Rees property, we need to understand the existence of elements that uniformly multiply $\overline{I^{n+k}}$ into I^n. We codify such elements in the following definition:

Definition 13.4.3 *Let R be a Noetherian ring. We define $T_k(R) = \cap_{n,I}(I^n : \overline{I^{n+k}})$, where the intersection is over all positive integers n and all ideals I of R. We set $T(R) = \cup_k T_k(R)$. (It is easy to see that as k increases, the ideals $T_k(R)$ also increase.)*

An element $c \in T(R)$ if and only if there is an integer k such that for all ideals I, $c\overline{I^{n+k}} \subseteq I^n$. When considering the existence of such elements, it is natural to assume that R is reduced: the nilradical $\sqrt{0}$ of R is contained in the integral closure of every ideal, so that $c \in T(R)$ would imply that $c\sqrt{0}$ is in every ideal of R. This is not a useful condition to work with if R is not reduced. We use these elements as follows:

Theorem 13.4.4 *Let R be a reduced ring. Assume that $T(R)$ is not contained in any minimal prime of R and in addition assume that R has the uniform Artin–Rees property. Then there exists a positive integer k such that for all ideals I of R, $\overline{I^{n+k}} \subseteq I^n$.*

Proof: Let I be an ideal of R and let m be an integer such that $T_m(R)$ is not in any minimal prime ideal. Choose an element c in $T_m(R)$ not in any minimal prime. Let l be an integer such that for all $n \geq 1$ and all ideals I, $I^{n+l} \cap (c) \subseteq I^n(c)$. Let $y \in \overline{I^{n+m+l}}$ for some $n \geq 1$. Then $cy \in (c) \cap \overline{I^{n+l}} \subseteq I^n(c)$ by the uniform Artin–Rees property. Therefore y is in I^n, which gives the theorem with $k = m + l$. $\qquad\square$

This simple proof shows that there are two crucial components involved in proving a general Briançon–Skoda Theorem: the existence of elements in $T(R)$ that are in no minimal prime ideal and a uniform Artin–Rees property.

We next give some criteria for $T(R)$ to not be zero. We first consider the case of positive characteristic; our results rest on the theory of test elements. See the latter part of Section 13.1 for the background on test elements.

Proposition 13.4.5 *Let R be Noetherian reduced ring of finite Krull dimension and of prime characteristic p with an infinite residue field. If R is F-finite, then $T(R) \cap R^o \neq \emptyset$.*

Proof: By Corollary 13.2.2, setting $d = \dim(R)$, we know that for every ideal I of R that is primary to a maximal ideal, $\overline{I^{n+d}}$ is contained in $(I^{n+1})^*$. By Theorem 13.1.6, R has a test element c. Then $c(I^{n+1})^* \subseteq I^{n+1}$ and hence c is in $T_{d-1}(R)$. Consequently $T(R) \neq 0$. $\qquad\square$

We next prove several results concerning the existence of elements in $T(R)$ in the case R contains the rational numbers. These results rely on the theorem of Lipman and Sathaye, Theorem 13.3.3.

Lemma 13.4.6 *Let R be a Noetherian ring of dimension d. Let k be a positive integer and let c be an element c not in any minimal prime ideal of R. Assume that for every ideal I primary to a maximal ideal and that for all $n \geq 1$, $c\overline{I^{n+k}} \subseteq I^n$. Then $T_k(R) \cap R^o \neq \emptyset$. In particular, $T(R) \cap R^o \neq \emptyset$.*

Proof: Let J be an arbitrary ideal of R. For all positive l and all maximal ideals $\mathfrak{m}$, $c(\overline{J^{n+k}}) \subseteq c(\overline{(J + \mathfrak{m}^l)^{n+k}}) \subseteq (J + \mathfrak{m}^l)^n \subseteq J^n + \mathfrak{m}^l$. Since this is true for all l and all $\mathfrak{m}$, it follows that $c(\overline{J^{n+k}}) \subseteq J^n$. $\qquad\square$

Theorem 13.4.7 *Let R be a regular domain of finite Krull dimension with quotient field K and having infinite residue fields. Assume that the character-*

istic of K is 0. *Furthermore, let S be a finitely generated R–algebra containing R that is a domain. Then $T(S) \neq 0$.*

Proof: Write $S = R[x_1, \ldots, x_s, u_1, \ldots, u_k]$. We may assume that $x_1, \ldots, x_s$ form a transcendence basis of the quotient field L of S over K and that u_i are algebraic over $K(x_1, \ldots, x_s)$. Each u_i satisfies a polynomial $Q_i(U_i)$ of minimal degree (though not necessarily monic) whose coefficients are in $B = R[x_1, \ldots, x_s]$, which is a regular ring. As the characteristic of K is 0 and the polynomials Q_i are minimal degree, the derivative of $Q_i(U_i)$ evaluated at u_i is non–zero. Set c equal to the image of $\prod_{i=1}^{k} \frac{\partial Q_i}{\partial U_i}$ in S. Note that $c \neq 0$. We will show that $c \in T(S)$.

Let I be any ideal of S that is primary to a maximal ideal M of S. Set $h = \dim(S)$, so by Theorem 8.7.3, I is integral over an ideal J generated by at most $h + 1 = l$ elements, say $a_1, \ldots, a_l$. Consider the ring $T = S[t^{-1}, a_1 t, \ldots, a_l t]$, the extended Rees algebra of J. The field of fractions $L(t)$ of T is finite and separable over the fraction field of $B[t^{-1}]$ and T is a finitely generated $B[t]$–algebra contained in $K(t)$. Furthermore, $B[t^{-1}]$ is a regular Noetherian domain. Thus we may apply Theorem 13.3.3 to obtain that $J_{T/B[t^{-1}]}\overline{T} \subseteq T$, where $\overline{T}$ is the integral closure of T. Write T as a homomorphic image of $B[t^{-1}][U_1, \ldots, U_k, X_1, \ldots, X_l]$, where U_i goes to u_i and X_i goes to $a_i t$. Call this map f. The kernel of f contains the polynomials $Q_i(U_i)$ and the polynomials $W_i(X_i, U_1, \ldots, U_k) = t^{-1}X_i - A_i(U_1, \ldots, U_k)$, where $A_i(U_1, \ldots, U_k)$ is any polynomial in $B[U_1, \ldots, U_k]$ that maps to a_i under the map f.

Consider the Jacobian $\frac{\partial(Q_1, \ldots, Q_k, W_1, \ldots, W_l)}{\partial(U_1, \ldots, U_k, X_1, \ldots, X_l)}$. The image in T of the determinant of this lower triangular matrix is ct^{-l} and is contained in J. In particular, this element multiplies $\overline{T}$ into T.

By Proposition 5.2.4, $\overline{T} = \overline{S}[t^{-1}, \overline{IS}t, \ldots, \overline{I^n S}t^n, \ldots]$, where $\overline{S}$ is the integral closure of S. Since $ct^{-l}\overline{T} \subseteq T$, we obtain that $ct^{-l}(\overline{I^n}t^n) \subseteq T$, so that for $n \geq 1$, $c(\overline{I^{n+l}}) \subseteq J^n \subseteq I^n$. We note that c does not depend upon the ideal I. Hence for every ideal I in S that is primary to a maximal ideal, $c(\overline{I^{n+l}}) \subseteq I^n$ for all $n \geq l$. Using Lemma 13.4.6 we obtain that $T(S) \neq 0$. $\qquad\square$

Theorem 13.4.8 (Huneke [138]) *Let R be a Noetherian reduced ring. Then there exists a positive integer k such that for all ideals I of R and all $n \in \mathbb{N}$, $\overline{I^{n+k}} \subseteq I^n$ if R satisfies either of the following conditions:*
(1) R is an analytically unramified Noetherian local ring containing $\mathbb{Q}$.
(2) R is of prime characteristic p and R is F-finite.

Proof: We first prove (2). By Proposition 13.4.5, $T(R) \cap R^o \neq \emptyset$, so by Theorem 13.4.4, we need only to see that R satisfies the uniform Artin–Rees property. This follows from Theorem 13.4.2, proving (2).

We next assume that R is an analytically unramified Noetherian local ring containing the rational numbers. We first make some observations. To prove the result for a ring R we may replace R by any faithfully flat reduced extension S of R, for if we can prove that the theorem holds in S, then since ideals are contracted from S and the integral closure of an ideal remains in the integral closure of the same ideal under any homomorphism, we may simply descend the statement of the theorem from S to R. We may thus replace R by its completion. Note that R remains reduced as R is analytically unramified. R satisfies the uniform Artin–Rees condition by the previous theorem, so it only remains to prove that $T(R) \cap R^o \neq \emptyset$, as above.

First consider $S = R/P$, where P is a minimal prime of R. We may apply Theorem 13.4.7: S satisfies the assumptions of that theorem because S is module–finite over a regular local ring by the Cohen Structure Theorem. Theorem 13.4.7 then gives that for all minimal primes P, $T(R/P) \neq 0$. Since there are only finitely many minimal primes, there exist an integer k and elements $c_P \notin P$ such that $c_P(\overline{I^{n+k}}) \subseteq I^n + P$ for all such P. Choose elements $d_P \notin P$ such that $d_P P = 0$. This choice is possible since R is reduced. Set $c = \sum c_P d_P$, where the sum ranges over all minimal prime ideals of R. Note that $c \in R^o$. Moreover, $c_P d_P(\overline{I^{n+k}}) \subseteq d_P(I^n + P) \subseteq I^n$. This holds for each minimal prime P, and hence $c(\overline{I^{n+k}}) \subseteq I^n$, proving that $T(R) \cap R^o \neq \emptyset$. $\qquad \square$

Huneke proved in [138, Theorem 4.13], that the conclusion of the theorem above also holds if R is reduced and essentially of finite type over an excellent Noetherian local ring or over $\mathbb{Z}$.

13.5. Exercises

13.1 Let R be a Noetherian ring, let $x_1, \ldots, x_r$ form a regular sequence in R and let $I = (x_1, \ldots, x_r)$ be an ideal. Prove that $((x_1, \ldots, x_r)^{q(n+1)} + (x^{qB} \mid B_1 + \cdots + B_r = n, B \neq A)) : x^{qA} \subseteq (x_1, \ldots, x_r)^q$, where the notation is as in Theorem 13.2.4.

13.2 Let R be a Noetherian local ring of characteristic p and let I be an ideal generated by a regular sequence of length d. Suppose that I^t is contained in the ideal generated by all the test elements of R. Prove for all $k \geq 0$, $\overline{I^{d+k+t}} \subseteq I^{k+1}$. (See [146] for generalizations of this exercise.)

13.3 Let $(R, \mathfrak{m})$ be a Noetherian reduced d–dimensional ring of characteristic p. Suppose that the set of all test elements generates an $\mathfrak{m}$–primary ideal (this is called **the test ideal**). Assume that the residue field of R is infinite. Prove that there exists an integer t such that if $I \subseteq \mathfrak{m}^t$, then $\overline{I^{n+d}} \subseteq I^n$ for all n.

13.4 Let R be a regular local ring containing an infinite field of positive characteristic. Let I be an ideal of R having analytic spread ℓ and let J be an arbitrary reduction of I. Set $h = \text{bight}(I)$. Prove that for all $n \geq 0$, $\overline{I^{n+\ell}} \subseteq J^{n+1}(J^{\ell-h})^{min}$. (Hint: Prove that there exists an element $f \neq 0$ such that for all $q = p^e$, $f J^{(\ell-1)q} \subseteq (J^{\ell-h})^{[q]}$.)

14
Two–dimensional regular local rings

This chapter presents the theory of integrally closed ideals in regular local rings of dimension two. This theory was begun by Zariski in his classic paper [320], and can be found in [322, Appendix 5]. Zariski's motivation was to give algebraic meaning to the idea of complete linear systems of curves defined by base conditions in which the curves pass through prescribed base points with given multiplicities. The set of base points includes infinitely near points; we will discuss these notions toward the end of the chapter. The work of Zariski can be broken into two parts: first the study of a class of ideals called **full** ideals, and secondly the study of **quadratic transformations**. The latter allows changing of rings and usage of induction. All integrally closed ideals are full, but not vice versa. Zariski's work was considerably enhanced and extended by Lipman. We present a reciprocity formula due to Lipman, which compares different divisorial valuations on a two–dimensional regular local ring. We also prove the famous Hoskin–Deligne Formula, which expresses the co–length of an integrally closed ideal in terms of the orders of the transforms of the ideal at infinitely near points. Our treatment follows that of [136]. For different approaches and generalizations see [181], [50], [56], [51], [189], [190], and [235].

We begin with a definition and the statements of the two main results of Zariski regarding the structure of integrally closed ideals in two–dimensional regular local rings. These results are proved later in this chapter.

Definition 14.0.1 *An ideal I is* **simple** *if $I \neq J \cdot K$, for any proper ideals J, K. We write $J \mid I$ if $I = JK$ for some ideal K.*

Theorem 14.4.4 (Zariski) *In a two–dimensional regular local ring products of integrally closed ideals are closed.*

Theorem 14.4.9 (Zariski) *Let $(R, \mathfrak{m})$ be a two–dimensional regular local ring. Then every non–zero integrally closed ideal I in R can be written uniquely (except for ordering) as*

$$I = a_1^{l_1} \cdots a_m^{l_m} I_1^{k_1} \cdots I_n^{k_n},$$

where $I_1, \ldots, I_n$ are simple $\mathfrak{m}$–primary integrally closed ideals, $a_1, \ldots, a_m$ are pairwise relatively prime irreducible elements of R, and $l_1, \ldots, l_m, k_1, \ldots, k_n$ are positive integers.

14.1. Full ideals

Let $(R, \mathfrak{m}, k)$ be a two–dimensional regular local ring with $\mathfrak{m} = (x, y)$. Recall that the **order** of $\mathfrak{m}$, denoted $\mathrm{ord}_{\mathfrak{m}}(\_)$, is the function that assigns to each non–zero $r \in R$ the largest integer k such that $r \in \mathfrak{m}^k$. For any non–zero ideal I, $\mathrm{ord}_{\mathfrak{m}}(I) = \min\{\mathrm{ord}(r) \,|\, r \in I\} = \max\{k \,|\, I \subseteq \mathfrak{m}^k\}$. In this chapter, by abuse of notation we will denote $\mathrm{ord}_{\mathfrak{m}}(\_)$ also as $\mathrm{ord}_R(\_)$. When R is known from the context, we will simply write $\mathrm{ord}(\_)$.

Remarks 14.1.1 We will use freely the facts below.
(1) R is a unique factorization domain.
(2) Every finitely generated R–module M has finite projective dimension at most two. In particular, if I is $\mathfrak{m}$–primary, then $\mathrm{depth}(R/I) = 0$, and the Auslander–Buchsbaum Formula shows that $\mathrm{pd}_R(R/I) = 2$. In this case, the resolution looks like:
$$0 \longrightarrow R^{n-1} \longrightarrow R^n \longrightarrow R \longrightarrow R/I \longrightarrow 0.$$
(To see that the last rank is $n-1$, tensor with the field of fractions of R.) The resolution is minimal if and only if $n = \mu(I)$.
(3) $\mathrm{gr}_{\mathfrak{m}}(R) \cong k[X, Y]$ is a polynomial ring.
(4) The order function ord is a valuation by Theorem 6.7.9 (because $\mathrm{gr}_{\mathfrak{m}}(R)$ is a domain).
(5) If $\mathrm{ord}(f) = n$, then by f^* we denote $f + \mathfrak{m}^{n+1} \in \mathfrak{m}^n / \mathfrak{m}^{n+1} \in gr_{\mathfrak{m}}(R)$. We call f^* the **leading form** of f.
(6) The greatest common divisor of the elements in
$$\frac{I + \mathfrak{m}^{\mathrm{ord}(I)+1}}{\mathfrak{m}^{\mathrm{ord}(I)+1}} \subseteq \frac{\mathfrak{m}^{\mathrm{ord}(I)}}{\mathfrak{m}^{\mathrm{ord}(I)+1}} \subseteq gr_{\mathfrak{m}}(R) \cong k[X, Y],$$
is called the **content** of I, and is denoted $c(I)$. Note that $c(I)$ is a homogeneous polynomial in two variables, of degree at most $\mathrm{ord}(I)$. It is uniquely determined up to a unit multiple.
For example, $c(x, y) = 1$, $c(x^2, xy^3, y^4) = X^2$, and $c(xy^2, x^2y, y^4, x^4) = XY$.
(7) If I and J are non–zero ideals in R, then $c(IJ) = c(I) \cdot c(J)$ (as $\mathrm{gr}_{\mathfrak{m}}(R)$ is a unique factorization domain).
(8) Let $x \in \mathfrak{m} \setminus \mathfrak{m}^2$, and let I be an $\mathfrak{m}$–primary ideal. Then $R' = R/(x)$ is a one–dimensional regular local ring, so that $\lambda(\frac{R}{I + (x)}) = \mathrm{ord}_{R'}(I')$, where $I' = IR'$. Consider the exact sequences
$$0 \longrightarrow \frac{I : \mathfrak{m}}{I} \longrightarrow \frac{I : x}{I} \longrightarrow \frac{I : x}{I : \mathfrak{m}} \longrightarrow 0,$$
$$0 \longrightarrow \frac{I : x}{I} \longrightarrow \frac{R}{I} \overset{x}{\longrightarrow} \frac{R}{I} \longrightarrow \frac{R}{I + (x)} \longrightarrow 0.$$
Because length is additive on exact sequences, we get
$$\mathrm{ord}(I) \leq \mathrm{ord}_{R'}(I') = \lambda\Big(\frac{R}{I + (x)}\Big) = \lambda\Big(\frac{I : x}{I}\Big)$$
$$= \lambda\Big(\frac{I : \mathfrak{m}}{I}\Big) + \lambda\Big(\frac{I : x}{I : \mathfrak{m}}\Big). \tag{14.1.2}$$

Lemma 14.1.3 *Let $(R, \mathfrak{m}, k)$ be a two–dimensional regular local ring, and let I be an $\mathfrak{m}$–primary ideal of R. Then*
(1) $\lambda(\frac{I:\mathfrak{m}}{I}) = \mu(I) - 1$.
(2) $\mu(I) \leq \operatorname{ord}(I) + 1$.
(3) *For any $x \in \mathfrak{m} \setminus \mathfrak{m}^2$,*
$$\mu(I) - 1 \leq \operatorname{ord}(I) \leq \operatorname{ord}_{R'}(I') = \lambda\left(\frac{R}{I + (x)}\right) = \mu(I) - 1 + \lambda\left(\frac{I : x}{I : \mathfrak{m}}\right).$$

Proof: We compute $\operatorname{Tor}_2^R(k, R/I)$ first using the resolution of k, then using the resolution of R/I. Let $\mathfrak{m} = (x, y)$. Then

$$0 \longrightarrow R \xrightarrow{\begin{bmatrix} y \\ -x \end{bmatrix}} R^2 \xrightarrow{[x \; y]} R \longrightarrow k \longrightarrow 0$$

is the projective resolution of k. Tensoring this resolution with R/I shows that $\operatorname{Tor}_2^R(k, R/I) \cong (I : \mathfrak{m})/I$. On the other hand, Remark 14.1.1 (2) shows that the k–vector space dimension of $\operatorname{Tor}_2^R(k, R/I)$, which is the second Betti number of R/I, is exactly one less than the minimal number of generators of I. This proves (1).

A minimal free resolution of R/I has the form

$$0 \longrightarrow R^{n-1} \xrightarrow{A} R^n \longrightarrow R \longrightarrow R/I \longrightarrow 0,$$

where $n = \mu(I)$. The Hilbert–Burch Theorem A.4.2 gives that $I = I_{n-1}(A)$. In particular, $\operatorname{ord}(I) \geq n - 1$, since the entries of A are contained in $\mathfrak{m}$. This proves (2).

Then by Inequalities (14.1.2), for any $x \in \mathfrak{m} \setminus \mathfrak{m}^2$, $\mu(I) - 1 \leq \operatorname{ord}(I) \leq \operatorname{ord}_{R'}(I') = \lambda(\frac{R}{I+(x)}) = \mu(I) - 1 + \lambda(\frac{I:x}{I:\mathfrak{m}})$, which proves (3). $\square$

Theorem 14.1.4 *Let $(R, \mathfrak{m})$ be a two–dimensional regular local ring and let I be an $\mathfrak{m}$–primary ideal.*
(1) If $x \in \mathfrak{m} \setminus \mathfrak{m}^2$, then $I : x = I : \mathfrak{m}$ if and only if $\mu(I) - 1 = \lambda(\frac{R}{I+(x)}) = \operatorname{ord}(I)$. If $x \in \mathfrak{m}^2$, then $I : x = I : \mathfrak{m}$ if and only if $I = \mathfrak{m}$.
(2) If $x \in R$ is such that $I : \mathfrak{m} = I : x$, then $\mu(I) = \operatorname{ord}(I) + 1$.
(3) If $\mu(I) = \operatorname{ord}(I) + 1$ and the residue field of R is infinite, then there exists an element $x \in \mathfrak{m} \setminus \mathfrak{m}^2$ such that $I : \mathfrak{m} = I : x$.

Proof: By Lemma 14.1.3 (3), if $x \in \mathfrak{m} \setminus \mathfrak{m}^2$, $I : x = I : \mathfrak{m}$ if and only if $\mu(I) - 1 = \lambda(\frac{R}{I+(x)}) = \operatorname{ord}(I)$.

Suppose that $x \in \mathfrak{m}^2$ and that $I : \mathfrak{m} = I : x$. Since I is $\mathfrak{m}$–primary, there exists $e \in \mathbb{N}$ such that $\mathfrak{m}^e \subseteq I$. Then $I : \mathfrak{m}^2 \subseteq I : x = I : \mathfrak{m} \subseteq I : \mathfrak{m}^2$, proving that $I : \mathfrak{m} = I : \mathfrak{m}^2$. Hence by induction, $I : \mathfrak{m}^c = I : \mathfrak{m} = R$, which occurs only if $I = \mathfrak{m}$. Conversely, if $I = \mathfrak{m}$, then $I : x = R = I : \mathfrak{m}$, which finishes the proof of (1).

Suppose that $x \in R$ such that $I : \mathfrak{m} = I : x$. If $x \in R \setminus \mathfrak{m}$, then $I : \mathfrak{m} = I$, whence by induction for all n, $I : \mathfrak{m}^n = I$, which contradicts the assumption that I is $\mathfrak{m}$–primary. So necessarily $x \in \mathfrak{m}$, and then the two parts of (1) prove that $\mu(I) = \text{ord}(I) + 1$. This proves (2).

To finish the proof, suppose that the residue field of R is infinite. Then there are infinitely many linear forms in $\text{gr}_{\mathfrak{m}}(R)$ that do not divide $c(I)$. Choose any $x \in \mathfrak{m} \setminus \mathfrak{m}^2$ whose leading form is one of these linear forms. Then there is $f \in I$ such that $f \notin \mathfrak{m}^{\text{ord}(I)+1}$ and such that the leading form of x does not divide the leading form of f. Then $\text{ord}(I) = \text{ord}(f) = \text{ord}_{R'}(f + I) = \text{ord}_{R'}(I')$, where $R' = R/(x)$ and $I' = IR'$. By Lemma 14.1.3 (3) then $\mu(I) = \text{ord}(I) + 1$ implies that $I : x = I : \mathfrak{m}$, proving (3). $\qquad\square$

Definition 14.1.5 *Let $(R, \mathfrak{m})$ be a two–dimensional regular local ring. If I is an $\mathfrak{m}$–primary ideal such that there exists an element $x \in R$ with $I : \mathfrak{m} = I : x$, we say that I is $\mathfrak{m}$–full. For brevity, in this chapter we call such ideals **full**.*

Full ideals play an important role throughout the rest of this chapter, so understanding equivalent conditions for an ideal to be full is critical. The next proposition summarizes some of these equivalent conditions. We then prove that integrally closed ideals are full when the residue field is infinite.

Proposition 14.1.6 *Let $(R, \mathfrak{m})$ be a two–dimensional regular local ring and let I be an $\mathfrak{m}$–primary ideal. Then the following are equivalent for $x \in \mathfrak{m} \setminus \mathfrak{m}^2$:*
(1) $I : \mathfrak{m} = I : x$.
(2) $\mathfrak{m}I : x = I$.

Proof: Assume (1). Clearly $I \subseteq \mathfrak{m}I : x$. To prove the opposite inclusion let $u \in \mathfrak{m}I : x$. Write $\mathfrak{m} = (x, y)$ for some y and $ux = xi_1 + yi_2$ for some $i_1, i_2 \in I$. Then $x(u - i_1) = yi_2$. Since x, y form a regular sequence, there exists an element $t \in R$ such that $xt = i_2$ and $yt = u - i_1$. But then $t \in I : x = I : \mathfrak{m}$, giving that $yt \in I$ and therefore $u \in I$, proving (2). Next assume (2). Since $I : \mathfrak{m} \subseteq I : x$ in any case, let $u \in I : x$. Clearly $\mathfrak{m}u \subseteq \mathfrak{m}I : x$, so that (2) implies $\mathfrak{m}u \subseteq I$, and thus $u \in I : \mathfrak{m}$, proving (1). $\qquad\square$

If the residue field of R is infinite, then the existence of an element $x \in \mathfrak{m} \setminus \mathfrak{m}^2$ such that I satisfies the two conditions of Proposition 14.1.6 is equivalent to the condition that $\mu(I) = \text{ord}(I) + 1$, by Theorem 14.1.4 (2) and (3). An important point to understand is which x satisfy these conditions.

Proposition 14.1.7 *Let $(R, \mathfrak{m})$ be a two–dimensional regular local ring. Let I be an $\mathfrak{m}$–primary ideal such that $\mu(I) = \text{ord}(I) + 1$, and let $x \in \mathfrak{m} \setminus \mathfrak{m}^2$. Then the following are equivalent:*
(1) $\lambda(\frac{R}{I+(x)}) = \text{ord}(I)$.
(2) $I : x = I : \mathfrak{m}$.
(3) $\mathfrak{m}I : x = I$.
(4) I is not contained in $(x) + \mathfrak{m}^{\text{ord}(I)+1}$.
(5) The leading form x^* of x in $\text{gr}_{\mathfrak{m}}(R)$ does not divide $c(I)$.
(6) $\mathfrak{m}^{\text{ord}(I)} \subseteq I + (x)$.
(7) For all $n \geq 0$, $\mathfrak{m}^{n+\text{ord}(I)} = I\mathfrak{m}^n + x^{n+1}\mathfrak{m}^{\text{ord}(I)-1}$.

Proof: (1) and (2) are equivalent by Theorem 14.1.4 (1). (2) and (3) are equivalent by Proposition 14.1.6. Observe that x^* does not divide $c(I)$ if and only if I is not contained in $(x) + \mathfrak{m}^{\mathrm{ord}(I)+1}$, which holds if and only if $\mathrm{ord}(I) = \lambda(\frac{R}{I+(x)})$ since $R/(x)$ is a principal ideal domain, which holds if and only if $\mathfrak{m}^{\mathrm{ord}(I)}$ and I have the same image in $R/(x)$. This establishes the equivalence of (1), (4), (5) and (6). Clearly (7) with $n = 0$ implies (6). Now assume that (6) holds. Let $r = \mathrm{ord}(I)$. Since R is a regular local ring, $xR \cap \mathfrak{m}^r = x\mathfrak{m}^{r-1}$, hence $\mathfrak{m}^r \subseteq (I + xR) \cap \mathfrak{m}^r = I + xR \cap \mathfrak{m}^r = I + x\mathfrak{m}^{r-1} \subseteq \mathfrak{m}^r$, and so equality holds. This proves (7) in the case $n = 0$. If (7) holds for n, then $\mathfrak{m}^{n+r+1} = I\mathfrak{m}^{n+1} + x^{n+1}\mathfrak{m}^r \subseteq I\mathfrak{m}^{n+1} + x^{n+1}(I + x\mathfrak{m}^{r-1}) = I\mathfrak{m}^{n+1} + x^{n+2}\mathfrak{m}^{r-1}$, which proves (7). $\qquad\square$

Theorem 14.1.8 *Let $(R, \mathfrak{m})$ be a two–dimensional regular local ring with infinite residue field and let I be an $\mathfrak{m}$–primary ideal. If $I = \bar{I}$, then I is full.*

Proof: (Cf. Exercise 10.11.) By the valuative criterion of integral closure, Theorem 6.8.3, or by the existence of Rees valuations, $I = \cap IV \cap R$, as V varies over all Noetherian $Q(R)$–valuation domains containing R. Since I is $\mathfrak{m}$–primary, R/I is Artinian, so a finite intersection suffices. Namely, $I = \cap_{i=1}^{t} IV_i \cap R$, where $V_1, \ldots, V_t$ are Noetherian valuation rings as above. Without loss of generality $\mathfrak{m}V_i \neq V_i$ for all i. By Lemma 6.4.3 (2) there exists $x \in \mathfrak{m}$ such that $xV_i = \mathfrak{m}V_i$ for all $i = 1, \ldots, t$. We claim that $I : x = I : \mathfrak{m}$.

That $I : \mathfrak{m} \subseteq I : x$ is obvious. Let $u \in I : x$, so that $xu \in I$. Then $\mathfrak{m}uV_i = xuV_i \subseteq IV_i$ for all i, so that $\mathfrak{m}u \subseteq \cap_i(IV_i) \cap R = I$, whence $I : x = I : \mathfrak{m}$. Thus I is full. $\qquad\square$

Example 14.1.9 Let $\mathfrak{m} = (x, y)$.

(1) $I = (x^2 y, x^5, y^5)$ is *not* full (and hence not integrally closed) since $\mu(I) = 3 \neq 4 = \mathrm{ord}(I) + 1$.

(2) Integrally closed ideals are full, but the converse does not hold. For example, let $I_1 = (x^2, xy^4, y^5)$ and $I_2 = (x^2, xy^3, y^5)$. Then $\mu(I_j) = \mathrm{ord}(I_j) + 1 = 3$, for $j = 1, 2$, so that by Theorem 14.1.4 (3), at least if the residue field is infinite, both I_j are full. However, the integral closures of I_1 and I_2 are the same, and so at least one of the two ideals is not integrally closed.

We are now in a position to start a proof of Zariski's two theorems by proving analogs for full ideals. The first theorem that products of integrally closed ideals remain integrally closed is true even when we replace "integrally closed" by "full". The second theorem concerning unique factorization is no longer valid for full ideals, but a part of it does remain true if we simply concentrate on divisibility by the maximal ideal. The uniqueness of factorization of integrally closed ideals into simple integrally closed ideals has two parts. The first part is cancellation: if R is a two–dimensional regular local ring and I, J, K are integrally closed ideals, then $IJ = IK$ implies that $J = K$. This part is easy to prove: because I is faithful, by the Determinant Trick $IJ = IK$ implies that J and K have the same integral closure, and as J and K are both integrally closed, they are equal. The second and far more difficult part is divisibility: if a simple integrally closed ideal I divides the product of two integrally closed ideals, it must divide one of them. We will prove that this is true for $I = \mathfrak{m}$, and even for the product of full ideals.

Theorem 14.1.10 *Let $(R, \mathfrak{m})$ be a two–dimensional regular local ring and let I and J be full $\mathfrak{m}$–primary ideals. Let $x \in \mathfrak{m} \setminus \mathfrak{m}^2$. If $I : x = I : \mathfrak{m}$ and $J : x = J : \mathfrak{m}$, then $IJ : x = IJ : \mathfrak{m}$.*
If the residue field is infinite and if I and J are full, then IJ is full.

Proof: Assume that $I : x = I : \mathfrak{m}$ and $J : x = J : \mathfrak{m}$. Since $x \notin \mathfrak{m}^2$, $R/(x)$ is a discrete valuation ring of rank one, and so there exist $f \in I$ and $g \in J$ such that $(I + (x))/(x) = (f, x)/(x)$ and $(J + (x))/(x) = (g, x)/(x)$. Therefore $I \subseteq (f, x)$, whence $I = (f) + (x) \cap I = (f) + x(I : x)$, and similarly, $J = (g) + x(I : x)$. It follows that $IJ = (fg, gx(I : x), fx(J : x), x^2(I : x)(J : x))$. Let $u \in IJ : x$. Then $xu \in IJ$, which implies that there is an equation

$$xu = fga + gxb + fxc + x^2 d$$

with $a \in R$, $b \in (I : x)$, $c \in (J : x)$, and $d \in (I : x)(J : x)$. Then $x(u - gb - fc - xd) \in (fg)$ which implies that $u - gb - fc - xd \in (fg)$. As $(I : x) = (I : \mathfrak{m})$ and $(J : x) = (J : \mathfrak{m})$ we obtain that $u\mathfrak{m} \subseteq gI + fJ + IJ \subseteq IJ$, proving that $IJ : x = IJ : \mathfrak{m}$.

Now assume that the residue field is infinite and that I and J are full. There exists $x \in R$ such that $I : x = I : \mathfrak{m}$. By Theorem 14.1.4 (2), $\mu(I) = \mathrm{ord}(I) + 1$. Similarly, $\mu(J) = \mathrm{ord}(J) + 1$. As the residue field is infinite, there exists $z \in \mathfrak{m} \setminus \mathfrak{m}^2$ whose leading form in $\mathrm{gr}_{\mathfrak{m}}(R)$ does not divide $c(I)$ and $c(J)$. Then by Proposition 14.1.7, $I : z = I : \mathfrak{m}$ and $J : z = J : \mathfrak{m}$, whence by the first part IJ is full. $\qquad\square$

Corollary 14.1.11 *Let $(R, \mathfrak{m})$ be a two–dimensional regular local ring and let I be an $\mathfrak{m}$–primary full ideal. Then $\mathfrak{m}I : \mathfrak{m} = I$. In particular, if I and J are $\mathfrak{m}$–primary full ideals and $\mathfrak{m}I = \mathfrak{m}J$, then $I = J$.*

Proof: Since I is full, there is an element $x \in \mathfrak{m}$ such that $I : x = I : \mathfrak{m}$. If $x \in \mathfrak{m}^2$, then by Theorem 14.1.4 (1), $I = \mathfrak{m}$, so clearly $\mathfrak{m}I : \mathfrak{m} = I$. Thus we may assume that $x \in \mathfrak{m} \setminus \mathfrak{m}^2$. Then by Proposition 14.1.6, $I = \mathfrak{m}I : x$. Hence $I \subseteq \mathfrak{m}I : \mathfrak{m} \subseteq \mathfrak{m}I : x = I$ gives the required equality.

If $\mathfrak{m}I = \mathfrak{m}J$, then $J \subseteq \mathfrak{m}J : \mathfrak{m} = \mathfrak{m}I : \mathfrak{m} = I$, and by symmetry $I \subseteq J$, whence $I = J$. $\qquad\square$

Proposition 14.1.12 *Let $(R, \mathfrak{m})$ be a two–dimensional regular local ring and let I be an $\mathfrak{m}$–primary full ideal. Then the following are equivalent.*
(1) $I = \mathfrak{m}(I : \mathfrak{m})$.
(2) $I = \mathfrak{m}J$ for some ideal J.
(3) $\deg(c(I)) < \mathrm{ord}(I)$.

Proof: Obviously (1) implies (2). Assume (2). By Remark 14.1.1 (7), $c(I) = c(\mathfrak{m}) \cdot c(J) = c(J)$, forcing $\deg(c(I)) = \deg(c(J)) \leq \operatorname{ord}(J) < \operatorname{ord}(I)$, which proves (3). Assume that $\deg c(I) < \operatorname{ord}(I)$. Since I is full, there exists $x \in \mathfrak{m}$ such that $I : x = I : \mathfrak{m}$. By Theorem 14.1.4 (3) and by Lemma 14.1.3, $\lambda(\frac{I:\mathfrak{m}}{I}) = \mu(I) - 1 = \operatorname{ord}(I)$. The exact sequence $0 \to \frac{I}{\mathfrak{m}(I:\mathfrak{m})} \to \frac{I:\mathfrak{m}}{\mathfrak{m}(I:\mathfrak{m})} \to \frac{I:\mathfrak{m}}{I} \to 0$ then shows that

$$\lambda\left(\frac{I}{\mathfrak{m}(I:\mathfrak{m})}\right) + \operatorname{ord}(I) = \lambda\left(\frac{I:\mathfrak{m}}{\mathfrak{m}(I:\mathfrak{m})}\right) = \mu(I:\mathfrak{m}) \leq \operatorname{ord}(I:\mathfrak{m}) + 1,$$

where the last inequality is by Lemma 14.1.3 (2) and the equality before that is by Nakayama's Lemma. If $\mathfrak{m}(I : \mathfrak{m}) \neq I$, then since in any case $\operatorname{ord}(I) \geq \operatorname{ord}(I : \mathfrak{m})$, necessarily $\lambda(\frac{I}{\mathfrak{m}(I:\mathfrak{m})}) = 1$ and $\operatorname{ord}(I) = \operatorname{ord}(I : \mathfrak{m})$. From $\lambda(\frac{I}{\mathfrak{m}(I:\mathfrak{m})}) = 1$ we deduce that there is $f \in I$ such that $I = (f) + \mathfrak{m}(I : \mathfrak{m})$. From $\operatorname{ord}(I) = \operatorname{ord}(I : \mathfrak{m})$ we deduce that $\operatorname{ord}(\mathfrak{m}(I : \mathfrak{m})) = \operatorname{ord}(\mathfrak{m}) + \operatorname{ord}(I : \mathfrak{m}) = 1 + \operatorname{ord}(I)$, and so $\mathfrak{m}(I : \mathfrak{m}) \subseteq \mathfrak{m}^{\operatorname{ord}(I)+1}$. It follows that $c(I) = f^*$, which means that $\deg c(I) = \operatorname{ord}(f) = \operatorname{ord}(I)$, which contradicts the assumption and proves (1). $\qquad\square$

In the example $I = (x^2y, xy^2, x^4, y^4)$, $\operatorname{ord}(I) = 3$, $\mu(I) = 4$. By Theorem 14.1.4 (3) this implies, if the residue field is infinite, that I is full. Since $c(I) = (XY)$, $\deg c(I) = 2 < 3 = \operatorname{ord}(I)$. Therefore $I = \mathfrak{m}(I : \mathfrak{m})$, and this further factors as $I = \mathfrak{m} \cdot (xy, x^3, y^3) = \mathfrak{m} \cdot (x, y^2) \cdot (y, x^2)$.

Corollary 14.1.13 *Let $(R, \mathfrak{m})$ be a two–dimensional regular local ring with infinite residue field. Let I and J be full $\mathfrak{m}$–primary ideals. If $\mathfrak{m} \mid IJ$, then $\mathfrak{m} \mid I$ or $\mathfrak{m} \mid J$.*

Proof: By Theorem 14.1.10, IJ is full. By Proposition 14.1.12, $\mathfrak{m} \mid IJ$ implies that $\deg(c(IJ)) < \operatorname{ord}(IJ)$. Hence $\deg(c(I)) + \deg(c(J)) < \operatorname{ord}(I) + \operatorname{ord}(J)$. Either $\deg(c(I)) < \operatorname{ord}(I)$ or $\deg(c(J)) < \operatorname{ord}(J)$, and another application of Proposition 14.1.12 shows that either $\mathfrak{m} \mid I$ or $\mathfrak{m} \mid J$. $\qquad\square$

14.2. Quadratic transformations

In this section we discuss the ideal theory of quadratic transformations. Geometrically we are blowing up the spectrum of the ring at a closed point and looking at various affine pieces of this blowup. We have used this in past chapters, most extensively in Chapter 10. However, in the special case we are considering of a two–dimensional regular local ring, this process can be described quite simply.

Throughout this section we let $(R, \mathfrak{m})$ be a two–dimensional regular local ring and $x \in \mathfrak{m} \setminus \mathfrak{m}^2$. We set $S = R[\frac{\mathfrak{m}}{x}] = \bigcup_{n \geq 0} \frac{\mathfrak{m}^n}{x^n}$. We refer to S as a **quadratic transformation** of R. The localization of S at a height two maximal ideal that contains $\mathfrak{m}S$ is called a **local quadratic transformation** of R. This notation is not totally consistent with existing literature, which often refers to the localization as a quadratic transformation of R, and may also allow localization at a height one prime ideal of S.

We begin with some easy observations. First, if t is an indeterminate over R, then $S = R[\frac{y}{x}] \cong \frac{R[t]}{P}$ for some height one prime ideal P of $R[t]$. Since R is a unique factorization domain, so is $R[t]$, therefore P is principal. As $xt - y \in P$ and is irreducible, $P = (xt - y)$. (See Corollary 5.5.9 for a more general result.) Clearly $xS = \mathfrak{m}S$ and for any $n \in \mathbb{N}$, $x^n S \cap R = \mathfrak{m}^n S \cap R = \mathfrak{m}^n$.

Ideals crucial for proofs of the theorems in this chapter are those that are contracted from some quadratic transformation.

Definition 14.2.1 *Let $(R, \mathfrak{m})$ be a two–dimensional regular local ring, and I an ideal of R. If for some $x \in \mathfrak{m} \setminus \mathfrak{m}^2$, $IR[\frac{\mathfrak{m}}{x}] \cap R = I$, then we say that I is **contracted** from the quadratic transformation $R[\frac{\mathfrak{m}}{x}]$. We refer to I as a **contracted ideal** if there exists a quadratic transformation from which I is contracted.*

Proposition 14.2.2 *Let $(R, \mathfrak{m})$ be a two–dimensional regular local ring and let I be an $\mathfrak{m}$–primary ideal. The following are equivalent.*
(1) I is full.
(2) I is contracted.
   Moreover, if I satisfies either of these conditions and if $x \in \mathfrak{m} \setminus \mathfrak{m}^2$, then I is contracted from $R[\frac{\mathfrak{m}}{x}]$ if and only if $I : x = I : \mathfrak{m}$, and this holds if and only if x^* does not divide $c(I)$.

Proof: First assume that I is contracted, i.e., that there exists $x \in \mathfrak{m} \setminus \mathfrak{m}^2$ such that $IR[\frac{\mathfrak{m}}{x}] \cap R = I$. To prove that I is full, it suffices to prove that $I : x \subseteq I : \mathfrak{m}$. Suppose that $ax \in I$ for some $a \in R$. Then $a\mathfrak{m} \subseteq a\mathfrak{m}R[\frac{\mathfrak{m}}{x}] \cap R = axR[\frac{\mathfrak{m}}{x}] \cap R \subseteq IR[\frac{\mathfrak{m}}{x}] \cap R = I$, which proves that I is full.

Conversely suppose that I is full. By definition there exists an element $x \in R$ such that $I : \mathfrak{m} = I : x$. Necessarily $x \in \mathfrak{m}$. If $x \in \mathfrak{m}^2$, then by Theorem 14.1.4, $I = \mathfrak{m}$, whence I is contracted from the quadratic transformation $R[\frac{\mathfrak{m}}{z}]$, for every $z \in \mathfrak{m} \setminus \mathfrak{m}^2$. So we may assume that $x \notin \mathfrak{m}^2$. Let $u \in IR[\frac{\mathfrak{m}}{x}] \cap R$. Since $R[\frac{\mathfrak{m}}{x}] = \cup_n \frac{\mathfrak{m}^n}{x^n}$, there exists a non–negative integer n such that $u \in I \cdot \frac{\mathfrak{m}^n}{x^n}$. Choose the smallest n with this property. If $n = 0$, then $u \in I$, which is what we need to prove. By way of contradiction, suppose that $n > 0$ and choose y such that $\mathfrak{m} = (x, y)$. We have that $x^n u \in \mathfrak{m}^n I = x\mathfrak{m}^{n-1}I + y^n I$ and we can write $x(x^{n-1}u - v) = y^n w$ for some $w \in I$ and $v \in \mathfrak{m}^{n-1}I$. As x, y^n are a regular sequence, there exists an element t such that $x^{n-1}u - v = y^n t$ and $xt = w \in I$. Hence $t \in I : x = I : \mathfrak{m}$, so that $y^n t = y^{n-1}(yt) \in \mathfrak{m}^{n-1}I$. Consequently, $x^{n-1}u \in \mathfrak{m}^{n-1}I$, implying that $u \in I \cdot \frac{\mathfrak{m}^{n-1}}{x^{n-1}}$ and contradicting the choice of n.

This also proves the first equivalence in the last part. By Theorem 14.1.4, the two equivalent assumptions on I imply that $\mu(I) = \operatorname{ord}(I) + 1$, so the last equivalence follows from Proposition 14.1.7. $\qquad\square$

Corollary 14.2.3 *Let $(R, \mathfrak{m})$ be a two–dimensional regular local ring and let $I_1, \ldots, I_n$ be $\mathfrak{m}$–primary ideals. If $x \in \mathfrak{m} \setminus \mathfrak{m}^2$ and if $I_1, \ldots, I_n$ are all contracted from $R[\frac{\mathfrak{m}}{x}]$, then $I_1 \cdots I_n$ is contracted from $R[\frac{\mathfrak{m}}{x}]$. If the residue field is infinite and if $I_1, \ldots, I_n$ are all contracted, then $I_1 \cdots I_n$ is contracted.*

Proof: By Proposition 14.2.2, an ideal is contracted if and only if it is full. By Theorem 14.1.10, the product of full ideals $I_1, \ldots, I_n$ in both cases is full, and hence contracted. $\qquad\square$

We determine the structure of a quadratic transformation in more detail. Let k be the residue field of R and let $S = R[\frac{\mathfrak{m}}{x}]$ for some $x \in \mathfrak{m} \setminus \mathfrak{m}^2$. Then $\operatorname{Spec} S$ is the disjoint union of $\operatorname{Spec}(S/xS)$ and $\operatorname{Spec}(S_x)$, i.e., the set of prime ideals containing x and the set of prime ideals not containing x.

Since $S \cong \frac{R[t]}{(xt-y)}$, then $S/xS \cong k[t]$, so that $xS = \mathfrak{m}S$ is a prime ideal in S and the maximal ideals in S contracting to $\mathfrak{m}$ are in one–to–one correspondence with irreducible polynomials in $k[t]$. If $f \in S$ has an irreducible image in $k[t]$, then (x, f) is a maximal ideal in S, of height 2, so that S localized at (x, f) is a regular local ring of dimension two. This proves that for all prime ideals Q in S that contain x, S_Q is regular.

Next consider $\operatorname{Spec}(S_x)$: since $R \subseteq S \subseteq R_x$, we see that $S_x = R_x$, and $\operatorname{Spec}(S_x) = \operatorname{Spec}(R_x)$. But R_x is a principal ideal domain, so S_x is also regular.

In particular, for all $Q \in \operatorname{Spec} S$, S_Q is regular, so that S is regular. It is also a domain since it is contained in the field of fractions of R. It follows that S is a unique factorization domain.

If $p \in \operatorname{Spec} R$, $\operatorname{ht} p = 1$, and $x \notin p$, then $p_x \in \operatorname{Spec}(R_x) = \operatorname{Spec}(S_x)$. The prime ideal in S corresponding to this prime is $pR_x \cap S$. We analyze which maximal ideals of S contain this height one prime.

Since R is a unique factorization domain, $p = zR$ for some z. Of course, S is also a unique factorization domain and we wish to find the generator of $pR_x \cap S$. Fix $\operatorname{ord}(z) = r$, write $z = z_r(x, y) + z_{r+1}$, where $\operatorname{ord}(z_{r+1}) \geq r + 1$ and $z_r(x, y)$ is a homogeneous polynomial in x, y over R of degree r with every non–zero coefficient a unit. Hence $\frac{z}{x^r} \in pR_x \cap S$. Let $\alpha \in pR_x \cap S$. Then $\alpha = \frac{a}{x^n}$ for some $a \in \mathfrak{m}^n$. Then $x^n \alpha = a \in pR_x \cap R = p = (z)$. Thus $a \in (z) \cap \mathfrak{m}^n = z\mathfrak{m}^{n-r}$, so that $\alpha = \frac{a}{x^n} \subseteq \frac{z\mathfrak{m}^{n-r}}{x^n} S = \frac{z}{x^r} S$. This proves that $\frac{z}{x^r} S = pR_x \cap S$.

All the height two maximal ideals of S contain xS (as $\dim S_x = \dim R_x = 1$), and $S/xS = S/\mathfrak{m}S \cong k[t]$. In S we may write

$$\frac{z}{x^r} = \frac{z_r(x, y)}{x^r} + \frac{z_{r+1}}{x^r} = z_r\left(1, \frac{y}{x}\right) + x\left(\frac{z_{r+1}}{x^{r+1}}\right).$$

As $S \cong \frac{R[t]}{(xt-y)}$ under identification of $\frac{y}{x}$ with t, then $\frac{S}{pS+xS}$ is a homomorphic image of

$$\frac{S}{(\frac{z}{x^r}, x)S} = \frac{S}{(z_r(1, \frac{y}{x}), x)S} \cong \frac{k[t]}{(z_r(1, t))}.$$

Hence there exist only finitely many maximal ideals containing $(\frac{z}{x^r}, x)S$, which correspond to the irreducible factors of $z_r(1, t)$.

However, $z_r(1, t)$ could be a unit – this occurs exactly when $z_r(x, y) = x^r$. In this case, $pR_x \cap S$ is a height one maximal ideal of S.

For a specific example, consider $S = R[\frac{y}{x}]$, $z = x^2 + y^3$. Let $y' = \frac{y}{x}$. Then in S, $z = x^2 + (y')^3 x^3 = x^2 \cdot (1 + (y')^3 x)$, so $(z)R_x \cap S = (1 + (y')^3 x)S$ is a height one maximal ideal of S.

The discrete valuation ring of rank one obtained by localizing S at the height one prime ideal xS is exactly the order valuation ord_R of R, by Example 10.3.1.

This example illustrates a very important theme — discrete valuations of rank one centered on the maximal ideal of R and contained in the field of fractions K of R are obtained by finitely many successive quadratic transformations. They are the discrete valuation rings of rank one associated to the adic valuation with respect to the maximal ideal in the last quadratic transformation in the sequence. In Section 14.5 we prove a correspondence between two–dimensional regular local rings T with $R \subseteq T \subseteq K$ and discrete K–valuations of rank one centered on the maximal ideal of R.

14.3. The transform of an ideal

Let S be a unique factorization domain. Every ideal J in S can be written in the form $J = aK$, where K is an ideal of height at least two (if $K = S$, we say that the height of K is infinity). The element a is unique up to associates, i.e., the ideal generated by a is unique since it is exactly the intersection of the primary components of J having height one. The ideal K is also clearly unique. When S is a unique factorization domain, such as a regular local ring, then a factors uniquely (up to associates and order) into irreducibles. Unique factorization of J occurs only rarely; the case of two–dimensional regular local rings is proved in the following section.

Definition 14.3.1 *Let R be a two–dimensional regular local ring and let I be an $\mathfrak{m}$–primary ideal of R. Let S be a unique factorization domain containing R. The **transform** of I in S is the unique ideal I^S of height at least two such that $IS = aI^S$ for some $a \in S$.*

Example 14.3.2 Let R be a two–dimensional regular local ring, and let I be an $\mathfrak{m}$–primary ideal of R. Write $\mathfrak{m} = (x, y)$ and let S be the quadratic transformation $S = R[\frac{y}{x}]$. Suppose that $\mathrm{ord}(I) = r$. We can write $IS = x^r I'$, where I' is an ideal of S. We claim that I' has height at least two, and therefore is the transform of I in S. For if I' has height one, then it is contained either in xS or in a height one prime Q of S of the form $qR_x \cap S$ where q is a height one prime of R. If $I' \subseteq xS$, then $I \subseteq x^{r+1}S \cap R = \mathfrak{m}^{r+1}$, a contradiction. But if $I' \subseteq Q$ then $I \subseteq Q \cap R = q$, which contradicts the assumption that I is $\mathfrak{m}$–primary.

Example 14.3.3 Let $I = (xy, x^3, y^3) \subseteq k[x, y]_{(x,y)} = R$. We let $\mathfrak{m}$ be the maximal ideal of R and identify $\mathrm{gr}_{\mathfrak{m}}(R)$ with $k[X, Y]$ as usual. The leading form of $x + y$ does not divide $c(I) = XY$, $\mu(I) = 3 = \mathrm{ord}(I) + 1$, so by Proposition 14.1.7, $I : (x + y) = I : \mathfrak{m}$. Hence by Proposition 14.2.2, I is contracted from $S = R[\frac{\mathfrak{m}}{x+y}]$. Write $x' = \frac{x}{x+y}$. Then $IS = (x + y)^2(x' - (x')^2, (x')^3(x+y), (x+y)(1-x')^3)S$, hence $I^S = (x' - (x')^2, (x')^3(x+y), (x+y)(1 - x')^3)$. The maximal ideals of S that contain I^S bifurcate into two groups — those containing x' and those containing $1 - x'$. There are exactly two such maximal ideals, $(x', x + y)$ and $(1 - x', x + y)$.

A crucial point in our arguments is that the co–length of the transform of an integrally closed $\mathfrak{m}$–primary ideal drops under quadratic transformations:

Lemma 14.3.4 *Let $(R, \mathfrak{m})$ be a two–dimensional regular local ring and let I be an integrally closed $\mathfrak{m}$–primary ideal of R. Suppose that $x \in \mathfrak{m} \setminus \mathfrak{m}^2$ such that I is contracted from $S = R[\frac{\mathfrak{m}}{x}]$. Then there is an R–module isomorphism*

$$\mathfrak{m}^{\mathrm{ord}(I)}/I \cong S/I^S.$$

In particular, $\lambda_R(R/I) > \lambda_T(T/I^T)$ for every local quadratic transformation T that is a localization of S.

Proof: Let $r = \mathrm{ord}(I)$. We know that $\mathfrak{m}^r S = x^r S$ and that $IS = x^r I^S$, so to prove the displayed isomorphism, it is enough to prove that $\mathfrak{m}^r/I \cong \mathfrak{m}^r S/IS$. We prove that the natural map from $\mathfrak{m}^r/I$ to $\mathfrak{m}^r S/IS$ is an isomorphism. It is injective because $IS \cap \mathfrak{m}^r = IS \cap R \cap \mathfrak{m}^r = I \cap \mathfrak{m}^r = I$. By Proposition 14.2.2, since I is contracted from S, $I : x = I : \mathfrak{m}$, hence by Theorem 14.1.4 (2) and Proposition 14.1.7, $\mathfrak{m}^{n+r} = I\mathfrak{m}^n + x^{n+1}\mathfrak{m}^{r-1}$ for all $n \geq 0$. An element in $\mathfrak{m}^r S$ is of the form a/x^n for some n and some $a \in \mathfrak{m}^{r+n}$. Then we may write $a = \sum i_j b_j + x^{n+1}c$, where $i_j \in I$, $b_j \in \mathfrak{m}^n$ and $c \in \mathfrak{m}^{r-1}$. Then $\frac{a}{x^n} = \sum i_j \frac{b_j}{x^n} + xc \in IS + \mathfrak{m}^r$, which proves surjectivity of the natural map $\mathfrak{m}^r/I \to \mathfrak{m}^r S/IS$.

The last statement of the lemma follows as $\lambda_S(S/I^S) \leq \lambda_R(S/I^S) = \lambda_R(\mathfrak{m}^r/I) = \lambda_R(R/I) - \lambda_R(R/\mathfrak{m}^r) = \lambda_R(R/I) - \binom{r+1}{2} < \lambda_R(R/I)$. $\square$

Before applying this theory to integrally closed ideals, we need one more fact about the transform of an ideal. This requires the concept of the inverse transform of an ideal.

Definition 14.3.5 *Let $(R, \mathfrak{m})$ be a two–dimensional regular local ring and let $S = R[\frac{y}{x}]$ be a quadratic transformation of R, where $\mathfrak{m} = (x, y)$. Let J be an ideal of S. The **inverse transform** of J is defined as follows. There is a least integer a such that $x^a J$ is extended from an ideal in R. We set $I = x^a J \cap R$ and call it the **inverse transform** of J.*

In the definition, I is the largest ideal of R whose extension is $x^a J$, as we explain next. By assumption, $x^a J = KS$ for some ideal K of R. Then $K \subseteq x^a J \cap R = I$, and $x^a J = KS \subseteq IS \subseteq x^a J$ gives equality throughout. Note in particular that the inverse transform of an ideal in S is then contracted. Moreover, if the height of J is two, then the height of the inverse transform of J is also two.

Example 14.3.6 $S = R[\frac{\mathfrak{m}}{x}], \mathfrak{m} = (x, y)$, let $J' = ((y')^2 + x^3, x^4)$, where $y' = \frac{y}{x}$. We compute the inverse transform of J': $x^2 \cdot J' = (y^2 + x^5, x^6)S$. The inverse transform is $x^2 J' \cap R \supseteq (y^2 + x^5, x^6)$. These ideals are not equal as $(y^2 + x^5, x^6)$ is *not* full. In fact, $x^2 J' \cap R = (y^2 + x^5, x^6, x^5 y) = (y^2 + x^5, xy^2, y^3)$.

Proposition 14.3.7 *Let $(R, \mathfrak{m})$ be a two–dimensional regular local ring and I a simple full $\mathfrak{m}$–primary ideal of R that is different from $\mathfrak{m}$. Let $x \in \mathfrak{m} \setminus \mathfrak{m}^2$ such that I is contracted from $S = R[\frac{y}{x}]$. Then I^S is contained in a unique maximal ideal N of S and $I^S S_N$ is simple.*

Proof: We first prove that I^S is simple. Suppose that there are proper ideals J' and K' in S such that $I^S = J'K'$. Let J (respectively K) be the inverse transform of J' (respectively K'), and choose integers a and b least such that $x^a J' = JS$ and $x^b K' = KS$. Then $x^{a+b} I^S = (x^a J')(x^b K') = JKS$. It follows that $\mathfrak{m}^{a+b} IS = JK\mathfrak{m}^r S$, where $r = \text{ord}(I)$ (since $IS = \mathfrak{m}^r I^S$). Since I, J, K, and $\mathfrak{m}$ are contracted from S, it follows by Proposition 14.2.2 and Theorem 14.1.10, that the products of any of these ideals are contracted from S as well. Thus $\mathfrak{m}^{a+b} I = JK\mathfrak{m}^r$. By Proposition 14.2.2, any products of I, J, K and $\mathfrak{m}$ are full as well. By Corollary 14.1.11, we can cancel the powers of the maximal ideal from this equation. Hence either $I = JK\mathfrak{m}^c$ or $I\mathfrak{m}^c = JK$ for some non–negative c. In the first case or if $c = 0$ in the second case, the simplicity of I gives that either J or K is R, and in the remaining case $\mathfrak{m}$ divides either J or K by Corollary 14.1.13. If, for example, $\mathfrak{m}$ divides K, say $K = \mathfrak{m}L$, then $x^a J' = \mathfrak{m}LS = xLS$ implying that $x^{a-1} J' = LS$, contradicting the least choice of a. Hence I^S is simple.

Since I^S has height two, there are only finitely many maximal ideals of S that contain it, say $N_1, \ldots, N_t$. Let $q_1, \ldots, q_t$ be the primary components of I^S corresponding to these primes. The Chinese Remainder Theorem shows that $I^S = q_1 \cdots q_t$. Since I^S is simple, necessarily $t = 1$, and so there is a unique maximal ideal N containing I^S. But if I_N^S were a product of two ideals, so would I^S be a product since I^S is N–primary. Thus I^S simple implies that I_N^S is simple. □

14.4. Zariski's theorems

In this section we prove the two Zariski's theorems stated in the introduction to the chapter. The proof of the second theorem is simpler when the residue field is infinite, so we prove that case first. The general case is proved at the end of the section. Namely, the proofs rely on full ideals, and full ideals can be manipulated more easily if the residue field is infinite. By Lemma 8.4.2, if X is a variable over a two–dimensional regular local ring $(R, \mathfrak{m})$, then $R(X)$ is a two–dimensional regular local ring with maximal ideal $\mathfrak{m}R(X)$ and with infinite residue field $R(X)/\mathfrak{m}R(X) = (R/\mathfrak{m})(X)$. Furthermore, by Lemma 8.4.2, for any ideal I in R, I is integrally closed if and only if $IR(X)$ is integrally closed.

Here is an easy consequence of the fact that integrally closed ideals are full:

Lemma 14.4.1 *Let I be an integrally closed $\mathfrak{m}$–primary ideal in a two–dimensional regular local ring $(R, \mathfrak{m})$. Then $I\mathfrak{m}^n$ is integrally closed for all $n \geq 0$.*

Proof: As above, via Lemma 8.4.2, we may assume that $R/\mathfrak{m}$ is infinite. It suffices to prove the theorem for $n = 1$ by induction. We know that I is full by Theorem 14.1.8, and Theorem 14.1.10 then shows that $\mathfrak{m}I$ is also full. Since the residue field is infinite, we may choose $x \in \mathfrak{m} \setminus \mathfrak{m}^2$ such that x^* does not divide $c(I)$. By Exercise 14.13, $c(\overline{\mathfrak{m}I}) = c(\mathfrak{m}I)$, so $c(\overline{\mathfrak{m}I}) = c(\mathfrak{m})c(I) = c(I)$, and x^* also does not divide $c(\overline{\mathfrak{m}I})$. But $\overline{\mathfrak{m}I}$ is full by Theorem 14.1.8, hence by Proposition 14.2.2, $\overline{\mathfrak{m}I} : x = \overline{\mathfrak{m}I} : \mathfrak{m}$. By Corollary 6.8.7 then $\overline{\mathfrak{m}I} : x = \overline{I} = I$.

Let $u \in \overline{\mathfrak{m}I}$. Going modulo xR we obtain a discrete valuation ring of rank one so that $u \in \mathfrak{m}I + xR$, as every ideal is integrally closed in a discrete valuation ring of rank one. Hence

$$u \in (\mathfrak{m}I + xR) \cap \overline{\mathfrak{m}I} = \mathfrak{m}I + xR \cap \overline{\mathfrak{m}I} = \mathfrak{m}I + x(\overline{\mathfrak{m}I} : \mathfrak{m}) = \mathfrak{m}I + xI = \mathfrak{m}I. \qquad \square$$

Example 14.4.2 Let $I = (x^2, xy^4, y^5)$. Then I is full but $I \neq \overline{I}$, since $xy^3 \in \overline{I} \setminus I$. Consider the quadratic transformation $S = R[\frac{\mathfrak{m}}{y}]$. Set $x' = \frac{x}{y}$. Then $IS = ((x')^2y^2, x'y^5, y^5) = y^2 \cdot I^S$, so that $I^S = ((x')^2, y^3)$ and I^S is not full, let alone integrally closed. However, in the next proposition we will prove that the transform of an integrally closed ideal is still integrally closed.

Proposition 14.4.3 *Let R be a two–dimensional regular local ring, and let I be an $\mathfrak{m}$–primary ideal of R. Suppose that I is integrally closed and $x \in \mathfrak{m} \setminus \mathfrak{m}^2$ such that I is contracted from $S = R[\frac{\mathfrak{m}}{x}]$. Then I^S is integrally closed.*

Proof: By Proposition 1.5.2, since S is integrally closed (being regular), to check $I^S = \overline{I}^S$, it is enough to prove that $IS = x^{\mathrm{ord}(I)}I^S$ is integrally closed. Hence it suffices to prove that $\overline{IS} = IS$. We observe that $\mathfrak{m}^n I$ is also integrally closed and contracted from S for all n. If $\frac{u}{x^n} \in \overline{IS}$, where $u \in \mathfrak{m}^n$, then $u \in x^n\overline{IS} \subseteq \overline{x^nIS} = \overline{\mathfrak{m}^nIS}$ and therefore $u \in \overline{\mathfrak{m}^nIS} \cap R$. We claim that this forces $u \in \mathfrak{m}^n I$. More generally, suppose that J is integrally closed and contracted from S. We claim that $\overline{JS} \cap R = J$. To prove this, let $w \in \overline{JS} \cap R$. There is an equation $w^n + s_1w^{n-1} + \cdots + s_n = 0$ for some $s_i \in J^iS = \cup_t \frac{J^i\mathfrak{m}^t}{x^t}$. By clearing denominators, there exists $t \geq 0$ such that $x^tw \in \overline{\mathfrak{m}^tJ} = \mathfrak{m}^tJ$, the last equality coming from Lemma 14.4.1. Since J is contracted from S, we know that $\mathfrak{m}^tJ$ is also contracted from S, and then using Propositions 14.2.2, 14.1.6, and induction on t, that $w \in \mathfrak{m}^tJ : x^t = J$. This proves the claim. Hence $u \in \mathfrak{m}^n I$ and $\frac{u}{x^n} \in \frac{\mathfrak{m}^n}{x^n}I \subseteq IS$. Therefore $\overline{IS} = IS$. $\qquad \square$

We can now prove the first main theorem of Zariski that products of integrally closed ideals are integrally closed. Conceptually the proof is fairly straightforward. We induct on the co–length of one of the ideals and pass to a quadratic transformation of the ring.

Theorem 14.4.4 (Zariski) *Let* $(R, \mathfrak{m})$ *be a two–dimensional regular local ring and let* I *and* J *be integrally closed ideals in* R. *Then* IJ *is integrally closed.*

Proof: By Lemma 8.4.2, without loss of generality the residue field of R is infinite. By unique factorization we may write $I = aI'$ and $J = bJ'$ for some $a, b \in R$ and some ideals I', J' in R that are either $\mathfrak{m}$–primary or R. By Proposition 1.5.2 I' and J' are integrally closed ideals, and if $I'J'$ is integrally closed, so is IJ. Thus it suffices to prove that $I'J'$ is integrally closed. By changing notation, we may thus assume that ht I, ht $J \geq 2$. If either I or J is R, then clearly the theorem follows. So we may assume that both I and J have height 2, so that they are both $\mathfrak{m}$–primary.

We use induction on $\lambda(R/I)$ to prove the theorem. If $\lambda(R/I) = 1$, then $I = \mathfrak{m}$. We proved in Lemma 14.4.1 that $\mathfrak{m}J$ is integrally closed. We may thus assume that $\lambda(R/I) > 1$.

By Theorem 14.1.8, the ideals I and J are full since they are integrally closed. As the residue field is infinite, we may choose $x \in \mathfrak{m} \setminus \mathfrak{m}^2$ such that x^* does not divide $c(IJ) = c(\overline{IJ})$ (Exercise 14.13). Proposition 14.2.2 then gives that IJ and $\overline{IJ}$ are contracted from $S = R[\frac{\mathfrak{m}}{x}]$. To prove $IJ = \overline{IJ}$, it suffices to prove $\overline{I^S J^S} = I^S J^S$, since contractions of integrally closed ideals are integrally closed. To prove this equality, it is enough to prove that for all maximal ideals N in S, setting $T = S_N$, $\overline{I^T J^T} = I^T J^T$.

Both I^S and J^S are integrally closed by Proposition 14.4.3, hence their localizations I^T and J^T are integrally closed. By Lemma 14.3.4, the co–length of I^T in T is strictly smaller than the co–length of I in R, so the induction gives $\overline{I^T J^T} = I^T J^T$.   □

This theorem fails in higher dimensions, see Exercises 1.14 and 1.13.

Example 14.4.5 Let $\mathfrak{m} = (x, y)$ and $I = (x^2 + y^3) + \mathfrak{m}^4$. We saw in Example 1.3.3 that I is integrally closed. This can be easily checked using the results above. First note that I is full since ord$(I) = 2$ and $\mu(I) = 3$. As $c(I) = X^2$ and Y does not divide X^2, we consider $S = R[\frac{\mathfrak{m}}{y}]$, and set $x' = \frac{x}{y}$, $x = x'y$. Then $IS = (y^2(x')^2 + y^3, (x')^2y^3) = y^2IS$, so that $I^S = ((x')^2 + y, (x')^2y)$. But $\sqrt{I^S} = (x', y)$, and we let $R_1 = S_{(x', y)}$, $I_1 = I^S R_1$. Set $y_1 = y + (x')^2$. Then $y_1 \notin \mathfrak{m}_1^2$, and $I_1 = (y_1, (x')^4)$ is integrally closed. Since I is contracted from S, it follows that I is integrally closed.

Before proving Zariski's second main theorem, that integrally closed ideals factor uniquely up to order into simple integrally closed ideals, there are several easy observations which make the proof below more transparent.

(1) In a Noetherian ring, every ideal I is a product of simple ideals. This is clear from the usual Noetherian argument of looking at a maximal counterexample.

(2) If I is integrally closed and $I = J_1 \cdots J_n$, then
$$I = J_1 \cdots J_n \subseteq \bar{J}_1 \bar{J}_2 \cdots \bar{J}_n \subseteq \overline{J_1 \cdots \cdots J_n} = \bar{I} = I$$
so $I = \bar{J}_1 \cdots \bar{J}_n$. We can repeat such factorization for each non–simple $\bar{J}_i$. If I is not zero, the procedure has to stop as I is not contained in all powers of $\mathfrak{m}$. Hence every integrally closed ideal is a product of integrally closed simple ideals.

(3) Cancellation holds for integrally closed ideals in a domain: If $IJ = IK$, $I \neq 0$, the determinant trick implies that $\bar{J} = \bar{K}$. If $J = \bar{J}$, $K = \bar{K}$, we then can cancel the I from $IJ = IK$.

We can now prove:

Theorem 14.4.6 (Zariski) *Let $(R, \mathfrak{m})$ be a two–dimensional regular local ring. Then every non–zero integrally closed ideal I in R can be written as*
$$I = a_1^{l_1} \cdots a_m^{l_m} I_1^{k_1} \cdots I_n^{k_n},$$
where $I_1, \ldots, I_n$ are simple $\mathfrak{m}$–primary integrally closed ideals, $a_1, \ldots, a_m$ are pairwise relatively prime irreducible elements of R, and $l_1, \ldots, l_m, k_1, \ldots, k_n$ are positive integers.

If the residue field is infinite, then this decomposition is unique up to ordering.

Proof: Let I be an integrally closed ideal in R. We may write $I = aJ$, where J is integrally closed and height two. The ideal generated by a is clearly unique, and the factors of a are unique up to unit multiples as R is a unique factorization domain. Hence it suffices to assume that I has height two. By Remark (2) above, we can factor I into a product of simple integrally closed ideals, so only the uniqueness needs to be proved. (In this theorem we prove it under the condition that the residue field is infinite, for a general proof see Theorem 14.4.9.)

Since we have cancellation, an easy induction shows that it suffices to prove:

Claim: If $I \mid J_1 \cdots J_n$ and $I, J_1, \ldots, J_n$ are simple integrally closed ideals, then $I = J_i$ for some i.

To prove this claim, we use induction on $\lambda(R/I)$. If $\lambda(R/I) = 1$ then $I = \mathfrak{m}$, and $\mathfrak{m} \mid J_1 \cdots J_n$. By Corollary 14.1.13, $\mathfrak{m} \mid J_i$ for some i, and then $\mathfrak{m} = J_i$ as J_i is simple. Hence we may assume that $\lambda(R/I) > 1$. Choose an $x \in \mathfrak{m} \setminus \mathfrak{m}^2$ such that x^* does not divide $c(IJ_1 \cdots J_n)$. This is possible since the residue field is infinite. Set $S = R[\frac{\mathfrak{m}}{x}]$. By Corollary 14.2.3 it follows that all of the ideals I, J_j and all possible products of these ideals are contracted from S. Write $IK = J_1 \cdots J_n$. Set $r = \mathrm{ord}(I)$, $r_i = \mathrm{ord}(J_i)$, and $\mathrm{ord}(K) = s$. Then $\mathrm{ord}(I) + \mathrm{ord}(K) = \mathrm{ord}(J_1 \cdots J_n) = \sum_{i=1}^{n} \mathrm{ord}(J_i)$, and $r + s = r_1 + \cdots + r_n$. Extending the ideals to S we see that $IS = x^r I'$, $KS = x^s K'$, $J_i S = x^{r_i} J_i'$, where

I', K', J_i' are the transforms of I, K, J_i, respectively. Then $I' \cdot K' = J_1' \cdots J_n'$ in S. We know that I' is a simple integrally closed ideal by Propositions 14.3.7 and 14.4.3, and that there exists a unique maximal ideal N containing I'. Localize at N to get $(I')_N \cdot (K')_N = (J_1')_N \cdots (J_n')_N$. Since I' is simple $(I')_N$ is also simple by Proposition 14.3.7, and likewise either $(J_i')_N$ is simple or $(J_i')_N = S_N$. But the co–length of $(I')_N$ is strictly less than the co–length of I by Lemma 14.3.4. By induction there exists an integer i such that $(I')_N \,|\, (J_i')_N$, and hence $(I')_N = (J_i')_N$ since $(J_i')_N$ is simple. Thus $I' = (I')_N \cap S = (J_i')_N \cap S = J_i'$ as N is the unique prime containing I' and J_i'. Multiplying by x^{r+r_i} yields $x^{r_i} \cdot (x^r I') = x^r \cdot (x^{r_i} J_i')$, so that $\mathfrak{m}^{r_i} \cdot IS = \mathfrak{m}^r \cdot J_i S$. Thus $\mathfrak{m}^{r_i} I = \mathfrak{m}^r IS \cap R = \mathfrak{m}^r J_i S \cap R = \mathfrak{m}^r J_i$. Since I and J_i are integrally closed, we can cancel powers of $\mathfrak{m}$ (using Corollary 14.1.11), and then by simplicity we get that $I = J_i$. $\qquad\square$

Both of Zariski's theorems are valid without the assumption of an infinite residue field, but removing this assumption necessitates technicalities that make the uniqueness of decomposition less transparent. The goal of the rest of this section is to prove Zariski's second theorem without the infinite residue field assumption.

Lemma 14.4.7 *Let $(R, \mathfrak{m})$ be a regular local ring and let I be a simple integrally closed $\mathfrak{m}$–primary ideal. If X is a variable over R, then $IR[X]_{\mathfrak{m}R[X]}$ is also simple.*

Proof: Let $v_1, \ldots, v_r$ be the Rees valuations of I. For each $i = 1, \ldots, r$, let w_i be the Gauss extension of v_i to the field of fractions of $R(X)$ as in Remark 6.1.3. Then w_i is positive on $\mathfrak{m}R[X]$ and zero on $R[X] \setminus \{0\}$, so it is non–negative on $R(X)$. By Proposition 10.4.9, $\mathcal{RV}(IR[X]) = \{w_1, \ldots, w_r\}$, and hence since Rees valuations localize (Proposition 10.4.1), $\mathcal{RV}(IR(X)) = \{w_1, \ldots, w_r\}$.

If $IR(X)$ is not simple, write $IR(X) = JK$ for some proper ideals J and K in $R(X)$. As in the reductions on page 270 we may assume that J and K are integrally closed. By Proposition 10.4.5, $\mathcal{RV}(J), \mathcal{RV}(K) \subseteq \mathcal{RV}(JK) = \mathcal{RV}(IR(X)) = \{w_1, \ldots, w_r\}$. Hence $J = \{u \in R(X) \,|\, w_i(u) \geq w_i(J)), i = 1, \ldots, r\}$, which is an ideal in $R(X)$ extended from R. In other words, $J = (J \cap R)R(X)$, and similarly, $K = (K \cap R)R(X)$. Thus $(J \cap R)(K \cap R)R(X) = JK = IR(X)$ and $(J \cap R)(K \cap R) \subseteq I$, so that by faithful flatness $(J \cap R)(K \cap R) = I$. Since I is simple, either $J \cap R = R$ or $K \cap R = R$, whence either $J = R(X)$ or $K = R(X)$. $\qquad\square$

Proposition 14.4.8 *Let $(R, \mathfrak{m})$ be a two–dimensional regular local ring and let I be a simple integrally closed $\mathfrak{m}$–primary ideal in R. Then I has only one Rees valuation.*

If $I \neq \mathfrak{m}$ and $x \in \mathfrak{m} \setminus \mathfrak{m}^2$ such that I is contracted from $S = R[\frac{\mathfrak{m}}{x}]$, then $\mathcal{RV}(I) = \mathcal{RV}(I^S)$.

Proof: Let X be a variable over R. By Lemma 14.4.7, $IR(X)$ is simple, by Lemma 8.4.2 it is integrally closed, it is primary to the maximal ideal of $R(X)$, and by Proposition 10.4.9, the number of Rees valuations of I is the same as the number of Rees valuations of $IR(X)$. Thus the hypotheses on I are satisfied also for $IR(X)$, and if we prove that $IR(X)$ has only one Rees valuation, the first part would be proved. Thus by possibly first passing to $R(X)$ we may assume that the residue field is infinite, in which case I is full (Theorem 14.1.8) and there exists $x \in \mathfrak{m} \setminus \mathfrak{m}^2$ such that the leading form of x does not divide $c(I)$, whence by Proposition 14.2.2 I is contracted from $S = R[\frac{\mathfrak{m}}{x}]$.

We proceed by induction on $\lambda(R/I)$. If this is 1, then $I = \mathfrak{m}$, and the only Rees valuation of $\mathfrak{m}$ is the $\mathfrak{m}$–adic valuation (established in Example 10.3.1). Now suppose that $\lambda(R/I) > 1$. By Proposition 14.3.7, I^S is contained in only one maximal ideal of S. Let N be that maximal ideal. Then also $I^S S_N$ is simple, and by Proposition 14.4.3, it is integrally closed. By Lemma 14.3.4, $\lambda(S_N/I_N^S) < \lambda(R/I)$. Thus by induction I^S has only one Rees valuation. Let V be the corresponding Rees valuation ring, and let $W = S_{xS}$. We already know that W is the unique Rees valuation ring of $\mathfrak{m}$ and of xS. By Proposition 10.4.8, the only Rees valuation rings of $IS = x^{\text{ord}(I)}I^S$ are V and W. For any $n \geq 1$, by Corollary 14.2.3 I^n is contracted from S, and by Theorem 14.4.4 I^n and $(I^S)^n$ are integrally closed. Thus

$$\overline{I^n} = I^n = I^n S \cap R = I^n V \cap I^n W \cap R = I^n V \cap \mathfrak{m}^n W \cap R = I^n V \cap \mathfrak{m}^n.$$

But $I^n V \cap S \subseteq (I^S)^n V \cap S = (I^S)^n \subseteq N^n$, whence $I^n V \cap R \subseteq N^n \cap R \subseteq \mathfrak{m}^n$ by the structure of height two ideals in S, which proves that for all n, $\overline{I^n} = I^n V \cap R$, whence by uniqueness of Rees valuations, Theorem 10.1.6, V is the only Rees valuation ring of I. $\qquad\square$

With this we can prove Zariski's uniqueness of factorization of integrally closed ideals in general:

Theorem 14.4.9 (Zariski) *Let $(R, \mathfrak{m})$ be a two–dimensional regular local ring. Then every non–zero integrally closed ideal I in R can be written uniquely (except for ordering)*

$$I = a_1^{l_1} \cdots a_m^{l_m} I_1^{k_1} \cdots I_n^{k_n},$$

where $I_1, \ldots, I_n$ are simple $\mathfrak{m}$–primary integrally closed ideals, $a_1, \ldots, a_m$ are pairwise relatively prime irreducible elements of R, and $l_1, \ldots, l_m, k_1, \ldots, k_n$ are positive integers.

Proof: By Theorem 14.4.6 it suffices to prove the uniqueness. As in the proof of that theorem, it suffices to prove the case where I is $\mathfrak{m}$–primary. Suppose that $I = I_1 \cdots I_n = J_1 \cdots J_m$ are two decompositions of I into products of simple $\mathfrak{m}$–primary integrally closed ideals $I_1, \ldots, I_n, J_1, \ldots, J_m$. Let X be an indeterminate over R, and let $R(X) = R[X]_{\mathfrak{m}R[X]}$. By Lemma 14.4.7, each $I_j R(X), J_j R(X)$ is simple, and also integrally closed and $\mathfrak{m}R(X)$–primary by

Lemma 8.4.2, and by faithful flatness, $IR(X) = (I_1 R(X)) \cdots (I_n R(X)) = (J_1 R(X)) \cdots (J_m R(X))$. By Theorem 14.4.6, since $R(X)$ has an infinite residue field, $n = m$, and after reindexing, for each $j = 1, \ldots, n$, $I_j R(X) = J_j R(X)$. But then $I_j = J_j$ for each j, which proves uniqueness. $\square$

This fundamental theorem of two–dimensional regular local rings also fails in higher dimension.

Example 14.4.10   (Lipman [189]) Let X, Y, Z variables over a field k and $R = k[X, Y, Z]$. One can show that the ideals $I = (X^3, Y^3, Z^3, XY, XZ, YZ)$, $J_1 = (X^2, Y, Z)$, $J_2 = (X, Y^2, Z)$ and $J_3 = (X, Y, Z^2)$ are integrally closed and simple, and that $(X, Y, Z)I = J_1 J_2 J_3$. However, these factors are definitely not equal, so that unique factorization fails in three–dimensional regular rings. However, Lipman has shown that some things can be recovered! See [189] for generalizations of the factorization of integrally closed ideals in higher dimension.

14.5. A formula of Hoskin and Deligne

Let K be a field. In this section we will study from a more abstract point of view the set of two–dimensional regular local rings with field of fractions K. Such objects will be called **points**. For any point T we denote its unique maximal ideals as $\mathfrak{m}_T$. A basic definition is the following one.

Definition 14.5.1   *A point T is **infinitely near** to R, written $T \succ R$, if there exists a sequence*

$$R = R_0 \subseteq R_1 \subseteq \cdots \subseteq R_n = T$$

such that for each $i = 1, \ldots, n - 1$, R_i is a local quadratic transformation of R_{i-1}.

In fact, *every* point T containing R with $\mathfrak{m}_R \subseteq \mathfrak{m}_T$ is infinitely near to R:

Theorem 14.5.2   (Abhyankar [6, p. 343, Theorem 3]) *Let $R \subseteq T$ be two–dimensional regular local rings with the same field of fractions. Then there exists a unique sequence $R = R_0 \subseteq R_1 \subseteq \cdots \subseteq R_n = T$ such that for each $i = 1, \ldots, n - 1$, R_i is a local quadratic transform of R_{i-1}.*

Proof: Without loss of generality $R \neq T$. Note that $\mathfrak{m}_R \subseteq \mathfrak{m}_T$ by the dimension inequality (Theorem B.2.5). A consequence of the conclusion is that there exists $x \in \mathfrak{m}_R$ such that $xT = \mathfrak{m}_R T$. Indeed in some sense this is the crucial point.

By Theorem 8.8.1, if $R \neq T$, the analytic spread of $\mathfrak{m}_R T$ is at most one. As T is integrally closed, all principal ideals are integrally closed. In particular, $\mathfrak{m}_R T$ is a principal ideal, and hence there exists $x_0 \in \mathfrak{m}_0$ such that $\mathfrak{m}_0 T = x_0 T$, where we have relabeled $\mathfrak{m}_0 = \mathfrak{m}_R$.

We have that $R[\frac{\mathfrak{m}_0}{x_0}] \subseteq T$. Let $\mathfrak{m}_1 = \mathfrak{m}_T \cap R[\frac{\mathfrak{m}_0}{x_0}]$. Set $R_1 = (R[\frac{\mathfrak{m}_0}{x_0}])_{\mathfrak{m}_1}$, so $R_1 \subseteq T$, and R_1 has a maximal ideal $\mathfrak{m}_1 R_1$ which by abuse of notation we call $\mathfrak{m}_1$. If $\mathfrak{m}_1$ has height 1, then R_1 is a Noetherian discrete valuation ring, hence $T = R_1$, a contradiction. So necessarily R_1 is again a two–dimensional regular local ring.

We claim that the ring R_1 just constructed is unique. Suppose not. Then there exists another local quadratic transformation $R[\frac{\mathfrak{m}_0}{y}]_Q \subseteq T$, where $Q = \mathfrak{m}_T \cap R[\frac{\mathfrak{m}_0}{y}]$. We will prove that $R_1 \subseteq R[\frac{\mathfrak{m}_0}{y}]_Q$, and by symmetry it will follow that these two rings are equal. Note that $\mathfrak{m}_0 T = x_0 T = yT$. Therefore $\frac{x_0}{y}$ is a unit in T. Then $\frac{x_0}{y}$ cannot be in $QR[\frac{\mathfrak{m}_0}{y}]_Q$, which means that $\frac{x_0}{y}$ is a unit in $R[\frac{\mathfrak{m}_0}{y}]_Q$. Hence $\frac{y}{x_0} \in R[\frac{\mathfrak{m}_0}{y}]_Q$, proving $R[\frac{\mathfrak{m}_0}{x_0}] \subseteq R[\frac{\mathfrak{m}_0}{y}]_Q$. Then $\mathfrak{m}_1 = \mathfrak{m}_T \cap R[\frac{\mathfrak{m}_0}{x_0}] = \mathfrak{m}_T \cap R[\frac{\mathfrak{m}_0}{y}]_Q \cap R[\frac{\mathfrak{m}_0}{x_0}] = QR[\frac{\mathfrak{m}_0}{y}]_Q \cap R[\frac{\mathfrak{m}_0}{x_0}]$, which forces $R_1 = R[\frac{\mathfrak{m}_0}{x_0}]_{\mathfrak{m}_1} \subseteq R[\frac{\mathfrak{m}_0}{y}]_Q$, as claimed. It follows that R_1 is the unique local quadratic transformation in T.

By repeating this construction we get a unique chain of rings $R = R_0 \subseteq R_1 \subseteq \cdots \subseteq R_n \cdots \subseteq T$ such that R_{i+1} is a local quadratic transformation of R_i. We need to prove $R_n = T$ for some n. If not, set $V = \cup_i R_i$. If V is not a valuation domain, there exists an element z in the common field of fractions of R and T such that $z \notin V$ and $\frac{1}{z} \notin V$. We can write $z = \frac{a}{b}$ with $a, b \in R_0$, and necessarily both $a, b \in \mathfrak{m}_0$. Hence $a = a_1 x_0$, $b = b_1 x_0$ with $a_1, b_1 \in R_1$. Again by necessity, $a_1, b_1 \in \mathfrak{m}_1$. Choose $x_1 \in R_1$ such that $\mathfrak{m}_1 T = x_1 T$ and repeat this process. We successively obtain elements $a_i, b_i \in \mathfrak{m}_i$ such that $z = \frac{a_i}{b_i}$ and $a_{i-1} = a_i x_{i-1}$, $b_{i-1} = b_i x_{i-1}$ with $\mathfrak{m}_{i-1} T = x_{i-1} T$ and $x_{i-1} \in \mathfrak{m}_{i-1}$, for all $i \geq 1$. But in T, this gives an increasing chain of ideals, $a_0 T \subseteq a_1 T \subseteq a_2 T \subseteq \cdots$ that cannot stop since at every step $a_i T \subseteq \mathfrak{m}_i a_{i+1} T \subseteq \mathfrak{m}_T a_{i+1} T$. This contradiction proves that V is a valuation domain. Any ring containing a valuation domain and sharing the same field of fractions is also a valuation domain, so in particular T is a valuation domain, which is a contradiction. So necessarily the chain must be finite. $\square$

Definition 14.5.3 *Let R be a point and I a non–zero ideal of R. The **point basis** of I, is the set $\{\mathrm{ord}_T(I^T)\}_{T \succ R}$ of non–negative integers. A **base point** of I is a point $T \succ R$ such that $\mathrm{ord}_T(I^T) \neq 0$, i.e., such that $I^T \neq T$.*

Observe that if $T \succ S$ and $S \succ R$, then for any ideal I in R, $(I^S)^T = I^T$.

Our main goal in this section to prove a classic formula of Hoskin and Deligne that gives the length of an $\mathfrak{m}$–primary ideal I in a point R in terms of the point basis of I. We will follow notes of Jugal Verma that are based on the thesis of Vijay Kodiyalam [171]. See also [154] and [153].

Theorem 14.5.4 (Hoskin–Deligne Formula) *Let $(R, \mathfrak{m})$ be a regular local ring of dimension two and with infinite residue field, and let I be an integrally closed $\mathfrak{m}$–primary ideal of R. Then (with $\binom{1}{2} = 0$)*

$$\lambda_R\left(\frac{R}{I}\right) = \sum_{T \succ R} \binom{\mathrm{ord}_T(I^T) + 1}{2} [\kappa(\mathfrak{m}_T) : \kappa(\mathfrak{m}_R)].$$

Proof: In Section 14.2 we showed that if $T \succ R$, then the residue field of T is a finite algebraic extension of the residue field of R. The degree of this field extension is denoted by $[\kappa(\mathfrak{m}_T) : \kappa(\mathfrak{m}_R)]$.

We first prove the case $I = \mathfrak{m}^n$. (Since $\mathfrak{m}$ is integrally closed, so is $\mathfrak{m}^n$, say by Theorem 14.4.4.) The only base point of I is just R, since if $T \succ R$ and $T \neq R$, then $\mathfrak{m}^T = T$. Thus the sum on the right side of the display is simply $\binom{\operatorname{ord}_R(I^R)+1}{2}[\kappa(\mathfrak{m}_R) : \kappa(\mathfrak{m}_R)]$, which is $\binom{n+1}{2} = \lambda(R/\mathfrak{m}^n)$.

We proceed with the general case by induction on $\lambda(R/I)$. If this is 1, then $I = \mathfrak{m}$, and this case has been proved.

Now let $\lambda(R/I) > 1$. Choose $x \in \mathfrak{m} \setminus \mathfrak{m}^2$ such that I is contracted from $S = R[\frac{\mathfrak{m}}{x}]$. By Proposition 14.4.3, I^S is integrally closed ideal, and by Example 14.3.2, I^S has height at least two. If $I^S = S$, then $IS = \mathfrak{m}^n S$ for some n, so since I is contracted from S, $I = \mathfrak{m}^n$, and this case has been proved. So we may assume that I^S is a proper ideal in S. Then necessarily it has height two. Let $Q_1, \ldots, Q_t$ be the maximal ideals of S that contain I^S. Set $T_i = S_{Q_i}$. Then $T_i \succ R$ and I^{T_i} is an integrally closed $\mathfrak{m}_{T_i}$–primary ideal in T_i. Moreover, $\lambda_R(T_i/I^{T_i}) = \lambda_{T_i}(T_i/I^{T_i}) \cdot [\kappa(\mathfrak{m}_{T_i}) : \kappa(\mathfrak{m}_R)]$. By the Chinese Remainder Theorem, $S/I^S \cong \prod_i T_i/I^{T_i}$, where $T_i = S_{Q_i}$ for $i = 1, \ldots, t$. Combining these remarks, Lemma 14.3.4, and induction, gives the following sequence of equalities:

$$
\lambda_R\left(\frac{R}{I}\right) = \lambda_R\left(\frac{R}{\mathfrak{m}^{\operatorname{ord}(I)}}\right) + \lambda_R\left(\frac{\mathfrak{m}^{\operatorname{ord}(I)}}{I}\right)
$$
$$
= \binom{\operatorname{ord}(I)+1}{2} + \lambda_R\left(\frac{S}{I^S}\right)
$$
$$
= \binom{\operatorname{ord}(I)+1}{2} + \sum_{T_i} \lambda_R\left(\frac{T_i}{I^{T_i}}\right)
$$
$$
= \binom{\operatorname{ord}(I)+1}{2} + \sum_{T_i} \lambda_{T_i}\left(\frac{T_i}{I^{T_i}}\right)[\kappa(\mathfrak{m}_{T_i}) : \kappa(\mathfrak{m}_R)]
$$
$$
= \binom{\operatorname{ord}(I)+1}{2} + \sum_{T_i}\sum_{T \succ T_i} \lambda_{T_i}\left(\frac{T}{I^T}\right)[\kappa(\mathfrak{m}_T) : \kappa(\mathfrak{m}_{T_i})] \cdot [\kappa(\mathfrak{m}_{T_i}) : \kappa(\mathfrak{m}_R)]
$$
$$
= \binom{\operatorname{ord}(I)+1}{2}[\kappa(\mathfrak{m}_R) : \kappa(\mathfrak{m}_R)] + \sum_{T_i}\sum_{T \succ T_i} \lambda_{T_i}\left(\frac{T}{I^T}\right)[\kappa(\mathfrak{m}_T) : \kappa(\mathfrak{m}_R)].
$$

Let $\mathcal{T}$ be the set of all T as above with $I^T \neq T$, together with the point R. It remains to prove that no $T \in \mathcal{T}$ appears twice in the last line in the display and that $\mathcal{T}$ equals the set of all base points of I. The non–repetition holds by Theorem 14.5.2— for each T that is infinitely near to R there is a unique sequence of quadratic transformations between R and T.

Certainly each $T \in \mathcal{T}$ is infinitely near to R. Now let $T \succ R$. We know from Theorem 14.5.2 that T dominates a unique local quadratic transformation of R. This local quadratic transformation is obtained as follows: choose

$z \in \mathfrak{m}$ such that $\mathrm{ord}_R(z) = \mathrm{ord}_R(\mathfrak{m})$. Then $R[\frac{\mathfrak{m}}{z}] \subseteq T$. The maximal ideal of T contracts to a prime ideal Q in $R[\frac{\mathfrak{m}}{z}]$, and the unique local quadratic transformation in question is exactly $(R[\frac{\mathfrak{m}}{z}])_Q$. Suppose that $\frac{x}{z} \in \mathfrak{m}_T$. As I is integrally closed, it is full, and as the residue field is integrally closed, by Theorem 14.1.4, $\mu(I) = \mathrm{ord}(I) + 1$. Then since I is contracted from S, by Propositions 14.2.2 and 14.1.7, $\mathfrak{m}^r = I + x\mathfrak{m}^{r-1}$. Thus we can write $z^r = a + xb$ for some $a \in I$ and $b \in \mathfrak{m}^{r-1}$. In T we may then write $1 = \frac{a}{z^r} + (\frac{x}{z})(\frac{b}{z^{r-1}})$, showing that $1 \in I^T + \mathfrak{m}_T$, a contradiction. So necessarily $\frac{x}{z} \notin \mathfrak{m}_T$. Then $\frac{z}{x} \in T$, which means that $S \subseteq T$. If $I^S \nsubseteq \mathfrak{m}_T \cap S$, then $I^T = T$, contradicting the assumption that T is a base point of I. So necessarily the maximal ideal of T must contract to one of the Q_i, so T is infinitely near to T_i for some i, whence $T \in \mathcal{T}$. $\square$

14.6. Simple integrally closed ideals

The purpose of this section is to elucidate the correspondence between divisorial valuations and simple integrally closed ideals in a two–dimensional regular local ring. For more information about this correspondence and related topics see [77], [78], [49], [51], and [93].

Throughout this section we work with a two–dimensional regular local ring $(R, \mathfrak{m})$ with an infinite residue field. By Proposition 14.4.8 we already know that every simple $\mathfrak{m}$–primary integrally closed ideal has a unique Rees valuation. If the ideal is I, we denote its unique Rees valuation by $\bar{v}_I$.

The goal of this section is to relate the values of $\bar{v}_I(J)$ and $\bar{v}_J(I)$ whenever I and J are simple $\mathfrak{m}$–primary integrally closed ideals. We need some more notation.

Since we assume that the residue field is infinite, I has a minimal two–generated reduction (Proposition 8.3.7 and Corollary 8.3.9). So let (a, b) be a minimal reduction of I. Set $A = R[\frac{b}{a}]$ and $B = R[\frac{I}{a}]$. Since I is integral over (a, b), it follows that B is integral over A. By Theorem 14.4.4, $B = \cup_n \frac{I^n}{a^n} = \cup_n \frac{\overline{I^n}}{a^n} = \overline{B}$, so B is integrally closed, and is thus the integral closure of A. We established that I has only one Rees valuation, namely $\bar{v}_I$. Thus by Theorem 10.2.2 (3), there is a unique prime ideal P in B that is minimal over aB. By the same theorem, B_P is the valuation ring corresponding to $\bar{v}_I$, i.e., B_P is the (unique) Rees valuation of I. The elements a, b form a regular sequence in R, hence $A \cong R[X]/(aX - b)$, where X is an indeterminate over R. In particular, $A/\mathfrak{m}A \cong (R/\mathfrak{m})[X]$ and $\mathfrak{m}A$ is a prime ideal. Thus $W = A \setminus \mathfrak{m}A$ is a multiplicatively closed subset of A and $W^{-1}A$ is a one–dimensional Noetherian local domain with integral closure $W^{-1}B$. Since P contains aB, it contains $b = a\frac{b}{a}$, hence it contains $\mathfrak{m}B$. But P is the only prime ideal in B minimal over aB, hence it is the only prime ideal in B minimal over $\mathfrak{m}B$, so that $W^{-1}B = B_P$, whence the integral closure of $W^{-1}A$ is exactly the valuation ring corresponding to $\bar{v}_I$. In particular, by Proposition 4.8.5, the residue field of $\overline{v}_I$ is finite over the residue field of $W^{-1}A$. We denote the degree of this extension by $\Delta(\overline{v}_I)$. Set $C = W^{-1}A$.

Proposition 14.6.1 (Lipman [185, 21.4])) *With notation as above, for all* $r \in C$, $\overline{v}_I(r)\Delta(\overline{v}_I) = \lambda_C(C/rC)$.

Proof: Let V be the Rees valuation ring of I. With notation above, $V = B_P = W^{-1}B$, and it is the integral closure of C. We know that $\overline{v}_I(r) = \lambda_V(V/rV)$. Since R is a regular local ring and C is a finitely generated R–algebra contained in the field of fractions of R, by Theorem 9.2.2, V is a finitely generated C–module. Thus we have the following exact sequence of finite–length C–modules:

$$0 \to \frac{C \cap rV}{rC} \to \frac{C}{rC} \to \frac{V}{rV} \to \frac{V}{C} \to 0.$$

If r is in the conductor of C, then the first module is $\frac{rV}{rC} \cong \frac{V}{C}$, so that $\lambda_C(C/rC) = \lambda_C(V/rV)$. It follows that for an arbitrary $r \in C$, if s is any non–zero element of the conductor of C,

$$\lambda_C\left(\frac{C}{rC}\right) = \lambda_C\left(\frac{C}{srC}\right) - \lambda_C\left(\frac{rC}{srC}\right) = \lambda_C\left(\frac{C}{srC}\right) - \lambda_C\left(\frac{C}{sC}\right)$$

$$= \lambda_C\left(\frac{V}{srV}\right) - \lambda_C\left(\frac{V}{sV}\right) = \lambda_C\left(\frac{V}{srV}\right) - \lambda_C\left(\frac{rV}{srV}\right)$$

$$= \lambda_C\left(\frac{V}{rV}\right).$$

Thus

$$\overline{v}_I(r) = \lambda_V(V/rV) = \lambda_C(V/rV)\text{tr.deg}_{\kappa(\mathfrak{m}_C)}(\kappa(\mathfrak{m}_V)) = \lambda_C(C/rC)\Delta(\overline{v}_I),$$

which finishes the proof. $\square$

With this we can present a proof of Lipman's reciprocity theorem:

Theorem 14.6.2 (Lipman [185, Proposition 21.4]) *Let* $(R, \mathfrak{m})$ *be a two–dimensional regular local ring with infinite residue field. Let I and J be two* $\mathfrak{m}$*–primary simple integrally closed ideals. Then*

$$\overline{v}_J(I)\Delta(\overline{v}_J) = \overline{v}_I(J)\Delta(\overline{v}_I).$$

Proof: Since the residue field is infinite, there are minimal reductions (a, b) of I and (c, d) of J.

We will prove the theorem by proving that $\overline{v}_I(J)\Delta(\overline{v}_I)$ and $\overline{v}_J(I)\Delta(\overline{v}_J)$ both equal the length of the ring $R(X, Y)/(bX - a, dY - c)$, where X, Y are variables and $R(X, Y)$ is the polynomial ring $R[X, Y]$ localized at the ideal $\mathfrak{m}R[X, Y]$. By symmetry it suffices to show that the length of this ring is equal to $\overline{v}_I(J)\Delta(\overline{v}_I)$.

Let v' be the Gauss extension of $\overline{v}_I$ as in Remark 6.1.3 to the field of fractions of $R[Y]$. Namely, if $a_0, \ldots, a_n \in R$, then $v'(a_0 + a_1Y + \cdots + a_nY^n) =$

$\min\{\overline{v}_I(a_0), \ldots, \overline{v}_I(a_n)\}$, and this extends to a valuation on the field of fractions. Let $R' = R[Y]_{\mathfrak{m}R[Y]}$. By Proposition 10.4.9, v' is the unique Rees valuation of IR'. By Lemma 14.4.7, IR' is a simple ideal. As in Lemma 8.4.2, R' is a two–dimensional regular local ring, its maximal ideal is $\mathfrak{m}R'$, IR' is integrally closed and primary to the maximal ideal, and $(a, b)R'$ continues to be a minimal reduction of IR'. The construction above Proposition 14.6.1, with IR' in place of I, gives a one–dimensional local domain $C' = (R'[X]/(bX - a))_{\mathfrak{m}R'[X]} = R(X, Y)/(bX - a) = C(Y)$ whose integral closure is the valuation ring corresponding to v'. Since $(c, d)R'$ is a reduction of JR', by Proposition 6.8.1, $v'(JR') = v'((c, d)R') = v'(cY - d)$, so by Proposition 14.6.1,

$$v(J)\Delta(v') = v'(J)\Delta(v') = v'(cY - d)\Delta(v') = \lambda_{C'}(C'/(cY - d)C')$$
$$= \lambda(R(X, Y)/(bX - a, cY - d)).$$

The theorem is then proved provided that $\Delta(v') = \Delta(v)$. But the valuation ring corresponding to v', namely the integral closure of $C' = C(Y)$, is $\overline{C}(Y)$, so that $\Delta(v') = \mathrm{tr.deg}_{\kappa(\mathfrak{m}_{C'})}(\kappa(\mathfrak{m}_{V'})) = \mathrm{tr.deg}_{\kappa(\mathfrak{m}_C)}(\kappa(\mathfrak{m}_V)) = \Delta(v)$. □

Discussion 14.6.3 This proof, while quite simple, does hide some of the basic geometry which makes the result more apparent. Namely, as Lipman explains in [185, (21.4)], Theorem 14.6.2 comes down to the fact that for the exceptional divisors D_1 and D_2 generated by I and J on a desingularization dominating the blowup of IJ, $D_1 \cdot D_2 = D_2 \cdot D_1$. The proof we give basically computes the intersection multiplicity of two "generic" curves, one defined by a generic element of a minimal reduction of I and one defined by a generic element of a minimal reduction of J.

14.7. Exercises

14.1 Let $(R, \mathfrak{m})$ be a two–dimensional regular local ring and let I be an $\mathfrak{m}$–primary ideal. Prove that I is $\mathfrak{m}$–full if and only if for all ideals J containing I, $\mu(J) \leq \mu(I)$.

14.2* Let $(R, \mathfrak{m})$ be a three–dimensional regular local ring. Prove or give a counterexample to the claim in Exercise 14.1. (This is open.)

14.3 Let $(R, \mathfrak{m})$ be a two–dimensional regular local ring and let I be a full $\mathfrak{m}$–primary ideal. Prove that $I : \mathfrak{m}$ is also full.

14.4 Let $(R, \mathfrak{m})$ be a two–dimensional regular local ring and let I be a full $\mathfrak{m}$–primary ideal. Prove that $I = \mathfrak{m}^{\mathrm{ord}(I)-\deg c(I)} \cdot J$ for some ideal J such that $\mathfrak{m}$ does not divide J.

14.5 Let $(R, \mathfrak{m})$ be a two–dimensional regular local ring. Give an example of three $\mathfrak{m}$–primary full ideals I, J, and K such that $IJ = IK$ but $J \neq K$.

14.6 Let R be a two–dimensional regular local ring. Suppose that I is a simple integrally closed ideal. Prove that $c(I)$ is a power of an irreducible polynomial.

14.7   Let R be a two–dimensional regular local ring with maximal ideal (x, y). Prove that (x^3, xy^3, x^2y, y^4) is integrally closed and find its factorization into simple integrally closed ideals.

14.8   Let $(R, \mathfrak{m})$ be a two–dimensional regular local ring and I an integrally closed ideal in R. Let (a, b) be a reduction of I. Prove that $(a, b)I = I^2$, i.e., that the reduction number of I is 1.

14.9   Let $(R, \mathfrak{m})$ be a two–dimensional regular local ring with $\mathfrak{m} = (x, y)$. For $n \in \mathbb{N}$ set $J = (x^n, y^n)$ and $I = (x^n, x^{n-1}y, y^n)$. Find the reduction number of $r_J(I)$.

14.10   Let $(R, \mathfrak{m}, k)$ be a two–dimensional regular local ring of characteristic p that contains a copy of its residue field k. Write $\mathfrak{m} = (x, y)$. Let $u \in k$. Set $I = (x^p + uy^p, \mathfrak{m}^{p+2})$. If $u^{1/p}$ is not in k, prove that I is integrally closed. If $u^{1/p}$ is in k, prove that I is not integrally closed.

14.11   Let $(R, \mathfrak{m})$ be a two–dimensional regular local ring. Let f be a non–zero element of R of order r. Prove that the ideal $(f) + \mathfrak{m}^{r+2}$ is integrally closed if and only if $(f) + \mathfrak{m}^n$ is integrally closed for all n.

14.12   Let R be a Cohen–Macaulay local ring with infinite residue field (of arbitrary dimension). Assume that R satisfies Serre's condition (R_2). Let I be a height two integrally closed ideal having analytic spread two. Prove that I^k is integrally closed for all $k \geq 1$.

14.13   Let I be an ideal in a two–dimensional regular local ring. Prove that $c(I) = c(\overline{I})$.

14.14   Let $(R, \mathfrak{m})$ be a two–dimensional regular local ring with infinite residue field, and let I be an $\mathfrak{m}$–primary integrally closed ideal. Prove that $e(I) = \sum_{T \succ R} \operatorname{ord}_T(I^T)^2 \cdot [\kappa(\mathfrak{m}_T) : \kappa(\mathfrak{m}_R)]$.

14.15   Let I be an $\mathfrak{m}$–primary ideal in a two–dimensional regular local ring $(R, \mathfrak{m})$. Prove that $c(I) = 1$ if and only if $\overline{I} = \mathfrak{m}^n$ for some n.

14.16   Let I be an $\mathfrak{m}$–primary ideal in a two–dimensional regular local ring $(R, \mathfrak{m})$. Let $x \in \mathfrak{m}$, and set $S = R[\frac{\mathfrak{m}}{x}]$. Prove that $IS \cap R = I\mathfrak{m}^n : x^n$ for all sufficiently large n.

14.17   Let $R = k[X, Y]_{(X,Y)}$, where k is a field and X and Y are variables over k. Set $I = ((Y^2 - X^3)^2, XY^4, X^2Y^3, X^5Y, X^7)R$. Prove that I is integrally closed, and find its factorization into simple integrally closed ideals. Do the same for the ideal $J = (X^2 - Y^4, X^2Y + Y^5, XY^3)$. Describe the valuations associated to the simple integrally closed ideals in the factorizations of I and J.

14.18   Let R be a two–dimensional regular local ring. Let v be a divisorial valuation with respect to R. Let $I = (a, b)$ be an ideal of height two such that $v(a) = v(b)$ and such that the image of a/b in $k(v)$ is transcendental over k. Prove that v is a Rees valuation of I.

14.19   Let $R = k[XYZ, X^3Z, Y^2Z] \subseteq k[X, Y, Z] = S$. Prove that R and S have the same field of fractions. Let $\mathfrak{m} = (XYZ, X^3Z, Y^2Z)R$. Prove that $\mathfrak{m}S = Z \cdot (XY, X^3, Y^2)S$ is not principal. Conclude that there is no $r \in \mathfrak{m} \setminus \mathfrak{m}^2$ such that $R[\frac{\mathfrak{m}}{r}]$ is contained in S.

15
Computing integral closure

Is the integral closure of rings and of ideals computable? In principle it is, but in practice often not due to the limitations of computers and the essential combinatorics that must enter into most calculations of integral closure. In this chapter we present some algorithms for computing the integral closure of ideals and rings. We do not require knowledge of the theory of Gröbner bases, but we make several claims in the chapter that can be either taken on faith or be proved using elementary knowledge of Gröbner bases. A recent book by Vasconcelos [307] is largely devoted to the theory of computations of integral closure.

In order to compute the integral closure of an ideal, by Proposition 1.1.5 it suffices to compute the integral closure of the ideal modulo each minimal prime, lift the closures to the ring, and intersect these finitely many ideals. Lifts, intersections, and finding minimal prime ideals are all computable operations: lifts and intersections are easily computable on any symbolic algebra computer package, but finding the minimal prime ideals is a harder, more time–consuming, task. See Exercise 15.17 for issues relating to finding minimal prime ideals. But in any case, the computation of the integral closure of an ideal reduces to the computation of the integral closure of an integral domain. Thus for the purposes of the computation of the integral closure of ideals we may assume that all rings are integral domains.

> The computation of the integral closure of a ring in an overring also reduces to the computation of integral closures of rings in overrings that are domains; however, a necessary step is also the computation of the intersection of two subrings, which is not so readily computable.

So let R be an integral domain and I an ideal. By Proposition 5.2.4, the integral closure of the Rees ring $R[It]$ is

$$\overline{R} \oplus \overline{I\overline{R}} \oplus \overline{I^2\overline{R}} \oplus \overline{I^3\overline{R}} \oplus \cdots,$$

where $\overline{R}$ is the integral closure of R. Reading off the degree one component $\overline{I\overline{R}}$ of this graded ring gives the integral closure of $I\overline{R}$. By Proposition 1.6.1, the desired result $\overline{I}$ is the intersection of $\overline{I\overline{R}}$ with R. For computable rings R, $\overline{R}$ is a module–finite extension of R (see Chapter 4). In this case the intersection of an ideal in $\overline{R}$ with R is a readily computable operation, say via an elimination order with Gröbner bases. Thus it is computationally no loss of generality if we compute the absolute integral closure of $R[It]$ in order to compute the integral closure of the ideal I. However, this method at the same time computes the integral closures of all the powers of I, much more than what may be asked, and so the method may not be the most efficient. This method is so slow that it is, in practice, often unusable. The search for more efficient general algorithms continues. For special ideals, such as monomial ideals, there do exist more efficient algorithms. In this chapter we present the general algorithm, as well as specialized algorithms for restricted classes of ideals.

Without the Noetherian assumption on the ring we cannot expect the computation to terminate (see Exercise 15.2), and naturally, to establish any algorithms we need to assume that the starting ring is computable. Examples of computable rings are $\mathbb{Q}$, finite fields, $\mathbb{Z}$, and finitely generated extensions of these. The algorithms proceed by building proper extensions contained in the integral closure, and then repeating the procedure on the extension. When the integral closure is a module–finite extension, which is the case for computable rings (see Chapter 4), this procedure must terminate. Some of the algorithms are also concerned with finding upper bounds on the number of steps needed to compute the integral closure. In particular, Stolzenberg's and Vasconcelos's second methods do this (see Sections 15.1 and the end of Section 15.3).

A Noetherian integral domain is integrally closed if and only if it satisfies Serre's conditions (R_1) and (S_2) (Theorem 4.5.3). Thus part of the computation of the integral closure means building extensions that satisfy at least one of these two conditions. Stolzenberg's procedure (see Section 15.1) first constructs a module–finite extension satisfying (R_1), and Vasconcelos's second algorithm, given in Section 15.3, first constructs a module–finite extension satisfying (S_2).

An algorithm for computing the integral closure of ideals and domains has been implemented in the symbolic algebra program Singular [100] and Macaulay2 [97]. We provide sample codes for Singular, for Macaulay2, and for CoCoA [35] in Section 15.2.

15.1. Method of Stolzenberg

Perhaps the oldest method for computing the integral closure is for cyclic extensions of $\mathbb{Z}$ and developed by Dedekind [58] in 1899. There are many explicit algorithms for computing special integral closures of this kind, for example by [11], [12]; [183]; [267]; [274]; [289]; or [290]. Our interest lies in providing algorithms for the integral closure of general computable integral domains, such as affine domains. In this section we briefly review the oldest such method. We do not give all the details, but point out the general idea and computational issues involved. This method was obtained by Stolzenberg [276] for affine domains that are separable over the base field. Stolzenberg's procedure is computable, but not effectively so (see comments below). His method was shortly after expanded on by Seidenberg [263], [264], with rudiments already in [261]: the separability assumption was removed, but the computational impracticality stayed. Both Stolzenberg and Seidenberg were also concerned with an a priori bound on the number of steps needed in the procedure, in the constructive spirit of Kronecker, and more concretely following Hermann [116]. We present an outline of these procedures, up to slight modifications. Another, computationally more practical modification, is due to Gianni and Trager [89].

The main steps of Stolzenberg's procedure for computing the integral closure of an affine domain R are as follows (details are explained further below):
(1) Find a Noether normalization A of R.
(2) Find a non–zero element $c \in A$ in the conductor of R.
(3) Compute a primary decomposition of cA.
(4) Find a module–finite extension of R that satisfies Serre's condition (R_1).
(5) Under the assumption that R satisfies (R_1), find the integral closure of R.

How does one carry out these steps? Noether normalization and primary decompositions are algorithmic (see exercises at the end of this chapter); and conductors and their computability are discussed in Chapter 12. Seidenberg's solutions to (4) and (5) are described below. The following lemma provides a method for solving (4):

Lemma 15.1.1 *Let R be a Noetherian domain, and P a prime ideal of height one in R such that R_P is not a Noetherian valuation ring. Let $a, b \in R$ be part of a minimal generating set of PR_P. Assume that R contains infinitely many units $u_1, u_2, \ldots$ such that for all $i \neq j$, $u_i - u_j$ is also a unit. Then there exists an integer i such that $a/(u_i b + a)$ is integral over R_P and is not in R_P. In fact, there exists an integer N such that whenever $\{u_i \,|\, i = 1, \ldots, N+1\} \cup \{u_i - u_j \,|\, 1 \leq i < j \leq N+1\}$ consists of units, then for some $i \in \{1, \ldots, N+1\}$, $a/(u_i b + a)$ is integral over R_P and is not in R_P.*

Proof: As R_P is local and one–dimensional, by Theorem 4.9.2 and Proposition 6.8.14, there exist only finitely many discrete valuations of rank one that determine the integral closure of R. Let N be the number of these valuations. For any such valuation v, if $v(a) \neq v(b)$, then $v(a/(u_i b + a)) \geq 0$. If instead $v(a) = v(b)$, either $v(a/(u_i b + a)) \geq 0$ or $v(a/(u_i b + a)) < 0$. Assume that there is no i such that $v(a/(u_i b+a)) \geq 0$ for all v. By the pigeonhole principle, since there are only finitely many valuations v, there exist $i \neq j$ such that for some v, $v(a/(u_i b + a)), v(a/(u_j b + a)) < 0$. Necessarily also $v(a) = v(b)$, and without loss of generality $v(u_j b + a) \geq v(u_i b + a) > v(a)$. Then there exists d in the field of fractions of R such that $v(d) > 0$ and $u_i b + a = da$. Then $u_j b = (u_j/u_i)(u_i b) = (u_j/u_i)a(d-1)$, whence $u_j b + a = a(1 + (u_j/u_i)(d-1))$. It follows that $v(1 + (u_j/u_i)(d - 1)) > 0$, whence $v(1 - u_j/u_i) > 0$. Hence $u_i - u_j$ is not a unit in the valuation ring of v, contradicting the assumption that there is no u_i such that $v(a/(u_i b + a)) \geq 0$ for all v.

Thus necessarily for some i, $v(a/(u_i b + a)) \geq 0$ for all v. By the choice of the v, this means that $a/(u_i b + a) \in \overline{R_P}$. However, $a/(u_i b + a)$ is not in R_P as otherwise for some $r \in R_P$, $a = r(u_i b + a)$. If r is a unit in R_P, then

$b \in ru_i bR_P = (1-r)aR_P \subseteq aR_P$, contradicting the choice of a, b, and if r is in PR_P, then $a \in (1-r)aR_P = ru_i bR_P \subseteq bR_P$, again contradicting the choice of a, b. $\qquad\square$

Thus if R does not satisfy (R_1), by the lemma above, after clearing denominators, there exists $r \in \overline{R} \setminus R$. Repeating the construction with $R[r]$ in place of R creates an ascending chain of domains between R and $\overline{R}$. Under the assumption that R is a localization of an affine domain, or more generally, whenever $\overline{R}$ is module–finite over R, the procedure has to stop, and it stops at a module–finite extension of R inside $\overline{R}$ that satisfies (R_1).

How can we use the lemma above? First of all, we can algorithmically compute the Jacobian ideal J of R over the underlying field k. If R is separable over k, as Stolzenberg assumed, then by Theorem 3.2.7 and by the Jacobian criterion 4.4.9, J is not zero, so that J must have height at least 1. Again by the Jacobian criterion, for any prime ideal P in R with $J \not\subseteq P$, R_P is regular. Thus there are at most finitely many prime ideals P of height one, all minimal over J, for which the condition (R_1) may fail. These prime ideals can be found algorithmically, and also one can algorithmically determine if they are locally principal. If each is locally principal, then by Proposition 4.1.2, R satisfies (R_1). Otherwise, if one has a way of finding enough units and of checking integrality of an element over R, then by the lemma above one can recursively construct a strictly larger module–finite extension of R in the field of fractions. A repetition of this construction either determines that this larger ring satisfies (R_1), or constructs a larger module–finite extension ring yet. As the integral closures of computable rings are module–finite extensions, this procedure must stop in finitely many steps. In short, if one has a way of finding enough units and of checking integrality of an element over R, by above one can construct a module–finite extension of R with the same field of fractions that satisfies (R_1).

Below is one way of checking integrality that applies in Stolzenberg's procedure with A being a Noether normalization of the given R:

Lemma 15.1.2 *Let A be a unique factorization domain, K its field of fractions, and L a finite field extension of K. Let S be a basis of L over K. Assume that one can effectively express elements of L as K–linear combinations of elements of S, the one can effectively factor elements in A and express elements of K as fractions of elements in A. Then there exists an effective criterion for determining integrality of elements of L over A.*

Proof: Let $\varphi : L \to \operatorname{Hom}_K(L, L)$ be the map taking each α to multiplication by α. This map is injective, so that the minimal polynomials of α and of $\varphi(\alpha)$ are the same. As for every $s \in S$, $\alpha \cdot s$ can be effectively written as a K–linear combination of the elements of S, the characteristic polynomial of $\varphi(\alpha)$ can thus also be effectively computed. Let f be a non–zero A–multiple of this polynomial all of whose coefficients are in A. Then $f(\alpha) = 0$. Since factorization of f is effective, one can effectively decide if α satisfies a monic polynomial with coefficients in A. □

Thus given an affine domain R with enough units, one can compute an affine domain module–finite over R and with the same field of fractions that satisfies (R_1). The integral closure can then be computed via primary decomposition:

Lemma 15.1.3 (Stolzenberg [276]; Seidenberg [263]) *Let R be a locally formally equidimensional Noetherian domain. Let c be a non–zero element in the conductor of the integral closure of R to R. Let I be the intersection of the minimal components of cR.*
(1) Then $I/c \subseteq \overline{R}$.
(2) Suppose that R satisfies Serre's condition (R_1). Then $\overline{R} = I/c$. In particular, if cR has no embedded prime ideals, then R is integrally closed.

Proof: (1) As $\overline{R} \subseteq R/c$, $\overline{R}$ is module–finite over R. Let Q be a prime ideal in $\overline{R}$ that is of height one. By Proposition 4.8.6 (formally equidimensional assumption and Theorem B.5.1), $P = Q \cap R$ has height one. If c is not in Q, then $I/c \subseteq \overline{R}_Q$. If $c \in Q$, then $c \in P$, and necessarily P is one of the minimal primes over cR. By assumption I is in the P–primary component of cR. Thus $I \subseteq cR_P \cap R \subseteq c\overline{R}_Q$. It follows that $I/c \subseteq \overline{R}_Q$ for every prime ideal Q of $\overline{R}$ of height one, so that as $\overline{R}$ is a Krull domain (Theorem 4.10.5), $I/c \subseteq \overline{R}$. This proves (1).

(2) Suppose that R satisfies (R_1). Let $\alpha \in \overline{R}$. Then $d = c\alpha \in R$. Let P be a prime ideal in R minimal over cR. By assumption, R_P is integrally closed, so $d/c = \alpha \in R_P$, and hence $d \in cR_P \cap R$. Thus d is in every minimal component of cR, which proves that $\overline{R} \subseteq I/c$. □

With this, a fuller description of Stolzenberg's procedure for deciding normality and computing the integral closure of affine domains, up to minor modifications, is as follows:

Description of Stolzenberg's procedure Let R be a finitely generated domain $k[x_1, \ldots, x_n]$ over a field k such that the field of fractions K of R is separable over k. By Theorem 4.2.2, there is a Noether normalization A of R such R is integral over A and K is separable over $Q(A)$. (This is computable; see Exercise 15.6.) There is also a constructively obtained element $r \in R$ and a polynomial f in one variable with coefficients in A such that $Q(R) = Q(A)(r)$ and such that f is a minimal polynomial of r over $Q(A)$. Let D be the discriminant of f. (Recall that the discriminant is defined to be $\prod_{i<j}(\alpha_i - \alpha_j)$, where $f(x) = \prod_i(x - \alpha_i)$ in an algebraic closure. Another way to compute D is as plus or minus the norm of $f'(r)$, the formal derivative of f at r; see [210].) By Theorem 12.1.1, $f'(r)\overline{R} \subseteq R$. As f is separable, $D/f'(r) \in \overline{R}$, so that $D\overline{R} \subseteq R$. If D is a unit in A, then R is integrally closed. So assume that D is not a unit in A. Let $P_1, \ldots, P_l$ be the prime ideals in A minimal over DA. Let S be the complement in A of $P_1 \cup \cdots \cup P_l$.

Then $S^{-1}A$ is a one–dimensional semilocal unique factorization domain, and $S^{-1}R$ is a module–finite extension of $S^{-1}A$. (Stolzenberg does not address the computability of localized rings.) By Lemma 15.1.2, checking integrality over A is effective, assuming that factorization in A is effective. But A is a polynomial ring over a field, and there are algorithms for factoring in polynomial rings (see the book [315] by von zur Gathen and Gerhard; or the tutorial on page 38 of [173]). Thus checking integrality over $S^{-1}A$ is effective. Under the additional assumption that R has sufficiently many units u_i as in Lemma 15.1.1, one can construct a module–finite extension of $S^{-1}R$ in the field of fractions that satisfies (R_1) by repeated application of Lemma 15.1.1. The constructive part comes from Lemma 15.1.2, and from the fact that the search for a unit u_i ends after a predetermined number of steps. Suppose that s is in the integral closure of $S^{-1}R$ but is not in $S^{-1}R$. By clearing denominators (which is computable) we may also assume that s is integral over R. This procedure builds a module–finite extension of R that is integrally closed after localization at all prime ideals of height 1. In other words, a module–finite extension of R contained in $\overline{R}$ that satisfies Serre's condition (R_1) is computable. This reduces to the case when R satisfies (R_1). Now compute the primary decomposition of DR (see Exercises 15.15–15.17), and set I to be the intersection of the minimal components of DR. By Lemma 15.1.3, I/D is the integral closure of R.

Without the additional assumption above on the existence of sufficiently many units, one first adjoins to k infinitely many variables u_i to form a new field $k(u_1, u_2, \ldots)$, then proceed with the procedure as above to compute the integral closure of this larger ring, and then obtain the integral closure of R by computable elimination theory. However, this adjunction of variables is not effective computationally, and makes the procedure unpractical.

Seidenberg's method is similar to Stolzenberg's, with the improvement that the separable assumption be removed. As before, write $R = k[x_1, \ldots, x_n]$. One first passes to a finite purely inseparable extension K of the base field k such that $K(x_1, \ldots, x_n)$ is separable over K. This extension K is computable: one simply analyzes the minimal polynomials of the generating set of R over k. Seidenberg then filters $k \subseteq K$ as $k = k_0 \subseteq k_1 \subseteq \cdots \subseteq k_l = K$ such that for each i, either $k_i \subseteq k_{i-1}(x_1, \ldots, x_n)$ or k_i and $k_{i-1}(x_1, \ldots, x_n)$ are linearly independent over k_{i-1}. Then the point is to construct an integral closure of $k_{i-1}[x_1, \ldots, x_n]$ from the integral closure of $k_i[x_1, \ldots, x_n]$, which can be done via elimination.

15.2. Some computations

In Sections 2.4 and 4.5 we proved some criteria for whether a computable ring is integrally closed. In this section we give simple sample codes in symbolic computer algebra systems CoCoA, Macaulay2, and Singular for determining normality of a ring.

We use the criterion of normality of domains from Corollary 4.5.8: if R is a finitely generated algebra over a perfect field and J is an ideal in R such that $V(J)$ is the singular locus of R, then R is normal if and only if J has grade at least two.

Grade of an (arbitrary) ideal J can be computed as the least integer l for which $\mathrm{Ext}^l_R(R/J, R)$ is non–zero. Thus J has grade at least two if and only if $\mathrm{Ext}^0_R(R/J, R)$ and $\mathrm{Ext}^1_R(R/J, R)$ are both zero, which is a computable problem in computable rings. Explicitly, $\mathrm{Ext}^0_R(R/J, R) = \mathrm{Hom}_R(R/J, R) = 0 : J$, which is a simple colon computation. Also, the computation of $\mathrm{Ext}^1_R(R/J, R)$ is feasible. Here is a way to determine if $\mathrm{Ext}^1_R(R/J, R)$ is zero after it has been established that $\mathrm{Ext}^0_R(R/J, R)$ is zero. Namely, if $\mathrm{Ext}^0_R(R/J, R) = 0$, then J contains a non–zerodivisor. The finding of a non–zerodivisor is computable: in an integral domain, choose any non–zero element x of J. Once a non–zerodivisor $x \in J$ is known, the following simple computation determines if J has grade at least two, i.e., if $\mathrm{Ext}^1_R(R/J, R)$ is zero: by Theorem A.4.1, $\mathrm{Ext}^1_R(R/J, R) \cong \mathrm{Ext}^0_{R/xR}(R/J, R/xR) \cong \frac{(x):J}{(x)}$, so that the grade of J is exactly one if and only if the ideal $(x) : J$ is strictly bigger than (x), and the grade of J is at least two if and only if $(x) : J = (x)$.

In the three symbolic computer packages CoCoA, Macaulay2, and Singular, such a computation can be carried out quite fast. Below is the very simple example $R = (\mathbb{Z}/3\mathbb{Z})[a, b, c, d]/(a^2d - b^2c)$ of an equidimensional affine domain worked out in each of the three packages. The three languages differ, but the code below is self–explanatory.

The first example is CoCoA code:

```
Use A ::= Z/(3)[abcd];
I := Ideal(a^2d-b^2c);
M := Jacobian([a^2d-b^2c]);
J := Minors(Dim(A) - Dim(A/I),M);
L := Ideal(J) + I;
L;
I : Ideal(a^2);
(Ideal(a^2) + I) : L;
```

The element a^2 is in L. As the answer to input `I : Ideal(a^2);` is the ideal I itself, the image $L(A/I)$ of L in A/I has grade at least one, but as the answer to the input `(Ideal(a^2) + I) : L;` contains abc, which is not in $(a^2) + I$, $L(A/I)$ has grade exactly one. Thus by the Jacobian criterion, the ring $(\mathbb{Z}/3\mathbb{Z})[a, b, c, d]/(a^2d - b^2c)$ is not integrally closed.

Macaulay2 code for the same example is as follows:

```
R = ZZ/3[a,b,c,d]/(a^2*d-b^2*c)
J = minors_1 jacobian R
ann a^2
ideal a^2 : J
```

and the Singular code for the same procedure is:

```
ring R =3,(a,b,c,d), dp;
ideal I = a^2*d-b^2*c;
ideal J = jacob(I);
ideal L = I + J;
print (L);
quotient(I, a^2);
quotient(ideal(a^2) + I, J);
```

Thus CoCoA, Macaulay2, and Singular can easily detect that the ring $(\mathbb{Z}/3\mathbb{Z})[a, b, c, d]/(a^2d - b^2c)$ is not integrally closed.

We have enough theory to compute the integral closure of this ring as well. Observe that $R = \frac{k[a,b,c,d]}{(a^2d-b^2c)} \cong A = k[x, y, x^2t, y^2t] \subseteq k[x, y, t]$, where x, y and t are variables over k. The Jacobian matrix of R over k is a 1×4 matrix $[2ad, -2bc, -b^2, a^2]$, so that the Jacobian ideal is $(2ad, -2bc, -b^2, a^2)R$. Alternately, $J = (2xy^2t, 2yx^2t, x^2, y^2)A$, and $\sqrt{J} = (x, y)A$. By Lemma 2.4.3,

$$\operatorname{Hom}_R(\sqrt{J}, \sqrt{J}) = \frac{1}{x}\left(x(x, y)R :_R (x, y)\right)$$

$$\cong \frac{1}{x}\left((x(x, y)A :_A x) \cap (x(x, y)A :_A y)\right)$$

$$= \frac{1}{x}\left((x, y)A \cap ((x) + (x^2A :_A y))\right)$$

$$= \frac{1}{x}\left((x, y)A \cap (x, yx^2t)\right) = \frac{1}{x}(x, yx^2t) = 1A + xytA.$$

Thus by Lemma 2.1.8, the integral closure of A in its field of fractions contains xyt, which is not in A. Set $S = A[xyt] = k[x, y, x^2t, y^2t, xyt]$. As S is integral over A and has the same field of fractions as A, the integral closure of A is the same as the integral closure of S. Thus it suffices to compute the integral closure of S. However, S is integrally closed. To prove this, we compute the radical of the Jacobian ideal. A presentation of S over k is as follows:

$$S \cong \frac{k[a, b, c, d, e]}{(a^2d - b^2c, ae - bc, be - da, e^2 - dc)} = \frac{k[a, b, c, d, e]}{(ae - bc, be - da, e^2 - dc)}.$$

As the dimension of S is the same as the dimension of A (which is 3), the height of $(ae - bc, be - da, e^2 - dc)$ is two. The Jacobian matrix of S over k is

$$\begin{bmatrix} e & -c & -b & 0 & a \\ -d & e & 0 & -a & b \\ 0 & 0 & -d & -c & 2e \end{bmatrix},$$

and the ideal of 2×2 minors of this contains $a^2, b^2, c^2, d^2, e^2 - cd$. The radical of this ideal is generated by a, b, c, d, e, which is also clearly the radical of the Jacobian ideal. Clearly a is a non–zerodivisor in S, and $S/aS = \frac{k[b,c,d,e]}{(bc,be,e^2-dc)}$, so that d is a non–zerodivisor on this ring. Thus a, d is a regular sequence in the radical of the Jacobian ideal, so that the Jacobian ideal has grade at least two. By Theorem 4.4.9, S satisfies (R_1) and (S_2), so that by Theorem 4.5.3, S is normal.

Another method to determine the normality of S is via Lemma 2.4.3:

$$\mathrm{Hom}_S(I, I) \cong \frac{1}{a}\left(aI :_S I\right)$$

$$= \frac{1}{a}\left(I \cap (aI :_S b) \cap (aI :_S c) \cap (aI :_S d) \cap (aI :_S e)\right)$$

$$= \frac{1}{a}\left(I \cap (a, c, e)S \cap (a, b)S\right) = \frac{1}{a}\,(a, bc, be)\,S = S,$$

so that by Theorem 15.3.2, S is integrally closed and so is the integral closure of A. As xyt in field of fractions of A corresponds to bc/a in the field of fractions of R, this proves that the integral closure of R is $R[bc/a]$.

Incidentally, by Proposition 5.2.4, from this example one can read off the integral closure of $(x^2, y^2)^n k[x, y]$: it is the component of S of t–degree n, which is the ideal $(x, y)^{2n} k[x, y]$.

Singular and Macaulay2 have automated such calculations to compute integral closures. The algorithms are based on the work of Grauert and Remmert [95], [96], and rediscovered by de Jong [59], which we describe in the next section. Singular was the first program to implement integral closure, and a description of the necessary algorithms as implemented in Singular is in [57] and [99]. Here is the Singular syntax for computing the integral closure of the ring $(\mathbb{Z}/3\mathbb{Z})[a, b, c, d]/(a^2 d - b^2 c)$:

```
LIB "normal.lib";
ring R =3,(a,b,c,d), dp;
ideal I = a2*d -b^2*c;
list J = normal(I);
def A = J[1]; setring A; norid;
```

To this Singular returns:

```
// 'normal' created a list of 1 ring(s).
// nor[1+1] is the delta-invariant in case of choose=wd.
// To see the rings, type (if the name of your list is nor):
      show( nor);
// To access the 1-such that ring and map
// (similar for the others), type:
      def R = nor[1]; setring R;  norid; normap;
// R/norid is the 1-such that ring of the normalization and
// normap the map from the original basering to R/norid
norid[1]=T(3)*T(4)-T(5)^2
norid[2]=T(1)*T(4)-T(2)*T(5)
norid[3]=T(2)*T(3)-T(1)*T(5)
```

The last lines say that the integral closure of $(\mathbb{Z}/3\mathbb{Z})[a, b, c, d]/(a^2 d - b^2 c)$ is

$$\frac{(\mathbb{Z}/3\mathbb{Z})[T_1, T_2, T_3, T_4, T_5]}{(T_3 T_4 - T_5^2, T_1 T_4 - T_2 T_5, T_2 T_3 - T_1 T_5)}.$$

Perhaps this is not too informative as the connections between the T_i and a, b, c, d are not obvious. However, normap; produces more information:

```
> normap;
normap[1]=T(1)
normap[2]=T(2)
normap[3]=T(3)
normap[4]=T(4)
```

from which one reads that a is identified with T_1, b with T_2, c with T_3, and d with T_4. Thus the integral closure of $(\mathbb{Z}/3\mathbb{Z})[a,b,c,d]/(a^2d-b^2c)$ is generated by one extra element, namely T_5, and from the defining relations of the integral closure we see that $T_5 = ad/b = bc/a$.

The command in Macaulay2 for computing the integral closure of rings is `integralClosure(R)`. For the ring above, Macaulay2 returns

```
i1 : R = ZZ/3[a,b,c,d]/(a^2*d-b^2*c)

o1 = R

o1 : QuotientRing

i2 : integralClosure(R)

            ZZ
            -- [w , a, b, c, d]
            3   0
o2 = ---------------------------------
                             2
            (w b - a*d, w a - b*c, w  - c*d)
              0           0        0

o2 : QuotientRing
```

In this example the integral closure can be immediately read off. We give another example in which the answer is not so immediate; one then needs to type in extra commands to get more information:

```
i11 : S = ZZ/3[x,y]/(x^2-y^3)

o11 = S

o11 : QuotientRing

i2 : integralClosure(S)

            ZZ
o12 = -- [w ]
            3   0

o12 : PolynomialRing
```

from which it is not obvious how w_0 was derived from the elements of S. However, the commands ICfractions and ICmap provide the extra information:

```
i13 : ICfractions S

o13 = | x/y |}
                    1                 1
o13 : Matrix frac(S)  <--- Frac(S)

i14 : ICmap S

            ZZ        3   2
o14 = map(-- [w ],S,{w , w })
           3    0        0   0

            ZZ
o14 : RingMap -- [w ] <--- S
              3    0
```

This says that $w_0 = x/y$.

Thus with Singular and Macaulay2 one can in principle compute the integral closures of affine domains, and in particular of Rees algebras of ideals in affine domains. This then constructs, as described at the beginning of this chapter, the integral closures of ideals in affine domains. As remarked earlier, passage to Rees algebras at the same time computes the integral closures of all the powers of the ideal, which is very time–consuming. Singular has also implemented a way to compute the integral closure of a few powers of one ideal. In particular, it is possible to compute the integral closure of the ideal itself without necessarily computing the integral closures of all the powers. This is correspondingly faster. However, as mentioned before, these computations tend to be very time– and memory–consuming.

Here is a worked example:

```
LIB "reesclos.lib";
ring R =0,(a,b,c), dp;
ideal I = a2-b5,ac2,b2c;
list J = normalI(I,2);
J[1];                    // J[1] is the integral closure of I
```

The function normalI(I [,n]]) above uses the following syntax: I is an ideal, and n is an optional integer. Then normalI(I [,n]]) computes the integral closures of $I, I^2, \ldots, I^n$. If n is not given or is zero, this command computes the integral closures of all the powers of I. For the example above, Singular returns

```
_[1]=-b5+a2
_[2]=ac2
_[3]=b2c
_[4]=abc
```

and for the integral closure of I^2 it returns:

```
> J[2];
_[1]=b10-2a2b5+a4
_[2]=-ab5c2+a3c2
_[3]=a2c4
_[4]=-b7c+a2b2c
_[5]=ab2c3
_[6]=b4c2
_[7]=-ab6c+a3bc
_[8]=a2bc3
_[9]=ab3c2
_[10]=a2b2c2
```

15.3. General algorithms

We present general algorithms for computing the integral closure of ideals. The first one is based on the work of Grauert and Remmert [95], [96], and is due to de Jong [59]. This algorithm is the basis of Singular's automation. The second algorithm is a modification of de Jong's algorithm due to Lipman's work. The last two algorithms are due to Vasconcelos.

An important ingredient for the computations in this section, as well as in the previous one, is Lemma 2.4.3: for any ideals I and J in a domain R, $\mathrm{Hom}_R(I, J)$ can be identified with with the R–submodule $\frac{1}{x}(xJ :_R I)$ of $Q(R)$, where x is an arbitrary non–zero element of I.

Lemma 15.3.1 *Let J be an integrally closed ideal in a Noetherian domain R. As submodules of the field of fractions K,*

$$\mathrm{Hom}_R(J, R) \cap \overline{R} = \mathrm{Hom}_R(J, J).$$

Proof: Let $f \in \mathrm{Hom}_R(J, R) \cap \overline{R}$. Write $f^n = r_0 + r_1 f + \cdots + r_{n-1} f^{n-1}$ for some $r_i \in R$. Then for all $x \in J$, $(fx)^n = r_0 x^n + r_1 x^{n-1}(fx) + \cdots + r_{n-1} x(fx)^{n-1}$ is an equation of integral dependence of the element $fx \in R$ over J. Thus $fx \in \overline{J} = J$, so $f \in \mathrm{Hom}_R(J, J)$. Conversely, if $f \in K$ such that $fJ \subseteq J$, then by Lemma 2.1.8, $f \in \overline{R}$, and certainly $fJ \subseteq R$.   □

We note that in the current computer implementations, this lemma is applied with J being a radical ideal.

With the lemma, Proposition 2.4.9 can be made computationally more reasonable: instead of computing $(I \cdot I^{-1})^{-1}$ for all ideals in R it suffices to compute it for one I:

Theorem 15.3.2 (Grauert and Remmert [95]) *Let R be a Noetherian integral domain. and J an ideal of R such that $V(J)$ contains the non–normal locus of R. Then the following are equivalent:*
(1) R is integrally closed.
(2) For all non–zero fractional ideals I of R, $\mathrm{Hom}_R(I, I) = R$.
(3) For all non–zero ideals I of R, $\mathrm{Hom}_R(I, I) = R$.
(4) Then $\mathrm{Hom}_R(J, J) = R$.

Proof: As for all I, $R \subseteq \operatorname{Hom}_R(I, I)$, by Lemma 2.1.8 or by Proposition 2.4.8, (1) implies (2). Certainly (2) implies (3) and (3) implies (4). Now assume condition (4). We will prove that R is integrally closed. Let $f \in \overline{R}$. Set $L = R :_R f$. A prime ideal P of R contains L if and only if $f \notin R_P$. Thus $V(L)$ is contained in the non–normal locus of R. Thus there exists an integer d such that $J^d \subseteq L$. Then $fJ^d \subseteq fL$, and by the definition of L, $fL \subseteq R$. Thus $fJ^d \subseteq R$. Let e be least non–negative integer such that $fJ^e \subseteq R$. If e is positive, there exists $r \in J^{e-1}$ such that fr is not in R. But $fr \in \overline{R} \cap \operatorname{Hom}_R(J, R)$, so that by the previous lemma, $fr \in \operatorname{Hom}_R(J, J) = R$, contradicting the choice of r. Thus necessarily e is not positive, so that $e = 0$, i.e., $f \in R$. $\qquad\square$

By the Jacobian criterion (Theorem 4.4.9), the non–normal locus is computable when R is a finitely generated algebra over a perfect field k. Namely, I as in (4) of the theorem above can be taken to be any non–zero integrally closed ideal contained in the radical of the Jacobian ideal of R/k, and then R is integrally closed if and only if $\operatorname{Hom}_R(I, I) = R$. Furthermore, by the theorem above, normality of a ring is computable also in other contexts, say if R is a finitely generated algebra over any regular ring, such as over $\mathbb{Z}$ (see Exercise 4.5).

Theorem 15.3.2 provides a criterion for determining whether a Noetherian domain with a known ideal defining the singular locus is integrally closed, and as observed by de Jong [59], it also indicates **an algorithm** for computing the integral closure of computable rings as follows:

(1) Compute $\operatorname{Hom}_R(I, I)$ for a single computable ideal I satisfying condition (4). Then R is integrally closed if and only if $\operatorname{Hom}_R(I, I)$ equals R, in which case stop the procedure.

(2) If R is not integrally closed, by Theorem 15.3.2, $\operatorname{Hom}_R(I, I) \neq R$. By Lemma 2.1.8, any element of $\operatorname{Hom}_R(I, I)$ is integral over R. (It may not be the case that the ring $\operatorname{Hom}_R(I, I)$ is the integral closure of R (see Exercise 15.5).) Choose any finite non–empty set of elements of $\operatorname{Hom}_R(I, I)$ that are not in R; adjoin those to R to obtain a new affine ring R'.

(3) Repeat the computation of $\operatorname{Hom}_{R'}(I', I')$ for the new ring R' and its corresponding new ideal I'.

By Theorem 4.6.3, the integral closure of R is a finitely generated R–module, so this procedure must stop eventually, and the final ring is necessarily the integral closure of R.

Here is one catch: to apply the algorithm and the theorem above, one has to compute an ideal I satisfying condition (4) of Theorem 15.3.2. If R is essentially of finite type over a perfect field, I can be taken to be the radical or the integral closure of the Jacobian ideal (see the Jacobian criterion 4.4.9). Once the Jacobian ideal is computed, one needs to find some integrally closed ideal between it and its radical before computing Hom. This adds a computational difficulty. To compute the integral closure of the ideal, in general one has to compute the integral closure of its Rees algebra and then read off the degree one component, but this makes the computation of the integral closure of an algebra successively more and more complex. Thus in general the theorem above should be applied with the radical of J rather than with the integral closure. Several methods of computing the radicals are in the exercises.

However, there are cases when we can avoid the computation of either the radical or the integral closure, as follows:

Theorem 15.3.3 (Lipman [184]) *Let R be an integral domain that is essentially of finite type over a field of characteristic 0. Let J be the Jacobian ideal for R over this field. Then the following are equivalent:*

(1) R is integrally closed.

(2) $\mathrm{Hom}_R(J^{-1}, J^{-1}) = R$.

(3) For all prime ideals P in R of grade 1, $\mathrm{Hom}_R(J^{-1}, J^{-1})R_P = R_P$.

Proof: By Proposition 2.4.9, the only new implication is that (3) implies (1). It suffices to prove that for every prime ideal P in R of grade 1, R_P is a discrete valuation ring of rank one. By Lemma 2.4.6 and the assumption, $(JJ^{-1})^{-1}R_P = R_P$. Thus by the same proof as in Proposition 2.4.9, $JJ^{-1}R_P = R_P$ and JR_P is principal. Now we apply Lipman's theorem from [184] whose proof we do not provide here: JR_P is principal if and only if R_P is regular. Hence R_P is a regular local ring. As it has depth one, it is a discrete valuation ring of rank one. $\square$

It follows that in characteristic zero this gives a simpler algorithm for computing the integral closure of ideals: the Jacobian ideal is computable, and there is no need to compute its integral closure or its radical.

We comment that the characteristic zero assumption is necessary. Matsumoto [199, page 404] gave the following example: $R = k[X, Y]/(X^p + Y^{p+1})$, where k is field of prime characteristic p. Here $J = J_{R/k} = Y^p R$, $J^{-1} = \frac{1}{Y^p}R$, $\mathrm{Hom}(J^{-1}, J^{-1}) = R$ does not contain R properly even though $\overline{R}$ is a proper extension of R (for example, $X/Y \in \overline{R}$ and is not in R).

A third algorithm for computing the integral closure of an affine domain R in characteristic zero was given by Vasconcelos [304]. Here are the steps:

(1) Compute a Noether normalization A of R (see Exercise 15.6).

(2) The process that yields A also yields a presentation of R as an A–module. Then compute $R^{**} = \mathrm{Hom}_A(\mathrm{Hom}_A(R, A), A)$. This is actually a subring of $\overline{R}$ that satisfies (S_2), see Exercises 15.9 and 15.12.

(3) If R satisfies (R_1), so does R^{**} (see Exercise 15.11), hence R^{**} is the integral closure of R. If, however, R does not satisfy (R_1), then in characteristic zero one can apply Theorem 15.3.3 to find a proper extension of R contained in $\overline{R}$. Then one repeats the procedure with this proper extension in place of R.

Note that Vasconcelos's algorithm does not require the computation of the radical of an ideal, but instead it requires the Noether normalization and the double dual rings. The characteristic zero assumption is necessary to guarantee the computation of $\overline{R}$.

In a later paper [306], Vasconcelos gave yet another algorithm for computing the integral closure of affine domains in characteristic zero, and at the same time gave an a priori upper bound on the number of steps needed to compute the integral closure. We give an outline of his method. Input is an integral domain R that is finitely generated over a field k of characteristic zero.

(1) Compute a Noether normalization A' of R (see Exercise 15.6), and an element $r \in R$ such that $Q(A')(r) = Q(R)$. (The last condition is the Primitive Element Theorem 3.1.1.) Set $A = A'[r]$. Then A is Gorenstein, $A \subseteq R$ is module–finite, and A and R have the same field of fractions.

(2) Set $R_1 = \operatorname{Hom}_A(\operatorname{Hom}_A(R, A), A)$. As before, R_1 is a subring of $\overline{R}$ that contains R and satisfies Serre's condition (S_2).

(3) If $R_1 \neq \overline{R}$, i.e., if R_1 is not integrally closed, compute a proper extension R_1' of R_1 contained in $\overline{R}$ (possibly use Theorem 15.3.3). Set $R_2 = \operatorname{Hom}_A(\operatorname{Hom}_A(R_1', A), A)$.

One gets a filtration

$$R \subsetneq R_1 \subsetneq R_2 \subsetneq \cdots \subseteq \overline{R},$$

where each R_i satisfies Serre's condition (S_2).

All the algorithms for computing the integral closure rely on the finite generation of $\overline{R}$ over R to claim that the procedure must stop at some $R_n = \overline{R}$. Vasconcelos [306] gives an a priori bound on the number of R_i needed in the construction above. Namely, let J be the Jacobian ideal of A over k. By Theorem 13.3.3, J is in the conductor ideal of R. (We may not know the conductor ideal prior to constructing $\overline{R}$, but we can readily compute the Jacobian ideal.) Note that as $A = k[x_1, \ldots, x_{d+1}]/(f)$, where $x_1, \ldots, x_{d+1}$ are variables over k, J is the ideal in A generated by all the partial derivatives of f. Vasconcelos proved [306, Theorem 2.2] that the number of R_i needed is at most $\sum_{\operatorname{ht} P = 1} \lambda(A_P / J A_P)$. (See Exercises 15.13 and 15.14.)

These algorithms above do not form an exhaustive list; a few more are presented through the exercises at the end of the chapters, and some other relevant papers on effective computation of the relative and absolute integral closure are listed here: [302]; [297]; [23], [24]; [42]; [61]; [300].

15.4. Monomial ideals

Recall Proposition 1.4.9: for a monomial ideal in $k[X_1, \ldots, X_d]$ generated by elements of degree at most N, the integral closure is generated by monomials of degree at most $N + d - 1$. There are only finitely many monomials in $k[X_1, \ldots, X_d]$ of degree at most $N + d - 1$, whence to compute the integral closure of monomial ideals it suffices to check if any of these finitely many monomials satisfy an equation of integral dependence.

So the problem of computing the integral closure of a monomial ideal I reduces to the problem of determining when a given monomial $X_1^{n_1} X_2^{n_2} \cdots X_d^{n_d}$ is integral over I. For this, if I is generated by elements $X_1^{v_{j1}} X_2^{v_{j2}} \cdots X_d^{v_{jd}}$, $j = 1, \ldots, l$, as in equation (1.4.5), one needs to find non–negative rational numbers $c_1, \ldots, c_d$ such that

$$(n_1, n_2, \ldots, n_d) \geq \sum c_j (v_{j1}, v_{j2}, \ldots, v_{jd}), \quad \sum c_j = 1.$$

This is now phrased as a linear^{j}programming problem,^{j}and algorithms exist for solving them.

In general, if d and N are large, the number of monomials to be tested for integrality over I can be very large. For a restricted class of monomial ideals there is a smaller number of monomials to be tested:

Proposition 15.4.1 *Let I be a monomial ideal in $k[X_1, \ldots, X_d]$ that is primary to $(X_1, \ldots, X_m)$, $m \leq d$. Then I is integrally closed if and only if none of the finitely many monomials in the set $((I : (X_1, \ldots, X_m)) \cap k[X_1, \ldots, X_m]) \setminus I$ is integral over I.*

Proof: Let S be the set of monomials in $((I : (X_1, \ldots, X_m)) \cap k[X_1, \ldots, X_m])$ that are not in I. This is a finite set. If I is integrally closed, none of the elements of S is integral over I. If $I \neq \bar{I}$, as integral closure of monomial ideals is monomial (see Proposition 1.4.2 or Corollary 5.2.3), there is a monomial $r \in \bar{I} \setminus I$. Without loss of generality $r \in k[X_1, \ldots, X_m]$. Some monomial multiple r' of r is in S and in $\bar{I}$, which proves the converse. $\qquad \square$

One can algorithmically and easily compute the finite set of monomials in $(I : (X_1, \ldots, X_m)) \cap k[X_1, \ldots, X_m]$ that are not in I. The number of these monomials is in general much smaller than the number of monomials of degree at most $N + d - 1$ that are not in I.

Another solution to the computation of integral closures of monomial ideals was provided by Bruns and Koch in [29] and implemented in Normaliz [30]. As in the beginning of this chapter, the computation is first converted to the computation of the integral closure of the Rees algebra of the ideal. We reiterate: for a monomial ideal I in $R = k[X_1, \ldots, X_d]$ generated by monomials $\underline{X}^{v_1}, \ldots, \underline{X}^{v_s}$, if t is a variable over R, the Rees algebra $R[It]$ of I is a k–algebra generated over k by the monomials $\underline{X}^{v_1}t, \ldots, \underline{X}^{v_s}t$, and by $X_1, \ldots, X_n$. The integral closure of I is the set of all elements of the integral closure of $R[It]$ whose t–degree is 1. By Theorem 2.3.2, the integral closure of $R[It]$, which is the integral closure of $R[It]$ in $R[t]$, is a monomial algebra. Thus the integral closure of I is generated by the monomial generators of $R[It]$ of t–degree 1. As before, monomials correspond to their exponent vectors, so the problem of computing the integral closure of $R[It]$ can be converted to a problem on exponent vectors. We explain next how Normaliz solves that, without linear programming.

Again, $v_1, \ldots, v_s$ are the exponent vectors of a monomial generating set of I. Let E be the subset of $\mathbb{N}^{d+1}$ consisting of the $d+s$ vectors $(v_{11}, \ldots, v_{1d}, 1), \ldots,$ $(v_{s1}, \ldots, v_{sd}, 1)$ and $(1, 0, \ldots, 0), (0, 1, 0, \ldots, 0), \ldots, (0, \ldots, 0, 1, 0)$. The monomial algebra $R[It]$ is generated over k by $\underline{X}^\mu$ as μ varies over elements of E.

In general, given a finite subset $E \in \mathbb{N}^{d+1}$ (or even in $\mathbb{Z}^{d+1}$), if $T = k[\underline{X}^\mu \,|\, \mu \in E]$, by Corollary 2.3.7, the exponent vectors of the monomials in the integral closure of T are precisely the elements of $\mathbb{Z}E \cap \mathbb{Q}_{\geq 0}E$.

By Carathéodory's Theorem (A.2.1), it suffices to find elements of $\mathbb{Z}E \cap (\mathbb{Q}_{\geq 0}\mu_1 + \cdots + \mathbb{Q}_{\geq 0}\mu_r)$, where $\mu_1, \ldots, \mu_r$ form a maximal linearly independent set in E. Naturally, $\mathbb{Z}_{\geq 0}\mu_1 + \cdots + \mathbb{Z}_{\geq 0}\mu_r$ are exponents of elements of T, so they do not add anything in constructing the integral closure of T, and we may ignore them. In fact, by subtracting the positive integer multiples of the monomials $\mu_1, \ldots, \mu_r$, it is easy to see that the integral closure of T is generated, in exponent form, by $\{c_1\mu_1 + \cdots + c_r\mu_r \in \mathbb{Z}E \,|\, c_i \in \mathbb{Q}, 0 \leq c_i < 1\}$, as $\{\mu_1, \ldots, \mu_r\}$ vary over the linearly independent subsets in E. By linear independence, these monomials are in one–to–one natural correspondence with $\mathbb{Z}E/(\mathbb{Z}\mu_1 + \cdots + \mathbb{Z}\mu_r)$. But by the fundamental theorem of finite abelian groups, the latter set is a computable finite set, and therefore $\mathbb{Z}E \cap \mathbb{Q}_{\geq 0}E$ and T are computable.

This is the theoretical basis of the program Normaliz by Bruns and Koch for computing the integral closure of monomial algebras and ideals. See also Exercise 2.18. The program Normaliz also provides optimizing features, such as finding maximally linearly independent subsets of $\{v_1, \ldots, v_s\}$ that correspond to a triangulation of the Newton polyhedron $\mathbb{Q}_{\geq 0}E$ into simplicial cones. The algorithm is described in more detail in [29].

15.5. Exercises

15.1 (Gianni and Trager [89]) Let R be a Noetherian domain such that $\overline{R}$ is module–finite over R. Let $t \in R :_R \overline{R}$, $t \neq 0$. Prove that if R is not integrally closed, then the ring $\operatorname{Hom}_R(\sqrt{tR}, \sqrt{tR})$ properly contains R. (This exercise, together with the conductor constructions in Section 12.3, gives a modified algorithm for computing the integral closure of affine algebras.)

15.2 Let R be the (non–Noetherian) ring $\mathbb{Z}[X_1, X_2, \ldots]/(X_1 - X_i^i \,|\, i \geq 1)$. Prove that the integral closure of X_2R is not finitely generated.

15.3 (Criterion for integral dependence. See Proposition 3.1.3 in [99]; background on Gröbner bases needed.) Let k be a field, $X_1, \ldots, X_n$ variables over k, I ideal in $k[\underline{X}]$, and $r, f_1, \ldots, f_m \in k[\underline{X}]$. Prove that r is integral over $k[f_1, \ldots, f_m]$ modulo I if and only if for variables $Y_1, \ldots, Y_m, Z$, under the lexicographic ordering $X_1 > \cdots > X_n > Z > Y_1 > \cdots > Y_m$, the Gröbner basis for the ideal $(Z - r, Y_1 - f_1, \ldots, Y_m - f_m) + I$ in $k[\underline{X}, \underline{Y}, Z]$ contains an element whose initial monomial is a power of Z.

15.4   (Criterion for integral dependence; background on Gröbner bases needed.) Let R be a Noetherian ring, $X_1, \ldots, X_n$ variables over R, and I an ideal in $R[\underline{X}]$. Assume that there is some term ordering on $R[\underline{X}]$ in which any non–trivial monomial on the X_i is greater than elements of R. Prove that $R[\underline{X}]/I$ is integral over R if and only if $(X_1, \ldots, X_n)$ lies in the radical of the initial ideal of I.

15.5   Let $k = \mathbb{Z}/2\mathbb{Z}$, and $R = \frac{k[x,y,z]}{(x^4+x^3y+(1+z)x^2y^2+zxy^3+z^2y^4)}$.

   (i) Prove that R is an integral domain.
   (ii) Find the Jacobian ideal J of R over k.
   (iii) Compute $S = \operatorname{Hom}_R(\sqrt{J}, \sqrt{J})$.
   (iv) Prove that $\frac{x}{y}$ is in the quotient field of R and is integral over R.
   (v) Prove that $\frac{x}{y}$ is not in S. (Thus Theorem 15.3.2 in general requires more than one step in the construction of the integral closure of the ring.)
   (vi) Compute $T = \operatorname{Hom}_R(J, J)$. Show that $\frac{x}{y}$ is in T.
   (vii) Compute $\overline{R}$.

15.6   Write an algorithm for finding a Noether normalization A of an affine domain R over a field k. In case R is separable over k, R should be separable algebraic over A.

15.7   Let R be a domain essentially of finite type over a field of characteristic 0. Let J be the Jacobian ideal for the extension R over this field. Then the following are equivalent:

   (i) R is integrally closed.
   (ii) $\operatorname{Hom}_R(J, R) = R$.
   (iii) For all prime ideals P in R of grade 1, $\operatorname{Hom}_R(J, R)R_P = R_P$.
   (iv) $\operatorname{Hom}_R(J, J) = R$, and for any/some/all non–zero $x \in J$, $xR :_R (xR :_R J) = R$.
   (v) $\operatorname{Hom}_R(J, J) = R$, and for any/some/all non–zero elements $x \in J$, $\operatorname{Hom}_R(xR :_R J, xR :_R J) = R$.

15.8   (Swanson and Villamayor [282]) Let R be a ring, d a positive integer, $a_1, \ldots, a_d \in R$, X a variable over R, and

$$S = R[X] \Big/ \left(X^d - \sum_{i=0}^{d-1} a_{d-i} X^i \right).$$

Then S is a finitely generated free R–module that is a ring, with basis $\{1, X, \ldots, X^{d-1}\}$.

   (i) For each $n \geq d$, find a closed form expression for X^n as an R–linear combination of the given basis elements.
   (ii) For any $f, g \in S$ written as R–linear combinations of the given basis elements, find a closed form expression for fg as an R–linear combination of the given basis elements.

Towards Vasconcelos's criterion

15.9 Let R be a Noetherian domain that satisfies Serre's condition (S_2), and let M be a finitely generated R–module. Prove that for every prime ideal P of R, either $(\operatorname{Hom}_R(M,R))_P$ is zero or the depth of M_P is greater than or equal to $\min\{2, \operatorname{ht} P\}$.

15.10 Let R be an integral domain and M a finitely generated R–module.

   (i) Suppose that M is a torsion–free R–module and that for all $P \in \operatorname{Spec} R$ depth $M_P \geq \min\{\operatorname{ht} P, 2\}$. Prove that $M = \cap_{\operatorname{ht} P=1} M_P$, where the intersection is taken in the vector space $M_{R\setminus\{0\}}$.

   (ii) Suppose that R satisfies (S_2) and is Gorenstein after localization at all height one prime ideals. Prove that if M is torsion–free, then $\operatorname{Hom}_R(\operatorname{Hom}_R(M,R),R) = \cap_{\operatorname{ht} P=1} M_P$, where the intersection is taken in the vector space $M_{R\setminus\{0\}}$.

15.11 Let R be a locally formally equidimensional Noetherian domain satisfying Serre's condition (R_1). Let S be a ring contained in the field of fractions and R and module–finite over R. Prove that S satisfies (R_1).

15.12 Let $A \subseteq R$ be Noetherian domains, with R module–finite over A. Prove that $\operatorname{Hom}_A(\operatorname{Hom}_A(R,A),A)$ is an algebra between R and $\overline{R}$.

15.13 Let A be a Noetherian integral domain that satisfies (S_2) and is Gorenstein after localization at any height one prime ideal. Assume that $\overline{A}$ is module–finite over A.

   (i) Let R be a ring between A and $\overline{A}$. Prove that
$$\operatorname{Hom}_A(\operatorname{Hom}_A(R,A),A) = \operatorname{Hom}_A(\operatorname{ann}_A(R/A),A).$$

   (ii) Let $A \subseteq R \subseteq R' \subseteq \overline{A}$ be rings such that R and R' satisfy (S_2). Prove that $R = R'$ if and only if $\operatorname{ann}(R/A) = \operatorname{ann}(R'/A)$.

15.14 Let R be a Noetherian integral domain.

   (i) Prove that the only proper ideals I in R satisfying $I = (I^{-1})^{-1}$ have grade at most 1.

   (ii) Assume that R is Cohen–Macaulay and that $\overline{R}$ is module–finite over R. Let $J = R :_R \overline{R}$. Prove that the only ideals I in R containing J and satisfying $I = (I^{-1})^{-1}$ lie in height one prime ideals associated to J. Prove that any chain of such ideals I has length at most $\sum_{\operatorname{ht} P=1} \lambda(R_P/JR_P)$.

Computation of radicals, associated primes, primary components

15.15 (Contraction of localization of an ideal, see Proposition 3.7 in Gianni, Trager and Zacharias [90]. Background on Gröbner bases needed.) Let A be an integral domain and $(p) \subseteq A$ a principal prime ideal in A. Write an algorithm that for any ideal I in the polynomial ring $A[X]$ in variable X over A computes an element $s \in A \setminus (p)$ such that $IA_{(p)}[X] \cap A[X] = IA_s[X] \cap A[X]$. In particular, $IA_{(p)}[X] \cap A[X]$ is computable.

15.16 (Determination of primality, following [90]) Let $A[X]$ be a Noetherian ring, X a variable over A, and P an ideal in $R = A[X]$.

    (i) Prove that P is a prime (respectively primary) ideal in R if and only if $P \cap A$ is a prime (respectively primary) ideal in A and the image of P in $A/P \cap A$ is a prime (respectively primary) ideal.

    (ii) Let A be an integral domain and $P \cap A = 0$. Let K be the field of fractions of A. Then P is a prime (respectively primary) ideal if and only if $PK[X]$ is a prime (respectively primary) ideal in $K[X]$ and $P = PK[X] \cap K[X]$.

    (iii) Assuming that irreducibility of polynomials in polynomial rings over fields or $\mathbb{Z}$ is algorithmic, write an outline of an algorithm to decide whether an ideal in such a ring is a prime ideal.

15.17 Let R be an Noetherian ring and I an ideal in R.

    (i) Let x be an element of R. Let n be such that for all $m \geq n$, $I : x^n = I : x^m$. Then $I = (I : x^n) \cap (I + (x^n))$. Thus a (possibly redundant) primary decomposition of I can be obtained from primary decompositions of $I : x^n$ and $I + (x^n)$.

    (ii) Prove that for any element $x \in R$, $\mathrm{Ass}(R/I) \subseteq \mathrm{Ass}(R/(I : x)) \cup \mathrm{Ass}(R/(I + (x)))$.

    (iii) Assuming that finding a zero–divisor modulo an ideal in R is algorithmic, write an outline of an algorithm to find all the minimal prime ideals.

    (iv) Assuming that finding a zero–divisor modulo an ideal in R is algorithmic, write an outline of an algorithm to find a primary decomposition of an ideal.

Another algorithm for computing some of the associated primes of an ideal is to first compute the radical of the ideal and then find the associated primes of the radical. Below are some methods for computing the radical. For more on computability of primary decompositions and radicals, see Hermann [116]; Gianni, Trager and Zacharias [90]; Eisenbud, Huneke and Vasconcelos [69]; and Shimoyama and Yokoyama [268].

15.18 (Krick and Logar [174]) Let k be a field, $X_1, \ldots, X_n$ variables over k, $R = k[\underline{X}]$. Let I be a zero–dimensional ideal in R. For each $j = 1, \ldots, n$, set $F_j k[X_j] = I \cap k[X_j]$. Let G_j be the square–free part of F_j. Prove that $\sqrt{I} = I + (G_1, \ldots, G_n)$.

15.19 Let k be a field, $X_1, \ldots, X_n$ variables over k, and I an ideal in $k[\underline{X}]$,

    (i) Prove that there is an algorithmic way of finding a subset S of $\{X_1, \ldots, X_n\}$ such that $I \cap k[S] = 0$, and the cardinality of S is maximal possible.

    (ii) (Background on Gröbner bases needed.) Let S be as above and set $T = \{X_1, \ldots, X_n\} \setminus S$. Prove that there is an algorithmic way of finding $h \in k[S]$ such that $I \cap k(S)[T] = I : h = I : h^2$.

    (iii) Prove that $\sqrt{I} = \sqrt{I \cap k(S)[T]} \cap \sqrt{I + hk[\underline{X}]}$.

    (iv) Conclude that $\sqrt{I}$ is computable in polynomial rings over fields.

15.20 (Eisenbud, Huneke, Vasconcelos [69, Theorem 2.1]) Let R be an equidimensional ring of dimension d that is finitely generated over a field k and that has no embedded primes. In case the characteristic p of k is positive, also assume that R, as a module over one of its Noether normalizations, has rank strictly less than p.

  (i) Let $X_1, \ldots, X_n$ be variables over k, and $f_1, \ldots, f_{n-d}$ a regular sequence in $k[\underline{X}]$. Suppose that $R = k[\underline{X}]/(f_1, \ldots, f_{n-d})$, and that $A = k[\underline{X}] \subseteq R$ is its Noether normalization. Let D be the determinant of the matrix $(\frac{\partial f_i}{\partial X_j})_{j>d}$. Prove that

$$\sqrt{(f_1, \ldots, f_{n-d})} = (f_1, \ldots, f_{n-d}) : D.$$

  (ii) Prove that if R is generically a complete intersection, then the nilradical of R equals $0 : J$, where J is the Jacobian ideal of R.

15.21 (Corso, Huneke, Vasconcelos [43]) Let k be a field of characteristic 0, X, Y, Z, W variables over k, $R = k[[X, Y, Z, W]]$, and $I = (X^2 - XY, Y^2 - XY, Z^2 - ZW, W^2 - ZW)$. Prove that $\bar{I} = I + (XZ - YZ - XW + YW)$ and that $\sqrt{I} = (X - Y, Z - W)$. Observe that $\bar{I}$ is not a binomial ideal.

15.22 ([43]) Let k be a field of characteristic 0, R a polynomial ring over k, and I an ideal in R all of whose minimal primes have the same height. Let J be the ideal in R that is the preimage of the Jacobian ideal of R/I.

  (i) Prove that if I is integrally closed, then $IJ : J = I$.

  (ii) Assume that I is generically a complete intersection. Prove that if $IJ : J = I$, then I is integrally closed. (Hint: you may need to use Exercise 15.20.)

15.23 (Matsumoto [199]) Suppose that radicals can be computed in $\mathbb{Z}$ and in polynomial rings over finite fields. Let $R = \mathbb{Z}[\underline{X}]$ be a polynomial ring over $\mathbb{Z}$, and I an ideal in R. The point of this exercise is to prove that the radical of I can be computed. Certainly $I \cap \mathbb{Z}$ can be computed with Gröbner bases. Let m generate the radical of $I \cap \mathbb{Z}$.

  (i) Suppose that $m \neq 0$. By simple exhaustion or by more sophisticated techniques, one can compute a prime decomposition of $m = p_1 \cdots p_l$ for some positive integers $p_1, \ldots, p_l$. Set $I_i = I : (p_1 \cdots \hat{p_i} \cdots p_l)^N$, where N is a very large (and computable) integer. Let $\varphi_i : R \to (\mathbb{Z}/p_i\mathbb{Z})[\underline{X}]$ be the natural map. Prove that $\sqrt{I} = \cap_{i=1}^l \sqrt{I_i}$ and that $\sqrt{I_i} = \varphi_i^{-1}\sqrt{(\varphi_i(I_i))}$.

  (ii) Suppose that $m = 0$. Let G be a Gröbner basis of I and g the product of the leading coefficients of elements of G. Thus g is a non-zero element of $\mathbb{Z}$. Prove that $\sqrt{I} = \sqrt{I : g^\infty} \cap \sqrt{I + gR}$, and that $\sqrt{I : g^\infty} = \mathbb{Z}[\underline{X}] \cap I\mathbb{Q}[\underline{X}]$. Thus $\sqrt{I}$ is computable.

15.24 Write an algorithm that computes the normalizations of domains that are finitely generated over $\mathbb{Z}$. Implement refinements: preserve homogeneity or separability.

16
Integral dependence of modules

Integral dependence of rings and ideals can be extended to modules. This was first done by Zariski and Samuel for modules contained in overfields in [322, Appendix 4], and later extended to general modules by Rees in [239] and even later by Eisenbud, Huneke and Ulrich [68]. The integral closures by Rees and by Eisenbud et al. do not agree on all rings and modules, e.g., if the ring is not an integral domain. A further discussion of the definitions is at the end of Section 16.2 and in the exercises.

We will be working with Rees's definition.

Once the definition is established, many properties of integral closure that we have established for ideals can also be proved for the integral closure of modules. In the first four sections we develop the basic notions. In Section 16.5, we define the Buchsbaum–Rim multiplicity, and use it to give a criterion for reduction of modules, generalizing Rees's Theorem 11.3.1. We follow the work in [269] in this section. Much of the chapter is based on the work of Rees [239], and Kirby and Rees [165]. We also work from Katz [162]. In Section 16.6, we present Rees's result that certain conditions on the heights appearing in a Koszul complex are equivalent to acyclicity of the Koszul complex up to integral closure. This can be viewed as a generalization of Serre's result on the depth sensitivity of the Koszul complex, with depth being replaced by height, and it can be also viewed as a generalization of Ratliff's Theorem 5.4.1. A more general acyclicity criterion is proved in Section 16.8, based on some results from tight closure.

16.1. Definitions

Definition 16.1.1 (Rees [239]) *Let R be a Noetherian ring and $N \subseteq M$ an inclusion of finitely generated R–modules. An element $x \in M$ is said to be* **integral** *over N if for every minimal prime ideal P in R and every valuation ring V between R/P and $\kappa(P)$, the image of x in M_P/PM_P can be written as $\sum_{i=1}^{m} v_i n_i$, where for each i, $v_i \in V$ and n_i lies in the canonical image of N inside M_P/PM_P.*

When R is a domain, then for any ring V between R and the field of fractions of R, the extension of scalars MV is well–defined: it is the V–submodule of $M_{R \setminus (0)}$ generated by elements vm, as v varies over V and m over M. In other words, $x \in M$ is integral over N if and only if for every $P \in \operatorname{Min} R$, the image of x in M_P/PM_P is in $\frac{N+PM}{PM}V$, as V varies over $\kappa(P)$–valuation domains containing R/P.

When R is an integral domain and M is the free R–module R^r, then for the integral dependence of $x \in M$ over N one has to check for every $Q(R)$–valuation ring V containing R whether $x \in NV$, where NV is the V–submodule of V^r generated by N. In particular, if $M = R$ is an integral domain, then N is an ideal, and by Proposition 6.8.2, the integral dependence of elements of the domain R over an ideal is a special case of the integral dependence of modules. If $N \subseteq M \subseteq L$ are finitely generated R–modules, an element $x \in M$ may be integral over N as an element of L but not as an element of M. This difference between the integral closure of ideals and the integral closure of modules is illustrated in the example below.

Example 16.1.2   Let k be a field, X a variable over k, and $R = k[X]/(X^2)$. Let I be the ideal generated by X. As I is nilpotent, all the elements of I are integral over the zero subideal. Thus $X \in I$ is integral over 0 as an element of the module R. However, X is not integral over 0 as an element of the module I, because in $I \otimes_R \kappa(XR) \cong k$, the image of I is non–zero, but the image of the zero submodule is of course zero.

It is easy to see that the set of all elements of M that are integral over N forms an R–submodule of M containing N. This module is called **the integral closure** of N in M.

There are other possible definitions of integral dependence; see Discussion 16.2.4 below. We have chosen to use Rees's original definition. Every choice of definition has its own problems.

To check whether an element x of M is integral over a submodule N reduces to checking whether for each minimal prime ideal P of R, the image of x in the (R/P)–module $\frac{M_P}{PM_P}$ is integral over the image of N. Thus it is no loss of generality to restrict the attention to the case where R is an integral domain, and to replace M and N by their images in $M_{R\backslash(0)}$, so that we may assume that M is torsion–free.

If R is an integral domain and M is a torsion–free R–module, then by the definition

$$\text{the integral closure of } N \text{ in } M = \bigcap_V (NV) \cap M,$$

as V varies over the valuation rings between R and the field of fractions of R.

Definition 16.1.3   *Let R be a Noetherian ring and $N \subseteq M \subseteq L$ inclusions of finitely generated R–modules. N is said to be **a reduction** of M in L if for all $P \in \text{Min } R$ and all $\kappa(P)$–valuation rings V containing R/P, the image of M in $L \otimes \kappa(P)$ lies in the V–span of the image of N. We say that N is **a reduction** of M if N is a reduction of M in M.*

Here is another indication that the integral closure of modules generalizes the integral closure of ideals:

Lemma 16.1.4   *Let R be a Noetherian ring, $N \subseteq M \subseteq L$ modules.*
(1) If $N \subseteq M$ is a reduction in M, then N is a reduction of M in L.

(2) If $N \subseteq M$ is a reduction in L, then the integral closure of N in L contains M.

(3) If $J \subseteq I$ are ideals in R such that $J \subseteq I$ is a reduction of modules, then the (ideal) integral closure of J (in R) contains I.

Proof: If the image of M in $M \otimes \kappa(P)$ lies in the V–span of the image of N, then also the image of M in $L \otimes \kappa(P)$ lies in the V–span of the image of N. This proves (1), and (2) follows. If $J \subseteq I$ is a reduction of modules, then by (1), $J \subseteq I$ is a reduction of modules in R, so that for all $P \in \operatorname{Min} R$, $J(R/P) \subseteq I(R/P)$ is a reduction of modules in R/P. This means that for all minimal prime ideals P in R and all $\kappa(P)$–valuation rings V containing R/P, $JV = IV$ in $\kappa(P)$. Thus by Proposition 6.8.2, $J(R/P) \subseteq I(R/P)$ is a reduction of ideals, whence by Proposition 1.1.5, $J \subseteq I$ is a reduction of ideals. □

16.2. Using symmetric algebras

In this section we connect the theory of the integral closure of modules and the theory of the integral closure of ideals (see Theorem 16.2.3).

Recall the definition of the tensor algebra of an R–module M:
$$T_R(M) = \sum_{n \geq 0} M^{\otimes n},$$

where $M^{\otimes n}$ denotes the n–fold tensor product of M (as R–module). By convention, $M^{\otimes 0} = R$, $M^{\otimes 1} = M$. For any $x \in M^{\otimes n}$ and $y \in M^{\otimes m}$, $x \otimes y \in M^{\otimes(n+m)}$. This makes $T_R(M)$ into an $\mathbb{N}$–graded ring that need not be commutative.

Fix a commutative ring R and an R–module M. Let I be the two–sided ideal in $T_R(M)$ generated by all elements of the form $m_1 \otimes m_2 - m_2 \otimes m_1$, as m_1, m_2 vary over elements of M. Define symmetric algebra $\operatorname{Sym}_R(M)$ of M to be $\operatorname{Sym}_R(M) = T_R(M)/I$. This is a commutative ring with identity, and is the largest commutative quotient of $T_R(M)$. As I is a homogeneous ideal, $\operatorname{Sym}_R(M)$ is $\mathbb{N}$–graded. Note that $\operatorname{Sym}_R(M)$ is generated over R by M, so that if R is Noetherian and M finitely generated, then $\operatorname{Sym}_R(M)$ is Noetherian as well. As an example, $\operatorname{Sym}_R(R^n)$ is a polynomial algebra in n variables over R. If M is an ideal of R, $\operatorname{Sym}_R(M)$ is the usual symmetric algebra of the ideal M.

If $N \subseteq M$, $\operatorname{Sym}_R(N)$ need not be contained in $\operatorname{Sym}_R(M)$.

Definition 16.2.1 *When M is an R–submodule of an R–module F, we denote by $S_F(M)$ the image of $\operatorname{Sym}_R(M)$ in $\operatorname{Sym}_R(F)$ under the natural map induced by the inclusion $M \subseteq F$. When F is understood, we simply write $S(M)$. When F is a free R–module we write $\mathfrak{R}(M) = S_F(M)$. If the free module F needs to be specified, then we will write $\mathfrak{R}_F(M)$.*

The ring $S_F(M)$ is $\mathbb{N}$–graded, with $[S_F(M)]_0 = R$, $[S_F(M)]_1 = M$. If $N \subseteq M \subseteq F$, then $S_F(N) \subseteq S_F(M) \subseteq \operatorname{Sym}_R(F)$. If F is free, then $\mathfrak{R}(M)$ does not have R–torsion as $\mathfrak{R}(M)$ is a subring of the polynomial ring $\operatorname{Sym}_R(F)$ over R.

There may be non–isomorphic $\mathfrak{R}_F(M)$ as F varies over finitely generated free R–modules containing M (see Exercise 16.16).

A special case of $\mathfrak{R}_F(M)$ is when $F = R$ and $M = I$ is an ideal. In this case $\mathfrak{R}(I)$ is the Rees algebra $R[It]$ of I. Thus $\mathfrak{R}(M)$ may be viewed as a **Rees algebra of the module M**. The following generalizes Theorem 5.1.4:

Lemma 16.2.2   *Let R be a Noetherian ring, $F = R^r$ a finitely generated free R–module and M an R–submodule of F.*
(1)   *There is a one–to–one inclusion–preserving correspondence between associated primes of $\mathfrak{R}_F(M)$ and associated primes of R, given by contraction.*
(2)   *For all $P \in \operatorname{Min} R$, $P \operatorname{Sym}_R(F) \cap \mathfrak{R}_F(M)$ is a minimal prime in $\mathfrak{R}_F(M)$, all the minimal primes of $\mathfrak{R}_F(M)$ are obtained in this way. Furthermore, $\mathfrak{R}_{F/PF}(\frac{M+PF}{PF}) \cong \mathfrak{R}_F(M)/(P \operatorname{Sym}_R(F) \cap \mathfrak{R}_F(M))$.*
(3) $\dim \mathfrak{R}_F(M) = \sup\{\operatorname{ht}(\mathfrak{m}/P) + \operatorname{rk}(M \otimes_R \kappa(P)) \mid P \in \operatorname{Min} R, P \subseteq \mathfrak{m} \in \operatorname{Spec} R\}$.
(4)   *If $\operatorname{ht}(\operatorname{ann}(F/M)) > 0$, then $\dim \mathfrak{R}_F(M) = \dim R + r$.*

Proof: As $R \subseteq \mathfrak{R}_F(M) \subseteq \operatorname{Sym}_R(F)$, every associated prime of R is contracted from an associated prime of $\mathfrak{R}_F(M)$, and every associated prime of $\mathfrak{R}_F(M)$ is contracted from an associated prime of $\operatorname{Sym}_R(F)$. But $\operatorname{Sym}_R(F)$ is a polynomial ring over R, so this correspondence is one–to–one. This proves (1). Under the natural surjection $\operatorname{Sym}_R(F) \to \operatorname{Sym}_{R/P}(F/PF)$, $\mathfrak{R}_F(M)$ maps onto $\mathfrak{R}_{F/PF}(\frac{M+PF}{PF})$ and the kernel is $P \operatorname{Sym}_R(F) \cap \mathfrak{R}_F(M)$. This proves (2).

As in the proof of (1) above, every prime ideal in R is contracted from a prime ideal in $\mathfrak{R}_F(M)$. Thus to prove (3), without loss of generality R is a Noetherian local integral domain. Let $\mathfrak{m}$ be its maximal ideal and K its field of fractions. Then it suffices to prove that $\dim(\mathfrak{R}_F(M)) = \operatorname{ht} \mathfrak{m} + \operatorname{rk}(M_{R \setminus (0)})$. Set $L = F_{R \setminus (0)}$, and $r = \operatorname{rk}(M_{R \setminus (0)})$. As $M_{R \setminus (0)}$ is a free finitely generated K–module contained in L, it follows that $\operatorname{ht}(M \mathfrak{R}_L(M_{R \setminus (0)})) = r$. Let $Q = M \mathfrak{R}_L(M_{R \setminus (0)}) \cap \mathfrak{R}_F(M)$. As $M \mathfrak{R}_F(M) \subseteq Q$, by Theorem B.2.2, $\operatorname{ht} Q \geq r$. Furthermore, $Q \cap R = 0$ and $\mathfrak{R}_F(M)/Q = R$, so that $\dim \mathfrak{R}_F(M) \geq \operatorname{ht} \mathfrak{m} + r$. By the Dimension Inequality B.2.5, $\dim \mathfrak{R}_F(M) = \max\{\operatorname{ht} Q' \mid Q' \in \operatorname{Spec}(\mathfrak{R}_F(M))\} \leq \dim R + \operatorname{tr.deg}_R \mathfrak{R}_F(M) = \operatorname{ht} \mathfrak{m} + r$. This proves (3).

The last claim follows from (3) since for every minimal prime P of R, P does not contain $\operatorname{ann}(F/M)$, and so $\operatorname{rk}(M \otimes_R \kappa(P)) = \operatorname{rk}(F \otimes_R \kappa(P)) = r$. $\square$

How does $\mathfrak{R}(M)$ behave under extensions by scalars? By definition, it is clear that if R is an integral domain, F is free, $M \subseteq F$, and V is any ring between R and the field of fractions of R, then $\mathfrak{R}_{FV}(MV) = (\mathfrak{R}(M))V$.

Theorem 16.2.3   (Rees [239]) *Let R be a Noetherian ring, and let $N \subseteq M$ be R–modules contained in a finitely generated free R–module F. Then the following are equivalent:*
(1) $N \subseteq M$ *is a reduction of modules in F.*
(2) $\mathfrak{R}(N) \subseteq \mathfrak{R}(M)$ *is an integral extension of rings.*
(3) $N \mathfrak{R}(M) \subseteq M \mathfrak{R}(M)$ *is an integral extension of ideals.*
(4) $M \mathfrak{R}(M) \subseteq \sqrt{N \mathfrak{R}(M)}$.

Proof: (2) and (3) are equivalent by Proposition 2.3.8, and (3) and (4) are equivalent by first passing modulo minimal prime ideals and then using Proposition 1.6.4.

Assume (1). By definition, for every $P \in \mathrm{Min}(R)$, the image of M in the (R/P)–module F/PF is in the integral closure of the image of N. Suppose that (1) implies (3) under the additional assumption that R is an integral domain. Then

$$\frac{N+PF}{PF}\mathfrak{R}_{F/PF}\left(\frac{M+PF}{PF}\right) \subseteq \frac{M+PF}{PF}\mathfrak{R}_{F/PF}\left(\frac{M+PF}{PF}\right)$$

is an integral extension of ideals. By Lemma 16.2.2 (2), for all minimal prime ideals Q of $\mathfrak{R}(M)$, $N(\mathfrak{R}(M)/Q) \subseteq M(\mathfrak{R}(M)/Q)$ is an integral extension of ideals, so that by Proposition 1.1.5, $N\mathfrak{R}(M) \subseteq M\mathfrak{R}(M)$ is an integral extension of ideals. Thus it suffices to prove that (1) implies (3) under the additional assumption that R be an integral domain. Let W be any valuation ring between $\mathfrak{R}(M)$ and its total ring of fractions. By Proposition 6.4.7, $V = W \cap K$ is a K–valuation ring containing R. Hence $\mathfrak{R}(N)W = \mathfrak{R}(N)VW = \mathfrak{R}_{FV}(NV)W = \mathfrak{R}_{FV}(MV)W = \mathfrak{R}(M)VW = \mathfrak{R}(M)W$, so that by Proposition 6.8.14, $\mathfrak{R}(N) \subseteq \mathfrak{R}(M)$ is an integral extension. This proves (2), and hence also (3) and (4).

Now assume that (2), (3), and (4) hold. Then $N\mathfrak{R}(M) \subseteq M\mathfrak{R}(M)$ is an integral extension of ideals, and the same is true after passing modulo each minimal prime ideal of $\mathfrak{R}(M)$. This means, by Lemma 16.2.2, that for all $P \in \mathrm{Min}\,R$, $\frac{N+PF}{PF}\mathfrak{R}_{F/PF}(\frac{M+PF}{PF}) \subseteq \frac{M+PF}{PF}\mathfrak{R}_{F/PF}(\frac{M+PF}{PF})$ is an integral extension of ideals. If (2) implies (1) in domains, it follows that $(N+PF)/PF \subseteq (M+PF)/PF$ is a reduction of modules in F/PF, whence $N \subseteq M$ is a reduction in F. Thus it suffices to prove the implication (2) $\Rightarrow$ (1) for domains. So let R be a domain with field of fractions K. Let V be a K–valuation ring containing R. Extend V trivially to a valuation ring on the field of fractions of $\mathrm{Sym}_R(F)$, and restrict to $\mathfrak{R}_F(M)$. Denote the obtained valuation ring by W. By assumption (3), $N\mathfrak{R}(M)W = M\mathfrak{R}(M)W$, whence $NV\mathfrak{R}_{FV}(MV)W = MV\mathfrak{R}_{FV}(MV)W$, so that by the structure theorem of finitely generated modules over valuation rings, necessarily $NV = MV$. This proves (1). □

Theorem 16.2.3 shows that questions of integral dependence and closure of modules can sometimes be converted to analogous questions for ideals, at least for submodules of finitely generated free modules. However, the intermediate construction of the rings $\mathfrak{R}(M)$ can obscure the otherwise concrete module M, so an explicit theory of the integral closure of modules can nevertheless be useful. Brennan and Vasconcelos [24] gave various criteria for integral closedness of the ring $\mathfrak{R}(M)$ under further assumptions on the ring and the module M.

Let $N \subseteq M \subseteq F$ be R–modules with F free. Theorem 16.2.3 shows that one could define M to be integral over N if and only if $\mathfrak{R}_F(M)$ is integral over $\mathfrak{R}_F(N)$. This would certainly cover most main cases and it seems to be a natural definition. However, in defining Rees algebra and integral closure of a module, one must pick one's poison; the problem is that this definition of the Rees algebra depends on how the module is embedded in a free module.

Discussion 16.2.4   This theme of finding a notion of a Rees algebra that does not depend on the embedding is addressed by Eisenbud, Huneke, and Vasconcelos in [68]. Define a map $f : M \to F$ from an R–module M to a free R–module F to be **versal** if every homomorphism from M to a free module factors through f. A versal map can be obtained if the dual $\mathrm{Hom}_R(M, R)$ is finitely generated, by taking the dual of a finitely generated free module mapping onto $\mathrm{Hom}_R(M, R)$, and mapping M to this dual via the natural map of M to its double dual. One could form the Rees algebra of M instead by using a versal map from M to F, i.e., define $R(M) = \mathfrak{R}_F(M)$. This is essentially the definition used in [68]. (See Exercise 16.13.) This definition can be used to give a notion of integrality that works well and agrees with our definition if R is a domain or if M has a rank. It can differ, however, in other cases. Among the results of [68] are that this definition of Rees algebra commutes with flat base change, that if R is torsion–free over $\mathbb{Z}$, then for every map of M to a free module G inducing an embedding on the torsionless quotient of M, $\mathfrak{R}_G(M) \cong R(M)$ (Exercise 16.17), that this is false in characteristic p (Exercise 16.16), and that it agrees with the usual definition of the Rees algebra of an ideal (Exercise 16.14). Other definitions of Rees algebras of modules are for example in Liu [194]; or in Simis, Ulrich, Vasconcelos [270].

16.3. Using exterior algebras

There is another characterization of integral closure of modules using exterior algebras (Rees [239]). Exterior algebras are non–commutative rings, but no theory of the integral closure in non–commutative rings is needed for the reformulation below of the integral closure of modules.

> For an R–module M, the exterior algebra $\Lambda_R(M)$ is the quotient of the tensor algebra $T_R(M) = \sum_{n \geq 0} M^{\otimes n}$ by the ideal generated by elements of the form $m_1 \otimes \cdots \otimes m_r$, with $m_1, \ldots, m_r \in M$, such that for some distinct $i, j \in \{1, \ldots, r\}$, $m_i = m_j$.
>
> The image of an element $m_1 \otimes \cdots \otimes m_r$ in $\Lambda_R(M)$ is denoted $m_1 \wedge \cdots \wedge m_r$. For all $m_1, m_2 \in M$,
>
> $$m_1 \wedge m_2 = (m_1 + m_2) \wedge (m_1 + m_2) - m_1 \wedge m_1 - m_2 \wedge m_2 - m_2 \wedge m_1 = -m_2 \wedge m_1.$$
>
> By definition, $\Lambda_R(M)$ is an $\mathbb{N}$–graded ring. An element of degree k is a finite R–linear combination of elements of the form $m_1 \wedge \cdots \wedge m_k$, with $m_i \in M$. The degree k component of $\Lambda_R(M)$ is denoted $\Lambda_R^k(M)$. Clearly $\Lambda_R^0(M) = R$, $\Lambda_R^1(M) = M$.
>
> If $M = R^r$, with basis $e_1, \ldots, e_r$, then $\Lambda_R^k(M)$ is a free R–module of rank $\binom{r}{k}$ whose natural basis consists of $e_{i_1} \wedge e_{i_2} \wedge \cdots \wedge e_{i_k}$ with $1 \leq i_1 < i_2 < \cdots < i_k \leq r$. In particular, if $k > r$, then $\Lambda_R^k(M) = 0$.

In the previous section we interpreted the integral closure of modules via (relative) symmetric algebras $\mathfrak{R}(M)$. For this we assumed that M be contained in a free module. In this section, we assume instead that M is a torsion–free module over a domain.

Let R be a Noetherian integral domain with field of fractions K, and let $N \subseteq M$ be finitely generated torsion–free R–modules. Set $L = M \otimes_R K$. Let $\mathcal{E}_R(N)$ (respectively $\mathcal{E}_R(M)$) be the canonical image of the exterior algebra of the R–module N (respectively M) in $\Lambda_K(L)$. Note that $\mathcal{E}_R(N) \subseteq \mathcal{E}_R(M) \subseteq \Lambda_K(L)$ are inclusions of $\mathbb{N}$–graded R–modules. By the definition, for any ring V between R and K, $\mathcal{E}_V(MV) = (\mathcal{E}_R(M))V$, $\mathcal{E}_V(NV) = (\mathcal{E}_R(N))V$.

Let r be the rank of M, i.e., the K–vector space dimension of L. Then for all $k > r$, $\mathcal{E}_R^k(M) = 0$, and the component $\mathcal{E}_R^r(M)$ of degree r is a non–zero torsion–free R–module of rank 1. From

$$\mathcal{E}_R^r(N) \subseteq \mathcal{E}_R^r(M) \subseteq \mathcal{E}_K^r(L) = K,$$

it follows that $\mathcal{E}_R^r(M)$ is canonically isomorphic to an ideal I of R, and under the same isomorphism, $\mathcal{E}_R^r(N)$ is canonically isomorphic to an ideal $J \subseteq I$. Using this notation, we get another characterization of module reductions:

Theorem 16.3.1   (Rees [239, Theorem 1.2]) *Let R be a Noetherian integral domain with field of fractions K, and $N \subseteq M$ an inclusion of finitely generated torsion–free R–modules. Then $N \subseteq M$ is a reduction in M if and only if $J \subseteq I$ is a reduction.*

Proof: If $N \subseteq M$ is a reduction in M, then for any K–valuation ring V containing R, $NV = MV$. Thus $\mathcal{E}_R(N)V = \mathcal{E}_R(M)V$, and since this is $\mathbb{N}$–graded, by reading off the degree r–part, so that $JV = IV$. Thus by Proposition 6.8.2, J is a reduction of I.

Conversely, suppose that N is not a reduction of M in M. Then there exists a valuation ring V between R and K such that $NV \subsetneq MV$. As N and M are torsion–free and finitely generated, NV and MV are both free V–modules. If their ranks are different, necessarily the rank of NV is strictly smaller than r so that $JV = 0$ and hence $J = 0$. But I is a non–zero ideal, so that J is not a reduction of I. Thus we may assume that the ranks of NV and MV are both r. We may choose a basis $u_1, \ldots, u_r$ of MV such that $p_1 u_1, \ldots, p_r u_r$ is a basis of NV, where $p_1, \ldots, p_r$ are elements of V. As NV is not equal to MV, necessarily $p_1 \cdots p_r$ is a non–unit in V, Hence $JV = p_1 \cdots p_r IV$ is not equal to IV, and so J is not a reduction of I. $\qquad\square$

Whenever $f : M \to R^r$ is an injective R–module homomorphism, then M can be represented as an $m \times r$ matrix (m_{ij}) of rank r and with entries in R. Let $\det_f(M)$ denote the ideal in R generated by the $r \times r$ minors of (m_{ij}). Theorem 16.3.1 above can be rephrased as follows:

Proposition 16.3.2   *Let R be a Noetherian integral domain. Let $N \subseteq M$ be torsion–free R–modules of rank r and $f : M \to R^r$ an embedding. Then $N \subseteq M$ is a reduction of modules (in M) if and only if $\det_f(N) \subseteq \det_f(M)$ is a reduction of ideals.*

By Proposition 6.8.2, for ideals $J \subseteq I$, $\bar{J} = \bar{I}$ if and only if for every valuation ring V containing R and in the field of fractions of R, $JV = IV$. Furthermore, by Theorem 10.2.2, $\bar{J} = \bar{I}$ if and only if for every Rees valuation ring of J, $JV = IV$. Thus an immediate consequence is:

Corollary 16.3.3 *Let R be a Noetherian integral domain, $N \subseteq M$ torsion–free R–modules of rank r, and $f : M \to R^r$ an embedding. Then $N \subseteq M$ is a reduction if and only if*

$$M = \bigcap_V (NV) \cap M,$$

as V varies over the Rees valuation rings of the ideal $\det_f(N)$ in R.

16.4. Properties of integral closure of modules

The theory of integral closure of modules has been used in many contexts. For example, Kodiyalam [172] developed the theory of integrally closed modules over two–dimensional regular local rings, in analogy with Zariski's theory of integrally closed ideals. (Zariski's theory of ideals is presented in Chapter 14.) Katz and Kodiyalam [164] proved that the symmetric algebra modulo torsion of an integrally closed module over a two–dimensional regular local ring is Cohen–Macaulay, with reduction number one. Mohan [207] explicitly computed the intersection of all reductions of an integrally closed module over a two–dimensional regular local ring, the so–called **core**, using a Fitting ideal of the module, generalizing the work of Huneke and Swanson [141] for ideals. Rees proved in [239] the existence of *Rees valuations* for a module. Mohan in [208] determined them more precisely for modules over two–dimensional regular local rings.

In this section we present a few of the properties of integral closures of modules, mostly analogs of the corresponding results for integral closures of ideals. Connections with Buchsbaum–Rim multiplicity and acyclicity of complexes are treated in subsequent sections.

The following generalization of Proposition 1.1.4 and Remark 1.3.2 (1) is easy to prove:

Proposition 16.4.1 *Let R be a Noetherian ring, $N \subseteq M \subseteq L$ all finitely generated R–modules. The following are equivalent:*
(1) $N \subseteq M$ is a reduction in L.
(2) For every multiplicatively closed subset $W \subseteq R$, $W^{-1}N \subseteq W^{-1}M$ is a reduction in $W^{-1}L$.
(3) For every prime ideal P in R, $N_P \subseteq M_P$ is a reduction in L_P.
(4) For every maximal ideal $\mathfrak{m}$ in R, $N_\mathfrak{m} \subseteq M_\mathfrak{m}$ is a reduction in $L_\mathfrak{m}$.

Persistence of integral closure fails for modules under tensoring:

Example 16.4.2 Let $R = k[X, Y]$ be a polynomial ring in variables X and Y over a field k, let $N = (X^2, Y^2)R \subseteq M = (X, Y)^2 R$. This is a reduction in

M (also as ideals). However, with the R–algebra $S = R/(X, Y)$, the image of $N \otimes_R S \cong k^2$ in $M \otimes_R S \cong k^3$ is not a reduction.

Persistence holds under extension of scalars (compare with persistence for ideals given on page 2):

Proposition 16.4.3 *Let R be a Noetherian ring, and $N \subseteq M \subseteq L$ finitely generated R–modules with $N \subseteq M$ a reduction in L. Let S be a Noetherian R–algebra, such that for every minimal prime Q in S, $Q \cap R$ is a minimal prime of R. Suppose that N, M, L extend by scalar extension to S–modules $NS \subseteq MS \subseteq LS$. (This holds, for example, if L is contained in a free R–module.) Then $NS \subseteq MS$ is a reduction of S–modules in LS.*

Proof: By passing to S modulo a minimal prime, we may assume that S is an integral domain. By assumption the kernel of the map $R \to S$ is a minimal prime, and by passing to R modulo it we may assume that R is a domain. Let V be a valuation ring between S and $Q(S)$. Set $V' = V \cap Q(R)$. By assumption $NV' = MV'$ in $L \otimes Q(R)$, whence $NV = MV$ in $LS \otimes Q(S)$. $\square$

Persistence also holds under projections:

Proposition 16.4.4 *Let R be a Noetherian ring, $M_1 \subseteq L_1$, $M_2 \subseteq L_2$ finitely generated R–modules, and $N \subseteq M_1 \oplus M_2$ a reduction in $L_1 \oplus L_2$. Let N_i be the image of N under the composition $N \subseteq M_1 \oplus M_2 \xrightarrow{\pi_i} M_i$, where π_i is the projection onto the ith component. Then $N_i \subseteq M_i$ is a reduction in L_i.*

Proof: By definition of integral closure, without loss of generality R is an integral domain and L_1, L_2 are torsion–free modules. Let V be a valuation ring between R and $Q(R)$. By assumption, NV equals $M_1 V \oplus M_2 V$ in $(L_1 \oplus L_2) \otimes Q(R)$, so that $N_i V = M_i V$ in $L_i \otimes Q(R)$. $\square$

Clearly if $N_1 \subseteq M_1$ and $N_2 \subseteq M_2$ are reductions, then so is $N_1 \oplus N_2 \subseteq M_2 \oplus M_2$. The following generalizes Propositions 1.6.1 and 1.6.2:

Lemma 16.4.5 *Let R be a Noetherian ring, and let $N \subseteq M \subseteq R^r = F$ be finitely generated R–modules.*
(1) Let $R \subseteq S$ be an integral extension of Noetherian rings. Then $NS \subseteq MS$ is a reduction in S^r if and only if $N \subseteq M$ is a reduction in R^r.
(2) If R is local with maximal ideal $\mathfrak{m}$ and if if and only if $N \subseteq M$ is a reduction in F.

Proof: Proposition 16.4.3 proves the necessity of both parts. Proofs of sufficiency of the two parts are similar, so we only prove it for (2). Assume that $\widehat{N} \subseteq \widehat{M}$ is a reduction in $\widehat{F}$. By Theorem 16.2.3, it follows that $\widehat{M} \subseteq \sqrt{\widehat{N} \mathfrak{R}_{\widehat{F}}(\widehat{M})}$. Hence $M \subseteq \sqrt{\widehat{N} \mathfrak{R}_{\widehat{F}}(\widehat{M})} \cap \mathfrak{R}_F(M) = \sqrt{N \mathfrak{R}_{\widehat{F}}(\widehat{M})} \cap \mathfrak{R}_F(M) = N \mathfrak{R}_F(M)$, where the last equality follows since $\widehat{R}$ is flat over R and $\mathfrak{R}_{\widehat{F}}(\widehat{M}) = \widehat{R} \otimes_R \mathfrak{R}_F(M)$, showing that $\mathfrak{R}_{\widehat{F}}(\widehat{M})$ is flat over $\mathfrak{R}_F(M)$. Another application of Theorem 16.2.3 finishes the proof. $\square$

The following generalizes the first part of Lemma 5.4.4:

Lemma 16.4.6 *Let R be a Noetherian ring, M a finitely generated R–module, and N a submodule of M. Let $\overline{N}$ be the integral closure of N in M and $P \in \mathrm{Ass}(M/\overline{N})$. Then there exists a minimal prime ideal $p \subseteq P$ such that P/p is associated to M'/N', where $M' = M/pM$ and N' is the integral closure in M' of the image of N.*

Proof: By localizing, without loss of generality P is the unique maximal ideal of R. Write $P = \overline{N} :_R z$ for some $z \in M$. So $z \notin \overline{N}$, and by the definition of $\overline{N}$, there exists a minimal prime ideal p of R such that z is not in the integral closure N' of the image of N in $M' = M/pM$. Hence $P/p \subseteq ((\overline{N}+pM)/pM :_{R/p} z) \subseteq (N' :_{R/p} z) \neq R/p$, whence P/p is associated to M'/N'. $\qquad\square$

An immediate corollary of Theorem 16.2.3 is:

Corollary 16.4.7 *Let $(R, \mathfrak{m})$ be a Noetherian local ring, and M a submodule of a finitely generated free R–module F. Then a reduction of M has at least $\dim(\mathfrak{R}(M)/\mathfrak{m}\mathfrak{R}(M))$ generators.*

With analogy with ideals, we define the **analytic spread** of a module M to be $\ell_F(M) = \ell(M) = \dim(\mathfrak{R}(M)/\mathfrak{m}\mathfrak{R}(M))$. If $R/\mathfrak{m}$ is infinite, then by Proposition 8.3.7, M has a reduction generated by $\ell(M)$ elements.

Remark 16.4.8 As reductions localize, for any prime ideal P in a Noetherian local ring $(R, \mathfrak{m})$, $\ell(M) \geq \ell(M_P)$.

The following easy and important corollary of Theorem 16.2.3 is in the direction towards constructing Rees valuations of a module. It shows that to check whether $N \subseteq M$ is a reduction one only needs to check discrete valuations of rank one:

Corollary 16.4.9 *Let $N \subseteq M \subseteq L$ be finitely generated modules over a Noetherian ring R. Then $N \subseteq M$ is a reduction in L if and only if for every minimal prime ideal P in R and every discrete valuation ring V of rank one or zero between R/P and $\kappa(P)$, the images of $N \otimes_{R/P} V$ and $M \otimes_{R/P} V$ in $L \otimes_{R/P} \kappa(P)$ agree.*

Furthermore, there exists a finite set of V as above such that $N \subseteq M$ is a reduction in L if and only if for each of these finitely many V, if P is the kernel of the map $R \to V$, then the images of $N \otimes_{R/P} V$ and $M \otimes_{R/P} V$ in $L \otimes_{R/P} \kappa(P)$ agree.

Proof: Without loss of generality one can replace R by R/P, as P varies over the minimal prime ideals, and assume that R is a domain. Let K be the field of fractions. Similarly we may assume that L is torsion–free. Thus for some non–zero $r \in R$, rL is contained in a free R–module F such that $L_r = F_r$.

Observe that $N \subseteq M$ is a reduction in L if and only if $rN \subseteq rM$ is a reduction in rL. Moreover, for every discrete valuation ring V of rank one or zero between R and $Q(R)$, the images of NV and MV in $L \otimes K$ agree if and only if the images of rNV and rMV in $rL \otimes K = F \otimes K$ agree. Hence we may replace N by rN, M by rM, and L by rL to assume that $N \subseteq M \subseteq L \subseteq F$, where F is a finitely generated free module.

By definition, if N is a reduction of M in L, the images of NV and MV in $L \otimes K$ agree. Conversely, assume that the images of NV and MV in $L \otimes_R K$ agree for every rank one discrete valuation ring V between R and K. If $N \subseteq M$ is not a reduction in L, then there exists a K–valuation ring V containing R such that the images of NV and MV in $L \otimes_R K = F \otimes_R K$ are not equal. Hence $N \subseteq M$ is not a reduction in F. By Theorem 16.2.3, $N\mathfrak{R}_F(M) \subseteq M\mathfrak{R}_F(M)$ is not a reduction, so that by Proposition 6.8.2 there exists a discrete valuation ring W containing $\mathfrak{R}_F(M)$ such that $N\mathfrak{R}_F(M)W \neq M\mathfrak{R}_F(M)W$. By Proposition 6.4.7, $V = W \cap K$ is a discrete K–valuation ring of rank one or zero. By assumption, the images of NV and MV in $F \otimes K$ agree. Hence $N\mathfrak{R}_F(M)W = N\mathfrak{R}_F(M)VW = M\mathfrak{R}_F(M)VW = M\mathfrak{R}_F(M)W$, which is a contradiction. Furthermore, if W are taken to vary over the Rees valuation rings of $N\mathfrak{R}_F(M)$, the corollary is proved. $\square$

Reductions can be tested also after passage to curves, i.e., after passing to one–dimensional quotient domains. The following theorem is a version for modules of a theorem of Böger [20] for ideals.

Theorem 16.4.10 (Katz [162, Theorem 2.6]) *Let $(R, \mathfrak{m})$ be a formally equidimensional Noetherian local ring, and $N \subseteq M \subseteq F$ finitely generated R–modules of rank r, with F free. Then N is a reduction of M in F if and only if for every $P \in \operatorname{Spec} R$ with $\dim(R/P) = 1$, the image of N in F/PF is a reduction of the image of M in F/PF.*

Proof: Without loss of generality R is a domain.

If N is a reduction of M in L, then by Theorem 16.2.3, $N\mathfrak{R}(M) \subseteq M\mathfrak{R}(M)$ is a reduction of ideals. Any equations of integral dependence remain equations of integral dependence in any quotient. In particular, if P is a prime ideal in R, then the passage to $\mathfrak{R}(M)/(P\operatorname{Sym}_R(F) \cap \mathfrak{R}(M))$ shows that the images of N and M still give a reduction of ideals. Then again by Theorem 16.2.3, the image of N in F/PF is a reduction of the image of M in F/PF.

Now assume that for every $P \in \operatorname{Spec} R$ with $\dim(R/P) = 1$, the image of N in F/PF is a reduction of the image of M in F/PF. Suppose for contradiction that $N \subseteq M$ is not a reduction. Then necessarily $\dim R \geq 2$. By Theorem 16.2.3, there exists a prime ideal in $\mathfrak{R}(M)$ that contains N but not $M\mathfrak{R}(M)$. By Krull's Height Theorem (Theorem B.2.1), there exists a homogeneous prime ideal Q' in $\mathfrak{R}(M)$ such that $\dim(\mathfrak{R}(M)/Q') = 1$ and Q' contains $N\mathfrak{R}(M)$ but not $M\mathfrak{R}(M)$. Similarly, there exists a homogeneous prime ideal Q contained in Q' such that $\dim(\mathfrak{R}(M)/Q) = 2$ and Q does not contain $\det(N)$. By Proposition 16.3.2, $\det(N)$ is a proper ideal in R. Set

$P = Q \cap R$. By the Dimension Formula (Theorem B.5.1), $\operatorname{ht} P = \operatorname{ht} Q - \operatorname{tr.deg}_R(\mathfrak{R}(M)) + \operatorname{tr.deg}_{\kappa(P)}(\kappa(Q)) = \dim R + r - 2 - r + \operatorname{tr.deg}_{\kappa(P)}(\kappa(Q))$. As Q does not contain $M\mathfrak{R}(M)$, $\operatorname{tr.deg}_{\kappa(P)}(\kappa(Q)) \geq 1$. Also, $\det(N) \not\subseteq P$, so that $\dim R > \operatorname{ht} P \geq \dim R - 2 + 1$, and so necessarily $\dim(R/P) = 1$. By assumption, the image of M in F/PF is integral over the image of N, whence by Theorem 16.2.3,

$$\frac{N + PF}{PF} \mathfrak{R}_{F/PF}\left(\frac{M + PF}{PF}\right) \subseteq \frac{M + PF}{PF}\mathfrak{R}_{F/PF}\left(\frac{M + PF}{PF}\right)$$

is a reduction of ideals. This is saying that

$$\frac{N\mathfrak{R}(M)}{P\operatorname{Sym}_R(F) \cap \mathfrak{R}(M)} \subseteq \frac{M\mathfrak{R}(M)}{P\operatorname{Sym}_R(F) \cap \mathfrak{R}(M)}$$

is a reduction of ideals. As $\det N \not\subseteq P$, also $\det M \not\subseteq P$. It follows that $\mathfrak{R}(M)_{R \backslash P}$ is a polynomial ring over R_P, so that $Q_{R \backslash P}$ contains the extension of P to this polynomial ring. Thus $P\operatorname{Sym}_R(F) \cap \mathfrak{R}(M) \subseteq Q \subseteq Q'$. Hence by passing to $\mathfrak{R}(M)/Q'$, the displayed result shows that the image of N is a reduction of the image of M, which contradicts the choice of Q'.   □

16.5. Buchsbaum–Rim multiplicity

In Chapter 11 we defined the multiplicity for modules over a Noetherian local ring. There is another multiplicity, due to Buchsbaum and Rim [32], which we present in this section. The results are largely due to Buchsbaum, Rim; Kirby and Rees [165]; and Katz [162]. A more geometric approach is in the work of [167] and [168]. Katz used the Buchsbaum–Rim multiplicity to prove a reduction criterion for modules, generalizing the theorem of Böger (Corollary 11.3.2). In this section we follow the treatment found in a paper of Simis, Ulrich and Vasconcelos [269]. One advantage of this treatment is that the theory of multigraded Hilbert polynomials is not needed, contrary to most approaches to this material. See also [22].

Discussion 16.5.1   Let $A = \oplus_{i \geq 0} A_i \subseteq B = \oplus_{i \geq 0} B_i$ be a homogeneous inclusion of graded Noetherian rings such that $R = A_0 = B_0$ is a local ring with maximal ideal $\mathfrak{m}$ and such that $A = R[A_1]$ and $B = R[B_1]$. Let I be the ideal in B generated by A_1, let $\mathfrak{R} = B[It]$ be the Rees algebra of I over B, and let $G = \operatorname{gr}_I(B)$.

Give $\mathfrak{R}$ a bigrading as follows: assign degree $(1, 0)$ to the image of B_1 in $\mathfrak{R}$, and $(0, 1)$ to the elements in $A_1 t$. (Although it is natural to assign elements in $A_1 t$ degrees $(1, 1)$, the theory works better with the given grading.) With this grading, $\mathfrak{R}_{(0,0)} = R$, and $\mathfrak{R} = R[\mathfrak{R}_{(1,0)}, \mathfrak{R}_{(0,1)}]$. The (i, j)th graded piece of $\mathfrak{R}$ is $B_i A_j t^j$.

The associated graded ring G inherits the bigrading. By $B_t G$ we mean the image of B_t in G via the composition of the maps from $B \to \mathfrak{R} \to G$.

The algebra G also admits a grading $G_n = \oplus_{i+j=n} G_{(i,j)}$ that makes G a non–negatively graded algebra over $G_0 = R$ such that $G = R[G_1]$.

Remark 16.5.2   Note that $G/B_1 G = \mathfrak{R}/B_1 \mathfrak{R} \cong A$, since $\mathfrak{R}/B_1 R = \oplus_{j \geq 0} \mathfrak{R}_{(0,j)} = \oplus_{j \geq 0} A_j t^j \cong A$. In general,

$$(B_t G)_n = \bigoplus_{j=1}^{n-t+1} (A_{j-1} B_{n-j+1}/A_j B_{n-j})$$

$$= (A_0 B_n/A_1 B_{n-1}) \oplus (A_1 B_{n-1}/A_2 B_{n-2}) \oplus \cdots \oplus (A_{n-t} B_t/A_{n-t+1} B_{t-1}).$$

Notice that since $A_j B_{n-j} = A_{j-1} A_1 B_{n-j} \subseteq A_{j-1} B_1 B_{n-j} = A_{j-1} B_{n-j+1}$, the quotient modules above make sense as R–modules. Using this isomorphism, we see that the $A = (G/B_1 G)$–module $B_1 G/(B_1 G)^2$ is isomorphic to $B_1 A/A_1 A \cong (A + B_1 A)/A$.

We begin with the following theorem (see Simis, Ulrich, Vasconcelos [269], Propositions 3.2, 3.4, and Corollary 3.5). If I is an ideal in a Noetherian ring R, then by $0 : I^\infty$ we denote the stable value of the ascending chain of ideals $0 : I \subseteq 0 : I^2 \subseteq \cdots$.

Theorem 16.5.3   *Let $A \subseteq B$ be as in Discussion 16.5.1. Set $d = \dim B$. Assume further that $\lambda_R(B_1/A_1) < \infty$. Fix an integer $t > 0$.*
(1) *For every $n \geq t - 1$, $\lambda_R(B_n/A_{n-t+1} B_{t-1}) = \lambda_R((B_t G)_n)$.*
(2) *For all n sufficiently large, $\lambda_R(B_n/A_{n-t+1} B_{t-1})$ is a polynomial function $p_t(n)$ of degree*

$$\dim(B_t G) - 1 = \dim(G/(0 :_G B_t G)) - 1 \leq \dim G - 1 = \dim B - 1 = d - 1.$$

(3) *The polynomial $p_t(n)$ has the form*

$$p_t(n) = \frac{e_t(A, B)}{(d - 1)!} n^{d-1} + O(n^{d-2}),$$

where $e_t(A, B) = 0$ if $\dim(G/(0 :_G B_t G)) < d$ and $e_t(A, B) = e(B_t G)$ if $\dim(G/(0 :_G B_t G)) = d$. In particular, $e_t(A, B)$ is a non–negative integer.
(4) *If $\dim(G/(0 :_G (B_1 G)^\infty)) = d$ and $(0 :_G B_t G) = (0 :_G (B_1 G)^\infty)$, then*

$$e_t(A, B) = e(G/(0 :_G (B_1 G)^\infty)).$$

In particular, $e_t(A, B)$ is constant for t sufficiently large; we denote the constant value by $e_\infty(A, B)$.
(5) *If B is integral over A, then $e_\infty(A, B) = 0$.*
(6) *If B is equidimensional, universally catenary, and $e_\infty(A, B) = 0$, then B is integral over A.*

Proof: By Remark 16.5.2,

$$\bigoplus_{j=1}^{n-t+1} (A_{j-1} B_{n-j+1}/A_j B_{n-j}) = (B_t G)_n.$$

Taking lengths on both sides, we get a telescoping sum on the left–hand side, giving (1).

As a module, $B_t G$ is a finitely generated graded module over the graded ring $G/(0 :_G B_t B)$, which is a non–negatively graded ring generated by its degree one piece over an Artinian local ring in degree 0. The degree 0 piece of this ring is Artinian since it is a homomorphic image of the degree 0 piece of $G/(0 :_G B_1 B)$, and this is R/J, where $J = \operatorname{ann}(B_1/A_1)$. Hence we may apply Remark 11.1.12 to conclude that $\lambda_R((B_t G)_n)$ is a polynomial in n for large n of degree equal to $\dim(B_t G) - 1$. The only remaining claim in (2) that is not immediate is that the dimension of G is the dimension of B, which follows from Proposition 5.1.6.

Claim (3) follows from Remark 11.1.12.

Every statement in (4) is immediate except for the claim that $e_t(A, B) = e(G/(0 :_G (B_1 G)^\infty))$. To prove this statement, use the Associativity Formula (Theorem 11.2.4):

$$e_t(A, B) = e(B_t G) = \sum_{q \in \operatorname{Min}(G), \dim(G/q) = d} \lambda_{G_q}((B_t G)_q) e(G/q).$$

By assumption, $(0 :_G B_t G) = (0 :_G B_1 G^\infty)$, which implies that $B_t G$ is not in q if q is minimal. Hence $(B_t)_q \cong G_q$, which gives us $e_t(A, B) = e(G/(0 :_G (B_1 G)^\infty))$, as required.

If B is integral over A, then $B_1 B \subseteq \sqrt{A_1 B}$, and hence $B_1 G \subseteq \sqrt{0}$. This implies that $\operatorname{ht}((0 :_G (B_1 G)^\infty)) > 0$. By (3) this implies that $e_t(A, B) = 0$ for all large t, proving (5).

To prove (6) we reverse these implications. By (3), the assumption that $e_\infty(A, B) = 0$ implies that $\dim G/(0 :_G (B_1 G)^\infty) < d$. Since B is equidimensional and universally catenary, by Proposition 5.4.8 the associated graded ring G is equidimensional and catenary. It follows that $(0 :_G (B_1 G)^\infty)$ is not contained in any minimal prime of G, and therefore $B_1 G \subseteq \sqrt{0}$. Hence $B_1 B \subseteq \sqrt{A_1 B}$ and by Proposition 1.6.4, B is integral over A. $\qquad\square$

Parts (5) and (6) of Theorem 16.5.3 give a numerical condition for the extension $A \subseteq B$ to be integral, in terms of the vanishing of $e_\infty(A, B)$. However, what we are most interested in is developing such a condition in terms of $e_1(A, B)$, since this number has to do with the behavior of the lengths of B_n/A_n. The next corollary gives such a condition.

Corollary 16.5.4 *Let $A \subseteq B$ be as in Discussion 16.5.1. Set $d = \dim B$. Assume further that $\lambda_R(B_1/A_1) < \infty$. If B is integral over A, and $B_q = A_q$ for every minimal prime ideal q in A, then $e_1(A, B) = 0$. Conversely, if B is equidimensional, universally catenary, and $e_1(A, B) = 0$, then B is integral over A and $B_q = A_q$ for every minimal prime ideal q in A.*

Proof: Setting $t = 1$ in Theorem 16.5.3 we obtain that for large n, there is a polynomial $p(n) = p_1(n)$ such that

$$p(n) = \lambda_R(B_n/A_n) = \lambda_R(B_1 G)$$

and

$$p(n) = \frac{e_1(A, B)}{(d - 1)!} n^{d-1} + O(n^{d-2}),$$

where $e_1(A, B) = 0$ if $\dim(G/(0 :_G B_1 G)) < d$ and $e_1(A, B) = e(B_1 G)$ if $\dim(G/(0 :_G B_1 G)) = d$.

First assume that B is integral over A, and $B_q = A_q$ for every minimal prime q in A. Since B is integral over A it follows that $B_1 B \subseteq \sqrt{A_1 B}$ or equivalently that $B_1 G \subseteq \sqrt{0}$ in G. The minimal primes in A are in one–to–one correspondence with the minimal primes in $G/B_1 G$ via the isomorphism of Remark 16.5.2. Since $B_1 G \subseteq \sqrt{0}$, this means that the minimal primes of A and the minimal primes of G are in one–to–one correspondence. Since $B_q = A_q$ for every minimal prime q in A, the isomorphism $B_1 G/(B_1 G)^2 \cong (A + B_1 A)/A$ of Remark 16.5.2 gives that for all minimal primes Q of G, $(B_1 G/(B_1 G)^2)_Q = 0$. By Nakayama's Lemma it follows that $(B_1 G)_Q = 0$, and hence that $(0 : B_1 G)$ is not in Q. This holds for every minimal prime Q of G, which implies that $\dim G/((0 : B_1 G) < \dim G$. By Theorem 16.5.3 (3), $e_1(A, B) = 0$.

Conversely, assume that $e_1(A, B) = 0$ and that B is equidimensional and universally catenary. From Theorem 16.5.3 (3) it follows that $\dim G/((0 : B_1 G) < \dim G$. As G is equidimensional (Proposition 5.1.6), this means that $(0 : B_1 G)$ is not contained in any minimal prime of G, and so for all minimal primes Q of G, $(B_1 G)_Q = 0$. Hence $B_1 G \subseteq \sqrt{0}$, which implies that $B_1 B \subseteq \sqrt{A_1 B}$. This also means that the minimal primes in A are in one–to–one correspondence with the minimal primes in G via the isomorphism of Remark 16.5.2. Moreover, B is integral over A since $B_1 B \subseteq \sqrt{A_1 B}$. Remark 16.5.2 shows that $((A + B_1 A)/A)_Q = 0$, which implies that $A_q = B_q$ for every minimal prime q of A as B is generated by B_1 over R (and hence over A as well). $\qquad\square$

We now specialize this situation to the Rees algebras of modules. Let $(R, \mathfrak{m})$ be a local Noetherian ring, and let $N \subseteq M \subseteq F = R^r$ be inclusions of R–modules. We apply the general results above by setting $A = \mathfrak{R}(N)$ and $B = \mathfrak{R}(M)$, the Rees algebras of N and M respectively coming from the embedding of N and M into the free module F. We write M^n (respectively N^n) to be the nth power of the image of M in B (respectively of N in A). In this case, G is the associated graded ring of the ideal $N\mathfrak{R}(M)$.

Definition 16.5.5 *Let $(R, \mathfrak{m})$ be a local Noetherian ring, and let $N \subseteq M \subseteq F = R^r$ be inclusions of R–modules. Suppose that $\lambda(F/M) < \infty$. We define the* **Buchsbaum–Rim multiplicity** *of M (with respect to this embedding), $\mathrm{br}(M)$, to be*

$$\mathrm{br}(M) = e_1(\mathfrak{R}(M), \mathfrak{R}(F)).$$

Similarly, if $\lambda(M/N) < \infty$ we define the **relative Buchsbaum–Rim multiplicity** *by $\mathrm{br}(N, M) = e_1(\mathfrak{R}(N), \mathfrak{R}(M))$.*

Theorem 16.5.6 *Let $(R, \mathfrak{m})$ be a local Noetherian ring of dimension d, and let $N \subseteq M \subseteq F = R^r$ be inclusions of R–modules such that $\lambda_R(M/N) < \infty$.*
(1) For all sufficiently large integers n, $\lambda_R(M^n/N^n)$ is a polynomial function $f(n)$ of degree at most $d + \max\{\mathrm{rk}(M \otimes_R \kappa(P)) \mid P \in \mathrm{Min}(R)\} - 1$.
(2) If $\mathrm{ht}(\mathrm{ann}\, F/M) > 0$, then
$$f(n) = \frac{\mathrm{br}(N, M)}{(d + r - 1)!} n^{d+r-1} + O(n^{d+r-2}).$$
(3) If N is a reduction of M, then $\mathrm{br}(N, M) = 0$ and the degree of $f(n)$ is strictly less than $d + r - 1$. The converse holds in case R is formally equidimensional and $\mathrm{ht}(\mathrm{ann}\, F/M) > 0$.

Proof: We set $A = \mathfrak{R}(N)$ and $B = \mathfrak{R}(M)$. Note that $\lambda_R(M^n/N^n)$ is the function studied in Theorem 16.5.3 when $t = 1$. By (2) and (3) of that theorem it follows that for all large n,

$$\lambda_R(M^n/N^n) = \frac{e_1(A, B)}{(\dim B - 1)!} n^{\dim B - 1} + O(n^{\dim B - 2})$$

is a polynomial function. By Lemma 16.2.2,

$$\dim B = \dim \mathfrak{R}(M) \le d + \max\{\mathrm{rk}(M \otimes_R \kappa(P)) \mid P \in \mathrm{Min}(R)\} - 1,$$

proving (1).

To prove (2), note that by Lemma 16.2.2, the condition $\mathrm{ht}(\mathrm{ann}\, F/M) > 0$ implies that $\dim B = d + r$. Then (2) follows at once Theorem 16.5.3 (in the case $t = 1$ in that theorem).

To prove (3), first suppose that N is a reduction of M. By Theorem 16.2.3, B is integral over A. By assumption the length of M/N is finite. Provided the dimension of R is positive, this implies that $A_q = B_q$ for every minimal prime q of A. Corollary 16.5.4 applies to conclude that $\mathrm{br}(N, M) = e_1(A, B) = 0$, proving that the degree of $f(n)$ is strictly less than $d + r - 1$. Suppose that the dimension of R is 0, i.e., R is Artinian. If $M = F$, then the condition that N is a reduction of M forces $N = F$ as well. Obviously $e_1(A, B) = 0$ in this case. If $M \ne F$, then $M \subseteq R^{r-1} \oplus \mathfrak{m} \subseteq F$, and then the degree of $f(n)$ is at most $r - 1$.

Conversely, if the degree of $f(n)$ is strictly less than $d + r - 1$, then because $\mathrm{ht}(\mathrm{ann}\, F/M) > 0$, the degree of $f(n)$ is strictly less than $\dim B$. Hence $\mathrm{br}(N, M) = e_1(A, B) = 0$. Corollary 16.5.4 gives the conclusion that B is integral over A, forcing N to be a reduction of M in F by Theorem 16.2.3.$\square$

With the set–up as in the theorem, the polynomial $f(n)$ is called the **Buchsbaum–Rim polynomial** of $N \subseteq M \subseteq F$.

With the notion of Buchsbaum–Rim multiplicity, the following is a reduction criterion for modules in the spirit of Rees's Theorem 11.3.1:

Corollary 16.5.7 *Let $(R, \mathfrak{m})$ be a formally equidimensional local Noetherian ring of dimension $d > 0$, and let $N \subseteq M \subseteq F = R^r$ be R–modules with*

$\lambda(F/N) < \infty$. Then N is a reduction of M in F if and only if $\mathrm{br}(N) = \mathrm{br}(M)$, which holds if and only if $\mathrm{br}(N, M) = 0$.

Proof: We have that $\lambda(M^n/N^n) = \lambda(\mathrm{Sym}_n(F)/N^n) - \lambda(\mathrm{Sym}_n(F)/M^n)$ is, for large n, a polynomial of degree strictly less than $d + r - 1$ if and only if $\mathrm{br}(N) = \mathrm{br}(M)$. By Theorem 16.5.6 this is equivalent to N being a reduction of M in F under the conditions of the corollary.

By Theorem 16.5.6 the assumption that N is a reduction of M in F implies that $\mathrm{br}(N, M) = 0$. The converse holds since R is formally equidimensional by the same theorem.   □

Observe that if I is an ideal in R and $\lambda(R/I) < \infty$, then $\mathrm{br}(I, R) = e(I)$. The corollary above then generalizes Rees's Theorem 11.3.1. More generally, the next theorem can be thought of as a generalization of Böger's theorem, Corollary 11.3.2.

Theorem 16.5.8 *Let $(R, \mathfrak{m})$ be a Noetherian local ring, F a finitely generated free R–module, and $N \subseteq M \subseteq F$.*
(1) Suppose that $N \subseteq M$ is a reduction in F. Then $\mathrm{br}_{R_Q}(N_Q, M_Q) = 0$ for every minimal prime ideal Q over $N :_R M$.
(2) Suppose that R is formally equidimensional and that there exists an integer r such that for all $P \in \mathrm{Min}(R)$, the image of M in $F \otimes_R \kappa(P)$ has rank r. Further suppose that $\ell(N) \leq \mathrm{ht}(N :_R M) + r - 1$. If for every minimal prime ideal Q over $N :_R M$, either $\ell(N_Q) < \mathrm{ht}(N_Q :_{R_Q} M_Q) + r - 1$ or $\mathrm{br}_{R_Q}(N_Q, M_Q) = 0$, then N is a reduction of M.

Proof: (1) Assuming that $N \subseteq M$ is a reduction in F, N_Q is also a reduction of M_Q. For Q minimal over $N :_R M$, the length over R_Q of M_Q/N_Q is finite. By Theorem 16.5.6 we obtain that $\mathrm{br}_{R_Q}(N_Q, M_Q) = 0$.

(2) Let $\overline{N}$ be the integral closure of N in F. We claim that for every $P \in \mathrm{Ass}_R(F/\overline{N})$, $\mathrm{ht}\, P \leq \ell(N_P) - r + 1$. To prove this, by Theorem B.5.2 and localization, without loss of generality we may assume that P is the unique maximal ideal of R. As R is formally equidimensional, by Lemma B.4.2, by passing to R modulo a minimal prime ideal, we do not change $\mathrm{ht}\, P$. By assumption this also does not change r, and $\ell(N_P)$ can only decrease. Thus we may assume that R is an integral domain. Write $P = \overline{N} :_R z$ for some $z \in F$. Then $Pz \subseteq \overline{N}$, so that if $\overline{z}$ is a linear form in $\mathrm{Sym}_R(F)$ corresponding to z, then $P\overline{z}$ is in the integral closure $\overline{\mathfrak{R}(N)}$ of $\mathfrak{R}(N)$ in $\mathrm{Sym}_R(F)$, and $\overline{z} \notin \mathfrak{R}(N)$. Hence there exists a prime ideal $\overline{Q}$ of height one in the Krull domain $\overline{\mathfrak{R}(N)}$ such that $\overline{z} \notin \overline{\mathfrak{R}(N)}_{\overline{Q}}$. Therefore $P \subseteq \overline{Q}$, and necessarily $P = \overline{Q} \cap R$. Set $Q = \overline{Q} \cap \mathfrak{R}(N)$. By Proposition 4.8.6, $\mathrm{ht}\, Q = 1$. By Theorem B.5.2, R satisfies the dimension formula, so

$$1 + \dim \mathfrak{R}(N)/P\mathfrak{R}(N) \geq \mathrm{ht}\, Q + \mathrm{tr.deg}_{\kappa(P)}\kappa(Q)$$
$$= \mathrm{ht}\, P + \mathrm{tr.deg}_R(\mathfrak{R}(N)) = \mathrm{ht}\, P + r,$$

which proves the claim.

Now let $P \in \operatorname{Ass}(F/\overline{N})$. By Remark 16.4.8, $\dim \ell(N) \geq \ell(N_P)$. Thus by assumption and the claim, $\operatorname{ht} P \leq \ell(N_P) - r + 1 \leq \ell(N) - r + 1 \leq \operatorname{ht}(N :_R M)$. Then either P does not contain $(N :_R M)$, or P is minimal over $(N :_R M)$. In the former case, $M_P \subseteq N_P \subseteq \overline{N}_P$. Suppose that P is minimal over $(N :_R M)$. We claim that $\operatorname{br}_{R_P}(N_P, M_P) = 0$. If not then by assumption $\ell(N_P) < \operatorname{ht}(N_P :_{R_P} M_P) + r - 1$. However in this case we would obtain that $\operatorname{ht} P \leq \ell(N_P) - r + 1 < \operatorname{ht}(N_P :_{R_P} M_P)$, which is impossible. Thus $\operatorname{br}_{R_P}(N_P, M_P) = 0$ is forced. By Theorems 16.5.6 and 16.2.3, $N_P \subseteq M_P$ is a reduction in F_P. Thus for all $P \in \operatorname{Ass}(F/\overline{N})$, $M_P \subseteq \overline{N}_P$, so that $M \subseteq \overline{N}$. This proves the theorem. □

16.6. Height sensitivity of Koszul complexes

Let $x_1, \ldots, x_n \in R$, and let M be an R–module. The standard construction of Koszul complex $K(x_1, \ldots, x_n; M)$ is given in Section A.4 in the appendix. It is well–known that the Koszul complex is depth–sensitive, i.e., that the depth of $(x_1, \ldots, x_n)$ on M equals

$$n - \max\{j \mid \text{the } j\text{th homology of } K(x_1, \ldots, x_n; M) \text{ is non–zero}\}.$$

In this statement, depth cannot be replaced by height. For example, in the ring $R = k[X, Y]/(X^2, XY)$, with X and Y variables over a field k, the ideal YR has height 1, the depth of YR on R is 0, and so $\operatorname{ht}(YR) \neq 1 - \max\{j \mid \text{the } j\text{th homology of } K(Y; R) \text{ is non–zero}\}$.

We prove that the Koszul complex is height–sensitive up to integral closure:

Theorem 16.6.1 (Rees [239]) *Let R be a formally equidimensional ring, and $I = (x_1, \ldots, x_n)$ an ideal in R of height h. Let $G_\bullet$ be the Koszul complex $K(x_1, \ldots, x_n; R)$. Then for all $i > n - h$, the module B_i of ith boundaries of $G_\bullet$ is a reduction in G_i of the module Z_i of ith cycles of $G_\bullet$.*

Proof: (This proof follows Katz [160].) By Proposition 16.4.1 and by Theorem B.5.2, the hypotheses and conclusion are unaffected by localization, so we may assume that R is a formally equidimensional local ring. By the definition of integral closure and by Corollary B.4.3 we may also tensor with R/P, for $P \in \operatorname{Min} R$, and thus assume that R is a formally equidimensional local domain. Koszul complexes on different n–element generating sets of I are isomorphic, so we may change the x_i by using prime avoidance to assume that whenever $1 \leq i_1 < i_2 < \cdots < i_j \leq n$, $1 \leq j \leq h$, then $(x_{i_1}, x_{i_2}, \ldots, x_{i_j})$ has height j.

We denote the maps in the complex $G_\bullet$ by $\varphi_\bullet$, and represent them as matrices. For each i, G_i is $\Lambda_R^i(R^n)$, so its standard basis consists of the elements e_J as J varies over all subsets of $\{1, \ldots, n\}$ of cardinality i. For $J = \{j_1, \ldots, j_m\} \subseteq \{1, \ldots, n\}$, with $j_1 < \cdots < j_m$, define $\operatorname{sgn}(j_i, J) = (-1)^{i-1}$.

Let $i > n - h$ and let $\alpha \in Z_i$. The standard columns of φ_i are labeled by sets $J \subseteq \{1, \ldots, n\}$, $|J| = i$. The column corresponding to J equals

$C_J = \sum_{j \in J} \text{sgn}(j, J) x_j e_{J \setminus \{j\}}$. Then $0 = \varphi_i(\alpha) = \sum_J \alpha_J C_J$, and for any set K with $|K| = i - 1$, this yields (row) equations

$$0 = \sum_{j \notin K} \alpha_{K \cup \{j\}} \text{sgn}(j, K \cup \{j\}) x_j.$$

Thus by Ratliff's Theorem 5.4.1, each α_J is in the integral closure of some $(n - i + 1)$-generated subideal of I, thus in particular $\alpha \in \bar{I} G_i$.

Multiplication by x_j on the Koszul complex $G_\bullet$ is homotopic to 0, and hence there is a homotopy $s : G_\bullet \to G_\bullet[1]$ such that $ds + sd = \mu_{x_j}$, the multiplication map by x_j. Hence $ds(\alpha) = x_j \alpha$. As j is arbitrary, and $\alpha \in Z_i$ is arbitrary, it follows that $I Z_i \subseteq \bar{I} B_i$. Therefore for any valuation ring V, $IV Z_i \subseteq IV B_i \subseteq IV Z_i$, which proves the theorem. □

A weak version of Ratliff's Theorem 5.4.1 follows from this theorem (although the proof above relies on it): if $x_1, \ldots, x_n$ of R generate an ideal of height n, then for any element $r = r_n$ of $(x_1, \ldots, x_{n-1}) : x_n$, by definition there exist $r_1, \ldots, r_{n-1}$ such that $(r_1, \ldots, r_n) \in Z_1$. Thus by the theorem, $(r_1, \ldots, r_n) \in \overline{B_1}$. By Proposition 16.4.4, r_n is integral over the ideal generated by the last row of the matrix φ_1. But the entries of this last row are $0, \pm x_1, \ldots, \pm x_{n-1}$, so that r_n is integral over $(x_1, \ldots, x_{n-1})$.

Definition 16.6.2 *Let $G_\bullet$ be the bounded complex of finitely generated free modules*

$$G_\bullet : \qquad 0 \longrightarrow G_n \xrightarrow{\varphi_n} G_{n-1} \xrightarrow{\varphi_{n-1}} \cdots \xrightarrow{\varphi_{i+1}} G_i \xrightarrow{\varphi_i} \cdots \xrightarrow{\varphi_2} G_1 \xrightarrow{\varphi_1} G_0 \longrightarrow 0.$$

*The kernel of φ_i is denoted as $Z_i = Z_i(G_\bullet)$, the image of φ_{i+1} is denoted $B_i = B_i(G_\bullet)$, and the ith homology of $G_\bullet$ as $H_i = H_i(G_\bullet) = Z_i/B_i$. We say that $G_\bullet$ is **acyclic up to integral closure** if for all $1 \le i \le n$, Z_i is in the integral closure of B_i in G_i.*

Thus we can rephrase Theorem 16.6.1:

Theorem 16.6.3 *Let (R, m) be a formally equidimensional Noetherian local ring, and let $x_1, \ldots, x_n$ be part of a system of parameters, i.e., the height of $(x_1, \ldots, x_n)$ is n. Then the Koszul complex of $x_1, \ldots, x_n$ is acyclic up to integral closure.*

Acyclicity of complexes up to integral closure can be formulated differently:

Proposition 16.6.4 *Let R be a Noetherian ring, and let $G_\bullet$ be a complex as in Definition 16.6.2. Then the following are equivalent:*
(1) $G_\bullet$ is acyclic up to integral closure.
(2) $G_\bullet \otimes_R (R/P)$ is acyclic up to integral closure for all minimal primes P of R.
(3) $G_\bullet \otimes_R R_{red}$ is acyclic up to integral closure, where R_{red} is R modulo its ideal of nilpotent elements.
If $(R, \mathfrak{m})$ is a Noetherian local ring, (1), (2), (3) are equivalent to:
(4) $G_\bullet \otimes_R \hat{R}$ is acyclic up to integral closure, where $\hat{R}$ is the completion of R.

Proof: The equivalence of (1)–(3) follows immediately from the definitions. The equivalence of (1) and (4) follows from Lemma 16.4.5.   □

16.7. Absolute integral closures

In this section we give a proof, due to Huneke and Lyubeznik [140], of a fundamental result due to Hochster and Huneke [125] that says that the absolute integral closure of a complete local domain in characteristic p is Cohen–Macaulay. Recall from Section 4.8 that the absolute integral closure of a commutative Noetherian domain R is the integral closure of R in a fixed algebraic closure of the field of fractions K of R. The original proof in [125] is quite difficult technically, while the proof given here is much simpler. We assume knowledge of local duality and local cohomology in this section. See [28], Section 3.5, for information concerning local duality.

We begin with a basic lemma that shows a way to kill cohomology classes in finite extensions. This lemma is closely related to the "equational lemma" in Hochster and Huneke [125] and its modification in Smith [272, 5.3].

Lemma 16.7.1 *Let R be a Noetherian domain containing a field of characteristic $p > 0$, let K be the field of fractions of R and let $\overline{K}$ be the algebraic closure of K. Let I be an ideal of R and let $\alpha \in H_I^i(R)$ be an element such that the elements $\alpha, \alpha^p, \alpha^{p^2}, \ldots, \alpha^{p^t}, \ldots$ belong to a finitely generated submodule of $H_I^i(R)$. Then there exists an R–subalgebra R' of $\overline{K}$ (i.e., $R \subseteq R' \subseteq \overline{K}$) that is finitely generated as an R–module and such that the natural map $H_I^i(R) \to H_I^i(R')$ induced by the natural inclusion $R \to R'$ sends α to 0.*

Proof: Let $A_t = \sum_{i=1}^t R\alpha^{p^i}$ be the R–submodule of $H_I^i(R)$ generated by $\alpha, \alpha^p, \ldots, \alpha^{p^t}$. The ascending chain $A_1 \subseteq A_2 \subseteq A_3 \subseteq \cdots$ stabilizes because R is Noetherian and all A_t sit inside a single finitely generated R–submodule of $H_I^i(R)$. Hence $A_s = A_{s-1}$ for some s, i.e., $\alpha^{p^s} \in A_{s-1}$. Thus there exists an equation $\alpha^{p^s} = r_1 \alpha^{p^{s-1}} + r_2 \alpha^{p^{s-2}} + \cdots + r_{s-1}\alpha$ with $r_i \in R$ for all i. Let T be a variable and let $g(T) = T^{p^s} - r_1 T^{p^{s-1}} - r_2^{p^{s-2}} - \cdots - r_{s-1}T$. Clearly, $g(T)$ is a monic polynomial in T with coefficients in R and $g(\alpha) = 0$.

Let $x_1, \ldots, x_d \in R$ generate the ideal I. If M is an R–module, the Čech complex $C^\bullet(M)$ of M with respect to the generators $x_1, \ldots, x_d \in R$ is

$$0 \to C^0(M) \to \cdots \to C^{i-1}(M) \overset{d_{i-1}}{\to} C^i(M) \overset{d_i}{\to} C^{i+1}(M) \to \cdots \to C^d(M) \to 0,$$

where $C^0(M) = M$ and $C^i(M) = \oplus_{1 \leq j_1 < \cdots < j_i \leq d} R_{x_{j_1} \cdots x_{j_i}}$, and the ith local cohomology module $H_I^i(M)$ is the ith cohomology module of $C^\bullet(M)$ [27, 5.1.19].

Let $\tilde{\alpha} \in C^i(R)$ be a cycle (i.e., $d_i(\tilde{\alpha}) = 0$) that represents α. The equality $g(\alpha) = 0$ means that $g(\tilde{\alpha}) = d_{i-1}(\beta)$ for some $\beta \in C^{i-1}(R)$. Since $C^{i-1}(R) =$

$\oplus_{1\leq j_1<\cdots<j_{i-1}\leq d}R_{x_{j_1}\cdots x_{j_{i-1}}}$, we may write $\beta = \sum_{1\leq j_1<\cdots<j_{i-1}\leq d}\frac{r_{j_1,\ldots,j_{i-1}}}{(x_{j_1}\cdots x_{j_{i-1}})^e}$ where $r_{j_1,\ldots,j_{i-1}}\in R$, $e\geq 0$, and $\frac{r_{j_1,\ldots,j_{i-1}}}{(x_{j_1}\cdots x_{j_{i-1}})^e}\in R_{x_{j_1}\cdots x_{j_{i-1}}}$.

Consider the equations $g\left(\frac{Z_{j_1,\ldots,j_{i-1}}}{(x_{j_1}\cdots x_{j_{i-1}})^e}\right) - \frac{r_{j_1,\ldots,j_{i-1}}}{(x_{j_1}\cdots x_{j_{i-1}})^e} = 0$, where $Z_{j_1,\ldots,j_{i-1}}$ is a variable. Multiplying such an equation by $((x_{j_1}\cdots x_{j_{i-1}})^e)^{p^s}$ produces a monic polynomial equation in $Z_{j_1,\ldots,j_{i-1}}$ with coefficients in R. Let $z_{j_1,\ldots,j_{i-1}}\in \overline{K}$ be a root of this equation and let R'' be the R–subalgebra of $\overline{K}$ generated by all the $z_{j_1,\ldots,j_{i-1}}$, i.e., by the set $\{z_{j_1,\ldots,j_{i-1}} | 1\leq j_1<\cdots<j_{i-1}\leq d\}$. Since each $z_{j_1,\ldots,j_{i-1}}$ is integral over R and there are finitely many $z_{j_1,\ldots,j_{i-1}}$, the R–algebra R'' is finitely generated as an R–module.

Let $\gamma = \sum_{1\leq j_1<\cdots<j_{i-1}\leq d}\frac{z_{j_1,\ldots,j_{i-1}}}{(x_{j_1}\cdots x_{j_{i-1}})^e}\in C^{i-1}(R'')$. The natural inclusion $R\to R''$ makes $C^\bullet(R)$ into a subcomplex of $C^\bullet(R'')$ in a natural way, and we identify $\tilde{\alpha}\in C^i(R)$ and $\beta\in C^{i-1}(R)$ with their natural images in $C^i(R'')$ and $C^{i-1}(R'')$ respectively. With this identification, $\tilde{\alpha}\in C^i(R'')$ is a cycle representing the image of α under the natural map $H_I^i(R)\to H_I^i(R'')$, and so is $\delta = \tilde{\alpha} - d_{i-1}(\gamma)\in C^i(R'')$. Since $g(\gamma) = \beta$ and $g(\tilde{\alpha}) = d_{i-1}(\beta)$, we conclude that $g(\overline{\alpha}) = 0$. Write $\delta = \sum\rho_{j_1,\ldots,j_i}$ where $\rho_{j_1,\ldots,j_i}\in R''_{x_{j_1}\cdots x_{j_i}}$. Each individual $\rho_{j_1,\ldots,j_i}$ satisfies the equation $g(\rho_{j_1,\ldots,j_i}) = 0$. Since $g(T)$ is a monic polynomial in T with coefficients in R, each $\rho_{j_1,\ldots,j_i}$ is an element of the field of fractions of R'' and is integral over R. Let R' be obtained from R'' by adjoining all the $\rho_{j_1,\ldots,j_i}$. Each $\rho_{j_1,\ldots,j_i}\in R'$ and the image of α in $H_I^i(R')$ is represented by the cycle $\delta = \sum\rho_{j_1,\ldots,j_i}\in C^i(R')$ that has all its components $\rho_{j_1,\ldots,j_i}$ in R'. Each $R'_{x_{j_1}\cdots x_{j_i}}$ contains a natural copy of R', namely, the one generated by the element $1\in R'_{x_{j_1}\cdots x_{j_i}}$. There is a subcomplex of $C^\bullet(R')$ that in each degree is the direct sum of all such copies of R'. This subcomplex is exact because its cohomology groups are the cohomology groups of R' with respect to the unit ideal. Since δ is a cycle and belongs to this exact subcomplex, it is a boundary, hence it represents the zero element in $H_I^i(R')$. $\square$

Theorem 16.7.2 *Let $(R,\mathfrak{m})$ be a Noetherian local domain containing a field of characteristic $p>0$, let K be the field of fractions of R and let $\overline{K}$ be the algebraic closure of K. Assume R is a surjective image of a Gorenstein local ring A. Let R' be an R–subalgebra of $\overline{K}$ (i.e., $R\subseteq R'\subseteq \overline{K}$) that is a finitely generated R–module. Let $i<\dim R$ be a non–negative integer. There is an R'–subalgebra R'' of $\overline{K}$ (i.e., $R'\subseteq R''\subseteq \overline{K}$) that is finitely generated as an R–module and such that the natural map $H_{\mathfrak{m}}^i(R')\to H_{\mathfrak{m}}^i(R'')$ is the zero map.*

Proof: As a first step, we will prove the existence of a finite extension where the image of the local cohomology has finite length, allowing us to apply Lemma 16.7.1. Let $n = \dim A$ and let $N_i = \operatorname{Ext}_A^{n-i}(R', A)$. Since R' is a finite R–module, so is N_i. Since we will be working with N_i for fixed i, we abuse notation and simply write $N = N_i$.

Set $d = \dim R$, and use induction on d to prove the theorem. For $d = 0$ there is nothing to prove, so we assume that $d > 0$ and that the theorem proven for all smaller dimensions. Since N is a finite R–module, the set of the associated primes of N is finite. Let $P_1, \ldots, P_s$ be the associated primes of N different from $\mathfrak{m}$. We make the following claim.

Claim: For each j there is an R'–subalgebra of $\overline{K}$, R^{P_j}, corresponding to P_j, such that R^{P_j} is a finite R–module and for every R^* such that $R^{P_j} \subseteq R^* \subseteq \overline{K}$ with R^* a finite R–module, the image $I \subseteq N$ of the natural map $\mathrm{Ext}_A^{n-i}(R^*, A) \to N$ induced by the natural inclusion $R' \to R^*$ vanishes after localization at P_j, i.e., $I_{P_j} = 0$.

Assuming this claim, let $\overline{R}' = R'[R^{P_1}, \ldots, R^{P_s}]$ be the compositum of all the R^{P_j}, $1 \leq j \leq s$. Clearly $R' \subseteq \overline{R}' \subseteq \overline{K}$. Since each R^{P_j} is a finite R–module, so is $\overline{R}'$. Clearly, $\overline{R}'$ contains every R^{P_j}. Hence the above claim implies that $I_{P_j} = 0$ for every j, where $I \subseteq N$ is the image of the natural map $\mathrm{Ext}_A^{n-i}(\overline{R}', A) \to N$ induced by the natural inclusion $R' \to \overline{R}'$. It follows that not a single P_j is an associated prime of I. But I is a submodule of N, and therefore every associated prime of I is an associated prime of N. Since $P_1, \ldots, P_s$ are all the associated primes of N different from $\mathfrak{m}$, we conclude that if $I \neq 0$, then $\mathfrak{m}$ is the only associated prime of I. Since I, being a submodule of a finite R–module N, is finite, and since $\mathfrak{m}$ is the only associated prime of I, we conclude that I is an R–module of finite length.

We next establish the claim. Let P denote one of the primes $P_1, \ldots, P_s$. Let $d_P = \dim(R/P)$. Since P is different from the maximal ideal, $d_P > 0$. As R is a surjective image of a Gorenstein local ring, it is catenary, hence the dimension of R_P equals $d - d_P$, and $i < d$ implies $i - d_P < d - d_P = \dim R_P$. By the induction hypothesis applied to the local ring R_P and the R_P–algebra R'_P, which is finitely generated as an R_P–module, there is an R'_P–subalgebra $\tilde{R}$ of $\overline{K}$ that is finite as an R_P–module, such that the natural map $H_P^{i-d_P}(R'_P) \to H_P^{i-d_P}(\tilde{R})$ is the zero map. Let $\tilde{R} = R'_P[z_1, z_2, \ldots, z_t]$, where $z_1, z_2, \ldots, z_t \in \overline{K}$ are integral over R_P. Multiplying, if necessary, each z_j by some element of $R \setminus P$, we can assume that each z_j is integral over R. We set $R^P = R'[z_1, z_2, \ldots, z_t]$ Clearly, R^P is an R'–subalgebra of $\overline{K}$ that is finite as R–module. Now let R^* be both an R^P–subalgebra of $\overline{K}$ (i.e., $R^P \subseteq R^* \subseteq \overline{K}$) and a finitely generated R–module. The natural inclusions $R' \to R^P \to R^*$ induce natural maps $\mathrm{Ext}_A^{n-i}(R^*, A) \to \mathrm{Ext}_A^{n-i}(R^P, A) \to N$. This implies that $I \subseteq J$, where J is the image of the natural map $\psi : \mathrm{Ext}_A^{n-i}(R^P, A) \to N$. Hence it is enough to prove that $J_P = 0$. Localizing this map at P we conclude that J_P is the image of the natural map $\psi_P : \mathrm{Ext}_{A_P}^{n-i}(\tilde{R}, A_P) \to \mathrm{Ext}_{A_P}^{n-i}(R'_P, A_P)$ induced by the natural inclusion $R'_P \to \tilde{R}$ (by a slight abuse of language we identify the prime ideal P of R with its full preimage in A). Let $D_P(\_) = \mathrm{Hom}_{A_P}(\_, E_P)$ be the Matlis duality functor in the category of R_P–modules, where E_P is the injective hull of the residue field of R_P in the category of R_P–modules. Local duality implies that $D_P(\psi_P)$ is the natural map $H_P^{i-d_P}(R_P') \to H_P^{i-d_P}(\tilde{R})$, which is the zero map by construction (note that $i - d_P = \dim A_P - (n-i)$). Since ψ_P is a map between finite R_P–modules and $D_P(\psi_P) = 0$, it follows that $\psi_P = 0$. This proves the claim.

We can write the natural map $\text{Ext}_A^{n-i}(\overline{R}', A) \to N$ as the composition of the two maps $\text{Ext}_A^{n-i}(\overline{R}', A) \to I \to N$, the first of which is surjective and the second injective, and applying the Matlis duality functor D, we get that the natural map $\varphi : H_{\mathfrak{m}}^i(R') \to H_{\mathfrak{m}}^i(\overline{R}')$ induced by the inclusion $R' \to \overline{R}'$ is the composition of two maps $H_{\mathfrak{m}}^i(R') \to D(I) \to H_{\mathfrak{m}}^i(\overline{R}')$, the first of which is surjective and the second injective. This shows that the image of φ is isomorphic to $D(I)$, which is an R–module of finite length since so is I. In particular, the image of φ is a finitely generated R–module. Let $\alpha_1, \ldots, \alpha_s \in H_{\mathfrak{m}}^i(\overline{R}')$ generate $\text{Im}(\varphi)$.

The natural inclusion $R' \to \overline{R}'$ is compatible with the Frobenius homomorphism, i.e., with the raising to the pth power on R' and $\overline{R}'$. This implies that φ is compatible with the action of the Frobenius f_* on $H_{\mathfrak{m}}^i(R')$ and $H_{\mathfrak{m}}^i(\overline{R}')$, i.e., $\varphi(f_*(\alpha)) = f_*(\varphi(\alpha))$ for every $\alpha \in H_{\mathfrak{m}}^i(R')$, which, in turn, implies that $\text{Im}(\varphi)$ is an f_*–stable R–submodule of $H_{\mathfrak{m}}^i(\overline{R}')$, i.e., $f_*(\alpha) \in \text{Im}(\varphi)$ for every $\alpha \in \text{Im}(\varphi)$. We finish the proof by applying the following lemma to a finite generating set of $\text{Im}(\varphi)$: let α_j generate this image. Applying Lemma 16.7.1 we obtain a $\overline{R}'$–subalgebra R_j' of $\overline{K}$ (i.e., $\overline{R}' \subseteq R_j' \subseteq \overline{K}$) such that R_j' is a finite R–module and the natural map $H_{\mathfrak{m}}^i(\overline{R}') \to H_{\mathfrak{m}}^i(R_j')$ sends α_j to zero. Let $R'' = R'[R_1, \ldots, R_s]$ be the compositum of all the R_j'. Then R'' is an R'–subalgebra of $\overline{K}$ and is a finite R–module since so is each R_j'. The natural map $H_{\mathfrak{m}}^i(\overline{R}') \to H_{\mathfrak{m}}^i(R'')$ sends every α_j to zero, hence it sends the entire $\text{Im}(\varphi)$ to zero. Thus the natural map $H_{\mathfrak{m}}^i(R') \to H_{\mathfrak{m}}^i(R'')$ is zero. $\square$

As an almost immediate corollary we obtain the main result of Hochster and Huneke [125], although our hypothesis that R be the homomorphic image of a Gorenstein is slightly different from the hypothesis in [125] that R be excellent.

Corollary 16.7.3 *Let $(R, \mathfrak{m})$ be a Noetherian local domain containing a field of characteristic $p > 0$. Assume that R is a surjective image of a Gorenstein local ring. Then the following hold:*
(1) $H_{\mathfrak{m}}^i(R^+) = 0$ for all $i < \dim R$.
(2) Every system of parameters of R is a regular sequence on R^+.

Proof: (1) R^+ is the direct limit of the finitely generated R–subalgebras R', hence $H_{\mathfrak{m}}^i(R^+) = \varinjlim H_{\mathfrak{m}}^i(R')$. But Theorem 16.7.2 implies that for each R' there is R'' such that the map $H_{\mathfrak{m}}^i(R') \to H_{\mathfrak{m}}^i(R'')$ in the inductive system is zero. Hence the limit is zero.

(2) Let $x_1, \ldots, x_d$ be a system of parameters of R. We prove that $x_1, \ldots, x_j$ is a regular sequence on R^+ by induction on j. The case $j = 1$ is clear since R^+

is a domain. Assume that $j > 1$ and that $x_1, \ldots, x_{j-1}$ is a regular sequence on R^+. Set $I_t = (x_1, \ldots, x_t)$. The short exact sequences

$$0 \to R^+/I_{t-1}R^+ \xrightarrow{x_t} R^+/I_{t-1}R^+ \to R^+/I_t R^+ \to 0$$

for $t \leq j - 1$ and the fact that $H_{\mathfrak{m}}^i(R^+) = 0$ for all $i < d$ imply by induction on t that $H_{\mathfrak{m}}^q(R^+/(x_1, \ldots, x_t)R^+) = 0$ for $q < d - t$. In particular, $H_{\mathfrak{m}}^0(R^+/(x_1, \ldots, x_{j-1})R^+) = 0$ since $0 < d - (j - 1)$. Hence $\mathfrak{m}$ is not an associated prime of $R^+/(x_1, \ldots, x_{j-1})R^+$. Suppose that P is an embedded associated prime ideal of $R^+/(x_1, \ldots, x_{j-1})R^+$. Then P is the maximal ideal of the ring R_P, $\dim(R_P) > j - 1$, and P is an associated prime of $(R^+/(x_1, \ldots, x_{j-1})R^+)_P = (R_P)^+/(x_1, \ldots, x_{j-1})(R_P)^+$, which is impossible by the above. Hence every element of $\mathfrak{m}$ not in any minimal prime ideal of $R/(x_1, \ldots, x_{j-1})R$, for example, x_j, is a regular element on $R^+/(x_1, \ldots, x_{j-1})R^+$.   □

16.8. Complexes acyclic up to integral closure

In this section we describe more generally when a complex is acyclic up to integral closure. Stronger results than what we present have been obtained by the theory of tight closure, see [124]. We need two results before proving the main theorem. The first result is Corollary 16.7.3, while the second is a generalization of the Buchsbaum–Eisenbud's Acyclicity Criterion.

Throughout this section, R is a Noetherian ring and $G_\bullet$ is a complex of finitely generated free modules:

$$G_\bullet: \quad 0 \longrightarrow G_n \xrightarrow{\varphi_n} G_{n-1} \xrightarrow{\varphi_{n-1}} \cdots \xrightarrow{\varphi_2} G_1 \xrightarrow{\varphi_1} G_0 \longrightarrow 0. \quad (16.8.1)$$

Definition 16.8.2 *Let R be a Noetherian ring and F and G finitely generated free R–modules. A map $\varphi : F \to G$ can be represented as a matrix relative to some chosen bases of F and G.*

(1) For any integer t, $I_t(\varphi)$ denotes the ideal in R generated by the $t \times t$ minors of a matrix expression for φ. It is easy to show that $I_t(\varphi)$ does not depend on the bases. By convention $I_0(\varphi) = R$, even if φ is the zero map. If t exceeds the rank of either F or G, then $I_t(\varphi) = (0)$.

(2) The rank of φ, $\mathrm{rk}(\varphi)$, is defined to be the largest integer r such that $I_r(\varphi) \neq (0)$.

*(3) More generally, if M is an R–module, the **rank** of the map $F \otimes_R M \to G \otimes_R M$ to be the largest integer r such that the induced map $\Lambda_R^r(F) \otimes_R M \to \Lambda_R^r(G) \otimes_R M$ is non–zero.*

Clearly $\mathrm{rk}(\varphi)$ equals the largest integer r for which $\Lambda_R^r(\varphi) : \Lambda_R^r(F) \to \Lambda_R^r(G)$ is non-zero.

Definition 16.8.3 *If M is a not necessarily finitely generated R–module, then whenever $I = (x_1, \ldots, x_n)$ is an ideal such that $IM \neq M$, the **Koszul***

depth *of* I *on* M, *denoted* K-depth$_I(M)$, *is* $n - j$, *where* $H_j(x_1, \ldots, x_n; M)$ *is the highest non–vanishing Koszul homology module on* M.

This agrees with the usual notion of depth in case R is Noetherian and M is a finitely generated R–module (see Section A.4 in the appendix). We use the convention that the depth and the height of an ideal I are $+\infty$ if $I = R$.

With these definitions we can state the generalized Buchsbaum–Eisenbud Acyclicity Criterion.

Theorem 16.8.4 (The Generalized Buchsbaum–Eisenbud Acyclicity Criterion Aberbach [1], original criterion in [31]) *Let* R *be a Noetherian ring. Let* $G_\bullet$ *be as in (16.8.1). Let* M *be a not necessarily finitely generated* R–module. *For* $1 \leq i \leq n$ *let* s_i *be the rank of the map* $G_i \otimes_R M \to G_{i-1} \otimes_R M$, *and let* $s_{n+1} = 0$. *Set* $I_i = I_{s_i}(\varphi_i)$. *Then* $G_\bullet \otimes_R M$ *is acyclic if and only if for all* $i = 1, \ldots, n$, K-depth$_{I_i}(M) \geq i$, *and* $s_i + s_{i+1} = \mathrm{rk}(G_i)$.

Definition 16.8.5 *Let* $G_\bullet$ *be as in display (16.8.1). Let* $b_i = \mathrm{rk}\, G_i$, *with the convention that* $b_i = 0$ *if* $i > n$ *or* $i < 0$. *Set* $r_i = \Sigma_{t=i}^{n}(-1)^{t-i}b_t$, $1 \leq i \leq n$, $r_{n+1} = 0$. *The* r_i *are the unique integers such that* $r_{n+1} = 0$ *and* $r_{i+1} + r_i = b_i$ *whenever* $1 \leq i \leq n$.

We say that $G_\bullet$ **satisfies the standard conditions on rank**, *if for* $1 \leq i \leq n$, $\mathrm{rk}\, \varphi_i = r_i$ (*equivalently,* $b_i = \mathrm{rk}\ \varphi_{i+1} + \mathrm{rk}\ \varphi_i$, $1 \leq i \leq n$).

We say that $G_\bullet$ **satisfies the standard conditions for depth** (*respectively the* **height**, *if for all* $i = 1, \ldots, n$, *the depth (respectively the height) of the ideal* $I_{r_i}(\varphi_i)$ *is at least* i.

Examples of complexes satisfying the standard conditions on height and rank can be found in the exercises.

With this set–up, the main theorem in this section is the following generalization of Theorem 16.6.3 (i.e., of Theorem 16.6.1):

Theorem 16.8.6 *Let* $(R, \mathfrak{m})$ *be a formally equidimensional Noetherian local ring containing a field, and let* $G_\bullet$ *be*

$$G_\bullet: \quad 0 \longrightarrow G_n \xrightarrow{\varphi_n} G_{n-1} \xrightarrow{\varphi_{n-1}} \cdots \xrightarrow{\varphi_{i+1}} G_i \xrightarrow{\varphi_i} \cdots \xrightarrow{\varphi_2} G_1 \xrightarrow{\varphi_1} G_0 \longrightarrow 0,$$

with G_i *finitely generated free* R–modules, *satisfying the standard conditions on height and rank. Then* $G_\bullet$ *is acyclic up to integral closure.*

As pointed out in Roberts [247, page 126], without the assumption that R contain a field, the result above would imply the monomial conjecture, which is not yet known. (Cf. Exercise 16.9, part (ii).) A few characteristic–free cases of complexes acyclic up to integral closure were proved by Katz in [160].

The proof of Theorem 16.8.6 proceeds by using the technique of reduction to characteristic p. This technique is beyond the scope of this book, but it is a standard technique in algebra, and especially in tight closure. Details are written in Hochster and Huneke [124]. Using reduction to characteristic p reduces Theorem 16.8.6 to the same statement in a ring of positive and prime characteristic:

Theorem 16.8.7 *Let $(R, \mathfrak{m})$ be a formally equidimensional Noetherian local ring of positive prime characteristic p, and let $G_\bullet$ be a complex of finitely generated free R–modules as in (16.8.1) satisfying the standard conditions on height and rank. Then $G_\bullet$ is acyclic up to integral closure.*

Proof: Let $\widehat{R}$ be the $\mathfrak{m}$–adic completion of R, and Q a minimal prime ideal in $\widehat{R}$. Then $\widehat{R}/Q$ is formally equidimensional, and $G_\bullet \otimes_R (\widehat{R}/Q)$ satisfies the standard conditions on height and rank. If we prove that $G_\bullet \otimes_R (\widehat{R}/Q)$ is acyclic up to integral closure, then by Proposition 16.6.4, $G_\bullet$ is acyclic up to integral closure. Thus without loss of generality we may assume that R is a complete local domain.

Choose $i \in \{1, \ldots, n\}$. As the height of $I_i = I_{\mathrm{rk}(\varphi_i)}(\varphi_i)$ is at least i, there exists a system of parameters $x_1, \ldots, x_i$ contained in I_i.

By Corollary 16.7.3, $x_1, \ldots, x_i$ form a regular sequence on R^+. Thus the K-depth$_{I_i}(R^+) \geq i$ for $1 \leq i \leq n$. Thus by Theorem 16.8.4, $G_\bullet \otimes_R R^+$ is acyclic.

Let $z \in Z_i$ be a cycle. Since $G_\bullet \otimes_R R^+$ is acyclic, it follows that $z \in B_i(G_\bullet \otimes_R R^+) = B_i \otimes_R R^+$. By writing z as a linear combination of the boundaries (over R) with coefficients in R^+, we may collect the finitely many coefficients appearing and adjoin them to R to get a complete local Noetherian ring S that is module finite over R. Then the theorem follows from Lemma 16.4.5. $\square$

16.9. Exercises

16.1   Let $N \subseteq M \subseteq T$ be inclusions of finitely generated modules over a Noetherian ring R. Prove or disprove: N is a reduction of T if and only if N is a reduction of M and M is a reduction of T.

16.2   Let $N \subseteq M$ be finitely generated modules over a Noetherian ring R. Let $\overline{N}$ be the integral closure of N in M. Prove that for any minimal prime ideal P in R and any Noetherian valuation ring V between R/P and $\kappa(P)$, $NV = \overline{N}V$ inside $M \otimes \kappa(P)$.

16.3   (Rees [239]) Let R be an integral domain, $\overline{R}$ its integral closure, and M a finitely generated torsion–free R–module. Let $x \in M_{(0)}$ be integral over M. Prove that x is an element of $\mathrm{Hom}_R(\mathrm{Hom}_R(M, \overline{R}), \overline{R})$, in a natural way.

16.4   Let R be a Noetherian domain with field of fractions K, and let $N \subseteq M$ be finitely generated R–modules such that $N \otimes_R K = M \otimes_R K$. Let $S'(M)$ (respectively $S'(N)$) denote the image of $\mathrm{Sym}_R(M)$ (respectively $\mathrm{Sym}_R(N)$) in $\mathrm{Sym}_K(M \otimes_R K)$. Prove that the following are equivalent:

   (i) $N \subseteq M$ is a reduction of modules.
   (ii) $S'(N) \subseteq S'(M)$ is an integral extension of rings.
   (iii) $NS'(M) \subseteq MS'(M)$ is an integral extension of ideals.

16.5   ([81]) Let R be a Noetherian ring, n a positive integer, and M a submodule of R^n. We think of M as generated by the columns of a matrix A. Suppose that there exists an integer k such that for all minimal primes P, the rank of $(M + PR^n)/PR^n$ is k. Let $m \in R^n$, and let B be the matrix whose columns generate $M + Rm$. Prove that m is in the integral closure of M in R^n if and only if $I_k(B) \subseteq \overline{I_k(A)}$.

16.6   Let R be a Noetherian local ring of positive dimension d. Let $G_\bullet$ be a complex of finitely generated free R–modules,
$$0 \to G_n \to G_{n-1} \to \cdots \to G_1 \to G_0 \to 0.$$
such that $H_i(G_\bullet)$ has finite length for $i \geq 1$ and such that $d \geq n$. Prove that $G_\bullet$ satisfies the standard conditions on height and rank.

16.7   Let $A = \mathbb{Z}[X_1, \ldots, X_d]$, a polynomial ring in variables $X_1, \ldots, X_d$ over $\mathbb{Z}$. Let J be an ideal in A generated by monomials in the X_i (with unit coefficients), and let $H_\bullet$ be an A–free resolution of A/J by finitely generated free A–modules. Let I_i denote the ideal of rank–order minors of the ith map in $H_\bullet$.
  (i) Prove that $\sqrt{I_i}$ is generated by monomials in $X_1, \ldots, X_d$.
  (ii) If $J = (X_1, \ldots, X_d)$, prove that $\sqrt{I_i} = (X_1, \ldots, X_d)$ for all $1 \leq i \leq d$, and $\sqrt{I_i} = A$ for $i \geq d + 1$. (We are not assuming that A be minimal.)

16.8   Let R be a locally formally equidimensional Noetherian ring, and $x_1, \ldots, x_n$ elements of R generating an ideal whose height at any localization is at least n. Let $G_\bullet$ be the Koszul complex $K_\bullet(x_1, \ldots, x_n; R)$. Prove that $G_\bullet$ satisfies the standard conditions on height and rank.

16.9   Let $A = \mathbb{Z}[X_1, \ldots, X_d]$, a polynomial ring in variables $X_1, \ldots, X_d$ over $\mathbb{Z}$. Let J be an ideal in A generated by monomials in the X_i (with unit coefficients), and let $H_\bullet$ be an A–free resolution of A/J by finitely generated free A–modules. Let R denote a complete local domain, and $x_1, \ldots, x_d$ a system of parameters of R. Let $f : A \to R$ be the ring map sending X_i to x_i, $J = f(I)R$, and $G_\bullet = H_\bullet \otimes_A R$.
  (i) Suppose that R contains a field. Prove that $G_\bullet$ is acyclic up to integral closure.
  (ii) Let $I = (x_1^{t+1}, \ldots, x_d^{t+1}, x_1^t \cdots x_d^t)$. Assume that $G_\bullet$ is acyclic up to integral closure. Prove that $(x_1^{t+1}, \ldots, x_d^{t+1}) : x_1^t \cdots x_d^t \subseteq \overline{(x_1, \ldots, x_d)}$.

16.10   Let R be an integral domain with field of fractions K, M a non–zero finitely generated torsion–free R–module. Let $\overline{M} = \cap_V MV$, where V varies over all K–valuation rings containing R. Prove that $\overline{M}$ is a finitely generated R–module if and only if the integral closure of the ring R is a finitely generated R–module.

16.11   Let $(R, \mathfrak{m})$ be a Noetherian local domain with infinite residue field. Let $M \subseteq R^r$ be modules of rank r. Assume that M is integrally closed. Prove that there exists $x \in \mathfrak{m}$ such that $\mathfrak{m}M :_{R^r} x = M$. (In other words, an integrally closed module is **$\mathfrak{m}$–full**.)

16.12 (Katz [162, Theorem 2.5]) Let $(R, \mathfrak{m})$ be a formally equidimensional Noetherian local ring, let $N \subseteq M \subseteq F$ be R–modules with F finitely generated and free. Assume that N, M, and F have rank r. Let $I(M)$ be the ideal of $r \times r$ minors of a presenting matrix of M, and similarly define $I(N)$. Suppose that $\dim S(M)/\mathfrak{m}S(M) = \mathrm{ht}(I(M)) + r - 1$, and that $\sqrt{I(N)} = \sqrt{I(M)}$. Prove that $N \subseteq M$ is a reduction if and only if for all $Q \in \mathrm{Min}(I(M))$, $\mathrm{br}_{R_Q}(M_Q) = \mathrm{br}_{R_Q}(N_Q)$.

16.13 (Eisenbud, Huneke, Ulrich [68]) Define the Rees algebra of an R–module M as $R(M) = \frac{\mathrm{Sym}\, M}{L}$, where $L = \cap_g \left(\mathrm{Sym}\, M \xrightarrow{\mathrm{Sym}\, g} \mathrm{Sym}\, F \right)$, as $g : M \to F$ varies over all R–module homomorphisms to free modules F. Prove that if $f : M \to G$ is versal, then $R(M) \cong R(f)$, where $R(f)$ is the image of $\mathrm{Sym}\, M$ in $\mathrm{Sym}\, G$.

16.14 ([68]) With the notation of Exercise 16.13, prove that if M is an ideal, then $R(M)$ is the usual Rees algebra of M.

16.15 ([68]) Let R be a ring, M an R–module, and $A \subseteq B$ submodules of M. Let A', B' be the images of A and B in the Rees algebra S of M, as defined in Exercise 16.14. Define B to be **integral** over A if the algebra extension $R[A'] \subseteq R[B']$ in S is an integral extension. Let k be a field, $R = k[x]/(x^2)$, and $A = (0) \subseteq B = M = (x)/(x^2)$. Prove that B is integral over A by the definition in this exercise, but not integral by the definition in 16.1.1.

16.16 ([68]) Let k be a field of characteristic $p > 0$, X, Y, Z variables over k, and $R = k[X, Y, Z]/((X^p, Y^p) + (X, Y, Z)^{p+1})$. Let $M = ZR$ and $f : M \to R^2$ the homomorphism $f(Z) = (X, Y)$.

  (i) Prove that $M \cong R/(X, Y, Z)^p$.

  (ii) Prove that f is injective.

  (iii) Let S be the Rees algebra of M defined by the embedding $M \subseteq R$. Prove that S is naturally isomorphic to the image of $\mathrm{Sym}_R(M) \to \mathrm{Sym}_R(R)$.

  (iv) Let S' be the image of $\mathrm{Sym}_R(M) \xrightarrow{\mathrm{Sym}\, f} \mathrm{Sym}_R(R^2)$. Prove that S' is not isomorphic to S.

  (Thus this possible definition of a Rees algebra of a module depends on the embedding of the module in a free module.)

16.17 ([68]) Let R be a Noetherian ring, let M be a finitely generated R–module, and let $g : M \to G$ be a homomorphism to a free R–module that induces an injection from the torsionless quotient of M. If R is torsion–free over $\mathbb{Z}$, prove that $R(M) \cong R(g)$.

16.18 (Gaffney [81] [82], [83], [84]; Gaffney and Kleiman [88] applied the definition of the integral closure of a module in the analytic setting to study the singularities). Let $\mathcal{X}, x$ be a complex analytic germ set, n a positive integer, and M a submodule of $\mathcal{O}_{\mathcal{X},x}^n$. Let $f \in \mathcal{O}_{\mathcal{X},x}^n$. Prove that f is in the integral closure of M in $\mathcal{O}_{\mathcal{X},x}^n$ if and only if for all analytic $\varphi : (\mathbb{C}, 0) \to (\mathcal{X}, x)$, $h \circ \varphi \in (\varphi^* M)\mathcal{O}_1$.

16.19 (Amao [13], Rees [238]) Let R be a Noetherian ring, M a finitely generated R–module, and $J \subseteq I$ ideals in R such that IM/JM has finite length.

    (i) Prove that $I^n M/J^n M$ has finite length for all $n \geq 0$, and that there exists a polynomial p such that for n sufficiently large, $p(n) = \lambda(I^n M/J^n M)$. Also prove that $p(n)$ has degree at most $\dim R$, and the coefficient of the term of degree $\dim R$ is of the form $a(J, I, M)/(\dim R)!$, where $a(J, I, M) \in \mathbb{N}$.

    (ii) Assume that R is a local formally equidimensional ring. Prove that J is a reduction of I if and only if $p(n)$ has degree strictly smaller than $\dim R$.

    (iii) Assume that R is a local formally equidimensional ring and that $\mathrm{ht}(J : I) \geq \ell(J)$. Prove that J is a reduction of I if and only if either $\mathrm{ht}(J : I) > \ell(J)$ or $\mathrm{ht}(J : I) = \ell(J)$ and for all prime ideals P minimal over $J : I$, $a(J_P, I_P, M_P) = 0$.

17
Joint reductions

Joint reductions were first introduced by Rees in [236]. Rees used joint reductions to prove that mixed multiplicities of zero–dimensional ideals, as defined by Teissier and Risler [292], are the usual multiplicities of ideals generated by joint reductions. Rees's definitions and existence results were generalized by O'Carroll in [216]. Here we prove basic properties of joint reductions and mixed multiplicities, and prove various connections between them.

In Chapter 8 we proved the existence of reductions in Noetherian local rings in several ways: Theorem 8.3.6 proved it without the assumption that the residue field be infinite and relied on the descending chain condition in Artinian modules; Proposition 8.3.7 relied on Noether Normalization; Theorem 8.6.3 used superficial elements, and Theorem 8.6.6 used Zariski–open sets. In Section 17.3 we generalize such existence results to joint reductions. The Rees–Sally proof of the existence of joint reductions via standard independent sets of general elements is in Section 17.8. In Section 17.4, we define mixed multiplicities, a generalization of multiplicities. The results of that section are mostly due to Teissier and Rees; they connect mixed multiplicities and joint reductions. The result that in formally equidimensional rings mixed multiplicities determine joint reductions is in Section 17.6. Section 17.7 contains the Teissier–Rees–Sharp results on Minkowski inequalities for mixed multiplicities. We end Section 17.8 with a brief discussion of the core of ideals.

There are many topics related to joint reductions that we do not go into in this book. For "joint reduction numbers" (cf. reduction numbers, Definition 8.2.3) and for Cohen–Macaulayness of multi–ideal Rees algebras, an interested reader may wish to read [309], [311], [312]; [291]; [117]; [118]. A recent interpretation of mixed multiplicities of monomial ideals in terms of mixed volumes of polytopes is in [299].

17.1. Definition of joint reductions

In this chapter, we deal with many ideals at the same time. For notational convenience, if $I_1, \ldots, I_k$ are ideals and $n_1, \ldots, n_k \in \mathbb{N}$, we write $\underline{I}^{\underline{n}}$ for $I_1^{n_1} \cdots I_k^{n_k}$.

Definition 17.1.1 Let $I_1, \ldots, I_k$ be ideals in a ring R, and for each $i = 1, \ldots, k$, let $x_i \in I_i$. The k–tuple $(x_1, \ldots, x_k)$ is **a joint reduction** of the k–tuple $(I_1, \ldots, I_k)$ if the ideal $\sum_{i=1}^{k} x_i I_1 \cdots I_{i-1} I_{i+1} \cdots I_k$ is a reduction of $I_1 \cdots I_k$.

In case $I = I_1 = \cdots = I_k$, a k–tuple $(x_1, \ldots, x_k)$ being a joint reduction of $(I_1, \ldots, I_k)$ says that $(x_1, \ldots, x_k)I^{k-1}$ is a reduction of I^k, so that there exists an integer n such that $(x_1, \ldots, x_k)I^{k-1}I^{kn} = I^{k(n+1)}$. In other words, a k–tuple $(x_1, \ldots, x_k)$ is a joint reduction of $(I, \ldots, I)$ if and only if the ideal $(x_1, \ldots, x_k)$ is a reduction of I. Thus joint reductions generalize reductions.

Example 17.1.2 For any elements $x, y \in R$, and any positive integers m, n, the pair (x, y) is a joint reduction of the pair $((x, y^n)R, (x^m, y)R)$ of ideals, because $x(x^m, y) + y(x, y^n) = (x, y^n)(x^m, y)$.

It is convenient to define reductions and joint reductions more generally:

Definition 17.1.3 *Let R be a ring, M an R–module, and $I \subseteq J$ ideals in R. We say that $I \subseteq J$ is a **reduction with respect to M** if there exists an integer l such that $IJ^lM = J^{l+1}M$. Similarly, if $I_1, \ldots, I_k$ are ideals in a ring R, and for each $i = 1, \ldots, k$, $x_i \in I_i$, then the k–tuple $(x_1, \ldots, x_k)$ is called a **joint reduction** of the k–tuple $(I_1, \ldots, I_k)$ **with respect to M** if the ideal $\sum_{i=1}^k x_i I_1 \cdots I_{i-1}I_{i+1} \cdots I_k$ is a reduction of $I_1 \cdots I_k$ with respect to M.*

*If for each i, I_i is a product of ideals $J_{i1} \cdots J_{il_i}$, and $x_i = y_{i1} \cdots y_{il_i}$ with $y_{ij} \in J_{ij}$, then the joint reduction $(x_1, \ldots, x_k)$ of $(I_1, \ldots, I_k)$ is called a **complete reduction**.*

O'Carroll [219] gave a generalization of the Eakin–Sathaye theorem (cf. Theorem 8.6.8) by using complete reductions. Jayanthan and Verma [151] used complete reductions to study the Cohen–Macaulay property of bigraded Rees algebras. Another application of complete reductions is in Viêt in [313].

Lemma 17.1.4 *Let R be a Noetherian ring and M a finitely generated R–module. Then any (joint) reduction with respect to M is also a (joint) reduction with respect to $R/\text{ann } M$, and vice versa.*

Proof: Suppose that $I \subseteq J$ is a reduction with respect to M. Let l be such that $IJ^lM = J^{l+1}M$. By Lemma 2.1.8, $I \subseteq J$ is a reduction with respect to $R/\text{ann } M$. If I is a reduction of J with respect to $R/\text{ann } M$, then there exists $l \in \mathbb{N}$ such that $J^{l+1} \subseteq IJ^l + \text{ann } M$, so that $J^{l+1}M \subseteq IJ^lM$, whence equality holds and I is a reduction of J with respect to M. A similar proof works for joint reductions. $\qquad\square$

A useful inductive tool for joint reductions is the following easy lemma:

Lemma 17.1.5 *Let $(x_1, \ldots, x_k)$ be a joint reduction of $(I_1, \ldots, I_k)$ with respect to M. Then there exists an integer l such that $(x_2, \ldots, x_k)$ is a joint reduction of $(I_2, \ldots, I_k)$ with respect to $(I_1^{l+1}M + x_1M)/x_1M$.*

Proof: By assumption there exists an integer l such that $(I_1 \cdots I_k)^{l+1}M = (\sum_{i=1}^k x_i I_1 \cdots I_{i-1}I_{i+1} \cdots I_k)(I_1 \cdots I_k)^lM$. This l works. $\qquad\square$

An important role in the theory of reductions is played by the Artin–Rees Lemma. For joint reductions, we need a multi–ideal version:

Theorem 17.1.6 (Artin–Rees Lemma) *Let R be a Noetherian ring, $I_1, \ldots,$ I_k ideals, and $M, N \subseteq T$ finitely generated R–modules. Then there exist* *integers $c_1, \ldots, c_k$ such that for all $n_i \geq c_i$, $i = 1, \ldots, k$,*

$$\underline{I}^{\underline{n}}M \cap N = \underline{I}^{\underline{n}-\underline{c}}(\underline{I}^{\underline{c}}M \cap N).$$

Proof: (Sketch) Let $t_1, \ldots, t_k$ be variables over R, S the finitely generated R–algebra $R[I_1 t_1, \ldots, I_k t_k]$, the so–called **multi–Rees algebra** of $I_1, \ldots, I_k$, G the finitely generated S–module $\oplus_{\underline{n}} \underline{I}^{\underline{n}} \underline{t}^{\underline{n}} T = T[I_1 t_1, \ldots, I_k t_k]$, and $H = \oplus_{\underline{n}} (\underline{I}^{\underline{n}} M \cap N) \underline{t}^{\underline{n}}$ an S–submodule of G. By the Noetherian properties, H is a finitely generated S–module. Let c_i be the maximal t_i–degree of elements in a generating set of H. These $c_1, \ldots, c_k$ work (as in the proof of the usual Artin–Rees Lemma). $\square$

17.2. Superficial elements

In Section 8.5, we defined and proved existence of superficial elements in the context needed for reductions. Analogously, we present here the definitions and existence for superficial elements for the multi–ideal version.

Definition 17.2.1 *Let R be a Noetherian ring, M a finitely generated R–module, and $I_1, \ldots, I_k$ ideals in R. An element $x \in I_1$ is **superficial** for $I_1, \ldots, I_k$ with respect to M if there exists a non–negative integer c such that for all $n_1 \geq c$ and all $n_2, \ldots, n_k \geq 0$,*

$$(\underline{I}^{\underline{n}}M :_M x) \cap I_1^c I_2^{n_2} \cdots I_k^{n_k} M = I_1^{n_1-1} I_2^{n_2} \cdots I_k^{n_k} M.$$

*A sequence of elements $x_1, \ldots, x_l$, with $x_i \in I_i$, is a **a superficial sequence** for $I_1, \ldots, I_l$ with respect to M if for $i = 1, \ldots, l$, $x_i \in I_i$ is superficial with respect to $M/(x_1, \ldots, x_{i-1})M$ for the images of $I_i, \ldots, I_k$ in $R/(x_1, \ldots, x_{i-1})$.*

Proposition 17.2.2 (Existence of superficial elements) *If $(R, \mathfrak{m})$ is a Noetherian local ring with infinite residue field, then superficial sequences for $I_1, \ldots, I_k$ with respect to a finitely generated module M exist. Explicitly, there exist a non–empty Zariski–open subset U of $I_1/\mathfrak{m}I_1$ and $c \in \mathbb{N}$ such that for any $x \in I_1$ with image in U, for all $i \geq 1$, all $n_1 \geq c + i - 1$, and all $n_2, \ldots, n_k \geq 0$,*

$$(\underline{I}^{\underline{n}}M :_M x^i) \cap I_1^c I_2^{n_2} \cdots I_k^{n_k} M = I_1^{n_1-i} I_2^{n_2} \cdots I_k^{n_k} M.$$

If $I_1 = J_1 \cdots J_r$, we can even take x to be of the form $y_1 \cdots y_r$ with $y_j \in J_j$. Also, if I_1 is not contained in the prime ideals $P_1, \ldots, P_s$ of R, x can be chosen to avoid the same prime ideals.

Proof: The module $I_1^{n_1-i} I_2^{n_2} \cdots I_k^{n_k} M$ is always contained in $(\underline{I}^{\underline{n}}M :_M x^i) \cap I_1^c I_2^{n_2} \cdots I_k^{n_k} M$. It suffices to prove the other inclusion for some x and c. If I_1 is nilpotent, let c be such that $I_1^{c-1} = 0$. Then for all $n_1 \geq c + i - 1$, $I_1^{n_1-i} = 0 = I_1^c$, and the proposition follows for all $x \in I_1$.

Now assume that I_1 is not nilpotent. Then $I_1/I_1^2 \neq 0$. Let A be the finitely generated R–algebra $\oplus_{\underline{n} \geq 0}(\underline{I}^{\underline{n}}/I_1 \underline{I}^{\underline{n}})$. We can make A an $\mathbb{N}^k$–graded ring by setting the degree of $\underline{I}^{\underline{n}}/I_1\underline{I}^{\underline{n}}$ to be $\underline{n} \in \mathbb{N}^k$. For each positive integer c define $A_c = \oplus_{\underline{n}\geq\underline{0},n_1=c}(\underline{I}^{\underline{n}}/I_1\underline{I}^{\underline{n}})$.

Let $G = \oplus_{\underline{n}\geq\underline{0}}(\underline{I}^{\underline{n}}M/I_1\underline{I}^{\underline{n}}M)$, a finitely generated graded A–module. Let $Q_1,\ldots,Q_t$ be the associated prime ideals of G that do not contain A_1. For each $i = 1,\ldots,t$, let W_i be the image of $Q_i \cap I_1$ in $I_1/\mathfrak{m}I_1$. Also, for each $i = 1,\ldots,s$, let V_i be the image of $P_i \cap I_1$ in $I_1/\mathfrak{m}I_1$. By Nakayama's Lemma, each W_i and each V_i is a proper $(R/\mathfrak{m})$–vector subspace of $I_1/\mathfrak{m}I_1$. Let U be the complement of the union of all these W_i and V_i. As $R/\mathfrak{m}$ is an infinite field, U is a non–empty Zariski–open subset of $I_1/\mathfrak{m}I_1$. Let $x \in I_1$ such that its image lies in U.

Suppose that I_1 is of the form $J_1\cdots J_r$. If all elements $y_1\cdots y_r$, with $y_i \in J_i$, have the image in $I_1/\mathfrak{m}I_1$ lie in some V_i or W_i, then since $R/\mathfrak{m}$ is infinite, for each $y_2 \in J_2,\ldots,y_r \in J_r$, the image of $J_1y_2\cdots y_r$ is contained in some V_i or W_i. Similarly, for each $y_3 \in J_3,\ldots,y_r \in J_r$, the image of $J_1J_2y_3\cdots y_r$ is contained in some V_i or W_i, etc., so that $I_1 = J_1\cdots J_r$ has the image contained in some V_i or W_i, contradicting that each V_i and W_i was proper.

It now remains to prove that $(0 :_G x^i) \cap A_cG = (0_G)$. Let $(0_G) = \cap_j G_j$ be an irredundant primary decomposition of the zero submodule of G. Each $\sqrt{G_j :_A G}$ is an associated prime ideal of G. If x is not contained in $\sqrt{G_j :_A G}$, then $\sqrt{G_j :_A G}$ is one of the Q_l and $G_j :_G x^i = G_j$ for all i. If instead x is contained in $\sqrt{G_j :_A G}$, then by the choice of x (avoiding the images of $Q_1,\ldots,Q_t$), A_1 is contained in $\sqrt{G_j :_A G}$, so that there exists a positive integer c_j such that $A_{c_j} \subseteq G_j :_A G$. Let c be the maximum of all such c_j (or $c = 0$ if there are no such c_j). Then

$$(0 :_G x^i) \cap A_cG = \bigcap_{x\in\sqrt{G_j:_A G}} (G_j :_A x^i) \cap \bigcap_{x\notin\sqrt{G_j:_A G}} (G_j :_A x^i) \cap A_cG$$

$$\subseteq \bigcap_{x\notin\sqrt{G_j:_A G}} G_j \cap A_cG \subseteq \bigcap_j G_j = (0_G). \qquad \square$$

Remark 17.2.3 Note that the proof shows that whenever the image of x in $I_1/I_1^2 \subseteq A = \oplus_{\underline{n}\geq\underline{0}}(\underline{I}^{\underline{n}}/I_1\underline{I}^{\underline{n}})$ is not in any associated prime ideal of the A–module $\oplus_{\underline{n}\geq\underline{0}}(\underline{I}^{\underline{n}}M/I_1\underline{I}^{\underline{n}}M)$ that does not contain $(I_1/I_1^2)A$, then x is superficial for $I_1,\ldots,I_k$ with respect to M.

Lemma 17.2.4 *Let R be a Noetherian ring, $I_1,\ldots,I_k$ ideals in R, and M a finitely generated R–module. Let i be a positive integer and $x \in I_1$ be such that for some $c \in \mathbb{N}$ and all large integers $n_1,\ldots,n_k$,*

$$(\underline{I}^{\underline{n}}M :_M x^i) \cap I_1^c I_2^{n_2}\cdots I_k^{n_k} M = I_1^{n_1-i}I_2^{n_2}\cdots I_k^{n_k}M.$$

Assume that $I_1 \subseteq \sqrt{I_2\cdots I_k}$, and that $\cap_n I_1^n M = 0$. Then for all sufficiently large $n_1,\ldots,n_m$,

$$\underline{I}^{\underline{n}}M :_M x^i = (0 :_M x^i) + I_1^{n_1-i}I_2^{n_2}\cdots I_k^{n_k}M,$$

and $(0 :_M x^i) \cap I_1^{n_1}I_2^{n_2}\cdots I_k^{n_k}M = 0$.

The hypothesis on x holds for example if $i = 1$ and $x^i \in I_1^i$ is superficial for $I_1^i, I_2, \ldots, I_k$ with respect to M.

Proof: By the multi–ideal version of the Artin–Rees Lemma (17.1.6), there exist $e_1, \ldots, e_k \in \mathbb{N}$ such that for all $n_j \geq e_j$, $\underline{I}^{\underline{n}} M \cap x^i M \subseteq x^i \underline{I}^{\underline{n}-\underline{e}} M$. Thus $\underline{I}^{\underline{n}} M :_M x^i \subseteq (0 :_M x^i) + \underline{I}^{\underline{n}-\underline{e}} M$. By the radical assumption, there exists a positive integer m such that $I_1^m \subseteq I_1^c I_2^{e_2} \cdots I_k^{e_k}$. Thus for $n_1 \geq m + e_1$, $n_2 \geq e_2, \ldots, n_k \geq e_k$, $\underline{I}^{\underline{n}} M :_M x^i \subseteq (0 :_M x^i) + I_1^c I_2^{n_2} \cdots I_k^{n_k} M$. It follows that for large $n_1, \ldots, n_k$,

$$\underline{I}^{\underline{n}} M :_M x^i = (\underline{I}^{\underline{n}} M :_M x^i) \cap \big((0 :_M x^i) + I_1^c I_2^{n_2} \cdots I_k^{n_k} M\big)$$

$$= (0 :_M x^i) + (\underline{I}^{\underline{n}} M :_M x^i) \cap I_1^c I_2^{n_2} \cdots I_k^{n_k} M$$

$$= (0 :_M x^i) + I_1^{n_1 - i} I_2^{n_2} \cdots I_k^{n_k} M.$$

In particular, for all sufficiently large n_i,

$$(0 :_M x^i) \cap I_1^{n_1} I_2^{n_2} \cdots I_k^{n_k} M \subseteq \bigcap_{n_1 \gg 0} (\underline{I}^{\underline{n}} M :_M x^i) \cap I_1^c I_2^{n_2} \cdots I_k^{n_k} M$$

$$\subseteq \bigcap_{n_1 \gg 0} I_1^{n_1 - i} I_2^{n_2} \cdots I_k^{n_k} M \subseteq \bigcap_{n_1 \gg 0} I_1^{n_1 - i} M = 0. \qquad \square$$

17.3. Existence of joint reductions

In this section we prove the existence of joint reductions of, for example, d ideals in a Noetherian d–dimensional local ring via superficial elements: an element–by–element construction is proved in Theorem 17.3.1 and global existence is proved more generally via Zariski–open sets in Proposition 17.3.2. Another proof of existence of joint reductions is given in Section 17.8. The first theorem below also proves existence of complete reductions.

Theorem 17.3.1 (Existence of joint reductions) *Let $(R, \mathfrak{m})$ be a Noetherian local ring with infinite residue field, let d be the dimension of R, and let $I_1, \ldots, I_d$ be ideals in R. Assume that for all i, $I_i \subseteq \sqrt{I_{i+1} \cdots I_d}$. Then there exists a joint reduction $(x_1, \ldots, x_d)$ of $(I_1, \ldots, I_d)$. If any I_i is a product of n ideals, the element x_i may be taken to be a product of n elements, the jth factor of x_i being taken from the jth factor of I_i.*

Proof: If $d = 0$, there is nothing to prove. So assume that $d > 0$. Suppose that $(x_1, \ldots, x_d)$ is a joint reduction of $(I_1, \ldots, I_d)$ modulo $0 : (I_1 \cdots I_d)^l$ for some positive integer l. Then there exists an integer n such that

$$(I_1 \cdots I_d)^n \subseteq \sum_{i=1}^d x_i I_1^n \cdots I_{i-1}^n I_i^{n-1} I_{i+1}^n \cdots I_d^n + (0 : (I_1 \cdots I_d)^l).$$

Hence $(I_1 \cdots I_d)^{n+l} \subseteq \sum_{i=1}^d x_i I_1^{n+l} \cdots I_{i-1}^{n+l} I_i^{n+l-1} I_{i+1}^{n+l} \cdots I_d^{n+l}$, which says that $(x_1, \ldots, x_d)$ is a joint reduction of $(I_1, \ldots, I_d)$. Thus it suffices to prove the theorem by first passing to the ring $R/(0 : (I_1 \cdots I_d)^l)$ for some l. If l is large enough such that for all n, $0 : (I_1 \cdots I_d)^n \subseteq 0 : (I_1 \cdots I_d)^l$, then the annihilator of the image of $I_1 \cdots I_d$ in $R/(0 : (I_1 \cdots I_d)^l)$ is zero. Thus by passing to $R/(0 : (I_1 \cdots I_d)^l)$, we may assume that $I_1 \cdots I_d$ has a zero annihilator, which implies that no I_i is contained in any associated prime ideal.

Let $x_1 \in I_1$ be superficial for $I_1, \ldots, I_d$ and not contained in any associated prime ideal of R. Such an element x_1 exists by Proposition 17.2.2, and if $I_1 = J_1 \cdots J_n$, then x_1 can be taken to be $y_1 \cdots y_n$ with $y_i \in J_i$. As $\dim(R/(x_1)) = \dim R - 1$, by induction there exist elements $x_i \in I_i$, $i \geq 2$, such that the image of $(x_2, \ldots, x_d)$ modulo (x_1) is a joint reduction of $(I_2, \ldots, I_d)$ modulo (x_1). Thus there exists a positive integer n such that $(I_2 \cdots I_d)^n$ is contained in $\sum_{i=2}^{d} x_i I_2^n \cdots I_{i-1}^n I_i^{n-1} I_{i+1}^n \cdots I_d^n + (x_1)$. Multiplication by I_1^n produces

$$(I_1 \cdots I_d)^n \subseteq \sum_{i=2}^{d} x_i I_1^n I_2^n \cdots I_{i-1}^n I_i^{n-1} I_{i+1}^n \cdots I_d^n + (x_1) \cap (I_1 \cdots I_d)^n.$$

As x_1 is superficial for $I_1, \ldots, I_d$, by Lemma 17.2.4, for sufficiently large n,

$$(I_1 \cdots I_d)^n \subseteq \sum_{i=1}^{d} x_i I_1^n \cdots I_{i-1}^n I_i^{n-1} I_{i+1}^n \cdots I_d^n \subseteq (I_1 \cdots I_d)^n,$$

so that equality holds throughout. This proves the theorem. □

This theorem proves the existence of a joint reduction, the first element of which is chosen from a Zariski–open subset of a homomorphic image of $I_1/\mathfrak{m}I_1$, after which the second element is chosen from a homomorphic image of $I_2/\mathfrak{m}I_2$, etc. The following proposition shows that when joint reductions exist, one can take the joint reduction d–tuple $(x_1, \ldots, x_d)$ from a Zariski–open subset of $(I_1/\mathfrak{m}I_1) \oplus \cdots \oplus (I_d/\mathfrak{m}I_d)$. The analogous result for reductions was proved by Northcott and Rees, see Theorem 8.6.6.

Proposition 17.3.2 *Let $(R, \mathfrak{m}, k)$ be a Noetherian local ring of dimension d. For $i = 1, \ldots, d$, let $I_i = (a_{11}, \ldots, a_{1l_i})$ be ideals in R, and let x_i be an element of I_i. Write $x_i = \sum_j u_{ij} a_{ij}$ for some $u_{ij} \in R$. Set $l = \sum_i l_i$. Then there exists a (possibly empty) Zariski open set $U \subseteq k^l$ such that $(x_1, \ldots, x_d)$ is a joint reduction of $(I_1, \ldots, I_d)$ if and only if the image of $(u_{i1}, \ldots, u_{il_i}, \ldots, u_{d1} \ldots, u_{dl_d})$ in k^l lies in U.*

Proof: Let S be the polynomial ring over k in variables A_{ij}, where i varies from 1 to d, and for each i, j varies from 1 to l_i. Let $T = \oplus_{\underline{n}} (\underline{I}^n / \mathfrak{m}\underline{I}^n)$, where $\underline{n}$ varies over $\mathbb{N}^d$. Both S and T are finitely generated R–algebras and are $\mathbb{N}^d$–graded, with $\deg A_{ij} = (0, \ldots, 0, 1, 0, \ldots, 0)$, with 1 in the ith spot and 0 elsewhere. For $i = 1, \ldots, d$, let $J_i = (A_{i1}, \ldots, A_{il_i})S$ and $X_i = \sum_j \overline{u}_{ij} A_{ij}$, where $\overline{u}_{ij}$ is the image of u_{ij} in $R/\mathfrak{m}$. Set $J = J_1 \cdots J_d$.

Let $\varphi : S \to T$ be the natural graded ring homomorphism with $\varphi(A_{ij}) = a_{ij} + \mathfrak{m}I_i$. Let K be the kernel of φ. This is a homogeneous ideal in S.

Claim: $(x_1, \ldots, x_d)$ is a joint reduction of $(I_1, \ldots, I_d)$ if and only if for some integer l, $J^l \subseteq K + (X_1, \ldots, X_d)S$.

If $(x_1, \ldots, x_d)$ is a joint reduction of $(I_1, \ldots, I_d)$, then there exists an integer l such that

$$(I_1 \cdots I_d)^l = \sum_{i=1}^d x_i I_1^l \cdots I_{i-1}^l I_i^{l-1} I_{i+1}^l \cdots I_d^l.$$

Let M be a monomial in J^l of degree $(l, \ldots, l)$. For each i there exists an element $r_i \in I_1^l \cdots I_{i-1}^l I_i^{l-1} I_{i+1}^l \cdots I_d^l$ such that $\varphi(M) = \sum r_i x_i + \mathfrak{m}(I_1 \cdots I_d)^l$. Let $R_i \in J_1^l \cdots J_{i-1}^l J_i^{l-1} J_{i+1}^l \cdots J_d^l$ be a preimage of r_i under φ. Then $M - \sum R_i X_i \in K$, which proves that $J^l \subseteq K + (X_1, \ldots, X_d)$.

For the other inclusion in the claim, assume that $J^l \subseteq K + (X_1, \ldots, X_d)$. By multi–homogeneity,

$$J^l \subseteq K + \sum_i X_i J_1^l \cdots J_{i-1}^l J_i^{l-1} J_{i+1}^l \cdots J_d^l.$$

After applying φ this says that $(x_1, \ldots, x_d)$ is a joint reduction of $(I_1, \ldots, I_d)$, proving the claim.

Fix $n \in \mathbb{N}$. If $M_1, \ldots, M_{c_n}$ are the monomials (in the A_{ij}) of degree $(n, \ldots, n)$, then the k–vector space $V_n \subseteq S$ that they generate has dimension c_n. Let W_n be the vector subspace of V_n generated by elements of $(K + (X_1, \ldots, X_d)S) \cap J^n$, with basis $\{B_1, \ldots, B_{s_n}\}$. Write $B_i = \sum_j c_{ij} M_j$ for some elements $c_{ij} \in k$. The X_i and the c_{ij} depend linearly on the u_{ij}. For each u_{ij}, let U_{ij} be a variable over k. Lift the dependence of the c_{ij} on the u_{ij} into a polynomial L_{ij} of degree at most 1 in $k[U_{11}, \ldots, U_{dl_d}]$. Set C_n to be the matrix (L_{ij}), and set L to be the ideal of all c_n–minors of C_n, as n varies over positive integers (L could be the zero ideal). Define the following (possibly empty) Zariski–open set:

$$U = \{\underline{u} \in k^l \mid F(\underline{u}) \ne 0 \text{ for some } F \in L\}.$$

If $\underline{u} \in U$, there exist a positive integer n and a c_n–minor F of C_n such that $F(\underline{u}) \ne 0$. In particular, $s_n = c_n$, so $J^n \subseteq K + (X_1, \ldots, X_d)$. Thus by the claim the corresponding $(x_1, \ldots, x_d)$ is a joint reduction of $(I_1, \ldots, I_d)$. The other implication is proved in a similar way. $\square$

The last two results showcase the role Zariski–open sets play in the theory of joint reductions. More results of this type are in Section 17.5.

Here is a twist on the existence of joint reductions: given elements, we can construct ideals for which the given elements form a joint reduction:

Proposition 17.3.3 *Let $(R, \mathfrak{m})$ be a Noetherian ring. Let $(x_1, \ldots, x_k)$ and $I_1, \ldots, I_k$ be ideals with $x_i \in I_i$, $i = 1, \ldots, k$, and with $(x_1, \ldots, x_k)$ being $\mathfrak{m}$–primary. Then for all large n, $(x_1, \ldots, x_k)$ is a joint reduction of $((x_1) + \mathfrak{m}^n, \ldots, (x_k) + \mathfrak{m}^n)$.*

Proof: There exists n such that $\mathfrak{m}^n \subseteq (x_1, \ldots, x_k)$. Let $J = \sum_{i=1}^{k} x_i((x_1) + \mathfrak{m}^n) \cdots ((x_{i-1}) + \mathfrak{m}^n)((x_{i+1}) + \mathfrak{m}^n) \cdots ((x_k) + \mathfrak{m}^n)$. Then

$$((x_1) + \mathfrak{m}^n) \cdots ((x_k) + \mathfrak{m}^n) = J + \mathfrak{m}^{kn} \subseteq J + (x_1, \ldots, x_k)\mathfrak{m}^{n(k-1)}$$
$$\subseteq J \subseteq ((x_1) + \mathfrak{m}^n) \cdots ((x_k) + \mathfrak{m}^n),$$

so that equality holds throughout. $\qquad\square$

17.4. Mixed multiplicities

In analogy with Hilbert polynomials and multiplicities for an ideal with respect to a module we develop the multi–graded Hilbert polynomials and mixed multiplicities for several ideals with respect to a module. An important ingredient is the exact sequence

$$0 \to \frac{\underline{I}^n M :_M x}{I_1^{n_1-1} I_2^{n_2} \cdots I_k^{n_k} M} \to \frac{M}{I_1^{n_1-1} I_2^{n_2} \cdots I_k^{n_k} M} \xrightarrow{x} \frac{M}{\underline{I}^n M} \to \frac{M}{xM + \underline{I}^n M} \to 0,$$

where M is a finitely generated R–module, $I_1, \ldots, I_k$ are ideals in R, and $x \in I_1$. If $I_1, \ldots, I_k$ have finite co–length with respect to M, then the modules in the sequence have finite length, and

$$\lambda\left(\frac{M}{\underline{I}^n M}\right) - \lambda\left(\frac{M}{I_1^{n_1-1} I_2^{n_2} \cdots I_k^{n_k} M}\right)$$
$$= \lambda\left(\frac{M}{xM + \underline{I}^n M}\right) - \lambda\left(\frac{\underline{I}^n M :_M x}{I_1^{n_1-1} I_2^{n_2} \cdots I_k^{n_k} M}\right). \qquad (17.4.1)$$

Theorem 17.4.2 (The multi–graded Hilbert polynomial) *Let $(R, \mathfrak{m})$ be a Noetherian local ring, M a finitely generated R–module, and $I_1, \ldots, I_k$ $\mathfrak{m}$–primary ideals in R. There exists a polynomial $P(n_1, \ldots, n_k)$ in k variables with rational coefficients and of total degree $\dim M$ such that for all sufficiently large $n_1, \ldots, n_k$,*

$$P(n_1, \ldots, n_k) = \lambda_R(M/I_1^{n_1} \cdots I_k^{n_k} M).$$

We call P the **multi–graded Hilbert** polynomial of $I_1, \ldots, I_k$ with respect to M. This theorem clearly generalizes Theorem 11.1.3, where $k = 1$.

Proof: The case $k = 1$ is proved in Theorem 11.1.3, and here we assume that $k > 1$. Using Lemma 8.4.2 we may assume that the residue field of R is infinite. We prove the theorem by induction on $d = \dim(M)$.

If $\dim M = 0$, then for all large $n_1, \ldots, n_k$, $I_1^{n_1} \cdots I_k^{n_k} M = 0$, and so $P(n_1, \ldots, n_k)$ is the constant polynomial $\lambda(M)$. The degree of the polynomial is 0, which is equal to the dimension of M.

Assume that $\dim M > 0$. By Proposition 17.2.2, there exists $x \in I_1$ that is superficial for $I_1, \ldots, I_k$ with respect to M and not contained in any prime ideal that is minimal over $\operatorname{ann} M$. For $n_1, \ldots, n_k$ sufficiently large, by Lemma 17.2.4 and Equation (17.4.1),

$$\lambda\left(\frac{M}{\underline{I}^n M}\right) - \lambda\left(\frac{M}{I_1^{n_1-1} I_2^{n_2} \cdots I_k^{n_k} M}\right) = \lambda\left(\frac{M}{xM + \underline{I}^n M}\right) - \lambda\left(0 :_M x\right).$$

Since x is not in any minimal prime ideal over ann M, $\dim(M/xM) = \dim M - 1$, so that by induction on dimension there is a polynomial $Q(n_1, \ldots, n_k)$ with rational coefficients and of total degree $\dim M - 1$ such that $Q(n_1, \ldots, n_k) = \lambda(M/(xM + \underline{I}^{\underline{n}}M))$ for all sufficiently large $n_1, \ldots, n_k$. If $Q(\underline{n}) = \lambda(0 :_M x)$ for all sufficiently large $n_1, \ldots, n_k$, then for such $n_1, \ldots, n_k$, by the displayed equation, $\lambda(M/\underline{I}^{\underline{n}}M)$ is constant, so that $\underline{I}^{\underline{n}}M/I_1\underline{I}^{\underline{n}}M = 0$, whence by Nakayama's Lemma, $\underline{I}^{\underline{n}}M = 0$, which contradicts the assumption that $\dim M > 0$. So necessarily $Q(\underline{n}) - \lambda(0 :_M x)$ is a polynomial of degree $\dim M - 1$, for large integers $n_1, \ldots, n_k$. By the displayed formula, for large $n_1, \ldots, n_k$, $\lambda(M/\underline{I}^{\underline{n}}M)$ equals $\sum_{i=1}^{n_1} Q(i, n_2, \ldots, n_k)$ plus an integer constant independent of $n_1, \ldots, n_k$, so that by Lemma 11.1.2, for large $n_1, \ldots, n_k$, $\lambda(M/\underline{I}^{\underline{n}}M)$ equals a polynomial in $n_1, \ldots, n_k$ with rational coefficients, of degree $\dim M$.   □

As was the case for one ideal, also the leading coefficients of the Hilbert polynomials of many ideals carry much information, as we prove below.

Definition 17.4.3 *Let $(R, \mathfrak{m})$ be a Noetherian local ring, M a finitely generated R–module, and $I_1, \ldots, I_k$ $\mathfrak{m}$–primary. Let P be the multi–graded Hilbert polynomial of $I_1, \ldots, I_k$ with respect to M. Write the homogeneous part of P of degree $\dim R$ as*

$$\sum_{d_1 + \cdots + d_k = \dim R} \frac{1}{d_1! \cdots d_k!} e(I_1^{[d_1]}, \ldots, I_k^{[d_k]}; M)\, n_1^{d_1} \cdots n_k^{d_k},$$

*where $e(I_1^{[d_1]}, \ldots, I_k^{[d_k]}; M) \in \mathbb{Q}$. (The notation $I_i^{[d_i]}$ is unrelated to Frobenius powers in tight closure.) We call this number the **mixed multiplicity** of M of type $(d_1, \ldots, d_k)$ with respect to $I_1, \ldots, I_k$. If $d_i = 1$, we also write I_i in place of $I_i^{[d_i]}$. Sometimes we also write the symbol $e(I_1^{[d_1]}, \ldots, I_k^{[d_k]}; M)$ as $e(I_1, \ldots, I_1, \ldots, I_k, \ldots, I_k; M)$, where each I_i is listed d_i times.*

Just as in the definition of multiplicity (Definition 11.1.5), we could have replaced the dimension of R with the dimension of M in the definition of mixed multiplicity above. However, for our purposes in this book, it is convenient to have the mixed multiplicities be 0 when the dimension of M is strictly less than the dimension of R. This stipulation is not a serious restriction: an arbitrary R–module M is a module over $R/\operatorname{ann} M$, $\dim M = \dim(R/\operatorname{ann} M)$, and Hilbert polynomials and joint reductions of M as an R–module are the same as the Hilbert polynomials and joint reductions of M as a module over $R/\operatorname{ann} M$.

By Theorem 17.4.2, $e(I_1^{[d_1]}, \ldots, I_k^{[d_k]}; M)$ is a rational number. We prove below that it is a non–negative integer.

The case $k = 1$ is of course the usual multiplicity, cf. Definition 11.1.5. The study of multi–graded Hilbert polynomials for bihomogeneous ideals in a bigraded polynomial ring over an Artinian ring was initiated by van der Waerden [303]. Bhattacharya [15] followed the methods of van der Waerden and developed the Hilbert polynomial for a pair of zero–dimensional ideals in a Noetherian local ring, i.e., Bhattacharya proved the case $k = 2$ above.

Mixed multiplicities are additive on exact sequences, just like multiplicities:

Lemma 17.4.4 *Let $(R, \mathfrak{m})$ be a Noetherian local ring, $I_1, \ldots, I_k$ $\mathfrak{m}$–primary ideals, and $0 \to M_1 \to M_2 \to M_3 \to 0$ a short exact sequence of finitely generated R–modules. Then for any $d_1, \ldots, d_k \in \mathbb{N}$ with $d_1 + \cdots + d_k = \dim R$,*

$$e(I_1^{[d_1]}, \ldots, I_k^{[d_k]}; M_2) = e(I_1^{[d_1]}, \ldots, I_k^{[d_k]}; M_1) + e(I_1^{[d_1]}, \ldots, I_k^{[d_k]}; M_3).$$

Proof: By the multi–ideal version of the Artin–Rees Lemma (Theorem 17.1.6), there exist $c_1, \ldots, c_k \in \mathbb{N}$ such that for all $n_i \geq c_i$, $\underline{I}^n M_2 \cap M_1 = \underline{I}^{n-c}(\underline{I}^c M_2 \cap M_1) \subseteq \underline{I}^{n-c} M_1$. Thus

$$\lambda\left(\frac{M_1}{\underline{I}^n M_1}\right) \geq \lambda\left(\frac{M_1}{\underline{I}^n M_2 \cap M_1}\right) \geq \lambda\left(\frac{M_1}{\underline{I}^{n-c} M_1}\right).$$

It follows that $\lambda(M_1/(\underline{I}^n M_2 \cap M_1))$ is

$$\sum_{d_1 + \cdots + d_k = \dim R} \frac{1}{d_1! \cdots d_k!} e(I_1^{[d_1]}, \ldots, I_k^{[d_k]}; M_1) \, n_1^{d_1} \cdots n_k^{d_k} + Q',$$

where Q' is dominated by a polynomial of degree at most $\dim R - 1$. The short exact sequence

$$0 \longrightarrow \frac{M_1}{\underline{I}^n M_2 \cap M_1} \longrightarrow \frac{M_2}{\underline{I}^n M_2} \longrightarrow \frac{M_3}{\underline{I}^n M_3} \longrightarrow 0$$

and Theorem 17.4.2 show that for sufficiently large $n_1, \ldots, n_k$, Q' is a polynomial, necessarily of degree at most $\dim R - 1$. The comparison of the leading coefficients of the polynomials of lengths of the modules in the last short exact sequence now proves the lemma. $\qquad \square$

An important inductive consequence is the following:

Lemma 17.4.5 *Let $(R, \mathfrak{m})$ be a Noetherian local ring, M a finitely generated R–module of dimension $d = \dim R > 1$, and $(x_1, \ldots, x_d)$ and $I_1, \ldots, I_d$ $\mathfrak{m}$–primary ideals, where $x_i \in I_i$ for each i. Set $M' = M/x_1 M$, and for any positive integer l, set $N_l = (I_1^l M + x_1 M)/x_1 M$. Then $\dim N_l = \dim M'$.*

Set $R' = R/x_1 R$ or $R' = R/\operatorname{ann} M'$, and suppose that $\dim R' = d - 1 = \dim M'$. Then $e_{R'}(I_2 R', \ldots, I_d R'; M') = e_{R'}(I_2 R', \ldots, I_d R'; N_l)$ and $e_{R'}((x_2, \ldots, x_d) R'; M') = e_{R'}((x_2, \ldots, x_d) R'; N_l)$.

Proof: From the short exact sequence

$$0 \longrightarrow N_l = \frac{I_1^l M + x_1 M}{x_1 M} \to \frac{M}{x_1 M} \to \frac{M}{I_1^l M + x_1 M} \to 0,$$

as $M/(I_1^l M + x_1 M)$ has dimension zero, it follows that $\dim N_l = \dim M'$ and by additivity of (mixed) multiplicities on the short exact sequence, the proposition follows. $\qquad \square$

With this notation, Theorem 17.4.2 can be made more precise:

Theorem 17.4.6 (Risler and Teissier [292]) *Let $(R, \mathfrak{m})$ be a Noetherian local ring, $I_1, \ldots, I_k$ $\mathfrak{m}$–primary ideals, and M a finitely generated R–module of dimension $d = \dim R$. Fix $c \in \mathbb{N}$. Let $x \in I_1$ be in the complement of all the minimal prime ideals of R, and suppose that for all sufficiently large $n_1, \ldots, n_k$, $(\underline{I}^{\underline{n}} M :_M x) \cap I_1^c I_2^{n_2} \cdots I_k^{n_k} M = I_1^{n_1 - 1} I_2^{n_2} \cdots I_k^{n_k} M$ (this holds for example if $x \in I_1$ is superficial for $I_1^i, I_2, \ldots, I_k$ with respect to M). Set $R' = R/xR$, $M' = M/xM$. Then for any integers $d_1, \ldots, d_k \in \mathbb{N}$ with $d_1 + \cdots + d_k = d$ and $d_1 > 0$,*

$$
e_R(I_1^{[d_1]}, \ldots, I_k^{[d_k]}; M) = \begin{cases} e_{R'}(I_1^{[d_1 - 1]} R', \ldots, I_k^{[d_k]} R'; M') & \text{if } d > 1; \\[2mm] \lambda(M/xM) - \lambda(0 :_M x) & \text{if } d = 1. \end{cases}
$$

Proof: From Equality (17.4.1) and Lemma 17.2.4 we get that

$$
\lambda\left(\frac{M}{\underline{I}^{\underline{n}} M}\right) - \lambda\left(\frac{M}{I_1^{n_1 - 1} I_2^{n_2} \cdots I_k^{n_k} M}\right) = \lambda\left(\frac{M}{xM + \underline{I}^{\underline{n}} M}\right) - \lambda(0 :_M x)
$$

for large $n_1, \ldots, n_k$. Both sides can be expressed in terms of the multi–graded Hilbert polynomials for such $n_1, \ldots, n_k$. The highest degree part on the left–hand side is then the highest degree in

$$
\sum_{d_1 + \cdots + d_k = d} \frac{1}{d_1! \cdots d_k!} e_R(I_1^{[d_1]}, \ldots, I_k^{[d_k]}; M) \left(n_1^{d_1} - (n_1 - 1)^{d_1}\right) n_2^{d_2} \cdots n_k^{d_k},
$$

which is

$$
\sum_{d_1 + \cdots + d_k = d} \frac{1}{(d_1 - 1)! d_2! \cdots d_k!} e_R(I_1^{[d_1]}, \ldots, I_k^{[d_k]}; M) n_1^{d_1 - 1} n_2^{d_2} \cdots n_k^{d_k}.
$$

On the right–hand side the highest degree then equals, in case $d > 1$,

$$
\sum_{d_1 + \cdots + d_k = d - 1} \frac{1}{d_1! \cdots d_k!} e_{R'}(I_1^{[d_1]} R', \ldots, I_k^{[d_k]} R'; M') n_1^{d_1} \cdots n_k^{d_k},
$$

and if $d = 1$, the highest degree on the right is the constant

$$
e_{R'}(I_1^{[0]} R', \ldots, I_k^{[0]} R'; M') - \lambda(0 :_M x) = \lambda(M/xM) - \lambda(0 :_M x).
$$

With this, the proposition follows by reading off the leading coefficients. $\square$

An immediate corollary is that mixed multiplicities are positive integers in important situations:

Corollary 17.4.7 (Risler and Teissier [292]) *Let $(R, \mathfrak{m})$ be a Noetherian local ring, $I_1, \ldots, I_k$ $\mathfrak{m}$–primary ideals, and M a non–zero finitely generated R–module with $\dim M = \dim R = d$. Let $d_1, \ldots, d_k \in \mathbb{N}$ with $d_1 + \cdots + d_k = d$. Let $x_1, \ldots, x_d$ be any superficial sequence for $I_1, \ldots, I_1, \ldots, I_k, \ldots, I_k$ with* *respect to M, with each I_i listed d_i times, and each x_i not in any minimal prime ideal over $(x_1, \ldots, x_{i-1})$. Then*

$$e(I_1^{[d_1]}, \ldots, I_k^{[d_k]}; M) = \lambda\Big(\frac{M}{(x_1, \ldots, x_d)M}\Big) - \lambda((x_1, \ldots, x_{d-1})M :_M x_d),$$

which equals the multiplicity of the ideal $(x_1, \ldots, x_d)$ on M. In particular, mixed multiplicities are positive integers when $\dim M = \dim R$.

Proof: Set $R_i = R/(x_1, \ldots, x_i)$ and $M_i = M/(x_1, \ldots, x_i)M$. By the previous theorem,

$$\begin{aligned}
e_R(I_1^{[d_1]}, \ldots, I_k^{[d_k]}; M) &= e_{R_1}(I_1^{[d_1-1]}R_1, \ldots, I_k^{[d_k]}R_1; M_1) \\
&= \cdots \\
&= e_{R_{d-1}}(I_1^{[0]}R_{d-1}, \ldots, I_{k-1}^{[0]}R_{d-1}, I_k^{[1]}R_{d-1}; M_{d-1}) \\
&= e_{R_d}(I_1^{[0]}R_d, \ldots, I_k^{[0]}R_d; M_d) - \lambda((x_1, \ldots, x_{d-1})M : x_d) \\
&= \lambda\Big(\frac{M}{(x_1, \ldots, x_d)M}\Big) - \lambda((x_1, \ldots, x_{d-1})M : x_d).
\end{aligned}$$

This proves the displayed formula of the statement, and that mixed multiplicities are integers. If $d = 0$, this number is just $\lambda(M)$, a positive integer (by assumption M is non–zero). If $d > 0$, set $I = (x_1, \ldots, x_d)$. By repeated use of Proposition 11.1.9, $e(I; M) = \lambda(M/(x_1, \ldots, x_d)M) - \lambda((x_1, \ldots, x_{d-1})M : x_d)$, which proves the corollary. $\qquad\square$

This corollary says for example that the mixed multiplicity $e(I^{[d-j]}, \mathfrak{m}^{[j]}; M)$ is the I–multiplicity of M after intersecting M with j **sufficiently general** hyperplanes in $\mathfrak{m}$ (and passing to the corresponding ring).

An Associativity Formula also holds for mixed multiplicities. One can prove it with brute force by using prime filtrations of modules and the additivity of mixed multiplicities on short exact sequences, or as follows:

Theorem 17.4.8 (Associativity Formula) *Let $(R, \mathfrak{m})$ be a Noetherian local ring, M a finitely generated R–module of dimension $d = \dim R$, and $I_1, \ldots, I_k$ $\mathfrak{m}$–primary ideals. Then for any $d_1, \ldots, d_k \in \mathbb{N}$ with $d_1 + \cdots + d_k = d$,*

$$e_R(I_1^{[d_1]}, \ldots, I_k^{[d_k]}; M) = \sum_p \lambda(M_p) e_{R/p}(I_1^{[d_1]}(R/p), \ldots, I_k^{[d_k]}(R/p); R/p),$$

where p varies over $\mathrm{Min}(R/\mathrm{ann}\,M)$ for which $\dim(R/p) = d$.

Proof: By passing to $R[X]_{\mathfrak{m}R[X]}$, where X is a variable over R, neither the hypotheses nor the conclusion change, so we may assume that $R/\mathfrak{m}$ is infinite. By Proposition 17.2.2, there is a superficial sequence $x_1, \ldots, x_d$, for $I_1, \ldots, I_1, \ldots, I_k, \ldots, I_k$ with respect to M and with respect to each of the finitely many R/p, with each I_i listed d_i times, and each x_i chosen to not be in any minimal prime ideal over $(x_1, \ldots, x_{i-1})$. Then if $I = (x_1, \ldots, x_d)$, by Corollary 17.4.7, $e(I_1^{[d_1]}, \ldots, I_k^{[d_k]}; M) = e(I; M)$, and for each

p, $e(I_1^{[d_1]}, \ldots, I_k^{[d_k]}; R/p) = e(I; R/p)$. Now the proposition follows from the Associativity Formula for multiplicities, Theorem 11.2.4.   □

Theorem 17.4.6 and Corollary 17.4.7 show a connection between joint reductions and mixed multiplicities: when a joint reduction is built by a superficial sequence, the corresponding mixed multiplicity is given by the multiplicity of the ideal generated by the superficial sequence. We next prove Rees's result that the ideal generated by a joint reduction has the same multiplicity as the corresponding mixed multiplicity:

Theorem 17.4.9   (Rees [237]) *Let $(R, \mathfrak{m})$ be a Noetherian local ring, M a finitely generated R–module of dimension $d = \dim R$, and $I_1, \ldots, I_k$ $\mathfrak{m}$–primary ideals in R. Let $d_1, \ldots, d_k \in \mathbb{N}$ with $d_1 + \cdots + d_k = d$, and $(x_1, \ldots, x_d)$ a joint reduction of $(I_1, \ldots, I_1, \ldots, I_k, \ldots, I_k)$ with respect to M, with each I_i listed d_i times. Then*

$$e(I_1^{[d_1]}, \ldots, I_k^{[d_k]}; M) = e((x_1, \ldots, x_d); M).$$

Proof: Set $I = (x_1, \ldots, x_d)$. Because $(x_1, \ldots, x_d)$ is a joint reduction of $(I_1, \ldots, I_1, \ldots, I_k, \ldots, I_k)$, I is $\mathfrak{m}$–primary, and therefore $e(I; M)$ makes sense. As (mixed) multiplicities and joint reductions are preserved under passage to the faithfully flat extension $R[X]_{\mathfrak{m}R[X]}$, where X is a variable over R, without loss of generality we may assume that the residue field is infinite. Since $e(I_1^{[d_1]}, \ldots, I_k^{[d_k]}; M) = e(I_1, \ldots, I_1, \ldots, I_k, \ldots, I_k; M)$, we may simplify notation and set $k = d$ and all $d_i = 1$.

If $d = 0$, then $e(I_1, \ldots, I_d; M) = \lambda(M) = e((0); M)$, so the theorem holds.

Let $d = 1$. Then (x_1) being a joint reduction of (I_1) with respect to M says that there exists an integer l such that $x_1 I_1^l M = I_1^{l+1} M$. Thus for all $n > l$, $x_1^n M \subseteq x_1^{n-l} I_1^l M = I_1^n M$ and $\lambda(M/x_1^n M) \geq \lambda(M/I_1^n M) \geq \lambda(M/x_1^{n-l} M)$. Thus the Hilbert polynomials of I and (x_1) with respect to M must have the same leading coefficients, namely $e((x_1); M) = e(I_1; M)$.

Now let $d > 1$. If $(x_1, \ldots, x_d)$ is a joint reduction of $(I_1, \ldots, I_d)$ with respect to M, then it is so with respect to each R/P, as P varies over the minimal prime ideals in R that contain $\operatorname{ann} M$. Then by the Associativity Formulas for (mixed) multiplicities (Theorems 11.2.4 and 17.4.8), it suffices to prove the theorem in case $M = R = R/P$ is an integral domain.

Set $l = \sum_i \mu(I_i)$. With notation as in Proposition 17.3.2, let $U \subseteq (R/\mathfrak{m})^l$ be a Zariski–open subset that determines all the joint reductions of $(I_1, \ldots, I_d)$; by assumption, U is non–empty. Let U' be a non–empty Zariski–open subset of $I_1/\mathfrak{m}I_1$ such that any preimage in I_1 of any element of U' is a non–zero superficial element for $I_1, \ldots, I_d$. Such U' exists and is non–empty by Proposition 17.2.2. By Lemma 8.5.12, there exists $y \in U'$ such that $(y, x_2, \ldots, x_d) \in U$, i.e., such that $(y, x_2, \ldots, x_d)$ is a joint reduction of $(I_1, \ldots, I_d)$. Set $R' = R/yR$. By Theorem 17.4.6, $e_R(I_1, \ldots, I_d; R) = e_{R'}(I_2 R', \ldots, I_d R'; R')$, and by Proposition 11.1.9, $e((y, x_2, \ldots, x_d); R) = e_{R'}((x_2, \ldots, x_d)R'; R')$.

Since $(y, x_2, \ldots, x_d)$ is a joint reduction of $(I_1, \ldots, I_d)$, by Lemma 17.1.5, for all large l, $(x_2, \ldots, x_d)$ is a joint reduction of $(I_2, \ldots, I_d)$ with respect to $(I_1^l + yR)/yR$. Choose one such large l and set $N = (I_1^l + yR)/yR$. By induction on d then $e_{R'}(I_2R', \ldots, I_dR'; N) = e_{R'}((x_2, \ldots, x_d)R'; N)$. By Lemma 17.4.5,

$$
\begin{aligned}
e_R(I_1, \ldots, I_d; R) &= e_{R'}(I_2R', \ldots, I_dR'; R') \\
&= e_{R'}(I_2R', \ldots, I_dR'; N) \\
&= e_{R'}((x_2, \ldots, x_d)R'; N) \\
&= e_{R'}((x_2, \ldots, x_d)R'; R') \\
&= e_R((y, x_2, \ldots, x_d); R).
\end{aligned}
$$

It remains to prove that $e((y, x_2, \ldots, x_d); R) = e((x_1, \ldots, x_d); R)$. Set $R'' = R/(x_d)$, $N'' = (I_d^l + x_dR)/x_dR$. By Proposition 11.1.9 and by Lemma 17.4.5,

$$
\begin{aligned}
e_R((y, x_2, \ldots, x_d); R) &= e_{R''}((y, x_2, \ldots, x_{d-1})R''; R'') \\
&= e_{R''}((y, x_2, \ldots, x_{d-1})R''; N'').
\end{aligned}
$$

Similarly,

$$
\begin{aligned}
e_R((x_1, \ldots, x_d); R) &= e_{R''}((x_1, \ldots, x_{d-1})R''; R'') \\
&= e_{R''}((x_1, \ldots, x_{d-1})R''; N'').
\end{aligned}
$$

By Lemma 17.1.5, for l sufficiently large, $(x_1, \ldots, x_{d-1})$ is a joint reduction of $(I_1, \ldots, I_{d-1})$ with respect to N''. Thus by induction on d, using R'' and N'', y can in addition be chosen sufficiently general in I_1 such that $e_{R''}((y, x_2, \ldots, x_{d-1})R''; N'') = e_{R''}((x_1, \ldots, x_{d-1})R''; N'')$. This proves the theorem. $\qquad\square$

In the next two sections we prove more connections between joint reductions and mixed multiplicities: Section 17.5 gives background on manipulations of joint reductions and superficial elements, and the main result generalizing Rees's Theorem connecting multiplicities and reductions is in Section 17.6.

17.5. More manipulations of mixed multiplicities

In this section we exhibit some connections between joint reductions, superficial elements, and mixed multiplicities. Most results were inspired by Böger's techniques in [19]. The main goal of this section is to prepare the background for the results in the subsequent section, so a reader may wish to skip the section and only read the parts as needed later.

Proposition 17.5.1 *Let* $(R, \mathfrak{m})$ *be a Noetherian local ring of dimension* d, $I_1, \ldots, I_d$ $\mathfrak{m}$-*primary ideals and* M *a finitely generated* R-*module. Then for any positive integer* l,

$$
e(I_1, \ldots, I_{d-1}, I_d^l; M) = le(I_1, \ldots, I_d; M).
$$

Proof: Without loss of generality $d_k > 0$. By using Lemma 8.4.2 we may assume that the residue field of R is infinite. By Proposition 17.2.2 there exist elements $x_1, \ldots, x_d$, the jth element taken from I_j, that form a superficial sequence for $I_1, \ldots, I_d$. We can even assume that for all positive integers l, $x_d^l \in I_d^l$ is superficial for I_d^l with respect to $M/(x_1, \ldots, x_{d-1})M$. By Theorem 17.4.7 and by Proposition 11.2.9,

$$e(I_1, \ldots, I_{d-1}, I_d^l; M) = e((x_1, \ldots, x_{d-1}, x_d^l); M)$$
$$= le((x_1, \ldots, x_{d-1}, x_d^l); M)$$
$$= le(I_1, \ldots, I_d; M). \qquad \square$$

Lemma 17.5.2 *Let $(R, \mathfrak{m})$ be a Noetherian local ring with infinite residue field, and $I_1, \ldots, I_k$ ideals in R such that $I_1 \subseteq \sqrt{I_2 \cdots I_k}$. Let Y be a variable over R and $x \in I_1$. Then there exist positive integers c and e and a non–empty Zariski–open subset U of $I_1/\mathfrak{m}I_1$ with the following property: whenever $y \in I_1$ has a natural image in U, and whenever $l \geq e$, and $n_1, \ldots, n_k$ are sufficiently large (depending on l), then*

$$\underline{I^n}R[Y] \cap (x^l + y^l Y)R[Y] = (x^l + y^l Y)I_1^{n_1-l}I_2^{n_2} \cdots I_k^{n_k}R[Y],$$

and

$$(\underline{I^n}R[Y] :_{R[Y]} (x^l + y^l Y)) \cap I_1^c I_2^{n_2} \cdots I_k^{n_k}R[Y] = I_1^{n_1-l}I_2^{n_2} \cdots I_k^{n_k}R[Y].$$

Proof: Let $(0) = q_1 \cap \cdots \cap q_s$ be a primary decomposition of (0). Assume that $I_1 \subseteq \sqrt{q_1}, \ldots, \sqrt{q_t}$ and that $I_1 \not\subseteq \sqrt{q_{t+1}}, \ldots, \sqrt{q_s}$. Let $e \in \mathbb{N}$ be such that $I_1^e \subseteq q_1 \cap \cdots \cap q_t$. By Proposition 17.2.2, we may assume that the preimages of elements of U do not lie in $\sqrt{q_{t+1}}, \ldots, \sqrt{q_s}$, and furthermore that for each y in the preimage of U and for each $i \geq 1$, y^i is superficial for $I_1^i, I_2, \ldots, I_k$. Fix y in the preimage of U. By construction as in Proposition 17.2.2, there exists $c \in \mathbb{N}$ such that for all $n_1 \geq c + i - 1$ and all $n_2, \ldots, n_k \geq 0$,

$$(\underline{I^n} : y^i) \cap I_1^c I_2^{n_2} \cdots I_k^{n_k} = I_1^{n_1-i}I_2^{n_2} \cdots I_k^{n_k}.$$

Let $l \geq e$ and $r \in (0 : y^l)$. Then $ry^l \in q_{t+1} \cap \cdots \cap q_s$, so by the choice of y, $r \in q_{t+1} \cap \cdots \cap q_s$, whence $rI_1^l = 0$. This proves that $(0 : y^l) = (0 : I_1^l)$. By Lemma 17.2.4, $\underline{I^n} : y^l = (0 : y^l) + I_1^{n_1-l}I_2^{n_2} \cdots I_k^{n_k}$ for all sufficiently large n_i (depending on l). Let $F = (x^l + y^l Y) \sum_{j=0}^a r_j Y^j \in \underline{I^n}R[Y]$, with $r_j \in R$. The coefficient $y^l r_a$ of Y^{a+1} is in $\underline{I^n}$, so $r_a \in (0 : I_1^l) + I_1^{n_1-l}I_2^{n_2} \cdots I_k^{n_k}$. Hence $(x^l + y^l Y)r_a Y^a$ and $(x^l + y^l Y)\sum_{j=0}^{a-1} r_j Y^j$ are both in $\underline{I^n}R[Y]$, and so by induction on a, $F \in (x^l + y^l Y)I_1^{n_1-l}I_2^{n_2} \cdots I_k^{n_k}$.

If in F above in addition each $r_j \in I_1^c I_2^{n_2} \cdots I_k^{n_k}$, one similarly shows that $(\underline{I^n}R[Y] :_{R[Y]} (x^l + y^l Y)) \cap I_1^c I_2^{n_2} \cdots I_k^{n_k}R[Y] = I_1^{n_1-l}I_2^{n_2} \cdots I_k^{n_k}R[Y]. \qquad \square$

Lemma 17.5.3 *Let $(R, \mathfrak{m})$ be a Noetherian local ring, M a finitely generated R–module of dimension $d = \dim R$, and $I_1, \ldots, I_d$ $\mathfrak{m}$–primary ideals.*
(1) If $J_1, \ldots, J_d$ are $\mathfrak{m}$–primary ideals and for $i = 1, \ldots, d$, $J_i \subseteq I_i$, then

$$e(J_1, \ldots, J_d; M) \geq e(I_1, \ldots, I_d; M).$$

(2) If for $i = 1, \ldots, d$, $x_i \in I_i$ and $(x_1, \ldots, x_d)$ is $\mathfrak{m}$–primary, then
$$e((x_1, x_2, \ldots, x_d); M) \geq e(I_1, \ldots, I_d; M).$$

Proof: (1) follows as for all $\underline{n} \in \mathbb{N}^d$, $\lambda(M/\underline{J}^{\underline{n}}M) \geq \lambda(M/\underline{I}^{\underline{n}}M)$, so that the multi–graded Hilbert polynomial for $J_1, \ldots, J_d$ with respect to M dominates the multi–graded Hilbert polynomial for $I_1, \ldots, I_d$ with respect to M, and hence the same holds for the leading coefficients.

To prove (2), first apply Proposition 17.3.3 to construct $\mathfrak{m}$–primary ideals $J_1, \ldots, J_d$, with $x_i \in J_i \subseteq I_i$ for all i, and $(x_1, \ldots, x_d)$ being a joint reduction of $(J_1, \ldots, J_d)$. Then (2) follows from (1) and Theorem 17.4.9.   □

Lemma 17.5.4 *Let $(R, \mathfrak{m})$ be a Noetherian local ring with infinite residue field, let $I_1, \ldots, I_k$ be ideals, and $x_i \in I_i$ $(i = 1, \ldots, k)$. Let Y be a variable over R. Assume that the ideal $(x_1, \ldots, x_k)$ and all the I_i have the same height k and the same radical. Let P be a prime ideal minimal over $(x_1, \ldots, x_k)$ such that $e((x_1, \ldots, x_k)R_P; R_P) = e(I_1 R_P, \ldots, I_k R_P; R_P)$. Set $S = R[Y]_{PR[Y]}$. Then there exists a non–empty Zariski–open subset U of $I_1/\mathfrak{m}I_1$ (actually of $(I_1 + (x_k))/(\mathfrak{m}I_1 + (x_k))$, but that can be lifted to a non–empty Zariski–open subset U of $I_1/\mathfrak{m}I_1$), such that for any preimage y of an element of U and for all sufficiently large integers l, $e_S((x_1^l + y^l Y, x_2, \ldots, x_k)S; S) = l\, e_{R_P}((x_1, x_2, \ldots, x_k)R_P; R_P).$*

Proof: We use induction on k. If $k = 1$, choose y as in Lemma 17.5.2. Then for all sufficiently large integers l, $x_1^l + y^l Y$ is superficial for $I_1^l R[Y]$, hence also for $I_1^l S$. By Corollary 17.4.7, $e_S((x_1^l + y^l Y)S; S) = e_S(I_1^l S; S) = e_{R_P}(I_1^l R_P; R_P)$. By Proposition 11.2.9, this equals $l\, e(I_1 R_P; R_P)$, which proves the case $k = 1$.

Now let $k \geq 2$. For $q \in \text{Min}(R_P)$, set $A = R_P/q$. By Lemma 17.5.3 (2), if $\dim(A) = k$, then $e_A((x_1, \ldots, x_k)A; A) \geq e_A(I_1 A, \ldots, I_k A; A)$. By the Associativity Formulas for multiplicity and mixed multiplicity (Theorems 11.2.4 and 17.4.8),

$$e_{R_P}((x_1, \ldots, x_k)R_P; R_P) = \sum_{A = R_P/q} \lambda(R_q) e_A((x_1, \ldots, x_k)A; A)$$
$$\geq \sum_q \lambda(R_q) e_A(I_1 A, \ldots, I_k A; A)$$
$$= e_{R_P}(I_1 R_P, \ldots, I_k R_P; R_P).$$

But then all terms in the display have to be equal, so that for each $A = R_P/q$, $e_A((x_1, \ldots, x_k)A; A) = e_A(I_1 A, \ldots, I_k A; A)$. Thus the hypotheses of the lemma hold for each R/p in place of R, with p varying over those minimal prime ideals of R for which $\dim(R_P/pR_P) = k$. If the conclusion holds with R/p in place of R, then there exists a Zariski–open non–empty subset U_p of $I_1/\mathfrak{m}I_1$ such that the conclusion of the lemma holds for R/p in place of R. Then by the Associativity Formula for multiplicities, Theorem 11.2.4, the conclusion holds also in R for y a preimage of any element of the non–empty

Zariski–open subset $\cap_p U_p$ of $I_1/\mathfrak{m}I_1$. Thus it suffices to prove the lemma in the case where R_P is an integral domain.

In this case, x_k is a non–zerodivisor on R_P. Set $T = R_P/x_k R_P$. Then

$$e(I_1 R_P, \ldots, I_k R_P; R_P) = e_{R_P}((x_1, \ldots, x_k) R_P; R_P) \quad \text{(by assumption)}$$
$$= e_T((x_1, \ldots, x_{k-1}) T; T) \quad \text{(by Proposition 11.1.9)}$$
$$\geq e_T(I_1 T, \ldots, I_{k-1} T; T) \quad \text{(by Lemma 17.5.3)}$$
$$= e_T((y_1, \ldots, y_{k-1}) T; T)$$
$$\text{(by Corollary 17.4.7, for some } y_i \in I_i)$$
$$\geq e(I_1 R_P, \ldots, I_k R_P; R_P) \quad \text{(by Proposition 11.1.9)},$$

so that equality has to hold throughout. In particular, $e((x_1, \ldots, x_{k-1}) T; T) = e(I_1 T, \ldots, I_{k-1} T; T)$. By induction on k, there exists a non–empty Zariski–open subset U of $I_1/\mathfrak{m}I_1$ such that for any preimage y of an element of U,

$$e_{S/x_k S}((x_1^l + y^l Y, x_2, \ldots, x_{k-1})(S/x_k S); S/x_k S) = l\, e((x_1, x_2, \ldots, x_{k-1}) T; T)$$

for all large l. Another application of Proposition 11.1.9 finishes the proof:

$$e_S((x_1^l + y^l Y, x_2, \ldots, x_k) S; S)$$
$$= e_{S/x_k S}((x_1^l + y^l Y, x_2, \ldots, x_{k-1})(S/x_k S); S/x_k S)$$
$$= l\, e((x_1, x_2, \ldots, x_{k-1}) T; T)$$
$$= l\, e((x_1, x_2, \ldots, x_k) S; S). \qquad \square$$

Lemma 17.5.5 *Let $(R, \mathfrak{m})$ be a formally equidimensional Noetherian local ring with infinite residue field, Y a variable over R and S the localization of $R[Y]$ at $\mathfrak{m}R[Y] + YR[Y]$. Let $I_1, \ldots, I_k$ be ideals in R, with $x_i \in I_i$ for $i = 1, \ldots, k$. Assume that the ideal $(x_1, \ldots, x_k)$ and all the I_i have height k and have the same radical. Let Λ be the set of prime ideals in R minimal over $(x_1, \ldots, x_k)$. Assume that for all $P \in \Lambda$, $e((x_1, \ldots, x_k) R_P; R_P) = e(I_1 R_P, \ldots, I_k R_P; R_P)$. Let $y \in I_1$ be a superficial element for $I_1, \ldots, I_k$ that is not in any prime ideal minimal over $(x_2, \ldots, x_k)$. Then for all sufficiently large integers l, the set of prime ideals of S minimal over $(x_1^l + y^l Y, x_2, \ldots, x_k) S$ equals $\{PS \mid P \in \Lambda\}$.*

Proof: Let $J_l = (x_1^l + y^l Y, x_2, \ldots, x_k) S$. By the choice of y, the height of J_l is k. Elements of Λ clearly extend to prime ideals in S minimal over J_l. Suppose that there exists a prime ideal Q in S minimal over J_l that is not extended from a prime ideal in Λ. By the Krull's Height Theorem, Theorem B.2.1, $\text{ht}\, Q \leq k$. As J_l has height k, necessarily $\text{ht}\, Q = k$. By Lemma B.4.7, S is formally equidimensional, so that by Lemma B.4.2, $\dim(S/Q) = \dim S - k$. Similarly, for each $P \in \Lambda$, $\dim(S/PS) = \dim S - k$. By the Associativity Formula, Theorem 11.2.4, for all $n \geq 1$,

$$e\left(\frac{S}{((x_1^l + y^l Y)^n, x_2^n, \ldots, x_k^n) S}\right) \geq e\left(\frac{S}{Q}\right) \cdot \lambda\left(\frac{S_Q}{((x_1^l + y^l Y)^n, x_2^n, \ldots, x_k^n) S_Q}\right)$$
$$+ \sum_{P \in \Lambda} e\left(\frac{S}{PS}\right) \cdot \lambda\left(\frac{S_{PS}}{((x_1^l + y^l Y)^n, x_2^n, \ldots, x_k^n) S_{PS}}\right).$$

By Lech's Formula 11.2.10 it follows that

$$
\lim_{n \to \infty} \frac{1}{n^k} e\Big(\frac{S}{((x_1^l + y^l Y)^n, x_2^n, \ldots, x_k^n)S}\Big)
$$
$$
\geq e(S/Q) \cdot e(J_l S_Q; S_Q) + \sum_{P \in \Lambda} e(S/PS) \cdot e(J_l S_{PS}; S_{PS})
$$
$$
> \sum_{P \in \Lambda} e(S/PS) \cdot e(J_l S_{PS}; S_{PS})
$$
$$
= \sum_{P \in \Lambda} e(R/P) l \cdot e((x_1, \ldots, x_k) R_P; R_P),
$$

the last equality by Lemma 17.5.4. By Lemma 11.1.7 and by the Associativity Formula,

$$
e\Big(\frac{S}{((x_1^l + y^l Y)^n, x_2^n, \ldots, x_k^n)S}\Big) \leq e\Big(\frac{S}{((x_1^l + y^l Y)^n, x_2^n, \ldots, x_k^n, Y)S}\Big)
$$
$$
= e\Big(\frac{R}{(x_1^{ln}, x_2^n, \ldots, x_k^n)R}\Big)
$$
$$
= \sum_{P \in \Lambda} e(R/P) \lambda\Big(\frac{R_P}{(x_1^{ln}, x_2^n, \ldots, x_k^n)R_P}\Big),
$$

so that by Lech's Formula and by Proposition 11.2.9,

$$
\lim_{n \to \infty} \frac{1}{n^k} e\Big(\frac{S}{((x_1^l + y^l Y)^n, x_2^n, \ldots, x_k^n)S}\Big) \leq l\, e(R/P) e((x_1, \ldots, x_k) R_P; R_P),
$$

contradicting the earlier inequality. Thus no such Q exists.   □

17.6. Converse of Rees's multiplicity theorem

Rees showed (see Theorem 17.4.9) that the ideal generated by a joint reduction has the same multiplicity as the corresponding mixed multiplicity, and Böger extended this result to non-$\mathfrak{m}$–primary ideals. The converse holds for formally equidimensional rings, and this we prove next. This converse generalizes Theorem 11.3.1 and Corollary 11.3.2 for multiplicities.

Theorem 17.6.1 (Swanson [279]) *Let $(R, \mathfrak{m})$ be a formally equidimensional Noetherian local ring, $I_1, \ldots, I_k$ ideals in R, and $x_i \in I_i$, $i = 1, \ldots, k$. Assume that the ideal $(x_1, \ldots, x_k)$ and the I_i have the same height k and the same radical. If $e((x_1, \ldots, x_k) R_P; R_P) = e(I_1 R_P, \ldots, I_k R_P; R_P)$ for all prime ideals P minimal over $(x_1, \ldots, x_k)$, then $(x_1, \ldots, x_k)$ is a joint reduction of $(I_1, \ldots, I_k)$.*

Proof: Let X be a variable over R, and $T = R[X]_{\mathfrak{m}R[X]}$. By work in Section 8.4, $R \to T$ is a faithfully flat extension, and radicals, heights, minimal prime ideals, multiplicities, and mixed multiplicities are preserved under passage to T, and some k–tuple of elements is a joint reduction of a k–tuple of ideals in R if and only if it is so after passage to T. Furthermore, by Lemma B.4.7, T is still formally equidimensional, so that by possibly switching to T we may assume that R has an infinite residue field.

Let Λ be the set of all prime ideals in R that are minimal over $(x_1, \ldots, x_k)$. Then Λ is a finite set, and by the Krull's Height Theorem, each prime ideal in Λ has height k. By Theorem B.5.2, each R_P is formally equidimensional.

If $k = 0$, there is nothing to prove. If $k = 1$, by assumption $e((x_1); R_P) = e(I_1 R_P; R_P) = e(I_1; R_P)$. As R_P is formally equidimensional, by Rees's Theorem 11.3.1, $(x_1)R_P \subseteq I_1 R_P$ is a reduction. Thus $I_1 \subseteq \cap_P \overline{(x_1)} R_P \cap R$, and by Ratliff's Theorem 5.4.1, $I_1 \subseteq \overline{(x_1)}$, so that $(x_1) \subseteq I_1$ is a reduction.

Now let $k > 1$. We first reduce to the case where R is a domain. Let $p \in \operatorname{Min} R$ and let $Q \in \operatorname{Spec} R$ be minimal over $p + (x_1, \ldots, x_k)$. Then since R is formally equidimensional and hence equidimensional and catenary, $\operatorname{ht} Q = \operatorname{ht}(Q/p) \leq k$, so necessarily $\operatorname{ht} Q = k$ and $Q \in \Lambda$. By Lemma 17.5.3,

$$e((x_1, \ldots, x_k)(R/p)_Q; (R/p)_Q) \leq e(I_1(R/p)_Q, \ldots, I_k(R/p)_Q; (R/p)_Q).$$

Then by the Associativity Formulas for multiplicities and mixed multiplicities (Theorems 11.2.4 and 17.4.8),

$$
\begin{aligned}
e((x_1, \ldots, x_k)R_Q; R_Q) &= \sum_{q \in \operatorname{Min} R, q \subseteq Q} \lambda(R_q)\, e((x_1, \ldots, x_k)(R/q)_Q; (R/q)_Q) \\
&\leq \sum_{q \in \operatorname{Min} R, q \subseteq Q} \lambda(R_q)\, e(I_1(R/q)_Q, \ldots, I_k(R/q)_Q; (R/q)_Q) \\
&= e(I_1(R/q)_Q, \ldots, I_k(R/q)_Q; (R/q)_Q) \\
&= e((x_1, \ldots, x_k)R_Q; R_Q),
\end{aligned}
$$

so that $e((x_1, \ldots, x_k)(R/q)_Q; (R/p)_Q) = e(I_1(R/p)_Q, \ldots, I_k(R/p)_Q; (R/p)_Q)$ for each $p \in \operatorname{Min} R$ and each $Q \in \Lambda$ such that $p \subseteq Q$. If we know the result for integral domains, since $\operatorname{ht}(Q/p) = \operatorname{ht} Q = k$, then $(x_1, \ldots, x_k)$ is a joint reduction of $(I_1, \ldots, I_k)$ with respect to R/p for each $p \in \operatorname{Min} R$. Then by Proposition 1.1.5, since the definition of joint reduction reduces to a reduction question, $(x_1, \ldots, x_k)$ is a joint reduction of $(I_1, \ldots, I_k)$. Thus it suffices to prove the theorem for integral domains.

Let $S = R[Y]_{\mathfrak{m}R[Y] + YR[Y]}$ and let y be as in the statements of Lemmas 17.5.2 and 17.5.4. Since both requirements are given by non-empty Zariski-open sets, y exists, and we may choose non-zero y. Thus $x_1^l + y^l Y$ is not zero for all l. Set $S' = S/(x_1^l + y^l Y)$, for some large integer l. By Lemma 17.5.4, if l is sufficiently large,

$$e((x_1^l + y^l Y, x_2, \ldots, x_k)S_{PS}; S_{PS}) = l\, e((x_1, x_2, \ldots, x_k)R_P; R_P)$$

for every $P \in \Lambda$. By Proposition 11.1.9,

$$e((x_1^l + y^l Y, x_2, \ldots, x_k)S_{PS}; S_{PS}) = e((x_2, \ldots, x_k)S'_{PS'}; S'_{PS'}).$$

By Lemma 17.5.2, there exists an integer c such that for all large $n_1, \ldots, n_k$,

$$(\underline{I^n}S :_S (x^l + y^l Y)) \cap I_1^c I_2^{n_2} \cdots I_k^{n_k} S = I_1^{n_1-l} I_2^{n_2} \cdots I_k^{n_k} S.$$

This in particular holds for all n_1 that are large multiples of l, and c replaced by a larger integer that is a multiple of l. Thus by Theorem 17.4.6,

$$e_{S'_{PS'}}(I_2 S'_{PS'}, \ldots, I_k S'_{PS'}; S'_{PS'}) = e(I_1^l S_{PS}, I_2 S_{PS}, \ldots, I_k S_{PS}; S_{PS})$$

$$= e(I_1^l R_P, I_2 R_P, \ldots, I_k R_P; R_P)$$

$$= l\, e(I_1 R_P, I_2 R_P, \ldots, I_k R_P; R_P).$$

By assumption and the derived equalities, $e(I_2 S'_{PS'}, \ldots, I_k S'_{PS'}; S'_{PS'})$ equals $e((x_2, \ldots, x_k)S'_{PS'}; S'_{PS'})$ for all $P \in \Lambda$. By Lemma 17.5.5, all the minimal prime ideals over $(x_1^l + y^l Y, x_2, \ldots, x_k)S$ are of the form PS, with $P \in \Lambda$.

Set $J = x_2 I_3 \cdots I_k + \cdots + x_k I_2 \cdots I_{k-1}$. By Lemma B.4.7, S is locally formally equidimensional, and by Proposition B.4.4, S' is formally equidimensional. By induction on k, $(x_2, \ldots, x_k)$ is joint reduction of $(I_2, \ldots, I_k)$ with respect to S', so JS' is a reduction of $I_2 \cdots I_k S'$. Thus for sufficiently large n, $(I_2 \cdots I_k)^{n+1}S' \subseteq J(I_2 \cdots I_k)^n S'$. Hence $(I_2 \cdots I_k)^{n+1}S$ is contained in $J(I_2 \cdots I_k)^n S + (x_1^l + y^l Y)S$. By the choice of y as in Lemma 17.5.2, for possibly larger n, if $J' = JI_1$,

$$(I_1 \cdots I_k)^{n+1}S \subseteq J'(I_1 I_2 \cdots I_k)^n S + (x_1^l + y^l Y)I_1^{n+1-l}(I_2 \cdots I_k)^{n+1}S.$$

Thus there exists $s \in R[Y] \setminus (\mathfrak{m}R[Y] + YR[Y])$ such that

$$s(I_1 \cdots I_k)^{n+1} \subseteq J'(I_1 I_2 \cdots I_k)^n R[Y] + (x_1^l + y^l Y)I_1^{n+1-l}(I_2 \cdots I_k)^{n+1}R[Y].$$

But the constant term u of s is a unit in R, so that by reading off the degree zero monomials in $R[Y]$ we get

$$(I_1 \cdots I_k)^{n+1} \subseteq J'(I_1 I_2 \cdots I_k)^n + x_1^l I_1^{n+1-l}(I_2 \cdots I_k)^{n+1}$$

$$\subseteq J'(I_1 I_2 \cdots I_k)^n + x_1 I_1^n (I_2 \cdots I_k)^{n+1},$$

which proves that $(x_1, \ldots, x_k)$ is a joint reduction of $(I_1, \ldots, I_k)$. $\qquad\square$

17.7. Minkowski inequality

In [292], Teissier conjectured a Minkowski–type inequality for mixed multiplicities:

$$e(IJ; M)^{1/d} \leq e(I; M)^{1/d} + e(J; M)^{1/d},$$

where R is a Noetherian local ring with maximal ideal $\mathfrak{m}$, I and J are $\mathfrak{m}$–primary ideals and M is a finitely generated R–module of dimension $d = \dim R$. In [293] Teissier proved the conjecture for rings R that are reduced Cohen–Macaulay and contain $\mathbb{Q}$. Rees and Sharp proved the conjectures in full generality in [241]. We present the proofs of Teissier, and Rees and

Sharp. In [294] Teissier also proved that if R is a Cohen–Macaulay normal complex analytic algebra, then equality holds above if and only if there exist positive integers a and b such that $\overline{I^a} = \overline{I^b}$. For a more general statement see Exercise 17.10.

Lemma 17.7.1 Let $(R, \mathfrak{m})$ be a two–dimensional Noetherian local ring, I and J $\mathfrak{m}$–primary ideals, and M a finitely generated R–module of dimension two. Then
$$e(I, J; M)^2 \le e(I; M)\, e(J; M).$$

Proof: By standard methods we may assume that R has an infinite residue field. There then exist $a, b \in I$ such that (a, b) is a reduction of I. For any positive integers r, s, n,

$$\lambda\Big(\frac{M}{I^{rn}J^{sn}M}\Big) \le \lambda\Big(\frac{M}{(a^{rn}, b^{rn})J^{sn}M}\Big)$$
$$= \lambda\Big(\frac{(a^{rn}, b^{rn})M}{(a^{rn}, b^{rn})J^{sn}M}\Big) + \lambda\Big(\frac{M}{(a^{rn}, b^{rn})M}\Big)$$
$$\le 2\,\lambda\Big(\frac{M}{J^{sn}M}\Big) + \lambda\Big(\frac{M}{(a^{rn}, b^{rn})M}\Big).$$

By multiplying through by $2!/n^2$ and taking the limit as $n \to \infty$, by definition of multiplicities and by Lech's Formula (Theorem 11.2.10), $e(I^r J^s; M) \le 2\,e(J^s; M) + 2\,e((a^r, b^r); M)$. By Proposition 8.1.5, (a^r, b^r) is a reduction of I^r, so that by Propositions 11.2.1 and 11.2.9,

$$e(I^r J^s; M) \le 2e(J^s; M) + 2e(I^r; M) = 2e(J; M)s^2 + 2e(I; M)r^2.$$

Furthermore,

$$e(I^r J^s; M) = \lim_{n \to \infty} \frac{2!}{n^2} \lambda\Big(\frac{M}{(I^r J^s)^n}M\Big)$$
$$= \lim_{n \to \infty} \frac{2!}{n^2} \Big(\frac{1}{2}e(I; M)(rn)^2 + e(I^{[1]}, J^{[1]}; M)(rn)(sn) + \frac{1}{2}e(J; M)(sn)^2\Big)$$
$$= e(I; M)r^2 + 2e(I, J; M)rs + e(J; M)s^2.$$

Thus $2\,e(I, J; M)rs \le e(I; M)r^2 + e(J; M)s^2$, for all positive integers r, s. In particular, the inequality holds for $r = e(I, J; M)$ and $s = e(I; M)$, which proves the lemma. $\square$

Theorem 17.7.2 Let $(R, \mathfrak{m})$ be a Noetherian local ring, M a finitely generated R–module of dimension $d = \dim R \ge 2$, and I and J $\mathfrak{m}$–primary ideals. Then for all $i = 1, \ldots, d-1$,
$$e(I^{[i]}, J^{[d-i]}; M)^2 \le e(I^{[i+1]}, J^{[d-i-1]}; M) \cdot e(I^{[i-1]}, J^{[d-i+1]}; M).$$

Proof: Without loss of generality we may assume that the residue field of R is infinite: this changes neither the hypotheses nor the conclusion.

First suppose that $d > 2$. By possibly switching i and $d - i$ we may assume that $i \geq 2$. By Theorem 17.4.6, there exists $x \in I$ such that $\dim(M/xM) = d - 1 = \dim R - 1$ and such that for all $j = 1, \ldots, d$, $e_R(I^{[j]}, J^{[d-j]}; M) = e_{R'}(I^{[j-1]}R', J^{[d-j]}R'; M')$, where $R' = R/xR$ and $M' = M/xM$. By induction on d,

$$
\begin{aligned}
e_R(I^{[i+1]}, &J^{[d-i-1]}; M) \cdot e_R(I^{[i-1]}, J^{[d-i+1]}; M) \\
&= e_{R'}(I^{[i]}R', J^{[d-i-1]}R'; M') \cdot e_{R'}(I^{[i-2]}R', J^{[d-i+1]}R'; M') \\
&\geq e_{R'}(I^{[i-1]}R', J^{[d-i]}R'; M')^2 \\
&= e_R(I^{[i]}, J^{[d-i]}; M)^2.
\end{aligned}
$$

Thus it suffices to prove the case $d = 2$. But then $i = 1$, and the conclusion follows from the lemma. $\qquad\square$

Corollary 17.7.3 (Minkowski inequality) *Let $(R, \mathfrak{m})$ be a Noetherian local ring, M a finitely generated R–module of dimension $d = \dim R \geq 1$, and I and J $\mathfrak{m}$–primary ideals. Then for all $i = 0, \ldots, d$,*
(1) $e(I^{[i]}, J^{[d-i]}; M)\, e(I^{[d-i]}, J^{[i]}; M) \leq e(I; M)\, e(J; M)$,
(2) $e(I^{[i]}, J^{[d-i]}; M)^d \leq e(I; M)^{d-i} e(J; M)^i$, and
(3) $e(IJ; M)^{1/d} \leq e(I; M)^{1/d} + e(J; M)^{1/d}$.

Proof: We switch to the compact notation $e_i = e(I^{[d-i]}, J^{[i]}; M)$. By Corollary 17.4.7, e_i is a positive integer for all i, and by Theorem 17.7.2, for all $i = 1, \ldots, d - 1$, $\frac{e_i}{e_{i-1}} \leq \frac{e_{i+1}}{e_i}$.

If $i = 0, d$, (1) and (2) hold trivially. If $d = 2$, (1) and (2) hold by Lemma 17.7.1. Now let $d > 2$ and $d > i > 0$.

To prove (1), by symmetry we may assume that $i \leq d/2$. By Theorem 17.7.2 and by induction on i,

$$
e_i e_{d-i} = \frac{e_i}{e_{i-1}} \frac{e_{d-i}}{e_{d-i+1}} e_{i-1} e_{d-i+1} \leq \frac{e_{d-i+1}}{e_{d-i}} \frac{e_{d-i}}{e_{d-i+1}} e_0 e_d = e_0 e_d,
$$

which proves (1). Furthermore, by Theorem 17.7.2,

$$
\left(\frac{e_i}{e_{i-1}}\right)^{d-i} \cdots \left(\frac{e_1}{e_0}\right)^{d-i} \leq \left(\frac{e_d}{e_{d-1}}\right)^i \cdots \left(\frac{e_{i+1}}{e_i}\right)^i,
$$

as there are $i \cdot (d - i)$ factors on each side. But the left side is e_i^{d-i}/e_0^{d-i}, and the right side is e_d^i/e_i^i, which proves (2).

From the definition of mixed multiplicities and (2),

$$
\begin{aligned}
e(IJ; M) = \sum_{i=0}^{d} \binom{d}{i} e_i &\leq \sum_{i=0}^{d} \binom{d}{i} e(I; M)^{(d-i)/d} e(J; M)^{i/d} \\
&= (e(I; M)^{1/d} + e(J; M)^{1/d})^d,
\end{aligned}
$$

which proves (3). $\qquad\square$

17.8. The Rees–Sally formulation and the core

We prove yet another form of the existence of joint reductions and superficial elements, after passing to generic extensions or to infinite residue fields.

Let $(R, \mathfrak{m})$ be a Noetherian local ring of dimension d. Let N be a sufficiently large integer, and $X_1, \ldots, X_N$ variables over R. By $R[X]$ we denote the ring $R[X_1, \ldots, X_N]$, and by $R(X)$ we denote the ring $R[X]$ localized at $\mathfrak{m}R[X]$. Then $R[X]$ and $R(X)$ are faithfully flat extensions of R and $R(X)$ is a d–dimensional Noetherian ring.

Let $I_1, \ldots, I_d$ be ideals in R. For each $i = 1, \ldots, d$, write $I_i = (a_{i1}, \ldots, a_{il_i})$.

Definition 17.8.1 (Rees and Sally [240]) *The **standard independent set of general elements** of $(I_1, \ldots, I_d)$ is a set $\{x_1, \ldots, x_d\} \subseteq R(X)$, where*

$$x_i = \sum_{j=m_i+1}^{m_i+l_i} a_{ij} X_{m_i+j}, \quad m_i = l_1 + \cdots + l_{i-1}.$$

Note that each x_i depends on distinct variables X_j.

Theorem 17.8.2 (Rees and Sally [240]) *With notation as above, $(x_1, \ldots, x_d)$ is a joint reduction of $(I_1 R(X), \ldots, I_d R(X))$.*

Proof: If $d = 0$, there is nothing to prove. So assume that $d > 0$. Choose a positive integer l such that for all n, $0 : (I_1 \cdots I_d)^n \subseteq 0 : (I_1 \cdots I_d)^l$. If $(x_1, \ldots, x_d)$ is a joint reduction of $(I_1 R(X), \ldots, I_d R(X))$ modulo the ideal $0 : (I_1 \cdots I_d)^l$ extended to $R(X)$, then there exists an integer n such that

$$(I_1 \cdots I_d)^n R(X) \subseteq \sum_{i=1}^{d} x_i I_1^n \cdots I_{i-1}^n I_i^{n-1} I_{i+1}^n \cdots I_d^n R(X) + (0 : (I_1 \cdots I_d)^l) R(X).$$

Hence $(I_1 \cdots I_d)^{n+l} R(X) \subseteq \sum_{i=1}^{d} x_i I_1^{n+l} \cdots I_{i-1}^{n+l} I_i^{n+l-1} I_{i+1}^{n+l} \cdots I_d^{n+l} R(X)$, or in other words, $(x_1, \ldots, x_d)$ is a joint reduction of $(I_1 R(X), \ldots, I_d R(X))$. Thus it suffices to prove the theorem in case $0 : (I_1 \cdots I_d)^l = 0$. This implies that each I_i contains a non–zerodivisor. Thus $\dim(R(X)/(x_1)) = \dim R(X) - 1 = \dim R - 1$.

Set $R' = R(X_1, \ldots, X_{m_1})/x_1 R(X_1, \ldots, X_{m_1})$. The images of $x_2, \ldots, x_d$ in $R'(X) = R(X)/x_1 R(X)$ form the standard independent set of general elements of $(I_2 R', \ldots, I_d R')$. Thus by induction on d, the image of $(x_2, \ldots, x_d)$ in $R'(X)$ is a joint reduction of $(I_2 R'(X), \ldots, I_d R'(X))$ and there exists a positive integer n such that

$$(I_2 \cdots I_d)^n R(X) \subseteq \sum_{i=2}^{d} x_i I_2^n \cdots I_{i-1}^n I_i^{n-1} I_{i+1}^n \cdots I_d^n R(X) + (x_1) R(X).$$

Multiplication by I_1^n shows that $(I_1 \cdots I_d)^n R(X)$ is contained in

$$\sum_{i=2}^{d} x_i I_1^n I_2^n \cdots I_{i-1}^n I_i^{n-1} I_{i+1}^n \cdots I_d^n R(X) + (x_1) \cap (I_1 \cdots I_d)^n R(X).$$

It suffices to prove that x_1 is a superficial for $I_1 R(X), \ldots, I_d R(X)$. Consider

$$A = \oplus_{\underline{n} \geq \underline{0}} (\underline{I}^{\underline{n}} / I_1 \underline{I}^{\underline{n}})$$
$$\subseteq A[X] = \oplus_{\underline{n} \geq \underline{0}} (\underline{I}^{\underline{n}} R[X] / I_1 \underline{I}^{\underline{n}} R[X])$$
$$\subseteq A(X) = \oplus_{\underline{n} \geq \underline{0}} (\underline{I}^{\underline{n}} R(X) / I_1 \underline{I}^{\underline{n}} R(X)).$$

All the associated primes of $A(X)$ are localizations of the associated primes of $A[X]$, which are extended from the associated primes of A. Thus the image of x_1 in $A[X]$ is not contained in any associated prime of $A[X]$ that does not contain I_1 / I_1^2. Thus by Remark 17.2.3, $x_1 \in I_1 R[X]$ is superficial for $I_1 R[X], \ldots, I_d R[X]$ and so $x_1 \in I_1 R(X)$ is superficial for $I_1 R(X), \ldots, I_d R(X)$. In particular, for all sufficiently large n, $(x_1) \cap (I_1 \cdots I_d)^n R(X)$ is contained in $x_1 I_1^{n-1} I_2^n \cdots I_d^n R(X)$, which proves the theorem. $\square$

The last part of the proof showed:

Lemma 17.8.3 *Any element of the standard independent set of general elements of $(I_1, \ldots, I_d)$ is a superficial element for $I_1 R(X), \ldots, I_d R(X)$.*

With Theorem 17.8.2 we get a new proof of the existence of joint reductions:

Theorem 17.8.4 *Let $(R, \mathfrak{m})$ be a Noetherian local ring with infinite residue field. Let d be the dimension of R. Then for any ideals $I_1, \ldots, I_d$ in R there exists a joint reduction of $(I_1, \ldots, I_d)$. In fact, almost any specialization of the X_i yields a joint reduction.*

Proof: Let $\{x_1, \ldots, x_d\} \subseteq R[X]$ be a standard independent set of general elements of $(I_1, \ldots, I_d)$. By Theorem 17.8.2, $(x_1, \ldots, x_d)$ is a joint reduction of $(I_1 R(X), \ldots, I_d R(X))$. Thus there exists a positive integer n such that

$$(I_1 \cdots I_d)^n R(X) \subseteq \sum_{i=1}^{d} x_i I_1^n \cdots I_{i-1}^n I_i^{n-1} I_{i+1}^n \cdots I_d^n R(X).$$

By definition of $R(X)$ there exists an element $f \in R[X] \setminus \mathfrak{m} R[X]$ such that

$$f \cdot (I_1 \cdots I_d)^n R[X] \subseteq \sum_{i=1}^{d} x_i I_1^n \cdots I_{i-1}^n I_i^{n-1} I_{i+1}^n \cdots I_d^n R[X].$$

As f is not in $\mathfrak{m} R[X]$ and $R/\mathfrak{m}$ is an infinite field, there exist $u_1, \ldots, u_N \in R$ such that f evaluated at $X_i \mapsto u_i$ is a unit in R. In fact, the "almost all" is taken to mean that the set of all $\underline{u} = (u_1, \ldots, u_N) \in R^N$ with $f(\underline{u}) \notin \mathfrak{m}$ satisfies the condition. Let $\varphi : R[X] \to R$ be the ring homomorphism defined by $\varphi(X_i) = u_i$. Set $a_i = \varphi(x_i)$. Under the image of φ we get that

$$(I_1 \cdots I_d)^n \subseteq \sum_{i=1}^{d} a_i I_1^n \cdots I_{i-1}^n I_i^{n-1} I_{i+1}^n \cdots I_d^n R \subseteq (I_1 \cdots I_d)^n,$$

which proves that $(a_1, \ldots, a_d)$ is a joint reduction of $(I_1, \ldots, I_d)$. $\square$

Lemma 17.8.5 *Let $(R, \mathfrak{m})$ be a Cohen–Macaulay local ring of dimension d and let $I_1, \ldots, I_d$ be $\mathfrak{m}$–primary ideals in R. Suppose that $(a_1, \ldots, a_d)$ is a joint reduction of $(I_1, \ldots, I_d)$. Let $\{x_1, \ldots, x_d\} \subseteq R[X]$ be a standard independent set of general elements of $(I_1, \ldots, I_d)$. Then for any $r = 0, \ldots, d$, $(a_1, \ldots, a_r, x_{r+1}, \ldots, x_d)$ is a joint reduction of $(I_1 R(X), \ldots, I_d R(X))$.*

Proof: If $r = d$, as $(a_1, \ldots, a_d)$ is a joint reduction of $(I_1, \ldots, I_d)$, then by extension to $R(X)$, $(a_1, \ldots, a_d)$ is a joint reduction of $(I_1 R(X), \ldots, I_d R(X))$. So we may assume that $r < d$. If $r = 0$, then the result follows from Theorem 17.8.2. So we may assume that $0 < r < d$.

By assumption there exists a positive integer l such that $(I_1 \cdots I_d)^l = \sum_i a_i I_1^l \cdots I_{i-1}^l I_i^{l-1} I_{i+1}^l \cdots I_d^l$. Let $K = \sum_{i<d} a_i I_1^l \cdots I_{i-1}^l I_i^{l-1} I_{i+1}^l \cdots I_{d-1}^l$. Then modulo K, $a_d I_d^{l-1}$ equals I_d^l. In particular, modulo K, by Theorem 17.8.2 the standard independent element x_d generates a reduction of $I_d R(X)$. Certainly $x_d I_d^{l-1} R(X) \subseteq (I_d^l + K) R(X)$. We want to prove that $x_d I_d^{l-1} R(X) + K R(X) = (I_d^l + K) R(X)$. Since $a_d I_d^{l-1} + K = I_d^l + K$, we get the following inequalities on vector space dimensions over $(R/\mathfrak{m})(X)$:

$$\dim\left(\frac{(I_d^l + K) R(X)}{(\mathfrak{m} I_d^l + K) R(X)} \right) = \dim\left(\frac{(a_d I_d^{l-1} + K) R(X)}{(\mathfrak{m} I_d^l + K) R(X)} \right)$$
$$\leq \dim\left(\frac{(x_d I_d^{l-1} + \mathfrak{m} I_d^l + K) R(X)}{(\mathfrak{m} I_d^l + K) R(X)} \right)$$
$$\leq \dim\left(\frac{(I_d^l + K) R(X)}{(\mathfrak{m} I_d^l + K) R(X)} \right),$$

so that equality holds throughout and by Nakayama's Lemma $x_d I_d^{l-1} R(X) + K R(X) = (I_d^l + K) R(X)$. This means that $(a_1, \ldots, a_{d-1}, x_d)$ is a joint reduction of $(I_1 R(X), \ldots, I_d R(X))$, proving the case $r = d - 1$. Then by Lemma 17.1.5, $(a_1, \ldots, a_{d-1})$ is a joint reduction of $(I_1 R(X), \ldots, I_{d-1} R(X))$ with respect to $(I_d^l + (x_d)) R(X) / (x_d) R(X)$ for some integer l, whence by Lemma 17.1.4, $(a_1, \ldots, a_{d-1})$ is a joint reduction of $(I_1 R(X), \ldots, I_{d-1} R(X))$ with respect to $R' = R(X)/x_d R(X)$. (We only need those variables X_i above that appear in x_d but by abuse of notation we still write $R(X)$.) Hence by induction on d, since $\{x_1, \ldots, x_{d-1}\}$ is a standard independent set of general elements of $(I_1 R', \ldots, I_{d-1} R')$ so that by induction $(a_1, \ldots, a_r, x_{r+1}, \ldots, x_{d-1})$ is a joint reduction of $(I_1 R', \ldots, I_d R')$. We write out an ideal equation in $R' = R(X)/x_d R(X)$ of what this means, then lift it to an ideal inclusion in $R(X)$, and then as in the proof of Theorem 17.8.2 we use the fact that x_d is superficial for the ideals to finish the proof of the lemma. $\square$

Theorem 17.8.6 *Let $(R, \mathfrak{m})$ be a Cohen–Macaulay local ring of dimension d and let $I_1, \ldots, I_d$ be $\mathfrak{m}$–primary ideals in R. Suppose that $(a_1, \ldots, a_d)$ is a joint reduction of $(I_1, \ldots, I_d)$. Let $\{x_1, \ldots, x_d\} \subseteq R[X]$ be a standard independent set of general elements of $(I_1, \ldots, I_d)$. Then $(x_1, \ldots, x_d) R(X) \cap R \subseteq (a_1, \ldots, a_d)$.*

Proof: We proceed by induction on d. First let $d = 1$. With the given generators $a_{11}, \ldots, a_{1l}$ of I_1, write $a_1 = \sum_i r_i a_{1i}$ for some $r_i \in R$. By Proposition 8.3.3, since $(a_1) \subseteq I_1$ is a reduction, there exists i such that $r_i \notin \mathfrak{m}$. Without loss of generality $r_1 \notin \mathfrak{m}$. By possibly rescaling a_1 without loss of generality $r_1 = 1$. By definition $x_1 = \sum_i a_{1i} X_i = (a_{11} + \sum_{i>1} r_i a_{1i}) X_1 + \sum_{i>1} a_{1i}(X_i - r_i X_1)$. For $i > 1$ set $X_i' = X_i - r_i X_1$ and then set $b = \sum_{i>1} a_{1i} X_i'$. It follows that $x_1 = a_1 X_1 + b$. As $(a_1) \subseteq I_1$ is a reduction, so is $a_1 R[X_2', \ldots, X_l'] \subseteq I_1 R[X_2', \ldots, X_l']$. In particular, b/a_1 is integral over $R[X_2', \ldots, X_l']$. Thus there exists a monic polynomial f in variable X_1 and coefficients in $R[X_2', \ldots, X_l']$ such that $f(-b/a_1) = 0$. In other words, $f \in R[X_1, X_2', \ldots, X_l'] = R[X]$ and is monic in X_1. By polynomial division after inverting a_1, $f \in (a_1 X_1 + b) R_{a_1}[X] = x_1 R_{a_1}[X]$. Since R is Cohen–Macaulay, so is $R[X]$, so every associated prime ideal of $x_1 R[X]$ is minimal. Thus f is in every minimal component of $x_1 R[X]$ whose prime ideal does not contain a_1. All other associated prime ideals of $x_1 R[X]$ contain a_1, hence I_1, hence $\mathfrak{m}$, so that by the minimality condition the only other associated prime ideal of $x_1 R[X]$ is $\mathfrak{m} R[X]$. Thus $f \cdot (x_1 R(X) \cap R)$ is contained in all the primary components of $x_1 R[X]$, hence $f \cdot (x_1 R(X) \cap R) \subseteq x_1 R[X] = x_1 R[X_1, X_2', \ldots, X_l']$. Since f is monic in X_1, by reading off the coefficient of this leading term we get that $x_1 R(X) \cap R \subseteq a_1 R$. This proves the case $d = 1$.

Now let $d > 1$. First pass to $R/(a_1)$: this is a $(d-1)$–dimensional Cohen–Macaulay ring. By Lemmas 17.1.5 and 17.1.4, $(a_2, \ldots, a_d)$ is a joint reduction of $(I_2, \ldots, I_d)$ with respect to $R/(a_1)$. By induction on d,

$$(x_2, \ldots, x_d)((R/(a_1))(X)) \cap (R/(a_1)) \subseteq (a_2, \ldots, a_d)(R/(a_1)).$$

In other words, $(a_1, x_2, \ldots, x_d) R(X) \cap R \subseteq (a_1, a_2, \ldots, a_d)$. By the previous lemma, $(a_1, x_2, \ldots, x_d)$ is a joint reduction of $(I_1 R(X), \ldots, I_d R(X))$. We may pass modulo $(x_2, \ldots, x_d)$ as above modulo (a_1) to conclude similarly that $(x_1, \ldots, x_d) R(X) \cap R(\text{variables in } x_2, \ldots, x_d) \subseteq (a_1, x_2, \ldots, x_d) R(X)$. It follows that

$$\begin{aligned}
(x_1, \ldots, x_d) R(X) \cap R &\subseteq (x_1, \ldots, x_d) R(X) \cap R(\text{variables in } x_2, \ldots, x_d) \cap R \\
&\subseteq (a_1, x_2, \ldots, x_d) R(X) \cap R \\
&\subseteq (a_1, a_2, \ldots, a_d). \qquad \square
\end{aligned}$$

Rees and Sally used these constructions in their proof of the Briançon–Skoda Theorem:

Theorem 17.8.7 (Rees and Sally [240]) *Let $(R, \mathfrak{m})$ be a regular local ring of dimension d and let $I_1, \ldots, I_d$ be $\mathfrak{m}$–primary ideals. Then the integral closure of $I_1 \cdots I_d$ is contained in every joint reduction of $(I_1, \ldots, I_d)$. In particular, for any $\mathfrak{m}$–primary ideal I, $\overline{I^d}$ is contained in every reduction of I.*

Proof: Let $\{x_1, \ldots, x_d\}$ be a standard independent set of general elements of $(I_1, \ldots, I_d)$. Set $L = R(X)/(x_2, \ldots, x_d)$. A linear change of variables

$X_{l_1+1}, \ldots, X_{m_2}$ that is invertible over R does not change L, but it enables us to assume that for each $i = 1, \ldots, l_2$, the coefficient a_{2i} of X_{i+l_1} in x_2 is a non–zerodivisor on L. Similarly we may assume that each a_{ji} is a non–zerodivisor on L, with $j = 2, \ldots, d$. For each $j = 2, \ldots, d$, choose $i \in \{m_{j-1}+1, \ldots, m_j\}$, let c_j be the coefficient of X_i in x_j, let $y_j = x_j - c_j X_i$, and let Y denote all the variables other than these chosen ones. Then the kernel of the natural map $R[X] \to R[Y][-y_2/c_2, \ldots, -y_d/c_d]$ is $(x_2, \ldots, x_d)R_{c_2 \cdots c_d}[X] \cap R[X]$. By the assumption on the a_{ji} then the kernel is $(x_2, \ldots, x_d)R[X]$. Thus L is isomorphic to a localization of $R[Y][-y_2/c_2, \ldots, -y_d/c_d]$. But $R[Y]$ is a regular domain, hence locally analytically unramified, so that by Theorem 9.2.2, $R[Y][-y_2/c_2, \ldots, -y_d/c_d]$ is locally analytically unramified. In particular, L is analytically unramified. This means that the integral closure $\overline{L}$ of L is module–finite over L. Let C be the conductor of $L \subseteq \overline{L}$. By Exercise 12.2, C and $x_1 C$ are integrally closed ideals in L. Since $x_1 L \subseteq I_1 L$ is a reduction, it follows that $\overline{I_1 C L} \subseteq \overline{x_1 C} = x_1 C \subseteq x_1 L$. From the given presentation of L we deduce that $c_2 \cdots c_d$ is in C. As the possible choices of the c_i generate I_i and since L is independent of these choices, it follows that $I_2 \cdots I_d \subseteq C$. Thus the image of $\overline{I_1 \cdots I_d}$ in L is contained in $x_1 L$, whence $\overline{I_1 \cdots I_d} \subseteq (x_1, \ldots, x_d)R(X) \cap R$, whence the conclusion follows from Theorem 17.8.6. □

Other versions of the Briançon–Skoda Theorem are in Section 13.3. More connections between joint reductions and the Briançon–Skoda Theorems are in [278], [280].

The results in the previous section and in Chapter 13 showed that the integral closures of powers of an ideal are contained in lower ordinary powers of the ideal. In particular, Theorem 17.8.7 and Corollary 13.3.4 each shows that in a regular local ring R of dimension d, for any ideal I, $\overline{I^d} \subseteq I$. If J is any reduction of I, then as $\overline{I^d} = \overline{J^d}$, this result proves that $\overline{I^d} \subseteq J$. Thus the integral closure of the dth power of I is contained in every reduction of I.

Definition 17.8.8 (Rees and Sally [240]) *For an ideal I, the* **core** *of I, denoted* core(I), *is the intersection of all reductions of I.*

Thus in a regular local ring of dimension d, for any ideal I, core(I) contains the integral closure of I^d. (Also by Lipman's adjoint results in Section 18.2, at least under additional assumptions on $\mathrm{gr}_I(R)$, core(I) contains the adjoint of I^d.) This seems to indicate that the radicals of core(I) and I are the same. Indeed, this is true quite generally:

Proposition 17.8.9 (Böger [18]) *Let R be a Noetherian local ring and let I be an ideal. Then $\sqrt{I} = \sqrt{\mathrm{core}(I)}$.*

Proof: Certainly core(I) $\subseteq I$. So it suffices to prove that if $r \in I$, then a power of r is in core(I). By Corollary 8.6.7, there exists an upper bound N on all reduction numbers of reductions of I. Thus for any reduction J of I, $I^{N+1} = JI^N \subseteq J$, so that $I^{N+1} \subseteq \mathrm{core}(I)$. □

As a consequence, if R/I is Artinian, then with notation as in the proof of the proposition, R/I^{N+1} is Artinian, so that by the descending chain condition the possibly infinite intersection of all the reductions of I may be pruned to be a finite intersection. This finiteness holds also more generally, see Exercise 17.13.

The core of I is a fascinating ideal associated to I that has been the subject of much work in recent years. Some easy results are outlined in the exercises. There are many more results in the literature, see [45], [46], [47]; [141]; [142]; [144], [145]; [222]; [80]; and [223] for additional information.

17.9. Exercises

17.1   Let R be a Noetherian ring, $I_1, \ldots, I_k$ ideals in R, and $x_i \in I_i$ for $i = 1, \ldots, k$. Prove that the following are equivalent:
   (i) $(x_1, \ldots, x_k)$ is a joint reduction of $(I_1, \ldots, I_k)$.
   (ii) For all integers $n_1, \ldots, n_k > 0$, $(x_1^{n_1}, \ldots, x_k^{n_k})$ is a joint reduction of $(I_1^{n_1}, \ldots, I_k^{n_k})$.
   (iii) For some integers $n_1, \ldots, n_k > 0$, $(x_1^{n_1}, \ldots, x_k^{n_k})$ is a joint reduction of $(I_1^{n_1}, \ldots, I_k^{n_k})$.

17.2   Let $x \in I_1$ be superficial for $I_1, \ldots, I_k$ with respect to M. Prove conditions that x be a non–zerodivisor. (Cf. Lemmas 8.5.3, 8.5.4.)

17.3   (O'Carroll [216]) Let $(R, \mathfrak{m})$ be a Noetherian local ring with infinite residue field and of dimension d, and let $I_1, \ldots, I_d$ be ideals in R. Prove that there exists a reduction $(y_1, \ldots, y_d)$ of $I_1 \cdots I_d$ such that each y_i is a product of d elements, the jth factor being taken from I_j.

17.4   (Verma [311]) Let $(R, \mathfrak{m})$ be a two–dimensional regular local ring with infinite residue field. Let I and J be $\mathfrak{m}$–primary integrally closed ideals and (a, b) a joint reduction of (I, J). Prove that $aJ + bI = IJ$. (We say that I and J have joint reduction number zero.)

17.5   (Cf. Corollary 8.6.7.) Let $(R, \mathfrak{m})$ be a Noetherian local ring, and let $I_1, \ldots, I_d$ be ideals in R. Prove that there exists an integer n such that $(I_1 \cdots I_d)^{n+1} = \sum_{j=1}^d a_j (I_1 \cdots I_{j-1} I_{j+1} \cdots I_d)(I_1 \cdots I_d)^n$ for any joint reduction $(a_1, \ldots, a_d)$ of $(I_1, \ldots, I_d)$. (In other words, joint reduction numbers are bounded.)

17.6   Let $(R, \mathfrak{m})$ be a Noetherian local ring with infinite residue field k. Let M be a finitely generated R–module of positive dimension $d = \dim R$. Let $I_1, \ldots, I_d$ be $\mathfrak{m}$–primary ideals, and $(x_1, \ldots, x_d)$ a joint reduction of $(I_1, \ldots, I_d)$ with respect to M. Prove that there exists a non–empty Zariski–open subset U of $I_1/\mathfrak{m}I_1$ such that if $y \in I_1$ is in the preimage of an element in U, then $e((x_1, \ldots, x_d); M) = e((y, x_2, \ldots, x_d); M)$.

17.7   Let $(R, \mathfrak{m})$ be a Noetherian local ring, M a finitely generated R–module, and $I_1, \ldots, I_k$ $\mathfrak{m}$–primary ideals in R. For any non–negative integers $n_1, \ldots, n_k$, find a formula for $e(I_1^{n_1} \cdots I_k^{n_k}; M)$ in terms of mixed multiplicities of M with respect to $I_1, \ldots, I_k$.

17.8 Let $(R, \mathfrak{m})$ be a two–dimensional regular local ring and I and J $\mathfrak{m}$–primary ideals. Use the Hoskin–Deligne Formula 14.5.4 to show that $e(I, J; R) = \lambda(R/IJ) - \lambda(R/I) - \lambda(R/J)$.

17.9 Let $(R, \mathfrak{m})$ be a d–dimensional Noetherian local ring, and I and J projectively equivalent $\mathfrak{m}$–primary ideals (i.e., there exist positive integers r and s such that the integral closures of I^r and J^s equal). Prove that $e(IJ; R)^{1/d} = e(I; R)^{1/d} + e(J; R)^{1/d}$.

17.10* (Rees and Sharp [241] for dimension 2; Katz [159] in general) Let $(R, \mathfrak{m})$ be a formally equidimensional d–dimensional Noetherian local ring, and I and J $\mathfrak{m}$–primary ideals such that $e(I; R)^{1/d} + e(J; R)^{1/d} = e(IJ; R)^{1/d}$. Prove that I and J are projectively equivalent.

Exercises about the core

17.11 Prove that if $(R, \mathfrak{m})$ is a two–dimensional regular local ring, then $\operatorname{core}(\mathfrak{m}^n) = \mathfrak{m}^{2n-1}$.

17.12 Let $I \subseteq J$ be ideals in R. Prove or find a counterexample: $\operatorname{core}(I) \subseteq \operatorname{core}(J)$.*

17.13 (Corso, Polini, Ulrich [45]) Let $(R, \mathfrak{m})$ be a Noetherian local ring with infinite residue field and I an ideal in R. Assume that there exists a finite set $U \subseteq \operatorname{Spec} R$ such that for any minimal reduction J of I, the set of associated primes of R/J is contained in U. Prove that $\operatorname{core}(I)$ can be written as a finite intersection of minimal reductions of I.

17.14 ([45]) Let R be an $\mathbb{N}^d$–graded Noetherian ring with maximal homogeneous maximal ideal $\mathfrak{M}$. Let I be a homogeneous ideal in R. Prove that $\operatorname{core}(IR_{\mathfrak{M}})$ is generated by homogeneous elements of R.

17.15 ([45]) Let k be a field of characteristic 0, X, Y, Z variables over k, and $R = k[X, Y, Z]_{(X,Y,Z)}$. Let $I = (X^2 - Y^2 + XZ, XY + XZ - YZ, XZ - 2YZ + Z^2, Y^2 + YZ - Z^2, Z^2 - 2YZ)$. Verify that $\operatorname{core}(I) = \mathfrak{m}^4$.

17.16 ([141]) (Review the techniques of Chapters 14 and 18.) Let $(R, \mathfrak{m})$ be a two–dimensional regular local ring with an infinite residue field. Let I be an integrally closed $\mathfrak{m}$–primary ideal.

   (i) Prove that $I \operatorname{adj}(I) \subseteq \operatorname{core}(I)$.

   (ii) (This part takes a lot of time.) Let x_1 be part of a minimal reduction of I. Prove that the intersection of all reductions (x_1, x_2) of I (as x_2 varies) equals $(x_1) + I \operatorname{adj}(I)$.

   (iii) Prove that $I \operatorname{adj}(I) = \operatorname{core}(I)$.

   (iv) Prove that for all $n \geq 1$, $\operatorname{core}(I^n) = I^{2n-1} \operatorname{adj}(I)$.

   (v) Prove that $\operatorname{adj}(\operatorname{core}(I)) = (\operatorname{adj}(I))^2$.

   (vi) Define core^n recursively: $\operatorname{core}^1(I) = \operatorname{core}(I)$ and $\operatorname{core}^n(I) = \operatorname{core}^{n-1}(\operatorname{core}(I))$ for $n > 1$. Prove that for all $n \geq 1$, $\operatorname{core}^n(I)) = I(\operatorname{adj}(I))^{2^n-1}$.

---

   * As the book is going to press, Kyungyong Lee found a counterexample in a four–dimensional regular local ring.

18
Adjoints of ideals

In this chapter we present adjoint ideals for regular local rings. Lipman defined and used them in [190], proving a generalized version of the Briançon–Skoda Theorem, and extending Zariski's theory on two–dimensional regular local rings. Our goal in this chapter is to cover the basic properties of adjoint ideals, present the generalized Briançon–Skoda Theorem, present Howald's work on monomial ideals, (and more generally, adjoint ideals of ideals generated by monomials in an arbitrary regular system of parameters), and to present special results for two–dimensional regular local rings. In the last section, we develop mapping cones in greater generality, and then apply them to adjoint construction in two–dimensional regular local rings.

The notion of adjoints is closely related to the notion of multiplier ideals. We discuss multiplier ideals and their characteristic p variants in Section 18.7. Any two of these notions agree whenever they are both defined, however, adjoint ideals are defined in arbitrary characteristics.

18.1. Basic facts about adjoints

Let R be a regular domain with field of fractions K. Recall from Definition 9.3.1 that the set $D(R)$ of divisorial valuations with respect to R consists of all K–valuation rings V such that the maximal ideal $\mathfrak{m}_V$ of V contracts to the prime ideal P in R and $\operatorname{tr.deg}_{\kappa(P)}\kappa(\mathfrak{m}_V) = \operatorname{ht} P - 1$. We proved in Theorem 9.3.2 that every such valuation ring V is essentially of finite type over R and hence Noetherian. Thus we can write V as a localization of $R[X_1, \ldots, X_n]/(f_1, \ldots, f_m)$ for some variables $X_1, \ldots, X_n$ and $f_1, \ldots, f_m \in R[X_1, \ldots, X_n]$. We can even choose $f_1, \ldots, f_m$ such that they generate a prime ideal, so that by Definition 4.4.1, $J_{V/R}$ is defined.

Definition 18.1.1 (Lipman [190]) *Let R be a regular domain with field of fractions K. The **adjoint** of an ideal I in R, denoted $\operatorname{adj} I$, is the ideal*

$$\operatorname{adj} I = \bigcap_{V \in D(R)} \{r \in K \mid r J_{V/R} \subseteq IV\}.$$

We will prove that $\operatorname{adj} I$ is an ideal in R. Since R is Noetherian and integrally closed, it is a Krull domain and therefore it is the intersection of its localizations at all the height one prime ideals. These localizations are elements of $D(R)$ for which the Jacobian ideal over R is a unit ideal, so that

$$\operatorname{adj} I \subseteq \bigcap_{\operatorname{ht} P = 1} \{r \in K \mid r J_{R_P/R} \subseteq I R_P\}$$

$$\subseteq \bigcap_{\text{ht } P=1} \{r \in K \,|\, r \in IR_P\} \subseteq \bigcap_{\text{ht } P=1} R_P = R.$$

Hence we can also write:

$$\text{adj}\, I = \bigcap_{V \in D(R)} \{r \in R \,|\, rJ_{V/R} \subseteq IV\}.$$

Furthermore, only those V are needed in this intersection for which $IV \neq V$.

Lemma 18.1.2 *Let I be an ideal in a regular domain R. Then $I \subseteq \bar{I} \subseteq \text{adj}\, I = \text{adj}(\bar{I})$, and $\text{adj}\, I$ is an integrally closed ideal.*

Proof: By Proposition 10.4.3, since R is locally formally equidimensional, each Rees valuation ring of I is in $D(R)$. Thus by Proposition 6.8.2 and the definition of Rees valuations, $\bar{I} = \cap_{V \in D(R)}\{r \in R \,|\, r \in IV\}$. This is clearly contained in $\text{adj}\, I$. Every ideal in a valuation ring is integrally closed, hence so is the contraction $\{r \in R \,|\, rJ_{V/R} \subseteq IV\}$ to R. The intersection of integrally closed ideals is integrally closed, so adj is an integrally closed ideal. Furthermore, for all $V \in D(R)$, $IV = \bar{I}V$ by Proposition 6.8.1, so $\text{adj}\, I = \text{adj}(\bar{I})$. $\qquad\square$

We make a notational convention: for an arbitrary valuation ring V, as in Section 6.2, one can construct a corresponding valuation v_V, or v for short, that defines the valuation ring V. Any two such corresponding valuations are equivalent, but there is a unique one whose value group is $\mathbb{Z}$ (not just isomorphic to $\mathbb{Z}$). Let $\widetilde{D}(R)$ denote the set of all (normalized) v_V obtained in this way. Then the definition of adjoint can be rephrased as follows:

$$\text{adj}\, I = \bigcap_{v \in \widetilde{D}(R)} \{r \in R \,|\, v(r) \geq v(I) - v(J_{R_v/R})\},$$

where R_v stands for the valuation ring determined by v.

Lemma 18.1.3 *Let R be a regular domain, I an ideal and x an element in R. Then $\text{adj}(xI) = x \cdot \text{adj}(I)$. In particular, the adjoint of every principal ideal is the ideal itself.*

Proof: Let K be the field of fractions of R. Then

$$\text{adj}(xI) = \bigcap_{v \in \widetilde{D}(R)} \{r \in K \,|\, v(r) \geq v(xI) - v(J_{R_v/R})\}$$

$$= \bigcap_{v \in \widetilde{D}(R)} \left\{r \in K \,\Big|\, v\left(\frac{r}{x}\right) \geq v(I) - v(J_{R_v/R})\right\}$$

$$= \bigcap_{v \in \widetilde{D}(R)} \{rx \in K \,|\, v(r) \geq v(I) - v(J_{R_v/R})\}$$

$$= x \bigcap_{v \in \widetilde{D}(R)} \{r \in K \,|\, v(r) \geq v(I) - v(J_{R_v/R})\}$$

$$= x \cdot \text{adj}\, I. \qquad\square$$

This says that the adjoint of any ideal in a zero– or one–dimensional regular domain is itself. Thus in the rest of this chapter we only consider two– or higher dimensional regular domains.

In general, adjoints are not easily computable. See Sections 18.3, 18.4 and 18.5 for more on computability.

18.2. Adjoints and the Briançon–Skoda Theorem

Briançon–Skoda–type results say that for a given ideal I there exists an integer l such that for all $n \geq l$, some ideal operation of I^n is contained in I^{n-l}. See Section 13.3 for results of this form involving integral closure. In this section we prove several such results involving adjoint ideals. We also prove various other basic results on adjoints. The results in this section are from Lipman [190].

Proposition 18.2.1 *For any ideals I and J in a regular domain R, $\mathrm{adj}(IJ) : \bar{I} = \mathrm{adj}(IJ) : I = \mathrm{adj}\, J$.*

In particular, for all $n \geq 0$, $\mathrm{adj}(I^{n+1}) : I = \mathrm{adj}(I^n)$.

Proof: For any $r \in \mathrm{adj}\, J$ and any $v \in \widetilde{D}(R)$, $v(r\bar{I}) \geq v(IJ) - v(J_{R_v/R})$, so $r\bar{I} \in \mathrm{adj}(IJ)$.

Let $r \in R$. Then $r \in \mathrm{adj}(IJ) : I$ if and only if $rI \in \mathrm{adj}(IJ)$, which holds if and only if for all $v \in \widetilde{D}(R)$, $v(rI) \geq v(IJ) - v(J_{R_v/R})$. This in turn holds if and only if $v(r) \geq v(J) - v(J_{R_v/R})$, i.e., if and only if $r \in \mathrm{adj}\, J$. $\qquad\square$

Partial information on adjoints can be obtained via the following:

Proposition 18.2.2 *Let R be a regular domain, with field of fractions K, and let I be an ideal of R. Let S be a finitely generated R–algebra contained in K, and let $\bar{S}$ be the integral closure of S. Assume that $I\bar{S}$ is principal. Then $\mathrm{adj}\, I \subseteq (I\bar{S} :_R J_{\bar{S}/R})$.*

Proof: By Theorem 9.2.3, $\bar{S}$ is module–finite over S and hence finitely generated over R. By Proposition 4.10.3, $I\bar{S}$ has a primary decomposition $\cap_{i=1}^s (I\bar{S}_{P_i} \cap \bar{S})$, where $P_1, \ldots, P_s$ are the prime ideals in $\bar{S}$ minimal over $I\bar{S}$. As R is locally formally equidimensional, by Theorem B.4.8 it satisfies the dimension formula, so the $\bar{S}_{P_i}$ are divisorial valuation rings with respect to R. It follows that

$$\mathrm{adj}\, I \subseteq \bigcap_{i=1}^s \{ r \in R \mid r J_{\bar{S}_{P_i}/R} \subseteq I\bar{S}_{P_i} \}$$

$$= \bigcap_{i=1}^s \{ r \in R \mid r J_{\bar{S}/R} \subseteq I\bar{S}_{P_i} \}$$

$$= \{ r \in R \mid r J_{\bar{S}/R} \subseteq I\bar{S} \}$$

$$= (I\bar{S} :_R J_{\bar{S}/R}). \qquad\square$$

It is clear that $\operatorname{adj} I$ is the intersection of all $(I\overline{S} :_R J_{\overline{S}/R})$, as S varies over all finitely generated R–algebras contained in the field of fractions of R.

This construction is used in the proof below of the following Briançon–Skoda–type result:

Theorem 18.2.3 *Let I be an l–generated ideal in a regular domain R. Then*
(1) $\operatorname{adj}(I^{n+l-1}) \subseteq I^n$ for all sufficiently large n.
(2) If $\operatorname{gr}_I(R)$ contains a homogeneous non–zerodivisor of positive degree, then $\operatorname{adj}(I^{n+l-1}) \subseteq I^n$ for all $n \geq 0$.

Proof: Let $I = (a_1, \ldots, a_l)$. For each $i = 1, \ldots, l$, set $S_i = R[\frac{I}{a_i}]$. By Proposition 18.2.2, $\operatorname{adj}(I^{n+l-1}) \subseteq \cap_{i=1}^l (I^{n+l-1}\overline{S_i} :_R J_{\overline{S_i}/R})$. Clearly $(I^{n+l-1}\overline{S_i} :_R J_{\overline{S_i}/R}) \subseteq a_i^{n+1-l}(\overline{S_i} :_R J_{\overline{S_i}/R})$. By Theorem 9.2.3, $\overline{S_i}$ is module–finite over S_i. Thus by Theorem 12.3.10, $(\overline{S_i} :_K J_{\overline{S_i}/R}) \subseteq (S_i :_K J_{S_i/R})$. It follows that

$$\operatorname{adj}(I^{n+l-1}) \subseteq a_i^{n+l-1}(\overline{S_i} :_K J_{\overline{S_i}/R}) \subseteq a_i^{n+l-1}(S_i :_K J_{S_i/R}).$$

Observe that $J_{S_i/R}$ contains $a_i^{l-1}S_i = I^{l-1}S_i$, so that $\operatorname{adj}(I^{n+l-1})$ is contained in $a_i^{n+l-1}(S_i :_K a_i^{l-1}) = a_i^n S_i = I^n S_i$. Thus $\operatorname{adj}(I^{n+l-1}) \subseteq \cap_{i=1}^l I^n S_i \cap R = \cup_{t\geq 0}(I^{n+t} :_R I^t)$.

It is well–known that for large n, $\cup_{t\geq 0}(I^{n+t} :_R I^t) = I^n$, which proves (1).

> Here are the details: By Proposition 8.5.7, there exist positive integers m and c and an element $x \in I^m$ such that for all $n \geq m + c$, $(I^n : x) \cap I^c = I^{n-m}$. By the Artin–Rees Lemma, for sufficiently large n, $I^n : x \subseteq I^c$, so that for all sufficiently large n, $I^n : x = I^{n-m}$. It follows that for any t and all sufficiently large n,
> $$I^n \subseteq I^{n+t} : I^t \subseteq I^{n+mt} : I^{mt} \subseteq I^{n+mt} : x^t = I^n.$$

If $\operatorname{gr}_I(R)$ has a non–zerodivisor of positive degree, by Lemma 8.5.8, for all $t \in \mathbb{N}$, $I^{n+t} : I^t = I^n$, which proves (2). $\qquad\square$

Proposition 18.2.4 *Let R be a regular domain and I an ideal. Then the R–module $\oplus_{n\geq 0}\operatorname{adj}(I^n)$ is a finitely generated module over the rings $\oplus_{n\geq 0}I^n$ and $\oplus_{n\geq 0}\overline{I^n}$ (the Rees algebra and its normalization). In particular, there exists n_0 such that for all $n \geq n_0$,*

$$\operatorname{adj}(I^{n+1}) = I\operatorname{adj}(I^n).$$

Proof: The result is trivial if $I = 0$. So we assume that $I \neq 0$. By Proposition 18.2.1, $\oplus_{n\geq 0}\operatorname{adj}(I^n)$ is a module over $\oplus_{n\geq 0}\overline{I^n}$, and by Theorem 18.2.3, for some positive integer l, $I^l(\oplus_{n\geq 0}\operatorname{adj}(I^n))$ is contained in $\oplus_{n\geq 0}I^n = R[It]$. Thus if x is a non–zero element in I, $\oplus_{n\geq 0}\operatorname{adj}(I^n)$ is contained in the module $\frac{1}{x^l}R[It]$, which is finitely generated over $R[It]$. This proves the proposition. $\square$

However, $\oplus_{n\geq 0}\operatorname{adj}(I^n)$ need not be a ring: Proposition 18.3.3 below shows that for a maximal ideal $\mathfrak{m}$ in a regular domain, $\operatorname{adj}(\mathfrak{m}^n \cdot \mathfrak{m})$ is not $\operatorname{adj}(\mathfrak{m}^n) \cdot \operatorname{adj}(\mathfrak{m})$ in general.

18.3. Background for computation of adjoints

In this section we indicate a general principle on how one can compute adjoints in some cases. At the end we also prove that the Jacobian ideals of elements in $D(R)$ are not "too big".

Principle 18.3.1 (Rhodes [244]) Fix elements $x_1, \ldots, x_l$ in R. For each $i = 1, \ldots, l$, set $S_i = R[\frac{x_1}{x_i}, \ldots, \frac{x_l}{x_i}]$. By Lemma 6.4.3, for each $V \in D(R)$, there exists i such that $S_i \subseteq V$. By Lemma 9.3.3, $V \in D(S_i)$, and furthermore, $D(S_i) \subseteq D(R)$. If S_i is regular after localization at the center of V, then by Lemma 4.4.8, $J_{V/R} = J_{V/S_i} J_{S_i/R}$. If $x_1, \ldots, x_l$ form a regular sequence, then by Corollary 5.5.9, $J_{S_i/R} = x_i^{l-1} S_i$. If moreover each S_i is regular at all the centers of elements of $D(R)$ that contain S_i, say if $x_1, \ldots, x_l$ form part of a regular system of parameters, then for any ideal I,

$$
\operatorname{adj} I = \bigcap_{i=1}^{l} \bigcap_{v \in \widetilde{D}(S_i)} \{r \in K \mid v(r) \geq v(I) - v(J_{R_v/R})\}
$$

$$
= \bigcap_{i=1}^{l} \bigcap_{v \in \widetilde{D}(S_i)} \{r \in K \mid v(r) \geq v(I) - v(J_{R_v/S_i}) - v(J_{S_i/R})\}
$$

$$
= \bigcap_{i=1}^{l} \bigcap_{v \in \widetilde{D}(S_i)} \{r \in K \mid v(r) \geq v(I) - v(J_{R_v/S_i}) - v(x_i^{l-1})\}
$$

$$
= \bigcap_{i=1}^{l} \frac{1}{x_i^{l-1}} \bigcap_{v \in \widetilde{D}(S_i)} \{r \in K \mid v(r) \geq v(I) - v(J_{R_v/S_i})\} \cap R
$$

$$
= \bigcap_{i=1}^{l} \frac{1}{x_i^{l-1}} \operatorname{adj}(IS_i) \cap R.
$$

This reduces the computation of $\operatorname{adj} I$ to the computation of the adjoints of each IS_i (and to the computation of a finite intersection).

There are cases when the ideals IS_i are simpler, so this principle can be successfully applied, but there are also cases where simplicity is not transparent (see Exercise 18.4).

The principle can be summarized as follows:

Proposition 18.3.2 *Let R be a regular domain. Let an ideal $\mathfrak{m}$ of height l be minimally generated by $(x_1, \ldots, x_l)$ and assume that $R/(x_1, \ldots, x_l)$ is regular. Then for any ideal I in R,*

$$
\operatorname{adj} I = \left(\bigcap_{i=1}^{l} \frac{1}{x_i^{l-1}} \operatorname{adj} \left(IR[\frac{\mathfrak{m}}{x_i}] \right) \right) \cap R.
$$

A special case when this principle can be applied is:

Proposition 18.3.3 *Let R be a Noetherian regular domain. Let $\mathfrak{m}$ be a d–generated maximal ideal in R of height d. Then*

$$\operatorname{adj}\mathfrak{m}^n = \begin{cases} R, & \text{if } n \le d-1, \\ \mathfrak{m}^{n-d+1}, & \text{otherwise.} \end{cases}$$

Proof: By Proposition 18.3.2,

$$\operatorname{adj}\mathfrak{m}^n = \bigcap_{i=1}^d \frac{1}{x_i^{d-1}} \operatorname{adj}\left(\mathfrak{m}^n R\left[\frac{\mathfrak{m}}{x_i}\right]\right) \cap R$$

$$= \bigcap_{i=1}^d \frac{1}{x_i^{d-1}} \operatorname{adj}\left(x_i^n R\left[\frac{\mathfrak{m}}{x_i}\right]\right) \cap R$$

$$= \bigcap_{i=1}^d x_i^{n-d+1} R\left[\frac{\mathfrak{m}}{x_i}\right] \cap R.$$

If $n \ge d$, this is $\bigcap_{i=1}^d \mathfrak{m}^{n-d+1} R[\frac{\mathfrak{m}}{x_i}] \cap R = \mathfrak{m}^{n-d+1}$, otherwise it is R. □

Similar techniques as in the proof of this proposition show the following:

Proposition 18.3.4 *Let R be a Noetherian regular domain, $\mathfrak{m}$ a d–generated maximal ideal in R of height $d > 1$, and r a positive integer. Then*

$$\operatorname{adj}(x_1^r, x_2, \ldots, x_d) = R.$$

In particular, with the set–up as in Proposition 18.3.3, for all $V \in D(R)$, $J_{V/R} \subseteq \mathfrak{m}^{d-1}V$, so that the Jacobian ideals are not "too big". Here is another testimony to this:

Lemma 18.3.5 *Let R be a regular domain, with $x_1, \ldots, x_d$ a regular sequence such that $R/(x_1, \ldots, x_d)$ is regular. Then for any divisorial valuation ring $V \in D(R)$, $\mathfrak{m}_V \cdot J_{V/R} \subseteq x_1 \cdots x_d V$.*

Proof: Let $v \in \widetilde{D}(R)$ corresponding to $V \in D(R)$. By possibly taking a subset of the x_i, without loss of generality all $v(x_i)$ are positive. Let $\mathfrak{m}$ be the contraction of the maximal ideal of V to R. After localizing at $\mathfrak{m}$, $x_1, x_2, \ldots, x_d$ are part of a regular system of parameters ($R/(x_1, \ldots, x_d)$ is regular, hence so is $R_\mathfrak{m}/(x_1, \ldots, x_d)_\mathfrak{m}$, whence $x_1, \ldots, x_d$ is part of a regular system of parameters in $R_\mathfrak{m}$). We may possibly extend them to a full regular system of parameters, so we may assume that R is a Noetherian local ring with maximal ideal $\mathfrak{m} = (x_1, \ldots, x_d)$.

If $d = 0$, the lemma holds trivially. If $d = 1$, then v is the $\mathfrak{m}$–adic valuation, in which case $V = R$, $J_{V/R} = R$. As v is normalized, $v(x_1) = 1$, and the lemma holds again. Let $S = R[\frac{x_1}{x_d}, \ldots, \frac{x_{d-1}}{x_d}]$. Then S is a regular ring contained in V, and locally, $\frac{x_1}{x_d}, \ldots, \frac{x_{d-1}}{x_d}, x_d$ form a regular system of parameters. By Corollary 5.5.9, $J_{S/R}$ equals $x_d^{d-1}S$. By induction on $\sum_i v(x_i)$, then $v(J_{V/S}) \ge v(\frac{x_1}{x_d} \cdots \frac{x_{d-1}}{x_d} x_d) - 1$, so that by Lemma 4.4.8, $v(J_{V/R}) = v(J_{S/R})v(J_{V/S}) \ge v(x_1 \cdots x_d) - 1$. □

18.4. Adjoints of monomial ideals

Monomial ideals typically mean ideals in a polynomial ring or in a power series ring over a field that are generated by monomials in the variables. We define monomial ideals more generally: let R be a regular ring, and $x_1, \ldots, x_d$ a permutable regular sequence in R such that for every $i_1, \ldots, i_s \in \{1, \ldots, d\}$, $R/(x_{i_1}, \ldots, x_{i_s})$ is a regular domain. By a **monomial ideal (in $x_1, \ldots, x_d$)** we mean an ideal in R generated by monomials in $x_1, \ldots, x_d$. For example, when R is regular local, $x_1, \ldots, x_d$ can be an arbitrary regular system of parameters. There are few results in the literature on such monomial ideals, see for example Kiyek and Stückrad [166] for results on the integral closure of such ideals. Most of this section is taken from Hübl and Swanson [133].

As in the usual monomial ideal case, we can define the Newton polyhedron:

Definition 18.4.1 *With the set–up for this section, let I be an ideal generated by monomials $\underline{x}^{\underline{a}_1}, \ldots, \underline{x}^{\underline{a}_s}$. Then the **Newton polyhedron of I** (relative to $x_1, \ldots, x_d$) is the set*

$$NP(I) = \{\underline{e} \in \mathbb{Q}_{\geq 0}^d \mid \underline{e} \geq \Sigma_i c_i \underline{a}_i, c_i \in \mathbb{Q}_{\geq 0}, \Sigma_i c_i = 1\} \subseteq \mathbb{Q}^d.$$

The set $NP(I)$ is the standard Newton polyhedron of the monomial ideal $(\underline{X}^{\underline{a}_1}, \ldots, \underline{X}^{\underline{a}_s})$ in the polynomial ring $\mathbb{Q}[X_1, \ldots, X_d]$. Thus we know that $NP(I)$ is a closed convex set in $\mathbb{Q}_{\geq 0}^d$, is bounded by coordinate hyperplanes and by hyperplanes of the form $p_1 X_1 + \cdots + p_d X_d = p$, with $p_i \in \mathbb{N}$ and $p \in \mathbb{N}_{>0}$. We will denote the interior of $NP(I)$ as $NP^\circ(I)$.

As expected, the integral closure of general monomial ideals is determined by their Newton polyhedra, just as for the usual monomial ideals (Proposition 1.4.9). We provide a proof for general monomial ideals for completeness:

Theorem 18.4.2 *Let R be a regular domain and let $x_1, \ldots, x_d$ be a permutable regular sequence in R such that for every $i_1, \ldots, i_s \in \{1, \ldots, d\}$, the ring $R/(x_{i_1}, \ldots, x_{i_s})$ is a regular domain. Let I be an ideal generated by monomials in $x_1, \ldots, x_d$. The integral closure $\overline{I}$ of I equals*

$$\overline{I} = (\underline{x}^{\underline{e}} \mid \underline{e} \in NP(I) \cap \mathbb{N}^d).$$

Thus the integral closure of an ideal generated by monomials is also generated by monomials.

Proof: Write $I = (\underline{x}^{\underline{a}_1}, \ldots, \underline{x}^{\underline{a}_s})$. Let $\alpha = \underline{x}^{\underline{e}}$ be such that $\underline{e} \in NP(I) \cap \mathbb{N}^d$. There exist $r_1, \ldots, r_s \in \mathbb{Q}_{\geq 0}$ such that $\sum r_i = 1$ and $\underline{e} \geq \sum r_i \underline{a}_i$ (component-wise). Write $r_i = m_i/n$ for some $m_i \in \mathbb{N}$ and $n \in \mathbb{N}_{>0}$. Then

$$\alpha^n = x_1^{n e_1 - \Sigma m_i a_{i1}} \cdots x_d^{n e_d - \Sigma m_i a_{id}} (\underline{x}^{\underline{a}_1})^{m_1} \cdots (\underline{x}^{\underline{a}_s})^{m_s} \in I^n,$$

so that $\alpha \in \overline{I}$. It remains to prove the other inclusion.

Let S be the set of hyperplanes that bound $NP(I)$ and are not coordinate hyperplanes. For each $H \in S$, if an equation for H is $p_1 X_1 + \cdots + p_d X_d = p$ with $p_i \in \mathbb{N}$ and $p \in \mathbb{N}_{>0}$, define $I_H = (\underline{x}^{\underline{e}} \mid \underline{e} \in \mathbb{N}^d, \sum_i p_i e_i \geq p)$. Clearly $I \subseteq I_H$.

Let $Y_1, \ldots, Y_t$ be variables over R and $R' = R[Y_1, \ldots, Y_t]/(Y_1^{p_1} - x_1, \ldots, Y_t^{p_t} - x_t)$. This is a free finitely generated R–module, and $Y_1, \ldots, Y_t$ form a regular sequence in R'. Set $P = (Y_1, \ldots, Y_t)R'$. Then $R'/P = R/(x_1, \ldots, x_t)$ is a regular domain, so P is a prime ideal, and for any prime ideal Q in R' containing P, R'_Q is a regular local ring. By construction, $I_H R'$ is contained in $(Y_1, \ldots, Y_t)^p = P^p$. By Theorems 6.7.9, 6.7.8 and Exercise 5.7, $P^p R'_P$ is integrally closed. As R' is finitely generated over a locally formally equi-dimensional (regular) ring, by Theorem B.5.2, R'_Q is locally formally equi-dimensional. By Ratliff's Theorem 5.4.1, the integral closure of $P^p R'_Q$ has no embedded prime ideals. It follows that the integral closure of $P^p R'_Q$ is $P^p R'_P \cap R'_Q$. As R'_Q is a regular domain and P is generated by a regular sequence, $\overline{P^p R'_P} \cap R'_Q = P^p R'_Q$. It follows that $\overline{P^p R'_P} \cap R' = P^p$ is the integral closure of P^p. Hence $\overline{I_H} \subseteq \overline{P^p} \cap R = P^p \cap R$, and by freeness of R' over R, the last ideal is exactly $(\underline{x}^{\underline{e}} \,|\, \underline{e} \in \mathbb{N}, \sum_i p_i e_i \geq p)$.

With this,

$$\overline{I} \subseteq \bigcap_{H \in S} \overline{I_H}$$

$$\subseteq \bigcap_{H \in S} (\underline{x}^{\underline{e}} \,|\, \underline{e} \in \mathbb{N}, \textstyle\sum_i p_i e_i \geq p, \ H \text{ is defined by } \sum_i p_i X_i = p)$$

$$= (\underline{x}^{\underline{e}} \,|\, \underline{e} \in \mathrm{NP}(I) \cap \mathbb{N}^d). \qquad \square$$

We proved that in the special case in Proposition 18.3.3, the adjoint of a monomial ideal is monomial. This is true in general. Howald [127] proved the precise characterization of adjoints of monomial ideals in a polynomial ring over a field of characteristic zero, the general form is due to Hübl and Swanson [133]:

Theorem 18.4.3 *Let R be a regular domain, and let $x_1, \ldots, x_d$ be a permutable regular sequence in R such that for every $i_1, \ldots, i_s \in \{1, \ldots, d\}$, the ring $R/(x_{i_1}, \ldots, x_{i_s})$ is a regular domain. Let I be an ideal generated by monomials in $x_1, \ldots, x_d$. Then the adjoint of I equals*

$$\mathrm{adj}\, I = (\underline{x}^{\underline{e}} \,|\, \underline{e} \in \mathbb{N}^d, \underline{e} + (1, \ldots, 1) \in NP^\circ(I)).$$

Thus the adjoint of an ideal generated by monomials in a regular system of parameters is also generated by monomials.

Proof: First we prove that whenever $\underline{e} \in \mathbb{N}^d$ with $\underline{e} + (1, \ldots, 1) \in \mathrm{NP}^\circ(I)$, then $\underline{x}^{\underline{e}} \in \mathrm{adj}(I)$. Let $v \in \widetilde{D}(R)$. By the definition of the Newton polyhedron, $v(x_1 \cdots x_d \underline{x}^{\underline{e}}) > v(I)$. As v is normalized, $v(\underline{x}^{\underline{e}}) \geq v(I) - v(x_1 \cdots x_d) + 1$. By Lemma 18.3.5, $v(J_{R_v/R}) \geq v(x_1 \cdots x_d) - 1$, so that $v(\underline{x}^{\underline{e}}) \geq v(I) - v(J_{R_v/R})$. As v was arbitrary, this proves that $(\underline{x}^{\underline{e}} \,|\, \underline{e} \in \mathbb{N}^d, \underline{e} + (1, \ldots, 1) \in \mathrm{NP}^\circ(I)) \subseteq \mathrm{adj}\, I$. It remains to prove the other inclusion.

Let S be the set of hyperplanes that bound $\mathrm{NP}(I)$ and are not coordinate hyperplanes. For each $H \in S$, if an equation for H is $p_1 X_1 + \cdots + p_d X_d = p$

with $p_i \in \mathbb{N}$ and $p \in \mathbb{N}_{>0}$, define $I_H = (\underline{x}^{\underline{e}} \,|\, \underline{e} \in \mathbb{N}^d, \sum_i p_i e_i \geq p)$. By the definition of Newton polytopes, $I \subseteq I_H$.

By possibly reindexing, without loss of generality $p_1, \ldots, p_t > 0$ and $p_{t+1} = \cdots = p_d = 0$. Let v_H be the valuation on $Q(R)$ defined by $v_H(x_i) = p_i$ with the property that for every $r = \sum_\nu r_\nu \underline{x}^\nu \in R$ with r_ν either zero or not in $(x_1, \ldots, x_t)$, $v_H(r) = \min\{v_H(\underline{x}^\nu) \,|\, r_\nu \neq 0\}$. Such valuations exist (first localize at $(x_1, \ldots, x_t)$, then adjoin a p_ith root y_i of x_i for $i \leq t$ to obtain a larger regular local ring in which the natural extension of v_H is a multiple of the valuation corresponding to the maximal ideal). By construction, $v_H(I) \geq v_H(I_H) \geq p$, and $\mathrm{adj}(I_H) \subseteq \{r \in R \,|\, v_H(r) \geq v_H(I_H) - v_H(J_{R_{v_H}/R})\}$. By the properties of v_H, the last ideal is generated by monomials in the x_i. The proof of Lemma 18.3.5 can be modified to show that $v_H(J_{R_{v_H}/R}) = v_H(x_1 \cdots x_d) - 1$. Thus

$$\mathrm{adj}(I_H) \subseteq (\underline{x}^{\underline{e}} \,|\, \underline{e} \in \mathbb{N}^d, v_H(\underline{x}^{\underline{e}} > v_H(I_H) - v_H(x_1 \cdots x_d))$$
$$\subseteq (\underline{x}^{\underline{e}} \,|\, \underline{e} \in \mathbb{N}^d, \Sigma_i p_i(e_i + 1) > v_H(I_H))$$
$$\subseteq (\underline{x}^{\underline{e}} \,|\, \underline{e} \in \mathbb{N}^d, \Sigma_i p_i(e_i + 1) > p),$$

whence

$$\mathrm{adj}\, I \subseteq \bigcap_{H \in S} \mathrm{adj}(I_H)$$

$$\subseteq \bigcap_{H \in S} (\underline{x}^{\underline{e}} \,|\, \underline{e} \in \mathbb{N}, \Sigma_i p_i(e_i + 1) > p, \ H \text{ is defined by } \textstyle\sum_i p_i X_i = p)$$

$$= (\underline{x}^{\underline{e}} \,|\, \underline{e} \in \mathbb{N}, \underline{e} + (1, \ldots, 1) \in \mathrm{NP}^\circ(I)). \qquad \square$$

Example 18.4.4 Let $R = k[X, Y]$, a polynomial ring in variables X and Y over R. We show here that the adjoint of $I = (X^4, XY, Y^2)$ is (X, Y). Observe that $(1, 0) + (1, 1)$ and $(0, 1) + (1, 1)$ are both in the interior of the Newton polytope of I, but $(0, 0) + (1, 1)$ is not, proving via Theorem 18.4.3 that $\mathrm{adj}(X^4, XY, Y^2)$ is the monomial ideal (X, Y). See the corresponding Newton polytope below.

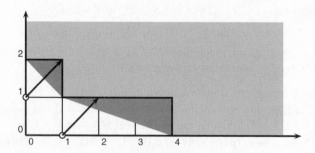

One can also compute the same adjoint by using the Computational Principle 18.3.1 step–by–step. In the next section we show yet another way of computing the adjoint of this ideal.

18.5. Adjoints in two–dimensional regular rings

Throughout this section R is a regular local ring of dimension two, with maximal ideal $\mathfrak{m}$. If I is integrally closed, we will prove that for any $\mathfrak{m}$–primary ideal I, $\mathrm{adj}\, I = I_{n-2}(A)$, where A is an $n \times (n-1)$ presenting matrix of $\bar{I}$. We will also prove that if I has a two–generated reduction (x_1, x_2), then

$$\mathrm{adj}\, I = I_{n-2}(A) = (x_1, x_2) : \bar{I}.$$

Thus in particular the adjoint of any ideal ($\mathfrak{m}$–primary or not) whose integral closure is known, is computable, and is independent of the reduction.

First some reductions: as $\mathrm{adj}(xI) = x \cdot \mathrm{adj}(I)$ (Lemma 18.1.3), it suffices to determine the adjoints of $\mathfrak{m}$–primary ideals. Also, as $\mathrm{adj}(I) = \mathrm{adj}(\bar{I})$, it suffices to determine the adjoints of integrally closed $\mathfrak{m}$–primary ideals. Any $\mathfrak{m}$–primary ideal I in R has a resolution of the form

$$0 \longrightarrow R^{n-1} \overset{A}{\longrightarrow} R^n \longrightarrow I \longrightarrow 0,$$

where n is not necessarily minimal possible. By the Hilbert–Burch Theorem A.4.2, $I = I_{n-1}(A)$.

We first assume that $R/\mathfrak{m}$ is an infinite field. Write $I = (x_1, \ldots, x_n)$, and assume that (x_1, x_2) is a reduction of I. As the residue field is infinite, such minimal reductions exist. Let B be the $(n-2) \times (n-1)$ submatrix of A consisting of the last $n-2$ rows. We will prove that $\mathrm{adj}\, I = I_{n-2}(B) = I_{n-2}(A)$. The reason for introducing B is purely technical – one can more readily prove that $I_{n-2}(B)$ contracts from a blowup. As x_1 and x_2 generate a minimal reduction of I, an easy observation about A and B is that if A contains a unit entry in some column, the same column contains a unit entry in B. Thus if A is an $n \times (n-1)$ presenting matrix of I and n is not minimal possible, by elementary row and column operations on A and its submatrix B and eliminating a column and one of the last $n-2$ rows, we get an $(n-1) \times (n-2)$ presenting matrix A' of I such that its submatrix B' consisting of its last $n-3$ rows satisfying the following:
(1) for all $j = 1, \ldots, n-2$, $I_j(A') = I_{j+1}(A)$,
(2) for all $j = 1, \ldots, n-3$, $I_j(B') = I_{j+1}(B)$.
Thus in order to determine the adjoint of I from its presenting matrix A or its submatrix B we may assume that all the entries in A are in $\mathfrak{m}$.

Theorems 14.1.4 and 14.1.8 show that when I is integrally closed, the minimal number of generators of I is one more than the order of I. By Proposition 14.2.2 and by Theorem 14.1.8, $I = IR[\frac{\mathfrak{m}}{x}] \cap R$ for infinitely many $x \in \mathfrak{m} \setminus \mathfrak{m}^2$.

Theorem 18.5.1 *Let I be an integrally closed $\mathfrak{m}$–primary ideal in a regular local ring $(R, \mathfrak{m})$ with an infinite residue field. Let*

$$0 \longrightarrow R^{n-1} \overset{A}{\longrightarrow} R^n \longrightarrow I \longrightarrow 0$$

be a short exact sequence. Then $\mathrm{adj}\, I = I_{n-2}(A)$.

If the images x_1, x_2 of the first two standard bases vectors of $R^n \to I$ form a reduction of I and if B is the submatrix of A consisting of the last $n - 2$ rows, then $\operatorname{adj} I = I_{n-2}(B) = I_{n-2}(A)$.

Proof: By reductions above we may assume that all the entries of A are in $\mathfrak{m}$. Thus the minimal number of generators of I is n. As I is integrally closed, by Theorems 14.1.4 and 14.1.8, the order of I is $n - 1$. By Proposition 18.3.4 we may assume that $n > 2$.

By the Hilbert–Burch Theorem A.4.2, each x_i is the determinant of of the $(n - 1) \times (n - 1)$ submatrix of A obtained by deleting the ith row. Thus $(x_1, x_2) \subseteq \mathfrak{m} I_{n-2}(B)$. It follows that the ideal $I_{n-2}(B)$ is $\mathfrak{m}$–primary or the whole ring. As the entries of A are all in $\mathfrak{m}$, the ideal $I_{n-2}(B)$ has order at least $n - 2$ and is thus not the whole ring. If its order is strictly bigger than $n - 2$, again as $(x_1, x_2) \subseteq \mathfrak{m} I_{n-2}(B)$, the order of (x_1, x_2) would be strictly larger than $n - 1$. But order is a valuation (Theorem 6.7.9) and (x_1, x_2) is a reduction of I, so that by Proposition 6.8.10, the order of I would be strictly larger than $n-1$, which is a contradiction. Thus the order of $I_{n-2}(B)$ is $n-2$.

As B is an $(n - 2) \times (n - 1)$ matrix with entries in $\mathfrak{m}$, by the Hilbert–Burch Theorem A.4.2 the transpose of B presents the $\mathfrak{m}$–primary ideal $I_{n-2}(B)$, and $I_{n-2}(B)$ is minimally generated by $n - 1$ elements.

Thus $\mu(I_{n-2}(B)) = \operatorname{ord}(I_{n-2}(B)) + 1$, so by Theorem 14.1.4 and Proposition 14.2.2 we can choose x, y in $\mathfrak{m}$ such that $(x, y) = \mathfrak{m}$ and such that $I_{n-2}(B)$ contracts from its extension to $R[\frac{y}{x}]$.

First assume that I is a power of $\mathfrak{m}$, i.e., that $I = \mathfrak{m}^{n-1}$. By Proposition 18.3.3, $\operatorname{adj} I = \mathfrak{m}^{n-2}$. Let $x_1 = x^{n-1}$, $x_2 = y^{n-1}$, and $x_i = x^{n-i+1} y^{i-2}$ for $i = 3, \ldots, n$. A presenting matrix A of I is

$$
A = \begin{bmatrix}
y & 0 & 0 & \cdots & 0 & 0 & 0 \\
0 & 0 & 0 & \cdots & 0 & 0 & -x \\
-x & y & 0 & \cdots & 0 & 0 & 0 \\
0 & -x & y & \cdots & 0 & 0 & 0 \\
 & & & \ddots & & & \\
0 & 0 & 0 & \cdots & -x & y & 0 \\
0 & 0 & 0 & \cdots & 0 & -x & y
\end{bmatrix}.
$$

(This is, up to reordering of the rows, the standard Koszul matrix of relations: by moving the second row to last row we get the standard Koszul matrix.)

Thus $I_{n-2}(A) = \mathfrak{m}^{n-2}$. If we take $x_1 = x^{n-1}$, $x_2 = y^{n-1}$, then B is the submatrix consisting of rows 2 through $n - 1$, so that $I_{n-2}(B) = \mathfrak{m}^{n-2}$. This proves the theorem in the case when I is a power of the maximal ideal.

Now assume that I is any $\mathfrak{m}$–primary integrally closed ideal that is not a power of $\mathfrak{m}$. By Proposition 18.3.2,

$$
\operatorname{adj} I = \frac{1}{x} \operatorname{adj} \left(IR[\tfrac{y}{x}] \right) \cap \frac{1}{y} \operatorname{adj} \left(IR[\tfrac{x}{y}] \right) \cap R
$$

$$
\subseteq x^{\operatorname{ord} I - 1} \operatorname{adj} \left(\frac{I}{x^{\operatorname{ord} I}} R[\tfrac{y}{x}] \right) \cap \cap R.
$$

In $R[\frac{y}{x}]$, the matrix $\frac{A}{x}$ is a presenting matrix of $\frac{I}{x^{\operatorname{ord} I}} R[\frac{y}{x}]$, and $\left(\frac{x_1}{x^{\operatorname{ord} I}}, \frac{x_2}{x^{\operatorname{ord} I}}\right)$ is a reduction of $\frac{I}{x^{\operatorname{ord} I}}$. Also, $\left(\frac{x_1}{x^{\operatorname{ord} I}}, \frac{x_2}{x^{\operatorname{ord} I}}\right)$ is contained in $I_{n-2}(\frac{B}{x})$.

As I is not a power of $\mathfrak{m}$, for any $x \in \mathfrak{m} \setminus \mathfrak{m}^2$ such that I contracts from $IR[\frac{\mathfrak{m}}{x}]$, $IR[\frac{\mathfrak{m}}{x}]$ is not a power of x, so that $IR[\frac{\mathfrak{m}}{x}] = x^{\operatorname{ord} I} I'$ for some ideal I' in $R[\frac{\mathfrak{m}}{x}]$ of height exactly two. By Lemma 14.3.4, with an appropriate choice of x, the co–length of I' at any maximal ideal is strictly smaller than the co–length of I, so we can proceed by induction on co–length. At least locally after localizing at all the maximal ideals of $R[\frac{y}{x}]$ containing $\frac{I}{x^{\operatorname{ord} I}}$,

$$\operatorname{adj}\left(\frac{I}{x^{\operatorname{ord} I}}\right) = I_{n-2}\left(\frac{A}{x}\right) = I_{n-2}\left(\frac{B}{x}\right).$$

Every prime ideal that contains $I_{n-2}(\frac{A}{x})$ or $I_{n-2}(\frac{B}{x})$ also contains the ideal $\left(\frac{x_1}{x^{\operatorname{ord} I}}, \frac{x_2}{x^{\operatorname{ord} I}}\right)$, and hence also $\frac{I}{x^{\operatorname{ord} I}}$. Thus the displayed equality holds globally.

Thus as $\operatorname{ord}(I) - 1 = n - 2$, it follows that

$$\operatorname{adj} I \subseteq x^{\operatorname{ord} I - 1} I_{n-2}\left(\frac{B}{x}\right) R\left[\frac{y}{x}\right] \cap R = I_{n-2}(B) R\left[\frac{y}{x}\right] \cap R = I_{n-2}(B),$$

the last equality by the choice of x. Finally,

$$I_{n-2}(B) \subseteq I_{n-2}(A) \subseteq I_{n-2} big(A) R\left[\frac{y}{x}\right] \cap R = I_{n-2}(B) R\left[\frac{y}{x}\right] \cap R = I_{n-2}(B),$$

so that $I_{n-2}(B) = I_{n-2}(A)$. $\square$

We next characterize the adjoints of integrally closed $\mathfrak{m}$–primary ideals.

Theorem 18.5.2 *Let I be an integrally closed $\mathfrak{m}$–primary ideal in a regular local ring $(R, \mathfrak{m})$. Let*

$$0 \longrightarrow R^{n-1} \stackrel{A}{\longrightarrow} R^n \longrightarrow I \longrightarrow 0$$

be a short exact sequence. Then $\operatorname{adj} I = I_{n-2}(A)$.

Proof: (We employ the standard reduction to infinite residue field, which is described in more detail in Section 8.4.) Let t be a variable over the field of fractions K over R. For any local algebra $(S, \mathfrak{n})$ contained in K, let $S(t)$ stand for $S[t]_{\mathfrak{n}S[t]}$. Note that if S is a d–dimensional local ring, so is $S(t)$. Furthermore, $S(t)$ is a faithfully flat S–algebra with an infinite residue field. In particular,

$$0 \longrightarrow R(t)^{n-1} \stackrel{A}{\longrightarrow} R(t)^n \longrightarrow IR(t) \longrightarrow 0$$

is a short exact sequence. Thus by the previous theorem, $\operatorname{adj}(IR(t)) = I_{n-2}(A)R(t)$.

First let $r \in I_{n-2}(A)$. Then $r \in I_{n-2}(A)R(t) = \operatorname{adj}(IR(t))$. For all $V \in D(R)$, $V(t) \in D(R(t))$. Furthermore, $J_{V/R} \subseteq J_{V(t)/R(t)}$. Thus by the definition of adjoints, $r J_{V/R} \subseteq r J_{V(t)/R(t)} \cap V \subseteq IV(t) \cap V = IV$. As this holds for all $V \in D(R)$, it follows that $r \in \operatorname{adj} I$.

Conversely, assume that $r \in \operatorname{adj} I$. We will prove that $r \in \operatorname{adj}(IR(t))$. For this, let $V \in D(R(t))$. As $IR(t)$ is $\mathfrak{m}R(t)$–primary, the center of V is without loss of generality $\mathfrak{m}R(t)$. Thus by Lemma 9.3.4, $V' = V \cap K \in D(R)$ By the definition of adjoints, $rJ_{V'/R} \subseteq IV'$. Let S be $V'[t]$ localized at the complement of $\mathfrak{m}R[t]$ in $R[t]$. Then $R(t) \subseteq S \subseteq V \subseteq K(t)$ are regular rings with the same field of fractions. By Lemma 4.4.8, $J_{V/R(t)} \subseteq J_{S/R(t)}V = J_{V'[t]/R[t]}V = J_{V'/R}V$, so that $rJ_{V/R(t)} \subseteq rJ_{V'/R}V \subseteq IV$. Thus for all $V \in D(R(t))$, $rJ_{V/R(t)} \subseteq IV$, whence $r \in \operatorname{adj}(IR(t)) \cap R = I_{n-2}(A)R(t) \cap R = I_{n-2}(A)$. $\qquad\square$

18.6. Mapping cones

In this section we develop some elementary facts about mapping cones over arbitrary commutative rings. A goal is to give yet another formulation of the adjoint of an ideal in a regular local ring.

Let $\mathbb{F} = \cdots \to F_2 \to F_1 \to F_0 \to 0$ and $\mathbb{G} = \cdots \to G_2 \to G_1 \to G_0 \to 0$ be two complexes with differential maps δ_F and δ_G, respectively. Let $\varphi : \mathbb{F} \to \mathbb{G}$ be a map of complexes. The **mapping cone** is the complex $\mathbb{M} = \mathbb{M}(\varphi)$ of φ whose ith module is $\mathbb{M}_i = \mathbb{F}_{i-1} \oplus \mathbb{G}_i$ and the differential δ_M is

$$\mathbb{F}_i \oplus \mathbb{G}_{i+1} \quad \longrightarrow \quad \mathbb{F}_{i-1} \oplus \mathbb{G}_i$$
$$(f, g) \quad \longmapsto \quad (-\delta_F(f), \varphi(f) + \delta_G(g)).$$

Then $\mathbb{M}$ is a complex because

$$\begin{aligned}
\delta_M^2(f, g) &= \delta_M(-\delta_F(f), \varphi(f) + \delta_G(g)) \\
&= (\delta_F^2(f), \varphi(-\delta_F(f)) + \delta_G(\varphi(f) + \delta_G(g))) \\
&= (0, \varphi(-\delta_F(f)) + \delta_G(\varphi(f))) = (0, 0),
\end{aligned}$$

as φ is a map of complexes.

Proposition 18.6.1 *With notation as above, there is a canonical short exact sequence* $0 \longrightarrow \mathbb{G} \longrightarrow \mathbb{M} \longrightarrow \mathbb{F}[-1] \longrightarrow 0$ *of complexes.*

We leave a straightforward proof to the reader.

A consequence of the proposition is the canonical long exact sequence on homology:

$$\cdots \to H_{i+1}(\mathbb{F}[-1]) \to H_i(\mathbb{G}) \to H_i(\mathbb{M}) \to H_i(\mathbb{F}[-1]) \to H_{i-1}(\mathbb{G}) \cdots,$$

or equivalently

$$\cdots \to H_i(\mathbb{F}) \to H_i(\mathbb{G}) \to H_i(\mathbb{M}) \to H_{i-1}(\mathbb{F}) \to H_{i-1}(\mathbb{G}) \cdots.$$

The connecting homomorphism $H_{i-1}(\mathbb{F}) \to H_{i-1}(\mathbb{G})$ is induced by φ.

The homology of the mapping cone carries some information about the homologies of $\mathbb{F}$ and $\mathbb{G}$, and vice versa. The following lemma illustrates this.

Lemma 18.6.2 *Assume that* $\mathbb{F}$ *and* $\mathbb{G}$ *are acyclic complexes. If* $\mathbb{M}$ *is the mapping cone of* $\varphi : \mathbb{F} \to \mathbb{G}$, *then for all* $i > 1$, $H_i(\mathbb{M}) = 0$, *and*

$$0 \to H_1(\mathbb{M}) \to H_0(\mathbb{F}) \to H_0(\mathbb{G}) \to H_0(\mathbb{M}) \to 0$$

is exact.

A proof is immediate from the long exact sequence above.

Mapping cones can be used for constructing a resolution of a module M from resolutions $\mathbb{F}$ and $\mathbb{G}$ of modules K, N, respectively, where $0 \to K \to N \to M \to 0$ is a short exact sequence. We will apply the mapping cone construction in the following situation:

Proposition 18.6.3 *Let* R *be a Cohen–Macaulay local ring,* $x_1, \ldots, x_d$ *a maximal regular sequence in* R, $J = (x_1, \ldots, x_d)$, *and* I *an ideal containing* J *with a finite free resolution. Then the mapping cone construction yields a free resolution of* $R/(J : I)$.

Proof: By assumption the projective dimension of R/I is finite, and by the Auslander–Buchsbaum Formula, it is equal to $\operatorname{depth} R - \operatorname{depth}(R/I) = d$. Since R is local, a projective resolution is free. Let $\mathbb{G}$ be the Koszul complex for $x_1, \ldots, x_d$ and $\mathbb{F}$ a free resolution of R/I. There is a map of complexes $\varphi : \mathbb{G} \longrightarrow \mathbb{F}$ with φ_0 being the identity lifting the surjection $R/J \to R/I$. If we write $I = (x_1, \ldots, x_d, \ldots, x_n)$, we can take $\mathbb{F}_1 = R^n$, and $\varphi_1 : \mathbb{G}_1 \to \mathbb{F}_1$ can be taken to be $\varphi_1(r_1, \ldots, r_d) = (r_1, \ldots, r_d, 0, \ldots, 0)$.

Applying the dual $\operatorname{Hom}_R(\_, R)$ gives the map of complexes $\varphi^* : \mathbb{F}^* \longrightarrow \mathbb{G}^*$. Let $\mathbb{M}$ be the mapping cone of φ^*. The long exact sequence of homologies is

$$0 \to H_{d+1}(\mathbb{M}) \to H_d(\mathbb{F}^*) \to H_d(\mathbb{G}^*) \to H_d(\mathbb{M}) \to H_{d-1}(\mathbb{F}^*)$$
$$\cdots \to H_1(\mathbb{F}^*) \to H_1(\mathbb{G}^*) \to H_1(\mathbb{M}) \to H_0(\mathbb{F}^*) \to H_0(\mathbb{G}^*) \to H_0(\mathbb{M}) \to 0.$$

But $H_i(\mathbb{F}^*) = \operatorname{Ext}_R^{d-i}(R/I, R)$ and $H_i(\mathbb{G}^*) = \operatorname{Ext}_R^{d-i}(R/J, R)$. As $x_1, \ldots, x_d$ is a regular sequence, $\operatorname{Ext}_R^i(R/I, R) = \operatorname{Ext}_R^i(R/J, R) = 0$ for all $i = 0, \ldots, d-1$. Thus the long exact sequence above shows that $H_i(\mathbb{M}) = 0$ for $i > 1$, and gives the exact sequence

$$0 \to H_1(\mathbb{M}) \to H_0(\mathbb{F}^*) \to H_0(\mathbb{G}^*) \to H_0(\mathbb{M}) \to 0.$$

All the maps are canonical. By Rees's Theorem A.4.1, there are canonical isomorphisms

$$H_0(\mathbb{F}^*) = \operatorname{Ext}_R^d(R/I, R) \cong \operatorname{Hom}_R(R/I, R/J) = \frac{J : I}{J},$$

$$H_0(\mathbb{G}^*) = \operatorname{Ext}_R^d(R/J, R) \cong \operatorname{Hom}_R(R/J, R/J) = \frac{R}{J},$$

and one can verify that in

$$0 \to H_1(\mathbb{M}) \to \frac{J : I}{J} \to \frac{R}{J} \to H_0(\mathbb{M}) \to 0$$

all the maps are canonical. In particular, $H_1(\mathbb{M}) = 0$ and $H_0(\mathbb{M}) = \frac{R}{J:I}$. Thus $\mathbb{M}$ is a free resolution of $\frac{R}{J:I}$. $\square$

Note that $\mathbb{M}$ in the proof is a resolution of length $d + 1$. It is possible to trim this resolution to a resolution of length d because the tail end of $\mathbb{M}$ is as follows:

$$0 \to \mathbb{F}_0^* \xrightarrow{\begin{bmatrix} -x_1 \\ \vdots \\ -x_n \\ 1 \end{bmatrix}} \mathbb{F}_1^* \oplus \mathbb{G}_0^* \xrightarrow{\left[\begin{array}{c|c|c} & & 0 \\ -(\delta_F)_2^* & & \vdots \\ & & 0 \\ \hline & & x_1 \\ I_d & 0 & \vdots \\ & & x_d \end{array}\right]} \mathbb{F}_2^* \oplus \mathbb{G}_1^*.$$

To change the leftmost non–zero map to $(0, \ldots, 0, 1)$ we need to perform a change of basis: adding x_i times row $n + 1$ from the ith row in the leftmost non–zero map above necessitates at the same time subtracting x_i times column i to column $n + 1$ in the rightmost map above. After repeating this for all $i = 1, \ldots, n$, the last column of the rightmost map above is zero. Thus the complex $\mathbb{M}$ at the tail end is the direct sum of the two complexes below as follows:

$$0 \to R \xrightarrow{=} R \xrightarrow{\hspace{3cm}} 0$$

$$0 \to \frac{\mathbb{F}_1^* \oplus \mathbb{G}_0^*}{R^*} \xrightarrow{\left[\begin{array}{c|c} -(\delta_F)_2^* \\ \hline I_d & 0 \end{array}\right]} \mathbb{M}_{d-1} \to \mathbb{M}_{d-2} \to \cdots.$$

Now we restrict to the case $d = 2$, and return to the computation of adjoints of $\mathfrak{m}$–primary ideals in a two–dimensional regular local ring $(R, \mathfrak{m})$. Then $(\delta_F)_2$ is the presentation matrix A of I. By the Hilbert–Burch Theorem A.4.2, as the second complex above presents $R/(J : I)$, then $J : I$ is the ideal of the maximal minors of

$$\left[\begin{array}{c|c} -(\delta_F)_2^* \\ \hline I_2 & 0 \end{array}\right] = \left[\begin{array}{c|c} -A^T \\ \hline I_2 & 0 \end{array}\right],$$

where A is the presentation matrix of I. But the ideal of the maximal minors of this matrix is the ideal of the maximal minors of the matrix obtained from A^T by deleting the first two columns. Thus by Theorem 18.5.1, we get the following characterization of adjoints:

Proposition 18.6.4 *If I is an $\mathfrak{m}$–primary integrally closed ideal in a two–dimensional regular local ring $(R, \mathfrak{m})$, then for any reduction (x_1, x_2) of I,*

$$\mathrm{adj}(I) = (x_1, x_2) : I.$$

18.7. Analogs of adjoint ideals

In this section, we briefly mention some related notions, but prove no theorems. For undefined terms see Lazarsfeld's book [179].

Definition 18.7.1 *If X is a smooth complex variety, D an effective $\mathbb{Q}$–divisor on X, and $\mu : X' \to X$ a log resolution of D, then the **multiplier ideal** associated to D is defined as $\mu_*(\mathcal{O}_{X'}(K_{X'/X} - \lfloor \mu^* D \rfloor)) \in \mathcal{O}_X$, where $K_{X'/X}$ is the relative canonical divisor. It is denoted $\mathcal{J}(D)$ or $\mathcal{J}(X, D)$.*

When adjoint ideals and multiplier ideals are both defined, the two notions agree. However, one disadvantage of multiplier ideals is that they are restricted to special rings in characteristic zero, whereas adjoint ideals do not have such a restriction.

There is a rich theory of multiplier ideals, partly due to their access to analytic techniques. For example, multiplier ideals satisfy the **subadditivity** property. Namely, for any two effective $\mathbb{Q}$–divisors D_1 and D_2, $\mathcal{J}(D_1 + D_2) \subseteq \mathcal{J}(D_1) \cdot \mathcal{J}(D_2)$, or if I and J denote ideal sheaves in $\mathcal{O}_X$, then $\mathcal{J}(I \cdot J) \subseteq \mathcal{J}(I) \cdot \mathcal{J}(J)$. This was proved by Demailly, Ein, and Lazarsfeld [62]. The analytic techniques needed to prove subadditivity are beyond the scope of this book. Excellent references are [179] and [17].

In contrast, subadditivity is not known in general for adjoint ideals, i.e., it is not known whether $\mathrm{adj}(IJ) \subseteq (\mathrm{adj}\, I)(\mathrm{adj}\, J)$. Equality definitely fails, see Proposition 18.3.3. Special cases of subadditivity on adjoint ideals are known: Takagi and Watanabe [287] proved it for two dimensional regular local rings, and Hübl and Swanson [133] proved it for (generalized) monomial ideals.

In positive characteristic, an analog of adjoint ideals (and multiplier ideals) has been established by N. Hara and K. Yoshida [107]. It has been further studied in Hara [104], [105]; Smith [273]; Takagi [285]; Hara and Takagi [106]; and Hara and Watanabe [108]. The definition comes from the idea of test elements from tight closure.

Definition 18.7.2 *Let R be a Noetherian ring of positive prime characteristic p. Fix an ideal J in R. The J–**tight closure of** I, denoted I^{*J}, is defined to be the set of all $z \in R$ for which there exists a $c \in R^o$ such that for all large $q = p^e$, $cJ^q z^q$ is in the ideal generated by the qth powers of elements of I. (Compare with definition of tight closure in 13.1.1.)*

Definition 18.7.3 *Let R be a Noetherian ring of positive prime characteristic p. Fix an ideal J in R. The J–**test ideal**, $\tau(J)$, is the unique largest ideal such that $\tau(J)I^{*J} \subseteq I$ for all ideals I.*

When $J = R$, this ideal is the usual test ideal coming from tight closure theory. In general, the J–test ideal measures subtle properties of how J sits in R. The following is Theorem 3.4 in Hara and Yoshida [107]:

Theorem 18.7.4 *Let R be a normal $\mathbb{Q}$–Gorenstein local ring essentially of finite type over a field, and let J be a non–zero ideal. Assume that $J \subseteq R$ is reduced from characteristic 0 to characteristic $p \gg 0$, together with a log resolution of singularities $f : X \to Y = \operatorname{Spec}(R)$ such that $J\mathcal{O}_X = \mathcal{O}_X(-Z)$ is invertible. Then*

$$\tau(J) = H_0(X, \mathcal{O}_X(K_X - \lceil f^* K_Y(-Z) \rceil)).$$

In characteristic 0, the multiplier ideal of J is the right–hand side of the equality asserted in the above theorem, so one can think of the J–test ideal as the analog of the multiplier ideal of J. Takagi has made this even more precise in his paper [286] dealing with singularities of pairs (see also [288]). Once again, as in Section 13.3, this shows the close connection between characteristic p methods and birational algebraic geometry in characteristic 0. In particular, subadditivity was proved for two dimensional regular local rings by Takagi and Watanabe [287], and via J–tight closure in some special cases by Hara and Takagi [106].

18.8. Exercises

18.1   Compute the adjoint of (X^3, X^2YZ^3, Y^5) in $k[X, Y, Z]$.

18.2   Compute the adjoint of $(X^2 + Y^3, XY^3, Y^4)$ in $k[X, Y]$.

18.3   Let R be $k[X, Y, Z]$. Prove that for all non–negative integers a, the adjoint of $(X^{a+1}, X^a Z^3, Y)$ is R.

18.4   (Rhodes [244]) Let k be a field, X, Y and Z over k, and $R = k[X, Y, Z]$. Let $I = (X^3, X^2YZ^3, Y^4)$.

    (i) Compute $\operatorname{adj}(I)$ using Proposition 18.3.2 and $\mathfrak{m} = (X, Y)$.

    (ii) Similarly set up the computation of $\operatorname{adj}(I)$ by applying Proposition 18.3.2 with $\mathfrak{m} = (X, Y, Z)$. Observe that the first step "reduces" the computation of adjoints to a seemingly more difficult computation.

18.5   Let R be a regular local ring domain of dimension d. For any ideal I in R and any positive rational number c, define (in analogy for multiplier ideals, and no meaning is attached to $c \cdot I$):

$$\operatorname{adj}(c \cdot I) = \bigcap_{v \in \widetilde{D}(R)} \{ r \in K \mid v(r) \geq c \cdot v(I) - v(J_{R_v/R}) \}.$$

    (i) Prove that $\operatorname{adj}(c \cdot I) \subseteq R$.

    (ii) Prove that $\operatorname{adj}(c \cdot \mathfrak{m}^n) = \mathfrak{m}^{\max\{0, \lceil cn \rceil - d + 1\}}$.

18.6   ([141]) Let R be a two–dimensional regular local ring, I an $\mathfrak{m}$–primary integrally closed ideal in R and (a, b) a reduction of I. Prove that for all $n \geq 1$, $(a, b)^n : I = (a, b)^{n-1} \operatorname{adj}(I)$.

18.7   Let R be a two–dimensional regular local ring. Prove that for every integrally closed ideal I in R and for every $n \in \mathbb{N}$, $\mathrm{adj}(I^n) = I^{n-1} \, \mathrm{adj}\, I$.

18.8   (Lipman [190]) Let R be a regular domain, and I an ideal. Prove that for any height h prime ideal P such that R/P is regular, $\bar{I} : P^{h-1} \subseteq \mathrm{adj}\, I$.

18.9   Let $(R, \mathfrak{m})$ be a two–dimensional regular local ring. Let I be an integrally closed $\mathfrak{m}$–primary ideal in R and (a, b) a reduction of I. Prove that for all $n \in \mathbb{N}_{>0}$, $(a, b)^n : I = (a, b)^{n-1} \mathrm{adj}\, I$.

18.10   (Eliahou and Kervaire [71]) Let $R = k[X_1, \ldots, X_n]$ be a polynomial ring in n variables over a field k. A monomial ideal I is **strongly stable** if for any monomial $m \in I$, if $X_i | m$ and $j < i$, then $X_j m / X_i \in I$. Construct a free resolution of I by the mapping cone construction. (See [120] for more general resolutions by mapping cone constructions.)

18.11   ([133]) Let R be a regular ring, and $x_1, \ldots, x_d$ a permutable regular sequence in R such that for every $i_1, \ldots, i_s \in \{1, \ldots, d\}$, $R/(x_{i_1}, \ldots, x_{i_s})$ is a regular domain. Let I and J be ideals in R generated by monomials in the x_i. Prove that $\mathrm{adj}(IJ) \subseteq \mathrm{adj}(I)\mathrm{adj}(J)$ ("subadditivity").

18.12   (Takagi and Watanabe [287], Hübl and Swanson [133]) Let R be a two–dimensional regular local ring, and I and J ideals in R. Prove that $\mathrm{adj}(IJ) \subseteq \mathrm{adj}(I)\mathrm{adj}(J)$ ("subadditivity").

19
Normal homomorphisms

In this chapter we present how integral closure behaves under homomorphisms. As we have already seen, if $R \to S$ is a homomorphism and I is an ideal of R, then it is obvious from the definition that $\bar{I}S \subseteq \overline{IS}$. We are interested in conditions that guarantee equality. It is not difficult to find an example where the integral closure of an ideal does not commute with flat base change. For example, let $R = k[X^2, Y^2] \subseteq S = k[X, Y]$, where k is a field and X, Y variables over k. This map is flat as S is free over R, but the maximal ideal $\mathfrak{m} = (X^2, Y^2)$ of R is not integrally closed upon extension to S.

Definition 19.0.1 *A ring homomorphism $f : R \to S$ is said to be* **normal** *if it is flat and if for every $P \in \mathrm{Spec}(R)$ and every field extension L of $\kappa(P)$, the ring $L \otimes_R S$ is normal.*

Theorem 19.4.2 in Section 19.4 proves that if $g : R \to R'$ is a normal homomorphism of locally Noetherian rings, then for every normal ring S containing R, the ring $S' = S \otimes_R R'$ is also normal. One of the most general theorem known concerning normality of rings under base change is proved in Section 19.5.

In Corollary 19.5.2 we prove that if $R \to S$ is a normal ring homomorphism of Noetherian rings, then for any ideal I of R, $\bar{I}S = \overline{IS}$. We follow the treatment in [101, Chapter IV]. The difficulty comes from the fact that the approach taken is to transfer the problem to the Rees algebra of a given ideal, where the problem can be rephrased in terms of integral closures of algebras instead of ideals. In the case R_{red} is not locally analytically unramified, the integral closure of the Rees algebra is not necessarily Noetherian, and this adds considerable difficulty. The easier case where R_{red} is locally analytically unramified is proved in Theorem 19.2.1. In particular, if locally the formal fibers are normal, then the integral closure of an ideal commutes with completion for a Noetherian local ring.

Without the restriction on formal fibers being normal, the integral closure of ideals need not commute with completion. In Section 19.2 we present an example by Heinzer, Rotthaus, and Wiegand of a three–dimensional regular local ring with a height two prime ideal P such that upon completing, the image of P is not integrally closed.

19.1. Normal homomorphisms

In this section we prove basic properties of normal homomorphisms.

Proposition 19.1.1 *Let k be a field and ℓ a k–algebra. If $k \subseteq \ell$ is normal, it is separable. If ℓ is a field that is separable over k, then $k \to \ell$ is normal.*

Proof: A normal ring is reduced, so normality of the field extension implies separability.

Now assume that $k \subseteq \ell$ is separable. An arbitrary field extension is flat, and it remains to show that for any field extension K of k, $\ell \otimes_k K$ is normal. By the separable assumption, $\ell \otimes_k K$ is reduced.

First suppose that ℓ is finitely generated over k. By Theorem 3.1.3, there exist algebraically independent elements $z_1, \ldots, z_d$ in ℓ such that ℓ is separable algebraic over $k(z_1, \ldots, z_d)$. Then $\ell \otimes_k K = \ell \otimes_{k(\underline{z})} k(\underline{z}) \otimes_k K$. The ring $k(\underline{z}) \otimes_k K$ is a localization of a finitely generated polynomial ring over the field K, hence regular. Let P be any prime ideal in this ring. Then $\ell \otimes_{k(\underline{z})} \kappa(P)$ is a module–finite extension of $\kappa(P)$, and reduced because ℓ is separable over $k(\underline{z})$. Thus $\ell \otimes_{k(\underline{z})} \kappa(P)$ is a direct sum of fields, so regular. Thus we have a map $k(\underline{z}) \otimes_k K \longrightarrow \ell \otimes_k K$ from a regular ring, with all fibers regular, so that by Theorem 4.5.5 $\ell \otimes_k K$ is regular as well. In particular, $\ell \otimes_k K$ is normal.

Now let ℓ be an arbitrary separable field extension of k. Write $\ell = \cup_i \ell_i$, where i varies over an index set and each ℓ_i is finitely generated field extension of k. By Proposition 3.2.2, each ℓ_i is separable over k. By the finitely generated case, each $\ell_i \otimes_k K$ is normal. Clearly $\ell \otimes_k K = \cup_i (\ell_i \otimes_k K)$. If x is in the integral closure of $\ell \otimes_k K$, there exists i such that x is in the total ring of fractions of $\ell_i \otimes_k K$ and is integral over it. Hence $x \in \ell_i \otimes_k K \subseteq \ell \otimes_k K$. $\square$

Proposition 19.1.2 *Let $f : R \to S$ be a normal homomorphism of Noetherian rings. Suppose that A is a finitely generated R–algebra. Then $A \to A \otimes_R S$ is a normal homomorphism of Noetherian rings.*

Proof: $A \otimes_R S$ is Noetherian because it is finitely generated over S. Let $Q \in \operatorname{Spec}(A)$. Let L be an extension field of $\kappa(Q)$. Then $L \otimes_A (A \otimes_R S) \cong L \otimes_R S$ and the latter is normal as L is an extension field of $\kappa(f^{-1}(Q))$. $\square$

We prove some basic facts concerning normal homomorphisms:

Proposition 19.1.3 *Let R be a Noetherian domain with field of fractions K. Let $\varphi : R \to R'$ be a normal homomorphism to a Noetherian ring R'. The following statements hold:*

(1) R' is reduced.

(2) If R is a field, then R' is a finite direct product of integrally closed domains whose fields of fractions are separable over R.

(3) Let $Q_1, \ldots, Q_r$ be the minimal prime ideals of R'. Set $L = K \otimes_R R'$. Then L is normal, and L is the direct product of integrally closed domains between R'/Q_i and $K_i = \kappa(Q_i)$. All K_i are separable over K.

(4) If R is local with maximal ideal $\mathfrak{m}$ and R' is local, then $\mathfrak{m}R'$ is a prime ideal. Moreover, $\dim R'_{\mathfrak{m}R'} = \dim R$. We can identify $R'_{\mathfrak{m}R'}$ as a subring of the product of those K_i for which $Q_i \subseteq \mathfrak{m}R'$.

(5) Assume that $(R, \mathfrak{m})$ is local of dimension one and that R' is local. Then $R' = L \cap R'_{\mathfrak{m}R'}$, where we identify both L and $R'_{\mathfrak{m}R'}$ as subrings of the total ring of fractions of R'. (Note that by (4), $R'_{\mathfrak{m}R'}$ is identified as a subring of the total ring of fractions of R'.)

Proof: The fiber of φ over 0 is $K \otimes_R R'$, and by assumption this is normal and therefore reduced. Since R' is flat over R, the inclusion $R \to K$ tensors to the inclusion $R' \to K \otimes_R R'$, so R' is reduced, proving (1).

Assume that R is a field. By Proposition 19.1.1, R' is separable over R. By Theorem 3.2.8 (5), for every minimal prime ideal P of the reduced ring R', $\kappa(P)$ is separable over R. Since R' is normal, it is a finite direct product of integrally closed domains whose field of fractions are $\kappa(P)$ as P ranges over the minimal prime ideals of R' (see Corollary 2.1.13 and Lemma 2.1.15). This proves (2).

Let L be as in (3). Then L is a localization of R', hence Noetherian. By the normal assumption on $R \to R'$, L is normal. By Corollary 2.1.13 and Lemma 2.1.15, L is a direct product of the integral closures of its quotients by minimal prime ideals. The normal assumption on $R \to R'$ localizes, so $K \to L$ is normal, whence separable by Proposition 19.1.1. Then (2) shows that each K_i is separable over L.

The ring $R'/\mathfrak{m}R'$ is the fiber of φ over the maximal ideal of R, and is therefore normal and hence reduced. It is also local since R' is local, and therefore cannot be properly a direct product of other rings. It follows that $R'/\mathfrak{m}R'$ is an integrally closed domain, and consequently $\mathfrak{m}R'$ is a prime ideal. Since $R \to R' \to R'_{\mathfrak{m}R'}$ is a composition of flat and faithful homomorphisms, $R'_{\mathfrak{m}R'}$ is faithfully flat over R, and by Proposition B.2.3 it follows that $\dim(R) = \dim(R'_{\mathfrak{m}R'})$ (note that $\mathfrak{m}R'$ is necessarily minimal among the prime ideals that contract to $\mathfrak{m}$).

We have identified R' as a subring of $\prod_i K_i$. It follows that $R'_{\mathfrak{m}R'}$ is contained in $\prod (R' \setminus \mathfrak{m}R')^{-1}(K_i)$, which is the product of those fields K_i for which every element in $R' \setminus \mathfrak{m}R'$ is non–zero. This set is exactly the set of those K_i for which $Q_i \subseteq \mathfrak{m}R'$.

Lastly, it is clear that $R' \subseteq L \cap R'_{\mathfrak{m}R'}$. Let $y \in L \cap R'_{\mathfrak{m}R'}$, and fix a parameter $a \in R$. Every element of K can be written in the form b/a^n for some $n \geq 1$ and some $b \in R$. It follows that every element of L can be written in the form s/a^n for some $s \in R'$ and $n \geq 1$, so we may write $y = s/a^n$. Since R' is flat over R and $\mathfrak{m}R'$ is prime, it follows that $\mathfrak{m}R'$ is the only associated prime ideal of $a^n R'$, as $\mathfrak{m}$ is the only associated prime ideal of $a^n R$. This forces $a^n y = s \in a^n R'_{\mathfrak{m}R'} \cap R' = a^n R'$, and hence $y \in R'$. $\square$

19.2. Locally analytically unramified rings

The main theorem of this section is the following:

Theorem 19.2.1 *Let R be a Noetherian ring such that R_{red} is locally analytically unramified. Suppose that $R \to S$ is a normal homomorphism of R to a Noetherian ring S. Then for every ideal I in R, $\bar{I}S = \overline{IS}$.*

Proof: Fix an ideal I in R. Since the nilradical of R is contained in both $\bar{I}S$ and $\overline{IS}$, we may factor it out and assume that R is reduced. Suppose we have shown that for every prime ideal p in R such that $I \subseteq p$, $\overline{I_p}S_p = (\overline{IS})_p$, where the subscript p denotes localization at the multiplicatively closed set $R \setminus p$. If $\overline{IS} \neq \bar{I}S$, choose a prime ideal P in S that is minimal in the support of $\overline{IS}/\bar{I}S$. There exists an element $r_P \in \overline{IS} \setminus \bar{I}S$ such that $P = \overline{IS} :_S r_P$. As the same holds after inverting all elements of $R \setminus (P \cap R)$, this is a contradiction. Thus to prove the theorem we may replace R by its localization at $P \cap R$, and assume that R is local.

We next reduce to the case in which R is a domain. Let $P_1, \ldots, P_n$ be the minimal prime ideals of R. Since R is local and analytically unramified, the intersection of the P_i is 0. Let $R_i = R/P_i$, $I_i = IR_i$, and $S_i = R_i \otimes_R S = S/P_iS$. By Proposition 9.1.3 R_i is analytically unramified and by Proposition 19.1.2 $R_i \to S_i$ is a normal homomorphism for every i. Assuming the result for analytically unramified domains, then $\bar{I_i}S_i = \overline{I_iS_i}$. By Proposition 1.1.5 $r \in \bar{I}$ if and only if $r_i \in \bar{I_i}$ for each $i = 1, \ldots, n$, where r_i is the image of r in R_i. This implies that

$$0 \to R/\bar{I} \to R_1/\bar{I_1} \times \cdots \times R_n/\bar{I_n}$$

is exact. By tensoring with S and using the flatness of S over R,

$$0 \to S/\bar{I}S \to S_1/\bar{I_1}S_1 \times \cdots \times S_n/\bar{I_n}S_n \tag{19.2.2}$$

is also exact. Let $s \in \overline{IS}$. Then s_i, the image of s in S_i is in $\overline{I_iS_i}$ for each $i = 1, \ldots, n$, and hence by assumption, $s_i \in \bar{I_i}S_i$ for each $i = 1, \ldots, n$. The exactness of the sequence (19.2.2) then shows that $s \in \bar{I}S$ as required.

We have reduced to the case in which R is a local domain. We next reduce to the case in which R is integrally closed. Let R' be the integral closure of R in its field of fractions. By Corollary 4.6.2 since R is analytically unramified, R' is a finitely generated R–module. Set $S' = R' \otimes_R S$. By Proposition 19.1.2, the map $R' \to S'$ is still normal. Suppose we have proved the theorem in the case in which R is integrally closed. Then

$$\overline{IR'}S' = \overline{IS'}. \tag{19.2.3}$$

Since by Proposition 1.6.1, $\overline{IR'} \cap R = \bar{I}R$, there is an embedding $R/\bar{I} \to R'/\overline{IR'}$, and upon tensoring with S we obtain an embedding

$$S/\bar{I}S \to S'/\overline{IR'}S'. \tag{19.2.4}$$

Now let $s \in \overline{IS}$. Then the image of s in S' is in $\overline{IS'}$ and by (19.2.3) it follows that $s \in \overline{IR'S'}$. Hence (19.2.4) implies that $s \in \overline{I}S$, proving the theorem.

We have now reduced to proving the theorem in the case in which R is a local integrally closed analytically unramified domain. Let $A = \oplus_{n \geq 0} \overline{I^n} t^n$, which by Proposition 5.2.4 is the integral closure of the Rees algebra $R[It]$ in $R[t]$. By Corollary 9.2.1, A is a finitely generated R–algebra, and it is integrally closed by Proposition 5.2.4. By Proposition 19.1.2, the map from $A \to A \otimes_R S$ is normal. In particular, all the fibers of the map $A \to A \otimes_R S$ normal. Thus by Corollary 4.5.6, $A \otimes_R S$ is locally normal, hence normal. However, $A \otimes_R S$ is isomorphic to the ring $\oplus_{n \geq 0} \overline{I^n} S t^n$, and its normality then implies that $\overline{I^n}S = \overline{I^n S}$ for all n, in particular for $n = 1$. $\square$

Corollary 19.2.5 *Let $(R, \mathfrak{m})$ be a local Noetherian ring such that $R \to \widehat{R}$ is normal (e.g., if R is excellent). Then for all ideals $I \subseteq R$,*

$$\overline{I}\widehat{R} = \overline{I\widehat{R}}.$$

Proof: The parenthetical remark of the corollary follows as excellent local rings by definition have geometrically regular formal fibers.

By Theorem 19.2.1 it suffices to prove that R_{red} is analytically unramified. Theorem 4.5.2 shows that the reduced property of a ring is equivalent to the ring satisfying Serre's conditions (S_1) and (R_0). Let $P \in \mathrm{Spec}(R_{red})$. By the normal assumption $\kappa(P) \otimes_{R_{red}} \widehat{R_{red}} = \kappa(P) \otimes_R \widehat{R}$ is normal, hence reduced. Thus by Theorem 4.5.5, since R_{red} satisfies (S_1) and (R_0), so does $\widehat{R_{red}}$, whence $\widehat{R_{red}}$ is reduced and R_{red} is analytically unramified. $\square$

The conclusion of Theorem 19.2.1 need not hold without the normal hypothesis, as shown by the following example:

Example 19.2.6 (Heinzer, Rotthaus, S. Wiegand [114]) Let k be a field, and X, Y, Z variables over k. By Exercise 3.13 there exist α and β in $Xk[[X]]$ that are algebraically independent over $k(X)$. Set $f = (Y - \alpha)^2$, $g = (Z - \beta)^2$. Write $f = \sum_{i \geq 0} a_i X^i$ for some $a_i \in k[Y]$, and $g = \sum_{i \geq 0} b_i X^i$ for some $b_i \in k[Z]$. Define $F_n = \sum_{i > n} a_i X^{i-n}$, $G_n = \sum_{i > n} b_i X^{i-n}$. Set $R' = k[X, Y, Z, F_0, G_0, F_1, G_1, F_2, G_2, \ldots]$. From the equations $XF_{n+1} = F_n - a_{n+1}X$, $XG_{n+1} = G_n - b_{n+1}X$, it follows that $(X, Y, Z)R'$ is a maximal ideal and that $R'/XR' = k[Y, Z]$. Define $R = R'_{(X,Y,Z)R'}$. Heinzer, Rotthaus and Wiegand developed a machinery that manipulates such constructions and determines if R is Noetherian. We need that R is Noetherian. Here are steps towards proving it. By Proposition 6.4.7, $V = k(X, \alpha, \beta) \cap k[[X]]$ is a Noetherian discrete valuation ring. Clearly the maximal ideal is XV, so $V_X = k(X, \alpha, \beta)$ and $XV = Xk[[X]] \cap V$. Thus $k \subseteq V/XV \subseteq k[[X]]/Xk[[X]] = k$, so that equality holds. Thus by Exercise 19.1, the completion of V is $k[[X]]$. Hence the ring $S = V[Y, Z]_{(X,Y,Z)}$ is Noetherian with completion $k[[X, Y, Z]]$, $S/(X^n) = k[X, Y, Z]_{(X,Y,Z)}/(X^n)$, and so the (X)–adic completion of S is

$k[Y, Z]_{(Y,Z)}[[X]]$. Let $A = k[X, Y, Z]_{(X,Y,Z)}$, and A^* its (X)–adic completion. Then $A^* = k[Y, Z]_{(Y,Z)}[[X]]$, and $S \to A^*$ is flat, and *a fortiori* also faithfully flat. Observe that

$$k[X, Y, Z, f, g] \hookrightarrow k[X, Y, Z, f, g]_X = R'_X \hookrightarrow k[X, Y, Z, \alpha, \beta]_{(X,Y,Z,\alpha,\beta)}[X^{-1}],$$

and that $k[X, Y, Z, \alpha, \beta]$ is free of rank four over $k[X, Y, Z, f, g]$. Hence the last ring in the display, $k[X, Y, Z, \alpha, \beta]_{(X,Y,Z,\alpha,\beta)}[X^{-1}]$, is flat over R_X, and thus over R. But S_X is a localization of $k[X, Y, Z, \alpha, \beta]_{(X,Y,Z,\alpha,\beta)}[X^{-1}]$, so that $R \to A^*_X$ factors through flat maps $R \to S_X \to A^*_X$, so that A^*_X is flat over R. As $R \subseteq A^*$, X is a non–zerodivisor on R and A^*, and $R/XR = A^*/XA^*$, so by Exercise 19.1, A^* is flat over R, and hence faithfully flat as R has only one maximal ideal, and this maximal ideal extends to a proper ideal in A^*. Thus R is Noetherian. Moreover, Exercise 19.1 shows that the (X)–adic completion of R is A^*, whence the (X, Y, Z)–adic completion of R is $k[[X, Y, Z]]$. In particular, as R is Noetherian, it is even regular local, with X, Y, Z a regular system of parameters. Let $P = (Y - \alpha, Z - \beta)k[[X, Y, Z]] \cap R$. Then P is not the maximal ideal, it contains f, g, so that P is a prime ideal of height exactly two. But $P\widehat{R}$ is not integrally closed as $(Y - \alpha)(Z - \beta)$ is in its integral closure, but $(Y - \alpha)(Z - \beta)$ is not in $Pk[[X, Y, Z]]$.

19.3. Inductive limits of normal rings

Throughout this section, let Λ be a partially ordered directed set, and let $\{R_i, \varphi_{ij}\}$ be a direct system of rings on Λ, with $\varphi_{ij} : R_i \to R_j$ a ring homomorphism whenever $i \leq j$. We set $R = \varinjlim R_i$, and by φ_j we denote the canonical homomorphism $R_j \to R$.

Proposition 19.3.1

(1) For every $i, j \in \Lambda$ with $i \leq j$, let J_i be an ideal of R_i satisfying $\varphi_{ij}(J_i) \subseteq J_j$. Then $\{J_i\}$ form a direct system with the restrictions of the maps φ_{ij}, and $J = \varinjlim J_i$ can be canonically identified with an ideal J of R. Moreover, R/J is canonically isomorphic to $\varinjlim(R_i/J_i)$.

(2) Conversely, for every ideal J of R, setting $J_i = \varphi_i^{-1}(J)$, the $\{J_i\}$ form a direct system and $J = \varinjlim J_i$.

(3) If the J_i are prime for all i, then J is also prime. Moreover, if $J_j = \varphi_{ji}^{-1}(J_i)$, then the localization R_J is canonically identified with $\varinjlim(R_i)_{J_i}$.

(4) If R_i are integrally closed domains for all i, and if all φ_{ij} are injective, then R is an integrally closed domain.

(5) If R_i is normal for every i, and if minimal prime ideals of R_i contract to minimal prime ideals of R_j whenever $j \leq i$, then R is normal.

Proof: The first part is easy to prove and we leave it to the reader. The second part follows directly from (1). To prove (3), since $R/J \cong \varinjlim(R_i/J_i)$, the first claim of (3) follows if we prove that whenever R_i is a domain for all i, so is R. Suppose that $x, y \in R$ such that $xy = 0$. Choose $i \in \Lambda$ such that $\varphi_i(x_i) = x$ and $\varphi_i(y_i) = y$. Then $\varphi_i(x_i y_i) = 0$, and hence there is a $j \geq i$ such that with $x_j = \varphi_{ij}(x_i)$ and $y_j = \varphi_{ij}(y_i)$ we have that $x_j y_j = 0$. As R_j is a domain, this forces either x_j or y_j to be 0, and hence either $x = 0$ or $y = 0$. The hypothesis that $J_j = \varphi_{ji}^{-1}(J_i)$ forces the sets $R_j \setminus J_j$ to form a direct limit of multiplicatively closed subsets of R_j with direct limit $R \setminus J$. Thus $(R_i)_{J_i}$ form a direct limit system of local rings (and local homomorphisms) and clearly $\varinjlim (R_i)_{J_i}$ maps onto R_J. We need to prove injectivity. If $x_i \in R_i$ and $s_i \in R_i \setminus J_i$ are elements such that $\varphi_i(x_i)/\varphi_i(s_i) = 0$ in R_J, then there exists an element $u \in R \setminus J$ such that $u\varphi_i(x_i) = 0$, and by replacing i by a possibly larger index, we can assume there is a $u_i \in R_i \setminus J_i$ such that $\varphi_i(u_i) = u$. Then $\varphi_i(u_i x_i) = 0$, and hence there exists a $j \geq i$ such that $\varphi_{ij}(u_i)\varphi_{ij}(x_i) = 0$, which forces $x_j/s_j = 0$, proving (3).

We prove (4). Let K_i be the field of fractions of R_i. By (3), we can identify $K = \varinjlim K_i$ with the field of fractions of R. Suppose that $z \in K$ is integral over R, say $z^n + r_1 z^{n-1} + \cdots + r_n = 0$ for some $r_1, \ldots, r_n \in R$. There exist $j \in \Lambda$ and elements $r_{kj}, z_j \in K_j$ such that $\varphi_j(r_{kj}) = r_k$ and such that $\varphi_j(z_j) = z$. Then $\varphi_j(z_j^n + r_{1j} z_j^{n-1} + \cdots + r_{nj}) = 0$, and hence for some $i \geq j$, $z_i^n + r_{1i} z_i^{n-1} + \cdots + r_{ni} = 0$ where $z_i = \varphi_{ji}(z_j)$ and $r_{ki} = \varphi_{ji}(r_{kj})$. This shows that $z_i \in R_i$ as R_i is integrally closed, and it follows that $z \in R$.

Finally we prove (5). Let P be a prime ideal of R, and set $P_i = \varphi_i^{-1}(P)$. Since R_i is normal, $(R_i)_{P_i}$ is an integrally closed domain. By (3), $R_P = \varinjlim (R_i)_{P_i}$, so by (4), R_P is an integrally closed domain. This proves that R is normal. $\qquad\square$

19.4. Base change and normal rings

In this section we prove the main results concerning base change and normal homomorphisms. Our treatment follows that of EGA IV [101, Chapter IV, Section 6.14].

Lemma 19.4.1 *Let R be a Noetherian ring, and $R \to R'$ a flat map to a Noetherian ring R'. Suppose that $S \subseteq T$ are domains that are R–algebras. Set $S' = S \otimes_R R'$ and $T' = T \otimes_R R'$. Then $S' \subseteq T'$ and minimal prime ideals of T' contract to minimal prime ideals of S'.*

Proof: That $S' \subseteq T'$ is immediate from the assumption that $R \to R'$ is flat and $S \subseteq T$. Observe that $T' \cong T \otimes_S (S \otimes_R R') = T \otimes_S S'$, and that $S \to S'$ is flat. Every non–zero element of S is a non–zerodivisor on S, and hence also on S' and on T'. Likewise every non–zero element of T is a non–zerodivisor on T'. Thus every minimal prime ideal of T' contracts to 0 in T, and we may therefore invert all non–zero elements of both S and T to assume that S

and T are fields. But then T is flat over S, and therefore T' is flat over S'. By Proposition B.1.2 the homomorphism $S' \to T'$ satisfies the Going–Down property. Let Q be a minimal prime ideal in T', and set $q = Q \cap S'$. If q is not minimal, then there is a prime ideal $p \subsetneq q$, and the Going–Down property shows that there is a prime ideal $P \subsetneq Q$ in T' such that $P \subseteq S' = p$. This contradicts the minimality of Q.   □

A first step in our goal of understanding base change and integral closures is taken in the following theorem (see [101, Chapter IV, Proposition 6.14.1]):

Theorem 19.4.2   *Let $g : R \to R'$ be a normal homomorphism of locally Noetherian rings. For every normal ring S that is an R–algebra, the ring $S' = S \otimes_R R'$ is also normal.*

Proof: We first reduce this proof of the theorem to the case in which R and R' are local Noetherian, R is a domain, and S is the integral closure of R. Suppose we are given R, R', S as in the statement of the theorem. To prove that $S' = S \otimes_R R'$ is normal, it suffices to prove that for all prime ideals Q' of S', $(S')_{Q'}$ is an integrally closed domain. Set $P' = Q' \cap R'$, $Q = Q' \cap S$, and $P = Q' \cap R$. Then $(S')_{Q'}$ is a localization of $S_Q \otimes_{R_P} (R')_{P'}$, so it suffices to prove that the latter tensor product is normal. As the map $R_P \to (R')_{P'}$ is normal and S_Q is a normal R_P–algebra, without loss of generality we may assume that R and R' are local Noetherian and that S is local. But S local and normal implies that S is an integrally closed domain.

Some general observations facilitate the proof below. Suppose that CS is a finitely generated R–algebra contained in S. Set $C' = C \otimes_R R'$. By Proposition 19.1.2, $C \to C'$ is normal. Furthermore, S is a normal C–algebra, and $S \otimes_C C' \cong S \otimes_C (C \otimes_R R') \cong S \otimes_R R' = S'$. To prove the theorem we may replace R by C. Furthermore, whenever we make such a reduction, we can repeat the additional reductions of the first paragraph with C in place of R and C' in place of R' to reduce to the case where C and C' are local Noetherian, S is a local integrally closed domain, and then also necessarily C is a domain.

Consider all finitely generated R–subalgebras R_i of S, and let S_i be their integral closures, so that $S = \cup S_i$ is written as a direct limit of such rings. We have that $S' = (\cup S_i) \otimes_R R' = \cup(S_i \otimes_R R')$, the second equality following from the flatness of $R \to R'$. Set $S'_i = S_i \otimes_R R'$. By flatness, tensoring with R' preserves all inclusions among the S_i. By Proposition 19.3.1, a direct limit of normal algebras is again normal provided that every minimal prime ideal of S'_i contracts to a minimal prime ideal of S'_j when $S'_j \subseteq S'_i$. But since the S_i are domains, and $R \to R'$ is flat, this is true by Lemma 19.4.1. Thus without loss of generality we may assume that S is the integral closure of a finitely generated R–algebra C. As in the second paragraph of this proof, we can replace R by C, and again localize to assume that R and R' are local Noetherian, R is a domain, and S is the integral closure of R.

We next reduce to the case in which R is a one–dimensional local Noetherian domain with integral closure S a Noetherian valuation domain, and R' is local Noetherian. As R is Noetherian, S is a Krull domain (see Theorem 4.10.5). In particular, S is the intersection of Noetherian valuation rings V_i with the same field of fractions K of R such that every $x \in R$ is in at most finitely many of the V_i. Set $K' = K \otimes_R R'$ and $V_i' = V_i \otimes_R R'$. As K' is the fiber over 0 of the normal map $R \to R'$, K' is normal, and is contained in the total ring of fractions of R'. (Here note that R' sits inside K' and is thus reduced. The non–zero elements of R are non–zerodivisors on R, and hence are also non–zerodivisors on R' by flatness.) Moreover, the total ring of fractions of each V_i' is also the total ring of fractions of R', and each V_i' sits inside K', again by flatness.

There is an exact sequence $0 \to S \to K \to \oplus_i(K/V_i)$, which upon tensoring with R' gives the exact sequence

$$0 \to S' \to K' \to \oplus_i(K'/V_i').$$

Hence $S' = \cap_i V_i'$. Since S' and V_i' have the same total ring of fractions, by Lemma 2.1.15 if each V_i' is normal, then S' is normal.

Fix $V = V_i$. By Lemma 4.9.4 V is the integral closure of C_p, where C is a finitely generated R subalgebra of K and p is a height one prime ideal of C. Let $D = C_p$. By base change, the map $D \to D' = D \otimes_R R'$ is normal, and $V' = V \otimes_R R' \cong V \otimes_D (D \otimes_R R') = V \otimes_D D'$. Replacing R by D and R' by D', we may assume that R is a one–dimensional local Noetherian ring with integral closure S a Noetherian valuation domain, and R' is local Noetherian (by again localizing R' at the contraction of a given prime ideal in V').

We wish to make R as close as possible to S; ideally S would be finite as an R–module, in which case the proof is much easier. In any case, we may write S as a union of a nested sequence of finitely generated R–algebras R_i contained in S. In fact, since S is integral over R, each R_i is a finitely generated R–module. We are free to replace R by any one of the R_i by the discussion in the second paragraph of this proof. We wish to reduce to the case in which a uniformizing parameter of S is in R and the residue field of S is the same as the residue field of R. To see that this reduction is possible, consider passing to the completion of R. The injections of $R_i \to S$ stay injections upon tensoring with $\widehat{R}$, the completion of R. Moreover, $S \otimes_R \widehat{R}$ is contained in $K \otimes_R \widehat{R}$, where K is the field of fractions of R. In turn, $K \otimes_R \widehat{R}$ is contained in the total ring of fractions of $\widehat{R}$. Thus we have that $R_i \otimes_R \widehat{R}$ form a nested set of finitely generated $\widehat{R}$–algebras contained in the total ring of fractions of $\widehat{R}$. In particular, since $(S \otimes_R \widehat{R})_{red}$ must be a finitely generated $(\widehat{R})_{red}$–module (Proposition 19.1.3), there must exist an index i such that $(R_i \otimes_R \widehat{R})_{red} = (S \otimes_R \widehat{R})_{red}$. We can replace R by R_i and assume that $(\widehat{R})_{red} = (S \otimes_R \widehat{R})_{red}$. (Note that $\hat{R}_i = R_i \otimes_R \widehat{R}$ as R_i is a finitely generated R–module.) We can choose an element $t \in R$ that is a uniformizing parameter in S, i.e., such that tS is the maximal ideal of S. If not, then the isomorphism

$(\hat{R})_{red} = (S \otimes_R \hat{R})_{red}$ would give a contradiction upon further tensoring with $R/\mathfrak{m}$. Moreover, as above we may assume that R' is also local. Furthermore, $S/tS = R/\mathfrak{m} = k$ (from the isomorphism $(\hat{R})_{red} = (S \otimes_R \hat{R})_{red}$), so that in addition $S'/tS' = (R' \otimes_R S)/t(R' \otimes_R S) = R' \otimes_R k = R'/\mathfrak{m}R'$.

To summarize, we have reduced this proof of the theorem to the case in which R is a one–dimensional Noetherian local domain with integral closure S, a rank one discrete valuation domain, and R' is local. Furthermore, R contains a uniformizing parameter t of S, and the residue fields of R and S are the same, implying that $S'/tS' \cong R'/\mathfrak{m}R'$. We now prove that $S' = R' \otimes_R S$ is normal. We do this by proving that $S'[t^{-1}]$ is normal, that tS' is a prime ideal Q, that S'_Q is a Noetherian discrete valuation domain, and finally that $S' = S'[t^{-1}] \cap S'_Q$.

Let K be the field of fractions of R, which is just $R[t^{-1}]$. Note that $R \subseteq S \subseteq K$ induces containments $R' \subseteq S' \subseteq K \otimes_R R'$, proving that $R'[t^{-1}] = S'[t^{-1}] = K \otimes_R R'$, the fiber of the map $R \to R'$ over 0. In particular, $K \otimes_R R'$ is normal, and thus so is $S'[t^{-1}]$. This ring is even Noetherian.

We next prove that tS' is a prime ideal of S'. As noted above, $S'/tS' = R'/\mathfrak{m}R'$, and this latter ring is an integrally closed domain as it is the fiber of the normal homomorphism from R to R' over the maximal ideal $\mathfrak{m}$ of R. It follows that tS' is prime, and we denote it by Q. Let $P = \mathfrak{m}R'$.

We prove that S'_Q is a Noetherian discrete valuation domain. First note that the equality of S'/tS' and $R'/\mathfrak{m}R'$ proves that S' is local and $S'_Q = W^{-1}S'$, where $W = R' \setminus \mathfrak{m}R'$. The latter claim holds since Q is the unique prime ideal of S' lying over $\mathfrak{m}R'$. Set $A = R'_{\mathfrak{m}R'}$. By Proposition 19.1.3, A is a one–dimensional reduced Noetherian ring. Since $S' \subseteq R' \otimes_R K$, S' is contained in the total ring of fractions of R' (note that the non–zero elements of R are non–zerodivisors on R' as R' is flat over R), and hence $S'_Q = W^{-1}S'$ is contained in the total ring of fractions of A. An application of Theorem 4.9.2 now proves that S'_Q is Noetherian. Since the maximal ideal of S'_Q is generated by the image of t, it follows that S'_Q is a Noetherian discrete valuation domain.

By applying Lemma 2.1.15, to finish this proof of the theorem it suffices to prove that $S' = S'[t^{-1}] \cap S'_Q$. Suppose that $a, b, s \in S'$ are such that $a/t^n = b/s$ and $s \notin Q$. Then in S', $as = t^n b$, as this intersection is taking place in the total ring of fractions of S'. Since tS' is prime and t is a non–zerodivisor on S' (as $t \in R$), the fact that $s \notin tS' = Q$ shows that we may cancel all powers of t from a to obtain that $b = a's$ for some $a' \in S'$. In this case $b/s = a' \in S'$, proving the claim and finishing the proof of Theorem 19.4.2.□

Via Proposition 19.1.1, a consequence of the last theorem is:

Theorem 19.4.3 Let $F \subseteq K$ be a separable field extension. Let R be a Noetherian normal F–algebra. Then $R \otimes_F K$ is a normal K–algebra.

In particular, the complexification $R \otimes_{\mathbb{R}} \mathbb{C}$ of any affine normal $\mathbb{R}$–algebra R is normal. One can also prove this fact directly from the Jacobian criterion.

19.5. Integral closure and normal maps

Consider a map of rings $R \to S$ and the integral closure C of R in S. In this section we prove that if $R \to R'$ is a normal homomorphism of Noetherian rings, then C' is the integral closure of R' in S', where $(\_)'$ denotes tensoring with R' over R. That this is non–trivial can be seen already by considering the case in which R, S and R' are all fields. In this case, what needs to be proved is that if R' is a normal extension of R (or, equivalently, a separable extension by Proposition 19.1.1), and if C is the algebraic closure of R in R', then $C \otimes_R R'$ is the integral closure of $R \otimes_R R'$ in $S \otimes_R R'$. We use the results proved in Chapter 3 to handle this case.

A consequence is that under a normal map $R \to R'$ of Noetherian rings, the image in R' of an integrally closed ideal in I is integrally closed.

Theorem 19.5.1 ([101, Chapter IV, Proposition 6.14.4]) *Let R be a Noetherian ring, and let R' be a Noetherian R–algebra such that the map $R \to R'$ is normal. Let S be an R–algebra and let C be the integral closure of R in S. Set $S' = S \otimes_R R'$ and $C' = C \otimes_R R'$. Then C' can be identified with a subring of S', and under this identification, C' is the integral closure of R' in S'.*

Proof: The proof below proceeds by several reductions. Let J be the nilradical of S. Clearly $J \subseteq C$. Let $T = S_{red} = S/J$, and let D be the integral closure of R in T. If $c \in D \subseteq T$, then for some positive integer n and some $r_1, \ldots, r_n \in R$, $c^n + r_1 c^{n-1} + \cdots + r_n$ is in J, hence some power of it is zero. This means that the lift of c to S is in C, and proves that $D \subseteq C/J$. Assume that we have proved the theorem with T in place of S. Let $x \in S'$ be in the integral closure of R'. Under the natural map $S' \to T'$, the image of x is integral over R', hence by assumption it is in $D \otimes_R R'$. Since $D \subseteq C/J$, there exists $c \in C'$ such that $x - c \in J \otimes_R R' \subseteq C \otimes_R R' = C'$, whence $x \in C'$. Thus it suffices to prove the theorem in case S is reduced.

Next we reduce to the case in which S is a domain and is finitely generated over R. Write S as a directed union of finitely generated R–subalgebras S_i, and let C_i be the integral closure of R in S_i. Clearly $C = \cup C_i$. Set $C'_i = C_i \otimes_R R' \subseteq S'_i = S_i \otimes_R R'$. Then $C' = \cup C'_i$ and $S' = \cup S'_i$. If we prove that C'_i is the integral closure of R' in S'_i, then it easily follows that C' is the integral closure of R' in S'. Hence without loss of generality we may assume that S is reduced and finitely generated over R; in particular it is Noetherian with finitely many minimal prime ideals $P_1, \ldots, P_r$. Let $S_0 = \prod_i S/P_i$; as S is reduced, we can identify S as a subring of S_0. Moreover, letting C_0 be the integral closure of R in S_0, we have that $C = C_0 \cap S$. Use $(\_)'$ to denote tensoring with R' over R. By flatness $C' = C'_0 \cap S'$, so to prove that C' is the integral closure of R' in S', it suffices to prove that C'_0 is the integral closure of R' in S'_0. (Note that S' is identified as a subring of S'_0.) Set $S_i = S/P_i$ and let D_i be the integral closure of R in S_i. If we prove that D'_i is the integral closure of R' in S'_i then it easily follows that $C'_0 = D'_1 \times \cdots \times D'_r$ is the integral closure of C_0 in in S_0. Hence we can assume that S is a domain, finitely generated over R.

We may assume that $R \subseteq S$. For let p be the kernel of the map from R to S. Then $S' = S \otimes_{R/p} (R'/pR')$. Moreover the map from $R/p \to R'/pR'$ is normal by Proposition 19.1.2. Thus we may assume that $R \subseteq S$.

Let K be the field of fractions of R and let L be the field of fractions of S, which is a finitely generated field extension of K. Let F be the algebraic closure of K in L; F is a finite algebraic extension of K by Proposition 3.3.2. Let C_0 be the integral closure of R in F, which is the same as the integral closure of R in L. Thus $C = C_0 \cap S$. Again we let $(\_)'$ denote tensoring with R' over R. By flatness, $C' = C_0' \cap S'$. We claim that C_0' is the integral closure of R' in F', and moreover that F' is a reduced Noetherian ring such that the map $F \to F'$ is normal. R' is Noetherian and F is a localization of an R–algebra A that is a finitely generated R–module. Then F' is a localization of A', which is finite as an R'–module, hence Noetherian. $F \to F'$ is normal by Proposition 19.1.2. The harder part is to prove the claim that C_0' is the integral closure of R' in F', and we prove this in the following paragraph.

As in the above paragraph, F' must be a reduced Noetherian ring with total ring of fractions L, a product of fields, all of them geometrically normal. As F is the field of fractions of C_0, F' is a subring of the total ring of fractions of C_0' and therefore the total ring of fractions of C_0' must be L. By Theorem 19.4.2 applied to the normal ring C_0 and the normal homomorphism $R \to R'$, it follows that C_0' is normal. By Lemma 2.1.15 it follows that C_0' is integrally closed in L, and thus also is integrally closed in F', and is therefore the integral closure of R' in F'.

By flatness, $C' = C_0' \cap S'$, so to prove that C' is integrally closed in S', it suffices, using the work in the above paragraph, to prove that F' is integrally closed in L'. Now we may replace R by F, S by L, and R' by $R' \otimes_R K = F'$. We have shown above that F' is Noetherian and $F \to F'$ is normal. To summarize, without loss of generality, we may assume that F is a field, $F \subseteq L$ is a finitely generated field extension such that F is algebraically closed in L, and $F \to F'$ is a normal homomorphism such that F' is Noetherian. It remains to prove that F' is integrally closed in $L' = F' \otimes_F L$.

Because $F \to F'$ is a normal homomorphism, F' is normal. Hence $F' = F_1' \times \cdots \times F_r'$ where F_i' is an integrally closed domain whose field of fractions K_i is separable over F (Proposition 19.1.3). It follows that F' is integrally closed in $L' = F' \otimes_F L$ if and only if F_i' is integrally closed in $L_i' = F_i' \otimes_F L$ for all $1 \leq i \leq r$, and we may therefore assume that F' is an integrally closed domain with field of fractions K that is separable over F. As L is flat over F, we can identify $L' = F' \otimes_F L$ as a subring of $K \otimes_F L$. Since F is algebraically closed in L and K is separable over F, by Theorem 3.3.3 we conclude that K is algebraically closed in $K \otimes_F L$. Any element of $F' \otimes_F L$ that is integral over F' is therefore in K, and since F' is integrally closed, it is in F'. $\square$

We can now prove the main application of this theorem to the behavior of integral closures of ideals under normal homomorphisms.

Corollary 19.5.2   *Let $f : R \to R'$ be a normal ring homomorphism of Noetherian rings. If I is an ideal in R, then $\bar{I}R' = \overline{IR'}$.*

Proof: Let t be a variable. Set $S = R[t]$. The integral closure C of $R[It]$ in S is the $\mathbb{N}$–graded ring whose nth graded piece is $\overline{I^n}t^n$ by Proposition 5.2.1. By Theorem 19.5.1, the integral closure of $R[It] \otimes_R R'$ in $S \otimes_R R' \cong R'[t]$ is $C \otimes_R R'$. There is a surjection $R[It] \otimes_R R' \to R'[IR't]$. Hence $C \otimes_R R'$ can be identified with the integral closure of $R'[IR't]$ in $R'[t]$, giving that for all n, $\overline{I^n}R' = \overline{I^nR'}$. $\square$

A notion related to normal homomorphisms is the following:

Definition 19.5.3 (Lipman [186]) *A ring R and an R–algebra R' **satisfy the condition** $(N_{R,R'})$ if whenever C is an R–algebra and S is a C–algebra in which C is integrally closed, then also $C \otimes_R R'$ is integrally closed in $S \otimes_R R'$. If R' is flat over R and $(N_{R,R'})$ holds, then R' called a **quasi–normal R–algebra**.*

Many examples of quasi–normal R–algebras R' are given in the exercises.

Lipman proved [186, Lemma 2.4]: if R is a ring and R' an R–algebra such that $(N_{R,R'})$ holds, then for every ideal I in R, $\bar{I}R' = \overline{IR'}$.

Theorem 19.5.1 guarantees that if $R \to R'$ is normal, then $(N_{R,R'})$ holds, and so R' is quasi–normal over R. Our proof of Corollary 19.5.2 is then basically the same as that of Lipman.

Lipman's work on quasi–normal rings and condition $(N_{R,R'})$ arose from his study of **Lipschitz saturation**, which was introduced by Pham and Teissier in [221]. The definition of relative Lipschitz saturation is based on the concept of integral closure of "diagonal ideals" and is related to a notion extensively studied by Zariski in connection with his theory of equisingularity.

19.6. Exercises

19.1   (Heinzer, Rotthaus, Wiegand [113]) Let $A \subseteq R$ be rings, not necessarily Noetherian. Let $x \in A$ be a non–zerodivisor in A and in R such that $\frac{A}{xA} = \frac{R}{xR}$.

   (i) Prove that for all integers n, $x^n A = x^n R \cap A$ and $\frac{A}{x^n A} = \frac{R}{x^n R}$.
   (ii) Prove that the xA–adic completion of A equals the xR–adic completion of R.
   (iii) Prove that $A = A_x \cap R$ and $R_x = A_x + R$.
   (iv) Prove that $0 \to A \xrightarrow{\varphi} A_x \oplus R \xrightarrow{\psi} R_x \to 0$ is a short exact sequence, where $\varphi(a) = (\frac{a}{1}, a)$ and $\psi(\frac{a}{x^n}, r) = \frac{a}{x^n} - r$.
   (v) Prove that R is flat over A if and only if R_x is flat over A.

19.2   Let $k \subseteq \ell$ be a separable field extension and $k \subseteq K$ a field extension. Suppose that either ℓ or K is finitely generated over k. Prove that $\ell \otimes_k K$ is a regular ring.

Quasi–normal algebras, Lipman [186]

19.3   Let R be a Noetherian ring and W a multiplicatively closed set in R. Prove that $R' = W^{-1}R$ is a quasi–normal R–algebra.

19.4   Let R be a Noetherian ring and set $R' = R[X_1, \ldots, X_n]$. Prove that R' is a quasi–normal R–algebra.

19.5   If R' is a quasi–normal R–algebra and R'' is a quasi–normal R'–algebra, prove that R'' is a quasi–normal R–algebra.

19.6   If R' is a quasi–normal R–algebra and S is any R–algebra, prove that $S' = S \otimes_R R'$ is a quasi–normal R'–algebra.

19.7   Prove that a filtered inductive limit of quasi–normal R–algebras is a quasi–normal R–algebra.

19.8*   Prove that the completion of an excellent ring R at an ideal I is a quasi–normal R–algebra.

19.9*   Prove that if R' is finitely presented, flat and unramified over R, (i.e., R' is étale over R), then R' is a quasi–normal R–algebra.

19.10*   Let k be a field of characteristic zero and K any field extension of k. Prove that for any variables $X_1, \ldots, X_d$ over K, $k[[X_1, \ldots, X_d]] \to K[[X_1, \ldots, X_d]]$ is normal.

A
Some background material

In this appendix we list several well–known results in commutative algebra that we need in the book. We provide proofs only for some results.

A.1. Some forms of Prime Avoidance

Theorem A.1.1 (Prime Avoidance) *Let R be a ring, not necessarily Noetherian. Let $P_1, \ldots, P_s$ be ideals of R, at most two of which are not prime ideals. Assume that $I \subseteq P_1 \cup \cdots \cup P_s$. Then there exists $i \in \{1, \ldots, s\}$ such that $I \subseteq P_i$.*

Theorem A.1.2 (Prime Avoidance [201]) *Let R be Noetherian $\mathbb{N}$–graded ring such that R_0 has infinite residue fields. Let $I, P_1, \ldots, P_s$ be homogeneous ideals of R, such that I is generated by elements of degree n. Suppose that I is contained in $P_1 \cup \cdots \cup P_s$. Then there exists $i \in \{1, \ldots, s\}$ such that $I_n \subseteq P_i$.*

Theorem A.1.3 (Prime Avoidance) *Let R be Noetherian $\mathbb{N}$–graded ring. Let $P_1, \ldots, P_s$ be homogeneous ideals in R, at most two of which are not prime ideals. If I is a homogeneous ideal generated by elements of positive degree and not contained in $P_1 \cup \cdots \cup P_s$, then there exists a homogeneous element $x \in I$ that is not in any P_i.*

A.2. Carathéodory's theorem

Theorem A.2.1 (Carathéodory's theorem) *Let n be a positive integer, and $v_1, \ldots, v_r \in \mathbb{R}^n$. Let S be the set of all linearly independent subsets of $\{v_1, \ldots, v_r\}$. Then if F stands for $\mathbb{R}$ or $\mathbb{Q}$,*

$$v_1 F_+ + \cdots + v_r F_+ = \bigcup_{\{v_{i_1}, \ldots, v_{i_s}\} \in S} (v_{i_1} F_+ + \cdots + v_{i_s} F_+),$$

Proof: If $v_1, \ldots, v_r$ are linearly independent, there is nothing to show. So assume that there exist $c_i \in F$, not all zero, such that $\sum_i c_i v_i = 0$. We may assume that $c_1 > 0$.

Let $v = a_1 v_1 + \cdots + a_r v_r$ for some non–negative $a_1, \ldots, a_r \in F$. It suffices to show that v can be written as a linear combination of $r - 1$ of the v_i, with non–negative coefficients in F. For contradiction we assume the contrary. Note that for any i such that $c_i \neq 0$, $v = \sum_{j \neq i} (a_j - a_i \frac{c_j}{c_i}) v_j$. If v cannot be written as a linear combination of the v_j with $j \neq i$ and non–negative coefficients, then there exists $j \neq i$ such that $a_j < a_i \frac{c_j}{c_i}$. In particular, $a_i > 0$.

We first start with $i = 1$. By reindexing j if necessary, we may then assume that $a_2 < a_1 \frac{c_2}{c_1}$. Thus $c_2 > 0$. Thus we get that $a_2/c_2 < a_1/c_1$. By repeating this step with $i = 2$, by possible reindexing of $v_3, \ldots, v_r$, we may assume that $a_3/c_3 < a_2/c_2$. By repeated applications of this step for $i = 3, \ldots, r - 1$, we then get that all c_i are positive and $a_r/c_r < \cdots < a_3/c_3 < a_2/c_2 < a_1/c_1$. Hence $v = \sum_{j \neq r} (a_j - a_r \frac{c_j}{c_r}) v_j$ gives v as a linear combination of $v_1, \ldots, v_{r-1}$ with non–negative coefficients. If these $r - 1$ vectors are linearly independent, we are done, otherwise we repeat the procedure. $\quad\square$

A.3. Grading

Proposition A.3.1 *Let G be a well–ordered abelian monoid (e.g., $\mathbb{N}^d \times \mathbb{Z}^e$). Let R be a G–graded ring and M a G–graded module. Let I be a homogeneous ideal in R, $x \in M$ (not necessarily homogeneous), and P a prime ideal such that $P = (0 :_R x)$. Then P is G–homogeneous.*

Proof: Let r be a non–zero element of P. Write $r = \sum_{i=1}^n r_i$ for some non–zero homogeneous elements r_i in R, with $\deg(r_1) > \cdots > \deg(r_n)$. We want to show that each $r_i \in P$. Let s be the sum of those r_i that are contained in P. Then $r - s \in P$, and it suffices to prove that each homogeneous summand of $r - s$ is in P. Thus we may assume (by possibly renaming r) that no $r_i \in P$.

Write $x = \sum_{i=1}^m x_i$ for some non–zero homogeneous elements x_i in M, with $\deg(x_1) > \cdots > \deg(x_m)$. Since $0 = rx$, by reading off the component of degree $\deg(r_1 x_1)$ we get that $r_1 x_1 = 0$. Suppose that we have proved that for some $e > 2$, $r_1^i x_i = 0$ for all $i = 1, \ldots, e - 1$. Then

$$0 = r_1^{e-1} r x = r_1^{e-1} \sum_{i=1}^n \sum_{j=1}^m r_i x_j = r_1^{e-1} \sum_{i=1}^n \sum_{j=e}^m r_i x_j$$

yields that the leading term $r_1^e x_e$ is zero. Thus induction shows that $r_1^m x_i = 0$ for all i. In particular, $r_1^m x = 0$, so that $r_1^m \in (0 :_R x) = P$, whence $r_1 \in P$. This contradicts the assumption, and hence proves the proposition. $\quad\square$

Corollary A.3.2 *Let G be a well–ordered abelian monoid (e.g., $\mathbb{N}^d \times \mathbb{Z}^e$), R a G–graded Noetherian ring, and M a finitely generated G–graded module. Then every G–graded submodule of M has a G–homogeneous primary decomposition and that the associated prime ideals are G–homogeneous.*

Proof: In a Noetherian ring, all associated prime ideals of finitely generated modules are expressible as annihilators of one element. Thus by Proposition A.3.1, all associated prime ideals are G–homogeneous.

Let N be a G–graded submodule of M. We know that all associated prime ideals of M/N are homogeneous. Let P be associated to M/N and be maximal among associated prime ideals.

First suppose that P is the only associated prime ideal of M/N. Then N is P–primary and also G–graded, hence there is nothing to prove.

So we may assume that P is not the only associated prime ideal of M/N. Choose a homogeneous element r in P that is not in some other associated prime ideal. By maximality assumption on P, such r exists. As M is Noetherian, the ascending chain $N \subseteq N :_M r \subseteq N :_M r^2 \subseteq \cdots$ terminates, say at $N :_M r^e$. Then $N = (N + r^e M) \cap (N :_M r^e)$ (one inclusion is trivial; if $x = n + r^e m$ for some $n \in N$ and $m \in M$, and if $r^e x \in N$, then $r^{2e} m \in N$, whence by the choice of e, $r^e m \in N$ and $x \in N$). As r is homogeneous, so are $N :_M r^e$ and $N + r^e M$. By the choice of r, P is not associated to $N :_M r^e$, so in particular $N :_M r^e$ is strictly bigger than N. Also, r is not nilpotent on M/N, so that $N + r^e M$ is strictly bigger than N. By Noetherian induction we can find homogeneous primary decomposition of each of these strictly larger G–graded modules. The intersection of all of their components is a possibly redundant homogeneous primary decomposition of N and then we simply remove the redundant intersectands. $\square$

Theorem A.3.3 *Let R be $\mathbb{Z}$–graded ring. Let $r \in R_0$ and let p be the prime ideal in R_0 of the form $(rR_0 :_{R_0} s)$ for some $s \in R_0$. Then there exists a homogeneous element $s' \in R$ such that $P = rR :_R ss'$ is a prime ideal such that $P \cap R_0 = p$.*

A.4. Complexes

A **complex** of R–modules is a countable collection $\{M_i\}$ of R–modules together with a collection of maps $\{\delta_i : M_i \to M_{i-1}\}$ such that for all i, $\delta_{i-1} \circ \delta_i = 0$. We encode all this information into saying that $(\mathbb{M}, \delta)$ is a complex. This complex is said to be **exact** if for all i, $\ker \delta_i = \operatorname{Im} \delta_{i+1}$. The ith **homology** of the complex is the R–module $(\ker \delta_i)/(\operatorname{Im} \delta_{i+1})$. A complex is **bounded from the right** if for all large i, $M_{-i} = 0$. A complex with $0 = M_0 = M_{-1} = M_{-2} = \cdots$ is **acyclic** if for all $i > 1$, the ith homology is zero.

Some standard examples of complexes are projective, free and injective **resolutions**. We assume familiarity with projective and free resolutions. Let R be a ring and M and N modules over R.

(1) If $\mathbb{F}$ is a free or projective resolution of M, then $\mathbb{F} \otimes N$ is a complex whose ith homology is denoted $\operatorname{Tor}_i^R(M, N)$. We use freely the fact that $\operatorname{Tor}_i^R(M, N) \cong \operatorname{Tor}_i^R(N, M)$.

(2) If $\mathbb{F}$ is an injective resolution of M, then $\operatorname{Hom}_R(N, \mathbb{F})$ is a complex whose ith homology is denoted $\operatorname{Ext}_R^i(N, M)$.

Another standard complex is the Koszul complex, which we define explicitly. Let R be a ring, let $x_1, \ldots, x_n \in R$ and let M be an R–module. The **Koszul complex** of $(x_1, \ldots, x_n)$ on M is the complex whose ith module is $\Lambda_R^i(R^n) \otimes_R M$ (so all except finitely many modules are zero). If $\{e_1, \ldots, e_n\}$ is a basis of R^n and $1 \le n_1 < \cdots < n_i \le n$, the ith map of the Koszul complex is defined by

$$e_{n_1} \wedge \cdots \wedge e_{n_i} \otimes m \mapsto \sum_{j=1}^{i} (-1)^j x_j e_{n_1} \wedge \cdots \wedge \widehat{e_j} \wedge \cdots \wedge e_{n_i} \otimes m.$$

Another way to view Koszul complexes is as follows:

$$K(x_1; M) := 0 \longrightarrow M \xrightarrow{x_1} M \longrightarrow 0,$$

where multiplication by x_1 is a map from $K(x_1; M)_1$ to $K(x_1; M)_0$, and for $n > 1$, $K(x_1, \ldots, x_n; M)$ is the tensor product of the complexes $K(x_1; R)$ and $K(x_2, \ldots, x_n; M)$.

The homology of $K(x_1, \ldots, x_n; M)$ is naturally isomorphic to the homology of the Koszul complex on the reordered $x_1, \ldots, x_n$, and similarly it is naturally isomorphic to the homology of the Koszul complex of the elements obtained from $x_1, \ldots, x_n$ by adding a multiple of one x_i to another x_j. Thus the homology of $K(x_1, \ldots, x_n; M)$ is independent of the generating set $x_1, \ldots, x_n$ of the ideal $(x_1, \ldots, x_n)$.

If $(\mathbb{M}, \delta)$ and $(\mathbb{M}', \delta')$ are two complexes of R–modules, then $\varphi : \mathbb{M} \to \mathbb{M}'$ is a **map of complexes** if for all i, $\varphi_i : \mathbb{M}_i \to \mathbb{M}'_i$ is an R–module homomorphism and $\delta'_i \circ \varphi_i = \varphi_{i-1} \circ \delta_i$. This map φ is **injective**, respectively **surjective**, if for all i, φ_i is injective, respectively surjective. By standard construction, a short exact sequence

$$0 \longrightarrow \mathbb{M} \longrightarrow \mathbb{M}' \longrightarrow \mathbb{M}'' \longrightarrow 0$$

of complexes yields a long exact sequence on homology:

$$\cdots \to H_{i+1}(\mathbb{M}'') \to H_i(\mathbb{M}) \to H_i(\mathbb{M}') \to H_i(\mathbb{M}'') \to H_{i-1}(\mathbb{M}) \cdots .$$

One can practice this on the short exact sequence $0 \to K(x_2, \ldots, x_n; M) \to K(x_2, \ldots, x_n; M) \otimes_R K(x_1; R) \to K(x_2, \ldots, x_n; M)(-1) \to 0$ (shift of labelling of graded parts of the complex; explain how this is a short exact sequence of complexes) to prove that the grade (depth) of $(x_1, \ldots, x_n)$ on M equals

$$n - \max\{j \,|\, \text{the } j\text{th homology of } K(x_1, \ldots, x_n; M) \text{ is non–zero}\}.$$

In other words, the Koszul complex is depth–sensitive.

Theorem A.4.1 (Rees) *Let R be a ring, M and N R–modules and $x \in R$ a non–zerodivisor on R and on M for which $xN = 0$. Then for all $n \geq 0$,*

$$\operatorname{Ext}_R^{n+1}(N, M) = \operatorname{Ext}_{R/xR}^n(N, M/xM).$$

In a few places in the book we assume the Buchsbaum–Eisenbud's Acyclicity Criterion [31]. A more general statement is in 16.8.4. A simple version of it is the following result:

Theorem A.4.2 (Hilbert–Burch Theorem) *Let R be a Noetherian ring. Let A be an $(n-1) \times n$ matrix with entries in R and let d_i be the determinant of the matrix obtained from A by deleting the ith column. Let I be the cokernel of the matrix A. Then $I = a(d_1, \ldots, d_n)$ for some non–zerodivisor $t \in R$.*

A.5. Macaulay representation of numbers

We use the convention that for any integers $a < b$, $\binom{a}{b} = 0$. It is easy to prove by induction on n that for all d,

$$\binom{n+d}{n} - 1 = \binom{n}{1} + \binom{n+1}{2} + \binom{n+2}{3} + \cdots + \binom{n+d-1}{d}.$$

Lemma A.5.1 *Let d be a positive integer. Any positive integer n can be expressed uniquely as*

$$n = \binom{c_d}{d} + \binom{c_{d-1}}{d-1} + \cdots + \binom{c_2}{2} + \binom{c_1}{1},$$

where $c_1, \ldots, c_d$ are non–negative integers satisfying $0 \le c_1 < c_2 < \cdots < c_{d-1} < c_d$.

Proof: This is trivially true if $d = 1$. Now assume that $d > 1$. Let c_d be the largest integer such that $\binom{c_d}{d} \le n$. If $n = \binom{c_d}{d}$, we are done. So suppose that $n > \binom{c_d}{d}$. By induction on d, there exist non–negative integers $c_1, \ldots, c_{d-1}$ such that $0 \le c_1 < c_2 < \cdots < c_{d-1}$ and such that $n - \binom{c_d}{d} = \binom{c_{d-1}}{d-1} + \cdots + \binom{c_2}{2} + \binom{c_1}{1}$. If $c_d \le c_{d-1}$, then $n \ge \binom{c_d}{d} + \binom{c_{d-1}}{d-1} \ge \binom{c_d}{d} + \binom{c_d}{d-1} = \binom{c_d+1}{d}$, contradicting the choice of c_d. Thus necessarily $c_d > c_{d-1}$.

It remains to prove that $c_1, \ldots, c_d$ are unique. Let $c'_1, \ldots, c'_d$ be non–negative integers satisfying $0 \le c'_1 < \cdots < c'_d$ and $n = \binom{c'_d}{d} + \cdots + \binom{c'_1}{1}$. Let i be the maximum integer such that $c_i \ne c'_i$. By possibly modifying n and d, without loss of generality $i = d$. By the choice of c_d, necessarily $c'_d < c_d$. Then

$$n = \binom{c'_d}{d} + \binom{c'_{d-1}}{d-1} + \cdots + \binom{c'_2}{2} + \binom{c'_1}{1},$$
$$\le \binom{c'_d}{d} + \binom{c'_d - 1}{d-1} + \cdots + \binom{c'_d - d + 2}{2} + \binom{c'_d - d + 1}{1}$$
$$= \binom{c'_d + 1}{d} - 1 \le \binom{c_d}{d} - 1,$$

which contradicts the choice of c_d. $\qquad\square$

Definition A.5.2 *Let n and d be positive integers. The dth **Macaulay representation** of n is the writing of n as*

$$n = \binom{c_d}{d} + \binom{c_{d-1}}{d-1} + \cdots + \binom{c_2}{2} + \binom{c_1}{1},$$

where $c_1, \ldots, c_d$ are non–negative integers satisfying $0 \le c_1 < c_2 < \cdots < c_{d-1} < c_d$. With this define

$$n_{\langle d \rangle} = n = \binom{c_d - 1}{d} + \binom{c_{d-1} - 1}{d-1} + \cdots + \binom{c_2 - 1}{2} + \binom{c_1 - 1}{1}.$$

B
Height and dimension formulas

B.1. Going–Down, Lying–Over, flatness

Proposition B.1.1 *A faithfully flat algebra homomorphism $R \to S$ satisfies the Lying–Over condition (meaning that every prime ideal in R is contracted from a (prime) ideal in S). In fact, any prime ideal P in R is the contraction of every prime ideal in S that is minimal over PS.*

Proof: Let $P \in \operatorname{Spec} R$. By faithful flatness, $PS \neq S$, so there exist prime ideals in S containing PS. Let $r \in R \setminus P$. Multiplication by r on R/P is an injective map, so that by flatness, multiplication by r on S/PS is an injective map. Then r cannot be in any minimal prime ideal over PS, which proves that P is a contraction of any prime ideal in S that is minimal over PS. $\square$

Proposition B.1.2 *A flat homomorphism $R \to S$ of rings satisfies the Going–Down condition, i.e., for any prime ideals $P_1 \subseteq P_2$ in R such that P_2 is contracted from a prime ideal Q_2 in S there exists a prime ideal $Q_1 \subseteq Q_2$ in S such that $Q_1 \cap R = P_1$.*

Proof: By localizing we may assume that R is local with maximal ideal P_2 and S is local with maximal ideal Q_2. Then the ring map is even faithfully flat. It suffices to prove that every prime ideal in R contracts from a prime ideal in S. But that holds by Proposition B.1.1. $\square$

Lemma B.1.3 *Let $R \to S$ be a ring homomorphism that satisfies the Going–Down and Lying–Over conditions. Let I be an ideal in R. Then every prime ideal in R minimal over I is contracted from a prime ideal in S minimal over IS, and conversely, every prime ideal in S minimal over IS contracts to a prime ideal in R minimal over I.*

Proof: Let P be a prime ideal in R minimal over I. By Lying–Over, there exists $Q' \in \operatorname{Spec} S$ such that $Q' \cap R = P$. Necessarily Q' contains IS. Let Q be a prime ideal in S that is contained in Q' and is minimal over IS. Then $I \subseteq Q \cap R \subseteq Q' \cap R = P$. But P is minimal over I, so that $Q \cap R = P$. Thus every prime ideal P in R minimal over I is contracted from a prime ideal Q in S minimal over IS.

Now let Q be a prime ideal in S minimal over IS. Let $P = Q \cap R$. Then P contains I. Let P' be a prime ideal in R contained in P and minimal over I. By Going–Down, there exists $Q' \in \operatorname{Spec} S$ such that $Q' \subseteq Q$ and $Q' \cap R = P'$.

As P' contains I, then Q' contains IS. But Q is minimal over IS, so that $Q = Q'$, whence $P = P'$. It follows that every prime ideal Q in S minimal over IS contracts to a prime ideal P in R minimal over I. $\square$

B.2. Dimension and height inequalities

Theorem B.2.1 (Krull's Height Theorem) *Let R be a Noetherian ring and I an ideal generated by d elements. Then every prime ideal that is minimal over I has height at most d.*

Theorem B.2.2 *Let $R \to S$ be a ring homomorphism of Noetherian rings, $Q \in \operatorname{Spec} S$ and $P = Q \cap R$. Then $\operatorname{ht} Q \leq \operatorname{ht} P + \dim(S \otimes_R \kappa(P))$.*

If the Going–Down condition is satisfied and Q is the unique maximal ideal of S, then $\operatorname{ht} Q = \operatorname{ht} P + \dim(S \otimes_R \kappa(P))$.

Proof: Without loss of generality R and S are local with P the maximal ideal of R and Q the maximal ideal of S. Let $m = \operatorname{ht} P$, $n = \dim(S/PS)$. Then there exist $x_1, \ldots, x_m \in P$ such that P is minimal over $(x_1, \ldots, x_m)$, and there exist $y_1, \ldots, y_n \in Q$ such that the images of $y_1, \ldots, y_n$ form a system of parameters in S/PS. Thus there exist integers u and v such that $P^v \subseteq (x_1, \ldots, x_m)$ and $Q^u \subseteq (y_1, \ldots, y_n) + PS$. Hence $Q^{uv} \subseteq (y_1, \ldots, y_n, x_1, \ldots, x_m)$. Thus by Krull's Height Theorem (Theorem B.2.1), $\operatorname{ht} Q \leq m + n$.

Now assume the Going–Down condition is satisfied. Let $Q' \in \operatorname{Spec} S$ such that $PS \subseteq Q'$ and $\dim(S/Q') = \dim(S/PS)$. Necessarily Q' contracts to P in R. By the Going–Down condition then $\operatorname{ht} P \leq \operatorname{ht} Q'$. Hence $\operatorname{ht} P + \dim(S/PS) \leq \operatorname{ht} Q' + \dim(S/Q') \leq \dim S = \operatorname{ht} Q$. $\square$

Proposition B.2.3 *Let $R \to S$ be an extension of Noetherian rings that satisfies the Going–Down condition. (For example, the extension could be flat, by Proposition B.1.2.) Let $P \in \operatorname{Spec} R$, and $Q \in \operatorname{Spec} S$ be minimal over PS. Then $\operatorname{ht} P = \operatorname{ht} Q$.*

Proof: The fiber $S \otimes_R \kappa(P)$ localized at the complement of Q is a zero–dimensional ring. Thus by Theorem B.2.2, $\operatorname{ht} Q = \operatorname{ht} P$. $\square$

Proposition B.2.4 *Let $R \subseteq S$ be an extension of Noetherian rings that satisfies the Going–Down and Lying–Over conditions. Then for every ideal I of R, $\operatorname{ht}(I) = \operatorname{ht}(IS)$.*

Proof: Let P be a prime ideal in R minimal over I. By Lemma B.1.3, there exists a prime ideal Q in S that is minimal over IS and contracts to P. By Proposition B.2.3, $\operatorname{ht} Q = \operatorname{ht} P$. If P is chosen so that $\operatorname{ht} I = \operatorname{ht} P$, this proves that $\operatorname{ht} I \geq \operatorname{ht}(IS)$.

Now let Q be a prime ideal in S minimal over IS. By Lemma B.1.3, $Q \cap R$ is minimal over I. By Proposition B.2.3, $\operatorname{ht} Q = \operatorname{ht}(Q \cap R)$. If Q is chosen so that $\operatorname{ht}(IS) = \operatorname{ht} Q$, this proves that $\operatorname{ht} I \leq \operatorname{ht}(IS)$ and hence that $\operatorname{ht} I = \operatorname{ht}(IS)$.$\square$

Theorem B.2.5 (Dimension Inequality) *Let R be a Noetherian ring and S an extension of R. Assume that both R and S are integral domains. (Note: we do not require that S be finitely generated over R. This version is due to Cohen [40].) Let Q be a prime ideal in S and $P = Q \cap R$. Then*

$$\operatorname{ht} Q + \operatorname{tr.deg}_{\kappa(P)}\kappa(Q) \le \operatorname{ht} P + \operatorname{tr.deg}_R S.$$

Proof: Without loss of generality R is local with maximal ideal P. As S is not necessarily Noetherian, $\operatorname{ht} Q$ and $\operatorname{tr.deg}_{\kappa(P)}\kappa(Q)$ need not be finite. Let h and t be non–negative integers such that $h \le \operatorname{ht} Q$ and $t \le \operatorname{tr.deg}_{\kappa(P)}\kappa(Q)$. Then there exist a chain of prime ideals in S: $Q_0 \subsetneq Q_1 \subsetneq \cdots \subsetneq Q_h = Q$. For each $i = 1, \ldots, h$, let $y_i \in Q_i \setminus Q_{i-1}$. Also, there exist $z_1, \ldots, z_t \in S$ such that their images in $\kappa(Q)$ are transcendental over $\kappa(P)$. Set S' be the finitely generated subalgebra of S generated over R by $y_1, \ldots, y_h, z_1, \ldots, z_t$, and set $Q' = Q \cap S'$. If the result is known for finitely generated algebras over a Noetherian domain, then $\operatorname{ht} Q' + \operatorname{tr.deg}_{\kappa(P)}\kappa(Q') \le \operatorname{ht} P + \operatorname{tr.deg}_R S'. \le \operatorname{ht} P + \operatorname{tr.deg}_R S$. By construction, $\operatorname{ht} Q' \ge h$, and $\operatorname{tr.deg}_{\kappa(P)}\kappa(Q') \ge t$, so that $h + t \le \operatorname{ht} P + \operatorname{tr.deg}_R S$. As this holds for all $h \le \operatorname{ht} Q$ and all $t \le \operatorname{tr.deg}_{\kappa(P)}\kappa(Q)$, the theorem follows.

Thus it suffices to prove the case S is finitely generated over R. Write $S = R[x_1, \ldots, x_n]$. If we know the result for $i = 1, \ldots, n-1$, then

$$\operatorname{ht} Q + \operatorname{tr.deg}_{\kappa(Q \cap R[x_1])}\kappa(Q) \le \operatorname{ht}(Q \cap R[x_1]) + \operatorname{tr.deg}_{R[x_1]} S,$$
$$\operatorname{ht}(Q \cap R[x_1]) + \operatorname{tr.deg}_{\kappa(P)}\kappa(Q \cap R[x_1]) \le \operatorname{ht} P + \operatorname{tr.deg}_R R[x_1],$$

which also proves the general case. Thus it suffices to prove the case $n = 1$.

If x_1 is transcendental, then Q is either PS or of height exactly one more than $\operatorname{ht}(PS) = \operatorname{ht} P$. In the former case, $\operatorname{tr.deg}_{\kappa(P)}\kappa(Q) = 1$, in the latter case, $\operatorname{tr.deg}_{\kappa(P)}\kappa(Q) = 0$. Thus equality holds.

Now assume that x_1 is not transcendental. Then $\operatorname{tr.deg}_R S = 0$. Write $S = R[X]/I$ for a variable X and a non–zero prime ideal $I \subseteq R[X]$. As $R \subseteq S$, by inverting all non–zero elements of R we see that $\operatorname{ht} I = 1$. Let Q' be the lift of Q to $R[X]$. Then $Q' \cap R = P$. By the transcendental case, $\operatorname{ht} Q' + \operatorname{tr.deg}_{\kappa(P)}\kappa(Q') = \operatorname{ht} P + \operatorname{tr.deg}_R R[X]$. But $\operatorname{ht} Q + 1 = \operatorname{ht} Q + \operatorname{ht} I \le \operatorname{ht} Q'$ and $\kappa(Q) = \kappa(Q')$, so that $\operatorname{ht} Q + \operatorname{tr.deg}_{\kappa(P)}\kappa(Q) \le \operatorname{ht} P$. □

B.3. Dimension formula

In the proof of Theorem B.2.5, if we could guarantee that $\operatorname{ht} Q + \operatorname{ht} I = \operatorname{ht} Q'$ for all situations as in the proof, the inequality in the statement of the theorem could be replaced by equality. The following condition guarantees it:

Definition B.3.1 *A ring R is* **catenary** *if for any prime ideals $P \subseteq Q$ in R, every saturated chain of prime ideals starting with P and ending with Q has the same length. A ring R is* **universally catenary** *if every finitely generated R–algebra is catenary.*

A modification of the proof of Theorem B.2.5 gives the following:

Theorem B.3.2 *Let R be a universally catenary Noetherian ring and S a finitely generated extension of R. If S and R are both integral domains and $Q \in \operatorname{Spec} S$, then $\operatorname{ht} Q + \operatorname{tr.deg}_{\kappa(Q \cap R)} \kappa(Q) = \operatorname{ht}(Q \cap R) + \operatorname{tr.deg}_R S$.*

Definition B.3.3 *A Noetherian domain R is said to* **satisfy the dimension formula** *if for every finitely generated extension S of R that is an integral domain, and for every $Q \in \operatorname{Spec} S$,*

$$\operatorname{ht} Q + \operatorname{tr.deg}_{\kappa(Q \cap R)} \kappa(Q) = \operatorname{ht}(Q \cap R) + \operatorname{tr.deg}_R S.$$

It is easy to prove the following:

Lemma B.3.4 *If R satisfies the dimension formula, then every localization of a finitely generated R–algebra that is a domain also satisfies the dimension formula.*

Lemma B.3.5 *Suppose that R is a Noetherian ring such that for all prime ideals $P \subseteq Q$, $\operatorname{ht}(PR_Q) + \dim(R_Q/PR_Q) = \operatorname{ht} Q$. Then R is catenary.*

Proof: Suppose that R is not catenary. Then there exist prime ideals $P \subseteq Q$ such that there are two saturated chains of prime ideals $P = P_0 \subsetneq P_1 \subsetneq P_2 \subsetneq \cdots \subsetneq P_n = Q$ and $P = Q_0 \subsetneq Q_1 \subsetneq Q_2 \subsetneq \cdots \subsetneq Q_m = Q$ with $m \neq n$. Without loss of generality $m < n$. Furthermore, we may assume that m is the smallest integer for which there exist prime ideals $P \subseteq Q$ in R and saturated chains of prime ideals between P and Q of unequal lengths, one of the lengths being m. Necessarily $m > 1$. By localizing without loss of generality R is local with maximal ideal Q. It follows by assumption that

$$\operatorname{ht} P + \dim(R/P) = \operatorname{ht} Q = \operatorname{ht}(P_1) + \dim(R/P_1)$$
$$= \operatorname{ht}(P) + \dim(R_{P_1}/PR_{P_1}) + \dim(R/P_1),$$

so that $\dim(R/P) = \dim(R_{P_1}/PR_{P_1}) + \dim(R/P_1) = 1 + \dim(R/P_1)$. Similarly, $\dim(R/P) = 1 + \dim(R/Q_1)$. By assumption on m, R/Q_1 is catenary, of dimension $m - 1$, so that $m = 1 + \dim(R/Q_1) = \dim(R/P) \geq n$, which contradicts the choice of m. $\square$

The hypotheses above are satisfied for Cohen–Macaulay rings, which gives the following:

Corollary B.3.6 *Every Cohen–Macaulay ring is catenary. Hence an algebra essentially of finite type over a Cohen–Macaulay ring is catenary.*

Corollary B.3.7 *A complete local ring is universally catenary. A complete local domain satisfies the dimension formula.*

Proof: By the Cohen Structure Theorem 4.3.3, a complete local ring is a quotient of a regular local ring. Every regular local ring is Cohen–Macaulay. Thus by Corollary B.3.6, a complete local ring is universally catenary. The rest follows from Theorem B.3.2. $\square$

B.4. Formal equidimensionality

Definition B.4.1 *A Noetherian local ring is said to be* **formally equidi-mensional** *if its completion in the topology defined by the maximal ideal is equidimensional. (Another name for it is* **quasi–unmixed***.)*

Lemma B.4.2 *Let R be a Noetherian local ring. If R is formally equidi-mensional, then for every $P \in \operatorname{Spec} R$, R/P is formally equidimensional and $\operatorname{ht} P + \dim(R/P) = \dim R$. Also, R is catenary.*

Proof: By the standard facts on completion, the completion of R/P is $\widehat{R}/P\widehat{R}$ and its dimension is $\dim(R/P)$. Let Q be a prime ideal in $\widehat{R}$ minimal over $P\widehat{R}$. By Proposition B.1.1, $\operatorname{ht} Q = \operatorname{ht} P$. By assumption $\widehat{R}$ is equidimensional. By Corollary B.3.7, it is also catenary, so that $\dim(\widehat{R}/Q) = \dim \widehat{R} - \operatorname{ht} Q = \dim R - \operatorname{ht} P$ is independent of Q. Thus R/P is formally equidimensional.

Furthermore, $\dim(R/P) = \dim(\widehat{R}/P\widehat{R}) = \dim(\widehat{R}/Q) = \dim R - \operatorname{ht} P$.

By Proposition B.1.2, every chain of prime ideals in R is contracted from a chain of prime ideals in $\widehat{R}$. But $\widehat{R}$ is catenary, hence so is R.   □

Corollary B.4.3 *Let R be a Noetherian local ring. Then R is formally equidimensional if and only if for every $P \in \operatorname{Min} R$, R/P is formally equidi-mensional and $\dim(R/P) = \dim R$.*

Proof: By Lemma B.4.2, if R is formally equidimensional, then for every $P \in \operatorname{Min} R$, R/P is equidimensional and $\dim(R/P) = \dim R$.

Now assume that for every $P \in \operatorname{Min} R$, R/P is formally equidimensional and $\dim(R/P) = \dim R$. Let $Q \in \operatorname{Min} \widehat{R}$. By Proposition B.1.2, $P = Q \cap R$ is a minimal prime ideal in R. It follows that the image of Q is the minimal prime ideal in the completion of R/P. By assumption that R/P is formally equidi-mensional and $\dim(R/P) = \dim R$, $\dim R/Q = \dim(\widehat{R}/P\widehat{R}) = \dim(R/P) = \dim R = \dim \widehat{R}$, which proves that $\widehat{R}$ is equidimensional.   □

Proposition B.4.4 *Let $(R, \mathfrak{m})$ be a formally equidimensional Noetherian local ring and let $x \in \mathfrak{m}$ be part of a system of parameters. Then R/xR is also formally equidimensional.*

Proof: Let $d = \dim R$. For any prime ideal P in R containing x, as x is part of a system of parameters of length d, $\dim(R/P) < d$. Hence by Lemma B.4.2, $\operatorname{ht} P > 0$. Thus necessarily x is not contained in any minimal prime ideal of R. Let Q be a prime ideal in $\widehat{R}$ minimal over $x\widehat{R}$. Then $\operatorname{ht} Q = 1$, so as $\widehat{R}$ is equidimensional and catenary, $\dim(\widehat{R}/Q) = d - 1$. Thus the completion $\widehat{R}/x\widehat{R}$ of R/xR is equidimensional.   □

Lemma B.4.5 *Let $(R, \mathfrak{m}) \to (S, \mathfrak{n})$ be a faithfully flat ring homomorphism of Noetherian rings. If S is equidimensional and catenary, then R is equidi-mensional.*

Proof: Let $p \in \operatorname{Min} R$. By Proposition B.1.1, p contracts from a prime ideal q in S that is minimal over pS. By Proposition B.2.3, $\operatorname{ht} q = \operatorname{ht} p = 0$. Then $\dim(S/q) = \dim S$, and so S/pS is equidimensional and catenary of dimension $\dim S$. By Theorem B.2.2 and by flatness of $R/pR \to S/pS$, $\operatorname{ht}(\mathfrak{m}/p) = \operatorname{ht}(\mathfrak{n}/pS) - \dim((S/pS) \otimes_R \kappa(\mathfrak{m})) = \dim S - \dim(S \otimes_R \kappa(\mathfrak{m}))$, which is independent of p. $\qquad\square$

Lemma B.4.6 *Let $(R, \mathfrak{m})$ be a Noetherian universally catenary ring. Then $\operatorname{gr}_{\mathfrak{m}}(R) = (R/\mathfrak{m}) \oplus (\mathfrak{m}/\mathfrak{m}^2) \oplus (\mathfrak{m}^2/\mathfrak{m}^3) \oplus \cdots$ is equidimensional if and only if R is equidimensional.*

Proof: (In this proof we use some elementary facts about Rees algebras and the associated graded rings from Section 5.1.) Let $S = R[\mathfrak{m}t, t^{-1}]$. Since $\operatorname{gr}_{\mathfrak{m}}(R) \cong S/t^{-1}S$, there is a one–to–one correspondence between the minimal prime ideals in $\operatorname{gr}_{\mathfrak{m}}(R)$ and the prime ideals in S minimal over $t^{-1}S$. Let Q be a prime ideal in S minimal over $t^{-1}S$. Let q be a minimal prime ideal in S contained in Q. As on page 93, $p = q \cap R$ is a minimal prime ideal of R and S/q is the extended Rees algebra of $I(R/p)$. Furthermore, every $p \in \operatorname{Min} R$ is $q \cap R$ for some $q \in \operatorname{Min} S$. By Theorem 5.1.4, $\dim(S/q) = \dim(R/p) + 1$. Furthermore, if M is the maximal ideal in S generated by $\mathfrak{m}$, It and t^{-1}, then $\operatorname{ht}(M/q) = \dim(R/p) + 1$. As Q is minimal over a principal ideal, by Theorem B.2.1, $\operatorname{ht}(Q) \le 1$. But $t^{-1} \notin q$, so that $\operatorname{ht}(Q/q) = 1$. By Proposition 5.1.6, $\dim(S/Q) = \operatorname{ht}(M/Q)$. By assumption on R, S is catenary, so that $\dim(S/Q) = \operatorname{ht}(M/Q) = \operatorname{ht}(M/q) - \operatorname{ht}(Q/q) = \dim(S/q) - \operatorname{ht}(Q/q) = \dim(R/p)$. Now, R is equidimensional if and only if $\dim(R/p)$ is independent of p, which by the last equalities holds if and only if $\dim(S/Q)$ is independent of Q, which in turn means that $\operatorname{gr}_{\mathfrak{m}}(R)$ is equidimensional. $\qquad\square$

Lemma B.4.7 *Let R be a locally formally equidimensional Noetherian ring. Let X be a variable over R. Then $R[X]$ is locally formally equidimensional.*

Proof: We have to prove that for any $Q \in \operatorname{Spec} R[X]$, $S = R[X]_Q$ is formally equidimensional. Without loss of generality R is local with maximal ideal $P = Q \cap R$. Then $\widehat{R}$ is equidimensional, and $\widehat{S}$ is the completion of a localization of $\widehat{R}[X]$. So it suffices to prove that $\widehat{R}[X]$ is locally formally equidimensional. Thus without loss of generality R is a complete equidimensional local ring. Let $p_1, \ldots, p_s$ be the minimal prime ideals of R. Then $p_1 R[X], \ldots, p_s R[X]$ are the minimal prime ideals of $R[X]$. By Corollary B.4.3 it suffices to prove that each $(R/p_i)[X]$ is locally formally equidimensional. Thus without loss of generality R is a complete local domain. As P is the maximal ideal of R, write $Q = PR[X] + yR[X]$ for some $y \in R[X]$. Without loss of generality y is either 0 or a monic polynomial in X.

First suppose that $y = 0$. Since R is complete, by Corollary B.3.7, R is universally catenary. By Lemma B.4.6, $\operatorname{gr}_P(R)$ is equidimensional. Hence $(\operatorname{gr}_P(R))[X] = \operatorname{gr}_{PR[X]}(R[X])$ and any of its localizations are equidimensional. In particular, $\mathrm{gr}_{PR[X]_Q}(R[X]_Q) = \mathrm{gr}_{PS}(S)$ is equidimensional. Hence $\mathrm{gr}_{Q\widehat{S}}(\widehat{S}) = \mathrm{gr}_{QS}(S) = \mathrm{gr}_{PS}(S)$ is equidimensional. As $\widehat{S}$ is universally catenary by Corollary B.3.7, by Lemma B.4.6, $\widehat{S}$ is equidimensional. It follows that $S = R[X]_Q$ is formally equidimensional.

Now assume that y is a monic polynomial in X. Set $T = R[y]_{PR[y]+yR[y]}$. As y is a variable over R, the completion of T is $R[[y]]$. The extension $T \subseteq S$ is module–finite, even free: $S = T[Z]/(f(Z) - y)$ for some monic polynomial $f(Z) \in R[Z]$ with $f(X) = y$. Then $\widehat{S} \cong \widehat{T} \otimes_T S \cong R[[y]][Z]/(f(Z) - y)$. We prove that $\widehat{S}$ is an integral domain. Let $g, h \in R[[y]][Z]$ such that $gh \in (f(Z) - y)$. As f is monic in Z, without loss of generality g and h have degrees in Z strictly smaller than $\deg f$. Let K be the field of fractions of R. As $f(Z) - y$ is irreducible in $K[[y, Z]]$, it is also irreducible in $K[[y]][Z]$, so that by possibly switching g and h, we may assume that $g \in (f(Z) - y)K[[y]][Z]$. But then by the Z–degree count, $g = 0$ in $K[[y]][Z]$, and hence also in $R[[y]][Z]$. This proves that $\widehat{S}$ is an integral domain, so that $R[X]_Q$ is formally equidimensional. $\square$

Theorem B.4.8 *A locally formally equidimensional Noetherian ring is universally catenary. A locally formally equidimensional Noetherian domain satisfies the dimension formula.*

Proof: By Lemma B.4.7, every finitely generated R–algebra is locally formally equidimensional, so it is catenary by Lemma B.4.2. Thus R is universally catenary. The last statement follows from Theorem B.3.2. $\square$

B.5. Dimension Formula

In the proof of the following important theorem we use parts of this book. However, the arguments are not circular.

Theorem B.5.1 (Dimension Formula, Ratliff [226, Theorem 3.6]) *Let R be a Noetherian integral domain. The following are equivalent:*
(1) R is universally catenary.
(2) R satisfies the dimension formula. In other words, for any finitely generated extension S of R that is a domain, and for any prime ideal Q in S,
$$\mathrm{ht}\, Q + \mathrm{tr.deg}_{\kappa(Q \cap R)}\kappa(Q) = \mathrm{ht}(Q \cap R) + \mathrm{tr.deg}_R S.$$
(3) R is locally formally equidimensional.

Proof: By Theorem B.3.2, (1) $\Rightarrow$ (2) and by Theorem B.4.8, (3) $\Rightarrow$ (1).

Assume (2). We will show that R is locally formally equidimensional. Without loss of generality we may assume that R is local and it suffices to prove that R is formally equidimensional. By Theorem 5.4.5 it suffices to prove that for every parameter ideal I and every integer n, $\overline{I^n}$ has no embedded primes. Let q be an associated prime ideal of $\overline{I^n}$ for some $n \geq 1$. Let $S = R[It, t^{-1}]$ and let $\bar{S}$ be its integral closure. By Proposition 5.3.2, $\overline{I^n} = t^{-n}\bar{S} \cap R$. As primary decompositions contract to possibly redundant primary decompositions, there exists a prime ideal Q in $\bar{S}$ that is associated to $t^{-n}\bar{S}$ and contracts to q. By Theorem 4.10.5, $\bar{S}$ is a Krull domain, so by Proposition 4.10.3, ht $Q = 1$. Let $P = Q \cap S$. As S is a finitely generated R–algebra, by Lemma B.3.4, S satisfies the dimension formula. Hence by Proposition 4.8.6, ht $Q =$ ht P. By assumption (2), $1 + \text{tr.deg}_{\kappa(q)}\kappa(P) = \text{ht}\, P + \text{tr.deg}_{\kappa(q)}\kappa(P) = \text{ht}\, q + \text{tr.deg}_R S = \text{ht}\, q + 1$. Observe that $\text{tr.deg}_{\kappa(q)}\kappa(P)$ is the dimension of the ring $R_q[I_q t, t^{-1}]/P_{R \backslash q}$, which is at most ht I since this ring is generated by at most ht I elements over $\kappa(q)$. Hence ht $q \leq$ ht I. As every prime ideal containing I has height at least ht I, q must be minimal over I, showing that $\overline{I^m}$ has no embedded primes. This proves $(2) \Rightarrow (3)$. □

Theorem B.5.2 *Let R be a formally equidimensional Noetherian local ring. Then every localization of R is formally equidimensional and every localization of a finitely generated R–algebra that is a domain satisfies the equivalent conditions of Theorem B.5.1.*

Proof: We will prove that R is locally formally equidimensional. The last part then follows from Theorem B.5.1 and Lemma B.4.2.

Let $P \in \text{Spec}\, R$. By Proposition B.1.1 there exists $Q \in \text{Spec}\, \widehat{R}$ such that $Q \cap R = P$. By assumption, $\widehat{R}$ is equidimensional. By Corollary B.3.7, $\widehat{R}$ is universally catenary, so every quotient of $\widehat{R}_Q$ is universally catenary. Let p be a minimal prime ideal in $\widehat{R}_Q$ and S the completion of $\widehat{R}_Q$. By Theorem B.5.1, S/pS is equidimensional of dimension equal to $\dim(\widehat{R}_Q/p)$. By assumption each p has the same dimension, so S is equidimensional.

Observe that S is flat over R_P.

Claim: S is flat over $\widehat{R}_P$. Every finitely generated $\widehat{R}_P$–module M can be written as the completion of a finitely generated R_P–module M': if $m_1, \ldots, m_n$ generate M, let M' be the R_P–submodule of M generated by $m_1, \ldots, m_n$. Then the completion $\widehat{M'}$ of M' is contained in the completion of M, which is M, and the map $\widehat{M'} \to M$ is also onto by the construction of M'. Thus if $N \subseteq M$ are finitely generated $\widehat{R}_P$–modules, there exist finitely generated R_P–modules $N' \subseteq M'$ such that $N' \otimes_{R_P} \widehat{R}_P = N$ and $M' \otimes_{R_P} \widehat{R}_P = M$. As $R_P \to S$ is flat, the claim follows from the following natural maps:

$$N \otimes_{\widehat{R}_P} S \cong N' \otimes_{R_P} \widehat{R}_P \otimes_{\widehat{R}_P} S \cong N' \otimes_{R_P} S$$

$$\subseteq M' \otimes_{R_P} S \cong M' \otimes_{R_P} \widehat{R}_P \otimes_{\widehat{R}_P} S \cong M \otimes_{\widehat{R}_P} S.$$

Now, S is equidimensional and faithfully flat over $\widehat{R}_P$, so $\widehat{R}_P$ is also equidimensional by Lemma B.4.5. □

References

1. I. Aberbach, Finite phantom projective dimension. *Amer. J. Math* **116** (1994), 447–477.

2. I. Aberbach and C. Huneke, An improved Briançon–Skoda Theorem with applications to the Cohen–Macaulayness of Rees algebras. *Math. Annalen* **297** (1993), 343–369.

3. I. Aberbach and C. Huneke, A theorem of Briançon–Skoda type for regular rings containing a field. *Proc. Amer. Math. Soc.* **124** (1996), 707–713.

4. I. Aberbach and C. Huneke, F-rational rings and the integral closure of ideals. *Michigan Math. J.* **49** (2001), 3–11.

5. S. Abhyankar, Local uniformization on algebraic surfaces over ground fields of characteristic $p \neq 0$. *Annals of Math.* **63** (1956), 491–526.

6. S. Abhyankar, On the valuations centered in a local domain. *Amer. J. Math.* **78** (1956), 321–348.

7. S. Abhyankar and T. T. Moh, On analytic independence. *Trans. Amer. Math. Soc.* **219** (1976), 77–87.

8. S. Abhyankar, Polynomial maps and Zariski's main theorem. *J. Algebra* **167** (1994), 142–145.

9. S. Abhyankar and O. Zariski, Splitting of valuations in extensions of local domains. *Proc. National Academy of Sciences* **41** (1955), 84–90.

10. R. Achilles and M. Manaresi, Multiplicity for ideals of maximal analytic spread and intersection theory. *J. Math. Kyoto Univ.* **33** (1993), 1029–1046.

11. A. A. Albert, A determination of the integers of all cubic fields. *Ann. of Math.* **31** (1930), 550–566.

12. A. A. Albert, Normalized integral bases of algebraic number fields. *Ann. of Math.* **38** (1937), 923–957.

13. J. Amao, On a certain Hilbert polynomial. *J. London Math. Soc.* **14** (1976), 13–20.

14. M. Artin, On the joins of Hensel rings. *Adv. Math.* **7** (1971), 282–296.

15. P. B. Bhattacharya, The Hilbert function of two ideals. *Proc. Cambridge Phil. Soc.* **53** (1957), 568–575.

16. C. Bivià-Ausina, The analytic spread of monomial ideals. *Comm. Algebra* **31** (2003), 3487–3496.

17. M. Blickle and R. Lazarsfeld, An informal introduction to multiplier ideals. In *Trends in Commutative Algebra*, Math. Sci. Res. Inst. Publ., **51**, Cambridge, Cambridge University Press, 2004, pp. 87–114.

18. E. Böger, Minimalitätsbedingungen in der Theorie der Reduktionen von Idealen. *Schr. Math. Inst. Univ. Münster* **40** (1968).

19. E. Böger, Eine Verallgemeinerung eines Multiplizitätensatzes von D.

Rees. *J. Algebra* **12** (1969), 207–215.

20. E. Böger, Einige Bemerkungen zur Theorie der ganz–algebraischen Abhängigkeit von Idealen. *Math. Ann.* **185** (1970) 303–308.

21. R. Bondil, Geometry of superficial elements. *Ann. Fac. Sci. Toulouse Math.* **14** (2005), 185–200.

22. J. Brennan, B. Ulrich, and W. V. Vasconcelos, The Buchsbaum–Rim polynomial of a module. *J. Algebra* **241** (2001), 379–392.

23. J. P. Brennan and W. V. Vasconcelos, Effective computation of the integral closure of a morphism. *J. Pure Appl. Algebra* **86** (1993), 125–134.

24. J. P. Brennan and W. V. Vasconcelos, Effective normality criteria for algebras of linear type. *J. Algebra* **273** (2004), 640–656.

25. J. Briançon and H. Skoda, Sur la clôture intégrale d' un idéal de germes de fonctions holomorphes en un point de $\mathbb{C}^n$. *C.R. Acad. Sci. Paris Sér. A.* **278** (1974), 949–951.

26. M. Brodmann, Asymptotic stability of $\mathrm{Ass}(M/I^n M)$. *Proc. Amer. Math. Soc.* **74** (1979), 16–18.

27. M. P. Brodmann and R. Y. Sharp, *Local Cohomology.* Cambridge Studies in Advanced Mathematics, **60**, Cambridge, Cambridge University Press, 1998.

28. W. Bruns and J. Herzog, *Cohen–Macaulay Rings.* Cambridge Studies in Advanced Mathematics, **39**, Cambridge, Cambridge University Press, 1993.

29. W. Bruns and R. Koch, Computing the integral closure of an affine semigroup. *Univ. Iagel. Acta Math.* **39** (2001), 59–70.

30. W. Bruns and R. Koch, Normaliz, a program for computing normalizations of affine semigroups (1998). Available via anonymous ftp from `ftp://ftp.mathematik.uni--osnabrueck.de/pub/osm/kommalg/software/`

31. D. Buchsbaum and D. Eisenbud, What makes a complex exact? *J. Algebra* **25** (1973), 259–268.

32. D. A. Buchsbaum and D. S. Rim, A generalized Koszul complex, II. Depth and multiplicity. *Trans. Amer. Math. Soc.* **111** (1964), 197–224.

33. L. Burch, On ideals of finite homological dimension in local rings. *Proc. Cambridge Phil. Soc.* **64** (1968), 941–948.

34. L. Burch, Codimension and analytic spread. *Proc. Cambridge Phil. Soc.* **72** (1972), 369–373.

35. A. Capani, G. Niesi, and L. Robbiano, CoCoA, a system for doing Computations in Commutative Algebra. Available via anonymous ftp from `cocoa.dima.unige.it`.

36. G. Caviglia, A theorem of Eakin and Sathaye and Green's hyperplane intersection theorem. In *Commutative Algebra*, Lecture Notes in Pure and Appl. Math., **244**, Boca Raton, FL, Chapman and Hall, 2006,

pp. 1–6.

37. S. Choi, Betti numbers and the integral closure of ideals. *Math. Scand.* **66** (1990), 173–184.

38. C. Ciupercă, Generalized Hilbert coefficients and the S_2–ification of a Rees algebra. Thesis, University of Kansas, (2001).

39. C. Ciupercă, First coefficient ideals and the S_2–ification of a Rees algebra. *J. Algebra* **242** (2001), 782–794.

40. I. S. Cohen, Lengths of prime ideal chains. *Amer. J. Math.* **76** (1954), 654–668.

41. I. S. Cohen and A. Seidenberg, Prime ideals and integral dependence. *Bull. Amer. Math. Soc.* **52** (1946), 252–261.

42. H. Cohen, *A Course in Computational Algebraic Number Theory.* Graduate Texts in Mathematics, **138**. Berlin, Springer-Verlag, 1993.

43. A. Corso, C. Huneke and W. Vasconcelos, On the integral closure of ideals. *Manuscripta Math.* **95** (1998), 331–347.

44. A. Corso, C. Huneke, D. Katz and W. Vasconcelos, Integral closure of ideals and annihilators of local cohomology. In *Commutative Algebra*, Lecture Notes in Pure and Appl. Math., **244**, Boca Raton, FL, Chapman and Hall, 2006, pp. 33–48.

45. A. Corso, C. Polini and B. Ulrich, The structure of the core of ideals. *Math. Ann.* **321** (2001), 89–105.

46. A. Corso, C. Polini and B. Ulrich, Core and residual intersections of ideals. *Trans. Amer. Math. Soc.* **354** (2002), 2579–2594.

47. A. Corso, C. Polini and B. Ulrich, Core of projective dimension one modules. *Manuscripta Math.* **111** (2003), 427–433.

48. V. Crispin Quiñonez, Integral Closure and Related Operations on Monomial Ideals. Thesis, Stockholm University, 2006.

49. S. D. Cutkosky, Factorization of complete ideals. *J. Algebra* **115** (1988), 144–149.

50. S. D. Cutkosky, On unique and almost unique factorization of complete ideals. *Amer. J. Math.* **111** (1989), 417–433.

51. S. D. Cutkosky, Complete ideals in algebra and geometry. In *Commutative algebra: Syzygies, Multiplicities, and Birational Algebra (South Hadley, MA, 1992)*. Contemp. Math., **159**, Providence, RI, 1994, American Mathematical Society, 1994, pp. 27–39.

52. S. D. Cutkosky, On unique and almost unique factorization of complete ideals II. *Invent. Math.* **98** (1989), 59–74.

53. S. D. Cutkosky, *Resolution of Singularities. Graduate Studies in Mathematics*, **63**. Providence, RI, American Mathematical Society, 2004.

54. S. D. Cutkosky and L. Ghezzi, Completions of valuation rings. In *Recent progress in arithmetic and algebraic geometry*, Contemp. Math., **386**, Providence, RI, Amer. Math. Soc., 2005, pp. 13–34.

55. E. Dade, Multiplicities and monoidal transformations. Thesis, Princeton University, 1960.

408   References

56.   C. D'Cruz, Quadratic transform of complete ideals in regular local rings. *Comm. Algebra* **28** (2000), 693–698.

57.   W. Decker, T. de Jong, G.-M. Greuel and G. Pfister, The normalization: a new algorithm, implementation and comparisons. In *Computational methods for representations of groups and algebras, (Essen, 1997)*, Progr. Math., **173**, Basel, Birkhäuser, 1999, pp. 177–185.

58.   R. Dedekind, Ueber die Anzahl der Idealklassen in rein kubischen Zahlkörpern. *J. für Math.* **121** (1899), 40–123.

59.   T. de Jong, An algorithm for computing the integral closure. *J. Symbolic Comput.* **26** (1998), 273–277.

60.   D. Delfino, On the inequality $\lambda(\overline{R}/R) \leq t(R)\lambda(R/G)$ for one–dimensional local rings. *J. Algebra* **169** (1994), 332–342.

61.   D. Delfino, A. Taylor, W. Vasconcelos, R. Villarreal, N. Weininger, Monomial ideals and the computation of multiplicities. In *Commutative Ring Theory and Applications (Fez, 2001)*, Lecture Notes in Pure and Appl. Math., **231**, New York, Dekker, 2003, pp. 87–106.

62.   J.-P. Demailly, L. Ein and R. Lazarsfeld, A subadditivity property of multiplier ideals. *Michigan Math. J.* **48** (2000), 137–156.

63.   A. J. Duncan and L. O'Carroll, A full uniform Artin–Rees Theorem. *J. Reine Angew. Math.* **394** (1989), 203–207.

64.   P. Eakin and A. Sathaye, Prestable ideals. *J. Algebra* **41** (1976), 439–454.

65.   L. Ein, R. Lazarsfeld, and K. E. Smith, Uniform approximation of Abhyankar valuation ideals in smooth function fields. *Amer. J. Math.* **125** (2003), 409–440.

66.   D. Eisenbud, *Commutative Algebra With A View Toward Algebraic Geometry*. Graduate Texts in Mathematics, **150**, New York, Springer–Verlag, 1995.

67.   D. Eisenbud and J. Harris, *The Geometry of Schemes*. Springer Graduate Texts in Mathematics, **197**, New York, Springer–Verlag, 2000.

68.   D. Eisenbud, C. Huneke and B. Ulrich, What is the Rees algebra of a module. *Proc. Amer. Math. Soc.* **131** (2002), 701–708.

69.   D. Eisenbud, C. Huneke and W. Vasconcelos, Direct methods for primary decomposition. *Invent. Math.* **110** (1992), 207–235.

70.   D. Eisenbud and B. Mazur, Evolutions, symbolic squares, and Fitting ideals. *J. Reine Angew. Math.* **488** (1997), 189–201.

71.   S. Eliahou and M. Kervaire, Minimal resolutions of some monomial ideals. *J. Algebra* **129** (1990), 1–25.

72.   J. Elias, Depth of higher associated graded rings. *J. London Math. Soc.* **70** (2004), 41–58.

73.   O. Endler, *Valuation Theory*. To the memory of Wolfgang Krull (26 August 1899–12 April 1971). *Universitext*, New York–Heidelberg, Springer–Verlag, 1972.

74.   E. G. Evans, A generalization of Zariski's main theorem. *Proc. Amer.*

Math. Soc. **26** (1970), 45–48.

75.   H. Flenner and M. Manaresi, A numerical characterization of reduction ideals. *Math. Z.* **238** (2001), 205–214.

76.   H. Flenner, L. O'Carroll, W. Vogel, *Joins and Intersections.* Springer Monographs in Mathematics, Berlin, Springer–Verlag, 1999.

77.   C. Favre and M. Jonsson, *The valuative tree.* Lecture Notes in Math., 1853. Berlin, Springer–Verlag, 2004.

78.   C. Favre and M. Jonsson, Valuations and multiplier ideals. *J. Amer. Math. Soc.* **18** (2005), 655–684.

79.   M. Fiorentini, On relative regular sequences. *J. Algebra* **18** (1971), 384–389.

80.   L. Fouli, C. Polini and B. Ulrich, The core of ideals in arbitrary characteristic. Preprint.

81.   T. Gaffney, Integral closure of modules and Whitney equisingularity. *Invent. Math.* **107** (1992), 301–322.

82.   T. Gaffney, Equisingularity of plane sections, t_1 condition and the integral closure of modules. In *Real and Complex Singularities (São Carlos, 1994)*, Pitman Res. Notes Math. Ser., **333**, Harlow, Longman, 1995, pp. 95–111.

83.   T. Gaffney, Aureoles and integral closure of modules. In *Stratifications, singularities and differential equations, II (Marseilles, 1990; Honolulu, HI, 1990)*, Travaux en Cours, **55**, Paris, Hermann, 1997, pp. 55–62.

84.   T. Gaffney, The theory of integral closure of ideals and modules: applications and new developments. With an appendix by Steven Kleiman and Anders Thorup. NATO ASI/EC Summer School: *New developments in singularity theory (Cambridge, 2000).* NATO Sci. Ser. II Math. Phys. Chem., **21**, Dordrecht, Kluwer Acad. Publ., 2001, pp. 379–404.

85.   T. Gaffney, Polar methods, invariants of pairs of modules and equisingularity. In *Real and Complex Singularities (São Carlos, 2002)*, *Contemp. Math.*, **354**, Providence, RI, Amer. Math. Soc., 2004, pp. 113–136.

86.   T. Gaffney, The multiplicity of pairs of modules and hypersurface singularities. Accepted by *Workshop on Real and Complex Singularities (São Carlos, 2004)*, (27 pages) math.AG/0509045.

87.   T. Gaffney and R. Gassler, Segre numbers and hypersurface singularities. *J. Algebraic Geom.* **8** (1999), 695–736.

88.   T. Gaffney and S. L. Kleiman, Specialization of integral dependence for modules. *Invent. Math.* **137** (1999), 541–574.

89.   P. Gianni and B. Trager, Integral closure of Noetherian rings. In *Proceedings of the 1997 International Symposium on Symbolic and Algebraic Computation (Kihei, HI)*, New York, ACM Press, 1997, pp. 212–216.

90.   P. Gianni, B. Trager and G. Zacharias, Gröbner bases and primary

decompositions of polynomial ideals. *J. Symbolic Comput.* **6** (1988), 149–167.

91. R. Gilmer, *Multiplicative Ideal Theory.* Queen's Papers in Pure and Appl. Math., No. 12. Kingston, Ont., Queen's University, 1968.

92. R. Gilmer and R. Heitmann, On $\mathrm{Pic}(R[X])$ for R seminormal. *J. Pure Appl. Algebra* **16** (1980), 251–257.

93. H. Göhner, Semifactoriality and Muhly's condition (N) in two–dimensional local rings. *J. Algebra* **34** (1975), 403–429.

94. S. Goto, Integral closedness of complete intersection ideals. *J. Algebra* **108** (1987), 151–160.

95. H. Grauert and R. Remmert, *Analytische Stellenalgebren.* Berlin, Springer–Verlag, 1971.

96. H. Grauert and R. Remmert, *Coherent Analytic Sheaves.* Berlin, Springer–Verlag, 1984.

97. D. Grayson and M. Stillman, Macaulay2. 1996. A system for computation in algebraic geometry and commutative algebra. Available via anonymous `ftp` from `math.uiuc.edu`.

98. M. Green, Restrictions of linear series to hyperplanes, and some results of Macaulay and Gotzmann. In *Algebraic Curves and Projective Geometry (Trento, 1988)*, Lecture Notes in Math., **1389**, Berlin, Springer, 1989, pp. 76–86.

99. G.-M. Greuel and G. Pfister, *A Singular Introduction to Commutative Algebra.* With contributions by O. Bachmann, C. Lossen and H. Schönemann. With 1 CD-ROM (Windows, Macintosh, and UNIX). Berlin, Springer, 2002.

100. G.-M. Greuel, G. Pfister and H. Schönemann, Singular. 1995. A system for computation in algebraic geometry and singularity theory. Available via anonymous `ftp` from `helios.mathematik.uni-kl.de`.

101. A. Grothendieck, Éléments de géométrie algébrique IV. *Inst. Hautes Études Sci. Publ. Math.* **24** (1965).

102. A. Guerrieri and M.E. Rossi, Estimates on the depth of the associated graded ring. *J. Algebra* **211** (1999), 457–471.

103. E. Hamann, On the R-invariance of R[X]. *J. Algebra* **35** (1975), 1–17.

104. N. Hara, Geometric interpretation of tight closure and test ideals. *Trans. Amer. Math. Soc.* **353** (2001), 1885–1906.

105. N. Hara, A characteristic p analog of multiplier ideals and applications. *Comm. Algebra* **33** (2005), 3375–3388.

106. N. Hara and S. Takagi, On a generalization of test ideals. *Nagoya Math. J.* **175** (2004), 59–74.

107. N. Hara and K. I. Yoshida, A generalization of tight closure and multiplier ideals. *Trans. Amer. Math. Soc.* **355** (2003), 3143–3174.

108. N. Hara and K. I. Watanabe, F-regular and F-pure rings vs. log terminal and log canonical singularities. *J. Algebraic Geom.* **11** (2002), 363–392.

109.   R. Hartshorne, *Algebraic Geometry*. New York, Springer, 1977.

110.   W. Heinzer, Minimal primes of ideals and integral extensions. *Proc. Amer. Math. Soc.* **40** (1973), 370–372.

111.   W. Heinzer and C. Huneke, A generalized Dedekind–Mertens lemma and its converse. *Trans. Amer. Math. Soc.* **350** (1998), 5095–5109.

112.   W. Heinzer, B. Johnston, D. Lantz, and K. Shah, The Ratliff–Rush ideals in a Noetherian ring: a survey. In *Methods in module theory (Colorado Springs, CO, 1991), Lecture Notes in Pure and Appl. Math.*, **140**, New York, Dekker, 1993, pp. 149–159.

113.   W. Heinzer, C. Rotthaus and S. Wiegand, Noetherian domains inside a homomorphic image of a completion. *J. Algebra* **215** (1999), 666–681.

114.   W. Heinzer, C. Rotthaus and S. Wiegand, Integral closures of ideals in completions of regular local domains. In *Commutative Algebra*, Lecture Notes in Pure and Appl. Math., **244**, Boca Raton, FL, Chapman and Hall, 2006, pp. 141–150.

115.   W. Heinzer and J. D. Sally, Extensions of valuations to the completion of a local domain. *J. Pure Appl. Algebra* **71** (1991), 175–185.

116.   G. Hermann, Die Frage der endlich vielen Schritte in der Theorie der Polynomideale. *Math. Ann.* **95** (1926), 736–788.

117.   M. Herrmann, E. Hyry and J. Ribbe, On the Cohen–Macaulay and Gorenstein properties of multi–Rees algebras. *Manuscripta Math.* **79** (1993), 343–377.

118.   M. Herrmann, E. Hyry, J. Ribbe and Z. Tang, Reduction numbers and multiplicities of multigraded structures. *J. Algebra* **197** (1997), 311–341

119.   M. Herrmann, S. Ikeda, and U. Orbanz, *Equimultiplicity and Blowing up. An Algebraic Study. With an appendix by B. Moonen.* Berlin, Springer–Verlag, 1988.

120.   J. Herzog and Y. Takayama, Resolutions by mapping cones. *Homology Homotopy Appl.* **4** (2002), 277–294.

121.   M. Hochster, Rings of invariants of tori, Cohen–Macaulay rings generated by monomials, and polytopes. *Ann. of Math.* **96** (1972), 491–506.

122.   M. Hochster, Presentation depth and the Lipman–Sathaye Jacobian theorem. *Homology, Homotopy and Applications* **4** (2002), 295–314.

123.   M. Hochster and C. Huneke, Tight closure and strong F-regularity. *Mém. Soc. Math. France* **38** (1989), 119–133.

124.   M. Hochster and C. Huneke, Tight closure, invariant theory, and the Briançon–Skoda Theorem. *J. Amer. Math. Soc.* **3** (1990), 31–116.

125.   M. Hochster and C. Huneke, Infinite integral extensions and big Cohen–Macaulay algebras. *Ann. of Math.* **135** (1992), 53–89.

126.   M. Hochster and C. Huneke, F-regularity, test elements, and smooth base change. *Trans. Amer. Math. Soc.* **346** (1994), 1–60.

127.   J. A. Howald, Multiplier ideals of monomial ideals. *Trans. Amer. Math. Soc.* **353** (2001), 2665–2671.

128. S. Huckaba and C. Huneke, Normal ideals in regular rings. *J. Reine Angew. Math.* **510** (1999), 63–82.

129. S. Huckaba and T. Marley, Hilbert coefficients and the depths of associated graded rings. *J. London Math. Soc.* **56** (1997), 64–76.

130. R. Hübl, Evolutions and valuations associated to an ideal. *J. Reine Angew. Math.* **517** (1999), 81–101.

131. R. Hübl, Derivations and the integral closure of ideals, with an appendix Zeros of differentials along ideals by I. Swanson. *Proc. Amer. Math. Soc.* **127** (1999), 3503–3511.

132. R. Hübl and I. Swanson, Discrete valuations centered on local domains. *J. Pure Appl. Algebra* **161** (2001), 145–166.

133. R. Hübl and I. Swanson, Adjoints of ideals. Preprint, 2006.

134. C. Huneke, On the symmetric and Rees algebra of an ideal generated by a d-sequence. *J. Algebra* **62** (1980), 268–275.

135. C. Huneke, The theory of d-sequences and powers of ideals. *Adv. in Math.* **46** (1982), 249–279.

136. C. Huneke, Complete ideals in two–dimensional regular local rings. In *Commutative Algebra (Berkeley, CA, 1987)* Math. Sci. Res. Inst. Publ., **15**, New York, Springer, 1989, pp. 325–338.

137. C. Huneke, Hilbert functions and symbolic powers. *Michigan J. Math.* **34** (1987), 293–318.

138. C. Huneke, Uniform bounds in Noetherian rings. *Invent. Math.* **107** (1992), 203–223.

139. C. Huneke, Desingularizations and the uniform Artin–Rees Theorem. *J. London. Math. Soc.* **62** (2000), 740–756.

140. C. Huneke and G. Lyubeznik, Absolute integral closure in positive characteristic. To appear in *Adv. Math.*. Preprint, 2006, arXiv:math.AC/0604046.

141. C. Huneke and I. Swanson, Cores of ideals in two dimensional regular local rings. *Michigan Math. J.* **42** (1995), 193–208.

142. C. Huneke and N. V. Trung, On the core of ideals. *Compos. Math.* **141** (2005), 1–18.

143. E. Hyry, Coefficient ideals and the Cohen–Macaulay property of Rees algebras. *Proc. Amer. Math. Soc.* **129** (2000), 1299–1308.

144. E. Hyry and K. E. Smith, On a non–vanishing conjecture of Kawamata and the core of an ideal. *Amer. J. Math.* **125** (2003), 1349–1410.

145. E. Hyry and K. E. Smith, Core versus graded core, and global sections of line bundles. *Trans. Amer. Math. Soc.* **356** (2004), 3143–3166.

146. E. Hyry and O. Villamayor, A Briançon–Skoda Theorem for isolated singularities. *J. Algebra* **204** (1998), 656–665.

147. S. Itoh, Integral closures of ideals of the principal class. *Hiroshima Math. J.* **17** (1987), 373–375.

148. S. Itoh, Integral closures of ideals generated by regular sequences. *J. Algebra* **117** (1988), 390–401.

149. S. Itoh, Coefficients of normal Hilbert polynomials. *J. Algebra* **150** (1992), 101–117.

150. A. S. Jarrah, Integral closures of Cohen–Macaulay monomial ideals. *Comm. Algebra* **30** (2002), 5473–5478.

151. A. V. Jayanthan and J. K. Verma, Grothendieck–Serre formula and bigraded Cohen–Macaulay Rees algebras. *J. Algebra* **254** (2002), 1–20.

152. A. V. Jayanthan and J. K. Verma, Fiber cones of ideals with almost minimal multiplicity. *Nagoya Math. J.* **177** (2005), 155–179.

153. B. Johnston, The higher–dimensional multiplicity formula associated to the length formula of Hoskin and Deligne. *Comm. Algebra* **22** (1994), 2057–2071.

154. B. Johnston and J. Verma, On the length formula of Hoskin and Deligne and associated graded rings of two–dimensional regular local rings. *Math. Proc. Cambridge Phil. Soc.* **111** (1992), 423–432.

155. D. Katz, Asymptotic primes and applications. Thesis, University of Texas, (1982).

156. D. Katz, A note on asymptotic prime sequences. *Proc. Amer. Math. Soc.* **87** (1983), 415–418.

157. D. Katz, Prime divisors, asymptotic R–sequences and unmixed local rings. *J. Algebra* **95** (1985), 59–71.

158. D. Katz, On the number of minimal prime ideals in the completion of a local ring. *Rocky Mountain J. of Math.* **16** (1986), 575–578.

159. D. Katz, Note on multiplicity. *Proc. Amer. Math. Soc.* **104** (1988), 1021–1026.

160. D. Katz, Complexes acyclic up to integral closure. *Math. Proc. Cambridge Phil. Soc.* **116** (1994), 401–414.

161. D. Katz, Generating ideals up to projective equivalence. *Proc. Amer. Math. Soc.* **120** (1994), 79–83.

162. D. Katz, Reduction criteria for modules. *Comm. Algebra* **23** (1995), 4543–4548.

163. D. Katz, On the existence of maximal Cohen–Macaulay modules over pth root extensions. *Proc. Amer. Math. Soc.* **127** (1999), 2601–2609.

164. D. Katz and V. Kodiyalam, Symmetric powers of complete modules over a two–dimensional regular local ring. *Trans. Amer. Math. Soc.* **349** (1997), 747–762.

165. D. Kirby and D. Rees. Multiplicities in graded rings. I. The general theory. In *Commutative algebra: Syzygies, Multiplicities, and Birational Algebra (South Hadley, MA, 1992)*. Contemp. Math., **159**, Providence, RI, 1994, American Mathematical Society, 1994, pp. 209–267.

166. K. Kiyek and J. Stückrad, Integral closure of monomial ideals on regular sequences. *Rev. Math. Iberoamericana* **19** (2003), 483–508.

167. S. Kleiman and A. Thorup, A geometric theory of the Buchsbaum–Rim

414   References

multiplicity. *J. Algebra* **167** (1994), 168–231.

168. S. Kleiman and A. Thorup, Mixed Buchsbaum–Rim multiplicities. *Amer. J. Math.* **118** (1996), 529–569.

169. A. Knutson, Balanced normal cones and Fulton–MacPherson's intersection theory. Preprint, 2005.

170. A. Knutson and V. Alexeev, Complete moduli spaces of branch varieties. Preprint, 2005.

171. V. Kodiyalam, *Syzygies, Multiplicities and Birational Algebra.* Thesis, Purdue University, 1993.

172. V. Kodiyalam, Integrally closed modules over two–dimensional regular local rings. *Trans. Amer. Math. Soc.* **347** (1995), 3551–3573.

173. M. Kreuzer and L. Robbiano, *Computational Commutative Algebra. 1.* Berlin, Springer–Verlag, 2000.

174. T. Krick and A. Logar, An algorithm for the computation of the radical of an ideal in the ring of polynomials. *Applied algebra, algebraic algorithms and error–correcting codes (New Orleans, LA, 1991)*, Lecture Notes in Comput. Sci., **508**, Berlin, Springer, 1991, pp. 195-205.

175. W. Krull, Zum Dimensionsbegriff der Idealtheorie. *Math. Zeit.* **43** (1937), 745–766.

176. N. Mohan Kumar, On two conjectures about polynomial rings. *Invent. Math.* **40** (1978), 225–236.

177. E. Kunz, Characterizations of regular local rings of characteristic p. *Amer. J. Math.* **91** (1969), 772–784.

178. E. Kunz, *Kähler Differentials.* Advanced Lectures in Mathematics. Braunschweig, Friedr. Vieweg & Sohn, 1986.

179. R. Lazarsfeld, *Positivity in Algebraic Geometry. II.* Positivity for vector bundles and multiplier ideals. Ergebnisse der Mathematik und ihrer Grenzgebiete. 3. Folge. A Series of Modern Surveys in Mathematics, **49**. Berlin, Springer–Verlag, 2004.

180. C. Lech, On the associativity formula for multiplicities. *Arkiv Math.* **3** (1956), 301–314.

181. M. Lejeune-Jalabert, Linear systems with infinitely near base conditions and complete ideals in dimension two. In *Singularity Theory, (Trieste, 1991)*, River Edge, NJ, World Sci. Publishing, 1995, pp. 345–369.

182. M. Lejeune-Jalabert and B. Teissier, Clôture intégrale des idéaux et équisingularité. *Séminaire Lejeune–Teissier, Centre de Mathématiques École Polytechnique*, 1974. Unpublished.

183. D. A. Leonard and R. Pellikaan, Integral closures and weight functions over finite fields. *Finite Fields Appl.* **9** (2003), 479–504.

184. J. Lipman, On the Jacobian ideal of the module of differentials. *Proc. Amer. Math. Soc.* **21** (1969), 422–426.

185. J. Lipman, Rational singularities with applications to algebraic surfaces and unique factorization. *Inst. Hautes Études Sci. Publ. Math.*

36 (1969), 195–279.

186.   J. Lipman, Relative Lipschitz saturation. *Amer. J. of Math.* **97** (1973), 791–813.

187.   J. Lipman, Equimultiplicity, reduction, and blowing up. In *Commutative algebra (Fairfax, Va., 1979)*, Lecture Notes in Pure and Appl. Math., **68**, New York, Dekker, 1982, pp. 111–147.

188.   J. Lipman, Cohen–Macaulayness in graded algebras. *Math. Res. Lett.* **1** (1994), 149–15.

189.   J. Lipman, On complete ideals in regular local rings. In *Algebraic Geometry and Commutative Algebra, Vol. I*, Tokyo, Kinokuniya, 1988, pp. 203–231.

190.   J. Lipman, Adjoints of ideals in regular local rings, with an appendix by S. D. Cutkosky. *Math. Research Letters* **1** (1994), 739–755.

191.   J. Lipman, Proximity inequalities for complete ideals in two–dimensional regular local rings, *Contemp. Math.* **159** (1994), 293–306.

192.   J. Lipman and A. Sathaye, Jacobian ideals and a theorem of Briançon–Skoda. *Michigan Math. J.* **28** (1981), 199–222.

193.   J. Lipman and B. Teissier, Pseudo–local rational rings and a theorem of Briançon–Skoda about integral closures of ideals. *Michigan Math. J.* **28** (1981), 97–112.

194.   J. C. Liu, Rees algebras of finitely generated torsion–free modules over a two–dimensional regular local ring. *Comm. Algebra* **26** (1998), 4015–4039.

195.   J. C. Liu, Ratliff–Rush closures and coefficient modules. *J. Algebra* **201** (1998), 584–603.

196.   G. Lyubeznik, A property of ideals in polynomial rings. *Proc. Amer. Math. Soc.* **98** (1986), 399–400.

197.   S. MacLane, A construction for absolute values in polynomial rings. *Trans. Amer. Math. Soc.* **40** (1936), 363–395.

198.   S. MacLane, Modular fields I. Separating transcendence bases. *Duke Math. J.* **5** (1939), 372–393.

199.   R. Matsumoto, On computing the integral closure. *Comm. Algebra* **28** (2000), 401–405.

200.   H. Matsumura, *Commutative Ring Theory.* Cambridge Studies in Advanced Mathematics, **8**, Cambridge, Cambridge University Press, 1986.

201.   S. McAdam, Finite covering by ideals. In *Ring Theory (Proc. Conf., Univ. Oklahoma, Norman, Okla., 1973)*, Lecture Notes in Pure and Appl. Math., **7**, New York, Dekker, 1974, pp. 163–171.

202.   S. McAdam, Asymptotic prime divisors and analytic spread. *Proc. Amer. Math. Soc.* **90** (1980), 555–559.

203.   S. McAdam, *Asymptotic Prime Divisors.* Lecture Notes in Math., **1023**, Berlin, Springer–Verlag, 1983.

204.   S. McAdam and P. Eakin, The asymptotic Ass. *J. Algebra* **61** (1979), 71–81.

205. S. McAdam and L. J. Ratliff, Sporadic and irrelevant prime divisors. *Trans. Amer. Math. Soc.* **303** (1987), 311–324.

206. R. Mines and F. Richman, Valuation theory: a constructive view. *J. Number Theory* **19** (1984), 40–62.

207. R. Mohan, The core of a module over a two–dimensional regular local ring. *J. Algebra* **189** (1997), 1–22.

208. R. Mohan, The Rees valuations of a module over a two–dimensional regular local ring. *Comm. Algebra* **27** (1999), 1515–1532.

209. M. Morales, Polynôme d'Hilbert–Samuel des clôtures intégrales des puissances d'un idéal m–primaire. *Bull. Soc. Math. France* **112** (1984), 343–358.

210. P. Morandi, *Field and Galois Theory*. Graduate Texts in Mathematics, **167**. New York, Springer–Verlag, 1996.

211. H. Muhly and M. Sakuma, Asymptotic factorization of ideals. *J. London Math. Soc.* **38** (1963), 341–350.

212. M. Nagata, An example of normal local ring which is analytically unramified. *Nagoya Math. J.* **9** (1955), 111–113.

213. M. Nagata, *Local Rings*. New York, John Wiley and Sons, 1962.

214. S. Noh, Adjacent integrally closed ideals in dimension two, *J. Pure and Appl. Algebra* **85** (1993), 163–184.

215. D. G. Northcott and D. Rees, Reductions of ideals in local rings. *Proc. Cambridge Phil. Soc.* **50** (1954), 145–158.

216. L. O'Carroll, On two theorems concerning reductions in local rings. *J. Math. Kyoto U.* **27** (1987), 61–67.

217. L. O'Carroll, A note on Artin–Rees numbers. *Bull. London Math. Soc.* **23** (1991), 209–212.

218. L. O'Carroll, Addendum to: "A note on Artin–Rees numbers". *Bull. London Math. Soc.* **23** (1991), 555–556.

219. L. O'Carroll, Around the Eakin–Sathaye theorem. *J. Algebra* **291** (2005), 259–268.

220. A. Ooishi, Reductions of graded rings and pseudo–flat graded modules. *Hiroshima Math. J.* **18** (1988), 463–477.

221. F. Pham, Fractions lipschitziennes et saturation de Zariski des algèbres analytiques complexes. Exposé d'un travail fait avec Bernard Teisser. Fractions lipschitziennes d'un algèbre analytique complexe et saturation de Zariski. Actes du Congrès International des Mathématiciens (Nice, 1970), Tome 2, Gauthier–Villars, Paris, 1971, pp. 649–654.

222. C. Polini and B. Ulrich, A formula for the core of an ideal. *Math. Ann.* **331** (2005), 487–503.

223. C. Polini, B. Ulrich and M. Vitulli, The core of zero–dimensional monomial ideals. Preprint.

224. H. Prüfer, Untersuchungen über die Teilbarkeitseigenschaften in Körpern. *J. Reine Angew. Math.* **168** (1932), 1–36.

225. K.N. Raghavan, A simple proof that ideals generated by d-sequences

are of linear type. *Comm. Alg.* **19** (1991), 2827–2831.

226. L. J. Ratliff, Jr., On quasi–unmixed local domains, the altitude formula, and the chain condition for prime ideals (I). *Amer. J. of Math.* **91** (1969), 508–528.

227. L. J. Ratliff, Jr., On prime divisors of I^n, n large. *Michigan Math. J.* **23** (1976), 337–352.

228. L. J. Ratliff, Jr., Locally quasi–unmixed Noetherian rings and ideals of the principal class. *Pacific J. Math.* **52** (1974), 185–205.

229. L. J. Ratliff, Jr. and D. E. Rush, Two notes on reductions of ideals. *Indiana Univ. Math. J.* **27** (1978), 929–934.

230. D. Rees, Valuations associated with ideals II. *J. London Math. Soc.* **31** (1956), 221–228.

231. D. Rees, Valuations associated with a local ring II. *J. London Math. Soc.* **31** (1956), 228–235.

232. D. Rees, $\mathcal{A}$–transforms of local rings and a theorem on multiplicities of ideals. *Proc. Cambridge Phil. Soc.* **57** (1961), 8–17.

233. D. Rees, A note on analytically unramified local rings. *J. London Math. Soc.* **36** (1961), 24–28.

234. D. Rees, Rings associated with ideals and analytic spread. *Math. Proc. Camb. Phil. Soc.* **89** (1981), 423–432.

235. D. Rees, Hilbert functions and pseudorational local rings of dimension two. *J. London Math. Soc.* **24** (1981), 467–479.

236. D. Rees, Multiplicities, Hilbert functions and degree functions. In *Commutative algebra: Durham 1981 (Durham, 1981*, London Math. Soc. Lecture Note Ser., **72**, Cambridge–New York, Cambridge Univ. Press, 1982, pp. 170–178.

237. D. Rees, Generalizations of reductions and mixed multiplicities. *J. London Math. Soc.* **29** (1984), 397–414.

238. D. Rees, Amao's theorem and reduction criteria. *J. London Math. Soc.* **32** (1985), 404–410.

239. D. Rees, Reduction of modules. *Math. Proc. Cambridge Phil. Soc.* **101** (1987), 431–449.

240. D. Rees and J. Sally, General elements and joint reductions. *Michigan Math. J.* **35** (1988), 241–254.

241. D. Rees and R. Y. Sharp, On a theorem of B. Teissier on multiplicities of ideals in local rings. *J. London Math. Soc.* **18** (1978), 449–463.

242. L. Reid, L.G. Roberts, and B. Singh, On weak subintegrality, *J. Pure Appl. Algebra* **114** (1996), 93–109.

243. L. Reid, L. G. Roberts, M. A. Vitulli, Some results on normal homogeneous ideals. *Comm. Algebra* **31** (2003), 4485–4506.

244. M. J. Rhodes, An algorithm for the computation of adjoints of monomial ideals in polynomial rings. Thesis, New Mexico State University, 2001.

245. P. Ribenboim, *The Theory of Classical Valuations.* Springer Mono-

graphs in Mathematics, New York, Springer–Verlag, 1999.

246. L. G. Roberts and B. Singh, Subintegrality, invertible modules and the Picard group. *Compositio Math.* **85** (1993), 249–279.

247. P. Roberts, The homological conjectures. In *Free resolutions in commutative algebra and algebraic geometry* (Sundance, UT, 1990), Res. Notes Math., **2**, Boston, MA, Jones and Bartlett, 1992, pp. 121–132.

248. G. Rond, Fonction de Artin et théorème d'Izumi. Thesis, University of Toulouse III, 2005.

249. P. Roquette, History of valuation theory, I. In *Valuation theory and its applications, Vol. I (Saskatoon, SK, 1999)*, Fields Inst. Commun., **32**, Providence, RI, Amer. Math. Soc., 2002, pp. 291–355.

250. M.E. Rossi, A bound on the reduction number of a primary ideal. *Proc. Amer. Math. Soc.* **128** (2000), 1325–1332.

251. M.E. Rossi and G. Valla, A conjecture of J. Sally. *Comm. Algebra* **24** (1996), 4249–4261.

252. M.E. Rossi and G. Valla, Cohen–Macaulay local rings of embedding dimension $e + d - 3$. *Proc. London Math. Soc.* **80** (2000), 107–126.

253. K. Saito, Quasihomogene isolierte Singularitäten von Hyperflächen. *Invent. Math.* **14** (1971), 123–142.

254. J. Sally, Fibers over closed points of birational morphisms of nonsingular varieties. *Amer. J. Math.* **104** (1982), 545–552.

255. J. Sally, Cohen–Macaulay local rings of embedding dimension $e + d - 2$. *J. Algebra* **83** (1983), 393–408.

256. J. Sally, One–fibered ideals. In *Commutative Algebra (Berkeley, CA, 1987)*, Math. Sci. Res. Inst. Publ., **15**, New York, Springer–Verlag, 1989, pp. 437–442.

257. J. B. Sancho de Salas, Blowing–up morphisms with Cohen–Macaulay associated graded rings. In *Géométrie algébrique et applications, I (La Rábida, 1984)*, Travaux en Cours, **22**, Paris, Hermann, 1987, pp. 201–209.

258. D. Schaub, Multiplicité et dépendance intégrale sur un idéal. *Deuxième Colloque d'Algèbre Commutative (Rennes, 1976)*, Exp. No. 1, Rennes, Univ. Rennes, 1976.

259. G. Scheja, Über ganz–algebraische Abhängigkeit in der Idealtheorie. *Comment. Math. Helv.* **45** (1970), 384–390.

260. P. Schenzel, Symbolic powers of prime ideals and their topology. *Proc. Amer. Math. Soc.* **93** (1885), 15–20.

261. A. Seidenberg, The hyperplane sections of normal varieties. *Trans. Amer. Math. Soc.* **69** (1950), 357–386.

262. A. Seidenberg, Derivations and integral closure. *Pacific J. of Math.* **16** (1966), 167–173.

263. A. Seidenberg, Construction of the integral closure of a finite integral domain. *Rendiconti Sem. Matematico e Fisico, Milano* **40** (1970), 101–120.

264. A. Seidenberg, Construction of the integral closure of a finite integral domain. II. *Proc. Amer. Math. Soc.* **52** (1975), 368–372.

265. J. P. Serre, *Algèbre locale. Multiplicités.* Lecture Notes in Math., **11**. Berlin–New York, Springer–Verlag, 1965.

266. K. Shah, Coefficient ideals. *Trans. Amer. Math. Soc.* **327** (1991), 373–384.

267. H. N. Shapiro and G. H. Sparer, Minimal bases for cubic fields. *Comm. Pure Appl. Math.* **44** (1991), 1121–1136.

268. T. Shimoyama and K. Yokoyama, Localization and primary decomposition of polynomial ideals. *J. of Symbolic Comput.* **22** (1996), 247–277.

269. A. Simis, B. Ulrich and W. V. Vasconcelos, Codimension, multiplicity and integral extensions. *Math. Proc. Camb. Phil. Soc.* **130** (2001), 237–257.

270. A. Simis, B. Ulrich and W. V. Vasconcelos, Rees algebras of modules. *Proc. London Math. Soc.*, **87** (2003), 610–646.

271. P. Singla, Minimal monomial reductions and the reduced fiber ring of an extremal ideal. `arXiv:math.AC/0512456`.

272. K. E. Smith, Tight closure of parameter ideals. *Invent. Math.* **115** (1994), 41–60.

273. K. E. Smith, The multiplier ideal is a universal test ideal. *Special issue in honor of Robin Hartshorne. Comm. Algebra* **28** (2000), 5915–5929.

274. B. K. Spearman, An explicit integral basis for a pure cubic field. *Far East J. Math. Sci.* **6** (1998), 1–14.

275. M. Spivakovsky, Valuations in function fields of surfaces. *Amer. J. Math.* **112** (1990), 107–156.

276. G. Stolzenberg, Constructive normalization of an algebraic variety. *Bull. Amer. Math. Soc.* **74** (1968), 595–599.

277. R. G. Swan, On seminormality. *J. Algebra* **67** (1980), 210–229.

278. I. Swanson, Joint reductions, tight closure and the Briançon–Skoda Theorem. *J. Algebra* **147** (1992), 128–136.

279. I. Swanson, Mixed multiplicities, joint reductions, and a theorem of Rees. *J. London Math. Soc.* **48** (1993), 1–14.

280. I. Swanson, Joint reductions, tight closure and the Briançon–Skoda Theorem, II. *J. Algebra* **170** (1994), 567–583.

281. I. Swanson, A note on analytic spread. *Comm. in Algebra* **22(2)** (1994), 407–411.

282. I. Swanson and O. Villamayor, On free integral extensions generated by one element. In *Commutative Algebra*, Lecture Notes in Pure and Appl. Math., **244**, Boca Raton, FL, Chapman and Hall, 2006, pp. 239–257.

283. M. E. Sweedler, A units theorem applied to Hopf algebras and Amitsur cohomology. *Amer. J. Math.* **92** (1970), 259–271.

284. M. E. Sweedler, When is the tensor product of algebras local? *Proc.*

Amer. Math. Soc. **48** (1975), 8–10.

285. S. Takagi, An interpretation of multiplier ideals via tight closure. *J. Algebraic Geom.* **13** (2004), 393–415.

286. S. Takagi, F-singularities of pairs and inversion of adjunction of arbitrary codimension. *Invent. Math.* **157** (2004), 123–146.

287. S. Takagi and K.-I. Watanabe, When does subadditivity theorem for multiplier ideals hold? *Trans. Amer. Math. Soc.* **356** (2004), 3951–3961.

288. S. Takagi and K.-I. Watanabe. On F-pure thresholds, *J. Algebra* **282** (2004), 278–297.

289. S.-L. Tan, Integral closure of a cubic extension and applications. *Proc. Amer. Math. Soc.* **129** (2001), 2553–2562.

290. S.-L. Tan and D.-Q. Zhang, The determination of integral closures and geometric applications. *Adv. Math.* **185** (2004), 215–245.

291. Z. Tang, A note on the Cohen–Macaulayness of multi–Rees rings. *Comm. Algebra* **27** (1999), 5967–5974.

292. B. Teissier, Cycles évanescents, sections planes, et conditions de Whitney. Singularités à Cargèse (Rencontre Singularités Géom. Anal., Inst. Études Sci., Cargèse, 1972). Asterisque, Nos. 7 et 8, Paris, Soc. Math. France, 1973, pp. 285–362.

293. B. Teissier, Sur une inégalité pour les multiplicités, (Appendix to a paper by D. Eisenbud and H. Levine). *Ann. Math.* **106** (1977), 38–44.

294. B. Teissier, On a Minkowski–type inequality for multiplicities. II. In *C. P. Ramanujam — a tribute*, Tata Inst. Fund. Res. Studies in Math., **8**, Berlin–New York, Springer, 1978, pp. 347–361.

295. B. Teissier, Résolution simultané –II. Résolution simultanée et cycles évanescents. In *Séminaire sur les singularités des surfaces, 1976–1977*, Lecture Notes in Math., **777**, Berlin, Springer, 1980. pp. 82–146.

296. C. Traverso, Seminormality and Picard group. *Ann. Scuola Norm. Sup. Pisa* **24** (1970), 585–595.

297. C. Traverso, A study on algebraic algorithms: the normalization. Conference on algebraic varieties of small dimension (Turin, 1985). *Rend. Sem. Mat. Torino* (1987), pp. 111–130.

298. N. V. Trung, Constructive characterization of the reduction numbers, *Compositio Math.* **137** (2003), 99–113.

299. N. V. Trung and J. Verma, Mixed multiplicities of ideals versus mixed volumes of polytopes, to appear in *Trans. Amer. Math. Soc.*

300. B. Ulrich and W. V. Vasconcelos, On the complexity of the integral closure. *Trans. Amer. Math. Soc.* **357** (2005), 425–442.

301. G. Valla, On the symmetric and Rees algebras of an ideal. *Manu. Math.* **30** (1980), 239–255.

302. L. van den Dries and K. Schmidt, Bounds in the theory of polynomial rings over fields. A nonstandard approach. *Invent. Math.* **76** (1984), 77–91.

303.   B. L. van der Waerden, On Hilbert's function, series of composition of ideals and a generalization of a theorem of Bezout. *Proc. K. Acad. Wet. Amst.* **3**, (1928), 49–70.

304.   W. Vasconcelos, Computing the integral closure of an affine domain. *Proc. Amer. Math. Soc.* **113** (1991), 633–639.

305.   W. Vasconcelos, *Arithmetic of blowup algebras*. London Mathematical Society Lecture Note Ser., 195, Cambridge, Cambridge University Press, 1994.

306.   W. Vasconcelos, Divisorial extensions and the computation of the integral closures. *J. Symbolic Computation* **30** (2000), 595–604.

307.   W. Vasconcelos, *Integral Closure. Rees algebras, multiplicities, algorithms.* Springer Monographs in Mathematics. Berlin, Springer-Verlag, 2005.

308.   J. K. Verma, On ideals whose adic and symbolic topologies are linearly equivalent. *J. Pure and Appl. Algebra* **47** (1987), 205–212.

309.   J. K. Verma, Rees algebras and mixed multiplicities. *Proc. Amer. Math. Soc.* **104** (1988), 1036–1044.

310.   J. K. Verma, On the symbolic topology of an ideal. *J. Algebra* **112** (1988), 416–429.

311.   J. K. Verma, Joint reductions of complete ideals. *Nagoya J. Math.* **118** (1990), 155–163.

312.   J. K. Verma, Joint reductions and Rees algebras. *Math. Proc. Cambridge Phil. Soc.* **109** (1991), 335–342.

313.   D. Q. Viet, Mixed multiplicities of arbitrary ideals in local rings. *Comm. Algebra* **28** (2000), 3803–3821.

314.   M. A. Vitulli and J. V. Leahy, The weak subintegral closure of an ideal. *J. Pure Appl. Algebra* **141** (1999), 185–200.

315.   J. von zur Gathen and J. Gerhard, *Modern Computer Algebra.* Second edition, Cambridge, Cambridge University Press, 2003.

316.   H.-J. Wang, On Cohen–Macaulay local rings with embedding dimension $e + d - 2$. *J. Algebra* **190** (1997), 226–240.

317.   H.-J. Wang, A note on powers of ideals. *J. Pure Appl. Algebra* **123** (1998), 301–312.

318.   J. Watanabe, m–full ideals. *Nagoya Math. J.* **106** (1987), 101–111.

319.   K. I. Watanabe, Chains of integrally closed ideals. In *Commutative algebra (Grenoble/Lyon, 2001), Contemp. Math.*, **331**, Providence, RI, Amer. Math. Soc., 2003, pp. 353–358.

320.   O. Zariski, Polynomial ideals defined by infinitely near base points. *Amer. J. Math.* **60** (1938), 151–204.

321.   O. Zariski, The reduction of the singularities of an algebraic surface. *Annals of Math.* **40** (1939), 639–689.

322.   O. Zariski and P. Samuel, *Commutative Algebra. II.* Reprint of the 1960 edition. Graduate Texts in Mathematics, **29**. New York–Heidelberg, Springer–Verlag, 1975.

Index

An underlined number marks a defining page.

A

Aberbach 244, 245, 326
Abhyankar 89, 92, 122, 126, 127, 142, 186, 274
absolute integral closure $\underline{77}$–79, 89, 321–325
acceptable sequence $\underline{238}$
acyclicity 325, 326
  up to integral closure 320
adic valuation $\underline{131}$, 132, 196
  for maximal ideal 131
adj I: see adjoint ideal
adjacent ideals 142
adjoint ideal 359, 360–377
  computation 364–365
  in dimension two 369, 375
  monomial ideal 366–368, 376
  powers of the maximal ideal 365
algebraic closure 77
almost finite 79
almost integral $\underline{20}$, 44, $\underline{45}$
Amao 330
analytic spread $\underline{94}$, 175
  bounds 155, 160
  Burch 103, 176
  determinantal ideal 174
  Eakin–Sathaye Theorem 168
  equimultiple 228, 229
  formally equidimensional 103, 201, 210, 228, 233
  integral closure 280
  modulo minimal primes 95
  monomial ideals 158, 174
  of module 311
  reductions 157, 158, 166, 174
  Sally's Theorem 171

analytically independent 175
analytically irreducible $\underline{123}$, 185
  criterion 211
analytically unramified $\underline{177}$–186
  and conductor 236, 242
  and normal maps 381
  criterion 178, 185
  list 177
  uniform Artin–Rees 255
Archimedean valuation 116, 126
Artin 77
Artin–Rees Lemma 332
  also referred to as Theorem
  uniform $\underline{253}$, 254, 255
associated graded ring $\underline{94}$
  depth 166
  dimension 94, 103, 110
  has a non–zerodivisor 163, 363
  integral closure 110
  quotient ring 165
  Rees valuations 209, 210
  reflects the ring 95, 110, 402
  superficial sequence 166
  valuations 131, 258
associated primes
  grading 393
  integral closure 22, 153
  powers of an ideal 135
  principal ideals 56
asymptotic Samuel function: see $\overline{v}_I$
asymptotic sequence 111

B

Böger 344, 348
base point $\underline{275}$
basic ideal 155

Bhattacharya 340
bight (big height) 256
Bivià-Ausina 158
$Bl_I(R)$: see blowup
blowup 108, 110, 112, 192
  of blowup 110
Böger 223, 312, 317, 318, 357
Brennan 306
Briançon–Skoda Theorem 244–256
  Rees and Sally 356
  uniform 254, 256
  with adjoints 362
Brodmann 97, 194
Bruns 296, 297
Buchsbaum 313
Buchsbaum–Eisenbud's Acyclicity
  Criterion 325, 326
Buchsbaum–Rim
  multiplicity 313–319
  polynomial 317
  relative multiplicity 316
Burch 16, 176
  analytic spread 103, 176

C
C_R: see conductor
canonical module 236, 237, 243
Carathéodory's theorem 392
catenary 399
Caviglia 169
Čech cohomology 108, 321
Chevalley's Theorem 186
Choi 16
$c(I)$: see content of an ideal
Ciupercă 224
CoCoA 282, 287
coefficient ideal 225
Cohen 30, 61, 117, 399
Cohen Structure Theorem 61
Cohen–Macaulay ring 70, 142, 209
  and conductor 243
  and core 355
  and flat maps 71
  and multiplicity 216, 217, 233

and normal ideals 280
catenary 400
Lipman–Sathaye Theorem 237
mapping cone 373
complete integral closure 20, 45
complete reduction 332
complex 394
  map of 395
complex analytic space 144
complexification 387
computation 281–301
  grade 287
  primality test 299
  primary decomposition 299
  radical 300
conductor 43, 234–243, 285, 295
content of a polynomial 17, 18, 21
content of an ideal 258, 262
contracted ideal 264, 269
contraction (integral closure) 3, 15
core 309, 357, 359
Corso 21, 243, 301, 359
Crispin Quiñonez 142, 158
Cutkosky 124, 202

D
Dade 233
de Jong 289, 293
Dedekind 234, 235, 282
Dedekind domain 57, 90
Dedekind–Mertens formula 17
Delfino 243
$\Delta(v)$ 278
Demailly 375
depth xii, 395
derivation 44, 147, 243
$\det_f(M)$ 308
determinantal ideal 112, 142, 325
  analytic spread 174
  associated graded ring 175
determinantal trick 4, 25
dimension formula 400
  theorem 403
Dimension Inequality 399

discrete valuation <u>124</u>, 125, 127, 139
  approximation 141
  example 130–132, 139, 140
  integral closure of ideals 133, 134
discriminant 285
divisorial valuation <u>183</u>, 186, 277,
    280, 360
  essentially of finite type 183
  example 196
  first and second kind 183
  Rees valuation 201, 202
D+M construction 141
$\widetilde{D}(R)$ (normalized valuations) <u>361</u>
$D(R)$ <u>183</u>, 360
d-sequence <u>104</u>, 105, 106, 112
dual basis 243
dual $(\mathrm{Hom}_R(\_, R))$ 299
  double 91, 299
DVR 125

E
Eakin 97
Eakin–Sathaye Theorem 168, 332
$e(I)$, $e_R(I; M)$: see multiplicity
Ein 375
Eisenbud 149, 325, 326
  primary decomposition 300
  radical 301
  Rees algebra of a module 307, 329
Eliahou–Kervaire 377
equation of integral dependence 2,
    4, 17, 110
  monomial ideals 9
  over a ring 23, 29
  uniqueness of 23, 29
equimultiple <u>226</u>, 228, 229
essential (prime) divisor 111
essentially of finite type <u>xi</u>
Euler's formula 146
Evans 92
excellent ring 177
exponent set and vector <u>10</u>
extended Rees algebra: see
    Rees algebra

extension of scalars 302, 310
exterior algebra 307

F
$\mathcal{F}_I(R), \mathcal{F}_I$: see fiber cone
faithful <u>xi</u>
Faridi 21
F-finite <u>246</u>–255
fiber cone <u>94</u>, 154
Fiorentini 105
Flenner 224, 225
formally equidimensional <u>401</u>–404
  Amao 330
  and height 82
  Buchsbaum–Rim 317, 318, 329
  criterion 111
  dimension formula 403
  equimultiplicity 228
  mixed multiplicity 348
  multiplicity 222, 223, 232, 233
  Ratliff's Theorem 100, 102, 137
  reduction criterion 312, 318, 327
  Rees valuations 195, 201, 204
  universally catenary 403
fractional ideal <u>40</u>, 292
Frobenius homomorphism <u>245</u>, 248
  flat 246
full ideal: see m–full

G
Gaffney 224, 225, 231, 328, 329
Galois extension 43
Gassler 224
Gauss's Lemma <u>17</u>, 21
generalized monomial ideal 366–368
Ghezzi 124
Gianni 283, 297, 299, 300
Gilmer 46, 141
Going–Down <u>30</u>, 32, 42, 397
  and height 398
  flatness 397
Going–Up <u>30</u>, 31
Gorenstein ring 236, 299
Goto 16, 209

$\mathrm{gr}_I(R)$: see associated graded ring
Gröbner bases 281, 297, 298, 299
grade (depth) 72, 287
grading 34, 393
Grauert 289, 292
Green 169
  hyperplane restriction 169
g_X 238

H
Hamann 46
Hara 375, 376
$\mathcal{E}_R(M)$ 308
Heinzer 17, 79, 123, 382, 390
Heitmann 46
Hermann 283, 300
Hilbert polynomial, multi–graded 339
Hilbert–Burch Theorem 369, 395
Hilbert–Samuel function 159, 213
Hilbert–Samuel polynomial 212–217
Hochster 77, 321
holomorphic function 143
$\mathrm{Hom}_R(I, J)$ 40
homogeneous 34
Hom–tensor adjointness 41
Hoskin–Deligne Formula 274, 359
Howald 360, 367
Hübl 147, 211, 366, 375, 377
Huckaba 21, 112
Huneke 21
  absolute integral closure 77, 321
  associated primes 22
  binomial ideal 301
  Briançon–Skoda Theorem 244, 256
  conductor 243
  Dedekind–Mertens formula 17
  d-sequence 105
  Huneke–Itoh Theorem 249
  normal ideal 21, 112
  primary decomposition 300
  radical 301
  Rees algebra of a module 307, 329
  uniform Artin–Rees Theorem 253
Hyry 244

I
I^{-1} xii, 40, 243
idempotent 60
Incomparability Theorem 30
infinitely near points 274
integral 2
  over a ring 23
integral closure 2
  and exponent set 11
  and grading 15, 21, 96, 153
  and products 8, 135, 210
  and Rees algebra 95
  and sums 8
  associated primes 22, 153
  binomial ideal 301
  cancellation 20, 261
  colon ideals 7, 20, 135, 137, 301
  compute 281
  contraction 3, 15
  discrete valuations 133
  examples 8
  ideals and modules 305, 308, 327
  is an ideal 6
  localization 3
  modulo minimal prime ideals 3
  modulo nilradical 2, 3
  monomial ideal 9, 12, 21
  Newton polyhedron 12
  persistence 2
  powers of a maximal ideal 22
  powers of an ideal 22, 97–104, 111,
    135, 194
  primary decomposition 140
  reduction: see reduction
  valuation 133–138
integral closure of modules 302–330
  integral 302, 329
  persistence 309, 310
  projection 310
  reduction 303
integral closure of rings 13, 24
  algebra over field 73–77
  algebra over $\mathbb{Z}$ 73–77
  and grading 34–39

and height 33
complete rings 60–63
completion reduced 74
dimension one 83, 283
dimension two 88
dimension 31
direct summand 44
localization 24, 25
module–finite 48, 56, 60, 62, 73–
   77, 180, 181, 185
modulo minimal primes 27, 28
Noetherian 83
not module–finite 89
not Noetherian 89
of a field 25
of monomial algebra 44
of Rees algebra 95, 96
polynomial extension 29
power series 44
primary decomposition 285
principal ideals 14, 56
rings of homomorphisms 41, 91
rings of integers 43
separable 48
transitivity of 27
integral dependence $\underline{2}$
   criterion 297, 298
   modules 302
   monomial ideal 9
integral dependence of rings
   and height 31, 82
   residue fields 82
integrally closed $\underline{2}$
   criterion 292, 294, 298
   rings 13, 24, 42
invertible fractional ideal 40
isolated subgroup 125
Itoh 244, 249
$I_t(\varphi)$ 325

J
Jacobian criterion $\underline{68}$, 89, 287
Jacobian ideal $\underline{64}$–69
   examples 64, 250, 288

localization 66
normality 294, 298
power series 144, 149, 244
special generators 238
transitivity 67
valuations 360, 365
well–defined 64
Jacobian matrix $\underline{64}$
Jacobson radical $\underline{31}$, 73, 153
Jarrah 22
Jayanthan 332
j-multiplicity 224, 225, 233
Jockusch 21
Johnson 205
joint reduction 331, $\underline{332}$–359
   complete joint reduction $\underline{332}$
joint reduction number 358
$J_{R/A}$: see Jacobian ideal

K
Kähler differentials 67
$\kappa(v), \kappa(V)$ $\underline{115}$, $\underline{124}$, 127
Katz 21, 313
   acyclic complex 319, 326
   associated primes 102
   asymptotic sequence 97
   Buchsbaum–Rim multiplicity 329
   conductor 243
   formally equidimensional 111,
      210, 312, 329
   integrally closed modules 309
   projectively equivalent 359
   reduction criterion 312
   reductions 169, 170
   Rees valuations 210
   rings of homomorphisms 39
Kirby 313
Kiyek 366
Kleiman 329
Koch 296, 297
Kodiyalam 275, 309
Koszul complex 21, 319, 328, $\underline{394}$
Koszul depth 326
Krick 300

Kronecker 283
Krull 30
Krull domain 85–88, 90, 192
Krull–Akizuki 83, 84
Krull's Height Theorem 398
$K_{S/R}$ 240
Kunz 246
Kürschák 113

L
Lazarsfeld 375
leading form 258
Leahy 46
Lech's Formula 220
Lee, Kyungyong 359
Lejeune-Jalabert 144, 146
lexicographic ordering 124
lifting idempotents 60
linear type 104, 106, 110, 174
linearly disjoint 50
Lipman 146, 221
  adjacent ideals 142
  adjoints 244, 357, 360–363, 377
  analytic spread 210, 233
  Briançon–Skoda 244, 245, 252
  conductor 242, 254
  dimension two 278
  Jacobian ideal 294
  multiplicity 233
  principle of specialization 225, 232
  quasi–normal 390–391
  reciprocity theorem 278
  unique factorization of ideals 274
Lipman–Sathaye Theorem 242, 244,
    252
Liu 225, 307
$\ell(I)$: see analytic spread
local xi
locally ringed space 144
Logar 300
Lying–Over 30, 42, 397
  and height 398
  finite 79
Lyubeznik 169, 321

M
m–full 209
  dimension two 260
  examples 261, 269
  modules 328
Macaulay representation 169, 396
Macaulay2 282, 287
MacLane's criterion 51
Manaresi 224, 225
mapping cone 372–375, 377
Mather 244
Matsumoto 72, 294, 301
Mazur 149
McAdam 97, 103
minimal equation 23
minimal polynomial 29, 235
minimal reduction 155
Minkowski inequality 331, 350, 352
mixed multiplicity 331, 338, 339
  associativity formula 342
module of homomorphisms 40
Moh 89
Mohan Kumar 170
Mohan, R. 309
monoid 34
monomial conjecture 326
monomial ideal 9
  adjoints 366
  and integral closure 9, 12, 295
  compute 295
  equation of integral dependence 9
  generalized 366–368
  normal 13
  Rees valuations 200
monomial reduction 158
Mori 86, 88
Mori–Nagata Theorem 86
$\mu(M)$, number of generators xii
Muhly 211
multiplicity 212, 214, 217–233
  additive 218
  associativity formula 218, 342
  Buchsbaum–Rim 313
  Cohen–Macaulay 216, 217, 233

extensions 219
hypersurface 219, 224
localization 218
mixed: see mixed multiplicity
one 216, 233
reduction 217, 222
semicontinuity 226
multiplier ideal 360, 375
multi–Rees algebra 332, _333_

N
Nagata 140–141
dimension two 88
examples 31, 89, 90
Lying–Over 78, 81
Mori–Nagata Theorem 86
multiplicity 233
Nagata ring: see pseudo–geometric
Newton polyhedron _12_
analytic spread 158
examples 11, 198, 368
generalized monomials _366_
Rees valuations 200, 208
Nishimura 88
Noether normalization 57–60, 221,
    283, 284, 285, 298
Noh 142
normal homomorphism _378_–391
normal ideal _2_, 21, 112, 210
monomial ideal 13
normal locus 72, 292
normal ring _27_, 69
base change 384
criterion 72
flat maps 71
integrally closed 28
Serre's conditions 70
Normaliz 296, 297
Northcott 150, 166, 336
$N_{R,R'}$ _390_

O
O'Carroll 225, 331, 332, 358
ω_R: see canonical module

Ooishi 155
ord_I: see order of an ideal
$\mathrm{ord}_R(\_)$ _258_
order of an ideal 131, 138, 196
overring _xi_

P
parameter ideal 100–102, 137
Pham 390
$P_{I,M}(n)$ 213
point basis _275_
point (ring) _274_
Polini 359
power series ring 24, 45, 145
convergent 144
generalized 131
Puiseux 45
powers of an ideal 97–104
associated primes 135, 194
integral closure 111, 135, 194
rational powers _205_
see Ratliff's Theorem
primary decomposition 86, 140, 299
prime avoidance 392
prime divisors: see divisorial valua-
    tions
Primitive Element Theorem 47
primitive polynomial _17_
principle of specialization of integral
    dependence 225, 231
Proj _108_
projectively equivalent _210_, 359
Prüfer 5
pseudo–geometric 46, 177, _186_
Puiseux series 45

Q
$Q(R)$ _xi_
quadratic transformation _263_–280
quadratic type 110
quasi–homogeneous 146
quasi–normal _390_, 391
quasi–unmixed: see formally equidi-
    mensional

R

$\mathfrak{R}_F(M)$ 304

R^+: see absolute integral closure

Raghavan 105

rank of a map of free modules 325

Ratliff 97, 101, 194, 320, 403

Ratliff–Rush closure 225

Ratliff's Theorem 17, 100, 102

reciprocity theorem 278

reduction 5, 150–175

   and grading 153

   and integral closure 6

   associated primes 153

   minimal 155

   modulo minimal primes 152

   monomial ideals 153, 158

   multiplicity 222

   non–local 169

   of modules 303

   persistence 151

   transitivity 6, 21

   with respect to 332

reduction number 154, 173, 176

   bound 166, 167

   one 250, 280

   see joint reduction number

reduction to characteristic p 326

reduction to infinite residue field
      159, 213, 222, 269, 338, 345, 371

Rees 185, 244

   analytically unramified 177–182

   Ext 395

   integral closure of modules 302–
      313, 319–320, 327

   joint reductions 331–359

   multiplicity 222, 317, 330, 348

   reductions 150–169

   Sally 353

Rees algebra 93–112

   and reductions 154

   defining equations 104–108, 112

   dimension 93

   integral closure 95, 180

   modulo minimal prime 93

   of module 305, 329

   symmetric algebra 305

Rees valuation 187–210, 277, 280

   construction 191–196, 208

   divisorial valuation 202

   examples 197, 203

   extension 205

   monomial ideals 197, 198, 199

   of $I \cdot J$ 203

   of modules 309, 311

   uniqueness 190

$\mathcal{RV}(I)$: see Rees valuations

reflexive module 240

regular ring

   adjoints 360–377

   and flat maps 71

   and valuations 132, 139, 186, 196

   are integrally closed 24

   Briançon–Skoda 244–252, 356

   dimension two 257–280, 359, 369

   extensions 181

   formal fiber not normal 378

   Jacobian criterion 68

   Jacobian ideal 67–72

   multiplicity 216, 219, 233

   Sally's Theorem 171

Reid 13, 46, 210

relation type 110

Remmert 289, 292

Rhodes 364, 376

Rim 313

rings of homomorphisms 39–42

   integral closure 41, 91

Risler 341

$r_J(I)$: see reduction number

(R_k): see Serre's conditions

rk(φ) 325

R^o xi, 136

Roberts, L. 13, 46, 210

Roberts, P. 326

Rotthaus 382, 390

R_{red} xi

Rush 101

$R(X)$ 159

S

Saito 146

Sakuma 211

Sally 123, 171, 195, 210, 244, 353, 357

Sally's machine 166, 175

Samuel 302

Sancho de Salas 109

Sathaye 242, 244, 252, 254

Schaub 222

Scheja 224

Schenzel 111

segment (of a group) 125

Seidenberg 30, 44, 117, 282, 286

seminormal <u>46</u>

separable 47, <u>48</u>–55, 285

  implies normal 379

  minimal polynomial 49

  module–finite integral closure 48

separably generated <u>50</u>

separated <u>xi</u>

Serre 216, 232

Serre's conditions <u>70</u>, 72–73, 209, 280, 285, 299

  and flat maps 71

  for normality 70

  (R_1) 70, 91, 283, 286

  (S_2) 70, 294, 295

sgn <u>319</u>

Shah 225

Sharp 331, 351, 359

Shimoyama 300

Simis 313

simple ideal <u>257</u>, 277

Singh, B. 46

Singla 13

Singular 282, 287, 289

singular locus 71, 72, 89, 293

(S_k): see Serre's conditions

$S(M)$ 304

Smith 321, 375

socle elements 16

standard conditions: on rank, height, depth 326, 327, 328

standard independent set of general elements 331, 353

Stolzenberg 282, 285

strongly stable monomial ideal 377

Stückrad 366

subadditivity 376, 377

superficial element <u>160</u>, 173, <u>333</u>, 354

  existence 162, 333

superficial sequence <u>164</u>, 165, <u>333</u>

Swan 46

Swanson 21, 175, 211, 298, 348, 366, 377

Sweedler 54

$\mathrm{Sym}_R(M)$ 304

symmetric algebra 110, 304

  integral closure of module 305, 327

  torsion 304

T

Takagi 375, 376, 377

Tate 91

$\tau(J)$: see test ideal

Teissier 144, 146

  Briançon–Skoda 245

  Lipschitz saturation 390

  Minkowski inequality 350

  mixed multiplicity 331, 341

  principle of specialization 225, 232

tensor algebra <u>304</u>, 307

test element <u>246</u>, 247, 256

test ideal <u>256</u>

  J–test ideal 375

tight closure 245–250

  J–tight closure 375

torsion–free <u>xi</u>

total ring of fractions <u>xi</u>

$T(R)$, $T_k(R)$ <u>253</u>–255

trace 48, 235, 243

Trager 283, 297, 299, 300

transcendence degree 55, 399, 403

  of a valuation 124, 127, 130

transform <u>266</u>

  inverse <u>267</u>

Traverso 46
tr.deg: see transcendence degree
Trung 166

U
Ulrich 91, 313, 329, 359
unique factorization domain 24, 90,
    258
Units Theorem 54
universally catenary 399, 403
    formally equidimensional 403
universally Japanese: see pseudo-
    geometric

V
Valla 104, 105
valuation 113–140
    adic: see adic valuation
    and completion 121–124
    approximation 141
    Archimedean 116
    center 122
    discrete: see discrete valuation
    divisorial: see divisorial valuation
    equivalent 115
    essentially of finite type 183
    examples 114, 130–133
    existence 117, 118
    extensions 201
    Gauss extension 114, 205
    independent 141
    integral closure 133–138
    intersection 121
    monomial 114, 198
    Noetherian 118, 120, 125
    (rational) rank 124–129
    Rees: see Rees valuation
    transcendence degree 124, 185
    values on minimal primes 189
valuation ring 115
valuative criterion 134, 144
value group 115, 116
    restriction and tensoring 121
van der Waerden 339

Vasconcelos 21, 91, 281, 313
    algorithm 294, 299
    binomial ideal 301
    conductor 243
    integral closure of modules 306
    primary decomposition 300
    radical 301
    Rees algebra of a module 307
Verma 275, 332, 358
    formally equidimensional 111
Veronese 208
versal 307, 329
Viêt 332
Villamayor 244, 298
Vitulli 13, 46, 210
Vogel 225
$\bar{v}_I$ 139, 189, 210, 277

W
Watanabe, J. 209
Watanabe, K. I. 142, 375, 376, 377
Wiegand 382, 390

Y
Yao 44
Yokoyama 300
Yoshida 375, 376

Z
Zacharias 299, 300
Zariski 113, 257, 270, 302
    topology 336
Zariski's Main Theorem 91, 92